T0190253

Lecture Notes of the Institute for Computer Sciences, Social Informatics and Telecommunications Engineering 416

More information about this series at https://link.springer.com/bookseries/8197

Shuai Liu · Xuefei Ma (Eds.)

Advanced Hybrid Information Processing

5th EAI International Conference, ADHIP 2021
Virtual Event, October 22–24, 2021
Proceedings, Part I

Editors
Shuai Liu
Hunan Normal University
Changsha, China

Xuefei Ma
Harbin Engineering University
Harbin, China

ISSN 1867-8211 ISSN 1867-822X (electronic)
Lecture Notes of the Institute for Computer Sciences, Social Informatics
and Telecommunications Engineering
ISBN 978-3-030-94550-3 ISBN 978-3-030-94551-0 (eBook)
https://doi.org/10.1007/978-3-030-94551-0

This Springer imprint is published by the registered company Springer Nature Switzerland AG
The registered company address is: Gewerbestrasse 11, 6330 Cham, Switzerland

Preface

We are delighted to introduce the proceedings of the second edition of the European Alliance for Innovation (EAI) International Conference on Advanced Hybrid Information Processing (ADHIP 2021), which took place online during October 22–24, 2021. This conference brought together researchers, developers, and practitioners around the world who are leveraging and developing hybrid information processing technology for smarter and more effective research and application. The theme of ADHIP 2021 was "social hybrid data processing".

The technical program of ADHIP 2021 consisted 94 full papers selected out of 254 submissions, with an acceptance rate of approximately 37.0%. The conference tracks were as follows: Track 1—Industrial Application of Multi-modal Information Processing; Track 2—Industrialized big data processing; Track 3—Industrial automation and intelligent control; and Track 4—Visual information processing. Aside from the high-quality technical paper presentations, the technical program also featured two keynote speeches. The two keynote speakers were David Camacho from the Universidad Politécnica de Madrid (UPM), Spain, who is currently a Full Professor and the head of the Applied Intelligence and Data Analysis (AIDA: https://aida.etsisi.uam.es) research group, and also acts as an Associate Editor of several journals including Information Fusion, Ambient Intelligence and Humanized Computing, Expert Systems, and Cognitive Computation, etc., and Muhammad Sajjad, who is an ERCIM Research Fellow with Norges Teknisk-Naturvitenskapelige Universitet (NTNU), Norway, the Head of the Digital Image Processing Laboratory at Islamia College University Peshawar, Pakistan, and an editor of a number of journals, such as IEEE Access and IEEE Trans ITS.

Coordination with the steering committee, comprising Imrich Chlamtac and Guanglu Sun, was essential for the success of the conference. We sincerely appreciate their constant support and guidance. It was also a great pleasure to work with such an excellent organizing committee team for their hard work in organizing and supporting the conference. In particular, we are grateful to the Technical Program Committee who completed the peer-review process for technical papers and helped to put together a high-quality technical program. We are also grateful to Conference Manager Natasha Onofrei for her support and all the authors who submitted their papers to the ADHIP 2021 conference and workshops.

We strongly believe that the ADHIP conference provides a good forum for all researchers, developers, and practitioners to discuss all science and technology aspects that are relevant to hybrid information processing. We also expect that the future ADHIP conferences will be as successful and stimulating as this year's, as indicated by the contributions presented in this volume.

December 2021

Shuai Liu
Yun Lin
Chen Cen
Danda Rawat
Zheng Ma

Organization

Steering Committee

Imrich Chlamtac (Chair)	University of Trento, Italy
Guanglu Sun	Harbin University of Science and Technology, China

Organizing Committee

General Chair

Zhongxun Wang	Yantai University, China

General Co-chairs

Wei Xue	Harbin Engineering University, China
Joey Tianyi Zhou	Institute of High Performance Computing, A*STAR, Singapore
Guan Gui	Nanjing University of Posts and Telecommunications, China
Shifeng Ou	Yantai University, China
Bo Nørregaard Jørgensen	University of Southern Denmark, Denmark

Technical Program Committee Chair

Shuai Liu	Hunan Normal University, China

Technical Program Committee Co-chairs

Chen Cen	Institute for Infocomm Research A*STAR, Singapore
Danda Rawat	Howard University, USA
Yun Lin	Harbin Engineering University, China
Zheng Ma	University of Southern Denmark, Denmark

Sponsorship and Exhibit Chairs

Jinwei Wang	Yantai University, China
Ali Kashif	Manchester Metropolitan University, UK

Local Chairs

Zhuoran Cai	Yantai University, China
Yidong Xu	Harbin Engineering University, China

Workshops Chairs

Zhuoran Cai	Yantai University, China
Ming Yan	Agency for Science, Technology and Research (A*STAR), Singapore
Ting Wu	Beihang University, China
Ya Tu	Harbin Engineering University, China

Publicity and Social Media Chairs

Pengfei He	Yantai University, China
Lei Chen	Georgia Southern University, USA

Publications Chairs

Weina Fu	Hunan Normal University, China
Wenjia Li	New York Institute of Technology, USA
Meiyu Wang	Harbin Engineering University, China

Web Chairs

Gang Jin	Yantai University, China
Ruolin Zhou	University of Massachusetts Dartmouth, USA

Posters and PhD Track Chair

Liyun Xia	Hunan Normal University, China

Panels Chair

Peng Gao	Hunan Normal University, China

Demos Chair

Jingyi Li	Hunan Normal University, China

Technical Program Committee

Adam Zielonka	Silesian University of Technology, Poland
Amin Taheri-Garavand	Lorestan University, Iran

Arun Kumar Sangaiah	Vellore Institute of Technology, India
Ashutosh Dhar Dwivedi	Technical University of Denmark, Denmark
Chen Cen	Institute for Infocomm Research A*STAR, Singapore
Chunli Guo	Inner Mongolia University, China
Dan Sui	Changchun University of Technology, China
Danda Rawat	Howard University, USA
Dang Thanh	Hue Industrial College, Vietnam
Dongye Liu	Inner Mongolia University, China
Fanyi Meng	Harbin Institute of Technology, China
Feng Chen	Xizang Minzu University, China
Fida Hussain Memon	Jeju National University, South Korea
Gautam Srivastava	Brandon University, Canada
Guanglu Sun	Harbin University of Science and Technology, China
Hari Mohan Pandey	Edge Hill University, UK
Heng Li	Henan Finance University, China
Jerry Chun-Wei Lin	Western Norway University of Applied Sciences, Norway
Jian Feng Cui	Xiamen University of Technology, China
Keming Mao	Northeastern University, China
Khan Muhammad	Sejong University, South Korea
Lei Ma	Beijing Polytechnic, China
Marcin Woźniak	Silesian University of Technology, Poland
Mu-Yen Chen	National Cheng Kung University, Taiwan
Norbert Herencsar	Brno University of Technology, Czech Republic
Ping Yu	Jilin University, China
Shuai Liu	Hunan Normal University, China
Shuai Wang	Hunan Normal University, China
Shuai Yang	Xinyang Vocational and Technical College, China
Shui-Hua Wang	Loughborough University, UK
Tenghui He	Hunan Normal University, China
Thippa Reddy Gadekallu	Vellore Institute of Technology, India
Uttam Ghosh	Vanderbilt University, USA
Wuxue Jiang	The Education University of Hong Kong, Hong Kong
Xiaochun Cheng	Middlesex University, UK
Xiaogang Zhu	Nanchang University, China
Xin Chun Zhou	Baoji University of Arts and Sciences, China
Xinyu Liu	Hunan Normal University, China
Xuan Yue Tong	Nanyang Institute of Technology, China
Yan Ning Zhang	Beijing Polytechnic, China

Yong Jun Qin	Guilin Normal College, China
Yun Lin	Harbin Engineering University, China
Zheng Ma	University of Southern Denmark, Denmark

Contents – Part I

Contents – Part II

Method Research on Internet of Things Technology

Intelligent Algorithms in Complex Environment

Human Physiological Behavior Understanding and Parameter Tracking Based on Complex Network Theory

Han Li[1,2](✉) and Peng Du[1,2]

[1] School of Electronic and Information Engineering, Liaoning University of Technology, Jinzhou 121000, China
lihan562323@yeah.net

[2] LiaoNing Wish Information Technology Co., Ltd., Jinzhou 121000, China

Abstract. In order to accurately track, analyze and understand human physiological behavior parameters, a method of human physiological behavior understanding and parameter tracking based on complex network theory is established. By defining the complex network theory, the statistical properties of complex networks are studied, and the central index of correlation degree of nodes is combined to analyze the structural characteristics of complex networks. On this basis, the human skeleton modeling conditions are set up, and the final result of parameter tracking behavior recognition is obtained according to the multi-scale feature extraction process. Experimental results show that the parameter tracking method based on complex network theory can accurately record the changes of human physiological behavior compared with traditional multi-scale human physiological behavior analysis method.

Keywords: Complex network · Human physiological behavior · Parameter tracking · Multi-scale feature

1 Introduction

With rapid economic development, increasingly fierce social competition, people's lives is getting faster and faster, the pressure of work and live increases. Common symptoms such as insufficient sleep, obesity, and immunity disorders threaten human health. With the development of science and technology, medical aids play an increasingly important role. Physiological parameters are the most basic life indicators of the human body, and medical workers can timely understand the disease condition through monitoring the human life indicators and parameters. The collection of physiological parameters and human wireless monitoring belong to biological telemetry. Through the method of radio magnetic signal transmission, the continuous collection of various important physiological parameters and intelligent monitoring is a part of the modern medical system. At present, technologies such as wireless communication and mobile Internet have been applied to small and medium-sized mobile communication devices, while the

S. Liu and X. Ma (Eds.): ADHIP 2021, LNICST 416, pp. 3–14, 2022.
https://doi.org/10.1007/978-3-030-94551-0_1

application of wireless communication and Internet technologies to medical equipment is still in its infancy. At present, there are some defects in the telemedicine monitoring system on the market, such as the monitoring results are not accurate enough, complex operation, and users can not get their own health status anytime and anywhere. To solve the above problems, this paper designs the human physiological behavior understanding and parameter tracking system based on complex network theory. Complex network theory focuses on the relationship between the complexity of the network structure and the network function, providing a powerful tool to reveal the nonlinear dynamic behavior of complex systems. The physiological behavior of the human body is very complex, dynamic and changeable, without real-time monitoring and analysis with a simple linear tracking system, so the complex network theory must be used. This paper presents a system of this theoretical design that can make users monitor their own physiological parameters in real time, and the data is more accurate, can better meet the health needs of people, so that everyone can understand their own health status at any time, prevent and monitor the occurrence of disease in advance [1].

Complex network is a kind of network with some or all properties of self-organization, self-similarity, attractor, small world and scale-free. Characteristics: Small world, cluster is the concept of degree of agglomeration, degree distribution of power law. Complex networks are, in short, networks of high complexity. Its complexity is mainly shown in the following aspects:

1) The complexity of the structure is mainly manifested in the huge number of nodes. network structure presents a variety of different characteristics.
2) Network evolution is mainly manifested in the emergence and disappearance of nodes or connections. For example, world-wide network, web pages or links may appear or break at any time, resulting in constant changes in the network structure.
3) Connection diversity: there are differences in connection weights among nodes, and there may be directionality.
4) Dynamical complexity: the node set may be a nonlinear dynamical system, for example, the node state changes complicated with time.
5) Node diversity: A node in a complex network can represent anything. For example, a node in a complex network composed of human relationships can represent an individual, and a node in a complex network composed of the World Wide Web can represent a different web page.
6) Multiple complexity fusion: that is, the above multiple complexities interact with each other, leading to more unpredictable results. For example, to design a power supply network, we need to consider the evolution process of the network, which determines the topology of the network. When two nodes transmit energy frequently, the connection weights between them will be increased, and the network performance will be improved gradually through continuous learning and memory.

2 Analysis of Structural Characteristics of Complex Network Theory

2.1 Overview of Complex Network Theory

Complex network is a large scale network with complex topological structure and complex dynamic behavior. It is a graph composed of a large number of nodes connected by edges, which is highly abstract. The complex network theory focuses on the relationship between the complexity of network structure and the function of network, which provides a powerful tool for revealing the nonlinear dynamic behavior of complex systems. The development of the complex network stems from the famous eighteenth-century Swiss mathematician Leonhard Euler's Seven Bridges at Cornisberg. That is, how to start from any of the four continents and cross each bridge exactly once when you return to the starting point. Euler uses abstraction analysis to turn the problem into a graphical problem: each land is replaced by a point, and each bridge is replaced by a line connecting the corresponding two points. So you get a simple "map".

Euler's scientific explanation of the seven-bridge problem initiated the study of graph theory, which is in line with the study of complex networks in some way. When people use stochastic networks to describe real networks, it is found that most of the real complex networks are not completely random in structure, which arouses the interest of many scholars. At the end of the nineteenth century, the discovery of the small-world and scale-free characteristics of complex networks led to a historic change in the scientific exploration of complex networks, which led to an upsurge of research on complex networks theory in many disciplines, marking the arrival of the era of "new science of networks" and the birth of a new interdisciplinary discipline of network science and engineering.

With the rapid development of complex network theory, many structural parameters in the sense of network statistics have been proposed by scholars. These parameters can be used to describe the importance of nodes or edges in complex networks from different perspectives. For example, the degree centrality index is the most basic evaluation method of node importance, which reflects the direct influence of nodes in their neighborhood. However, key nodes cannot be accurately identified because the indicator does not take into account the global importance of nodes [2]. In contrast, the central index can evaluate the importance of nodes in the whole network from a global point of view, which shows the control ability of nodes or edges in providing the shortest available routing. These methods use a single centrality index to measure the importance of nodes from different angles, but it is difficult to evaluate the importance of nodes accurately because of the large scale and complexity of real networks and the diversity of different topologies.

2.2 Statistical Properties of Complex Networks

For a complex network with small-world characteristics, it has a large clustering coefficient and a small average path length. Here, "a large clustering coefficient" is relative to a regular network, while "a small average path length" is relative to a random network. Clustering coefficients and average path length are proposed to characterize the topological characteristics of complex networks [3]. In addition, many other concepts of statistical properties have been proposed, such as degree and its distribution, medium number, efficiency, etc.

The degree value l_i of a node refers to the number of connected edges between node s_i and adjacent nodes (where i represents the actual statistical measurement coefficient of the complex network). Node degree is one of the most simple and important concepts in the statistical characteristics of complex networks. It is used to reflect the richness of nodes in the network and the direct influence of nodes in the network. According to the definition of node degree, we can get the average degree of the network, that is, the average value of all node degrees in the network, which is represented by l. Furthermore, the degree centrality index of node s_i refers to the ratio of the number of nodes directly associated with node s_i to the maximum number of nodes possibly associated with node 1.

$$DC(i) = \frac{l_i}{i-1} \tag{1}$$

Among them, $DC(i)$ reflects the proportion of the number of nodes with degree value of l_i in the total number of nodes in the network. The degree distribution of different network topologies is shown in Table 1.

Table 1. Comparison of characteristics of different network types

	Degree distribution	Feature path length	Agglomeration coefficient
Rule network	δ distribution	Large	Large
Random network	Poisson distribution	Small	Large
Small world network	Exponential distribution	Small	Large
Ba scale free network	Power law distribution	Small	Small
Actual network	Approximate power law distribution	Small	Large

Because of the individual differences in network connection and network function among the nodes that make up the basic unit of complex network, the strategies of node protection should be different. The structural attributes of complex networks are described by the central index of the statistics which characterizes the topological structure, while the functional attributes are described by the characteristic attributes which have specific meanings and functions according to the operation mode of networks. When a system is abstracted into a complex network, the importance of a node must be determined by its structural and functional attributes in the network. Accurately identifying the importance of complex network nodes is not only the premise for network managers to defend the network security, but also the effective way for attackers to lock the attack target.

The basic idea of multi-attribute node importance recognition algorithm based on information entropy is to take each node in complex network as a decision scheme, take the selected centrality index as the attribute of each decision scheme, get the direct value of each attribute and calculate the weight of each attribute according to information entropy, and the node importance is the weighted sum of the attribute values of each scheme [4]. For a decision attribute, the smaller its direct value, the more uneven the distribution of the value of the attribute, the corresponding greater its weight value; otherwise, the smaller the weight value. In particular, when the equal probability of the value of an attribute appears, the distribution of the value of the attribute is the most uniform, the direct value is the largest, and the weight of the attribute for evaluating the node importance is the least, that is, the attribute has the least impact on the evaluation results.

2.3 Node Correlation Degree Central Index

In 1865, the German physicist Rudolf Clausius first proposed the concept of the entropy and applied it to the study of human physiology to show the uniformity of the distribution of any energy in space. Then, Boltzmann and Planck gave the microscopic statistical formula of the entropy, which is used to measure the disorder degree of the system and lay the foundation for the application of the entropy in multidisciplinary. In 1948, Shannon, the American mathematician and founder of information theory, first introduced the concept of entropy into information theory when he solved the problem of quantifying information in communication. Information entropy is a concept used in information theory to measure the degree of uniformity and orderliness of information.

For a discrete random signal source $X = \{x_1, x_2, \cdots, x_n\}$ and its probability distribution, the definition of information entropy $P(X) = \{P(x_1), P(x_2), \cdots, P(x_n)\}$ is as follows

$$H = \frac{-\sum_{i=1}^{m} P(x_i) \ln P(x_i)}{DC(i)} \tag{2}$$

Among them, m represents the definition coefficient of correlation degree, and $P(x_i)$ represents the network microscopic statistics of signal source x_i, i has the same meaning as formula (1).

In complex networks, there are complex interdependencies among nodes. How to describe the strength of the interdependencies is closely related to the accuracy of node importance identification. Based on this, a centrality index of association degree is proposed in this section. This index is mainly considered from the following three aspects:

First, the correlation range of importance between nodes. It is generally believed that the most direct form of correlation between nodes lies in the adjacent nodes. In practice, does the interaction between nodes only exist between adjacent nodes? If the correlation strength between non adjacent nodes is much lower than that between adjacent nodes, the correlation between non adjacent nodes can be ignored. However, if the degree of correlation between non adjacent nodes is significantly higher than that of adjacent nodes, then the correlation between non adjacent nodes can not be ignored. The relationship between nodes can be considered as interdependence, and can be described by the contribution of node importance. In addition, due to the inherent importance of the node itself, the important contribution of the node is bound to be uneven.

3 Human Physiological Behavior Understanding and Parameter Tracking Method

3.1 Human Skeleton Modeling

When the clothing of human body is close to the background, the detected human body will appear holes, and due to the interference of noise and shadow, the detected foreground image often has isolated points. In order to eliminate these adverse effects, we use morphological methods to deal with noise and holes, so as to get a better binary image, as shown in Fig. 2.

First, all the pixels in the image are added with the corresponding tag array, and the array is initialized to 0; then, a point on the background area is used as the seed point to grow the region, and all the pixels scanned in the growth process are marked as 1; finally, through the operation of step 2, all the points marked as 1 are the background points, and the points marked as 0 are the detected objects and holes. If the gray value of all the pixels marked as 0 is set as the foreground color, the hole can be filled, and a better binary image effect can be obtained. Although the principle of background subtraction is simple, the construction and update method of background image is very important, because it directly affects the adaptability of background model to scene change and foreground target granularity [5, 6]. There are many methods for background update, such as selective update and blind update.

In view of the fact that expansion and corrosion are not a pair of reciprocal operations, we can do some special processing on the image through the operation of expansion first and then corrosion. After several times of expansion operation, the tiny fracture of binary image can be connected, and then several times of corrosion operation can be carried out, so that the redundant information generated by the expansion of the object periphery can be eliminated. In this way, not only the image information is complete, but also the noise is not introduced (Fig. 1).

Fig. 1. Human skeleton modeling form

In the field of complex network, the preprocessing of human physiological behavior image usually includes the main steps of foreground detection, denoising and filling holes. If the human body is modeled to extract features, the known 2D modeling method is a star skeleton model based on the distance from the center point to the contour point, or a 2D model based on the mapping of human body structure to the skeleton [7]. Firstly, the gray image is denoised by Gaussian filter, then the moving object is detected by background subtraction method, and the moving object image is post-processed, such as denoising and filling holes. The skeleton is extracted by morphological method, and then the key points of the skeleton, including limb end point, bifurcation point and upper limb inflexion point, are obtained by using chain code and curvature.

3.2 Multi-Scale Feature Extraction

For different tracking occasions and understanding purposes, human physiological behavior maintains different motion characteristics on the rough layer, the middle layer and the fine layer of the complex network. In the middle layer, people's motion-related behaviors reflect people's actions at every moment, such as bending down, stretching out hands, etc., and the features are usually represented by the outline of a human being or binary silhouette. In the details layer, the movement of a person's limb needs to be accurately expressed, the state changes quickly, and the features are usually represented by point coordinates. The features of different levels reflect the motion on different scales. We extract the features from different scales, which are called multi-scale features. These three scales are referred to as large scale, mesoscale and small scale [8]. Motion details on different scales play different roles in behavior recognition. How to combine them properly is of great significance to behavior recognition.

On a large scale, the speed of human action has the characteristics of this scale.

(1) Get the outline of the human body's binary silhouette;
(2) Finding the center point of the outline;

The speed of human motion is determined from the distance and time of the centroid movement in the continuous frame image.

If there is an obvious reference object in the background, the distance between the reference object and the human being is another large-scale feature; if there is no reference object, only the speed is selected as the feature, so the large-scale feature in the complex network environment only selects the tracking speed of human physiological behavior.

On the mesoscale, the body contour is used for feature extraction. The features of this scale include three levels:

(1) Scanning contours from top to bottom, taking the width of each line of contours as the first mesoscale feature, which can reflect the amplitude of upper and lower limb movement;
(2) The second mesoscale feature is the number of intersections between the line and the contour. The position of limb can be determined by the number of intersections. If the number of intersections is more, the position of limb can be determined.

(3) The cycle of upper limb swings is taken as a third characteristic dimension, as it reflects the speed of upper limb movement.

On the small scale, we use the position of human limbs, head and upper limb inflection point to depict human motion, i. e. extract the coordinated position of the human head, two hands and two feet, bifurcation point and upper limb inflection point as the small scale feature, altogether 8 points and 16 dimensional feature. Firstly, the background is extracted from the motion sequence, and then the foreground region is extracted by background subtraction to obtain binary silhouette. Many behavioral recognition algorithms based on multiple viewpoints choose binary silhouette as feature input. Features of different scales describe different behavioral characteristics, which can combine features of different scales together to express the behavioral performance more completely. A simplified 2-D model of human body is established by calculating the local maximum distance from body mass center to limbs and head, and the human body action sequence is represented as a continuous 2-D star skeleton, which is matched with the feature template for behavior recognition [9]. The location of these key parts is of great significance to the analysis of more subtle and subtle movements in human behavior. Based on the human skeleton model obtained in the previous chapter, the position of head, limb, bifurcation point and inflection point can be obtained accurately, which can more accurately depict the state of human in motion.

3.3 Parameter Tracking Behavior Recognition

Clustering is a collection of data objects. In the same class, data objects are similar, but objects between different classes are not similar. Cluster analysis is the grouping of data sets into several clusters. Clustering is an unsupervised classification, that is, there is no predefined class. Typical applications of clustering are: as an independent tool to look at data distribution; as a preprocessing step for other algorithms. Tracking behavior is a multilinear mapping over a set of vector spaces. Multilinear analysis is an extension of traditional linear methods such as PCA or matrix singular value decomposition. Tracking behavior is a generalization of the concepts of vector and matrix. Because the tracking behavior has such good characteristics, it can play a very good role in pattern recognition under the influence of multiple factors. Each factor is placed on a dimension of a high dimensional tensor, so that each factor can be decomposed independently on the corresponding dimension, and the components decomposed independently are controlled by a unified kernel tensor.

Feature extraction is to map the data set to a space, and take the basis of the space as a variable to model from the data set; Behavior classification classifies the behavior sequence into a concrete class. In the research of parameter tracking behavior recognition, first of all, the current feature extraction methods are introduced, mainly based on non-model and model-based methods. The non-model based methods are mainly based on contour or binary silhouette as feature, and the method of extracting the optical flow vector from the optical flow field to reflect the movement of the object by the change of the optical flow vector, which is mainly applied to the detection and tracking of the moving object, or combined with the time characteristic of the movement, use the moving historical image as feature to carry out template matching to identify the object.

Model-based methods mainly include two-dimensional models and three-dimensional models [10].

Under the support of complex network theory, several common behavioral recognition methods, such as template-matching method, probabilistic network method, hidden Markov model and dynamic Bayesian network are the most commonly used models. Other methods such as the method based on clustering analysis, neural network -based methods. Recently, the behavior recognition method based on tensor quantum space analysis has been paid more attention to, and it has been applied to the field of behavior recognition from the initial face recognition.

Let m_0 represent the minimum human physiological behavior understanding parameter, m_n represent the maximum human physiological behavior understanding parameter, and n represent the parameter tracking coefficient. With the support of the above physical quantities, the recognition result of parameter tracking behavior can be expressed by simultaneous formula (2)

$$C = H \cdot \sqrt{\frac{\sum_{m_0}^{m_n} (A_2 - A_1)^2}{f \times n^2}} \tag{3}$$

Among them, f represents the human physiological behavior understanding permission value based on complex network, A_1 and A_2 represent two different human physiological behavior index parameters respectively. So far, the calculation and processing of each coefficient application index have been completed. With the support of complex network theory, the method of human physiological behavior understanding and parameter tracking has been successfully applied.

4 Comparative Experimental Analysis

24 consecutive images are selected to form a behavior sequence, which is used as a sample to describe a behavior. The 24 consecutive images constitute a feature space, and the feature extraction from these 24 images is used as the input feature of behavior recognition. The continuous frame image is shown in Fig. 2. Figure 2 (1) shows a continuous action image of a behavior, Fig. 2 (2) is a corresponding binary image, and Fig. 2 (3) shows a continuous frame image of each corresponding behavior.

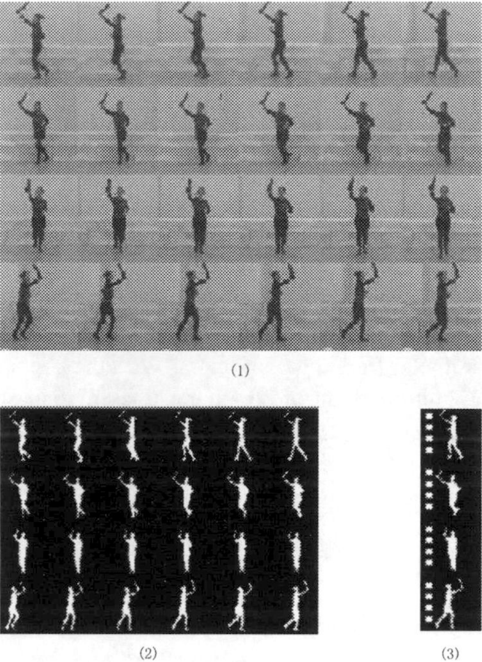

(1)

(2) (3)

Fig. 2. Continuous frame image

Many pose recognition algorithms are characterized by a single binary silhouette image. But the behavior is composed of a continuous action sequence, so the STB index is used to determine the recognition accuracy of the change form of human physiological behavior. Generally, the larger the STB index value is, the higher the recognition accuracy of the complex network for the change form of human physiological behavior is, and vice versa. During the experiment, the experimental group was equipped with the method of human physiological behavior understanding and parameter tracking based on complex network theory, while the control group was equipped with the traditional multi-scale human physiological behavior analysis and understanding method.

According to the analysis of Table 2, with the extension of experimental time, the STB index of the experimental group kept rising, and the maximum value of the whole experimental process reached 88.8%. The STB index of the control group began to rise slightly and steadily after a period of time. The maximum value of the whole experiment was only 54.3%, which decreased by 34.5% compared with the maximum value of the experimental group. To sum up, after the application of human physiological behavior understanding and parameter tracking method based on complex network theory, the value of STB index shows a significant upward trend, which meets the practical application needs of enhancing the accuracy of complex network for human physiological behavior change recognition.

Table 2. Comparison of STB index values

Experiment time/(min)	STB index value/(%)	
	Experience group	Control group
5	81.6	40.2
10	82.1	40.2
15	82.4	43.1
20	82.7	43.1
25	82.9	45.8
30	83.2	45.8
35	83.8	48.3
40	84.0	48.3
45	84.3	48.9
50	84.5	49.4
55	84.7	49.8
60	84.9	50.3
65	85.3	50.9
70	85.8	51.4
75	86.2	51.8
80	86.8	52.3
85	87.4	52.8
90	87.8	53.4
95	88.3	53.7
100	88.8	54.3

5 Conclusion

Compared with the traditional multi-scale human physiological behavior analysis and understanding method, the human physiological behavior understanding and parameter tracking method based on complex network theory can model the human skeleton under the support of the centrality index of node association degree, and then combine the multi-scale feature extraction results to realize the accurate recognition of parameter tracking behavior. From the practical point of view, the increase of STB index value can enhance the recognition accuracy of complex network for human physiological behavior changes, and has strong practical application ability.

Fund Projects. The Doctoral Scientific Research Foundation of Liaoning Province (2019-BS-121), the Central Govermment Guided Local Special Funds for Scientific and Technological Development (2020JH6/10500067).

References

1. Zhang, Y.: Simulation design of human physiological parameters remote monitoring system based on single chip microcomputer system. J. Changchun Normal Univ. (Nat. Sci.) **39**(4), 28–31 (2020)
2. Yang, Z., Wang, W., Kim, D.: On multi-path routing for reliable communications in failure interdependent complex networks. J. Comb. Optim. **41**(1), 170–196 (2021)
3. Kadokawa, J.I., Tanaka, K., Yamamoto, K.: Enzymatic preparation of supramolecular networks composed of amylosic inclusion complexes with grafted guest polymers. J. Electrochem. Soc. **166**(9), B3171–B3175 (2019)
4. Xiong, K., Yu, J., Hu, C., et al.: Synchronization in finite/fixed time of fully complex-valued dynamical networks via nonseparation approach. J. Franklin Inst. **357**(1), 473–493 (2020)
5. Yang, Y., Pan, J., Zhou, Z., et al.: Complex microbial nitrogen-cycling networks in three distinct anammox-inoculated wastewater treatment systems. Water Res. **168**(1), 115142.1-115142.11 (2020)
6. Zhou, W., Li, B., Zhang, J.E.: Matrix measure approach for stability and synchronization of complex-valued neural networks with deviating argument. Math. Probl. Eng. **2020**(8), 1–16 (2020)
7. Fu, W., Liu, S., Srivastava, G.: Optimization of big data scheduling in social networks. Entropy **21**(9), 902 (2019)
8. Liu, S., Li, Z., Zhang, Y., et al.: Introduction of key problems in long-distance learning and training. Mob. Netw. Appl. **24**(1), 1–4 (2019)
9. Liu, S., Bai, W., Zeng, N., et al.: A fast fractal based compression for MRI images. IEEE Access **7**, 62412–62420 (2019)
10. Dai, H., Jia, J., Yan, L., et al.: Event-triggered exponential synchronization of complex dynamical networks with cooperatively directed spanning tree topology. Neurocomputing **330**(FEB.22), 355–368 (2019)

Design of Fractal Image Coding Compression and Transmission Model Based on Wavelet Transform

Jie He(✉), Hai-xiao Gong, and Hong- yan Lu

School of Data Science and Software Engineering, Wuzhou University, Wuzhou 543002, China

Abstract. Coding and compression of image data is the key to high quality storage and transmission of image data. For this reason, this research designed a new fractal image coding compression and transmission model based on the wavelet transform process based on the analysis of the advantages and disadvantages of the basic fractal coding compression algorithm. Firstly, the image is decomposed by two-level wavelet transform, and then the decomposed high-level sub-images are subjected to basic fractal coding, and according to the similarity between the sub-image structures of different layers, the low-level sub-image fractal coding is constructed from the high-level fractal coding to realize the image High-quality transmission. Experimental results show that the model has achieved good application effects in shortening the image encoding time and improving the compression ratio.

Keywords: Wavelet transform · Fractal image · Image coding · Image compression · Transmission model

1 Introduction

The development of information technology brings more opportunities and challenges to image compression and transmission technology. Therefore, it is necessary to explore some new image coding and compression methods [1]. Fractal image coding has been paid more and more attention for its potential high compression ratio, but fractal coding can not solve the problem of obtaining higher compression ratio, especially shortening the coding time. Therefore, the combination of fractal coding and compression technology with other image processing technology may be an important method to solve these problems [2].

At present, fractal images are usually encoded, compressed and transmitted based on Gaussian model or Wiener filter. However, in practical application, it is found that the traditional image coding model has a long time and low application efficiency.

Wavelet transform is a kind of global transform, which has good positioning ability in time domain and frequency domain at the same time. It can focus on any detail of the processed image by gradually fine step size in time domain and spatial domain. Because

S. Liu and X. Ma (Eds.): ADHIP 2021, LNICST 416, pp. 15–25, 2022.
https://doi.org/10.1007/978-3-030-94551-0_2

of these advantages of wavelet transform, it has been widely used in the field of image processing [3].

Based on the above analysis, a new image coding compression and transmission model combining wavelet transform and fractal theory is proposed based on fractal coding theory. Firstly, the multi-scale image decomposition is carried out, and then the fractal coding of the high level sub-image is carried out according to the similarity of the sub-image structures of different frequency bands, and then the fractal coding of the high level sub-image is constructed to construct the fractal coding of the low level sub-image. Using this coding method, higher compression ratio can be obtained, and the coding time can be greatly shortened, so as to improve the efficiency of image transmission to a certain extent.

2 Fractal Image Coding and Compression Based on Wavelet Transform

Wavelet is a wave with limited energy and very concentrated in the time domain. The use of wavelet characteristics to analyze instantaneous variable signals can well extract faults. The wavelet analysis method can show the time domain characteristics and frequency domain characteristics of the signal well, so as to more accurately describe the signal and extract the fault [4]. For any function $f(t) \in L^3(R)$, a basic wavelet can be used to obtain a formula with the inner product of $f(t)$ after scaling and translation operations. The calculation formula is as follows: (1):

$$W_f(a,b) = \left[f, \psi(t)\right] = |a|^{\frac{1}{2}} \int f(x)\psi(\frac{1-b}{a})dx \tag{1}$$

In formula (1), $\psi(t)$ is the mother wavelet function, a is the expansion factor, and b is the parallel factor. The smaller ${}^1/_a$ is, the higher the corresponding frequency is, and the larger ${}^1/_a$ is, the lower the corresponding frequency is. Through the time-frequency analysis of the wavelet transform of the signal, observe the component of t corresponding to a a at a certain time [5].

The signal is divided into high-frequency and low-frequency sub-band by using the filter of frequency-domain averaging. The high-frequency part is retained every time, and then the low-frequency part is continued to be equally divided until the low-pass filter represented by the function. After analyzing the wavelet, the wavelet packet is decomposed. Wavelet packet is proposed to solve the two problems of poor frequency resolution in high frequency band and poor time resolution in low frequency band. Wavelet packet analysis can provide a more precise method. On the basis of multi-resolution analysis, the high frequency part is further subdivided, and the combination can be selected adaptively Appropriate frequency band, matching with the signal spectrum, so as to improve the time-frequency resolution [6]. The process is shown in Fig. 1.

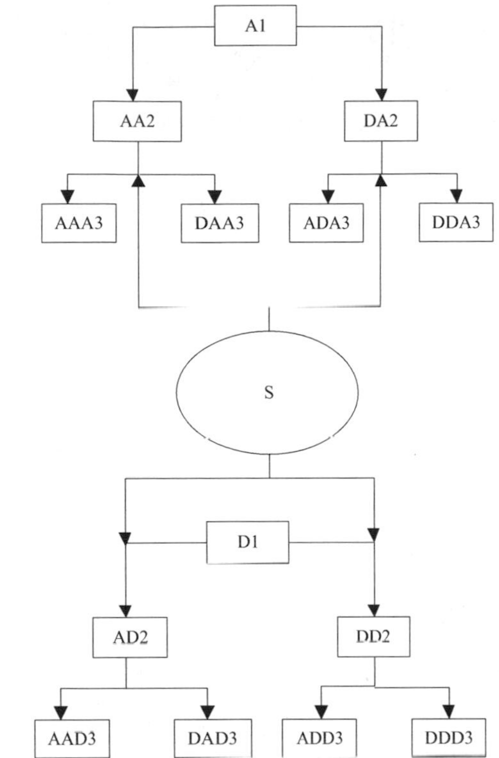

Fig. 1. Schematic diagram of the three-layer decomposition process of wavelet packet

The extracted signals are respectively represented as aaa3, daa3, ADA3, DDA3, aad3, dad3, add3, ddd3. Where a is the low frequency and B is the number after the high frequency letter, which is the level of wavelet packet decomposition. The formula of wavelet packet decomposition is s = aaa3 + daa3 + ADA3 + DDA3 + aad3 + dad3 + add3 + ddd3. According to the decomposition relation, the fault characteristics of the circuit are extracted [7].

Image coding based on fractal transform is a model coding technology developed on the basis of fractal theory. The main steps are as follows:

(1) The image to be encoded is divided into non overlapping small blocks, namely range block, referred to as R-block, while the image is divided into larger overlapping blocks, namely domain block, referred to as D-block. The combination of all the classified blocks is the original image, and the combination of all the overlapped blocks is just the "source" to find the affine compressed image blocks.
(2) The R-block and D-block are classified. Because an image can be divided into smooth region with smooth transformation, edge region with abrupt transformation and middle region with gentle transformation, and the matched blocks have the same regional properties. Therefore, in order to speed up the coding process and realize the matching search process between blocks with the same regional properties,

the segmented blocks must be divided into flat r block and D-block, edge R-block and D-block, and transition R-block and D-block according to the classification principle.

(3) For each R-block in the same region after classification, the matching D-block is found, so that D-block can approximate R through affine function. In this way, a group of affine transformation groups can be found, which is the segmentation iterative system. As long as the transformation of the system is contractive and simpler than the original system, the fractal compression can be realized.

To sum up, the basic fractal compression algorithm is the process of searching and matching R-block and D-block after image segmentation. This algorithm is characterized by high compression rate (1–2 orders of magnitude higher than the classical coding), and the operation speed has little to do with the improvement of image resolution, but the problem is that the amount of calculation is large and the coding compression time is long. Therefore, this paper proposes a new fractal image coding and compression technology based on wavelet transform.

After wavelet decomposition, the image is divided into four sub bands: low frequency sub-band ll which contains most of the image information, horizontal sub-band HL which contains texture features and edge information, vertical sub-band LH and diagonal sub-band HH.

The image decomposition algorithm based on wavelet multi-resolution analysis is to decompose the image by two-dimensional wavelet on the basis of selecting the appropriate orthogonal wavelet basis.

Assuming that signal $\{C_{j+1,m,n};\, m, n \in z\}$ is the approximate representation of two-dimensional image $f(x, y)$ at resolution $j+1$, the finite orthogonal wavelet decomposition formula of two-dimensional signal $\{C_{j+1,mt,n};\, m, n \in z\}$ is:

$$\begin{cases} c^1_{j,m_1,m_2} = \sum\limits_{k_1,k_2} h_{k_1-2m_1} h_{k_2-2m_2} c_{j+1,k_1k_2} \\ d^1_{j+1,m_1,m_2} = \sum\limits_{k_1,k_2} h_{k_1-2m_1} g_{k_2-2m_2} c_{j+1,k_1,k_2} \\ d^2_{j+1,m_1,m_2} = \sum\limits_{k_1+k_2} g_{k_1-2m_1} h_{k_2-2m_2} c_{j+1,k_1,k_2} \\ d_{j+1,m_1,m_2} = \sum\limits_{k_1,k_2} g_{k_1-2m_1} g_{k_2-2m_2} c_{j+1,k_1,k_2} \end{cases} \tag{2}$$

It can be seen that the image signal is decomposed into four parts c^1, d^1, d^2, d^3 through the two-dimensional wavelet decomposition of Eq. (2), corresponding to one low-frequency component and three high-frequency components of the image respectively. The main energy of the image signal is concentrated in the low frequency region, which reflects the average brightness of the image. The multi-level wavelet decomposition can be repeated for this part, and the detail edge information is mainly concentrated in the high frequency region, so it is necessary to code the high and low frequency components according to the human visual physiological characteristics.

The working principle of image acquisition designed in this paper is to call line synchronization signal, frame synchronization signal and pixel clock signal. The three signals cooperate with each other and work together to complete the collection of visually conveyed image information. The processor of the image clarification processing system disperses the image data. During the image data transmission, a data frame and a synchronization frame signal are carried at the same time, and a horizontal synchronization signal is output before and after the transmission. In the image acquisition, the rise of the frame synchronization signal represents the end of a visually communicated image data, and the decrease of the signal level represents the arrival of the visually communicated image data. The rise of the line sync signal strength represents the beginning of the image data frame data, and the module should immediately take the image information data; the fall of the line sync signal strength indicates the end of the image frame data, and the module will stop collecting information. The pixel clock signal in the collection represents the start of visually conveying the image pixel data, the rising of the pixel clock signal represents the beginning of the pixel data, and the falling of the pixel clock signal represents the end of the pixel data.

The basic idea of fractal compression based on wavelet transform is that the wavelet transform decomposes the image into sub images in different spatial frequency bands, and there is a great similarity between the corresponding sub image structures in different layers; and the fractal compression algorithm mainly uses the image spatial structure information for compression. In this way, the basic fractal compression can be carried out in the high-level sub image after wavelet transform, and according to the fractal characteristics, the similarity between different level sub images can be used to obtain higher compression ratio. At the same time, the number of image blocks in the classified r block domain and the corresponding overlapped D block domain after image segmentation can be greatly reduced by compression in the high-level sub image after wavelet transform The search time of matching is greatly reduced, so as to shorten the coding compression time. The process is as follows:

Step 1: wavelet decomposition of the image. Firstly, the original image is decomposed by one-stage wavelet transform to obtain four image components D10, D11, D12 and D13, and then the low-frequency component D10 is transformed by two-stage wavelet transform to obtain two-stage high and low-frequency components A0, D21, D22 and D23;

Step 2: segment and classify the wavelet transform secondary high frequency sub images D21, D22 and D23 respectively, and then compress them by using the block search matching method with the same regional property to obtain the fractal coding of the secondary sub images;

According to the fractal image's similarity, the first level coding is constructed.

Let a subblock $X = \left(x_{i,j}\right) \in R^{n \times n}$, then X be normalized as:

$$\hat{X} = \left(\hat{x}_{i,j}\right) = X - \overline{x}I \tag{3}$$

$$\hat{x}_{i,j} = \frac{x_{i,j} - \overline{x}}{\sqrt{\sum_{i,j}\left(x_{i,j} - \overline{x}\right)^2}} \tag{4}$$

Nine blocks and features are defined as:

$$q(x) = \begin{cases} \sum_{2}^{j=0} \sum_{2}^{i=0} \left|\hat{x}_{ik+1,j+1}\right| n = 2k+1 \\ \sum_{1}^{j=0} \sum_{1}^{i=0} \left(\left|\hat{x}_{ik+1,jk+1}\right| + \left|\hat{x}_{(i+1)k,(j+1)k}\right|\right) + \sum_{1}^{i=0} \left(\left|\hat{x}_{2k,ik+1}\right| + \right. \\ \left. \left|\hat{x}_{k,ik+1}\right| + \left|\hat{x}_{ik+1,2k}\right| + \left|\hat{x}_{ik+1,k}\right|\right) n = 2k \end{cases} \quad (5)$$

Obviously, the sum of nine blocks of sub block $\hat{X}$ is the sum of its four vertices, the middle points on the four edges, and the center points (for even order sub blocks, the middle points on the edges take the sum of the middle two points, and the center points take the average value of the sum of the four points in the center).

Let $R, D \in R^{n \times n}$, then the following inequality holds:

$$E(R, D) \geq \frac{n}{32} \sigma_R |q(R) - q(D)|^2 \quad (6)$$

Where: $\sigma_R = \frac{R - rI}{n}$. Sub block $Q = \left(q_{i,j}\right) \in R^{n \times n}$ is defined as follows:

$$q_{i,j} = \begin{cases} 1 \ i \in \{1, k+1, 2k+1\} \text{ and } j \in \{1, k+1, 2k+1\} \\ 0 \text{ other} \end{cases} \quad (7)$$

Through the two-level wavelet decomposition, the image is decomposed into 7 subbands, the wavelet coefficients of the low frequency subbands are retained, and the signs of the wavelet coefficients of the remaining 6 subbands are separately coded.

For these 6 subbands, the fractal image coding based on the nine-block sum feature is performed. For the first-level subband image, it is divided into R blocks (cannot overlap, 8 × 8 size)and D blocks (can overlap, 16 × 16 size). For the secondary sub-band image, it is divided into R blocks (can not overlap, 4 × 4 size) and D blocks (can overlap, 8 × 8 size).

Take the entire contracted block of each D block as a codebook, then set the standard deviation threshold, the standard deviation threshold of the code block, and the radius of the neighborhood, and filter the codebook to form a new codebook, which is ordered according to the nine blocks and the feature size. The fractal code of each subband is synthesized into the fractal code of the entire original image. According to the fractal code information, each subband is decoded to obtain the decoded image, and then the inverse wavelet transform is performed to generate the reconstructed image.

3 Fractal Transmission Model Based on Wavelet Transform

The fractal transmission model based on wavelet transform is set up. The fractal transmission model is the basic part of the normal operation of the whole work. The fractal transmission model based on wavelet transform is shown in Fig. 2.

It can be seen from Fig. 2 that the fractal transmission model based on wavelet transform designed in this paper is based on B/S structure and data source.

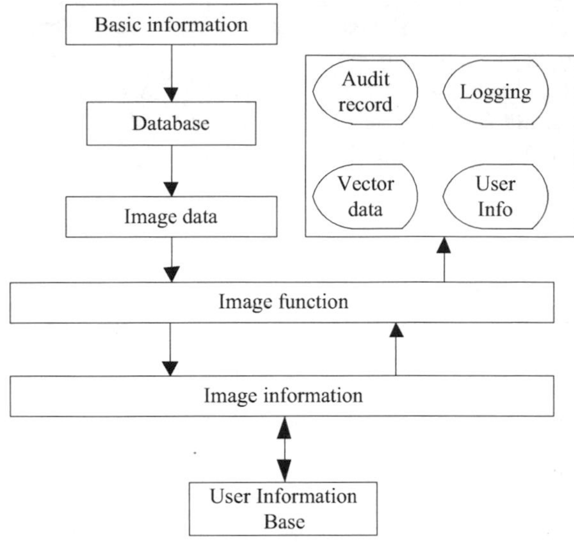

Fig. 2. Fractal transmission model based on wavelet transform

Vector data is the core part of collection, which establishes important parameters for the system. Topology data can facilitate users to find information and quickly find the target area.

The basic data source collection and calculation process is as follows:

$$u_{ad} = \sum_{n=1} (\varphi_{ad} - \varphi_s) \times \partial(x_1, y_1) \tag{8}$$

In formula (8), u_{ad} represents the collected data, φ_{ad} represents the vector data contained, φ_s represents the useless data, and $\partial(x_1, y_1)$ represents the coordinate position of the data node in the space vector data.

$$u_{bd} = \sum_{n=1} (\varphi_{bd} - \varphi_m) \times \partial(x_2, y_2) \tag{9}$$

In formula (9), u_{bd} is the collected data, φ_{bd} is the vector data containing data features, φ_m is the useless data, and $\partial(x_2, y_2)$ is the node coordinate position.

$$u_{cd} = \sum_{n=1} (\varphi_{cd} - \varphi_n) \times \partial(x_3, y_3) \tag{10}$$

In formula (10), u_{cd} represents the collected POI data, φ_{cd} represents the vector data containing the characteristics of the POI data, φ_n represents the useless data in the POI data, and $\partial(x_3, y_3)$ represents the coordinate position of the POI data node in the space vector data.

Integrate formula (8), (9), (10) into formula (4), and the result obtained is all the space vector data collected.

$$u_d = \varphi_{ad} + \varphi_{bd} + \varphi_{cd} \tag{11}$$

In formula (11), u_d is the space vector data collected in the network geographic information system [8].

The collected basic data source is applied to the transmission service, so as to better collect internal data and facilitate users to find information. The collected transmission data is divided into three types, and all kinds of transmission services present parallel relationship [9, 10], as shown in Fig. 3.

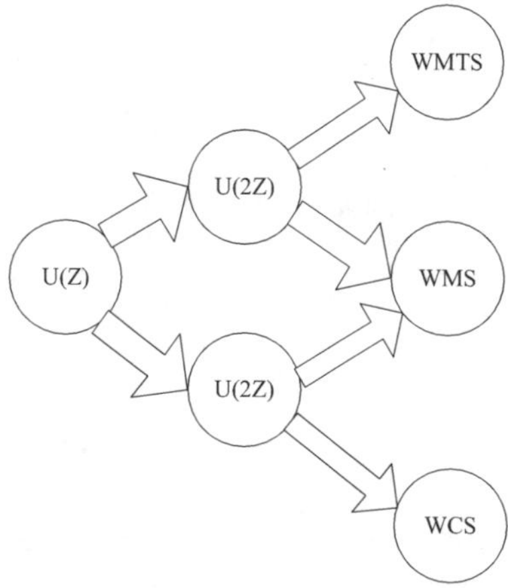

Fig. 3. Schematic diagram of the relationship between transmission vector data

Observing Fig. 3, it can be seen that the collected transmission data includes network map transmission (WMTS) data, network image transmission (WMS) data, and network information transmission (WCS) data. These three types of data run under the control of a user terminal, so the collected data The three kinds of data must be collected at the same time. If collected separately, it will seriously affect the working efficiency of the user side.

The collected transmission function data mainly includes 6 function data, which are display data, query data, collected data, coded data, audit data and management data [11, 12]. When collecting vector data, the established space vector database should be collected to improve collection efficiency. Including submitted vector data and unsubmitted vector data.

4 Experimental Study

In order to verify the effectiveness of the fractal image coding compression and transmission model based on wavelet transform, the performance of the fractal image coding compression and transmission model is compared with the traditional fractal image coding compression and transmission model.

Test image: couple; Lena; woman2; crowd, etc. (block segmentation), experimental platform: Windows 10 operating system (2.3 ghz i5-6300hq/8 GB memory), algorithm is implemented by MATLAB r2016a software. The experimental environment is shown in Fig. 4.

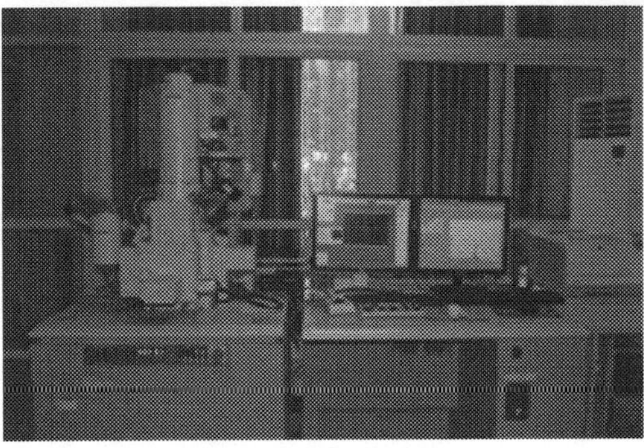

Fig. 4. Test environment

In the above detection experimental environment, in order to avoid the uniformity of experimental results, the traditional image equalization enhancement and transmission model in the compression domain based on MSR algorithm is taken as the comparison model, and it is compared with the fractal image coding compression and transmission model based on wavelet transform proposed in this paper. The number of selected images is 1200, and the size of each image is 50 cm × 50 cm, and the gray level is 256. The experimental results are as follows.

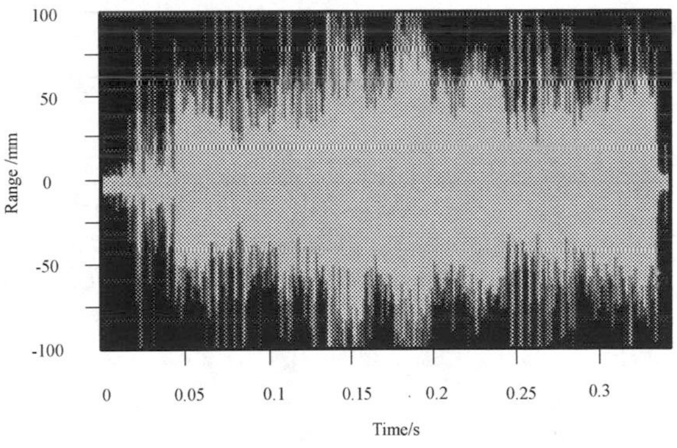

Fig. 5. Transmission waveform of traditional model

Figure 5 and Fig. 6 show the transmission waveforms of different models. By comparing Fig. 5 and Fig. 6, it can be seen that the time-frequency characteristics detected by the model in this paper are clearer than those detected by the traditional model.

The energy of the transmitted signal under normal conditions is mainly concentrated in the low frequency band. Under different damage levels, the emitted energy is distributed in the high frequency band with a wide frequency range. The entire time period shows obvious energy fluctuations. The non-stationary characteristics are obvious. There are also obvious differences in energy distribution between different degrees. In each frequency range, energy intensity and energy fluctuation elapsed time are obviously

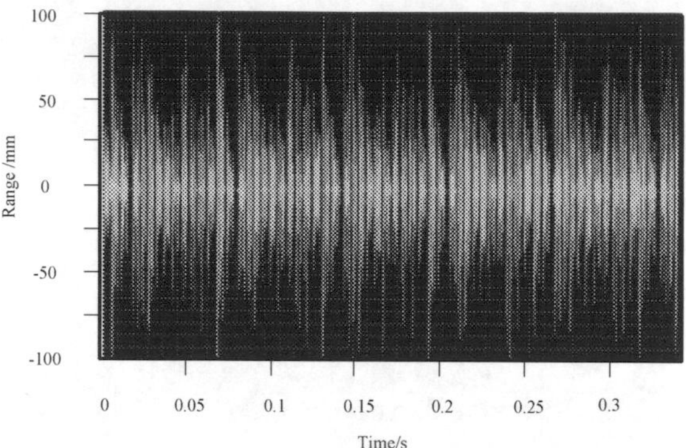

Fig. 6. Transmission waveform of this method

different. As the transmission degree increases, the vibration energy increases, and the fluctuation elapsed time decreases accordingly. It can be seen that the difference between different signals in the discrete wavelet time-frequency domain is obviously more prominent than the difference in the time domain, indicating that the discrete wavelet transform can fully display the time-frequency characteristics of the fault and better analyze the image.

It can be shown that the traditional model is difficult to judge the data encoding only by relying on the signal time domain waveform, but the model in this paper has better classification effect and can better realize the image transmission on the premise of ensuring the image quality.

5 Conclusion and Prospect

Of this study is to design a fractal image coding based on wavelet transform compression and transmission model, the secondary decomposition of the wavelet transform to image, on the basis of on the top of the decomposed sub image basic fractal coding, according to the similarity between different straton image structure, from senior fractal coding structure low straton fractal image coding, so as to realize the high quality of image transmission. Experimental results show that this model can achieve high quality image transmission on the premise of guaranteeing image quality.

However, due to the limitations of research time and equipment, the method in this paper still has a great scope for development. In the future research, the model will be further optimized from the perspective of improving transmission efficiency.

Fund Projects. National Natural Science Foundation of China: The Key Technologies about Fast Coding and Quality Controlling of Fractal Image Compression (61961036).

Natural Science Foundation of Guangxi:Research on the Key Technologies about Decoder for Reliable Transmission of HEVC for Microscopic Video(2020JJA170007).

Basic Ability Improvement Project for Young and Middle-aged Teachers in Guangxi:Research on Compound Optimization Strategy of Fractal Image Compression(2017KY0629).

Basic Ability Improvement Project for Young and Middle-aged Teachers in Guangxi: Research on fast image retrieval algorithm based on fractal coding (2018KY0537).

Basic Ability Improvement Project for Young and Middle-aged Teachers in Guangxi: Research on 3D Terrain Rendering for Large Scene Oblique Photography (2020KY17019).

Industry and Education and Research projects for Wuzhou University and Wuzhou High-tech zone: the portable remote control and HD Endoscope Camera System and cloud service platform Construction(2020G001).

Natural Science Foundation of Wuzhou University:Research on the Key Technologies of integrated navigation and positioning for GNSS receiver(2020C001) .

References

1. Zhu, S., Zong, X.: Fractal lossy hyperspectral image coding algorithm based on prediction. IEEE Access **9**(9), 1–8 (2017)
2. Cao, J., Zhang, A., Shi, L.: Orthogonal sparse fractal coding algorithm based on image texture feature. IET Image Proc. **13**(11), 1872–1879 (2019)
3. Cuiping, S., Liguo, W., Junping, Z., et al.: Remote sensing image compression based on direction lifting-based block transform with content-driven quadtree coding adaptively. Remote Sens. **10**(7), 999–1005 (2018)
4. Hualong, Y., Leihong, Z., Jian, C., et al.: Information transmission based on a Fourier transform and ascending coding temporal ghost imaging algorithm. Laser Phys. **30**(12), 125–135 (2020)
5. Wu, Y., Kampf, J.H., Scartezzini, J.L.: Performance assessment of the BTDF data compression based on wavelet transforms in daylighting simulation. Sol. Energy **190**(9), 329–336 (2019)
6. Wu, J., Jia, Y., Liu, Y.: Coverless information hiding algorithm based on image coding and stitching. Huanan Ligong Daxue Xuebao/J. South China Univ. Technol. (Nat. Sci.) **46**(5), 32–38 (2018)
7. Palaz, S., Ozer, Z., Mamedov, A.M., et al.: Ferroelectric based fractal phononic crystals: wave propagation and band structure. Ferroelectrics **557**(1), 85–91 (2020)
8. Liu, S., Lu, M., Li, H., et al.: Prediction of gene expression patterns with generalized linear regression model. Front. Genet. **10**, 120 (2019)
9. Liu, S., Liu, D., Srivastava, G., Połap, D., Woźniak, M.: Overview and methods of correlation filter algorithms in object tracking. Complex Intell. Syst. **7**(4), 1895–1917 (2020). https://doi.org/10.1007/s40747-020-00161-4
10. Liu, S., Bai, W., Zeng, N., et al.: A fast fractal based compression for MRI images. IEEE Access **7**, 62412–62420 (2019)
11. Zheng, Y.: Formal process virtual machine for smart contracts verification. Int. J. Performabil. Eng. **14**(8), 1–9 (2018)
12. Wand, L., Liu, Z.L.: A fractal image compression algorithm based on centroid features and important sensitive area classification. Comput. Eng. Sci. **42**(05), 869–876 (2020)
13. Niu, T.C., Zhang, A.H., Ji, H.F.: Sub-block average points features fast fractal image compression coding. Telecommun. Eng. **59**(03), 306–310 (2019)

Research on Intelligent Retrieval Model of Multilingual Text Information in Corpus

Ri-han Wu[1(✉)] and Yi-jie Cao[2]

[1] School of Chinese Language and Literature, Northwest Minzu University, Lanzhou 730030, China
wurihan21322@yeah.net

[2] School of Ethnology and Sociology, Northwest Minzu University, Lanzhou 730030, China

Abstract. Cross language information retrieval focuses on how to use the query expressed in one language to search the information expressed in another language. One of the key problems is to adopt different methods to establish bilingual semantic correspondence. In recent years, topic model has become an effective method in machine learning, information retrieval and natural language processing. This paper systematically studies the cross language retrieval model, cross language text classification method and cross language text clustering method. Without the help of cross language resources such as machine translation and bilingual dictionaries, it can effectively solve the many to many problem of Vocabulary Translation in CLIR and the problem of partial decomposition of unknown words. The experimental results on the cross language text classification evaluation corpus established in this paper show that the performance of cross language and single language text classification on the bilingual topic space constructed by this method is close to or better than that of single language classification on the original feature space, and the performance of cross language text clustering is close to or better than that of single language document clustering.

Keywords: Corpus · Language · Information retrieval

1 Introduction

Language and text information processing technology is a technology that transforms the language processing used by humans in interactive communication into machine language that computers can understand. It is a model and algorithm framework that uses language ability as the research object. It involves linguistics and computers. Cross-cutting research areas of science. In the "Internet+ " era, the innovation and breakthrough of this technology can not only promote the development of human-machine intelligence, bring about a revolution in computing technology, but also enable humans to further understand their own thinking and language, and pay more attention to language teaching and learning [1]. Information technology is rapidly penetrating into all levels of the economy and society and profoundly changing people's work and lifestyle. At the same time, it also brings new challenges to the improvement of the quality of information

S. Liu and X. Ma (Eds.): ADHIP 2021, LNICST 416, pp. 26–40, 2022.
https://doi.org/10.1007/978-3-030-94551-0_3

technology application of the entire nation. The competition in modern society is not only manifested in politics, economy, military, etc., but also in informatization, that is, the competition between information system and the right to speak [2]. Information technology has brought new development opportunities to the development of language itself. As the carrier of information, language information construction will be put on the national agenda, and the development of information technology will in turn promote the development of linguistics itself [3]. The current period is not only an important period of strategic opportunities for my country's modernization drive, but also an important period of strategic opportunities for the construction of language informatization. It is necessary to enhance the awareness of opportunities and the overall situation, and make full use of contemporary information technology to carry out language work in the new era.

The experimental results on the cross-language text classification evaluation corpus established in this paper show that the cross-language and single-language text classification performance completed on the bilingual topic space constructed by this method is close to or better than the single-language classification of the original feature space.

2 An Intelligent Retrieval Model of Multilingual Text Information in Corpus

2.1 Corpus Multilingual Text Information Feature Collection

A key issue in information retrieval is a variety of semantic representations of a description object, which are expressed as polysemous words, synonyms and synonyms in language form [4]. Even if the words or phrases contained in the query appear in the document, they may indicate another meaning because of the different context. The rich semantic representation of natural language [5] increases the difficulty of the IR system to retrieve and query related documents. In cross language information retrieval or multi language information retrieval, queries and documents are expressed in different languages. In addition to the semantic combination of words or phrases in a single language, there are also cross language semantic combinations, which makes it more difficult to retrieve related documents [6]. The overall strategic structure of language information construction is shown in the following figure. The implementation of the overall strategic plan relies on the construction of the information chemistry department and the construction of standards. The application support platform is based on the construction of a resource library, and the service system is built on the application support platform. It is not appropriate to pursue informatization construction in one step. It is necessary to combine advancement and practicability, make overall planning, and implement step by step. Based on this, the corpus multilingual text information database model is constructed, as follows (Fig. 1):

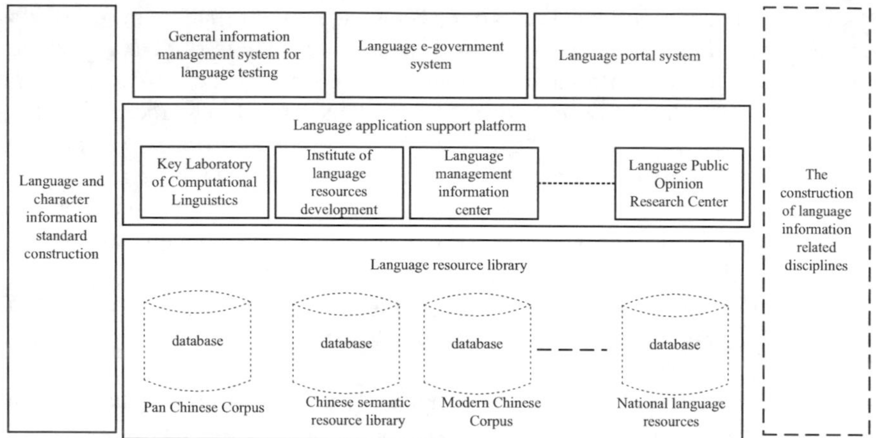

Fig. 1. Corpus multilingual text information database

The construction of language and text resource base is the foundation of information strategy research. The construction of resource base will change from word annotation database, tree database to semantic knowledge base; from text resource base to multimedia and multimodal resource base integrating voice, video and text [7]. It is not only necessary to establish a balanced resource database, but also to pay attention to the establishment of a multi-language simultaneity data database [8]. It is not only necessary to establish a domestic language resource monitoring corpus, but also to pay attention to the establishment of a language resource monitoring resource database in the entire Chinese-speaking region. It is not only necessary to establish a written language resource database, but also to pay attention to the establishment of a spoken language and dialect resource database. In addition, it is necessary to strengthen the sorting and inventory of existing databases, avoid duplication of construction, and take the road of intensive development.

Preserve, monitor, analyze and utilize language phenomena and language resources through the construction of a resource database, provide data services for social language and writing applications, provide experimental objects for scientific research, and provide decision-making basis for the formulation of national language and writing guidelines and policies [9, 10]. The object of language description "cross-language information retrieval" is expressed in five languages, and the form of expression is to describe the object using different language symbol systems endowed with rich semantics. Essentially, it is a multi-view representation formed by "cross-language information retrieval" on different language symbol systems [11, 12]. It can be seen from this that in natural language, for the same semantic object, using different languages (or languages) for text representation is essentially a different representation of the object, and different views formed from different sides (different languages). (Multilingual documentation). Based on this, the corresponding relations of different semantic granularities of the parallel corpus are displayed, as follows (Fig. 2):

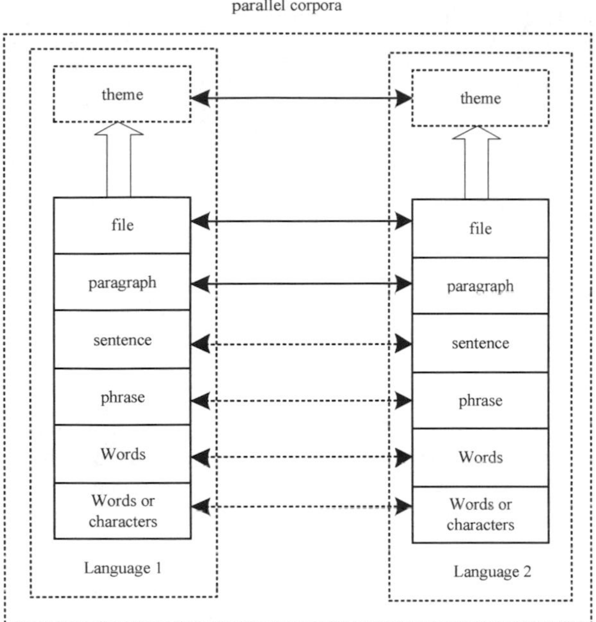

Fig. 2. Correspondence of different semantic granularity of parallel corpus

As shown in the figure, in the bilingual parallel corpus, it is divided according to the actual semantic granularity in the document set. From small to large, the granularity can be divided into words (or characters), words, phrases, sentences, paragraphs and documents. In the parallel corpus, there are semantic correspondences at each level of semantic granularity. Generally speaking, except for the document and paragraph levels, there is no strict bilingual semantic equivalence at the other levels. In the figure, there is no strict bilingual semantic equivalence relationship indicated by a dashed arrow, and a strict bilingual semantic equivalence relationship is indicated by a solid arrow [13]. For bilingual words or characters, the semantically equivalent forms include time, number, datè and other language-independent content. Because of the differences in language expression, it is more difficult to align words, phrases, and sentences. However, there are many identical expressions in the languagesof the same language family, such as English and French. They can achieve the translation correspondence between words and phrases relatively easily through homology matching methods. If the corpus is aligned with sentences or words (phrases), the semantic correspondence between words, phrases, and sentences in bilinguals can be easily extracted in cross-language information retrieval and machine translation.

2.2 Corpus Multilingual Text Information Evaluation Algorithm

There are many commonly used evaluation indicators for the performance of information retrieval systems. In addition to the use of these evaluation indicators, the cross-language information retrieval system also uses the ratio relative to the performance of a single language as an evaluation indicator. The most basic evaluation indicators are precision rate and recall rate. The accuracy rate is the ratio of the number of related documents in the number of retrieved documents:

$$\text{Accuracy} = \frac{\text{Number of related documents in search results}}{\text{Number of documents retrieved}} \tag{1}$$

The recall rate is the ratio of the number of documents retrieved in the number of related documents:

$$\text{Recall} = \frac{\text{Number of documents retrieved}}{\text{Number of related documents}} \tag{2}$$

The relationship between the precision rate and the recall rate is that when the precision rate increases, the recall rate decreases; conversely, when the recall rate increases, the precision rate decreases. In order to fully evaluate the performance of the model, the average precision rate at the 11-point recall rate is usually used. When the recall rate is 0, the precision rate is obtained by interpolation. Another widely used evaluation index is the average accuracy rate, MAP, whose calculation method is:

$$\text{MAP} = \frac{1}{M}\sum_{j=1}^{M}\frac{1}{N_j}\sum_{i=1}^{N_j}\text{pr}\left(d_{ij}\right) \tag{3}$$

$$\text{pr}\left(d_{ij}\right) = \begin{cases} \frac{r_i}{n_i} & \text{if } n_i \leq \text{MAX} \\ 0 & \text{otherwise} \end{cases} \tag{4}$$

Among them, n is the ranking of the Chinese documents in the search ranking results of the documents related to the query J, 1n is the number of related documents found with a ranking of n, M is the number of queries, and MAX is the ranking threshold (the TREC evaluation generally takes a value of 1000). Calculate the top n documents in the document retrieval ranking results to get the PN, which mainly reflects the direct evaluation of the user on the retrieval results. The correlation degree is obtained by calculating the similarity of the vector. The commonly used similarity calculation method is the angle cosine, which is defined as follows:

$$\text{sim}(\overline{D}, \overline{Q}) = \frac{\text{pr}\left(d_{ij}\right)\overline{D}\cdot\overline{Q}}{|\overline{D}| \times |\overline{Q}|} \tag{5}$$

Among them:

$$|\overline{D}| = \sqrt{\sum_{i=1}^{n}(d_i)^2} \tag{6}$$

Further construct a simple probability model and a binary independent retrieval model. The BIR model assumes that the terms are independent of each other. The documents are sorted according to the odds ratio:

$$\text{Score}(Q, D) = \log\frac{P(\text{rel}|D, Q)}{P(\text{irrel}|D, Q)}$$

$$= \log \frac{P(D|Q,\ \text{rel})P(\text{rel}|Q)}{P(D|Q,\ \text{irrel})P(\text{irrel}|Q)}$$
$$= \propto \log \frac{P(D|Q,\ \text{rel})}{P(D|Q,\ \text{irrel})} \tag{7}$$

The document D is represented as a collection of independent binary events xn, and $x = 1$ represents the term x, which does not appear in the document. Then, from the formula:

$$\text{Score}(Q, D) \propto \sum_{x_i \in D} \log \frac{P(x_i = 1|Q, rel)^{x_i}(1 - P(x_i = 1|Q, \text{rel}))^{1-x_i}}{P(x_i = 1|Q,\ \text{irrel})^{x_i}(1 - P(x_i = 1|Q,\ \text{irrel}))^{1-x_i}}$$
$$= \sum_{x_i \in D} x_i \log \frac{P(x_i = 1|Q, \text{rel})(1 - P(x_i = 1|Q,\ \text{irrel}))}{P(x_i = 1|Q,\ \text{irrel})(1 - P(x_i = 1|Q, rel))} + Const \tag{8}$$

Given such a document sample set, a contingency table for each term t can be calculated. Assuming that ni is the total number of documents in the sample set and R is the number of related documents, then (Table 1):

Table 1. Co-occurrence contingency table of information terms

Name	Relevant	Irrelevant	Total
Number of documents containing t	r_i	$n_i - r_i$	n_i
Number of documents without t	$R - r_i$	$N - n_i - (R - r_i)$	$N - n_i$
Total	R	N − R	N

The following probabilities can be derived from the table:

$$P(x_i = 1|Q, rel) = \frac{r_i}{R} \tag{9}$$

$$P(x_i = 1|Q,) = \frac{n_i - r_i}{N - R} \tag{10}$$

2.3 Realization of Multilingual Information Retrieval in Corpus

Based on the overall framework construction method described above, after we construct a bilingual topic space from a bilingual parallel corpus, we can achieve specific cross-language retrieval tasks. Without loss of generality, suppose the query expressed in language L1 retrieves the document set expressed in language L2. As shown in the figure, the general process of cross-language retrieval is (Fig. 3):

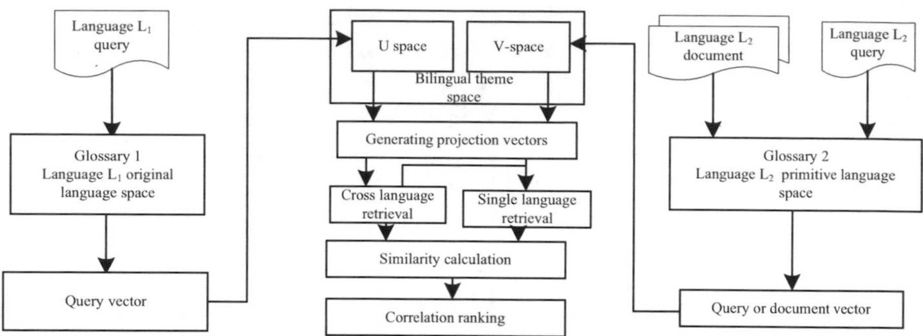

Fig. 3. The general process of cross-language retrieval under the overall framework

Perform simple similarity (such as angle cosine) or retrieval model calculation of similarity to query vector and document vector. According to the similarity calculated in the previous step, the retrieved documents are sorted based on the relevance ranking evaluation function, and the relevance ranking document list is returned. Information retrieval can be simply described as: According to the user's information needs, a query string is constructed and submitted to the information retrieval system. Then the system

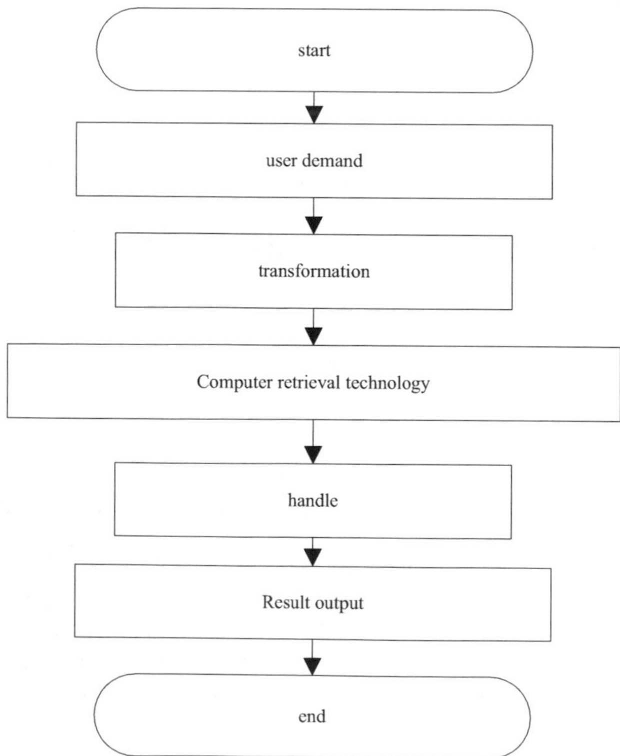

Fig. 4. Language and text information recognition steps

retrieves the document information related to the user's query from the relatively stable unstructured or semi-structured text data set, and sorts it according to the relevance from high to low. Finally, the sorted retrieval results are returned to the user, which is also called document retrieval. The identification steps of database information retrieval technology are shown in the figure (Fig. 4):

Information retrieval technology adopts the conversion technology of demand recognition between computer and user, involving retrieval structure, programming language, personalized demand, agent intelligence, information filtering, machine translation and other technologies. As can be seen from the figure, information retrieval is based on the user's information, enter the question in the computer, and then match the retrieved question with the identifier between the storage. The core is mainly to meet the needs of the retrieval subject, and then calculate the identified retrieval expression. The search expression can use the position operator and the restriction operator to combine and match the key words of the question, thereby determining the concept and position of the search key words, expressing the accurate content of the user's question, and ensuring the accuracy of the search accuracy.

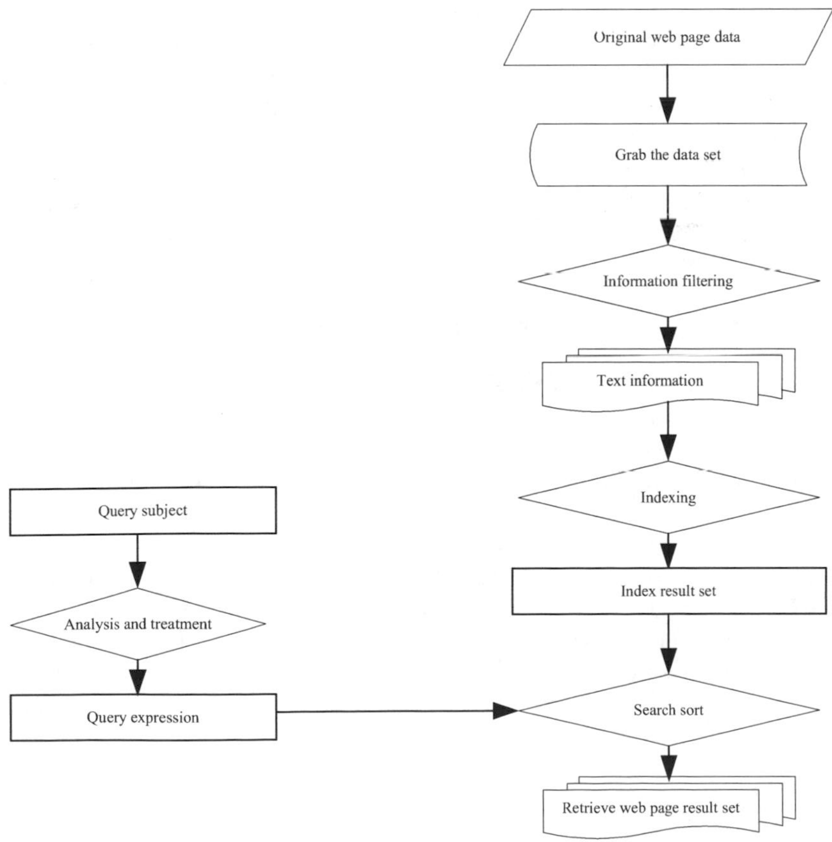

Fig. 5. Language and text information retrieval steps

Information retrieval includes clarifying users' information needs, information retrieval methods and techniques, and whether they can meet users' information needs. Among them, clarifying user information requirements, that is, user query, is a prerequisite for information retrieval, which is equivalent to when we perform a task, we must first clearly understand the requirements of the task, so as not to deviate from the direction. The method and technology of information retrieval is a means used in order to better achieve the goal during information retrieval, including some models and methods of information retrieval. Whether the information needs of users can be met is to evaluate the results of information retrieval to see how well it matches the needs of users. The higher the matching degree, the closer to the information needs of users. The basic process of general information retrieval is shown in the figure (Fig. 5):

In cross-language information retrieval, corpus is a very important basic data resource. As far as we know, there is no Chinese-English bilingual parallel corpus for cross-language information retrieval and evaluation. Regarding bilingual parallel documents as two views of the same semantic content, and assuming that they share the same semantic information, a cross-lingual information retrieval framework is proposed. The framework can extract the semantic representations of the same semantic object at various representation levels from the bilingual parallel corpus, construct a topic space representing linear or non-linear bilingual correspondence, and perform cross-language retrieval, multi-language retrieval, cross-language text classification, and cross-language text classification. Language and text clustering, etc., and lay the foundation for subsequent research content. When using the statistical language model for information retrieval, the query likelihood scoring method is defined as a conditional probability, that is, the statistical query Q is in the model M. Probability of generation under this condition. Therefore, the model of the query likelihood scoring method can be expressed as follows:

$$\text{Score}(Q, D) = p(Q|M_D) \tag{11}$$

In order to build a cross-language text classification corpus, firstly, some Chinese and English documents are clustered using K-Means algorithm to determine the document category. The SVM classification model is used to train the documents with the marked categories, and the unmarked documents are marked with categories. Perform manual category marking for proofreading and adjustment according to the content of the document. In order to establish a cross-language information retrieval corpus, the query content is determined according to the content description of the document set, and a standard query set is established. Based on the established query set, the probabilistic retrieval model is used to retrieve relevant documents to provide a preliminary basis for judging the relevance of documents. According to the retrieval results of the BM25 model, the document relevance is manually judged. Specific steps are as follows (Fig. 6):

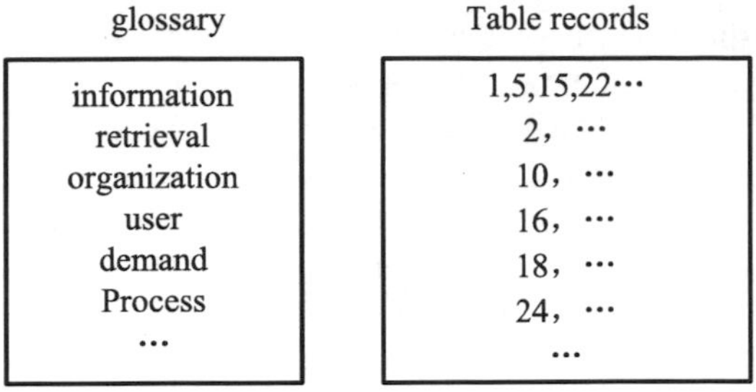

glossary	Table records
information	1,5,15,22···
retrieval	2, ···
organization	10, ···
user	16, ···
demand	18, ···
Process	24, ···
···	···

Fig. 6. Indexing process of text information files

Because a record table involves every word that appears in the text and its position in the text, the record table occupies a relatively large space. Retrieval using an inverted index is usually divided into three steps: vocabulary search, record table search and record table operation, which can do operations such as inserting, deleting, and updating documents. When searching with an inverted index, you only need to search the inverted table to obtain the documents required by the user. This search method is much faster than matching all documents linearly, thereby better guaranteeing the information retrieval effect.

3 Analysis of Results

In order to verify the actual application effect of the corpus multilingual text information intelligent retrieval model, the experimental environment is set up to ensure the detection effect. The operating environment of the experiment is:

Software Environment:
Operating system: Windows732 bit; Development tool: MicrosoftVisualStudio2008; Development language: C++; Search tool: Lemur.
Hardware environment:
System Model: LENOVOThinkCentreM8200T; processor: Intel(R)CoreM)Core 15650; Memory: 4 GBDDR2 experimental environment settings as shown in the table (Table 2).

Table 2. Experimental parameter settings

Parameter	Parameter
Host memory	4 GB
CPU frequency	3.50 GHz
Operating system	64 bit
Program running platform	Visual studio

The experimental reference database uses data collected randomly from 2000 networks, and the above data is repeatedly loaded 20 times to reach a database with a scale of 40,000. The experimental data are randomly selected from the database. The data used in the word sense disambiguation module of this article comes from the training corpus released by the 2009 National Statistical Machine Translation Conference. The corpus contains 67288 double sentences and their corpus. Through comparative experiments on the information retrieval system, a series of test data are obtained, and the reasons why the statistical language model is better than the traditional information retrieval model are analyzed, and the three common smoothing techniques in the statistical language model are also compared and analyzed. And lists the results of several common evaluation indicators (Table 3).

Table 3. IFIDF common index evaluation results

Evaluation index	TFIDF evaluation results
map	0.3309
ndcg	0.6064
R-prcc	0.3465
bpref	0.8732
P10	0.2904

In order to generate a more intuitive result image, Lemur's own GUI program is used to graphically represent the evaluation results, and the graphical interface part is developed with Java/Swing. The following figure shows the recall rate based on Lemur's own corpus, query subject and standard set-accuracy rate and the accuracy rate of the top N results returned by the system for the query (Fig. 7).

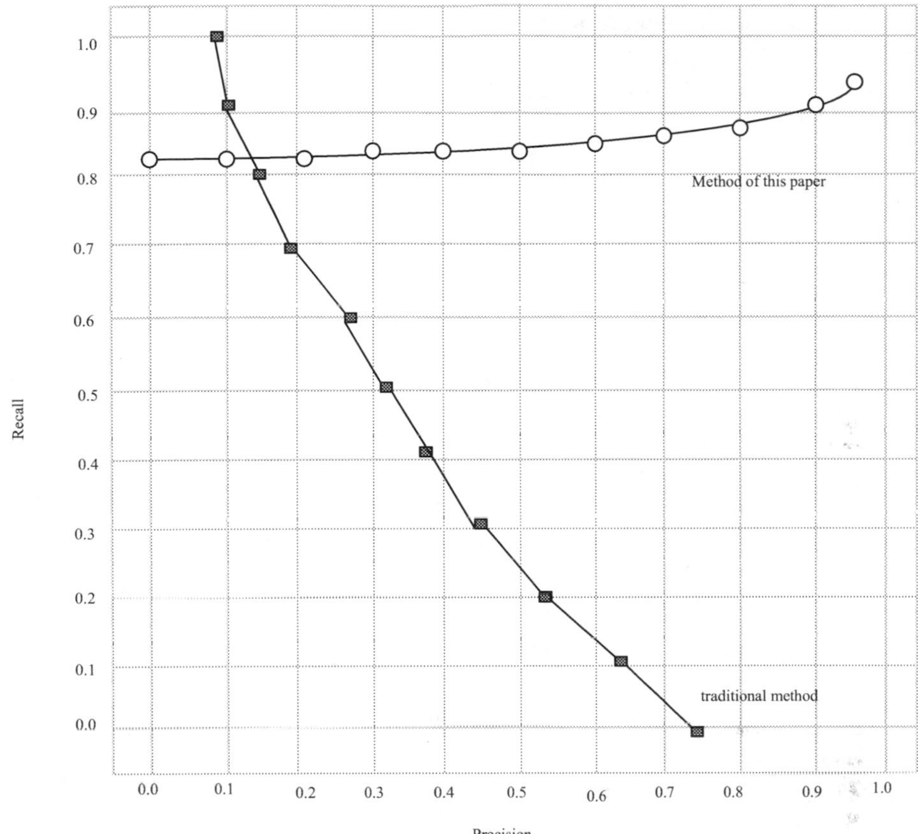

Fig. 7. Comparison of experimental monitoring results

Information retrieval evaluation can evaluate the pros and cons of different technologies and the impact of different factors on the system, thereby promoting the continuous improvement of the research level in this field. In the experiment, the three models of vector space model, probability model and statistical language model as well as the three smoothing methods of Jelinek-Mercer, Absolute Discounting and Dirichlet Prior in the statistical language model were retrieved in the corpus CWT200G, using MAP, P@N and R -Precision is the evaluation index. When searching, use the homepage as the entry point, and require the top ten results to include as many sites as possible. Therefore, the experimental results after treatment are shown in the table (Table 4):

Table 4. Retrieval model evaluation results

Evaluation model Evaluation index	TFIDF	OKPAI	KL-JM	KL-Abs	KL-Dir
MAP	0.1504	0.1387	0.1655	0.1728	0.1810
P@10	0.1182	0.1418	0.1768	0.1843	0.2098
P@20	0.0991	0.1136	0.1551	0.1693	0.1777
R-Precision	0.0446	0.0492	0.0603	0.0619	0.075

The traditional retrieval system is further compared with the recall rate of the system designed in this paper. In this experiment, segmentation, indexing and retrieval are all carried out by block processing. Therefore, in the retrieval process, each piece of data is retrieved, and the first 200 records are retained for each query, and then all the results are combined, and the results are sorted again according to the correlation score, and then the top 200 records are selected, so the most relevant results are basically included in the 200 records. The results are shown in the table (Table 5).

Table 5. Comparison results of recall rates of the two systems

Name	Traditional retrieval method	Article retrieval method
Decimation number	2165	2987
Correct number	1600	2987
Retrieval accuracy /%	52%	100%
System recall rate /%	60%	100%
Retrieval time / ms	2260	2100

For a large-scale corpus, it is impossible to list the accuracy of each query, so the calculation of the recall rate is not very accurate. Therefore, this article selects the accuracy as the evaluation index, which is simple and intuitive. For a given set of queries, the overall accuracy of the language model proposed in this article is higher than the previous two models. It is obvious that the retrieval performance of the model in this paper is better.

4 Concluding Remarks

The construction of language informatization is a basic work related to the long-term development of national informatization and information industry. Combined with the development status and practical application of language informatization, this paper makes a Macro Thinking on the construction of language informatization, and further discusses the key points, tasks and specific measures of the construction of language informatization in China. The construction of language informatization should closely focus on the information application needs of national economic and social development, adhere to the principle of demand traction and technology promotion, and constantly contribute to the construction of a harmonious language life, the improvement of national cultural quality, the enhancement of the soft power of national sustainable development, and the promotion of all-round economic and social progress.

There is a large semantic gap between the low-level data features and the high-level semantic features of multimodal data sets, which makes the retrieval accuracy and retrieval quality need to be further improved. Most of the research focuses on constructing the neighbor structure based on the category label information and the similarity matrix, and maintaining semantic consistency. Perhaps a more perfect algorithm suitable for this field can be proposed based on the particularity of cross-modal data.

Fund Projects. 1. This paper is supported by the Postgraduate Research and Innovation Project of Northwest Minzu University , the name of the project is "A study on the names and descriptions of the things of mongolian <qin ding li fan yuan ze li>" (project number is Yxm202009), which is the phased achievement of this project.

2. This paper is supported by "the Fundamental Research Funds for the Central Universities-A Study on the Influence of Ecological Protection Policy of Qilian Mountains on Surrounding Herdsmen from the Perspective of Ecological Safety", which number is 31920190133.

References

1. Zou, J., Kanoulas, E.: Towards question-based high-recall information retrieval: locating the last few relevant documents for technology-assisted reviews. ACM Trans. Inf. Syst. **38**(3), 1–35 (2020)
2. Kim, H., Cha, M., Kim, B.C., et al.: Part library-based information retrieval and inspection framework to support part maintenance using 3D printing technology. Rapid Prototyping J. **25**(3), 630–644 (2019)
3. Kanwal, S., Malik, K., Shahzad, K., et al.: Urdu named entity recognition: corpus generation and deep learning applications. ACM Trans. Asian Lang. Inf. Process. **19**(1), 8.1-8.13 (2020)
4. Rascon, C., Ruiz-Espitia, O., Martinez-Carranza, J.: On the use of the AIRA-UAS corpus to evaluate audio processing algorithms in unmanned aerial systems. Sensors **19**(18), 3902 (2019)
5. Rojc, M., Mlakar, I.: A new unit selection optimisation algorithm for corpus-based TTS systems using the RBF-based data compression technique. IEEE Access **7**(10), 1 (2019)
6. Mishra, S., Soni, D.: Smishing detector: a security model to detect smishing through SMS content analysis and URL behavior analysis. Futur. Gener. Comput. Syst. **108**(10), 803–815 (2020)
7. Oh, I., Kim, T., Yim, K., et al.: A novel message-preserving scheme with format-preserving encryption for connected cars in multi-access edge computing. Sensors **19**(18), 3869–3870 (2019)
8. Deng, N., Deng, S., Hu, C., et al.: An efficient revocable attribute-based signcryption scheme with outsourced unsigncryption in cloud computing. IEEE Access **8**(10), 42805–42815 (2020)
9. Ferdinando, D.M., Sabrina, S., Salvatore, S.: A lightweight clustering–based approach to discover different emotional shades from social message streams. Int. J. Intell. Syst. **34**(7), 1505–1523 (2019)

10. Jia, Z., Jafar, S.A.: On the asymptotic capacity of X-secure T-private information retrieval with graph-based replicated storage. IEEE Trans. Inf. Theory **66**(10), 6280–6296 (2020)
11. Bhattacharya, P., Goyal, P., Sarkar, S.: Using communities of words derived from multilingual word vectors for cross-language information retrieval in indian languages. ACM Trans. Asian Lang. Inf. Process. **18**(1), 1.1-1.27 (2019)
12. Liu, S., Bai, W., Liu, G., et al.: Parallel fractal compression method for big video data. Complexity **2018**, 2016976 (2018)
13. Liu, S., He, T., Dai, J.: A survey of CRF algorithm based knowledge extraction of elementary mathematics in Chinese. Mobile Netw. Appl. **26**, 1891–1903 (2021)

The Dynamic Monitoring System of Document Bibliography Information Based on Wireless Communication

Ri-han Wu(✉)

School of Chinese Language and Literature, Northwest Minzu University, Lanzhou 730030, China
wurihan21322@yeah.net

Abstract. The dynamic monitoring of changes in the bibliographic information is an important task for insight into the development of informatization in the field of science and technology. Based on this, a wireless communication-based dynamic monitoring system for document bibliographic information is designed. By optimizing the system hardware equipment, the system operation efficiency is improved, and the system software functions and operating procedures are further optimized by combining wireless communication technology. Through the collection of information and knowledge extraction, information analysis and other technologies can effectively reveal the important dynamic information of the target scientific research institution in terms of strategic planning, research layout, important research progress, etc., and in-depth monitoring of the dynamic changes of literature bibliographic information. Finally, it is confirmed by experiments that the dynamic monitoring system of bibliographic information based on wireless communication has high effectiveness in the actual application process and fully meets the research requirements.

Keywords: Wireless communication · Document bibliography · Information dynamic monitoring

1 Introduction

Dynamic monitoring of bibliographic information can effectively query the changes of science and technology in real time. The national science library requires each strategic information research team to continuously track and accumulate the changes of literature information in their respective fields, so as to effectively provide multi-level literature information services and grasp the major trends and strategic trends of scientific and technological development. Literature [1] designed a Scrapy-based network space information dynamic monitoring system, with the help of the information collection layer in the system, from various information sources and Scrapy web crawlers in the network, the required network space information was collected, and the information collection layer was constructed based on the Scrapy framework Establish the

S. Liu and X. Ma (Eds.): ADHIP 2021, LNICST 416, pp. 41–55, 2022.
https://doi.org/10.1007/978-3-030-94551-0_4

request response to the web page, analyze the web page layout and cyclically capture the required cyberspace information. Use regular expressions to remove the captured cyberspace information as formatted data, and store the collected data in the cyberspace information processing module database In. The network information processing module uses reasonable methods to purify the webpage where the information is located, and transfers the content extracted from the database to the network spatial information analysis module. The analysis module uses text clustering, feature selection and other technologies to explore the spatial information in the network and form a topic List and set up keyword information, and establish a cyberspace information lexicon in related fields. Use WM algorithm to process homophones, split words and interference symbols, to achieve accurate matching of required cyberspace information, and achieve the purpose of dynamic monitoring of cyberspace information.

However, the above-mentioned systems have low efficiency in network information monitoring. Based on this, combined with wireless communication technology, the literature bibliographic information dynamic monitoring system is optimized to optimize the functions of the automatic information monitoring service system, simplify the dynamic information monitoring process, and analyze through simulation experiments. The validity of the system in this paper is verified, and the problems existing in the traditional system are solved.

2 Literature Bibliographic Information Dynamic Monitoring System

2.1 The Hardware Configuration of the Dynamic Monitoring System for Bibliographic Information

The information collection layer of the document bibliographic information dynamic monitoring system mainly realizes the regular collection and harvesting of target resources. In the document bibliographic information dynamic monitoring system, it is necessary to build a series of distributed network directional collectors to achieve accurate collection of target resources [2]. The monitoring calculation and analysis layer is the main functional layer of the monitoring system. For each information resource collected by the network directional collector, the monitoring system first identifies new resources, and from the collected resources, it identifies new resources that were not in the original system [3]. For the identified new HTML page, the structure of this HTML page is analyzed, and the main text content of this page is extracted [4]. If this resource is not an HTML page, but a rich document resource such as PDF and WORD, the monitoring system will automatically realize the analysis of this rich document resource and extract the corresponding text content. After that, the monitoring system then implements title and abstract extraction, monitoring target object and object relationship extraction, and domain term extraction for this resource, transforming the original free text into structured data with certain semantic support, and incorporating it into the corresponding semantic knowledge base Stored in [5]. For these resources that have been expressed by structured monitoring objects and object relationships, the monitoring system can adopt corresponding wireless communication methods to realize the intelligence value judgment of these resources to reveal important intelligence resources; through automatic classification tools, the intelligence type can be realized. Recognition; through semantic similarity calculation, to reveal the clustering and distribution relationship between intelligence. The monitoring service provider layer is the layer that provides automatic monitoring services to strategic intelligence personnel [6]. For the new resources collected from the network, through various processing of the monitoring calculation and analysis layer, a specific service function is formed at the monitoring service providing layer, including the disclosure of important information content, the tracking of key monitoring targets, the disclosure of hot topics and hot objects, The overall situation of the organization is revealed, etc., through the friendly user interface, it serves the strategic intelligence monitoring [7]. Based on the business level of the above-mentioned network technology information automatic monitoring system, the corresponding system hardware structure framework is proposed, as shown in the figure:

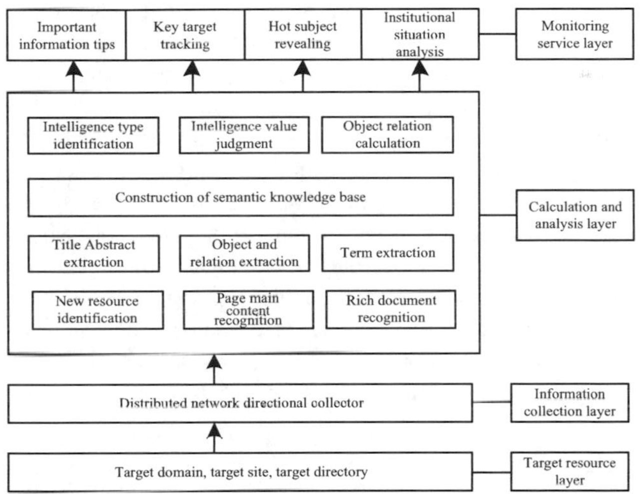

Fig. 1. Hierarchical structure of information automatic monitoring system

The main function of the monitoring information collection subsystem (as shown in Fig. 1) is to discover and collect new content from the target website in real time and accurately. The specific technical realization mainly has the following functional characteristics:

Distributed collection architecture realizes a scalable monitoring collection platform [8]. The central control server controls the collection strategy and collection cycle, and distributes the collection tasks to each collection node in a unified format, and the central control server performs load balancing control on each collection node at the same time.

Fine-grained collection methods are used to accurately collect target resources [9]. Different from the traditional full-site collection system, the monitoring subsystem starts from the particularity of the application, refines the monitoring collection resources from the site to the catalog, and conducts targeted collection according to different catalog page formats and URL characteristics.

Adopting efficient and accurate new resource identification technology. In addition to using the traditional MD5 code to check the text and extract page fingerprints, the monitoring collection subsystem also uses the scientific research objects extracted from the network information resources to calculate the similarity of the pages to realize the identification of new resources.

In network monitoring, URLs of monitored institutions and directories will change due to factors such as website revisions or adjustments to related modules. At this time, timely early warning and feedback mechanisms are particularly important, which can help strategic intelligence personnel to monitor the monitoring in a timely manner. The target is adjusted [10]. In the WSN data processing system, the upper application system mainly provides user services including model deployment, data playback, and situation analysis. These services are based on the validity and accuracy of the data in the data processing system. The core functions of data processing mainly include data fusion, data compression, data association, data completion, data compensation, data storage and query, etc. These core functions provide basic functional support for upper-layer applications. Among them, data fusion and data compression are necessary data processing to save hardware resources during data transmission; data association is to restore the data after fusion and compression.; Data completion and compensation is a kind of data post-processing to ensure the accuracy and effectiveness of the data in order to ensure that the data is accurate and effective; data storage and query are basic data services. According to requirements, the software in the data processing system adopts a modular design method, which integrates deployment model generation, data post-processing and data query and other related modules to encapsulate the large and complex data processing process based on WSN to form an organic whole. The software structure of the WSN-based dynamic monitoring data processing system is shown in the figure. The data processing system is composed of four parts: a system manager, a database,

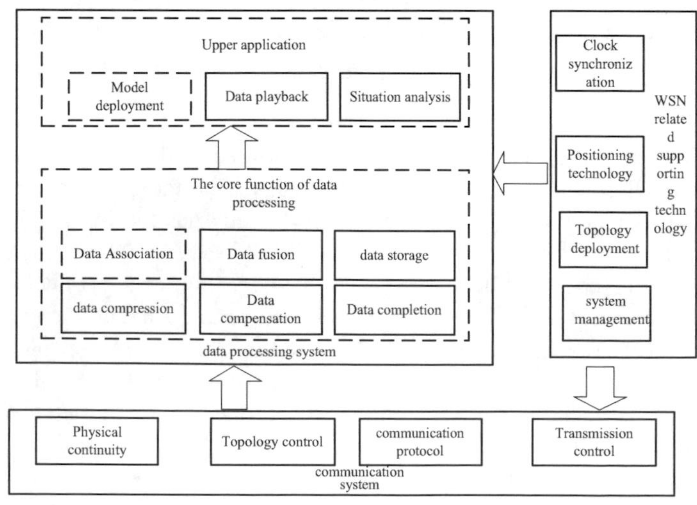

Fig. 2. The frame structure of the dynamic monitoring data processing system

a functional component and a user interaction interface, and each part completes the transmission and management of information through a public interface (Fig. 2).

The system structure design adopts a design model similar to MVC, which separates system management, functional components and user interaction interfaces, effectively guarantees the integration between modules, and provides convenience for system development and later maintenance. The functional components of the system adopt a modular design, which is divided into a deployment model generation module, a data post-processing module, and a data query module to realize data processing at different stages.

2.2 Literature Bibliographic Information Dynamic Monitoring System Software Function

The data processing process in WSN nodes and gateways is called data pre-processing. Compared with data pre-processing, the processing of data after the data is transmitted to the gateway is called data post-processing in this article. Each module is divided into more specific sub-modules according to its function to achieve high cohesion of the internal functions of the module. The model deployment generation module is composed of model data generation and model parameter setting modules; the data post-processing module is composed of data analysis and association, data compensation and data completion modules; the data query module is composed of integrated query and data playback modules.

The database of the system mainly includes original database, process database, result database, model database, algorithm database and rule database. Among them, the original database stores the original data transmitted by WSN, the process database and the effective database store some data in the data processing process and the effective data after analysis, and the model library stores the basic model and algorithm library of the large deformation flexible body of the airdrop equipment. Stored are the algorithms used in the data processing process, and the rule base is the rules used in the data parsing process.

The user interaction interface component of the system is to provide the user's operation interface for the user to use. When users use software, they pay more attention to the data in their domain and do not care about the process and logic of the software itself. Therefore, the user interface encapsulates the three functional modules, and only provides users with the data and processing processes they care about (Fig. 3).

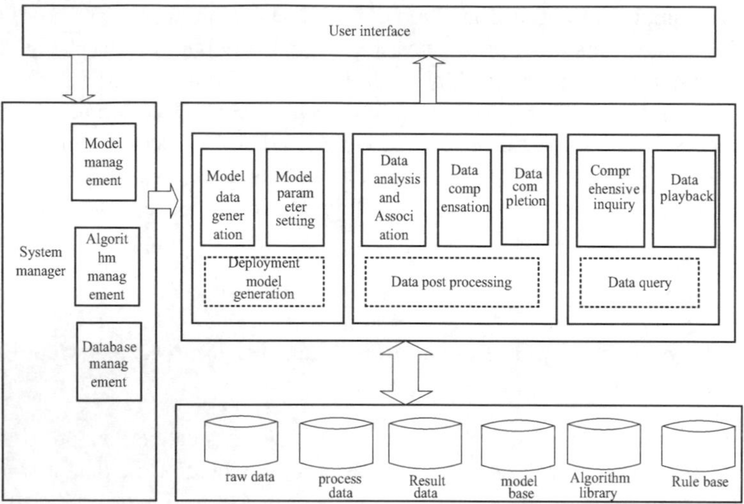

Fig. 3. Software structure of dynamic monitoring data processing system

The three parts of database, functional components and user interaction interface complete the interaction between each function through the control and coordination of the system management module. The user operates the system through the interaction interface and feeds it back to the system manager. The system manager determines the needs according to the user's operation. After obtaining the instructions of the manager, which functional component to use, which algorithm to choose, and which data is required, the functional component selects the relevant functional module and database as needed to complete the operation, and presents the operation result on the user interaction interface. The following will design and introduce the workflow of the data processing system according to the interactive process. The data post-processing module receives the monitoring data transmitted from the communication interface, and imports these data into each data table built to complete the storage of the local original data. Analyze and associate the original data, establish the internal index of the module, load the original data into the module through the index information, perform operations such as denoising and removing anomalies on the data, and store the parsed effective data. Call the core algorithm to effectively compensate and complete the data, and store the process data for processing and analysis. Analyze the compensated and supplemented data, and store the analysis results.

The data query module can replay the analyzed and processed raw data, help users understand the various states of the large-deformation flexible body of the airdrop and airborne equipment, and provide data support for the development of the large-deformation flexible body (Fig. 4).

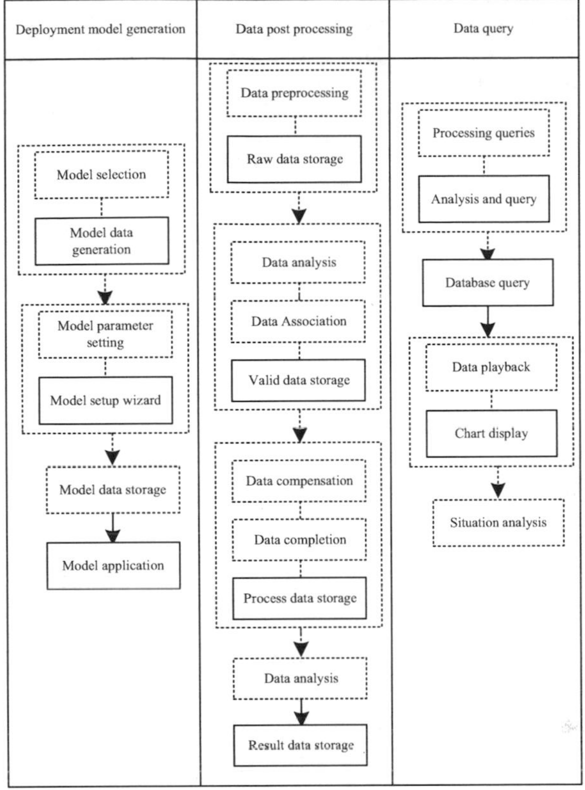

Fig. 4. The flow of the dynamic monitoring data processing system

When a certain data is not updated after multiple searches, it means that it has fallen into a local optimum, and a feasible solution is randomly searched for as the new dependent data, and a new search is restarted. Through the division of labor and cooperation, a better information source can be obtained. The specific algorithm is as follows.

$$\min\left(f(x) : x \in S \subset R^d\right) \tag{1}$$

Where $x = (x_1, x_2, \ldots, x_d)$ is the variable to be optimized, f(x) is the objective optimization function, and S is the solution space, then:

$$S = \{\left(x_j^{\min}, x_j^{\max}\right)\Big| j = 1, 2, \ldots, d)\Big\} \tag{2}$$

Further calculations can be:

$$v = r(-1, 1) \times \left(x_{ij} - x_{\text{neighberw}\ .j}\right) \tag{3}$$

$$x_{ij}^{\text{new}} = x_{ij} + v \tag{4}$$

After each round of searching, observe the information to judge the income of the information source, and make selections based on probability to speed up the convergence of the algorithm.

$$h_j = g\left(\sum_{i=1}^{I-1} w_{ij}t_i + Sx_{ij}^{\text{new}} - v\right) \tag{5}$$

Among them, w_{ij} is the connection weight between the i-th neuron in the input layer and the j-th neuron in the hidden layer, a is the threshold of the j-th hidden layer neuron, if h is the j-th neuron in the hidden layer The output of the excitation function is as follows:

$$g(x) = \frac{1}{1 + he^{-x}} \tag{6}$$

Then the output of wireless communication is:

$$p = \sum_{j=1}^{J} h_j c_j - b \tag{7}$$

The mean square error of data compensation wireless communication is:

$$E = \frac{1}{2}\sum_{k}^{s} e_k^2 \tag{8}$$

According to the gradient descent method in the feedback adjustment mechanism of wireless communication, the specific calculation process is as follows:

$$\Delta w_{\text{neighbour},j} = \begin{cases} \delta H_j(1 - H_j)t_{\text{neighbour}}\, c_j e \\ \delta H_{j-I*J} e \\ \delta H_{j-(I+1)*J}\left(1 - H_{j-(I+1)*J}\right)c_{j-(I+1)*J} e \\ \delta e \end{cases} \tag{9}$$

Based on the above algorithm, further use the document ontology storage query method to store the constructed ontology document. The specific steps are as follows (Fig. 5):

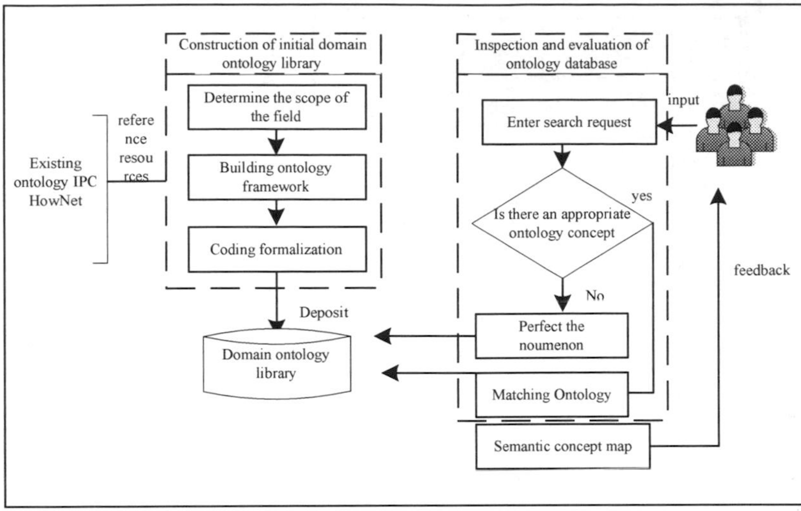

Fig. 5. The flow chart of the ontology construction of the document bibliographic information dynamic monitoring system

2.3 Realization of Dynamic Monitoring of Bibliographic Information

With the computerization of information management, people's use and reliance on databases has become more and more common. Although the database building software is becoming more intelligent and easier to use, many organizations can build a database system that meets its characteristics. However, the design of the database table is very important at the beginning of the development. If you do not pay attention to it, it will cause the later stage of the system development. There were some unsatisfactory points in the created table, and finally the table had to be modified. The program made before must also be modified due to the modification of the table structure. Therefore, when building a database, you must pay attention to the following:

(1) The data is complete, and the fields of the data table can fully reflect the document information required by the user.
(2) Reduce data redundancy, set reasonable field lengths, and avoid waste of storage space; at the same time, eliminate some unnecessary fields to prevent field duplication.
(3) Regularization of field naming. When creating a table, the name of the data table should not conflict with the keyword name in the table. Choose the appropriate field type. Based on this, the document information collection model is constructed as follows (Fig. 6):

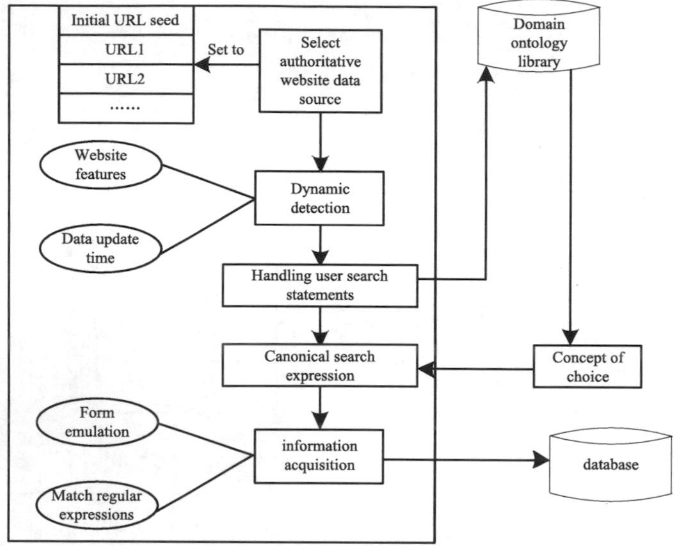

Fig. 6. Document information collection model

After the document database is formed, it needs to be maintained and managed, because in the document information collection process, due to network speed or other problems, the document information collection procedure may be interrupted, resulting in incomplete data records in the document database. If it is not timely Dealing with these incomplete data, the user's literature information analysis results will lack objectivity

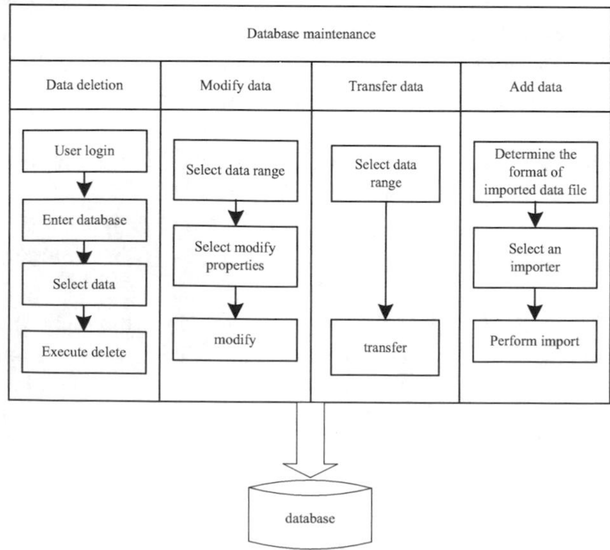

Fig. 7. Document information management model

and scientificity, so it is necessary to maintain the literature database from time to time. The document information management model is shown in the figure (Fig. 7).

According to the above-mentioned table building rules, design the document data table and the mapping relationship between it suitable for this system, and create the document database. The database should include data deletion, modification, and transfer. The system front-end development tool chooses PHPS. The PHPS development language is chosen among the many development tools mainly because the PHPS development language has advantages that other development tools do not have, such as fast speed, good openness and scalability, and support for a variety of mainstream and non-standard development tools. Mainstream databases, object-oriented programming, rich functions, etc. The realization principle of the user login module is to compare whether the user name and password entered by the user match the two fields of name and password in the login table in the literature data. If they match, enter the main module of the system. The flowchart of the user login module is shown in the figure (Fig. 8):

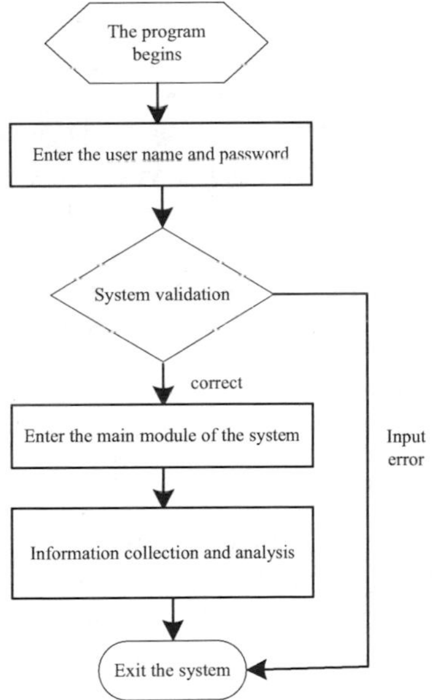

Fig. 8. Flow chart of user login module

Wireless sensor network node is mainly composed of four parts: sensor module, processing module, wireless communication module and energy supply module. Through the physical connection of Q-type sensor, the force information of large deformation flexible body and other measured target parameters are detected, and the circuit is adjusted. After a/D conversion, the sensing signal is processed and amplified, analyzed and transmitted by the processor, and stored by the memory. The wireless communication module is mainly responsible for the communication with other nodes, mainly including ZigBee wireless network, MAC layer and wireless transceiver. The energy supply module generally uses micro dry batteries to provide energy for the above three modules to ensure the normal operation of the node (Fig. 9).

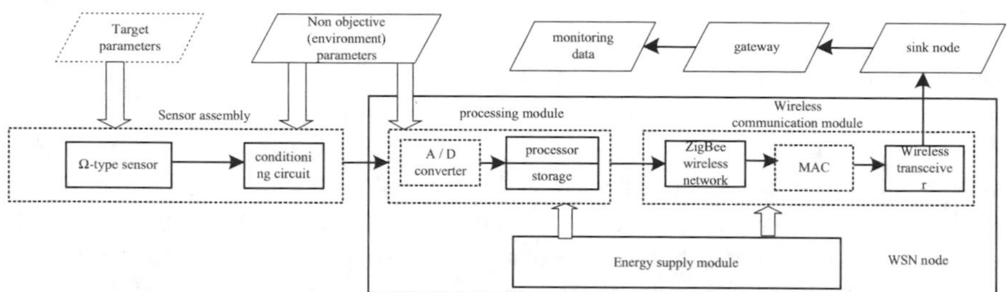

Fig. 9. WSN node and working mechanism structure

As can be seen in the figure, the sensor component will be affected by non-target parameters such as the environment during the data collection process. Therefore, the actual force of the measured object will deviate from the monitoring data obtained by the gateway node, and a certain amount must be used. The method to eliminate these deviations, this chapter gives the corresponding compensation model and algorithm to eliminate the deviation. After the data undergoes A/D conversion, the analog signal becomes a digital signal. The digital signal generally does not produce deviations, but abnormalities or missing during high-frequency transmission and processing. The processing of missing data is in Chapter 4. The corresponding model and algorithm are given.

3 Analysis of Results

In order to verify the actual application effect of the document bibliographic information dynamic monitoring system based on wireless communication, experimental testing was carried out. The experimental data processing system was developed in the VS2010 integrated development environment using C++ language, and the hardware and software environment is as follows:

Hardware environment:
CPU: Intel Core i54210M2.6 GHz/3200 MHz/3 MB
Memory: 4GDDR3L1600MHz*2
Hard Disk: SOOGHDDSATA
Communication interface: RS-232 interface

Software Environment:
Operating system: Windows7NT6.1 × 64
Database: Oracle11g
Integrated development environment: VisualStudio2010.

In order to facilitate users to observe the processing progress of data compensation, the processing progress bar is at the bottom of the interface. The figure shows the status of data processing (Fig. 10).

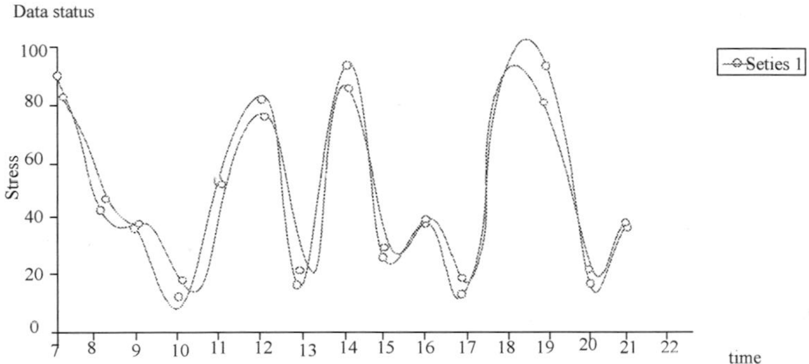

Fig. 10. System data status detection

The interface of data completion is similar to this interface, you can select the corresponding parameters, and you can view the corresponding processing progress (Table 1).

Table 1. Frequency table of main and subordinate classification numbers

Classification number	Frequency	Classification number	Frequency	Classification number	Frequency
A41D	5796	A61M	347	F21V	114
A41B	2212	A61H	267	A45C	100
A44B	1792	H05B	262	D06M	98
A41C	1294	A41G	261	A43B	91
A41H	654	A61K	186	A61L	91
A61F	510	A62B	146	B29C	85
A41F	479	B32B	134	D04B	79
A61N	408	A42B	117	A63B	77

In this system, in order to prevent suspended animation during data processing, the data processing system uses multi-threading technology. Several functions that require high computer resources such as CPU and memory are data association, data compensation, and data completion. When processing a large amount of data, it is necessary to test the time and space spent on these functions. At the same time, the growth trend of them can also be used to determine whether the data processing volume is within the acceptable range of the user. The figure below shows the data processing time of the system and related important algorithms (Fig. 11).

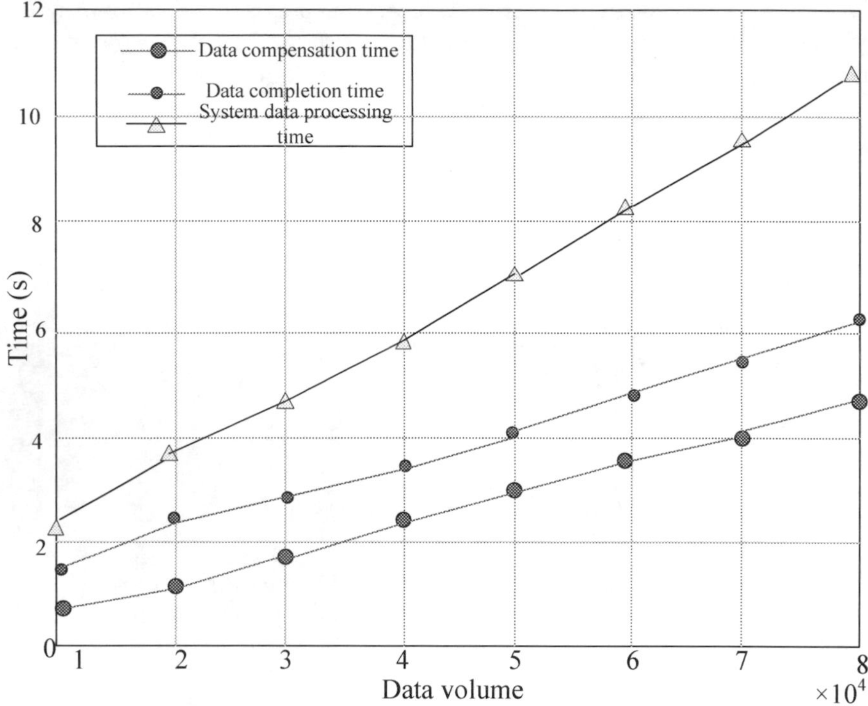

Fig. 11. System document detection time-consuming

The completion algorithm is tested when the data is missing at least 1%. It can be seen from the figure that as the amount of data increases, the running time of the system increases linearly, and the running time of the data compensation algorithm is significantly less than that of the data completion algorithm. The figure shows the running time of the system from 10,000 data to 80,000 data. In an experiment, the effective data monitoring time of the drop experiment and wind tunnel experiment is within a few seconds. If the data collection per second is 1000 times, The data volume of an experiment is basically within the scope of this performance test. Therefore, from the operating time of the entire system, the user is basically acceptable. This proves that the system in this paper has high monitoring accuracy and efficiency in the actual application process, and fully meets the research requirements.

4 Conclusion and Outlook

At present, digital resources have become an indispensable and important resource for social development, and the rational use and scientific analysis of network literature information resources can form literature information with high technical and commercial value, which is helpful to analyze the technology distribution situation and master related technologies. Historical origin, current conditions and future trends, so as to make a scientific and reasonable positioning for technological innovation activities, and provide important direction guidance and decision-making assistance for enterprises to improve independent innovation capabilities, optimize implementation effects, and increase development speed. The scientific analysis results of literature information rely on comprehensive and authoritative information sources and a reasonable analysis index system. However, the system in this paper did not consider the security of network literature information when designing it. In the following research, we will focus on information encryption.

FundProjects. This paper is supported by the Postgraduate Research and Innovation Project of Northwest Minzu University , the name of the project is "A study on the names and descriptions of the things of mongolian <qin ding li fan yuan ze li>" (project number is Yxm202009), which is the phased achievement of this project.

References

1. Li, C., Jian, M.: Scrapy-based agricultural network spatial information dynamic monitoring algorithm. J. Shandong Agric. Univ. (Nat. Sci. Ed.) **51**(02), 64–69 (2020)
2. Shen, S., Zhu, D., Rousseau, R., et al.: A refined method for computing bibliographic coupling strengths. J. Informet. **13**(2), 605–615 (2019)
3. Yang, J., Wen, J., Wang, Y., et al.: Fog-based marine environmental information monitoring toward ocean of things. IEEE Internet Things J. **7**(5), 4238–4247 (2020)
4. Shahshahani, A., Zilic, Z., Bhadra, S.: an ultrasound based biomedical system for continuous cardiopulmonary monitoring: a single sensor for multiple information. IEEE Trans. Biomed. Eng. **7**, 1–10 (2019)
5. Barua, A., Zhang, Z., Al-Turjman, F., et al.: Cognitive intelligence for monitoring fractured post-surgery ankle activity using channel information. IEEE Acccss **99**, 1 (2020)
6. Jiang, B., Luo, Y., Lu, Q.: Maximized mutual information analysis based on stochastic representation for process monitoring. IEEE Trans. Ind. Inf. **15**(3), 1579–1587 (2019)
7. Kodai, S., Tatsuya, O., Yuichi, O., Masayuki, M.: Traffic engineering and traffic monitoring in the case of incomplete information. IEICE Trans. Commun. **E102.B**(1), 111–121 (2019)
8. Lu, Z., Wang, M., Dai, W., et al.: In-process complex machining condition monitoring based on deep forest and process information fusion. Int. J. Adv. Manuf. Technol. **104**, 5–8 (2019)
9. Liu, S., Fu, W., He, L., Zhou, J., Ma, M.: Distribution of primary additional errors in fractal encoding method. Multimedia Tools Appl. **76**(4), 5787–5802 (2014). https://doi.org/10.1007/s11042-014-2408-1
10. Liu, S., Bai, W., Liu, G., et al.: Parallel fractal compression method for big video data. Complexity **2018**, 2016976–2016982 (2018)

The Method of Anomaly Location Data Recognition Based on Improved YOLO Algorithm

Chen-can Wang[1](✉), Yan Ge[2], and Yang Li[1]

[1] Information Engineering University, Zhengzhou 450000, China
[2] Zhengzhou Campus of Armyartillery Air Defense College, Zhengzhou 450000, China

Abstract. The existing anomaly location data recognition methods usually have poor accuracy due to the rough contour curve, so the anomaly location data recognition method is studied based on the improved YOLO algorithm. The improved YOLO algorithm is designed to judge the input and output residual error comparison in the normalization process. Based on the algorithm, the abnormal data location technology is studied, and the contour curve with low noise factor is obtained. Based on the improved YOLO algorithm, the abnormal location data recognition method is designed, and the accuracy of the method is optimized. The experimental results show that in the calculation of the first type error rate and the second type error rate, the slope of the method is gentle, and the value is small. It can be seen that the method will not produce large numerical changes under the changes of mathematical expectation and regression parameters, and can more accurately realize the anomaly location data recognition.

Keywords: Data exception · Improved Yolo algorithm · Exception recognition · Residual error

1 Introduction

For centralized data, if some data are very different from other data in structure, then they belong to abnormal data. Due to some errors occurring in the process of data collection and data transmission, abnormal data has little value, but for some special abnormal data, it has value that can be used, and some useful rules and knowledge can be obtained through their analysis. Therefore, how to remove these abnormal data has become an expert to solve the key problems [1]. Abnormal data greatly reduces the quality of data. If the quality of data is not guaranteed, there will be deviation in the follow-up data mining, and the analysis results such as clustering and regression can not be obtained. Therefore, a correct analysis report can not be provided to the decision-makers, which leads to deviation in prediction and decision-making, and ultimately leads to the loss of significance of the whole research. Therefore, more and more scholars are paying attention to this field, trying to find a systematic, complete and more adaptive anomaly data detection scheme.

S. Liu and X. Ma (Eds.): ADHIP 2021, LNICST 416, pp. 56–66, 2022.
https://doi.org/10.1007/978-3-030-94551-0_5

Reference [2] uses statistical concepts to construct a practical method, combined with simple formulas and visual representation of data to simplify the evaluation process. In reference [3], the structure of data existence is defined in the way of "object and attribute". Then the concept of similarity between theoretical distribution and attribute data distribution is proposed and analyzed, and the similarity algorithm of attribute data distribution is established. The advantage of using statistical methods to detect abnormal data in reference [4] is that it has enough statistical theory support, so the detected results are more reliable.

In this paper, through the above literature, based on the improved YOLO algorithm for anomaly location data recognition method is studied. Comparison of input and output residues during judging standardization. Based on this algorithm, the anomalous data localization technique is studied, and the profile curve of the low noise coefficient is obtained.

2 Research on Anomaly Location Data Recognition Method Based on Improved YOLO Algorithm

2.1 Design of Improved YOLO Algorithm

The YOLO algorithm, namely you only look once, means that you can identify the target you need at one glance. It is an end-to-end target detection algorithm, which has the huge advantage of fast detection speed. The main advantage of polo algorithm is that the monitoring effect is faster, and it can process video in real time in less than 25 ms. Compared with the traditional data location algorithm, this algorithm does not generate prediction box. In the case of convolution operation for each prediction box, a single regression method is designed for target detection, and the prediction box and confidence are obtained by deep convolution network processing directly from the input image. Instead of running on common frameworks, the algorithm uses Darknet framework, which uses C language and supports CPU and GPU. For target detection, Darknet can choose between two opencv and CUDA dependencies. Although Darknet is not as powerful as tensorflow and cafe, it is easy to install. Because it is written in C language, it is easy to check and debug the code. It can run without dependencies. It has Python interface, which is convenient to call the code and enhances the expansion ability of the framework. In addition, it can be well installed on personal computers, and has strong portability [5]. In addition, alexnet and vgg16 can also run on Darknet. Although there is no special team for maintenance, with these advantages, although they are relatively small, there is a lot of room for development. However, under the premise of extremely fast computing speed, the accuracy of this algorithm can not reach the ideal state, so this paper improves the algorithm, and designs the improved algorithm under different scales.

The improved YOLO algorithm mainly makes the following improvements: firstly, target detection of different scales, with 13×13, 26×26 and 52×52 as three different scale prediction selection values. In addition, the network structure is improved to some extent, using RESNET design method, adding residual blocks to reduce the possibility of gradient disappearance. The minimum scale feature map is responsible for the detection of large objects, the medium scale feature map is responsible for the detection of medium objects, and the maximum feature map is responsible for the detection of small objects. The introduction of residual block reduces the parameter redundancy of convolution, reduces the situation of gradient disappearance and gradient explosion, and optimizes the overall network structure. For improved algorithms at different scales, the data input/output process is shown at Fig. 1.

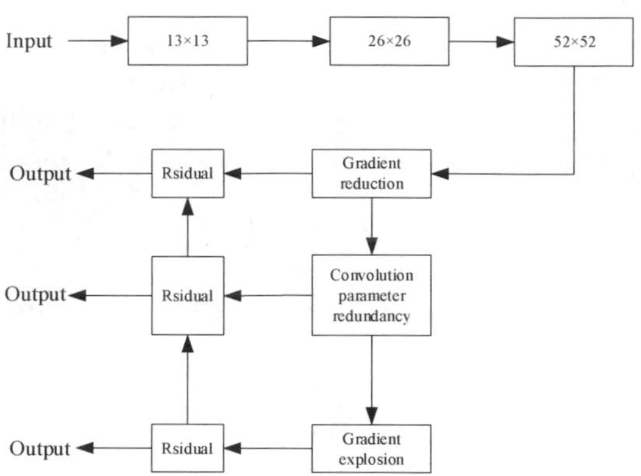

Fig. 1. Data input and output process

The overall structure of the improved algorithm is designed based on darknet-53 as the backbone network. In the network, 5 times of down sampling and the last 3 times of down sampling are used to get the feature map, as well as the prediction for three scales [6]. Finally, The feature pyramid is used to transfer the last two downsampled outputs to the previous one. Through up sampling and large-scale feature map, deep features and shallow features can be fused. Then, the residual network is used to output the three scales. Finally, through the non maximum suppression algorithm, the error and redundant prediction box is discarded, and the prediction box with accurate positioning and high confidence is retained as the final algorithm marker box. The comparison of input and output residuals is shown in Fig. 2.

The improved YOLO algorithm normalizes the input image to square 416 × 416 pixels, and then divides it into 8 × 8 grids. Each grid predicts 8 prediction boxes, and returns the confidence C and location information of each prediction box. The value of confidence C is calculated according to the following formula:

$$C = O_b \times IOU \frac{T}{pred} \tag{1}$$

Where, when the value of O_b is 1, it means there is an object, and its value is 0, indicating no object; IOU means the intersection and ratio between the prediction box and the actual region, and the prediction box returns five values of (x, y, w, h) and C values, so the last output tensor is $8 \times 8 \times (5 \times 8 + C)$. Category value is the value for each equal grid, and the category value C is for each prediction box. In the YOLO improved algorithm, the value of C is usually 7, so the image is normalized and divided

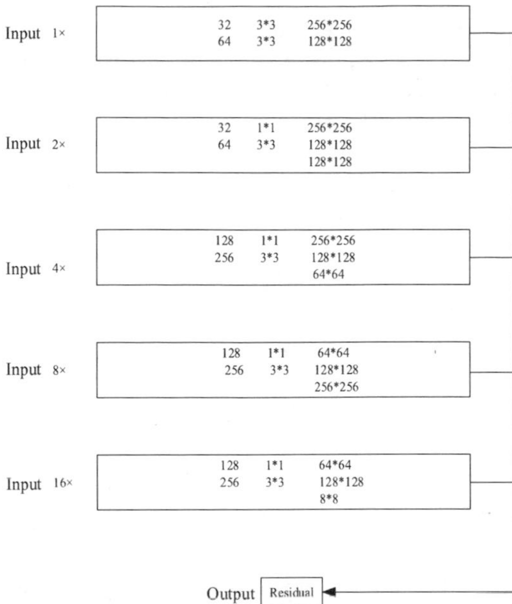

Fig. 2. Comparison of input-output residuals

into 7 × 7 mesh. O_b is usually 2, each grid is responsible for predicting two prediction boxes. In the YOLO improved algorithm, there are about 20 categories of data sets using the formula.

2.2 Noise Reduction for Abnormal Data Location

Because the shape of contour line is greatly affected by noise, a method of smoothing contour line is selected, that is, noise reduction. The improved YOLO algorithm is supported not by fitting the parameters of the contour line, but by fitting the regression contour line itself for smoothing, and the generalization error of the improved YOLO algorithm is small, which can avoid the phenomenon of "over fitting". Therefore, the improved YOLO algorithm is selected to smooth the nonlinear contour line, and the radial basis function is selected as the kernel function of the improved YOLO algorithm [7, 8].

When it is difficult to describe the functional form of a nonlinear contour with modeling method, a set of N contours with m points per contour can be regarded as an N × M matrix. A contour can be regarded as a point in high-dimensional space, so the contour data set can be regarded as n points in m-dimensional space. In the process of dimensionality reduction, the method of data depth can avoid the loss of data information in dimensionality reduction, and the data depth can judge the degree of data eccentricity when the data distribution is unknown or non normal distribution. Therefore, this paper selects the data depth to further process the smoothed data points. The improved YOLO algorithm is a common and effective data depth method. In this paper, the improved

YOLO algorithm is used to reduce the dimension of nonlinear contour data which is subject to normal distribution variation and non normal distribution variation.

In the YOLO improved algorithm, K-means clustering analysis can be used to classify data conveniently and efficiently. Therefore, the group of K-means is defined as 2 in the processing of identifying outliers. Assuming that outliers account for a small proportion in data, the data depth values transformed by YOLO algorithm are clustered and analyzed, and the few classes are distinguished from the data set as outliers, thus realizing the recognition of abnormal nonlinear contour [9]. Support vector regression has been widely used in various fields because of its small generalization error and can avoid the "over fitting" phenomenon. The basic operation idea is to find a real value function $f(x) = \frac{\omega x+b}{a}$ to minimize the expected risk. The expected risk can be expressed as follows:

$$R[f] = \int \frac{\omega x + b}{a} dP(x, y) \tag{2}$$

In the formula, $P(x, y)$ means to select the independent and identically distributed sample points to form the training set according to the probability distribution, and $\frac{\omega x+b}{a}$ means the loss function.

In this process, the support vector regression mechanism of the improved YOLO algorithm can be divided into linear and nonlinear regression models. Therefore, the nonlinear control problem is mainly studied. The basic principle of this method is to first map the original data set to Hilbert space through nonlinearity, and make the estimated regression function linear, and transform the training set into a linear one The following spatial coordinates:

$$\{(\phi(x_1 y_1)), (\phi(x_2 y_2)), \cdots, (\phi(x_3 y_3))\} \tag{3}$$

Then, the approximate linear regression is made in this high-dimensional feature space. By introducing the penalty factor c, and using Lagrange function and KKT condition to calculate the kernel function, the quadratic programming problem is solved

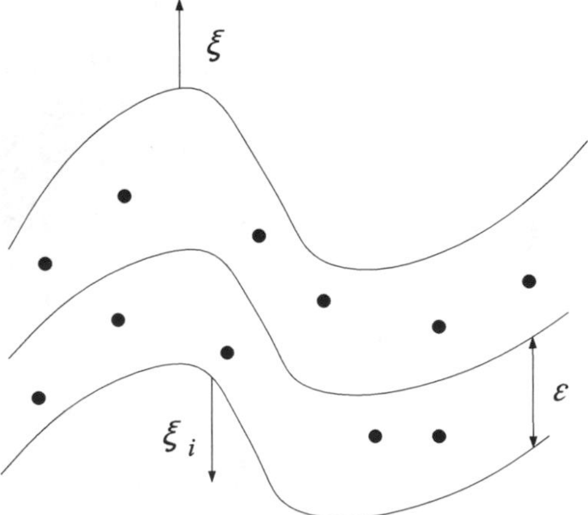

Fig. 3. Abnormal data positioning model

[10, 11]. Among them, the existence of relaxation variables ξ_i and ξ_i^* allows some data points to remain outside the confidence interval determined by c, as shown in Fig. 3, so as to obtain the optimized regression function.

Because the performance of Gaussian radial kernel function is better than other kinds of kernel functions, this paper uses RBF to check the nonlinear contour model for contour regression, so as to realize the abnormal data location based on the improved YOLO algorithm.

2.3 Design of Exception Location Data Recognition Based on Improved YOLO Algorithm

In the method of abnormal data recognition, neural network model training is mainly used to extract normal data samples and various types of abnormal data to establish the recognition model of abnormal data categories. In this process, the input information of the abnormal data identification model is mainly the load data collected by the collection system, and the output is the identification result. When it is normal data, the output result is the normal data type, and when it is abnormal data, the category of the abnormal data can be identified. In this paper, the abnormal data recognition method is mainly to predict the input data output through the improved YOLO algorithm calculation model completed by training. According to the output results, whether the input data is abnormal data can be judged, and the category of abnormal data can be judged [12].

Before using the recognition model, select the appropriate sample data, and optimize the weight attribute and threshold of the improved Yolo algorithm through genetic algorithm, and then use the sample data to train the improved YOLO algorithm. Therefore, the number and composition of the training sample data will have a great impact on the accuracy of the prediction output of the abnormal data recognition model, which plays a very important role effect. In general, the more sample data, the more knowledge the improved algorithm gets, the stronger its cognitive ability, and the more accurate the sample mapping. But in the current situation, due to the influence of environmental factors, it is impossible to obtain too much sample data. At the same time, too much training sample data not only can not improve the performance of the improved YOLO algorithm, but also can increase the network training time. The output accuracy of the abnormal data recognition model also depends on the parameter setting of the improved YOLO algorithm.

3 Experimental Research

3.1 Preparation of the Experiment

The whole experiment is implemented by Python language. IDE is eclipse pydev. All the data packages used in the process of data processing and algorithm implementation include numpy, pandas, Matplotlib, sciki learn, tensorflow, etc. Numpy is used to support a large number of dimensional array and matrix operations, and also provides a large number of mathematical function libraries, mainly for array operations, which belongs to an extension library of Python language. Pandas has been incorporated into a large number of databases and some standard data models, which provides the tools needed to operate large data sets efficiently. The tool is mainly used to solve the data analysis task. Maplotlib is a 2D drawing library in Python, which generates publishing quality level graphics in various hard copy formats and cross platform interactive environments [13, 14]. Sciki learn is a machine learning algorithm library implemented by python, which is used to realize data preprocessing, classification, regression, dimension reduction, model selection and other common machine learning algorithms. Tensorflow is an open source software library which uses data flow chart for numerical calculation. The universality of this system makes it widely used in other computing fields.

In this simulation experiment, MATLAB software is used to calculate. Firstly, the nonlinear contour data which obey normal distribution variation and non normal distribution variation are generated by MATLAB simulation through Monte Carlo simulation. The data are identified by SDC method, and compared with the two conventional methods.

Firstly, 500 nonlinear contour lines are generated, including some proportion of abnormal contour, and the function relation of the nonlinear contour set meets the following requirements:

$$\begin{cases} Y = 3\cos(x) + 5\sin(x) + \varepsilon_N \\ Y = 3\cos(x) + 5\sin(x) + \varepsilon_A \end{cases} \tag{4}$$

In the formula, $x \in (0, 2\pi)$; ε_N is the noise factor of normal contour; ε_A is the noise factor of abnormal contour; In this experiment, the performance of the new method and the comparison method when the proportion of abnormal contour in the total is different and ε_N and ε_A take different distribution.

In this paper, the first kind of error, the second kind of error and the running time of the program are used as the evaluation indexes of the performance of each method to identify the abnormal contour type of the target. The first kind of error refers to the case that the normal control line is judged as the abnormal contour line, and its calculation formula is as follows:

$$\alpha = \frac{FN}{TP + FN} \tag{5}$$

In the formula, *FN* is the number of normal contour lines determined as abnormal contour lines; *TP* is the number of normal contour lines determined as normal contour lines; α is the probability of the first type of error. Similarly, the probability of the second error is obtained.

$$\beta = \frac{FP}{TN + FP} \tag{6}$$

In the formula, *FP* represents the number of abnormal contours determined as normal contours; *TN* represents the number of abnormal contours determined as abnormal contours; β represents the probability of the second type of error. Next, this experiment will verify the performance of the proposed method and the conventional method under the condition that the proportion of abnormal contour lines in the total is 0.1, 0.15, 0.2, and the target contour data is distributed differently.

3.2 Performance Test for Algorithm

Because the proportion of abnormal contour in the total may affect the value of the first kind of error and the second kind of error, this experiment explores the influence of the proportion of abnormal contour in the total of the first kind of error and the second kind of error while exploring the different values of mathematical expectation under the lognormal distribution of noise factor. Matlab is used for data simulation and classification. The parameter C of support vector regression is set to 1, and the parameter g is set to 0.0625. According to perc = 0.1, 0.15, 0.2, 500 contour lines are randomly generated each time. According to the different mathematical expectation values of noise factor lognormal distribution, each value is repeated 1000 times, that is, a total of 2000 times, and the first type error and the second type error are calculated incidence. The probability of the occurrence of a type of error is shown in Table 1 when the proportion of abnormal contour in the total is different.

Table 1. Type I error incidence

Mathematical expectations	The algorithm in this paper	Conventional algorithms 1	Conventional algorithms 2	Conventional algorithms 3
0.1	0.965	0.398	0.978	0.967
0.2	0.869	0.684	0.857	0.759
0.3	0.586	0.489	0.594	0.475
0.4	0.369	0.569	0.674	0.578
0.5	0.852	0.875	0.965	0.954
0.6	0.967	0.987	0.578	0.477
0.7	0.598	0.965	0.415	0.488
0.8	0.496	0.859	0.552	0.955
0.9	0.756	0.598	0.485	0.789
1.0	0.975	0.785	0.758	0.958

At the same time, under the influence of different regression parameters, the first contrast effect picture of error rate is obtained, as shown in Fig. 4.

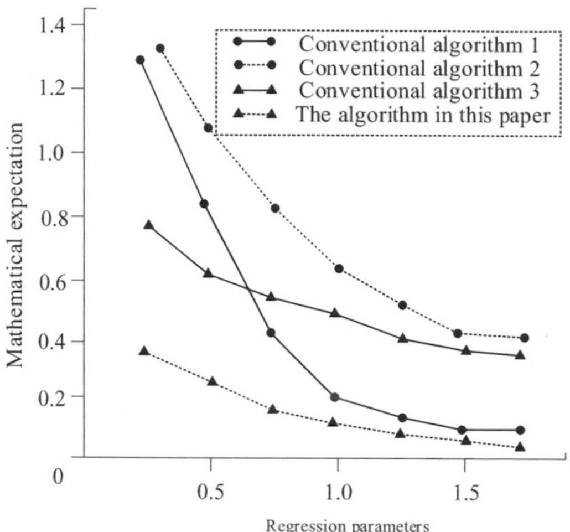

Fig. 4. Comparison of first error rate

As shown in Fig. 4, the image obtained by the algorithm in this paper is relatively flat, and has been given a relatively small value, because the error rate obtained by the algorithm is less affected during the change of regression parameters. The other three algorithms have very steep curves, especially the conventional algorithm 1 and the conventional algorithm 2. Therefore, this method has good performance in the first error rate test. The probability of the second type of error is shown in Table 2 when the proportion of abnormal contour is different.

Table 2. Type II error incidence

Mathematical expectations	The algorithm in this paper	Conventional algorithms 1	Conventional algorithms 2	Conventional algorithms 3
0.1	0.569	0.963	0.963	0.935
0.2	0.486	0.852	0.846	0.369
0.3	0.693	0.741	0.951	0.259
0.4	0.478	0.987	0.756	0.645
0.5	0.369	0.654	0.944	0.284
0.6	0.852	0.321	0.759	0.368
0.7	0.654	0.951	0.869	0.947
0.8	0.789	0.753	0.765	0.768
0.9	0.637	0.846	0.864	0.486
1.0	0.951	0.864	0.894	0.957

Under the influence of different regression parameters, the second effect chart of error rate comparison is obtained, as shown in Fig. 5.

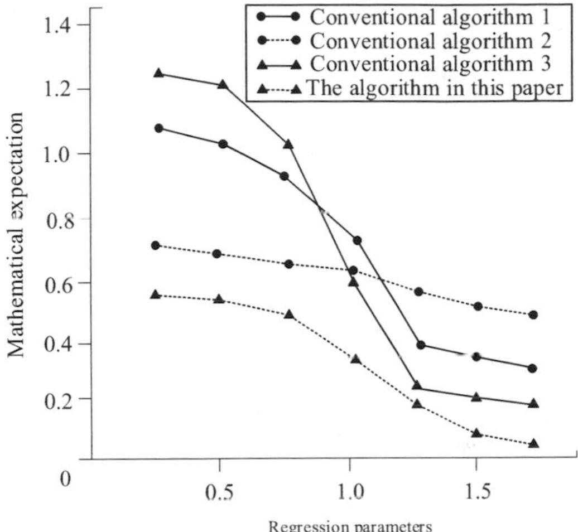

Fig. 5. Comparison of second error rate

As shown in Fig. 5, all four curves are of the "s" type structure with different slopes. The results show that the second error rate is least affected by the regression parameters. Among the other three algorithms, the slope of conventional algorithm 1 and conventional algorithm 3 is steepest, and the influence is the most. However, the conventional algorithm 2 has a relatively slow slope, but its value is larger. Therefore, according to the four curves, the method designed in this paper has the best performance in the second error rate test.

4 Conclusion

In order to improve the performance of the method in the case of non normal distribution, this paper proposes the comprehensive use of support vector regression, data depth and cluster analysis technology to identify the abnormal points of the first stage control chart of nonlinear contour. Then through the comparison between the new method and the conventional method, it is proved that this method has better data in the two evaluation indexes of the first type error rate and the second type error rate. The method of anomaly location data recognition is optimized. However, this study only identifies and detects for fixed datasets, and did not consider the characteristics of more data sets, such as nonstructural data. This will be deeply studied and analyzed in the next step.

References

1. Zhou, Z., Chen, Q., Ma, B., et al.: An improved YOLO target detection method with its application in cable device abnormal condition recognition. Electric. Measur. Instrum. **57**(2), 14–20 (2020)

2. Wang, L., Zheng, D.: anomaly identification of dam safety monitoring data based on convolutional neural network. J. Yangtze River Sci. Res. Inst. **38**(1), 72–77 (2021)
3. Wenli, J.I., Liutao, X.I., Bin, W.A.N.G.: Abnormal data recognition method of coal mine monitoring system based on imbalanced data set. Ind. Mine Autom. **46**(1), 18–25 (2020)
4. Xia, J., Liang, W., Wu, Z.: Research on automatic recognition algorithm of abnormal data in power monitoring based on mobile wavelet tree. Electron. Des. Eng. **28**(18), 148–152 (2020)
5. Zhang, H., Fan, Z., Chen, M. Application of isolated forest in abnormal identification of dam monitoring data. Yellow River **42**(8), 154–157, 168 (2020)
6. Lei, J., Chu, X., Jiang, Z., et al.: Abnormal automatic identification system data by visual analytics. J. Harbin Eng. Univ. **41**(6), 840–845 (2020)
7. Li, Y., Li, T.: Application of improved K-means algorithm in recognition of wind power abnormal data. Comput. Era **2**, 6–8 (2020)
8. Li, W.: Fast recognition and simulation of fuzzy anomaly data in nonlinear electronic networks. Comput. Simul. **36**(7), 351–354 (2019)
9. Xu, W.: Abnormal data recognition method based on power big data cleaning model. New Gener. Inf. Technol. **2**(17), 41–46 (2019)
10. Liu, S., Lu, M., Li, H., et al.: Prediction of gene expression patterns with generalized linear regression model. Front. Genet. **10**, 120 (2019)
11. Li, J., Zhang, R., Safonov, P., et al.: Outlier recognition method for spatio-temporal data based-on copula function and M-K test. Mod. Account. **39**(12), 3229–3236 (2019)
12. Xu, G., Hou, M., Xiong, H.: Moving target detection of remote tower based on improved YOLO algorithm. Sci. Technol. Eng. **19**(14), 377–383 (2019)
13. Liu, S., Liu, D., Srivastava, G., Połap, D., Woźniak, M.: Overview and methods of correlation filter algorithms in object tracking. Complex Intell. Syst. **7**(4), 1895–1917 (2020). https://doi.org/10.1007/s40747-020-00161-4
14. Fu, W., Liu, S., Srivastava, G.: Optimization of big data scheduling in social networks. Entropy **21**(9), 902 (2019)

A Factor Graph-Based Attitude Determination Approach for Microsatellite with Star Sensor and Gyros

Xiwei Wu[1], Dechao Ran[2], Lu Cao[2], and Bing Xiao[1](✉)

[1] School of Automation, Northwestern Polytechnical University, Xi'an 710072, China
xiaobing@nwpu.edu.cn

[2] National Innovation Institute of Defense Technology, Beijing 100071, China
randechao@nudt.edu.cn

Abstract. Attitude determination is an important part of microsatellite attitude control system. Its determination accuracy directly affects microsatellite's attitude control performance. Hence, a novel attitude determination approach is presented in this work for microsatellites with star sensor and gyros fixed. The factor graph representing the probabilistic graphical model as a bipartite graph is applied to solve the attitude estimation problem. This allows multi-rate, asynchronous, and possible delayed measurements to be incorporated in a natural way. An incremental smoothing algorithm is then proposed to achieve optimal estimation. Simulation results are finally presented to verify that the proposed approach can significantly improve the attitude determination accuracy.

Keywords: Microsatellite · Attitude determination · Factor graph · Smooth optimization · Star sensor · Gyro

1 Introduction

When designing attitude control system for microsatellites, attitude determination should be carried out. It provides the attitude controller design with feedback measurements of the attitude control system's states. Moreover, it directly affects the attitude control performance. Hence, it is essential to obtain high-precision attitude information. However, microsatellites are limited by cost, power consumption, and mass. The attitude sensors fixed in them are with low power consumption and low precision. The development of novel approaches to achieve high-precision attitude determination is therefore critical for microsatellites with low-accuracy sensors.

The main task of attitude determination is to develop methods to estimate the attitude and the angular velocity states of microsatellite attitude control system by using the measurements provided by attitude sensors with noise.

Supported by the National Natural Science Foundation of China under Grants 11902359, 11972373, and 61873207.

S. Liu and X. Ma (Eds.): ADHIP 2021, LNICST 416, pp. 67–80, 2022.
https://doi.org/10.1007/978-3-030-94551-0_6

The attitude determination mainly consists of the measurement sensors and the corresponding information processing algorithm, namely attitude determination algorithm [1]. Therefore, the accuracy of attitude determination depends on the measurement accuracy of attitude sensor and the accuracy of attitude determination algorithm. The attitude determination algorithm is to process the attitude sensor's measurements and filter or estimate the attitude of the microsatellite.

The attitude determination methods can be divided into the deterministic method and the state estimation method. For the deterministic method, the attitude matrix of a microsatellite is obtained only from a set of vector measurements. An algorithm to determine the three-axis attitude with two or more vector observations was presented in [2]. That algorithm has high calculation efficiency and was described as the problem of finding the orthogonal matrix which minimizes the non-negative loss function. More typical deterministic algorithms include the TRIAD [3], the QUEST [4], the SVD [5], the FOAM [6], and Euler-q [7]. The characteristic of the deterministic method that it does not require the prior knowledge of the attitude. However, this method can only estimate the attitude information of the microsatellite. Due to the uncertainty of the measured reference vector, higher attitude accuracy may not be obtained. Moreover, when there is only one measurement vector, all the deterministic algorithms are unable to determine the attitude of satellite.

The state estimation algorithm uses the attitude measurement information of continuous-time states, and combines the motion model and the estimation algorithm of the satellite attitude to obtain the optimal attitude estimation. Its advantage is that it can not only estimate the attitude, but also can estimate the uncertain parameters and system errors. The attitude determination accuracy is thus improved. For example, the extended Kalman filter (EKF) [8], the unscented Kalman filter (UKF) [9], the federated filter [10], nonlinear predictive filter [11], and particle filter [12] are available to state estimation methods design. Since, the performance of microsatellites' sensors are inferior, those filtering methods are not capable of achieving stability and real-time performance of the optimizer. To solve the stability problem, more accurate model may be required. For the real-time problem, it is quite necessary to design the optimal estimation algorithm consuming simple calculation and less time. Hence, it is necessary to develop a new optimal attitude determination methods and theoretically break through the nonlinear and real-time problems of current filtering methods.

Motivated by solving the above problems, a factor graph-based approach is presented for the microsatellite attitude determination system in this work. The proposed factor graph method has a plug-and-play characteristics that can directly solve the problems of nonlinear and sensor asynchronous, even the system uncertainties. The rest of paper is organized as follows. In Sect. 2, the factor graph and the correlation between the factor graph and the attitude determination system are introduced. In Sect. 3, the modeling of sensors and microsatellite attitude determination are given. In Sect. 4, the approach of smoothing optimization is proposed. Simulation results presented in Sect. 5. Finally, concluding remarks are drawn in Sect. 6.

2 Factor Graph

2.1 Factor Graphs Theory

The factor graph is a graphical representation of a function that visualizes the relationships between variables in a model and between variables and factors [13,14]. Its essence is the function factorization. A complex global function consisting of multiple variables is represented by the product of multiple simple local sub-functions. The relationship between subfunctions and the corresponding variables is reflected in the factor graph model.

The factor graph algorithm describes the integrated navigation information fusion problem as the connection factor node in the factor graph model and solves the integrated navigation result through a factor graph model inference [15]. Good scalability and flexibility of the factor graph model can efficiently and quickly integrate asynchronous measurement information. It has received more and more attention in multi-sensor integrated navigation systems.

2.2 Formulations of Factor Graphs

The factor graph is actually a bipartite graph. Its model is represented by a set $G = (X, F, E)$. $X = \{X_1, X_2, \ldots, X_n\}$ is the variable node, $F = \{f_1, f_2, \ldots, f_n\}$ is the factor node, and the undirected edge E connecting the two nodes represents a functional relationship between the factor node and the variable node. The necessary and sufficient condition for the existence of an edge between the factor node f_j and the variable node X_k is that $X_k \in S_j$ exists. Suppose that there is a function $g(X_1, X_2, \ldots, X_n)$ factored into m factors as

$$g(X_1, X_2, \ldots, X_n) = \prod_{j=1}^{m} f_j(S_j) \tag{1}$$

where $S_j \subseteq \{X_1, X_2, \ldots, X_n\}$ is the j-th variable subset of X and f is a real value function.

Define the state of microsatellite attitude determination set $\boldsymbol{X}_k = \{x_i\}_{i=1}^{k}$. At the time t_k, $\boldsymbol{Z}_k = \{z_i\}_{i=1}^{k}$ represents all the current measurement sets. Then, the joint probability density function of the attitude system is expressed as $p(\boldsymbol{X}_k/\boldsymbol{Z}_k)$. The maximum posteriori estimate of the system states can be obtained as

$$\boldsymbol{X}_k^* = \arg\max_{\boldsymbol{X}_k} p(\boldsymbol{X}_k/\boldsymbol{Z}_k) \tag{2}$$

Let $p(\boldsymbol{X}_k)$ be used to denote $p(\boldsymbol{X}_k/\boldsymbol{Z}_k)$ and the local function f_i be utilized to represent the local probability. Then, (2) can be rewritten as

$$p(\boldsymbol{X}_k) \propto \prod_{i} f_i(\boldsymbol{X}_k^i) \tag{3}$$

where $\boldsymbol{X}_k^i$ represents a subset of the variable nodes, *i.e.*, $\boldsymbol{X}_k^i \subseteq \boldsymbol{X}_k$ exists. Each factor node f_i in (2) represents a local function constructed from the states of the attitude measurements.

Note that the function representation of the factor nodes depends on the particular measurement model. For Gaussian noise distribution, the general factor node is given by

$$f_i(\boldsymbol{X}_k^i) = \exp(-\frac{1}{2}||err_i(\boldsymbol{X}_k^i, \boldsymbol{z}_i)||^2_{\boldsymbol{\Sigma}_i}) \tag{4}$$

Define the cost function g_i as

$$g_i(\boldsymbol{X}_k^i) = d(err_i(\boldsymbol{X}_k^i, \boldsymbol{z}_i)) \tag{5}$$

where $d(\cdot)$ is the square of the Mahalanobis distance for the Gaussian noise distribution, *i.e.*, $d(\boldsymbol{a}) = ||\boldsymbol{a}||^2_{\Sigma} = \boldsymbol{a}^{\mathrm{T}}\boldsymbol{\Sigma}^{-1}\boldsymbol{a}$, $\boldsymbol{\Sigma}$ is the covariance matrix of the estimated measurements. To this end, the maximum posteriori estimate can be established by minimizing the following global cost function

$$\sum_i g_i(\boldsymbol{X}_k^i) = \sum_i ||err_i(\boldsymbol{X}_k^i, \boldsymbol{z}_i)||^2_{\boldsymbol{\Sigma}_i} \tag{6}$$

3 Modeling of Mircosatellite Attitude Determination System

3.1 Modeling of Star Sensor Measurement

The output of a star sensor needs to convert the vector measurements to the quaternion information $\boldsymbol{q}_s$. The measurement noise $\boldsymbol{v}_s$ also needs to be converted to quaternion which is denoted as $\boldsymbol{v}_{sc} = [\boldsymbol{v}_s^{\mathrm{T}} \quad 1]^{\mathrm{T}}$. Moreover, $\boldsymbol{v}_s$ is assumed as Gaussian white noise and can be expressed as

$$E[\boldsymbol{v}_s(t)\boldsymbol{v}_s^{\mathrm{T}}(\tau)] = \sigma_s^2(t)\delta(t-\tau) \tag{7}$$

where σ_s^2 is the variance of $\boldsymbol{v}_s$. Hence, the attitude measurement information represented by quaternion of star sensor modeled as

$$\boldsymbol{q}_{sc} = \boldsymbol{q}_s \otimes \boldsymbol{v}_{sc} \tag{8}$$

For the case that the installation matrix of the star sensor is in the body-fixed coordinate system of the microsatellite, then the output measurement of the star sensor can be modeled as

$$\boldsymbol{q}_{sc} = (\boldsymbol{q}_s \otimes \boldsymbol{v}_{sc}) \otimes \boldsymbol{q}_{sb} \tag{9}$$

where $\boldsymbol{q}_{sb}$ is the equivalent quaternion representing the installation error.

When applying the method of factor graph, the star sensor's measurement model should be transformed into an factor graph, which is given by

$$\boldsymbol{z}^s = h^s(\boldsymbol{x}_k) + \boldsymbol{n}^s \tag{10}$$

where h^s is the measurement function, $\boldsymbol{x}_k$ is the measurement vector, and $\boldsymbol{n}^s$ is the measurement noise of the star sensor.

According to the measurement equation, the measurement factor of star sensor can be modeled as

$$\boldsymbol{f}^{\mathrm{STAR}}(\boldsymbol{x}_k) \triangleq d(\boldsymbol{z}^s - h^s(\boldsymbol{x}_k)) \tag{11}$$

Then, it Jacobian matrix is calculated as

$$\boldsymbol{J}^{\mathrm{STAR}}(\boldsymbol{x}_k) = \frac{\partial \boldsymbol{f}^{\mathrm{STAR}}}{\partial \boldsymbol{x}_k} = \boldsymbol{\Sigma}_s^{-\frac{1}{2}} \left\{ -\frac{\partial h^s}{\partial \boldsymbol{x}_k} \right\} \tag{12}$$

Moreover, the Jacobian matrix of the residual term can be expressed as

$$\boldsymbol{b}^{\mathrm{STAR}}(\boldsymbol{x}_k) = \boldsymbol{\Sigma}_s^{-\frac{1}{2}} \left\{ h^s(\boldsymbol{x}_k) - \boldsymbol{z}^s \right\} \tag{13}$$

3.2 Modeling of Gyro Measurement

The gyro directly measures the projection of the microsatellite's angular velocity in the inertial system. For simplicity, it is assumed that the gyro measurement coordinate system coincides with the body-fixed coordinate system of microsatellite. Then, the measurement model of the gyro is established as

$$\boldsymbol{\omega}_g = \boldsymbol{\omega} + \boldsymbol{b} + \boldsymbol{v}_g \tag{14}$$

$$\dot{\boldsymbol{b}} = \boldsymbol{v}_b \tag{15}$$

where $\boldsymbol{\omega} = [\omega_x \quad \omega_y \quad \omega_z]^{\mathrm{T}}$ is the angular velocity of the microsatellite and expressed in the body-fixed coordinate system. $\boldsymbol{b}$ is gyro drift and considered as a first-order random walk process driven by Gaussian white noise. $\boldsymbol{v}_g$ and $\boldsymbol{v}_b$ are irrelevant zero-mean Gaussian white noise. Moreover, the gyro error constantly satisfies

$$\begin{cases} E\{\boldsymbol{v}_g(t)\boldsymbol{v}_g{}^{\mathrm{T}}(\tau)\} = \sigma_g^2 \delta(t-\tau)\boldsymbol{I} \\ E\{\boldsymbol{v}_b(t)\boldsymbol{v}_b{}^{\mathrm{T}}(\tau)\} = \sigma_b^2 \delta(t-\tau)\boldsymbol{I} \end{cases} \tag{16}$$

where σ_g^2 and σ_b^2 are the variances of $\boldsymbol{v}_g$ and $\boldsymbol{v}_b$ respectively. $\boldsymbol{I}$ is the identity matrix with appropriate dimensions.

Given that the gyro's measurement value $\boldsymbol{z}_k$, the current state estimation value is $\boldsymbol{x}_k$, and the prediction value at the next time is $\boldsymbol{x}_{k+1}$, then the graph factor $\boldsymbol{f}^{\mathrm{GYRO}}$ of the gyro can be modeled as

$$\boldsymbol{f}^{\mathrm{GYRO}}(\boldsymbol{x}_{k+1}, \boldsymbol{x}_k) = d(\boldsymbol{x}_{k+1} - h(\boldsymbol{x}_k, \boldsymbol{z}_k)) \tag{17}$$

For the gyro factor $\boldsymbol{f}^{\mathrm{GYRO}}$, its Jacobian matrix is given by

$$\begin{cases} \boldsymbol{J}^{\mathrm{GYRO}}(\boldsymbol{x}_{k+1}) = \frac{\partial \boldsymbol{f}^{\mathrm{GYRO}}}{\partial \boldsymbol{x}_{k+1}} = \boldsymbol{\Sigma}^{-\frac{1}{2}} \left\{ -\frac{\partial h}{\partial \boldsymbol{x}_{k+1}} \right\} \\ \boldsymbol{J}^{\mathrm{GYRO}}(\boldsymbol{x}_k) = \frac{\partial \boldsymbol{f}^{\mathrm{GYRO}}}{\partial \boldsymbol{x}_k} = \boldsymbol{\Sigma}^{-\frac{1}{2}} \left\{ -\frac{\partial h}{\partial \boldsymbol{x}_k} \right\} \end{cases} \tag{18}$$

3.3 Modeling of Satellite Attitude Determination System

The star sensor, as the long-term attitude reference of the microsatellite, provides the three-axis attitude quaternion information of the satellite with a certain sampling frequency. Therefore, its measurement can be taken as the benchmark to correct the measurement information of the gyro. Using the measurements of the gyro and the star sensor, the attitude of satellite can be determined.

Define the quaternion as $\boldsymbol{Q} = [p_0\ \boldsymbol{p}^{\mathrm{T}}]^{\mathrm{T}} \in \Re^4, \boldsymbol{p} \in \Re^3, p_0^2 + \boldsymbol{p}^{\mathrm{T}}\boldsymbol{p} = 1$, where

$$\boldsymbol{p} = [p_1\ p_2\ p_3]^{\mathrm{T}}, [\boldsymbol{p}\times] = \begin{bmatrix} 0 & -p_3 & p_2 \\ p_3 & 0 & -p_1 \\ -p_2 & p_1 & 0 \end{bmatrix}$$

According to the multiplicative relationship among error quaternion, real attitude quaternion and estimated attitude quaternion, the attitude deviation quaternion is defined as follows

$$\boldsymbol{q} = \hat{\boldsymbol{q}} \otimes \boldsymbol{q}_e \tag{19}$$

The equation of satellite attitude motion is defined as

$$\dot{\boldsymbol{q}} = \frac{1}{2}\boldsymbol{q} \otimes [0\ \boldsymbol{\omega}] = \frac{1}{2}\begin{bmatrix} q_0 & -q_1 & -q_2 & -q_3 \\ q_1 & q_0 & -q_3 & q_2 \\ q_2 & q_3 & q_0 & -q_1 \\ q_3 & -q_2 & q_1 & q_0 \end{bmatrix}\begin{bmatrix} 0 \\ \omega_x \\ \omega_y \\ \omega_z \end{bmatrix} \tag{20}$$

According to the properties of quaternion multiplication, the derivative of formula (19) can be obtained as follows

$$\dot{\boldsymbol{q}} = \dot{\hat{\boldsymbol{q}}} \otimes \boldsymbol{q}_e + \hat{\boldsymbol{q}} \otimes \dot{\boldsymbol{q}}_e \tag{21}$$

Combining with formulas (20) and (21), we can obtain

$$\frac{1}{2}\boldsymbol{q} \otimes [0\ \boldsymbol{\omega}] = \frac{1}{2}\hat{\boldsymbol{q}} \otimes [0\ \hat{\boldsymbol{\omega}}] \otimes \boldsymbol{q}_e + \hat{\boldsymbol{q}} \otimes \dot{\boldsymbol{q}}_e \tag{22}$$

Substituting (19) into (22) yields

$$\dot{\boldsymbol{q}}_e = \frac{1}{2}\boldsymbol{q}_e \otimes [0\ \boldsymbol{\omega}] - \frac{1}{2}[0\ \hat{\boldsymbol{\omega}}] \otimes \boldsymbol{q}_e \tag{23}$$

Suppose the measured value of the gyros is $\boldsymbol{\omega}_{meas} = \boldsymbol{\omega} - \boldsymbol{b} - \boldsymbol{v}$, and the estimated value of the gyro is $\hat{\boldsymbol{\omega}} = \boldsymbol{\omega} - \hat{\boldsymbol{b}}$, then

$$\Delta\boldsymbol{\omega} = \boldsymbol{\omega}_{meas} - \hat{\boldsymbol{\omega}} = -(\boldsymbol{b} - \hat{\boldsymbol{b}}) - \boldsymbol{v} = -\Delta\boldsymbol{b} - \boldsymbol{v} \tag{24}$$

Combining with (23), simplifying and omitting higher-order terms yield

$$\dot{\boldsymbol{Q}}_e = -[\hat{\boldsymbol{\omega}}\times]\boldsymbol{Q}_e - \frac{1}{2}(\Delta\boldsymbol{b} + \boldsymbol{v}) \tag{25}$$

where $\boldsymbol{q}_e = [q_{e0}\boldsymbol{Q}_e^{\mathrm{T}}]^{\mathrm{T}}$, $\dot{q}_{e0} = 0$, and

$$\Delta\dot{\boldsymbol{b}} = \boldsymbol{v}_b \tag{26}$$

The star sensor is used to compensate the gyro drift. Let the estimate of the gyro drift be $\hat{\boldsymbol{b}}$, then the equation of state for the attitude determination is obtained by combining formulas (25) and (26) as follows

$$\begin{bmatrix} \dot{\boldsymbol{Q}}_e \\ \Delta\dot{\boldsymbol{b}} \end{bmatrix} = \begin{bmatrix} -[\boldsymbol{\omega}\times] & -\frac{1}{2}\boldsymbol{I}_3 \\ 0_{3\times3} & 0_{3\times3} \end{bmatrix} \begin{bmatrix} \boldsymbol{Q}_e \\ \Delta\boldsymbol{b} \end{bmatrix} + \begin{bmatrix} -\frac{1}{2}\boldsymbol{I}_3 & 0_{3\times3} \\ 0_{3\times3} & \boldsymbol{I}_3 \end{bmatrix} \begin{bmatrix} \boldsymbol{v}_g \\ \boldsymbol{v}_b \end{bmatrix} \tag{27}$$

Define the state vector as $\boldsymbol{X}_k = [\delta\boldsymbol{q}\ \Delta\boldsymbol{b}]^{\mathrm{T}} = [q_{e1}\ q_{e2}\ q_{e3}\ \Delta b_1\ \Delta b_2\ \Delta b_3]^{\mathrm{T}}$. The measurement equation for attitude determination can be obtained from the output of the star sensor

$$\boldsymbol{Z}_k = \boldsymbol{H}_k\boldsymbol{X}_k + \boldsymbol{D}_k\boldsymbol{V}_k \tag{28}$$

where $\boldsymbol{H}_k = [\boldsymbol{I}_{3\times3}\ 0_{3\times3}]$, $\boldsymbol{V}_k = [v_{sc1}\ v_{sc2}\ v_{sc3}]$.

To determine the attitude of satellite via factor graph, the Jacobian matrix needs to calculated to update the state. Since the gyro's measurement frequency is higher than that of the star sensor, a state connection is established with the star sensor, and the Jacobian matrix between time $t + \Delta t$ and time t can be expressed as

$$\boldsymbol{J}_{t+\Delta t} = (\boldsymbol{I} + \boldsymbol{F}\cdot\Delta t)\boldsymbol{J}_t \tag{29}$$

where $\boldsymbol{F} = \begin{bmatrix} -[\omega\times] & -\frac{1}{2}\boldsymbol{I}_3 \\ 0_{3\times3} & 0_{3\times3} \end{bmatrix}$, the initial value of the Jacobian matrix $\boldsymbol{J}_t$ is the identity matrix, and Δt is the time interval of the gyro measurement output.

The Jacobian matrix is used to establish the relationship between the state at time $k+1$ and time k as follows

$$\boldsymbol{X}_{k+1} = \boldsymbol{J}_{\mathrm{state}}\boldsymbol{X}_k \tag{30}$$

where $\boldsymbol{J}_{\mathrm{state}} = \begin{bmatrix} \boldsymbol{J}_{\Delta q}^{\Delta q} & \boldsymbol{J}_{\Delta b}^{\Delta q} \\ \boldsymbol{J}_{\Delta q}^{\Delta b} & \boldsymbol{J}_{\Delta b}^{\Delta b} \end{bmatrix}$.

Since $\boldsymbol{J}_{\Delta q}^{\Delta b} = 0$, $\boldsymbol{J}_{\Delta b}^{\Delta b} = \boldsymbol{I}$, we can obtain

$$\begin{cases} \delta\boldsymbol{q}_{k+1} = \boldsymbol{J}_{\Delta q}^{\Delta q}\delta\boldsymbol{q}_k + \boldsymbol{J}_{\Delta b}^{\Delta q}\Delta\boldsymbol{b}_k \\ \Delta\boldsymbol{b}_{k+1} = \Delta\boldsymbol{b}_k \end{cases} \tag{31}$$

Then we can get the gyro factor f^{GYRO} as

$$\boldsymbol{f}^{\mathrm{GYRO}} = \begin{bmatrix} \delta\boldsymbol{q}_{k+1} - \boldsymbol{J}_{\Delta q}^{\Delta q}\delta\boldsymbol{q}_k + \boldsymbol{J}_{\Delta b}^{\Delta q}\Delta\boldsymbol{b}_k \\ \Delta\boldsymbol{b}_{k+1} - \Delta\boldsymbol{b}_k \end{bmatrix} \tag{32}$$

Assuming that the current value of the star sensor is $\boldsymbol{Q}_{k+1}$, the star sensor factor $\boldsymbol{f}^{\mathrm{STAR}}$ can be established as

$$\boldsymbol{f}^{\text{STAR}} = \boldsymbol{Q}_{k+1} \otimes \left[\begin{bmatrix} \delta \boldsymbol{q}_{k+1} \\ 1 \end{bmatrix} \otimes \hat{\boldsymbol{Q}}_{k+1|k} \right]^{-1} \tag{33}$$

where $\hat{\boldsymbol{Q}}_{k+1|k}$ is the predicted value at the current time, and $\delta \boldsymbol{q}_{k+1}$ is the state at the current time.

Combining the proposed attitude measurement model of the star sensor and gyro, a star sensor/gyro attitude measurement system based on factor graph can be constructed. The factor graph representation is shown in Fig. 1.

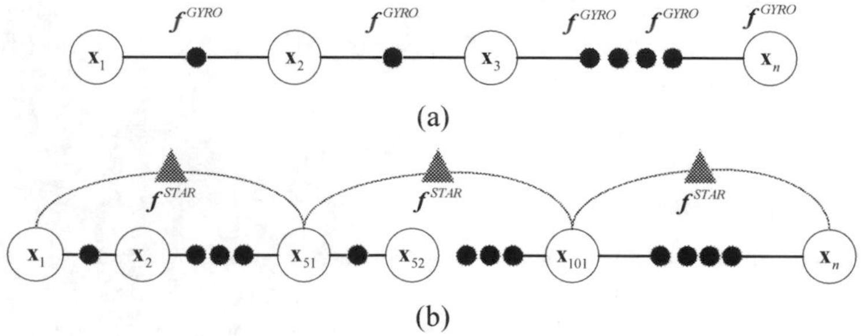

Fig. 1. Factor graph-based attitude determination system with star sensor and gyros.

4 Smoothing Optimization

Let $\boldsymbol{\Theta} = \boldsymbol{X}_k = [q_{e1}\ q_{e2}\ q_{e3}\ \Delta b_1\ \Delta b_2\ \Delta b_3]^{\text{T}}$, then for factor graph G define the factorization of function $f(\boldsymbol{\Theta})$ as

$$f(\boldsymbol{\Theta}) = \prod_i f_i(\boldsymbol{\Theta}_i) \tag{34}$$

The goal of attitude determination is to find the variable assignment $\boldsymbol{\Theta}^*$ of maximized formula (34) as follows

$$\boldsymbol{\Theta}^* = \arg\max_{\boldsymbol{\Theta}} f(\boldsymbol{\Theta}) \tag{35}$$

When assuming that the Gaussian measurement model is

$$f_i(\boldsymbol{\Theta}_i) \propto \exp(-\frac{1}{2}||h_i(\boldsymbol{\Theta}_i) - \boldsymbol{z}_i||^2_{\boldsymbol{\Sigma}_i}) \tag{36}$$

The decomposition objective function of the maximized (35) corresponds to the nonlinear least square criterion and can be expressed as

$$\arg\min_{\boldsymbol{\Theta}}(-\log f(\boldsymbol{\Theta})) = \arg\min_{\boldsymbol{\Theta}} \frac{1}{2} \sum_i ||h_i(\boldsymbol{\Theta}_i) - \boldsymbol{z}_i||^2_{\boldsymbol{\Sigma}_i} \tag{37}$$

where $h_i(\boldsymbol{\Theta}_i)$ is a measurement function, $\boldsymbol{z}_i$ is a measurement value, and $||\boldsymbol{e}||^2_{\boldsymbol{\Sigma}} \triangleq \boldsymbol{e}^{\mathrm{T}}\boldsymbol{\Sigma}^{-1}\boldsymbol{e}$ is defined as the square of the Mahalanobis distance of the covariance matrix $\boldsymbol{\Sigma}$.

Let $\boldsymbol{J}_i(\hat{\boldsymbol{X}}_k)$ and $\boldsymbol{b}_i$ represent the Jacobian matrix and the residual of a particular factor node respectively, the updated variable $\boldsymbol{\Delta}_i$ of the factor node is obtained as a form that can be normalized to a standard least squares

$$\begin{aligned}\boldsymbol{\Delta}_i{}^* &= \underset{\boldsymbol{\Delta}_i}{\arg\min}\,||\boldsymbol{J}_i\boldsymbol{\Delta}_i - \boldsymbol{b}_i||^2_{\boldsymbol{\Sigma}_i} \\ &= \underset{\boldsymbol{\Delta}_i}{\arg\min}\,||\boldsymbol{\Sigma}_i^{-\frac{\mathrm{T}}{2}}(\boldsymbol{J}_i\boldsymbol{\Delta}_i - \boldsymbol{b}_i)||^2\end{aligned} \tag{38}$$

According to the above formula, the linearization solution for updating the increment $\boldsymbol{\Delta}$ is a standard least squares problem. Generally, the Cholesky decomposition or the QR decomposition is needed. Hence, the QR decomposition is used to update the linear solution of the increment $\boldsymbol{\Delta}$. For the convenience of calculation, $\boldsymbol{A} \in \Re^{m\times n}(m \geq n)$ is employed to measure the Jacobian matrix.

The QR decomposition is firstly applied to the matrix $\boldsymbol{A}$, then one has

$$\boldsymbol{A} = \boldsymbol{Q}\begin{bmatrix}\boldsymbol{R}\\0\end{bmatrix}, \boldsymbol{A} \in \Re^{m\times n}, \boldsymbol{Q} \in \Re^{m\times m}, \boldsymbol{R} \in \Re^{n\times n} \tag{39}$$

where $\boldsymbol{Q}$ is an m-order unitary matrix and $\boldsymbol{R}$ is an n-order upper triangular matrix. Then, solving the least squares solution is equivalent to minimizing $||\boldsymbol{A}\boldsymbol{\Delta} - \boldsymbol{b}||^2_{\boldsymbol{\Sigma}}$

$$\begin{aligned}||\boldsymbol{A}\boldsymbol{\Delta} - \boldsymbol{b}||^2 &= \left\|\boldsymbol{Q}^{\mathrm{T}}\boldsymbol{Q}\begin{bmatrix}\boldsymbol{R}\\0\end{bmatrix}\boldsymbol{\Delta} - \boldsymbol{Q}^{\mathrm{T}}\boldsymbol{b}\right\|^2 \\ &= \left\|\begin{bmatrix}\boldsymbol{R}\\0\end{bmatrix}\boldsymbol{\Delta} - \begin{bmatrix}\boldsymbol{d}\\\boldsymbol{e}\end{bmatrix}\right\|^2 \\ &= ||\boldsymbol{R}\boldsymbol{\Delta} - \boldsymbol{d}||^2 + ||\mathbf{e}||^2\end{aligned} \tag{40}$$

Define $[\boldsymbol{d} \quad \boldsymbol{e}]^{\mathrm{T}} = \boldsymbol{Q}^{\mathrm{T}}\boldsymbol{b}$, $\boldsymbol{d} \in \Re^n$, $\boldsymbol{e} \in \Re^{m-n}$, then if and only if $\boldsymbol{R}\boldsymbol{\Delta} = \boldsymbol{d}$, the above formula is the smallest, i.e., $||\mathbf{e}||^2$ is the residual of the least squares solution. Therefore, the QR decomposition simplifies the least squares solution problem to solve linear equations with unique solutions, as shown below

$$\boldsymbol{R}\boldsymbol{\Delta}^* = \boldsymbol{d} \tag{41}$$

where $\boldsymbol{\Delta}^*$ is the updated increment of all state variables.

5 Simulation Results

To verify the effectiveness of the proposed factor graph method, this paper uses the EKF method for comparison. In the simulation, the initial quaternion of the satellite attitude is $[0 \quad 0.7071 \quad 0.7071 \quad 0]^{\mathrm{T}}$, The initial error covariance of the attitude is 3.0462×10^{-6}. The simulation parameters of star sensor are: The measurement of the star sensor takes the z-axis as the visual axis, the sampling period of the star sensor is 1 s, the field of view of the star sensor is 6 deg, the amplitude threshold of the star sensor is 6 deg, and the standard deviation of the star sensor is $\sigma_s = 2.9089 \times 10^{-5}$ rad. The simulation parameters of gyro are: The sampling frequency of the gyro 50 Hz, the standard deviation of the gyro bias noise $\sigma_u = 3.1623 \times 10^{-10}$, the standard deviation of the gyro noise is $\sigma_v = 3.1623 \times 10^{-7}$, and the initial error covariance of the gyro deviation is 9.4018×10^{-13}.

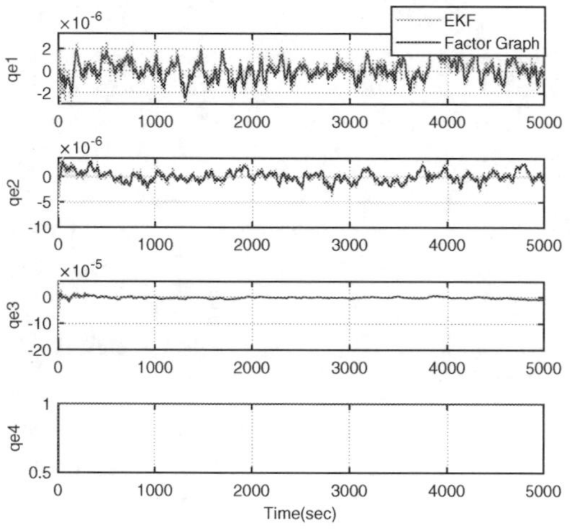

Fig. 2. The error of attitude quaternion.

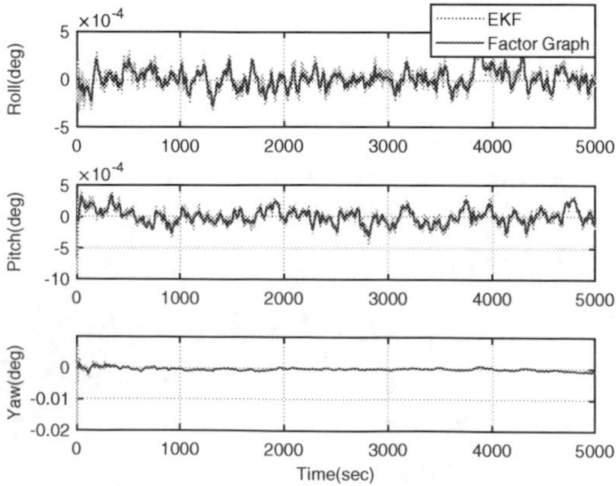

Fig. 3. The error of attitude angle.

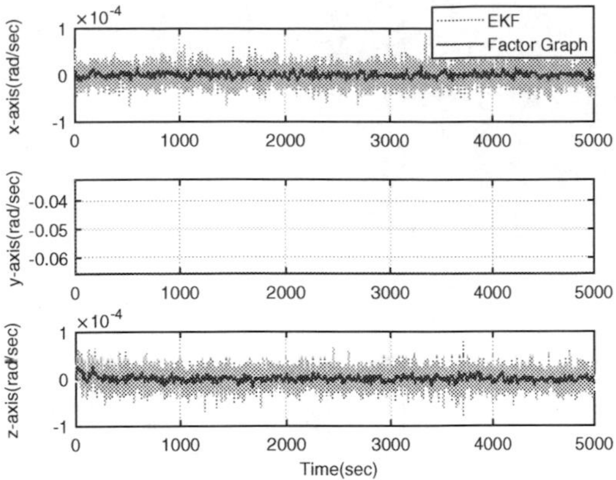

Fig. 4. The error of attitude angular velocity.

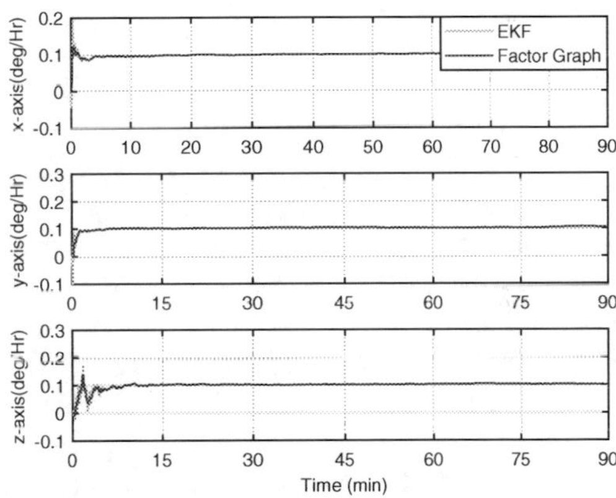

Fig. 5. The bias of gyro.

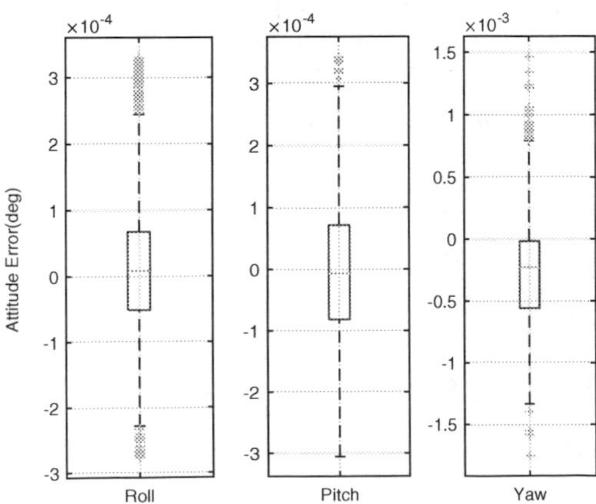

Fig. 6. The statistics of attitude angle erro.r

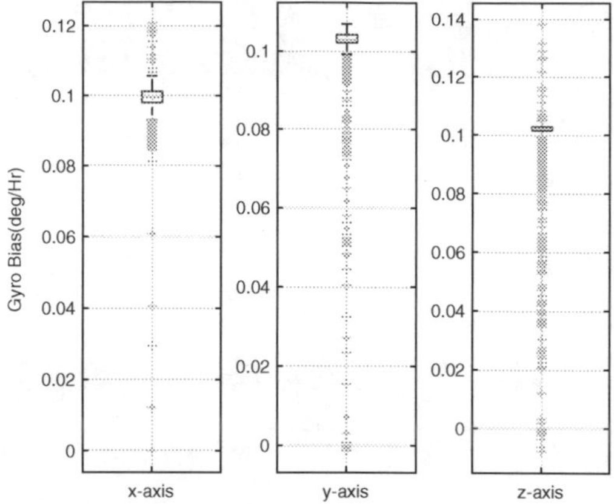

Fig. 7. The statistics of gyro bias.

It is seen in Fig. 2 that the magnitude of quaternion for the attitude error is 10^{-4}~10^{-6}. The mean value of the attitude quaternion is $[0 \quad 0 \quad 0 \quad 1.0000]^{\mathrm{T}}$. The mean square deviation of the attitude quaternion is $[8.3783\times10^{-7} \quad 1.0137\times10^{-6} \quad 3.4553\times10^{-6} \quad 1.2120\times10^{-11}]^{\mathrm{T}}$. Figures 3 and 4 show the attitude angle error and attitude angular velocity error, respectively. Figure 6 is the statistics of attitude angle error. According to the statistical characteristics of the three-axis attitude angle error shown in Table 1, it can be known that the mean square error (MSE) of the three-axis attitude angle error less than 4.2304×10^{-4} deg. Therefore, compared with EKF, the attitude determination accuracy of the proposed algorithm is higher than that of EKF. Figures 5, 6 and 7 are the gyro bias and the satistics of gyro bias, respectively. Combined with the MSE of gyro bias in Table 2, it can be seen that the cumulative error of the three-axis gyroscope is about 0.1 deg/Hr, and the mean square error is less than 0.0116 deg/Hr, the gyro deviation of the proposed method is less than EKF. It is shown that the proposed method ensures that the gyro bias also fluctuates within a certain range, and will not increase the error, thereby improving the accuracy of satellite attitude determination.

Table 1. The MSE of attitude angle

Value	Roll (deg)	Pitch (deg)	Yaw (deg)
EKF	1.0368×10^{-4}	1.2550×10^{-4}	5.3336×10^{-4}
FGs	9.4190×10^{-5}	1.1642×10^{-4}	4.2304×10^{-4}

Table 2. The MSE of gyro bias

Value	X-axis (deg/Hr)	Y-axis (deg/Hr)	Z-axis (deg/Hr)
EKF	0.0061	0.0099	0.0121
FGs	0.0052	0.0077	0.0116

6 Conclusion

In this paper, a factor graph-based attitude determination approach for satellites with star sensor and gyros was presented. The star sensor, gyro and satellite attitude dynamic model based on factor graph were given. The problem of satellite attitude determination was modeled as a factor graph optimization. Moreover, the attitude was determined through a smoothing optimization algorithm. The numerical simulations was developed to evaluate the optimization performance, and compared with the existing EKF method. The good scalability and flexibility of the proposed approach make it possible to efficiently and quickly integrate asynchronous measurement information. Mostly, it is possible to achieve plug and play of the sensor.

References

1. Lin, Y., Deng, Z.: Extended Kalman filtering for satellite orbital attitude estimation based on Gibbs vector. J. Guid. Control Dyn. **27**(3), 509–511 (2004)
2. Wahba, G.: A least squares estimate of satellite attitude. SIAM Rev. **7**(3), 409 (1965)
3. Shuster, M.D., Oh, S.D.: Three-axis attitude determination from vector observations. J. Guid. Control **4**(1), 70–77 (1981)
4. Shuster, M.: A simple Kalman filter and smoother for spacecraft attitude. J. Astronaut. Sci. **37**(1), 89–106 (1989)
5. Markley, F.L.: Attitude determination using vector observations and the singular value decomposition. J. Astronaut. Sci. **36**(3), 245–258 (1988)
6. Markley, F.L.: Attitude determination using vector observations: a fast optimal matrix algorithm. J. Astronaut. Sci. **41**(2), 261–280 (1993)
7. Mortari, D.: Euler-q algorithm for attitude determination from vector observations. J. Guid. Control Dyn. **21**(2), 328–334 (1998)
8. Kang, G., Xiaojing, D., Xinyuan, M.: Research on multi update rate method of precise satellite attitude determination based on gyro and star-sensor. In: AIAA Guidance, Navigation, and Control Conference, pp. 6645. AIAA, Portland, Oregon (2011)
9. Li, H., Sun, Z.-W., Zhang, S.-J.: Satellite attitude estimation UKF algorithm based on star-sensor. In: The Proceedings of the Multiconference on "Computational Engineering in Systems Applications", pp. 799–802. IEEE, Beijing (2006)
10. Cao, L., Yang, W., Chen, X., Huang, Y.: Application of multi-sensors data fusion based on improved federal filtering in micro-satellite attitude determination. In: 2011 International Workshop on Multi-Platform/Multi-Sensor Remote Sensing and Mapping, pp. 1–5. IEEE (2011)

11. Mook, D.J., Junkins, J.L.: Minimum model error estimation for poorly modeled dynamic systems. J. Guid. Control Dyn. **11**(3), 256–261 (1988)
12. Zhiru, F. and Jing, Y.: A research of gyro/star-sensor integrated attitude determination based on particle filter. In: 3th International Conference on Instrumentation, Measurement, Computer, Communication and Control, pp. 256–261, Shenyang (2013)
13. Frey, B.J., Kschischang, F.R., Loeliger, H.-A., Wiberg, N.: Factor graphs and algorithms. In: Proceedings of the Annual Allerton Conference on Communication Control and Computing, vol. 35, pp. 666–680. University of Illinois (1997)
14. Kschischang, F.R., Frey, B.J., Loeliger, H.-A.: Factor graphs and the sum-product algorithm. IEEE Trans. Inf. Theory **47**(2), 498–519 (2001)
15. Indelman, V., Williams, S., Kaess, M., and Dellaert, F.: Factor graph based incremental smoothing in inertial navigation systems. In: 15th International Conference on Information Fusion, pp. 2154–2161. IEEE, Singapore (2012)

Research on Tunable Laser Temperature Measurement Method Based on Spectral Absorption

Ya-ping Li(✉) and Ming-fei Qu

College of Mechatronic Engineering, Beijing Polytechnic, Beijing 100176, China

Abstract. Traditional laser temperature measurement method has a slow warning speed because of its poor temperature measurement scheme. To solve this problem, a new tunable laser temperature measurement method is designed based on the spectral absorption process. According to the number of atoms, the gas molecular types are divided, and then the absorption line broadening is set based on the absorption process. On this basis, according to the relationship between the output light intensity and the incident light intensity, a tunable laser temperature measurement scheme is designed. In the experimental part, the measurement model of gas concentration and temperature field was constructed according to the absorption area ratio, and two temperature environments of −70 °C and 210 °C were set. According to the test results, the warning speed of this method is 3.39 min, 4.13 min and 7.43 min faster than the three groups of traditional methods at −70 °C, and 2.91 min, 4.08 min and 7.31 min faster than the three groups of traditional methods at 210 °C. Therefore, this method is more timely.

Keywords: Spectral absorption · Tunable laser · Temperature measurement model · Light intensity · Temperature field · Early warning effect

1 Introduction

In view of the importance of temperature field to combustion diagnosis, advanced and effective temperature measurement method has been a hot spot in the field of combustion research. For this reason, a high-precision wavelength tuning method of modulated grating Y-branch (mGy) tunable laser is designed in reference [1]. The temperature measurement is realized through high-precision wavelength scanning calibration of coarse scanning combined with fine scanning. According to the characteristics of tunable laser grating in reference [2], a new type of tunable distributed Bragg reflection laser based on sampling grating is designed. SG is used to replace the uniform grating of traditional DBR laser, and the temperature measurement and adjustment of different DBR laser tuning range are realized in a monolithic integrated chip.

On the basis of the above traditional methods, this paper introduces the application method of spectral absorption technology to further optimize the temperature measurement method. The method divides gas molecules according to the number of atoms,

S. Liu and X. Ma (Eds.): ADHIP 2021, LNICST 416, pp. 81–93, 2022.
https://doi.org/10.1007/978-3-030-94551-0_7

and then sets absorption line broadening based on spectral absorption process. Then according to the relationship between the light intensity and the incident light intensity, a tunable laser temperature measurement scheme is designed, so as to provide more reliable technical support for the temperature measurement in dangerous and complex environments.

2 Design of Tunable Laser Temperature Measurement Method

2.1 Classification of Gas Molecules

According to the number of atoms, molecules can be divided into diatomic molecules and polyatomic molecules. In addition to a small number of monatomic and diatomic molecules, most of gas molecules are polyatomic molecules. Therefore, the rotation of polyatomic molecules is studied. According to the interaction of rotation, vibration and electron motion, three mutually orthogonal directions passing through the molecular center of mass are selected, which are called the principal axes, and the corresponding moment of inertia is the principal moment of inertia. According to the different cases of moment of inertia, the molecular rotation can be divided into four categories: three principal moment of inertia are not equal, called asymmetric gyro molecules; two principal moment of inertia are equal, called symmetric gyro molecules; three principal moment of inertia are completely equal, called spherical gyro molecules; if the rotational inertia of the axis direction of the symmetric gyro is very small, the inertia ellipsoid is close to a circle Cylindrical molecules are called linear molecules. According to the degree of symmetry, linear molecules are divided into two types: centrosymmetric and noncentrosymmetric. Except for a few exceptions, the angular momentum of the electrons of linear polyatomic molecules in the ground state around the inter nuclear axis is zero, and the pure rotational energy level of the molecules is as follows:

$$\frac{E}{ab} = ZJ(J+1) - CJ^2(J+1)^2 \tag{1}$$

In formula (1): E represents rotational energy; a represents Planck constant; b represents speed of light; Z represents rotational constant; J represents rotational term; C represents constant term. $CJ^2(J+1)^2$ in formula (1) comes from the non rigidity of the molecule, which represents the influence of centrifugal force. Compared with $ZJ(J+1)$, it is very small, so it can be omitted [3]. The selection rule of rotational quantum number J for pure rotational spectra of linear molecules is $\Delta J = \pm 1$. Assuming that J' is a higher state and J'' is a lower state, then according to formula (1), the rotational spectrum of linear molecule can be expressed as:

$$g = F(J') - F(J'') \tag{2}$$

Substituting $J' = J'' + 1 = J + 1$ into the above formula, we can get the following result:

$$g = 2Z(J+1) - 4C(J+1)^3 \tag{3}$$

According to the above formula, the pure rotational spectra of linear molecules are a series of spectral lines with approximately equal distance. For a symmetric gyro molecule, the two equal principal moment of inertia G_x and G_y are expressed by G_A, and the third principal moment of inertia G_z is expressed by G_B, then the molecular pure rotational energy level of the symmetric gyro molecule is:

$$F(J, W) = Z(J + 1) + (A - B)W^2 \tag{4}$$

In formula (4): A and B are the ratios of Planck's constants a and $8\pi^2$ and the speed of light to G_A and G_B; W represents the magnitude of the J component of the rotational quantum number. A symmetric gyroscope molecule has a symmetry axis, and the permanent dipole moment of the molecule is along the direction of the symmetry axis. The selection rule of W and J is:

$$\begin{cases} \Delta W = 0 \\ \Delta J = 0, \pm 1 \end{cases} \tag{5}$$

In pure rotational spectroscopy, the first set of conditions means that there is no transition, the transition at $\Delta J = J' - J'' = -1$ will not occur, and the transition will occur only at $\Delta J = +1$, that is, only those adjacent energy levels with the same W can merge with each other. The position of the spectral line of the infrared rotation spectrum of the symmetric gyro molecule is:

$$g = F\left(J', W'\right) - F\left(J'', W''\right) = 2Z(J + 1) \tag{6}$$

According to formula (6), the pure rotation spectrum of the symmetric gyro molecule is a set of equidistant spectral lines. If the molecule does not have a three-fold axis or a higher weight axis, all three principal moments of inertia are generally different. Such molecules are asymmetric gyroscopic molecules. Most polyatomic molecules are asymmetric gyroscopic molecules, which can be regarded as an intermediate situation between long symmetrical gyroscopic molecules and flat symmetrical gyroscopic molecules [4]. According to quantum mechanics, the energy level of asymmetric gyroscope molecules cannot be expressed as an explicit functional formula similar to the energy level formula of symmetric gyroscope molecules. The three main moments of inertia of the asymmetric gyro molecules are called G_A, G_B and G_C in the order of their increase, and are similar to the labeling method in the case of symmetric gyro molecules, and the following quantities are introduced:

$$A = \frac{a}{8\pi^2 bG_A}, B = \frac{a}{8\pi^2 bG_B}, C = \frac{a}{8\pi^2 bG_C} \tag{7}$$

If the molecule has a permanent dipole moment, in the case of dipole radiation, the choice of the rotational quantum number J is: $J = 0, \pm 1$. For asymmetric gyro molecules, transitions can occur between groups of energy levels belonging to different J values, and transitions can also occur between groups of energy levels belonging to a given J value at $\Delta J = 0$. Asymmetric gyro molecules have a large number of rotational transitions, and the rotational spectral lines are very rich. At present, only some very simple molecules have been studied in detail. The moments of inertia of the molecules of the ball spinning around all axes passing through the center of mass are completely equal. The molecules simply rotate around a fixed axis in space, and the fixed axis can have any orientation relative to the molecule. The energy level of the ball gyro molecule can be obtained on the basis of the symmetric gyro molecule energy level formula (4). Let $G_A = G_B$, namely $A = B$, obtain the molecular pure rotational energy level of the ball gyro molecule as:

$$F(J) = ZJ(J + 1) \tag{8}$$

In formula (8): J represents all values starting from 0, and the molecular energy depends on the quantum number J. Only when the molecule has a permanent dipole moment, the pure rotation spectrum can be produced. If the molecule has two or more symmetry axes, the permanent dipole moment of the molecule is zero, and the spherical gyro molecule does not display the infrared rotation spectrum. Divide the types of gas molecules according to the above formula.

2.2 Set Absorption Line Broadening Based on Spectral Absorption

Absorption line type can be used to describe the shape of absorption line in spectroscopy, which is a very important parameter when measuring gas concentration. The reason for the absorption line is that the energy of the gas molecule jumps to produce the absorption line. In an ideal state, the absorption line should be a straight line, but in actual situations, due to various interference factors, the absorption peak of the gas to be measured is a broad line type. The light absorption coefficient α of gas is related to the peak f of the linear function. Since the linear function is used to describe the change of the wavelength or frequency of the coefficient of the gas absorption spectrum, the peak f should be located at the center frequency of the function. Therefore, the light absorption coefficient α of gas can also be expressed as a normalized linear function centered on the center frequency of the spectral line, and its form is as follows:

$$\alpha(g) = Kf(g - g_0)N \tag{9}$$

The spread of spectral lines can be divided into three types of broadening: natural broadening, collision broadening and Doppler broadening. Among them, the natural broadening is determined by the nature of the molecule or atom itself, and it is caused by the instability of spontaneous radiation without being affected by the external environment [5]. The linear function of natural broadening is Lorentz type, and its linear function is:

$$f_L(g) = \frac{\Delta g_L}{2\pi} \frac{1}{(g - g_0)^2 + \left(\frac{\Delta g_L}{2}\right)^2} \tag{10}$$

In formula (10): g_0 represents the center frequency of different types; Δg_L represents the half-height width of the line type.

Collision broadening is caused by the collision between background molecules and radiation molecules. It is also caused by the instability of gas molecules in the excited state. Molecules will return to the ground state after a short residence time in the excited state. The residence time determines the characteristics of collision broadening. Therefore, the collision broadening is the same as the natural broadening, and their linear functions are Lorentz type. Doppler broadening is caused by the irregular motion of molecules or atoms in space, that is, the instability of molecules or atoms in space

$$f_G(g) = \frac{2}{g_G}\sqrt{\frac{\ln 2}{\pi}} \exp\left[-4\ln 2\left(\frac{g - g_0}{\Delta g_G}\right)^2\right] \tag{11}$$

In formula (11): Δg_G is the width at half height of Gaussian line [6]. So far, according to the characteristics of spectral absorption, the broadening of spectral line is set up.

2.3 Design of Tunable Laser Temperature Measurement Scheme

According to the Lambert Beer absorption law of spectrophotometry, the relationship between the intensity P of a beam of light passing through the absorption medium and the intensity P_0 of the incident light can be expressed as follows:

$$P = P_0 \exp(-\beta D) = P_0 \exp(-\eta ND) \tag{12}$$

In formula (12): β is the absorptivity; D is the absorption length; η is the absorption cross section; N is the particle number density of the absorption medium. The absorption cross section can be calculated by the following formula:

$$\eta = U(T)Qu(f) \tag{13}$$

In formula (13): $U(T)$ is the spectral line strength of the absorption medium, which is the inherent parameter of the absorption peak and is only related to the transition energy level and temperature; Q is the pressure; $u(f)$ is the linear function of $u(f)$ [7]. Substituting formula (13) into formula (12), the parameters after integration are obtained:

$$U(T)PNL = -\int \ln\left(\frac{P}{P_0}\right) dg \tag{14}$$

The right half of formula (14) is the integral absorptivity of the whole absorption peak. Because the absorption line strength of the medium is only a function related to temperature, the integral absorptivity of two different absorption peaks can be compared to eliminate the particle number density and pressure in the formula, and a new value of line strength a can be obtained. When the two selected absorption peaks are two different rotational transitions of the same vibrational band, the integral absorption ratio of formula (14) is finally obtained. The integral absorptivity of each absorption peak is $U(T)$ parameter that can be obtained by experimental measurement, and then the integral absorptivity θ can be obtained, so as to determine the relationship between the flame temperature and the measured value θ [8].

For the absorption spectrum temperature measurement technology, the selection of absorption spectrum line is very important, which should meet the following three basic principles: first, the interval between the two spectrum lines should not be too far or beyond the tuning range of the laser, and the spectrum line should be clean and free from the interference of other absorption spectrum lines; second, the selected absorption spectrum line should have a certain strength to ensure a good measurement signal-to-noise ratio; third, the selection of absorption spectrum line should have a good measurement signal-to-noise ratio The two absorption lines have proper lower level separation, which makes the measurement have enough temperature sensitivity.

2.4 Building the Measurement Model of Gas Concentration and Temperature Field

When the temperature distribution is known, the measurement model of gas concentration and temperature is constructed. In algebraic iterative reconstruction, the initial value of the iteration has a great influence on the reconstruction result. If the initial value of the iteration deviates too much, the reconstruction error will be large, and even the reconstruction result will be distorted. Therefore, the initial value close to the real value should be given as much as possible in the actual measurement reconstruction process. In order to give a reasonable two-dimensional distribution of gas concentration and temperature, tunable optical absorption signal is used to predict the gas concentration and temperature field. When two lasers with different wavelengths pass through the same test area, the path length and gas component concentration must be the same

$$\theta = \frac{S_1(T)}{S_2(T)} = \frac{S_1(T_0)}{S_2(T_0)} \exp\left[-\frac{ab}{m}(E_1 - E_2)\left(\frac{1}{T} - \frac{1}{T_0}\right)\right] \tag{15}$$

In formula (15): $S_1(T)$ and $S_2(T)$ denote the absorption area at the temperature of T; $S_1(T_0)$ and $S_2(T_0)$ denote the absorption area at the initial temperature of; m denotes the absorption strength; E_1 and E_2 denote the energy value under two conditions [9]. Then the actual value of absorption line wavelength is found in HITRAN database to determine the relationship between line strength and temperature. With the change of temperature, the line strength of the two absorption lines changes, and the relationship between the line strength ratio of the two absorption lines and the temperature is generated. According to the temperature relationship of line strength at different nodes, the range of line strength in the temperature field is determined, and the relationship between line strength and temperature in different intervals is analyzed [10]. As shown in Fig. 1, it is the schematic diagram of the measurement model for gas concentration and temperature field.

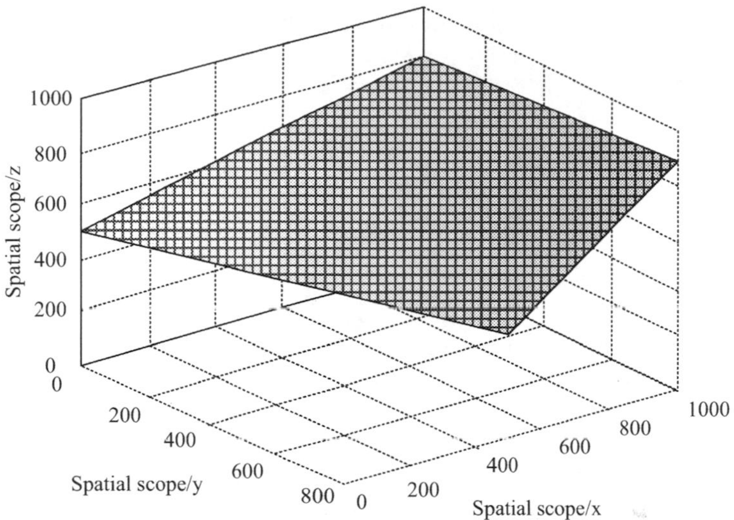

Fig. 1. Schematic diagram of gas concentration and temperature field measurement model

So far, the design of tunable laser temperature measurement method based on spectral absorption is completed.

3 Experimental Study

In order to verify the reliability of the tunable laser temperature measurement method based on spectral absorption, this method is compared with three groups of traditional temperature measurement methods, setting different temperature levels, comparing the temperature warning effect of the four groups of methods, and testing the temperature difference of the four methods.

3.1 Experimental Preparation

For this experimental test, an overall control circuit of the experimental operating system is constructed to perform the temperature measurement process of tunable laser. The accessories of the circuit mainly include: power adapter, RS-232 port, power indicator, single power level conversion chip, alarm, temperature sensor, matrix keyboard, memory, communication selection, function selection, trigger button, antenna, induction module, communication module, LCD, output indicator and power switch. On this basis, the serial port is debugged. Before debugging, connect the development board with the serial port line and power line, insert the SIM card on the back of the operating system circuit board, use the jumper cap to connect the communication, select the P and Q positions on the port, connect the power supply, press the trigger button and connect the wireless network. At this time, the system operation indicator starts to flash, which proves that the experimental system is established and can be applied to the temperature test. The parameters of the experimental platform are shown in Table 1.

Table 1. Experimental parameters

Serial number	Name	Parameter value
1	Temperature difference information	10 GB, 50 GB
2	System memory	32 GB
3	Working frequency	600 MHz
4	CPU main frequency	360 MHz

After setting the experimental parameters, the two groups of temperature field test environment were set as T1 and T2, and the temperature warning critical values of the two groups of environment were −70 °C and 210 °C respectively. The environmental temperature of the two groups was known, and reached the warning threshold value from the 5th min and the 9th min respectively. Four groups of temperature measurement methods were used to measure the temperature in the experimental environment.

3.2 Test Results at −70 °C

After debugging the experimental test system, real-time temperature measurement is carried out for the temperature change in the test background. The temperature measurement method designed in this paper is taken as experimental group A, and the three groups of traditional methods are taken as experimental group B, group C and group D respectively. Figure 2 shows the measurement results of four methods when the critical temperature is −70 °C.

As can be seen from Fig. 2, group a began to recognize abnormal temperature from the 5th, 32th min, and quickly reached the critical value of early warning. However, the three groups of traditional methods only warned the abnormal temperature from 8.45 min, 9.16 min and 12.5 min respectively. In order to ensure the authenticity, reliability and universality of the experimental test data, a total of 8 rounds of tests were carried out at this stage, and the results are shown in Table 2.

According to Table 2, the average values of the four groups of test data are 5.13 min, 8.52 min, 9.26 min and 12.56 min respectively. Comparing the four groups of test results, it is found that the warning speed of the temperature measurement method designed in this paper is 3.39 min, 4.13 min and 7.43 min faster than the three groups of traditional methods respectively.

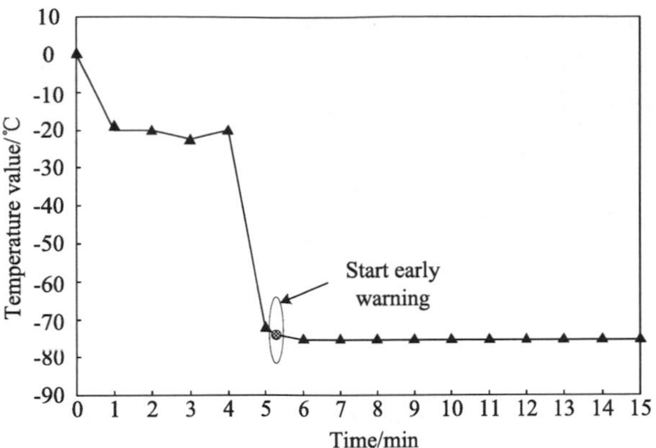

(a) Experimental group A

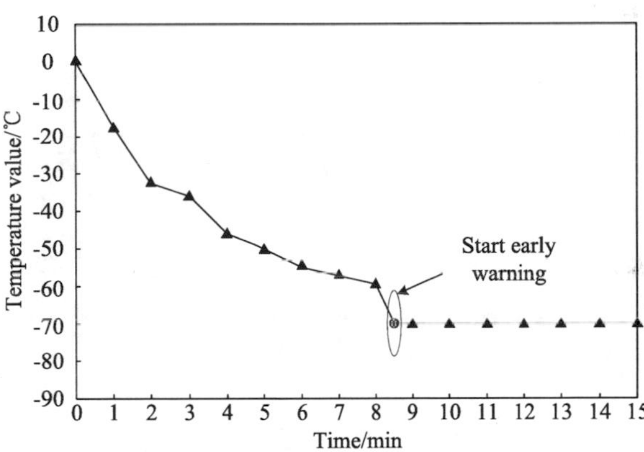

(b) Experimental group B

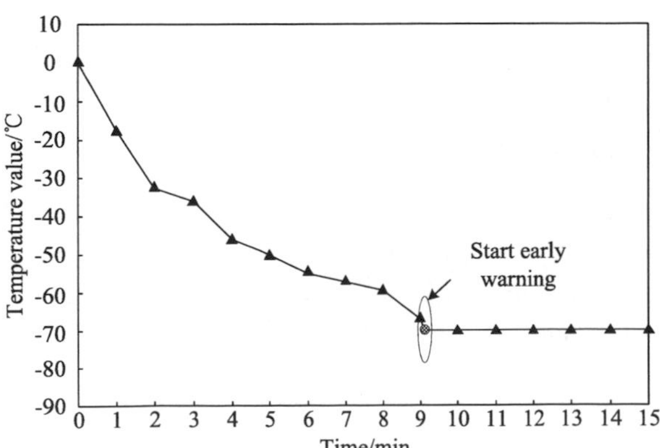

(c) Experimental group C

Fig. 2. Test results at −70 °C

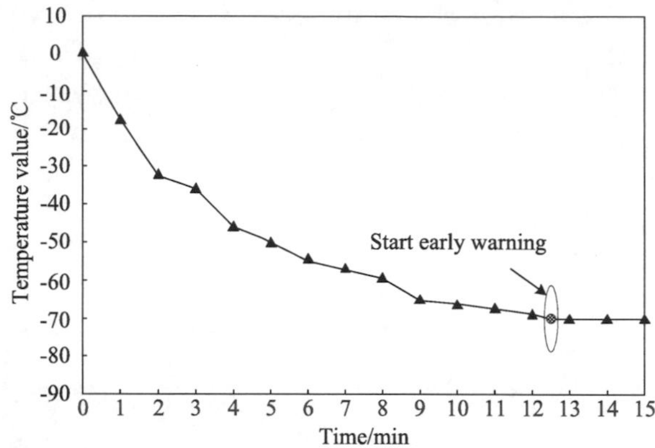

(d) Experimental group D

Fig. 2. continued

Table 2. Temperature warning speed test (min)

Test group	Group A	Group B	Group C	Group D
1	5.32	8.45	9.16	12.5
2	5.07	8.48	9.25	12.63
3	5.15	8.52	9.49	12.58
4	5.04	8.63	9.29	12.44
5	5.11	8.54	9.08	12.47
6	5.10	8.55	9.22	12.69
7	5.06	8.49	9.34	12.58
8	5.22	8.48	9.27	12.55

3.3 Test Results at 210 °C

Keep the basic test conditions unchanged and change the test environment to 210 °C. Figure 3 shows the measurement results of four methods when the critical temperature is 210 °C.

According to the curve change trend in Fig. 3, the designed temperature measurement method has the fastest warning speed in the face of high temperature test. Similarly, 8 rounds of tests were conducted, and the temperature warning results of each round are shown in Table 3.

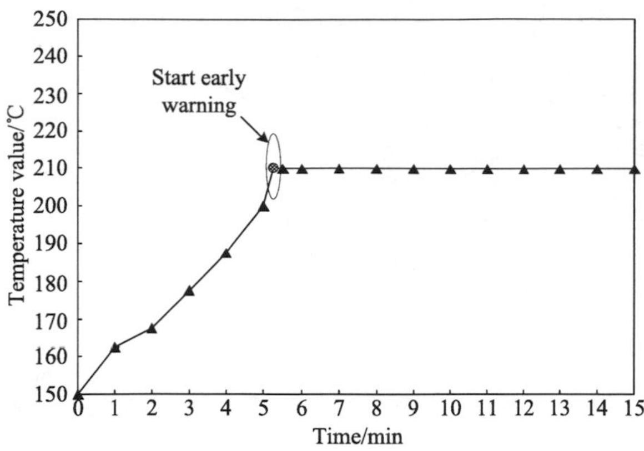

(a) Experimental group A

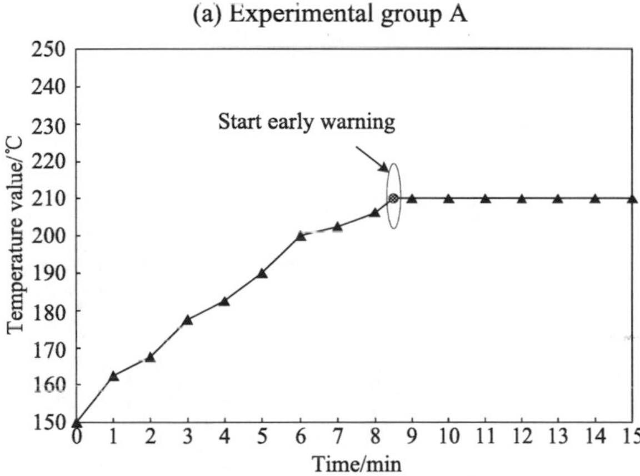

(b) Experimental group B

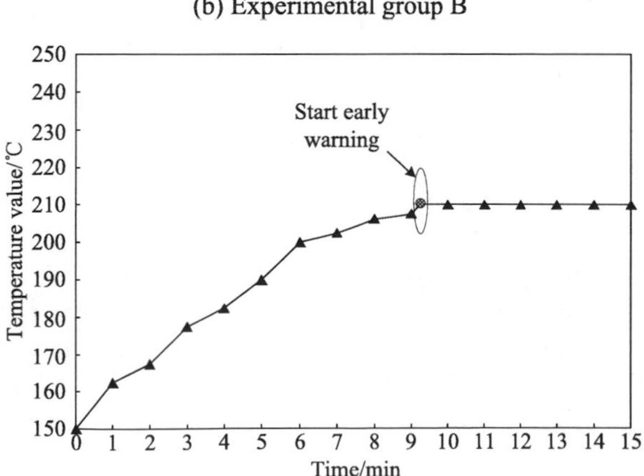

(c) Experimental group C

Fig. 3. Test results at 210 °C

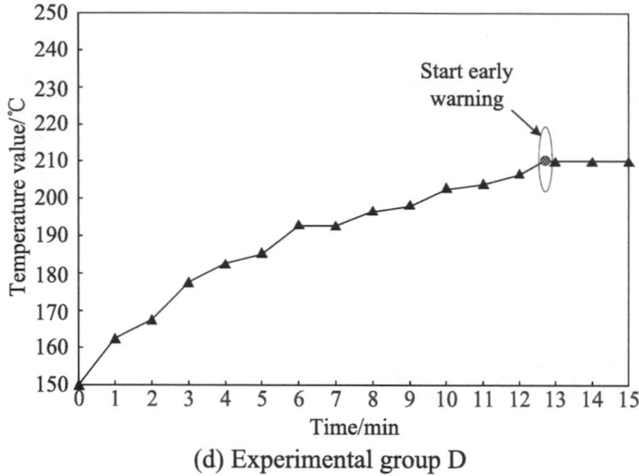

(d) Experimental group D

Fig. 3. continued

Table 3. Temperature warning speed test (min)

Test group	Group A	Group B	Group C	Group D
1	5.35	8.55	9.24	12.63
2	5.29	8.29	9.33	12.58
3	5.14	8.54	9.25	12.6
4	5.22	5.76	9.28	12.62
5	5.25	8.55	9.37	12.55
6	5.3	8.52	9.41	12.49
7	5.24	8.47	9.4	12.46
8	5.18	8.61	9.35	12.58

According to Table 3, the average warning time of the four groups was 5.25 min, 8.16 min, 9.33 min and 12.56 min, respectively. According to the above calculation results, when the critical temperature is 210 °C, the warning speed of the designed temperature measurement method is 2.91 min, 4.08 min and 7.31 min faster than the three methods respectively. Based on the above two sets of test results, it is verified that the temperature measurement method designed in this paper has good temperature warning effect in high temperature and low temperature environment.

4 Conclusion

On the basis of traditional temperature measurement methods, the tunable laser temperature measurement technology is optimized by means of spectral absorption, which provides a more reliable technical means for the current temperature measurement.

However, due to the limitation of research time, the method in this paper still has some deficiencies. For example, the amount of calculation is too large, and the calculation of various data is completed under the guidance of a large number of formulas, so there may be some calculation errors. Therefore, in the future research, it will be considered to design a supplementary detection module for real-time testing of various data to improve the reliability of temperature measurement results.

Fund Projects. Doctoral project of Beijing Polytechnic,A tunable laser temperature measurement method based on spectral absorption (2017Z004-006-KXZ).

References

1. Lv, X., Zhao, J., Xiong, Y., et al.: Design and study of a sampled gratings based tunable DBR laser. Study Opt. Commun. **12**(06), 47–51 (2019)
2. Zheng, S., Yang, Y.: High precision and accuracy wavelength tuning characteristics of modulated grating Y-branch tunable lasers. Chin. J. Lasers **46**(02), 9–16 (2019)
3. Zheng, W., Xie, R., Lu, Y., et al.: Study on crystal morphology and gas molecular transport mechanism of flammable ice generation process. Shanghai Energy Cons. **25**(02), 135–139 (2019)
4. Zhang, J., Ma, H., Ma, L., et al.: First-principles study on modulation mechanism of the electronic structures and magnetic properties of CdG by gas molecules adsorption. J. Synth. Cryst. **48**(06), 1030–1039 (2019)
5. Chen, S., Song, T., Yin, X.: Research on long-range hydrogen sulfide gas concentration detection system based on spectral absorption principle. Chin. J. Electron. Dev. **43**(04), 736–740 (2020)
6. Zhang, B., Xu, Z., Liu, J., et al.: absorption model of wavelength modulation spectroscopy in combustion flow field. Chin. J. Lasers **46**(07), 300–306 (2019)
7. Su, C., Yan, C., Wang, F., et al.: Research on error compensation method for infrared temperature measurement under laser irradiation. J. Appl. Opt. **40**(06), 1084–1090 (2019)
8. Liu, S., Liu, D., Srivastava, G., Połap, D., Woźniak, M.: Overview and methods of correlation filter algorithms in object tracking. Complex Intell. Syst. **7**(4), 1895–1917 (2020). https://doi.org/10.1007/s40747-020-00161-4
9. Liu, S., Lu, M., Li, H., et al.: Prediction of gene expression patterns with generalized linear regression model. Front. Genet. **10**, 120 (2019)
10. Liu, S., Bai, W., Zeng, N., et al.: A fast fractal based compression for MRI images. IEEE Access **7**, 62412–62420 (2019)

Thermal Zero Drift Compensation of Pressure Sensor Based on Data Mining and BP Neural Network

Ya-ping Li(✉) and Dan Zhao

College of Mechatronic Engineering, Beijing Polytechnic, Beijing 100176, China

Abstract. Due to the poor compensation accuracy, the traditional compensation algorithm for thermal zero shift of pressure sensor results in large error of pressure measurement. Therefore, this paper proposes a pressure sensor thermal zero drift compensation algorithm based on data mining and BP neural network. Combined with the data mining process, the characteristics of the thermal zero drift of the pressure sensor are analyzed, and the hysteresis and nonlinear characteristic curve of the pressure sensor is obtained to prepare for the compensation of the thermal zero drift. Then BP neural network is introduced to determine the parameter update mode, which is effectively combined with artificial fish swarm algorithm, and the compensation of pressure sensor thermal zero shift is realized by implementing the thermal zero shift compensation algorithm of pressure sensor. The experimental results show that the pressure measurement error range of the algorithm in this paper is 0.30 N–1.45 N. Compared with the three existing algorithms, the pressure measurement error of the algorithm in this paper is smaller, which indirectly shows that the algorithm in this paper has a higher thermal zero drift compensation accuracy, which fully shows that the algorithm in this paper compensates better performance.

Keywords: Data mining · BP neural network · Pressure sensor · drift compensation · Artificial fish swarm algorithm

1 Introduction

The three foundations of modern information technology are information collection, transmission and processing technology, namely sensor technology, communication technology and computer technology, which constitute the "sense", "nerve" and "brain" of information technology system respectively. The most important component of information acquisition system is sensor, which is placed in the front of the system. In this case, sensor can be called cutting-edge technology [1]. With the development of communication technology and computer technology, sensor technology has been improved. Sensor is a kind of detection device, which can feel the measured information and transform the detected information into electrical signal or other required information output according to certain rules, so as to meet the requirements of information transmission,

S. Liu and X. Ma (Eds.): ADHIP 2021, LNICST 416, pp. 94–104, 2022.
https://doi.org/10.1007/978-3-030-94551-0_8

processing, storage, display, recording and control. Sensor is the first step to realize automatic detection and control.

The pressure sensor is one of the key types of sensors. The core of the pressure sensor is the diffused silicon resistance bridge, and the single chip microcomputer technology is used to collect data, process and output the display results. The piezoresistive coefficient of diffused silicon is a function of temperature, so there is a temperature drift of sensitivity. There are many factors that affect the temperature, such as the change of measuring environment, the change of measuring circuit heat, etc.

The zero point of pressure sensor has thermal drift, electrical drift and time drift [2, 3]. The thermal zero drift of pressure sensor will lead to the increase of temperature variation and the change of internal circuit parameters, which will lead to the output of interference signal which has nothing to do with the measurement and greatly affect the performance of pressure sensor. The existing thermal zero drift compensation algorithm of pressure sensor mainly includes bridge arm series and parallel constant resistance algorithm, bridge arm thermistor compensation algorithm and bridge external series and parallel thermistor compensation algorithm. Through investigation and research, it is found that the existing algorithms have the problem of large measurement pressure error.

Based on data mining and BP neural network, a new compensation algorithm for thermal zero drift of pressure sensor is designed in this paper to overcome the shortcomings of traditional compensation algorithms. The design idea of this method is as follows:

(1) Combined with the data mining process, the thermal zero drift characteristics of the pressure sensor were analyzed, and the hysteresis nonlinear characteristic curves of the pressure sensor were obtained to prepare for the thermal zero drift compensation.
(2) BP neural network was introduced to determine the parameter update mode, which was effectively combined with artificial fish swarm algorithm, and then the compensation of pressure sensor thermal zero shift was realized by implementing the thermal zero shift compensation algorithm of pressure sensor.
(3) Experimental verification shows that: compared with the three existing algorithms, the proposed algorithm has a smaller error in measuring pressure, which indirectly indicates that the proposed algorithm has a higher compensation accuracy for thermal zero drift.

2 Research on Compensation Algorithm for Thermal Zero Drift of Pressure Sensor

2.1 Analyze the Characteristics of the Thermal Zero Drift of the Pressure Sensor Through Data Mining

The so-called zero point output of the pressure sensor refers to the output of the pressure sensor when there is no external pressure under a certain reference temperature and a certain excitation source. The zero output voltage can sometimes be larger than the output voltage at full scale. At this time, a resistor can be connected in series or in parallel with a certain bridge arm to make the bridge output zero, which is easy to achieve at a certain temperature [4]. The parameter to judge the quality of the zero point output of a pressure sensor is the ratio of the maximum zero point output to the upper range limit of the sensor. If the ratio is less than 0.8%, the zero point output of the sensor is considered to be small and can be ignored; if it is greater than 4%, the zero point output is equivalent big.

The main reasons for the zero-point output of the pressure sensor include: first, the difference between the design size of the force sensitive resistance bar and the actual size formed by lithography, and the inconsistency in the resistance value of the resistance bar due to the uneven doping concentration; Second, the stress introduced by the package is applied to the sensor to produce the output. Therefore, this study analyzes the characteristics of thermal zero drift of pressure sensors through data mining, and the process is as follows:

The relationship curve between the output voltage of the pressure sensor and the pressure is shown in Fig. 1.

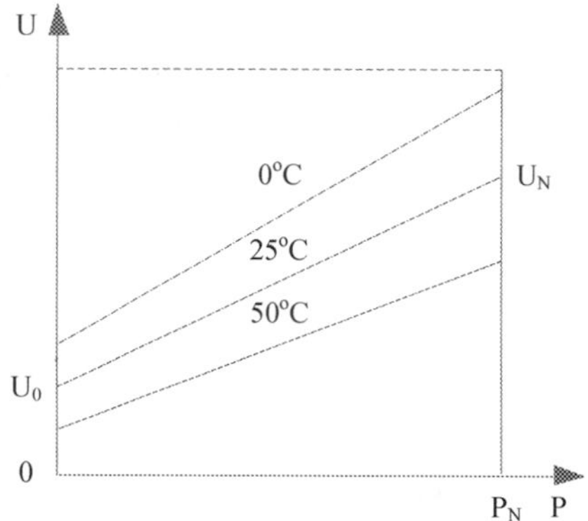

Fig. 1. Schematic diagram of relationship between output voltage and pressure of pressure sensor

In Fig. 1, U represents the output voltage of the pressure sensor; P represents the pressure value of the pressure sensor.

As shown in Fig. 1, it shows the relationship between the typical output voltage of the pressure sensor and the pressure under a certain excitation source. The three straight lines respectively represent the output conditions of the sensor at 0 °C, 25 °C and 50 °C. The zero point of the pressure sensor affected by temperature has changed significantly. This is because the doping of the force-sensitive resistor strips is inconsistent, resulting in a difference in the temperature coefficient of resistance. Considering only a single temperature situation, the zero-point output voltage offset can easily be offset by adjustment [5]. However, after the temperature changes, a new zero point output will appear. This phenomenon is called temperature drift of the zero point output voltage, or thermal zero point drift for short. When the three curves in Fig. 1 are subjected to the same initial pressure P_0, the output difference is caused by the thermal zero drift. Thermal zero drift is one of the important indicators of the performance of pressure sensors. Generally, the thermal zero drift coefficient α_0 is used to represent:

$$\alpha_0 = \frac{U_0^{\max}(T) - U_0^{\min}(T_0)}{\Delta T\left[\overline{U}_F - \overline{U}_0\right]} \times 100\%FS/^{\circ}\mathrm{C} \tag{1}$$

In Eq. (1), $U_0^{\max}(T)$ is the maximum zero point output at a reference temperature T; $U_0^{\min}(T_0)$ is the minimum zero point output at a reference temperature T_0; ΔT is the temperature working range of the sensor; $\overline{U}_F$ is the average value of the full-scale output at each reference temperature; $\overline{U}_0$ is the average value of the zero point output at each reference temperature; *FS* is the unit name of the thermal zero point drift coefficient α_0.

Full scale output refers to the maximum effective output of the pressure sensor within the required range under the specified reference temperature and certain excitation current or voltage conditions [6]. As shown in Fig. 1, the full scale output is:

$$U_{sig}^{N}(T_0) = U_N(T_0) - U_0(T_0) \tag{2}$$

In formula (2), $U_{sig}^{N}(T_0)$ is the maximum effective output at a reference temperature T_0; $U_N(T_0)$ is the upper limit of the standard full scale at a reference temperature T_0; $U_0(T_0)$ is the zero point output value at a reference temperature T_0.

When the force-sensitive resistor is distributed in the appropriate area of the silicon diaphragm, the output signal will be the largest when under pressure, and the full scale should be widened as much as possible. The sensitivity is usually used to describe the performance of the pressure sensor. The sensitivity S is defined as:

$$S = \frac{1}{V_B}\frac{U_N - U_0}{P_N} = \frac{1}{V_B}\frac{U_{sig}^{N}}{P_N} \tag{3}$$

In formula (3), V_B is the external excitation voltage; P_N is the upper limit of the sensor range. It can be seen from the formula (3) that the full-scale output U_{sig}^{N} is equal to the sensitivity, V_B multiplied by the upper limit of the range, that is, the full-scale output is:

$$U_{sig}^{N} = S \times V_B \times P_N \tag{4}$$

Pressure sensitivity S and full-scale output U_{sig}^{N} both decrease with increasing temperature. This is because the pressure sensitivity is proportional to the piezoresistive coefficient, and the piezoresistive coefficient decreases with increasing temperature. In addition, the thermal stress is caused by the difference between the two sides of the force sensitive resistor $p-n$ junction and the thermal expansion coefficient of the surface SiO_2 or the passivation layer Si_3N_4, resulting in an additional piezoresistive effect, which is also related to temperature. The thermal sensitivity drift is also a measure of the quality of the pressure sensor. One of the important indicators [7]. The expression of thermal sensitivity drift coefficient K_0 is:

$$K_0 = \frac{\Delta U_F^{\max}(T) - \Delta U_F^{\min}(T_0)}{\left[\overline{U}_F - \overline{U}_0\right]\Delta T} \times 100\%FS/^{\circ}\mathrm{C} \tag{5}$$

In Eq. (5), $\Delta U_F^{\max}(T)$ refers to the maximum value of full-scale output minus zero output at a reference temperature T; $\Delta U_F^{\min}(T_0)$ refers to the minimum value of full-scale output minus zero output at a reference temperature T_0.

2.2 Analysis of Hysteresis and Nonlinear Characteristics of Pressure Sensors Through Data Mining

Based on the analysis results of the thermal zero drift characteristics of the pressure sensor, the hysteresis nonlinear characteristic curve of the pressure sensor is obtained through the data mining process, as shown in Fig. 2.

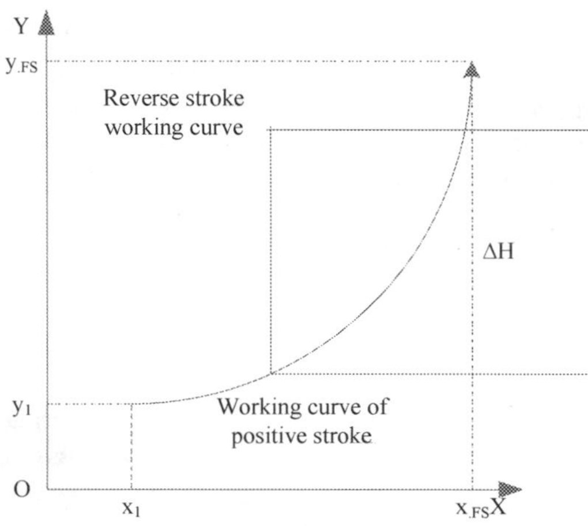

Fig. 2. Hysteresis and nonlinear characteristic curve of pressure sensor

As shown in Fig. 2, the hysteresis nonlinearity refers to the degree of inconsistency between the positive and negative stroke input and output curves of the sensor in the full range. The hysteresis nonlinear characteristics widely exist in sensors, piezoelectric ceramics, ferromagnets, semiconductor materials and smart materials, etc. field. As a pressure sensor with excellent performance, the pressure sensor also has hysteresis and nonlinear characteristics, which limits the accuracy of the measurement [8]. The magnitude of hysteresis nonlinearity can be determined by pressure calibration experiments. In the calibration experiment of the pressure sensor, first calculate the difference between the arithmetic mean of the forward output and the arithmetic mean of the reverse output at a set pressure measurement point:

$$\Delta y_{H_i} = \overline{y}_{+i} - \overline{y}_{-i} (i = 1, 2, 3, \cdots) \tag{6}$$

In Eq. (6), $\overline{y}_{+i}$ and $\overline{y}_{-i}$ refer to the arithmetic mean value of forward output and backward output respectively; the value range of i is $(1, +\infty)$.

The calculation formula of sensor hysteresis is:

$$\xi_H = \frac{|\Delta y_H|}{y_{FS}} \times 100\% \tag{7}$$

In formula (7), ξ_H is the hysteresis value of the sensor; y_{FS} is the full-scale output.

Solve the hysteresis and nonlinear problems of pressure sensors. Generally, data fusion algorithms are used to compensate. The pressure sensor designed in this research uses the BP neural network algorithm to compensate for the hysteresis. Formula (7) is used to calculate the change in the hysteresis of the sensor before and after compensation to verify the effect of the hysteresis correction.

2.3 BP Neural Network Algorithm Introduction

BP neural network algorithm is a multilayer feedforward network trained by error back propagation algorithm, which is usually used for data classification and prediction. The most important part of BP neural network algorithm is the learning part of its weight and threshold. Generally, the learning process is divided into two parts. One part is the forward transmission process, that is, the input samples are transferred from the input layer, and then processed layer by layer by each hidden layer, and then transferred to the output layer. The other part is the process of error reverse transmission, that is, if the actual output of the output layer is not the same as the expected output, the error is used as the adjustment signal to reverse the transmission layer by layer, and the connection weight matrix between neurons is processed to reduce the error. After repeated learning process, the error is finally reduced within the initial set range [9].

BP neural network is composed of input layer, output layer and intermediate layer between them. The middle layer can be single-layer or multi-layer. Because the middle layer is not connected with the external environment, it is also called hidden layer. The input layer, hidden layer and output layer are connected with each other, but the nodes of single layer are not connected. The structure of BP neural network is shown in Fig. 3.

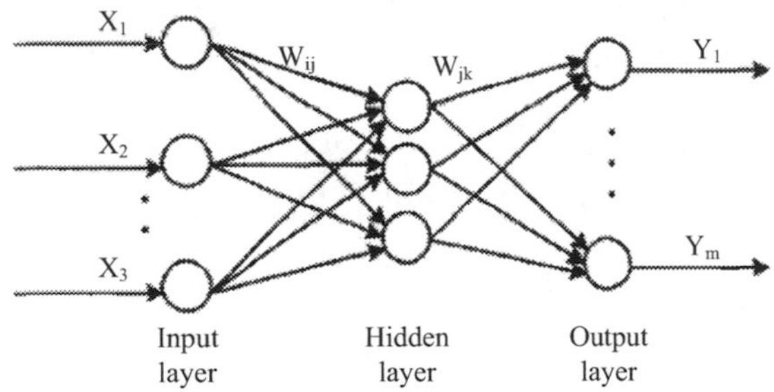

Fig. 3. BP neural network structure diagram

It can be seen from Fig. 3 that the BP neural network algorithm is composed of multiple input layers, hidden layers and output layers. Specifically, the input layer accepts external input by a series of input signals from X_1 to X_2. Correspondingly, the output layer is composed of a series of network output values, which is $Y_1, Y_2, \cdots, Y_m$ in the figure above. The hidden layer between the two is represented by the corresponding network weights W_{ij} and W_{jk}. In addition, θ_j and θ_k can be used to represent the network threshold between two adjacent layers. Then the output of each node can be obtained by the following formula:

$$\begin{cases} x_j^* = f\left(\sum_{i=1}^{n} w_{ij}x_j - \theta_j\right) \\ y_k = f\left(\sum_{j=1}^{l} w_{jk}x_j^* - \theta_k\right) \end{cases} \tag{8}$$

In Eq. (8), f represents the function transformation relationship between the input and output of neurons, which is called the excitation function and type S function:

$$f(x) = \frac{1}{1 + e^{-x}} \tag{9}$$

The prediction error E of BP neural network can be obtained, where o_k is the expected value of BP neural network:

$$E = \frac{1}{2}\sum_{k=1}^{m}(o_k - y_k) \tag{10}$$

If the error at this time is less than the previously set error, the calculation ends, otherwise it enters the error reverse transmission part.

Minute. In order to further reduce the error, the gradient descent method is used to adjust and update the weights and thresholds. The so-called gradient descent method is to adjust the weights and thresholds to the negative gradient direction, so that the adjustment amount of the two and the gradient of the error fall into direct ratio. The parameter adjustment formula is:

$$\begin{cases} W'_{ij} = w_{ij} - \eta\frac{\partial E}{\partial w_{ij}} \\ W'_{jk} = w_{jk} - \eta\frac{\partial E}{\partial w_{jk}} \\ \theta'_j = \theta_j - \eta\frac{\partial E}{\partial \theta_j} \\ \theta'_k = \theta_k - \eta\frac{\partial E}{\partial \theta_k} \end{cases} \tag{11}$$

In formula (11), η represents the learning rate, and the value range is [0, 1]. Train the BP neural network according to the result of formula (11). If the error is less than the set error, the learning ends, otherwise it will enter the next round of learning.

2.4 Design of Thermal Zero Drift Compensation Algorithm

Based on the above introduced BP neural network algorithm combined with artificial fish school algorithm, a pressure sensor thermal zero drift compensation algorithm is introduced.

Since the artificial fish school algorithm can iteratively optimize in the global scope, combining the artificial fish school algorithm with the BP neural network can combine the advantages of the two to make up for its own shortcomings. In other words, the artificial fish swarm algorithm can be used for global optimization to solve the problem of BP neural network being easily trapped in local minimums [10, 11]. Assuming that each artificial fish in the artificial fish school represents a BP neural network, the initial weights and thresholds of the BP neural network need to be optimized to correspond to the individual state of the artificial fish, and the best position of the BP neural network is obtained by finding the optimal artificial fish position. Optimal weights and thresholds, and then use the obtained optimal weights and thresholds to train the BP neural network. This study defines the fitness function F of the artificial fish school as the reciprocal of the mean square error value E of the BP neural network output result and the expected value result.

The specific steps of the pressure sensor thermal zero drift compensation algorithm are as follows:

Step 7: if *num* reaches the maximum number of iterations *trynumber* or the error of the solution is less than the set minimum error ε, the behavior ends and the calculation result is output, otherwise $num = num + 1$, go to step 4.

Through the above process, the thermal zero drift compensation of pressure sensor is realized, which provides effective means for the application and development of pressure sensor.

3 Simulation Experiment and Result Analysis

The above process realizes the design of thermal zero drift compensation algorithm of pressure sensor based on data mining and BP neural network, but whether it can solve the problems of existing algorithms is uncertain. Therefore, MATLAB software is used to design simulation experiments to verify the compensation performance of the proposed algorithm.

3.1 Simulation Experiment Data Acquisition and Experiment Preparation

The experiment of the temperature influence of the pressure sensor is carried out in an experiment box. The equipment used in the experiment is: two pressure sensors, a thermostat, a signal simulator, and a standard thermometer.

During the measurement, it needs to wait for the pressure value at each set temperature point before reading, and it needs to carry out 10 repeated experiments and read 10 times of data, and then it needs to average the 10 times of data corresponding to each observation point, so as to complete the acquisition of the measured value of the pressure sensor. For the calculation of the standard value of the pressure sensor, it is necessary to make further correction on the basis of the measured value of the pressure sensor, that is, to add the corresponding correction value on the basis of the average value of each test point. Some of the data are shown in Table 1.

The difference between the measured value of the pressure sensor and the standard value is taken as the measurement error value of the pressure sensor, the temperature value is taken as the abscissa, and the measurement error value is taken as the ordinate.

The temperature influence error curve of the pressure sensor corresponding to each temperature is drawn, as shown in Fig. 4. Based on the data in Fig. 4, BP neural network algorithm was trained to prepare for the simulation experiment.

Table 1. Part of simulation experiment data table

Interference amount (°C)	Measured value	Standard value	Error
20	23.08	23.55	0.47
25	22.57	25.05	2.48
30	45.89	48.25	2.36
35	45.59	48.00	2.41
40	68.65	71.20	2.55
45	67.65	71.25	3.60
50	88.77	90.35	1.58
55	85.40	92.30	6.90
60	92.50	96.50	4.00
65	91.88	97.60	5.72

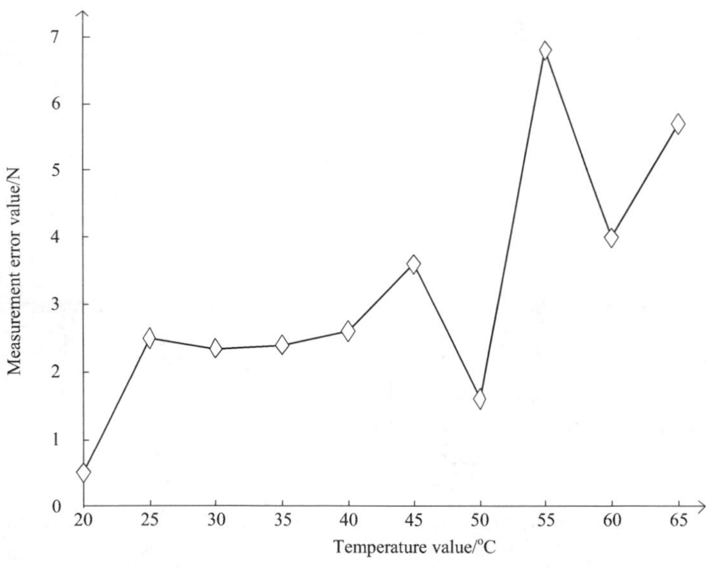

Fig. 4. Temperature influence error curve of pressure sensor

3.2 Analysis of Results

In order to clearly show the performance of the proposed algorithm, the bridge arm series and parallel appropriate constant resistance algorithm, the bridge arm thermistor compensation algorithm and the bridge outside series and parallel thermistor compensation algorithm and the proposed algorithm are used for simulation and comparative experiments. In order to facilitate the experiment, the above existing algorithms are called existing algorithm 1, existing algorithm 2 and existing algorithm 3, and the measurement pressure error is obvious after compensation It shows the performance of the algorithm.

The measured pressure error data obtained by simulation experiment is shown in Table 2.

Table 2. Data sheet of measurement pressure error

Interference amount (°C)	Measurement pressure error/N			
	The algorithm in this paper	Existing algorithm 1	Existing algorithm 2	Existing algorithm 3
20	0.30	2.13	2.56	2.48
25	1.23	2.00	2.01	2.44
30	1.20	2.10	3.10	2.69
35	1.11	2.11	3.00	2.58
40	1.01	2.05	2.99	2.15
45	1.00	2.56	2.01	2.12
50	1.03	2.48	2.55	2.30
55	1.22	2.55	2.41	3.10
60	1.45	2.97	3.01	3.00
65	0.98	3.01	3.12	3.05

According to the data in Table 2, the measurement error range of the existing algorithm 1 is 2.00–3.01; the measurement error range of the existing algorithm 2 is 2.01–3.12; the measurement error range of the existing algorithm 3 is 2.12–3.10; the measurement error range of the proposed algorithm is 0.30–1.45. Through the comparative study, it is found that the pressure measurement error of the proposed algorithm is far lower than the existing three algorithms.

The reasons for the above results are as follows: compared with the existing three algorithms, the proposed algorithm has smaller pressure measurement error, which indirectly indicates that the proposed algorithm has higher accuracy of thermal zero drift compensation, and fully indicates that the proposed algorithm has better compensation performance. The reason for the above results is that the algorithm in this paper combines the data mining process to analyze the thermal zero drift characteristics of the pressure sensor and obtain the hysteresis nonlinear characteristic curve of the pressure sensor, which makes sufficient preparation for the thermal zero drift compensation. Then, BP neural network is introduced to determine the parameter updating mode, which is combined with artificial fish swarm algorithm effectively, thus improving the accuracy of thermal zero drift compensation.

4 Conclusion

In this paper, data mining and BP neural network are applied to the process of pressure sensor thermal zero drift compensation, which greatly improves the accuracy of pressure sensor thermal zero drift compensation, reduces the pressure sensor measurement error, and provides a certain reference value for the research of pressure sensor thermal zero drift compensation. Although the existing experimental results can prove the effectiveness of the proposed algorithm, the explanability of the obtained experimental results is low due to the lack of experimental data in the simulation experiment, and there are also certain deviations. Therefore, in the following research, the compensation algorithm of thermal zero shift of pressure sensor will be further studied and discussed, the amount of simulation data will be increased, and the compensation results will be optimized.

Fund Projects. Research on temperature compensation method of silicon sapphire high temperature pressure sensor (CJGX2016-KY-YZK032).

References

1. Zhang, J., Soangra, R., Lockhart, T.: Automatic detection of dynamic and static activities of the older adults using a wearable sensor and support vector machines. Sci **2**(3), 56 (2020)
2. Song, H., Chen, W., Wang, X., et al.: Detection of methanol by a sensor based on rare-earth doped TiO 2 nanoparticles. J. Wuhan Univ. Technol.-Mater. Sci. Ed. **33**(5), 1070–1075 (2018)
3. Utama, D.T., Lee, S.G., Baek, K.H., et al.: Effects of high-pressure processing on taste-related ATP breakdown compounds and aroma volatiles in grass-fed beef during vacuum aging. Asian-Australas. J. Anim. Sci. **31**(8), 1336–1344 (2018)
4. Ogura, Y., Sato, K., Miyahara, S.I., et al.: Efficient ammonia synthesis over a Ru/La0.5Ce0.5O1.75 catalyst pre-reduced at high temperature. Chem. Sci. **9**(8), 2230–2237 (2018)
5. Zanos, T.P., Silverman, H.A., Levy, T., et al.: Identification of cytokine-specific sensory neural signals by decoding murine vagus nerve activity. Proc. Natl. Acad. Sci. U.S.A. **115**(21), 4843–4852 (2018)
6. Gao, X., Li, H., Wei, J.X.: MiR-4421 regulates the progression of preeclampsia by regulating CYP11B2. Eur. Rev. Med. Pharmacol. Sci. **22**(6), 1533–1540 (2018)
7. Bakis, A., Isik, E., El, A., Ülker, M.: Mechanical properties of reactive powder concretes produced using pumice powder. J. Wuhan Univ. Technol.-Mater. Sci. Ed. **34**(2), 353–360 (2019). https://doi.org/10.1007/s11595-019-2059-1
8. Liu, S., Lu, M., Li, H., et al.: Prediction of gene expression patterns with generalized linear regression model. Front. Genet. **10**, 120 (2019)
9. Liu, S., Bai, W., Zeng, N., et al.: A fast fractal based compression for MRI images. IEEE Access **7**, 62412–62420 (2019)
10. Liu, S., Liu, D., Srivastava, G., Połap, D., Woźniak, M.: Overview and methods of correlation filter algorithms in object tracking. Complex Intell. Syst. **7**(4), 1895–1917 (2020). https://doi.org/10.1007/s40747-020-00161-4
11. Wu, Q., Zhang, C., Zhang, M., et al.: A modified comprehensive learning particle swarm optimizer and its application in cylindricity error evaluation problem. Int. J. Performability Eng. **15**(3), 2553 (2019)

Clustering Mining Method of College Students' Employment Data Based on Feature Selection

Mei-bin Qi(✉)

Admission and Employment Office, Dongying Vocational Institute, Dongying 257090, China
hbdsfisdf545@sohu.com

Abstract. In order to further clarify the employment trend and employment status of college students, this paper proposes a clustering mining method for college students' employment data based on feature selection. This paper analyzes the employment market of college students, and gives the data structure of college Students' employment. On this basis, the employment data of college students are normalized and compressed. Based on the preprocessed employment data of college students, the sparse score method is used to select data features. And through online clustering strap algorithm deep Clustering Mining College Students' employment data, so as to realize the clustering mining of College Students' employment data. The experimental data show that the clustering accuracy of the existing methods ranges from 45.23% to 54.79%, and the clustering accuracy of the proposed methods ranges from 70.15% to 83.54%. Through data comparison, it is found that compared with the existing methods, the clustering accuracy of the proposed method is higher, which fully shows that the clustering mining effect of the proposed method is better.

Keywords: College student · Employment data · Clustering · Excavate · Feature selection

1 Introduction

With the rapid development of information technology, data mining technology has become one of the hot issues in the field of artificial intelligence. In this information age of rapid growth of data, data mining technology has become a powerful tool to solve the problem of big data explosion [1]. At present, this technology has been applied in many fields, and its application in the field of education has gradually developed, which is also an important research direction in the future. Colleges and universities generally have established a student management system, which saves a large amount of student employment history data. There are many important patterns and knowledge hidden behind these massive data, which are of great significance to the analysis and research of factors affecting student employment [2].

It is urgent to solve the employment problem of colleges and universities. In recent years, colleges and universities also attach great importance to the employment problem

S. Liu and X. Ma (Eds.): ADHIP 2021, LNICST 416, pp. 105–115, 2022.
https://doi.org/10.1007/978-3-030-94551-0_9

of graduates, and promote the employment problem to the strategic decision-making of school development. Many colleges and universities have set up employment guidance centers, which have specialized teachers who are responsible for opening up various employment channels, providing a lot of employment information and solving students' employment problems, including public institutions, postgraduate entrance examinations, entering enterprises, encouraging self employment, etc.

At the same time, colleges and universities have basically realized the digitization of students' personal information, and the student management system, educational administration management system and employment management system of colleges and universities have accumulated a lot of student data. These accumulated large amounts of student data have not been fully and effectively utilized. In terms of the employment of college students, the employment data of college graduates is still relatively simple and insufficient. According to the evaluation and evaluation of the school employment rate and professional employment rate, it is concluded that the majors, classes and colleges with high employment rate sometimes do not play the potential role and value of the data [3].

With the rapid development of the field of artificial intelligence, it is a new research direction in the education industry to process the massive data of colleges and universities with intelligent algorithms, and it is also one of the research hotspots in recent years. Based on feature selection, this study processes the employment information of graduates, classifies and forecasts the unemployed students, which not only solves the problems of the current data mining algorithm in the analysis of employment data in colleges and universities. It can also provide some reference for the employment of college graduates and serve for education and teaching.

2 Research on Clustering Mining Method of College Students' Employment Data

2.1 Data Analysis of College Students' Employment

The analysis of the employment market of college students mainly includes three main subjects, one of which is the employment of college students. The second main body is the employer (recruiter). The third main body is the talent market (intermediary). The third main body has various forms, including local talent market and online virtual talent market. It also includes the employment center of colleges and universities or the human resources department of large enterprises. At present, the data of College Students' employment market includes three main aspects: personal employment data of college students, recruitment data of recruitment units (including post data), and employment data of talent market. College Students' personal employment data mainly refers to the employment related data of college students, including knowledge, skills, humanistic spirit, job-hunting willingness and other information.Recruitment data of recruitment units mainly refers to the data related to recruitment units, including the size, nature, industry, salary and welfare, job description and requirements of recruitment units [4]. The employment data of the talent market refers to the relevant data generated by both sides in the talent market, including the employment agreement and the relevant agreements of the agreement. The data structure of College Students' employment market information is shown in Table 1, 2, 3 and 4.

Table 1. "Graduate" structure of college students' employment data table

Field name	Field meaning	Type definition
Stu_id	Graduate number	Decimal (10,0) (Main code)
Stu_name	Name of graduate	Varchar (30)
Stu_sex	Gender of graduates	Smallint [1/0]
Stu_region	Graduate nationality	Varchar (10)
Stu_year	Graduation year	Smallint
Stu_tel	Contact number	Varchar (50)
Stu_marks	Credit basis	Decimal (4,2)
Stu_school	Graduate School	Varchar (30)
Stu_major	Graduation major	Varchar (30)
Stu_in_marks	College entrance examination results	Smallint
Stu_talk	Putonghua score	Smallint
Stu_cer	Certification qualification	Smallint
Stu_honor	Awards	Smallint
Stu_cet	The passing of CET-4 and CET-6	Varchar (10)
Stu_job	Employment intention	Varchar (50)

Table 2. Recruitment data "enterprise" structure of recruitment units

Field name	Field meaning	Type definition
en_id	Enterprise number	Decimal (10,0) (Main code)
en_name	Enterprise name	Smallint [1/0]
en_num	Enterprise code certificate No	Varchar (10)
en_adr	Business address	Smallint
en_man	Business contacts	Varchar (50)
en_tel	Contact number	Smallint
en_tax	Enterprise tax amount of last year	Smallint
en_salary	Average salary of enterprises	Varchar (50)
en_cha	Nature of enterprise	Varchar (10)
en_scale	Enterprise scale	Varchar (30)
en_kind	Enterprise industry	Smallint

Table 3. "Employee" table of recruitment position data of recruitment unit

Field name	Field meaning	Type definition
em_id	Enterprise number	Decimal (10,0) (Main code)
em_man	Business contacts	Smallint
em_tel	Contact number	Varchar (50)
em_tax	Job description	Varchar (50)
em_salary	Post salary	Smallint
em_cha	Post nature	Varchar (50)
em_duty	Job requirements	Varchar (50)
em_ad	Post welfare	Varchar (50)

Table 4. Employment data sheet "agreement" structure

Field name	Field meaning	Type definition
ag_id	Agreement number	Decimal (10,0) (Main code)
ag_en	Enterprise number	Smallint
ag_stu	Graduate number	Smallint
ag_date	Date of agreement	Date
ag_salary	Salary of agreed position	Smallint

The employment information data of college students is increasing with the increase of the evaluation index of college students and employers. The data in Table 1, 2, 3 and 4 are relatively complete and accurate, covering most of the attributes of employment data.

2.2 Data Preprocessing of College Students' Employment

To preprocess the data in the database, the main tasks are as follows.

One is to clean up the data, mainly to deal with the empty value in the database. The second is the integration of data, which is mainly the standardized transformation of data [5, 6]. The standardization includes the binarization of graduates' gender, nationality and CET-4 passing, and the indexation of graduates' awards and certificates.

The calculation formula (1) is

$$r = \sum_{k=1}^{n} R_n \sigma_n \tag{1}$$

In formula (1), R_n is the award constant of graduates, and the value of R is assigned at the international, national, provincial and university levels. σ_n is the award coefficient, which reflects the weight of the award in the whole award system; n is the number of awards won by graduates.

This study uses the same algorithm to obtain the graduate certification index. Third, data specification is mainly to compress data sets and reduce the amount of data.

According to the needs of this research, the basic data is processed in Fig. 1 to get the data of Table 5 structure.

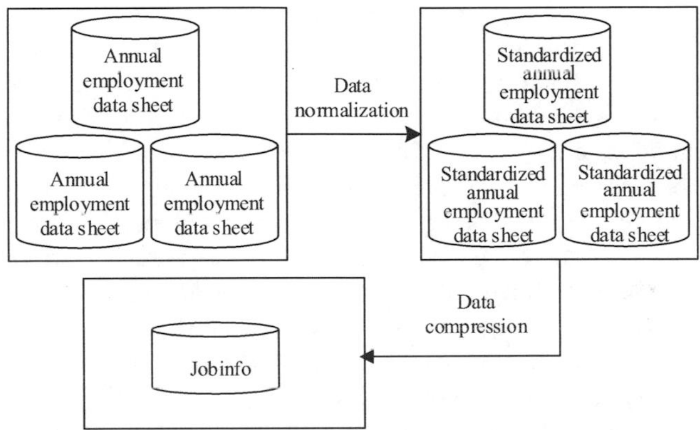

Fig. 1. Data preprocessing process

Table 5. "Jobinfo" structure of college students' employment data association analysis.

Field name	Field meaning	Type definition
gra_id	Graduate number	Decimal (10,0) (Main code)
gra_name	Name of graduate	Varchar (30)
gra_sex	Gender of graduates	Smallint [1/0]
gra_region	Graduate nationality	Varchar (10)
gra_ cet	The passing of CET-4 and CET-6	Smallint [1/0]
gra_year	Graduation year	Smallint
gra_empl	Employment situation of graduates	Smallint [1/0]
gra_marks	Base point credit graduate	Decimal (4,2)
gra_honor	Award winning index of graduates	Smallint
gra_cer	Graduate certification index	Smallint

2.3 Data Feature Selection

Although feature selection is a traditional research problem in the field of data mining and machine learning, it is still a research hotspot in these two fields. Especially with the emergence of large-scale data and high-dimensional feature data, the original feature selection algorithm has been difficult to meet the actual needs, and the research on feature selection presents a comprehensive and diversified trend [7]. On the one hand, feature selection technology began to expand to the field of machine learning. Different learning algorithms are used to select the features of the training samples and select a a reasonable feature subset. On the other hand, feature selection technology presents a systematic and structured trend. After analyzing a large number of feature selection methods, dash and Liu give a general framework of feature selection technology, which shows that a feature selection algorithm is mainly composed of four parts: generation strategy, evaluation criteria, The stop conditions and conclusion verification.

The generation strategy is to generate some feature subsets from the original feature set. The evaluation criterion is to evaluate the relevance of feature subset and judge the rationality of the generated feature subset. The stop condition is to judge whether the generated feature subset meets the requirements of the initial definition. Conclusion verification is to verify the effectiveness of the feature subset generated. The whole process of feature selection method extracts relevant feature subsets from the original feature set by generating strategy, and calculates the correlation of candidate feature subsets by evaluation criteria. If the candidate feature subset meets the The stop condition, it is submitted to the verification module for verification. If not, the generation strategy module continues to generate the next set of candidate feature subsets until the The stop condition is met. Obviously, the generation strategy and evaluation criteria are the core of the whole feature selection algorithm, which directly determines the performance of data clustering mining [8].

According to the label usage of the training data set, feature selection algorithms can be divided into three types: supervised, unsupervised and semi supervised. The supervised feature selection algorithm determines the importance of feature subset by using the correlation between feature and each data label. Unsupervised feature selection algorithm uses the variance or separation of data to judge the importance of each feature. Semi supervised feature selection algorithm uses a small number of labeled data samples as additional information to improve the performance of the unlabeled sample feature selection algorithm. According to the different evaluation methods of feature selection algorithm, feature selection can be roughly divided into the following three categories. The first is to separate the feature selection from the classification verification process, which is independent of the specific classifier. This kind of algorithm is called filtering method. The second is encapsulation method, which combines feature selection with classifier design. The pre specified classification learning algorithm is used to evaluate the classification performance of each feature or feature subset, so as to make a reasonable feature selection. The third type is embedded method, that is, feature selection method is a part of classifying the training, and the final feature selection subset is obtained by analyzing the results of the learning model. At present, most feature selection methods are based on filtering or encapsulation strategies. According to the idea of sparse fraction, this study selects data features, and the specific process is as follows.

Simultaneous interpreting is based on Sparse Score feature selection method, which is similar to the traditional fractional equation method. It is a filtering feature selection method. At the same time, it is an unsupervised feature selection algorithm because it does not use the class information of the training samples. Sparse score feature selection method calculates the corresponding sparse score size of all d-dimensional features respectively, and selects the first k features ($k < d$) in order from the smallest to the largest, so as to realize the purpose of feature selection.

The specific steps of feature selection method based on sparse fraction are as follows. Firstly, for the data set to be processed, based on the idea of sparse representation, the L1 norm minimization method is used to obtain the sparse representation reconstruction coefficient of each data in the whole data set. Specifically, for a given data set $\{X_i\}_{i=1}^n$, where $X_i \in R^d$, let each column in the data matrix $X = [x_1, x_2, \cdots, x_n] \in R^{d\times n}$ be a data vector of that data set. For each data vector x_i, formula (1) is used to obtain the reconstruction coefficient s_i of its corresponding sparse representation, that is, to solve the linear programming problem with the minimum L1 norm between the following data.

$$\begin{cases} \min\limits_{s_i} \|s_i\|_1 \\ s.t. x_i = X'_{s_i} \end{cases} \tag{2}$$

In formula (2), X' is the x_i data matrix X does not contain the ith column, and $s_i = \left[s_{i1}, \cdots, s_{ii-1}, 0, s_{ii+1}, \cdots, s_{in}\right]^T$ is an n maintenance number vector. Since x_i is not included in X when calculating s_i, the ith element in s_i is set to 0, and $s_{ij}, j \neq i$ represents the contribution of the jth data vector x_j in the dataset to the reconstruction of x_i. Thus, the reconstruction coefficient matrix $S = \left(\tilde{s}_{ij}\right)_{n\times n}$ of the sparse representation of the whole dataset $\{X_i\}_{i=1}^n$ is obtained. Where, each column vector $\tilde{s}_i$ is the sparse representation coefficient of the data vector x_i obtained by using formula (1).

Secondly, on the basis of the sparse representation, the sparse representation is obtained. The error between each feature of the original data and the reconstructed feature obtained by sparse representation reconstruction coefficient is calculated. By accumulating the reconstruction errors of the feature on all data samples, the sparsity preserving ability of each feature on the whole dataset is obtained. The specific criterion is: when the error between a feature and the reconstructed feature is small, it can be considered that the feature has good sparse representation retention ability in the whole dataset [9]. Thus, the sparse fractional objective function $S(r)$ is defined as follows.

$$S(r) = \frac{\sum\limits_{i=1}^{n} (x_{ir} - (X\tilde{s}_i)_r)^2}{Var(X(r, :))} \tag{3}$$

In formula (3), $\sum\limits_{i=1}^{n} (x_{ir} - (X\tilde{s}_i)_r)^2$ calculates the cumulative error between the r-th feature x_{ir} and the r-th reconstructed feature $(X\tilde{s}_i)_r$ of the whole sample set, and $Var(X(r, :))$ represents the variance of the r-th feature of the data set.

Finally, according to the sparsity score $S(r)$ of each feature in the data set, the sparsity preserving ability of each feature in the whole data set is sorted from small to large, and the smaller feature subset in $S(r)$ is selected to achieve the purpose of feature selection.

In a word, the feature selection method based on sparse fraction extends the idea of sparse representation to the field of feature selection. Assuming that the reconstruction error of one feature is less than that of another feature, it shows that the feature has better sparsity preserving ability and is considered as a good feature. The new method calculates the error between each feature of the original data and the feature reconstructed by sparse representation coefficient, and obtains the sparse fraction $S(r)$ of each feature. Finally, the sparsity preserving ability of all features is sorted.

2.4 Implementation of Data Clustering Mining

According to the selected data features, online clustering StrAP algorithm is used to mine the employment data of college students. The specific process is as follows.

StrAP is a relatively new algorithm. Based on AP clustering algorithm, this algorithm extends AP clustering to streaming data clustering. It is simple in theory but very practical and effective, and has been applied to some extent in grid computing. The detailed steps of StrAP algorithm are as follows.

Firstly, each data in a data stream is predefined to arrive in time order, and AP algorithm is used to cluster the first data points. So we get the initial model of StrAP.

Then, compare each data x_t that arrives successively with the center point of the StrAP model. Find out the center point x_t nearest to x_i and proceed to the next step.

Then, the distance $d(x_i, x_t)$ between x_i and x_t is calculated by Euclidean distance formula, and the size of S and threshold ε is compared. If the threshold ε is greater than the distance $d(x_i, x_t)$, the latest point x_t can be concluded from the StrAP model. In this case, only the class corresponding to the center point x_i in the StrAP model needs to be updated, which is very simple and efficient. If the threshold ε is less than the distance $d(x_i, x_t)$, it is considered that the latest point x_t is far away from the model and is not suitable for the StrAP model. Because the change of stream data is regular in time and space, point x_t may be a new class or noise, which cannot be accepted by the current model. Therefore, in order to update the data stream, the point x_t is put into a temporary storage box in memory.

When the algorithm meets one of the above two criteria, it is necessary to rebuild the StrAP model. At this time, the center points in the existing StrAP model and the points in the temporary storage box are clustered by WAP algorithm to get a new StrAP model. Repeat the above steps for new data points. From the process of the algorithm, we can see that the StrAP algorithm can meet the changing needs of stream data mining, that is, stream data change detection. At the same time, the complexity of the algorithm is much lower than that of the AP algorithm directly used in stream data clustering, and the processing efficiency is higher, which can meet the real-time processing requirements of large complex stream data and multiple data streams [10].

Through the above process, the cluster mining of college students' employment data is realized, which can provide some reference for college graduates' employment work and provide better service for education and teaching.

3 Experiment and Result Analysis

In order to verify the performance difference between the proposed method and the existing methods, a simulation experiment is designed. The specific experimental process is as follows.

3.1 Construction of Hadoop Platform

Hadoop is a cloud computing framework capable of distributed processing of big data. Its core modules are HDFS and MapReduce. HDFS is the underlying file system and MapReduce is the parallel computing mode. In order to enable users to easily develop distributed applications and make full use of the powerful computing and storage capabilities of Hadoop cluster, Hadoop designers cleverly designed the Hadoop architecture, so that users can not understand the underlying details of the distributed platform. With the understanding of MapReduce computing mode, we can develop distributed applications. Hadoop assumes that the failure of computing and storage is a normal state. In order to achieve the purpose of reliability, it maintains multiple copies of data in the cluster. HDFS can back up the data of failed nodes every time a failed node appears. Hadoop speeds up the data processing by parallelizing, and ensures the stability and reliability of the cluster through the underlying mechanism. So it is reliable and efficient. Hadoop is also scalable, and can meet the requirements of cluster dynamic growth with the increasing data volume and computing power. In addition, the Hadoop platform does not require high machine configuration, and the cost of building small-scale Hadoop is relatively low, which is suitable for small companies who want to own their own data center. And it can expand the cluster with the increase of data.

Therefore, this study uses the open source distributed software Hadoop to build the experimental cloud computing platform to test the clustering mining effect of the proposed method.

The platform consists of five PCs and a router. The software configuration of the cluster is shown in Table 6.

Table 6. Experimental software environment configuration

Software	Version
Linux	Ubuntu 9.10
Hadoop	Hadoop-0.20.2
JDK	Jdk-6u13-linux-i586
SSH	OpenSSH 5.8

3.2 Experimental Data Preparation

In this paper, we use the cluster ensemble toolbox of MATLAB to generate the experimental data. According to the requirements of the experiment, 10 data sets are generated, and the data sets are shown in Table 7.

Table 7. Experimental data set

Data set	Original document (MB)	Number of records (10^6)	Occupying HDFS space (MB)
1	4.02	0.01	64
2	19.56	0.05	64
3	39.07	0.1	64
4	40.12	0.3	64
5	44.71	0.5	64
A	389.91	1	448
B	1948.85	5	1984
C	3894.74	10	3904
D	5003	15	5016
E	7002	20	7012

3.3 Analysis of Experimental Results

Based on the Hadoop platform and the experimental data, the cluster mining experiment of College Students' employment data is carried out. The data of clustering accuracy obtained through experiments are shown in Table 8.

Table 8. Clustering accuracy data

Experimental data set	Existing methods	Proposed method
1	45.23%	70.15%
2	50.10%	80.14%
3	52.13%	78.54%
4	53.87%	80.12%
5	55.12%	83.01%
A	54.79%	78.95%
B	52.41%	83.54%
C	52.13%	81.09%
D	53.67%	82.01%
E	54.97%	83.79%

As shown in Table 8, the clustering accuracy of existing methods ranges from 45.23% to 54.97%, and the clustering accuracy of proposed methods ranges from 70.15% to 83.79%. Through data comparison, it is found that compared with the existing methods, the clustering accuracy of the proposed method is higher, which fully shows that the clustering effect of the proposed method is better.

4 Conclusion

In order to understand the employment trend of college graduates, this research introduces a feature selection algorithm and proposes a new cluster mining method for college graduates' employment data. On the basis of the employment data structure, the sparse score method is used to select data features. Then use the StrAP algorithm to intensively mine the employment data of college students. Comparative experiments show that the method in this paper greatly improves the accuracy of clustering, and the range of clustering accuracy is 70.15%–83.54%. Provide a certain theoretical reference for the research on employment data mining of college students.

References

1. Romero, C., Ventura, S.: Educational data mining and learning analytics: an updated survey. Wiley Interdiscip. Rev. Data Min. Knowl. Discov. **10**(3), e1355 (2020)
2. Zhou, F., Xue, L., Yan, Z., et al.: Research on college graduates employment prediction model based on C4. 5 algorithm. J. Phys. Conf. Ser. **1453**(1), 012033 (2020)
3. Agnihotri, D., Verma, K., Tripathi, P.: An automatic classification of text documents based on correlative association of words. J. Intell. Inf. Syst. **50**(3), 549–572 (2018)
4. Yi, Z., Rui, Z., Shi, X.W., et al.: Feature selection method based on high-resolution remote sensing images and the effect of sensitive features on classification accuracy. Sensors **18**(7), 2013 (2018)
5. Liu, P., Huang, W.: Incremental data mining-based software failure detection. Int. J. Performability Eng. **16**(8), 1279–1288 (2020)
6. Chang, G., Huo, H.: A method of fine-grained short text sentiment analysis based on machine learning. Neural Netw. World **28**(4), 325–344 (2018)
7. Liu, S., Liu, D., Srivastava, G., Połap, D., Woźniak, M.: Overview and methods of correlation filter algorithms in object tracking. Complex Intell. Syst. **7**(4), 1895–1917 (2020). https://doi.org/10.1007/s40747-020-00161-4
8. Kalaivani, A., Chitrakala, S.: An optimal multi-level backward feature subset selection for object recognition. IETE J. Res. **65**(4), 460–472 (2019)
9. Fu, W., Liu, S., Srivastava, G.: Optimization of big data scheduling in social networks. Entropy **21**(9), 902 (2019)
10. Vincenti, G., Bucciero, A., Helfert, M., Glowatz, M. (eds.): E-Learning, E-Education, and Online Training. LNICSSITE, vol. 180. Springer, Cham (2017). https://doi.org/10.1007/978-3-319-49625-2

High Precision Recognition Method of Basketball Dribbling Posture Based on Lightweight RFID Mobile Authentication Protocol

Qiang Huang(✉) and Yi-de Liao

Guangzhou Huali College, Guangzhou 511325, China
xqq5642@sina.cn

Abstract. The conventional high-precision recognition method of basketball dribbling posture has distance deviation and angle deviation when recognizing the three-dimensional position of bone points. To solve this problem, a high-precision recognition method of basketball dribbling posture based on lightweight RFID mobile authentication protocol is studied. The camera is placed in the basketball court, the RFID mobile authentication protocol label is arranged, and the basketball dribble posture image is collected. After compressing the image, the image is preprocessed by means of compression, noise suppression and graying, and the scale space of integral image is constructed to recognize the dribble posture edge and contour. Combined with convolution neural network, the precise characteristics of dribble posture are recognized. The results show that, compared with the two conventional methods, the distance deviation of the spatial position of the skeleton points is reduced by 0.007 m and 0.011 m, and the angle deviation is reduced by 2.9° and 4.11° respectively, which improves the attitude recognition accuracy.

Keywords: Lightweight RFID · Authentication protocol · Basketball · Dribbling posture · Image recognition

1 Introduction

With the progress of society and the improvement of living standards, people have begun to pay attention to the topic of physical health, and constantly improve their physical quality through sports. In the process of sports, the standard movement posture can not only determine the effect of sports to a certain extent, but also protect themselves from injury to the maximum extent in the process of sports. For basketball, using the basketball dribble mark Quasi movement for daily exercise can not only effectively exercise, but also improve the level of competition. However, in the process of motion, the definition of standard motion posture is mostly based on pictures or oral guidance, which leads to the lack of quantitative standard of standard motion posture. Therefore, it

S. Liu and X. Ma (Eds.): ADHIP 2021, LNICST 416, pp. 116–127, 2022.
https://doi.org/10.1007/978-3-030-94551-0_10

is of great significance to estimate human posture and recognize human action through the collected image or video sequence [1].

In reference [2], a recognition method based on image structure model is proposed. Heuristic local search technology is used for optimization processing to search reasonable initial solution and global optimal solution. Multiple limb components are used to represent the human body model. By counting the confidence of each component, human posture recognition and target detection are realized. In this method, each component can be represented as a joint or limb. If the components are connected with each other, they can represent the different postures of the human body model and qualitatively describe the meaning of the human body posture. In reference [3], a flexible hybrid model was used to capture the relationship between the limbs. At the same time, a star structure was used to represent the human body structure by using the variable component model DPM. Through rotation, scaling and size transformation, the position changes of different postures were displayed.

On this basis, a high-precision recognition method of basketball dribble posture based on lightweight RFID mobile authentication protocol is proposed. The efficiency of basketball dribble posture recognition is improved by using lightweight RFID mobile authentication protocol, which provides a theoretical basis for measuring standard movement posture.

2 Design of High Precision Recognition Method for Basketball Dribble Attitude Based on Lightweight RFID Mobile Authentication Protocol

2.1 Image of Basketball Dribble

Placement RFID Camera Based on Lightweight Mobile Authentication Protocol The camera is placed on the basketball court, and the angle of view is broadcast to collect the image of basketball dribbling posture. The standard basketball court is a rectangular court. In the actual field test, adjust the camera angle, let the camera hang down above the basketball court behind the basketball players, and form an angle of 45° to 60° with the horizontal plane, showing a kind of overlooking posture. You can see all the actions of the basketball players, synchronously collect the left and right camera images, and get two human posture coordinate systems, and then pass the test. The coordinated transformation of binocular vision transforms the posture of the human body to the world coordinated system. The linear distance between the camera and the basketball player is about 10.1 m, so that the shooting angle of the camera is the same as the height of the camera parallel to the horizontal plane and the height of the human body. The linear distance between the camera and the player is reduced, and the image features of the human body posture in the image are retained, so that the basketball player can drive the network to find the elbow covered by the trunk to reach the hand when he dribbles in order to obtain the feature representation information [4].

Due to the long distance of the camera, the human body is easily compressed to a very small area, and many image features are lost in the collected image. Therefore, the image features are perceived by using UHF RFID equipment combined with lightweight RFID mobile authentication protocol. RFID reader is selected. The RFID reader adopts R2000 imported RFID module, the kingerton R2000 chip, and four antennas are connected at the same time for data receiving, processing and command sending. The RFID antenna adopts Keller 12 dB high gain linear polarized UHF antenna, which has directivity and can detect long-distance RFID tags. The mobile authentication protocol tag adopts d68 long-distance electronic tag, which is a kind of RFID reader Passive tag, which has no harm to human health, is economical and practical, and can transmit data from lower computer to upper computer. Considering the rationality and economy, the distance between RFID mobile authentication protocol tags is half of the wavelength of RFID equipment.

$$A = \begin{cases} \left[\frac{2a}{\xi}\right] \times \left[\frac{2(b-0.2)}{\xi}\right] & b \leq 2.2\,\mathrm{m} \\ \left[\frac{2a}{\xi}\right] \times \left[\frac{2 \times 2}{\xi}\right] & b > 2.2\,\mathrm{m} \end{cases} \tag{1}$$

Where, A is the label spacing, ξ is the wavelength of RFID equipment, b is the width of basketball court, and a is the length of basketball court. Through formula (1), RFID tags are arranged to sense the change of tag return information caused by players in the range of tag array in the basketball court, assist the camera to capture the players' dribbling state, and retain more image features. So far, the camera position is placed.

Clear Camera Acquisition Parameters

Clear camera parameters, including interface type, pixel size, frame rate size, focal length size, color channel, the specific setting parameters are as follows: select USB3.0 as the interface between camera and computer, the theoretical maximum transmission bandwidth of USB3.0 is as high as 5 GB/s, in fact about 200 MB/s, the image storage size is about 1 MB, use USB3.0 to maintain the transmission speed of 200 frames/s, ensure the image quality The USB3.0 interface of holding camera is consistent, which makes it possible to use the interface layer to adjust and realize the compatibility of multiple cameras. Pixel is the basic unit of a picture. In the image presented by the computer, the image is composed of pixels. The image can be abstracted as a basic unit of matrix rate. The number of pixels in the source image determines the quality of the image. The more the number of pixels, the more feature information can be extracted. Select a high-pixel camera, take the cost control into account, and test a variety of different pixels The pixel parameters of the camera are given. Frame rate refers to the number of images collected per second. The frame rate itself does not affect the image quality, and mainly determines the delicacy of athletes' posture extraction. The higher the frame rate is, the more images of athletes' posture are collected per second. The posture changes of athletes' posture during dribbling will not be missed, that is, there will be no dynamic blur in the collected images, while the camera frame rate is higher If it is too low, there may be dynamic blurring of human limbs in the collected image, which is manifested by the diffusion of pixels in the dynamic blurring area to the surrounding area. Therefore, the frame rate is also an important parameter of the camera. Considering the cost control, test a number of cameras with different frame rates, and give the camera frame rate parameters that meet the requirements [5, 6]. The camera lens is divided into wide angle, standard and long focus. The focal length of the three lenses has their own range. When collecting images from the left side of the court to the athletes, it is found that the size of the camera focus will affect the acquisition angle. The effect test is carried out by using a variety of cameras with different focal lengths, and the camera with smaller focus is selected to make the acquisition angle include the whole half-time, so as to ensure the collected camera The image, including the basketball player in half-time, collect all the dribble posture of basketball players. The image has more than one color channel, which is used to save the color information of the image, and to overlay the color of all color channels in a certain position of the image, representing the color of the position in the image. Currently, the mainstream cameras have two color channels, one is RGB three channel camera, the other is gray single channel camera. The gray scale map is used to expand the single channel to three channels, and then RGB is selected Three channels are used to learn. The calculation formula of pixel Q at each position is:

$$Q = \frac{R \times 30 + G \times 59 + B \times 11 + 50}{100} \tag{2}$$

Among them, R, G and B are the pixel values of RGB channels of the camera. Using formula (2) to transform, the RGB three channels can be restored in single channel without external constraints, and the color features of the image can be extracted. So far, the camera acquisition parameters are clear, and the image data acquisition of basketball dribbling posture is realized.

2.2 Preprocessed Basketball Dribble Attitude Image Data

Preprocess the collected image to eliminate the irrelevant information in the image and ensure the image quality. Image preprocessing uses image compression, graying, geometric transformation, image enhancement and other means. Firstly, bilinear interpolation is used to compress the image, recover useful feature information, enhance the detectability of feature information, and simplify the image data to the maximum extent, so as to improve the reliability of feature extraction. The side length ratio of the source image and the compressed target image is calculated, and then the edge length ratio is calculated compared with the corresponding traceability image, the corresponding coordinate (X, Y) is calculated.

$$\begin{cases} X = \frac{\varpi \times \mathrm{u}}{\tau} \\ Y = \mathrm{n} \times \mathrm{m} \end{cases} \tag{3}$$

Where (ϖ, n) is the pixel position coordinates of the compressed target image, (u,m) is the size of the source image, and $(\tau, 1)$ is the size of the compressed target image [7]. After image compression, the cognitive region of the source image is preprocessed by noise suppression. The initial frame of the source image is used as the background, and compared with the subsequent image to eliminate the background points in the cognitive region of the image. Then, the source image is decomposed by using the wavelet domain Markov tree model. The cognitive region of the source image is transformed in the wavelet domain, and the wavelet coefficients are reconstructed to obtain the denoised image. The average filtering method is used to deal with the general noise. The given points of the pixels in the cognitive region are selected, and the neighboring pixels of the given points are weighted to calculate the average value and then replace the value of the given points. The average filtering formula is as follows.

$$B=\sum_{\alpha=1}^{C}\sum_{\beta=1}^{D}\mathrm{e}\left(\gamma_{\alpha\beta},\eta_{\alpha\beta}\right) \tag{4}$$

Among them, B is the given point value after mean filtering, α, β is the left neighborhood and right neighborhood of the given point in the image cognitive region, C, D is the left neighborhood threshold and right neighborhood threshold of the given point, e is the normalized weighting coefficient, $\gamma_{\alpha\beta}$ and $\eta_{\alpha\beta}$ are the horizontal neighborhood and vertical neighborhood of the given point [8]. For discrete noise, k-nearest neighbor filtering is used to denoise, establish the topological relationship between the pixels in the cognitive region, read in the pixels to be processed, establish KD tree, give a reference point in the pixels, and calculate the distance mean of k-nearest neighbor of the reference point. The calculation formula of distance mean E is as follows:

$$E=\frac{1}{\zeta}\sum_{\mathrm{s}\in H}\|\mathrm{s}-E\| \tag{5}$$

Where s is the given reference point of the pixels in the cognitive region, H is the k-nearest neighbor of s, and ζ is the KD tree nearest neighbor coefficient. The threshold of k-nearest neighbor distance is set. When the mean distance E is less than the threshold, the pixel is retained. When E is greater than or equal to the threshold, it is judged as discrete noise point and marked out. Finally, the image is transformed into gray image, and three primary colors R, G, B are selected as blue gray, green gray and red gray respectively. The maximum value of primary color component of each pixel is selected and multiplied by the corresponding weight to obtain the gray value after image graying. The calculation formula of gray value V is as follows:

$$V = \max(R \times W_R + G \times W_G + B \times W_B) \tag{6}$$

Among them, W_R, W_G, W_B are the different weights of R, G, B, which are converted into the gray image of quantization level 0–255 by formula (6). Then the gray-scale image is enhanced to highlight the key information of the image. The segmented linear enhancement method is used to select two lines with different slopes, so that the amplitude of gray-scale transformation of the image is different. The transformation function S is as follows.

$$S=\int_0^1 P_{\mathrm{r}}(\mathrm{w})\mathrm{dw} \tag{7}$$

Where, r is the gray level of the image, which is 0–255, P_{r} is the distribution function of the gray level of the image, and w is the integral variable of the gray level distribution function. After the transformation, the probability density of gray level of the image with uniform distribution is generated to expand the distribution range of pixel values and further enhance the background degree of the target and background region. Impulse noise and Gaussian noise are taken as the main noise of preprocessing, and linear mean filter is selected for denoising. For the pixels to be processed with noise, a 3 × 3 window is given to cover several adjacent pixels of the pixels to be processed. So far, the preprocessing of the image data of basketball dribble posture has been completed.

2.3 Recognition of Basketball Dribble Attitude

Recognition of Coarse Features of Basketball Dribble Posture

After preprocessing, the rough features of basketball dribble pose image are extracted, and the edge and contour of human body pose are preliminarily recognized. Hessian matrix is applied to the preprocessed image to select the scale. The scale σ of the matrix in image h(x,y) is defined as:

$$H=\begin{bmatrix} L_{xx}(h,\sigma) & L_{xy}(h,\sigma) \\ L_{xy}(h,\sigma) & L_{yy}(h,\sigma) \end{bmatrix} \tag{8}$$

Where H is the Hessian matrix of image h(x,y), $L_{xx}(h,\sigma)$ convolution h at x for Gaussian second order partial derivative, $L_{xy}(h,\sigma)$ is the convolution of h at x and y, and $L_{yy}(h,\sigma)$ is the convolution of h at y [9, 10]. The Hessian matrix template is selected to calculate the convolution, and all the continuous regions of the image are given the same weight. The integral image at any pixel position in the image is calculated, and the convolution of the image at the pixel position is obtained. Let the integral image corresponding to (x,y) be D(x,y), and the definition formula is as follows:

$$D(x,y)=\sum_{i=1}^{x}\sum_{j=1}^{y} I(i,j) \tag{9}$$

Where I(i,j) is the pixel value at position (x,y) for the image after scale selection. The scale space of the integrated image is constructed by using pyramid method, and the filter size is changed with the change of the construction size, so as to ensure the image size remains unchanged. In different scales of the integrated image, the target feature points are found. By using fitting method, the pixel points of the same scale in pyramid are compared, local extreme points are obtained, the position of feature points of target is initially located, and the spatial scale of contour and edge is determined. Using surf algorithm, the main direction of the target feature points is allocated. Firstly, a circular neighborhood is delineated, and the Harr wavelet features in the horizontal and vertical directions of all the image feature points in 60° fans are counted. Then, the sector area is rotated according to the interval standard of 0.2 rad, and the Harri small Porter sign is counted again until the whole circle area is traversed, and the longest vector direction is selected and taken as the target feature point The main direction of. At this point, the recognition of coarse features of the basketball dribble posture image is completed.

Identification of Fine Characteristics of Basketball Dribble Attitude
Combined with convolution neural network, the fine features of basketball dribbling attitude image are extracted, and each layer of features is used to identify human posture. Around the feature point, a square area block with a side length of 20 s is selected, which rotates to the main direction of the feature point, and the target is fixed with a square mesh so that it is not affected by the size of the spectral image. According to the fixed in the square grid, the convolution operation is carried out layer by layer. First, the weight function is used to weight the image. The formula is F as follows:

$$F = \int D(\mathrm{x,y})R(\mathrm{t})\mathrm{dt} \tag{10}$$

Where $R(\mathrm{t})$ is the weighting function of the integral image and t is the time interval from the current time after the convolution neural network training. Input the integral image, convolute the image in multiple dimensions, pool the maximum value, carry out nonlinear transformation on the weighted image, divide the contour image of rough target recognition into multiple sub regions, output the average value of each sub region, and search the local details of the target in small scale [12, 13]. Through the classification layer of neural network, the multi-dimensional vector of local features is output, and the vector is mapped to the probability space with sum of 1, so as to reduce the data dimension between layers in the connection layer of neural network, and constantly extract the front-end features above the search features. The first convolution layer output search feature is O_{z}, z is the search feature category, and the jump connection formula is expressed as follows:

$$O_{\mathrm{z}} = I\left(U + C_{1-\mu}\right) + C_{1-\mu} \tag{11}$$

In the formula, μ is the number of skip layers of the search feature in the neural network, I is the weighting function of the integral image, and C is the composite function of different layers. Finally, for the small-scale target features, the convolution kernel with step size of 2 is used to carry out 0-complement interpolation expansion to ensure the same size of the detail features in the connection layer. The output channels of different convolution layers are added, and the detail features extracted from each layer are used to output the fine features of the image target. The specific process is as follows: searching the feature points corresponding to the template image, matching the fine-features of the target image, and recognizing the basketball dribbling posture (Fig. 1).

Through the convolution neural network, the feature vectors of the reference image are trained repeatedly to get the feature vector set of the reference image, and the precise features extracted from the target are verified to realize the recognition of basketball dribbling posture. So far, the recognition of basketball dribble posture based on fine features is completed, and the design of high-precision recognition method of basketball dribble posture based on lightweight RFID mobile authentication protocol is completed.

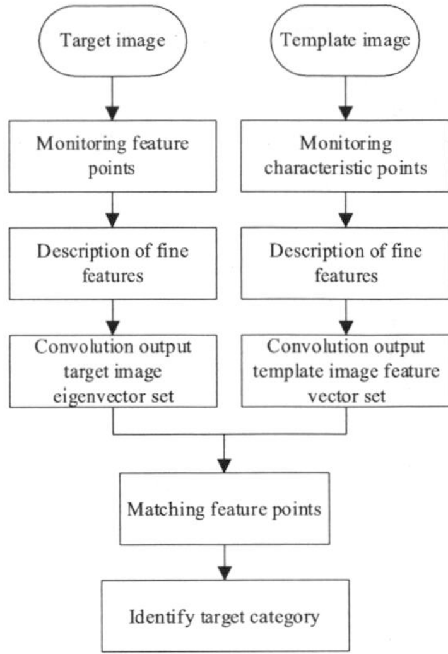

Fig. 1. Basketball dribble attitude recognition process

3 Experiment and Analysis

The design method is compared with two groups of high-precision recognition methods of basketball dribble posture, and the accuracy of three groups of methods is compared.

3.1 Preparation of the Experiment

Selection Microsoft. NET Platform, the programming language is matlab 2010a, the development environment is visual studio integrated development environment, the development tool is Visual C++ 7.0, the configuration of 4gb64 bit lpddr4 memory, Quad ARM® A57/2 MB L2 CPU graphics card. Through the camera, the original sequence image of basketball dribbling posture is collected, and the normalized image size is 50 * 165 pixels. The original image in the sequence image is selected as the experimental object of the three recognition methods. The technical parameters of camera and RFID equipment are shown in the following Table 1:

Table 1. Technical parameters of robot sensing equipment

Equipment	Parameter	Numerical value
Video camera	Pixel size	278 × 278
	Horizontal definition	Line 490
	Signal to noise ratio	>50 dB
	Minimum illumination	≥0.02lx

(*continued*)

Table 1. (*continued*)

Equipment	Parameter	Numerical value
	Scanning frequency	60 Hz
	Volume	78 × 78 × 63 mm
	Storage temperature range	−20 °C–+70 °C
RFID device	Working temperature	−10 °C–+55 °C
	Working humidity	20%–85%RH
	Range resolution	1.5 cm
	Power Supply	5 V
	Scanning frequency	30 Hz
	Angular resolution	0.36°
	Scanning frequency	5 Hz
	Measuring speed	5 kHz

The captured images are shown below (Fig. 2):

Fig. 2. Image of basketball dribble

Until the original image to be processed, the overall brightness of the image sequence is low, and multiple cognitive regions are arranged. In the cognitive region of the image, less detailed information can be obtained, some regions are fuzzy, and there is a certain degree of color distortion. After random pruning, noise disturbance, color dithering and rotation transformation for many times, the original image data is expanded, and a total of 1154 basketball dribbling posture images are obtained in the data set.

3.2 Distance Recognition Accuracy Test

Three groups of methods are used to identify the three-dimensional position of the main bone points in the basketball dribbling posture, and the distance deviation between the identified position and the actual position is compared. Firstly, the athlete's bone points are numbered as follows (Table 2):

Table 2. Number of bones in basketball posture

Serial number	Skeletal point	Serial number	Skeletal point
1	Right shoulder	2	Right elbow
3	Right wrist	4	Left shoulder
5	Left elbow	6	Left wrist
7	Right buttock	8	Right knee
9	Right ankle	10	Left hip
11	Left knee	12	Left ankle

The spatial position of the identified bone points is recorded, and the distance deviation between the identified position and the actual position of the bone points is measured (Table 3).

Table 3. Comparison of distance deviations between skeletal points (m)

Inspection point	Distance deviation of design method	Conventional method 1 distance deviation	Conventional method 2 distance deviation
1	0.016	0.023	0.029
2	0.018	0.025	0.031
3	0.016	0.025	0.026
4	0.017	0.024	0.027
5	0.018	0.023	0.026
6	0.016	0.024	0.026
7	0.016	0.022	0.027
8	0.019	0.023	0.027
9	0.018	0.024	0.027
10	0.018	0.024	0.027
11	0.016	0.025	0.030
12	0.015	0.024	0.028

It can be seen from the above table that the distance deviation of the three groups of methods to identify the spatial position of the bone points is less than 0.03 m. Within the control range of the distance accuracy standard index, the average distance deviation of the design method is 0.017 m, and the average distance deviation of the conventional method 1 and the conventional method 2 are 0.024 m and 0.028 m respectively. Compared with the two traditional methods, this design method can use the lightweight RFID mobile authentication protocol to identify the spatial position of bone points, and the distance deviation between the two positions is reduced by 0.007 m and 0.011 m respectively.

3.3 Precision Testing for Angle Recognition

Measure the angle deviation between the three groups of methods to identify the spatial position of the bone points and the actual position of the bone points.

Table 4. Comparison of Angle Dev of bone points(°)

Inspection point	Design system angle deviation	Angle deviation of conventional system 1	Angle deviation of conventional system 2
1	7.92	10.73	12.03
2	7.74	10.82	11.99
3	7.93	11.92	11.02
4	8.19	10.24	12.74
5	8.03	11.69	12.63
6	7.64	10.16	12.92
7	7.83	10.24	11.95
8	8.53	11.53	11.83
9	8.02	11.91	12.95
10	8.59	11.83	11.68
11	7.52	10.26	12.62
12	8.92	10.29	11.82

It can be seen from the above Table 4 that the angle deviation of the three groups of methods for identifying the spatial position of bone points is less than 13°. Within the control range of the angle accuracy standard index, the average angle deviation of the identification results of the design method is 8.07°, and the average angle deviation of the identification results of the conventional method 1 and the conventional method 2 are 10.97° and 12.18° respectively 9° and 4.11° respectively. To sum up, the design method uses lightweight RFID mobile authentication protocol, which can accurately identify the skeleton points of basketball dribble posture, reduce the distance deviation and angle deviation of recognition space position, and improve the accuracy of posture recognition.

4 Conclusion

In the field of basketball, the existing training plan mainly depends on the coach's artificial observation and personal experience, which is inevitably subjective. The application of body area network technology in the training of athletes can help coaches to assist in decision-making and greatly improve the competitive ability of athletes, Accurate recognition of basketball posture plays an important role in basketball competition and training. To sum up, the design method gives full play to the technical advantages of lightweight RFID mobile authentication protocol, and improves the accuracy of basketball dribble gesture recognition. However, there are still some shortcomings in this study. In the future research, human posture bone points will continue to be classified, and the influence of high priority bone points on low priority bone points will be considered to further improve the accuracy of posture recognition.

References

1. Tian, Y., Li, F.: Research review on human body gesture recognition based on depth data. Comput. Eng. Appl. **56**(4), 1–8 (2020)
2. Liu, J., Liu, Y., Jia, X., et al.: Research on human pose visual recognition algorithm based on model constraints. Chin. J. Sci. Instrum. **41**(4), 208–217 (2020)
3. Ma, Y., Hong, Z.: GMM-HMM human body posture recognition based on portable sensor data. Comput. Syst. Appl. **29**(11), 204–209 (2020)
4. Duan, J., Liang, M., Wang, R.: Human pose recognition based on human bone point detection and multi-layer perceptron. Electron. Meas. Technol. **43**(12), 168–172 (2020)
5. Hu, D., Ke, H., Zhang, W.: Research on human body attitude recognition based on Kinect and ROS. Chin. High Technol. Lett. **30**(2), 177–184 (2020)
6. Zhang, J., Yang, L., Wang, H., et al.: Posture recognition algorithm based on neural network. Softw. Guide **19**(11), 33–36 (2020)
7. Xue, T.: Improved human pose estimation based on hybrid component model. China Comput. Commun. **32**(10), 116–118 (2020)
8. Liu, S., Bai, W., Zeng, N., et al.: A fast fractal based compression for MRI images. IEEE Access **7**, 62412–62420 (2019)
9. Zhou, Y., Xu, Y.: Real-time human pose recognition in complex environment based on the bidirectional LSTM. Chin. J. Sci. Instrum. **41**(3), 192–201 (2020)
10. Cai, Y., Wang, X., Ma, J., et al.: 3D human pose estimation based on random forest misclassification processing mechanism. Acta Automatica Sinica **46**(7), 1457–1466 (2020)
11. Fu, W., Liu, S., Srivastava, G.: Optimization of big data scheduling in social networks. Entropy **21**(9), 902 (2019)
12. Jiao, Q., Xu, D.: Research on the design method of virtual human body maintenance posture examples. Comput. Simul. **36**(4), 324–329+444 (2019)
13. Liu, S., Liu, D., Srivastava, G., Połap, D., Woźniak, M.: Overview and methods of correlation filter algorithms in object tracking. Complex Intell. Syst. **7**(4), 1895–1917 (2020). https://doi.org/10.1007/s40747-020-00161-4

A Dynamic Monitoring Method for Marathon Athletes Based on Wireless Sensor Network

Yi-de Liao and Qiang Huang(✉)

Guangzhou Huali College, Guangzhou 511325, China

Abstract. Due to the continuous development of the Internet, mobile devices are widely used, and wireless sensor networks have new application significance in the dynamic monitoring of athletes' physical fitness. It is currently a new research hotspot in the field of physical fitness monitoring of marathon athletes. According to the prescribed evaluation indexes of athletes' competitive ability, use wireless sensing technology to monitor the real-time dynamics of athletes' physical training and obtain athlete data results. According to the monitoring data, the athlete's physical training implementation can be reasonably adjusted, so as to standardize the athlete's training method and improve the quality of training. The marathon athlete's dynamic monitoring method based on wireless sensor network proposed in this paper can better guide athletes to improve their physical fitness, improve sports perseverance and explosive ability, stimulate their own potential, and improve the actual training effect. It has a great role in improving the training efficiency of marathon athletes per unit time, and at the same time, it is of great significance to realize the precise and personalized physical training plan of elite athletes.

Keywords: Wireless sensor network · Marathon · Fitness monitoring · Exercise energy consumption

1 Introduction

The wireless sensor network for dynamic monitoring of physical fitness training is not a new thing. Part of the function of sports science is to quantify the process of training by data and summarize it as a rule. In competitive sports, the collection and analysis of human movement data is an indispensable core of all sports analysis. In order to reduce sports injuries caused by high-intensity sports, the state of sports load can be monitored in real time. Using wireless sensor network nodes to collect sports load status parameters, to build a detection statistical feature analysis model, to achieve sports load status feature parameter spectrum feature analysis and real-time exercise status monitoring, and to achieve real-time monitoring and monitoring of high-intensity sports based on the monitoring results Health level test [1]. Sports physiology and biochemistry quantify the internal load of athletes to realize the control of athletes' functional state and fatigue degree. Sports training science and sports biomechanics quantify the external

S. Liu and X. Ma (Eds.): ADHIP 2021, LNICST 416, pp. 128–141, 2022.
https://doi.org/10.1007/978-3-030-94551-0_11

performance of athletes in the process of training and realize the management of athletes' competitive performance. Due to the limitations of equipment and technology, it was formerly difficult to conduct real-time dynamic monitoring during training and immediately adjust the training content of athletes based on feedback. However, with the rapid development of the Internet, the dynamic monitoring system for athletes' physical fitness has been gradually improved. Athlete dynamic monitoring methods enable sports science practitioners to effectively collect athletes' physical fitness information and plan athletes' training load levels. Adjust the athlete's training method according to the actual competition needs to stimulate the athlete's own potential. Athlete dynamic monitoring can also help athletes reduce the risk of overload, optimize potential sports performance, promote physical function development, and improve actual training effects [2]. For example, a marathon athlete can accurately monitor the training load with a heart rate meter, and a cyclist can monitor the training load with a heart rate meter and a power meter. However, dynamic monitoring of physical fitness training involves different forms of movement and complex technical movements, so it is more difficult to achieve dynamic monitoring of wireless sensor networks [3]. Based on this, this paper discusses the dynamic monitoring method of wireless sensor network physical training, and enumerates a large number of elite athletes' real cases, expounds the important role of dynamic monitoring of wireless sensor network physical training in improving the quality of dynamic monitoring of elite athletes' physical training, and predicts that dynamic monitoring of wireless sensor network physical training will become an important direction of dynamic monitoring of physical training.

2 Dynamic Monitoring of Physical Fitness of Marathon Runners

2.1 Evaluation Index of Athlete's Athletic Fitness

Dynamic monitoring of physical fitness training can deeply excavate the potential of human movement. With the development of wireless sensor network technology, Dynamic monitoring of wireless sensor network has become possible [4]. The dynamic monitoring of wireless sensor network physical fitness training is an extension of traditional physical fitness training dynamic monitoring, which is a kind of training idea and concrete manifestation. The dynamic monitoring of physical fitness training in wireless sensor network is a two-way process, which is of great significance to improve the training efficiency per unit time and to realize the precise and personalized dynamic monitoring of elite athletes.

With the continuous improvement of modern sports science theory system, especially the development and improvement of sports training discipline theory, people have a more and more clear understanding of the essence of sports training and its internal laws [5]. The research on the structure and elements of athletic ability of athletes further clarifies the important position and function of dynamic monitoring of physical strength and physical training in modern sports training system [6]. At the same time, people have a more accurate understanding of the content and task of the dynamic monitoring of physical fitness training. Athlete's physical fitness, as a major component of competitive ability, plays an increasingly prominent role in sports competition [7]. Based on this, this paper analyzes the influencing factors of athletes' competitive ability, and the concrete structure is shown in Fig. 1.

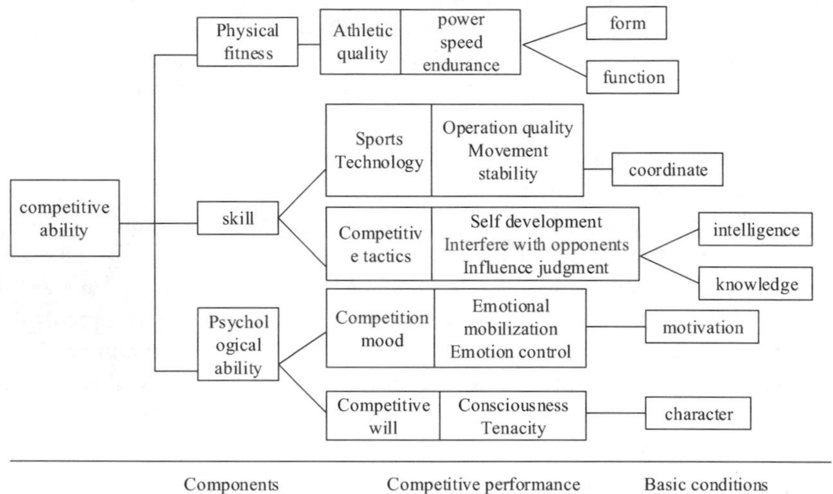

Fig. 1. The constitution of athlete's physical strength

According to the theory of sports training, physical fitness is an important part of athlete's competitive ability, and it is the basic ability of athlete's body. To establish the evaluation index system, the first thing is to determine the object and goal of the evaluation. This study chooses the special physical fitness of elite male marathon runners as the evaluation object, and then determines the index system, determines the weight

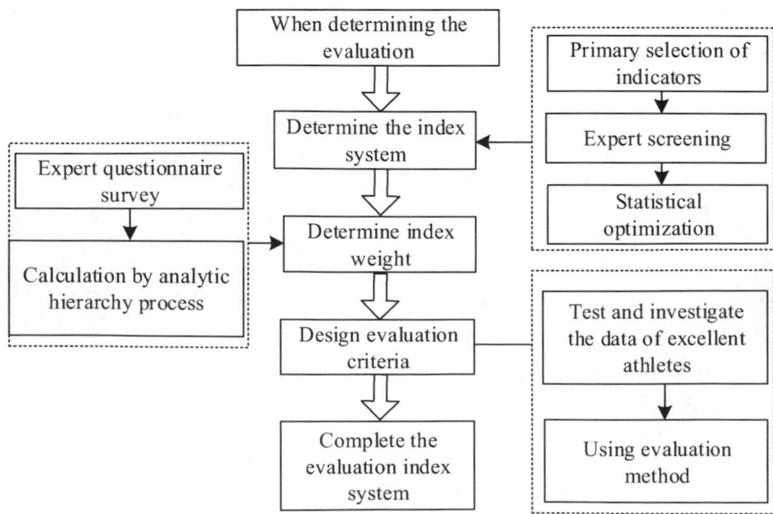

Fig. 2. Athlete fitness evaluation index system

of each index, sets the evaluation standard for physical fitness training, and finishes the construction of the whole system. The specific structure is shown in Fig. 2.

Because of the long-term nature of the dynamic monitoring process and the complexity of the influencing factors, coaches and researchers need to continuously monitor and evaluate the training state of athletes. After the construction of the evaluation index system is completed, the marathon athletes need to be put into training for trial, so as to guide the dynamic monitoring of marathon athletes' physical training. Provide more comprehensive monitoring and management for the dynamic monitoring level of marathon athletes' physical fitness training, improve the training effect, and finally achieve the goal of improving athletes' competitive level [8]. At the same time, according to the results of evaluation and some practical problems encountered in the evaluation process, constantly modify the evaluation index system, improve the evaluation function.

The dynamic monitoring of physical stamina training is a comprehensive project involving many subjects and fields. The important role and value of dynamic monitoring of physical fitness training on athletes' competitive ability and mass physical health have been re-recognized and widely recognized [9]. At present, the research on dynamic monitoring of physical fitness training in our country still lags behind the developed countries in Europe and America, and is still in the period of introducing and learning advanced theories and methods of dynamic monitoring of physical fitness training abroad. We should pay more attention to the research and application of specific practical problems when we are engaged in basic theoretical research. Generally speaking, the function of training monitoring is flexible and changeable. The main study is engaged in sports and competitive activities of special groups of people's physical fitness, the study is the physical function of athletes, so we must put the athlete's physical system in a larger system one athlete's competitive ability system for research [10]. Explore its function in this system, and the function of physical fitness system is always shown in the interaction with other factors, such as: technical ability, tactical ability, psychological ability, athletic intelligence, and the impact of health level and environmental factors. Based on this, the relationship between physical fitness training and competitive ability factors is analyzed, as shown in Fig. 3.

Physical fitness refers to physical ability, then dynamic monitoring of physical fitness training refers to the development of human physical fitness training activities of the general term. The general objective of dynamic monitoring of physical fitness training is to fully develop the physical activity ability of the human body, promote the improvement of morphological structure and physiological function, and improve the physical fitness level of the human body to meet the needs of sports [11]. In the field of competitive sports, dynamic monitoring of physical fitness training is a sub-goal under the general goal of sports training. It serves for the superior goal together with other subgoals, such as technical training goal, tactical training goal and psychological training goal. The dynamic monitoring of physical fitness training is composed of several subtargets. According to the four extensions of the physical fitness concept, the objectives of dynamic monitoring of physical fitness training include four aspects: to improve the physical structure; to comprehensively improve the physiological functions of various organs and systems of the human body; to fully develop various physical qualities; and to enhance the adaptability of the body.

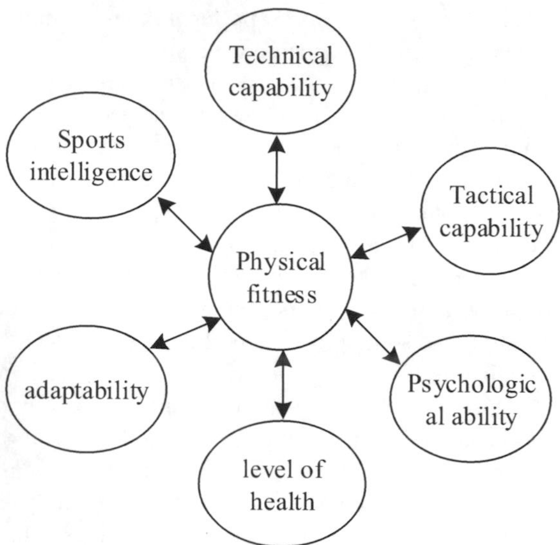

Fig. 3. Relationship between physical fitness training and competitive ability

Dynamic monitoring of physical fitness training is of great significance to both physical exercise and athletic competition. Without good physical fitness, people will not have overall health in the modern sense, and athletes will not have high level of technical and tactical performance. Dynamic monitoring of physical fitness training is the basis of other training, which is of great value to the improvement and maintenance of athletes' competitive ability.

2.2 Structural Elements of Dynamic Detection in Physical Fitness Training

It is recognized that the dynamic monitoring theory system of physical fitness training is an indispensable and important part of the discipline system of sports training. The dynamic monitoring theory of physical fitness training has experienced a long course of self-development and improvement. With the development of science and technology, people's understanding of physical fitness and dynamic monitoring of physical fitness training has been deepened, forming a multi-disciplinary, multi-disciplinary integration of emerging disciplines theoretical system [12]. The theory system of dynamic monitoring of physical fitness training is produced and developed in the field of sports training. Athlete's physical fitness, as an important sub-item of competitive ability, is highly valued and has a very strong meaning of specialization, which is the basic link in sports training. The fundamental goal of training monitoring is to develop the human body's exercise ability, and the basic material basis of exercise ability should be the human body's physical ability [13]. Therefore, in the large system of sports training, dynamic monitoring of physical fitness training is indispensable as a process of specialized transformation of the basic elements of athletic ability, and is also the basic condition of other training monitoring activities (Fig. 4).

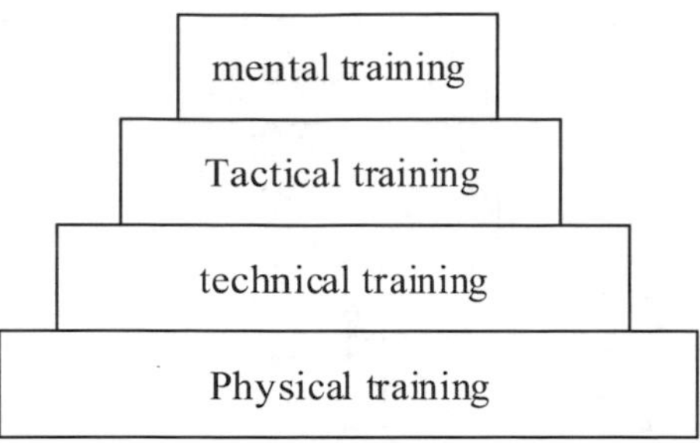

Fig. 4. Structural elements of sports training

In the practice of dynamic monitoring of physical fitness training, it involves the knowledge of many sports science, such as human anatomy, sports physiology, sports biomechanics, sports biochemistry, nutrition, psychology, and so on. Therefore, it can be said that the dynamic monitoring of high-level athletic training is the closest combination of scientific research on sports human body and training practice.

3 Dynamic Monitoring of Physical Fitness of Marathon Runners Based on Wireless Sensor Network

3.1 Athlete Fitness Data Collection

The rapid development of wireless sensor network technology makes it possible to monitor the athlete's state in marathon training. Through RFID and wireless sensor, the athlete's state can be fully sensed, captured and measured, and the acquired data can be transmitted to the data processing center via the Internet for analysis and processing, so as to provide the coaches with the best guidance scheme, thus helping them to monitor and guide the training of athletes in real time. The dynamic monitoring of marathon athletes' physical fitness is the design and grasp of training process. The whole process of dynamic monitoring of marathon athletes' physical fitness is essentially the process of athletes' competitive ability changing from the actual state to the target state. It is the precondition of making dynamic monitoring plan of physical training and the basis of evaluating training effect to test and evaluate athletes before they start dynamic monitoring of physical training. Based on this, the dynamic data of physical fitness training is first collected, as shown in Fig. 5.

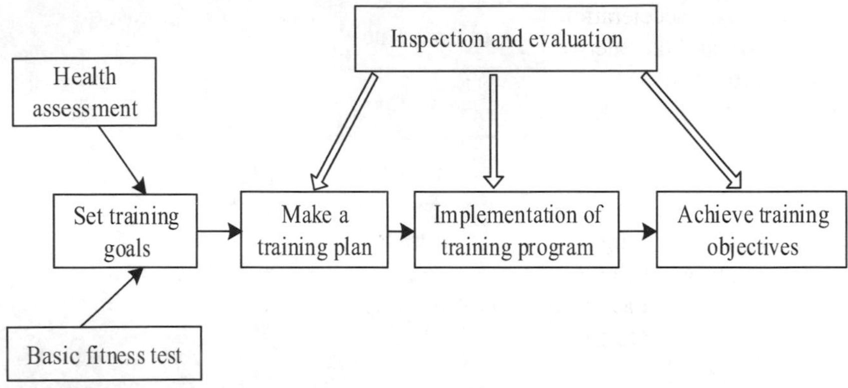

Fig. 5. Dynamic data acquisition process for physical fitness training

The energy consumption of human body should be considered when collecting physical energy data in wireless sensor network. However, the basic daily energy consumption (BEE) of the human body is subject to certain standards. Due to the physiological differences between men and women, the corresponding formula for calculating BEE is different, as follows:

$$\begin{aligned} &\text{MAN BEE (kcal)} \\ &= 66.5 + 13.7 \times \mathrm{W} + 5.0 \times \mathrm{H} - 6.8 \times Q \end{aligned} \tag{1}$$

$$\begin{aligned} &\text{WOMAN BEE (kcal)} \\ &= 65.1 + 9.56 \times \mathrm{W} + 1.85 \times \mathrm{H} - 4.68 \times Q \end{aligned} \tag{2}$$

In Eqs. (1) and (2), BEE takes kcal as the unit of calculation, 1 kcal – 4.184 kJ, W is weight (kg), H is height (cm), and Q is age (age). Since each index has different effects on physical stamina, it is necessary to define the judgment matrix by quoting the reciprocal x_{ij} as the scale, specifically as follows:

$$(x_{ij})_{n \times n} = \begin{bmatrix} x_{11}, & x_{12}, & x_{13} & \ldots, & x_{1n} \\ x_{21}, & x_{22}, & x_{23}, & \ldots, & x_{2n} \\ \ldots, & \ldots, & \ldots, & \ldots, & \ldots \\ x_{n1}, & x_{n2}, & x_{n3}, & \ldots, & x_{nn} \end{bmatrix} \tag{3}$$

In Formula (3), x_{ij} is the ratio of importance of the indicators, and n is the evaluation indicator. The index weight is calculated by arithmetic average method. The arithmetic average method is described by mathematical formulas as follows:

$$w = \frac{1}{m} \sum_{j=1}^{n} x_{ij} \cdot \alpha \tag{4}$$

In formula (4), m is the weighted average coefficient and α is the saturation weight. Using this formula, we can accurately calculate the weights of each index in Table 1 and make full preparation for the follow-up physical fitness assessment.

The three-axis acceleration generated by human motion is collected by wireless network sensor, and the energy consumption of human motion is calculated at E. The formula is as follows:

$$E = \frac{d}{2w} u \cdot g \tag{5}$$

In formula (5), u is the training frequency, d is the training intensity, and g is the acceleration in the training process (g = 9.8 m/s). Further, the human body energy information collected by the sensor is drawn, displayed in the mode of three-dimensional coordinates, and the characteristic value is output on X, Y, Z triaxial axis and converted to the geometric average generated after the acceleration of each axis. The calculation formula is as follows:

$$G = M\sqrt{\left(E_x^2 + E_y^2 + E_z^2\right)/3w} \tag{6}$$

In Formula (6), E_x^2, E_y^2, E_z^2 is the curve change of each axis after treatment respectively. According to the results of the evaluation and the actual problems encountered in the evaluation process, constantly modify the evaluation index system and improve the evaluation function.

3.2 Construction of Physical Fitness Monitoring System

Based on the results of training monitoring, a series of movement combination models are designed to get close to the characteristics of special sports techniques to the greatest extent. Functional training focuses on the training of the whole dynamic chain, emphasizes the integrity and control of the generation and transmission of the whole exercise force, and enhances the balance and the proprioception of neuromuscular system. Functional training is a dynamic monitoring method of physical training, which is based on basic physical ability and more similar to special technical movement characteristics. Based on this, the monitoring factors of athletes' physical fitness training are further analyzed, as shown in Fig. 6.

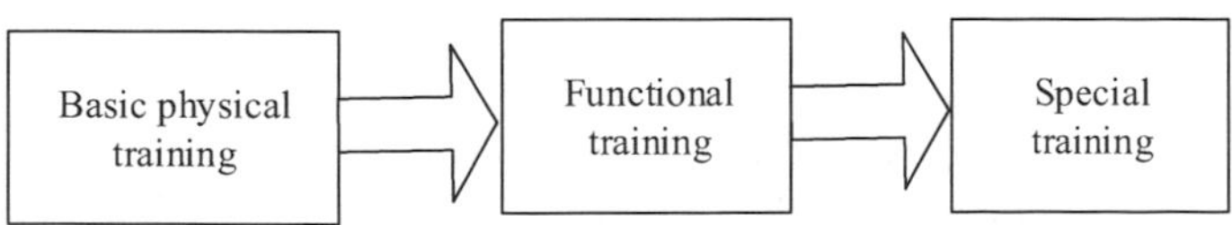

Fig. 6. Monitoring factors of athlete physical fitness training

As shown in Fig. 6, the dynamic monitoring process of modern physical fitness training is a scientific, rigorous and standardized operation process, which is divided into five closely connected processes: exercise capacity measurement, physical fitness assessment, scientific design of dynamic monitoring plan of physical fitness training, formulation of physical fitness development goals and implementation of dynamic monitoring of physical fitness training. How to define the nature of physical fitness coaches is the key problem to break down all layers of obstacles and establish a professional training system of physical fitness coaches as soon as possible.

Based on the analysis of the components of physical fitness structure, the dynamic monitoring of physical fitness training should include the transformation of body shape and structure, the improvement of physiological function, the development of sports quality and the enhancement of physical adaptability. Therefore, the dynamic monitoring of physical fitness training is to develop physical fitness by training the sports quality of physical performance, and to achieve the training goal of profoundly influencing and changing the physical form, physiological function and improving the adaptability of the body by developing the sports quality closely related to the special competitive ability of athletes, such as strength, endurance, speed, flexibility, agility and coordination. But with the development of training theory and practice, these six qualities can not reflect all the contents of dynamic monitoring of physical fitness training. Based on this, the methods of core stability training, balance ability training and rapid telescopic compound training are put forward. Supplementing the original dynamic monitoring system of physical stamina training, the dynamic monitoring system of physical stamina training with various physical stamina as the main body is formed (Fig. 7).

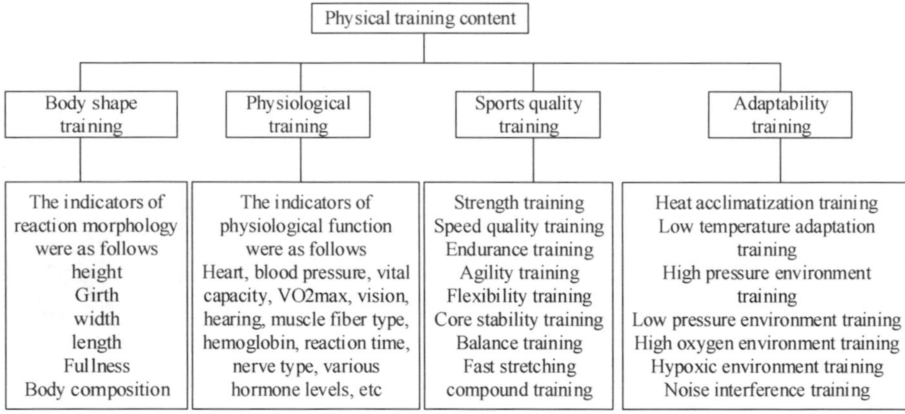

Fig. 7. Monitoring system of physical fitness training

Training monitoring is to achieve the goal of sports training, according to the specific training content and design and use of specialized training means or behavior of the general term. Training method is the core content in the theoretical system of special training, and other factors are the supporting conditions for the implementation of training method. The main factors of sports training methods are exercise and its combination, exercise load and its changing way, process arrangement and its changing way, information media and its transmitting way, external condition and its changing way. Different combinations of these elements and their changes can form a variety of training methods with different functions. The physical training process is also the process of making training plans and implementing them. According to the periodical division of training stages in sports training science, the training of athletes can be planned and organized by stages. Training plans of different length can be made, such as multi-year training plan, annual training plan, stage training plan and weekly training plan. Regardless of the length of training time, each training process should include basic contents such as athletes' health status and risk assessment, basic physical fitness assessment, determination of training objectives, formulation of training plans, implementation of training programs, and inspection and evaluation. In the dynamic monitoring system of physical fitness training, according to the principle of taxonomy, different classification systems can be established according to different classification standards (Fig. 8).

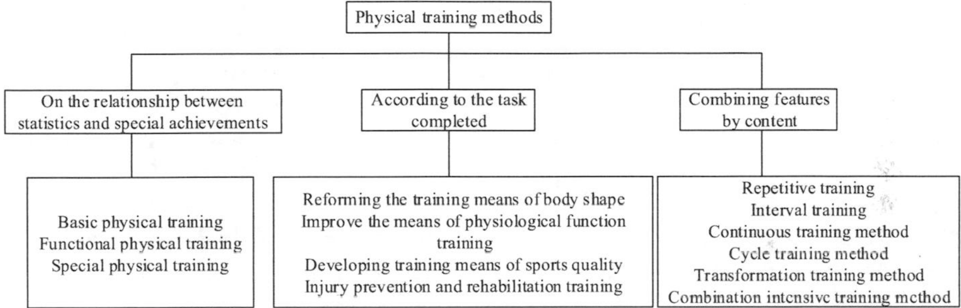

Fig. 8. Health information monitoring system for physical fitness training

According to the relationship between training and special achievements, it can be divided into dynamic monitoring of basic physical fitness training, which mainly refers to the dynamic monitoring of physical fitness training to ensure the health level and the physical quality and ability needed for daily activities. Dynamic monitoring of functional physical training is a kind of multi-plane compound movement training based on the theory of "dynamic chain of movement", which strengthens body posture, movement pattern and dynamic stability. Dynamic monitoring of special physical fitness training refers to the dynamic monitoring of physical fitness training, such as the dynamic monitoring of special physical fitness training. According to the training task, it can be divided into training means to reform body shape, training means to improve physiological function, training means to prevent injury and rehabilitation training means. According to the characteristics of training content combination, it can be divided into repeated training method, intermittent training method, continuous training method, cycle training method, change training method, combination strengthening training method, etc. Dynamic Monitoring of Marathon Athletes in Wireless Sensor Network Environment Can Better Guide Athletes to Improve Their Physical Fitness.

4 Analysis of Experimental Results

In order to ensure the effectiveness of the dynamic monitoring method for marathon runners in wireless sensor network environment, the expert validity test method is adopted. First of all, the questionnaire of this study is designed according to the design procedure of the questionnaire in the research content and the educational research method. The first draft of the questionnaire was designed on the basis of the consulted documents and expert interviews, and 12 experts and coaches were selected to make a logical judgment on the validity of the content and structure of the questionnaire by adopting the method of expert evaluation. The validity of the content and structure of the questionnaire was assessed on the basis of five grades. More than 90% of experts believed that the questionnaire designed was effective and reasonable. The specific situation was shown in Table 1, 2 and 3.

Table 1. Expert composition table

	Expert	Coach
Number of people	7	5
Title/grade	Professor	National level

It can be seen from Table 1 that the evaluation objects of this experiment are 7 professor-level experts and 5 national-level coaches, so as to fully ensure the objectivity of the evaluation.

Table 2. Evaluation of the validity of the questionnaire by experts (N = 12)

	Very effective	Effective	Commonly	Not very effective	Invalid
Physical coach questionnaire	8	3	1	0	0
Q & A of competitive coaches	9	2	1	0	0

Table 3. Expert evaluation of the structural validity of the questionnaire (N = 12)

	Very reasonable	Reasonable	Commonly	It's not reasonable	Unreasonable
Physical coach questionnaire	7	4	1	0	0
Q & A of competitive coaches	6	3	2	0	0

From Table 2 and Table 3, we can know that experts and coaches fully affirmed the content and structure of the questionnaire, and believe that the marathon athlete dynamic monitoring method has a certain effect on content improvement, and has a certain degree of rationality in the structure. Based on this, the results of further aerobic endurance training 3–4 times a week also show that planned dynamic monitoring of physical training is effective. The results are shown in Table 4.

Table 4. Statistics of test results

Number of subjects	Test index	Preliminary survey	Unmeasured	Improved value
3	Beeb-test	10–1	11–1	1.1*
	Repeat Sprint (6 times, seconds)	35.6	31.6	4*

It can be known from Table 4 that the preliminary test index values obtained by the athletes' training monitoring are relatively large, and there are many unmeasured data. Therefore, it is necessary to improve the overall data calculation ability of the system and optimize the test results. Further comparative analysis of athletes before and after training monitoring changes in physical fitness, the specific statistical results as shown in Fig. 9.

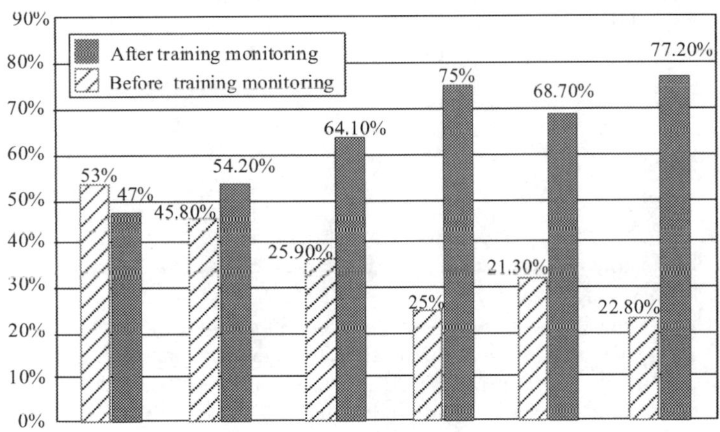

Fig. 9. Comparative analysis of monitoring results of athlete's physical fitness

Based on the testing results in Fig. 9, the dynamic monitoring method of marathon runner based on infinite sensor can improve athletes' stamina and explosive force. Improve the efficiency of training by stimulating the potential of the athletes themselves through enhanced feedback.

5 Concluding Remarks

The task of dynamic monitoring of physical fitness training is to carry out systematic, scientific and reasonable physical ability training according to the characteristics of athletes and the needs of sports competitions, so as to improve their special sports quality. As far as marathon runners are concerned, dynamic monitoring of physical fitness training can fully tap the potential of athletes and lay a solid physical foundation and key ability guarantee for their future racing and excellent sports results. Therefore, it is very important to monitor marathon athletes' physical fitness training in a scientific way. Marathon athletes should focus on the future, reasonable planning, strict implementation and scientific control, which is the fundamental path of dynamic monitoring of marathon athletes' physical training.

This paper uses wireless sensing technology to monitor athletes' physical training, obtains the athletes' effective training data results, and makes reasonable adjustments to the athletes' physical training implementation based on the monitoring data, so as to standardize the athletes' training methods. The research on the dynamic monitoring method of marathon athletes' physical fitness based on the wireless sensor network is beneficial to improve the athletes' physical fitness, enhance sports perseverance and explosive ability, so as to achieve the training effect. However, due to the limited time and research conditions, the method designed this time still has shortcomings, and the results still have limitations. The confidentiality of the athlete's dynamic monitoring data has not been analyzed in detail. In the future research process, data security technology will be applied to the physical fitness dynamic monitoring system to further improve the privacy security of the system, and provide athletes with more reliable training methods, with a view to monitoring the physical fitness training dynamics of Chinese marathon athletes and marathon reserve Talent training provides suggestions and references that can be used for reference.

References

1. Wang, H., Pei, Y.: Load monitoring system of high-intensity sports based on characteristic parameters. J. Qiqihar Univ. (Nat. Sci. Ed.) **37**(3), 76–79+94 (2021)
2. Duggan, J.D., Moody, J.A., Byrne, P.J., et al.: Training load monitoring considerations for female Gaelic team sports: from theory to practice. Sports **9**(6), 84 (2021)
3. Watanabe, F.: Wireless sensor network localization using AoA measurements with two-step error variance-weighted least squares. IEEE Access **9**, 10820–10828 (2021)
4. Tian, S., Li, Y., Kang, Y., Xia, J.: Multi-robot path planning in wireless sensor networks based on jump mechanism PSO and safety gap obstacle avoidance. Future Gener. Comput. Syst. **118**(5), 37–47 (2021)
5. Rouxia, C., Xiaodong, C., Shifang, T., et al.: Research on inverse simulation of physical training process based on wireless sensor network. Int. J. Distrib. Sens. Netw. **16**(4), 155014772091426 (2020)
6. Benson, L.C., Stilling, C., Owoeye, O.B.A., et al.: Evaluating methods for imputing missing data from longitudinal monitoring of athlete workload. J. Sports Sci. Med. **20**(2), 188–196 (2021)
7. Zhang, L.: Evaluation and simulation of sports balance training and testing equipment based on medical video image analysis. IEEE Sens. J. **20**(20), 12005–12012 (2020)
8. Iliadis, A., Tomovic, M., Dervas, D., et al.: A novel mHealth monitoring system during cycling in elite athletes. Int. J. Environ. Res. Public Health **18**(9), 4788 (2021)
9. Liu, S., Liu, D., Srivastava, G., Połap, D., Woźniak, M.: Overview and methods of correlation filter algorithms in object tracking. Complex Intell. Syst. **7**(4), 1895–1917 (2020). https://doi.org/10.1007/s40747-020-00161-4
10. McLean, S., Read, G.J.M., Ramsay, K., et al.: Designing success: applying cognitive work analysis to optimise a para sport system. Appl. Ergon. **93**, 103369 (2021)

11. Nuuttila, O.P., Nummela, A., Häkkinen, K., et al.: Monitoring training and recovery during a period of increased intensity or volume in recreational endurance athletes. Int. J. Environ. Res. Public Health **18**(5), 2401 (2021)
12. Zhang, L., Liang, F., Wagner, N., et al.: Monitoring and analysis of athletes' local body movement status based on BP neural network. J. Intell. Fuzzy Syst. **40**(2), 2325–2335 (2021)
13. Damji, F., MacDonald, K., Hunt, M.A., et al.: Using the VERT wearable device to monitor jumping loads in elite volleyball athletes. PloS One **16**(1), e0245299 (2021)

Research on Key Technologies of Analysis of User Emotion Fluctuation Characteristics in Wireless Network Based on Social Information Processing

Jia Yu(✉)

School of Information Engineering, East China Jiaotong University, Nanchang 330013, China
liuwenjun0213@sina.cn

Abstract. With the rapid development of the mobile Internet, people's living habits have been greatly changed, and the ways to express their feelings have been gradually extended from offline to online. The research in the field of Internet user sentiment analysis provides a new method and tool for the prediction of social behavior in specific fields. Based on this, this research proposes the key technology of user's emotion fluctuation analysis in wireless network based on social information processing. The affective dictionary with multiple affective types is used to identify the affective types, and the database of users, products and affective types is formed. Based on large-scale general knowledge base, sentence level emotion is detected. Through the automatic learning of emotion expression, the classification of sentence emotion is realized, and the analysis model of network user comment emotion is constructed. The confusion matrix of the classifier is introduced to analyze the user's emotion fluctuation. In order to evaluate the effect of text emotion analysis, the accuracy index is introduced to evaluate the analysis results. Experimental results show that the proposed technique has ideal accuracy and strong applicability.

Keywords: Social information · Wireless network · Emotion fluctuation · Emotion characteristics

1 Introduction

In recent years, many scholars have used social media user sentiment analysis to study the prediction methods of social behavior in specific fields [1]. Since different social activities have different sensitivity to affective factors, affective analysis factors play different roles in different prediction tasks. A sentiment analysis method with independent predictive ability in a prediction task in a certain field may prove to not have independent predictive ability in another field, and can only be used as an important supplementary factor. In addition, the sentiment analysis methods used by different scholars are also different, which objectively affects the researcher's judgment on the importance of sentiment analysis.

S. Liu and X. Ma (Eds.): ADHIP 2021, LNICST 416, pp. 142–154, 2022.
https://doi.org/10.1007/978-3-030-94551-0_12

Literature [2] proposes a method for predicting user's emotional behavior intentions in message push process of mobile terminals, and constructs a correlation matrix to obtain the correlation between user's emotional behavior intentions and push messages. Based on the relational relationship, the evaluation function of user's emotional behavior intention is constructed, and the weights of user's emotional behavior safety are calculated by combining the evaluation function of user's emotional behavior intention. In reference [3], based on the theory of emotional load, with the help of coagmento system, 52 users were recruited to complete three planning tasks in a group of two. The relationship between task difficulty, time pressure, collaboration division, search results and collaboration results and users' negative emotion perception was investigated. However, the above two traditional methods ignore the statistical processing of emotional words, resulting in a large deviation in the analysis results of online users' emotional characteristics.

Based on this, this paper expounds the emotional analysis of network users, analyzes the key technologies of the emotional fluctuation characteristics of wireless network users based on social information processing, analyzes the typical methods and key technologies of applying emotional analysis to predict, and summarizes and prospects the existing problems and future development of current related research.

2 Key Technologies for Analyzing User Emotion Fluctuation Characteristics in Wireless Networks

2.1 Collection of Emotion Fluctuation Characteristics of Wireless Network Users

Sentiment classification refers to the automatic analysis of the relevant comments on goods, services, people and other research objects to find the reviewers' attitudes and opinions towards the research object. Online user sentiment analysis mainly focuses on automatic sentiment analysis of comment information generated by social media [4]. At present, the research in the field of online user sentiment analysis is mainly carried out in one of three levels: document level, sentence level and attribute level. According to the sentiment tendency of the whole review, the document level sentiment analysis divides the review into three polarity categories: positive, negative or neutral. It mainly uses various machine learning techniques to analyze the sentiment of reviews. A method used to classify opinions in multilingual web forums. Emotion analysis at sentence level mainly focuses on identifying subjective sentences and judging their emotional polarity [5]. Based on the method of sentiment dictionary, the sentiment lexicon with multiple sentiment type labels is used to identify sentiment types, and the database of users, products and sentiment types is formed. The method of emotion type discrimination is based on the intensity of emotion words in comparative sentences. Assuming that there are n emotion words in the corpus, the discriminant formula of emotion type is as follows

$$\text{SentiType} = \max\left(\sum_{i=0}^{n} D_i T_i\right) \tag{1}$$

Among them, SentiType is the emotional type; I is the sequence number of all emotion words involved in a comment, Di is the corresponding sequence number, the degree score of adverbs in the first five bytes of emotion words, and Ti is the emotion score of emotion words with corresponding sequence number [6]. In the case of considering degree adverbs, the emotional type corresponding to the emotional word with the highest emotional score in the review is selected as the overall emotional type of the sentence by ranking. Based on the technology of network projection, the calculation formula of the single-vertex network of the emotional network projection is proposed, such as the formula:

$$T = \text{SentiType} \times \left(\frac{Q^2}{N} \right) \tag{2}$$

Among them, Q: Is the network projection weight composed of commentator or comment object and emotion tag, and further proposes the method of emotion information extraction, and gives the experimental results and analysis of each method. The extracted emotion information includes three types: User dictionary, domain related emotion words and evaluation collocation. The extraction of evaluation collocation depends on domain related emotion words and user dictionary. For the construction of user dictionary, this paper proposes a method based on statistics, covering five kinds of statistics [7]. Secondly, co-occurrence graph is proposed to extract domain related emotional words. Finally, evaluation collocation extraction is realized based on word2vec model and syntax analysis. Experimental results show that the proposed method can accurately and efficiently extract the emotional information from the comment text. Based on this, this paper first analyzes the dependence of emotional information. The details are as follows (Fig. 1):

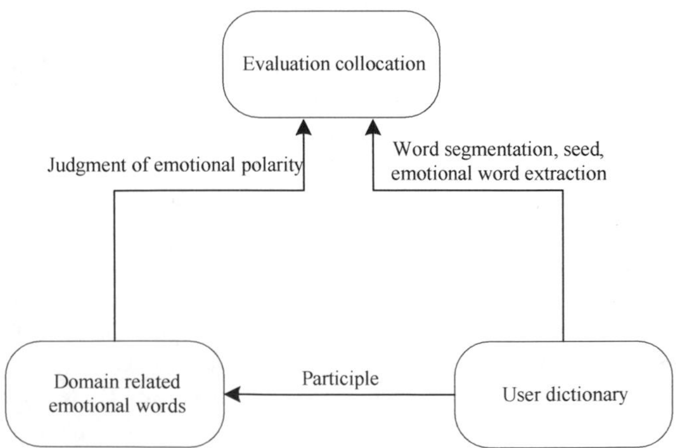

Fig. 1. Emotional information dependence

The emotional polarity of emotional words is not invariable, and the same emotional word may have polarity reversal when modifying different evaluation objects [8]. For example, the word “strong” is positive when modifying the atmosphere, and negative when modifying the smell of the room. The word “luxury” is negative in price decoration and positive in room decoration. In the task of sentiment analysis, it is very important to identify the emotion words whose polarity is not universal. This kind of emotion words is defined as "domain related emotion words". To solve this problem, a domain related emotion word recognition algorithm based on co-occurrence graph is proposed. "Co-occurrence graph" is a concept proposed, which refers to the undirected weighted graph based on the co-occurrence relationship of words in the corpus. The nodes in the graph are words in the corpus. If two words appear in the same sentence, they are connected.

2.2 The Relationship Between Interest and Passion of Wireless Network Users

Based on the large-scale general knowledge base to detect the sentence level emotion method, through the automatic learning of emotion expression to realize the classification of sentence emotion. The sentiment analysis of document layer and sentence layer is too rough to accurately determine the user's sentiment. To solve this problem, sentiment analysis at the attribute level, which extracts product-specific attribute opinions from reviews, is proposed. The part-of-speech tagging sequence rules are used to extract product attributes, and the polarity of the emotional description phrases of the attributes is determined based on the context information. In terms of emotional orientation, more related studies use positive and negative emotions to distinguish the emotional orientation in the text [9]. With the gradual deepening of research, some studies believe that such a simple emotional division may ignore many rich and multidimensional human emotional information. Therefore, some related studies try to further subdivide emotional orientation. In order to realize the organic combination of social information and emotion analysis, an online user comment emotion analysis model based on social information is constructed, as shown in Fig. 2.

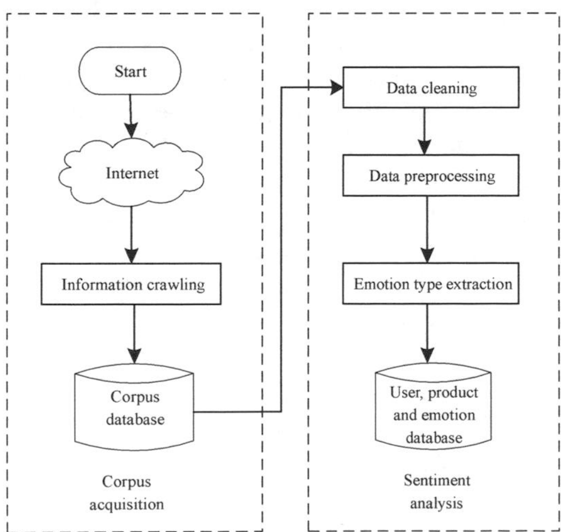

Fig. 2. Analysis model of online users' comments emotion

Sentiment analysis has been applied in many areas of prediction activities, and it has been proved that it can obtain better prediction results in most cases. However, due to the differences in application fields, emotion analysis and prediction target are not always closely related to each other [10]. In addition, sentiment analysis technology itself also has limitations in application fields or text processing objects, so in the research of prediction methods in different fields, sentiment analysis plays different roles. Let A be the behavior data set of social network users, and B be the data set to be recommended. The query history of a user is represented by a matrix.

$$S = T|A| \otimes |B| \tag{3}$$

Where, $a \in A$ represents the user, $b \in B$ represents the b in the set B, T represents the evaluation matrix, where each item $t_{ab} \in \{-1, 0, 1\}$ represents the search of user a for b, "–1" indicates that user a is not interested in b, "0" indicates that user a has not paid attention to b, "1" indicates that user a is interested in b.

Given a user's interest in $\{a, b\}$, let $Q\{a, b\}$ represent the degree of user a's interest in b, namely $Q \in \{A \otimes B \to T\}$. Given a data set list C to be recommended, it is expressed as follows:

$$R_a = \arg \underset{b \in C}{topN}\, SQ\{a, b\}, \forall a \in A \tag{4}$$

According to $Q\{a, b\}$, the score of each item in C is calculated, and then it is arranged in the order from big to small according to the score. The top N R_a with the highest score is recommended as the data cycle of the user's behavior. According to the different application ways of emotional factors in the prediction process, the prediction methods can be divided into the following two categories: The prediction method based on the results of emotional analysis as an auxiliary basis, and the prediction method based on the results of emotional analysis as the main basis. In the prediction method based on emotion analysis results, a combination of multiple indicators is used to predict. These indicators are usually proved to have a correlation with the predicted indicators, and the integration of multiple indicators can make the prediction effect optimal. Other prediction indicators integrated with the results of sentiment analysis include the number of mentions of the objects to be predicted in social media, the deformation of the number of mentions, and the historical data of authoritative institutions related to the research objects. Relevant studies believe that considering the emotional factors expressed by social media in the prediction process can effectively improve the prediction performance of existing indicators. The flow chart of typical prediction method based on the combination of emotion analysis results and other prediction indexes is shown in Fig. 3.

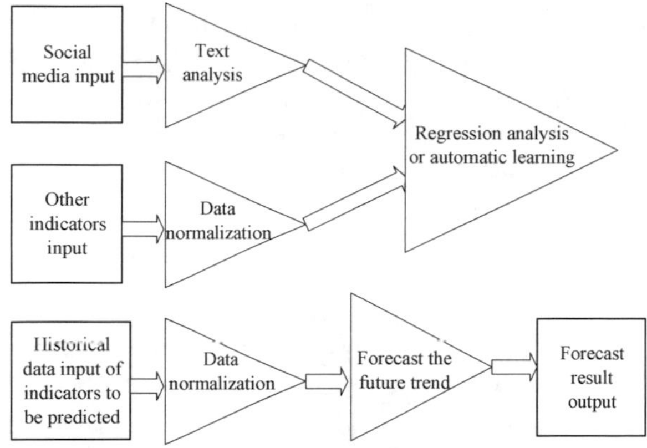

Fig. 3. Flow chart of emotion fluctuation analysis and prediction method

In the figure, the process of the prediction method based on the results of sentiment analysis mainly includes four key steps.

Text analysis of social media input. Text analysis can be divided into text preprocessing and text sentiment analysis. In the process of text preprocessing, it is necessary to standardize the social media content input in the form of natural text, so as to facilitate the subsequent text emotion analysis. The specific tasks of text preprocessing include: We should remove stop words, special characters other than exclamation marks and question marks, remove link addresses and user IDs, and use specific names to replace the names of the subjects (for example, the names of films with predicted box office revenue and books with predicted sales volume). In the process of text sentiment analysis, we can identify and count the sentiment words in the text with the help of the existing sentiment vocabulary to determine the text sentiment orientation. Machine learning method can also be applied to establish classification model through training to classify the emotion of the text. Because the third-party sentiment analysis tools are usually developed for general domain, it is difficult to make targeted adjustments when applied to specific domains, so the effect of sentiment analysis is not always satisfactory.

The input of other indicators and the input of historical data of indicators to be predicted are standardized. The input of other indicators to be integrated with the results of sentiment analysis and the historical data input of indicators to be predicted may have the problem of nonstandard data expression, which makes the subsequent indicators integration or comparison difficult. The range of data scale should be set, and the input of other indicators and historical data of indicators to be predicted should be standardized to a unified range of data scale.

Integrate the timing of sentiment indicators with the timing of other indicators. Regression analysis or automatic learning methods are used to integrate the standardized emotional indicator time series and other indicator time series to obtain the integrated indicator time series. Compared with the individual sentiment indicator timing and other indicator timings, the integrated indicator timing is more consistent with the indicator timing to be predicted when the timing advance is subtracted.

Use the integrated index trend to predict the future trend of the predictive index. The current integrated index trend reflects the future trend change of the index to be predicted, and the time advance n between the two can be calculated through the training text collection and historical index data to be predicted. Therefore, the trend of the integrated index near day t can be used to predict the trend of the index to be predicted on day t + n.

2.3 Realization of User Emotion Fluctuation Characteristic Analysis

Sentiment analysis has been applied in many areas of prediction activities, and it has been proved that it can obtain better prediction results in most cases. However, due to the differences in application fields, sentiment analysis and prediction target are not always closely related to each other. In addition, sentiment analysis technology itself also has limitations in application fields or text processing objects, so in the research of prediction methods in different fields, sentiment analysis plays different roles. According to the different ways in which affective factors are applied in the prediction process, the prediction methods can be divided into the following two categories: the prediction methods based on the results of affective analysis as the auxiliary basis, and the prediction methods based on the results of affective analysis as the main basis. The confusion matrix of classifier is introduced to explain the symbol, as shown in the Table 1:

Table 1. Confusion matrix and its symbol

Data category	Prediction example	Counter example of prediction
Practical examples	TP	FN
Actual counterexample	FP	TN

Different social media information sources have different update frequencies, and the performance of prediction methods based on these information sources is obviously affected by their update frequencies. Social network user behavior data cycle recommendation algorithm is carried out under the following premise, that is, if users score some items more similar, then they score other items more similar. Therefore, the first step of collaborative filtering is to establish a scoring matrix according to the user's interest in a certain item, that is, to score the user's interest items by using the user's browsing records, clicks, favorite content, etc. as the scoring index, and to establish a scoring matrix, as shown in Table 2 below.

Table 2. User project scoring matrix

Project	Project B1	Project B2	……	Project Bb	……	Project Bn
User A1	T11	T12	……	T1b	……	T1n
User A2	T21	T22	……	T2b	……	T2n
……	……	……	……	……	……	……
User Am	Tm1	Tm2	……	Tmb	……	Tmn

In the table, TMB is the score of user am on item BB, and a value in the table represents the score of a user on an item. Because wireless network has a limit of 140 characters, it is usually easier and more frequent to publish wireless network articles than blog articles. The timeliness of wireless network reflecting the emotional changes of network users is better than blog. Although the time advance of using wireless networks to predict epidemics is not obvious, shorter monitoring intervals, timely epidemic status information, and lower forecasting costs are fundamental to its existence.

The number of statements of related research objects in the information source should have a certain scale, and can more comprehensively and objectively reflect the category, proportion and intensity of people's emotional orientation towards research objects in the real world. Otherwise, the input of sparse data or false data will inevitably lead to the failure of prediction. The principle of richness of information source content and the principle of timeliness of information source update may need to make a trade-off under specific circumstances. For example, in the task of using sentiment analysis to predict the possible defects of automobile products, although wireless network information has the advantage of timely updating, its content length limit makes network users unable to describe the product experience and express emotions more deeply, so it is more suitable as a prediction information source.

Based on the quantitative rules of users' implicit emotion space, the corresponding relationships between the controllability, stability, transitivity, external force dependence of users on social platforms and the above characteristics are shown in the Table 3:

Table 3. Corresponding relationship between user characteristics and implicit emotion characteristics

User characteristics	Corresponding features
Controllability	Agree/disagree
Stability	Responsible/irresponsible, emotional/neurotic
Transitivity	Extroversion/introversion
External force dependence	Experience open/experience closed

As can be seen from the table, user controllability is measured by the "agree/disagree" feature. The more users tend to agree, the higher the controllability. For example, in a hot news event, if the user holds a negative view on the event, it can be used as a key monitoring and analysis object in public opinion control, and it needs to be properly guided and controlled by public opinion. User stability refers to the variability of user's emotional value for historical events, etc., which is measured by "responsible/irresponsible, emotional/neural sensitivity". User transitivity is measured by the "extroversion/introversion" feature. Outgoing users tend to have more attention, so they play an important role in the emotional communication of social networks. External force dependence refers to whether the user's emotion needs external force intervention. Experience open users need more external intervention to change their emotional tendency. This paper briefly introduces and discusses the method and model of text sentiment tendency analysis based on dictionary. Based on the emotion tendency calculation model of emotion group counting, the algorithm is designed and the emotion tendency of Chinese wireless network is analyzed. The main process of sentiment tendency analysis based on dictionary is shown in Fig. 4

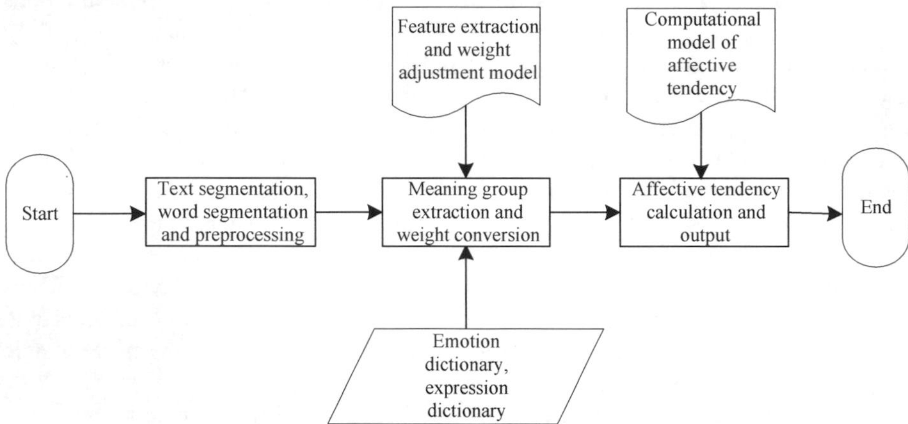

Fig. 4. Optimization of emotional tendency analysis process

Based on the above process, the analysis of user sentiment fluctuation can better guarantee the analysis effect, so as to effectively monitor the network environment.

3 Analysis of Experimental Results

In order to evaluate the effectiveness of the model, experiments and results analysis will be carried out according to the user emotion space model. Microblog is used as a test platform to query 1000 users' browsing records, clicks, and collections. Their interests and hobbies are classified as movie, food, beauty and movie stars. User project evaluation and scoring are conducted, and about 10 million scoring data are generated. As experimental data, social network user behavior data cycle is recommended (Table 4).

Use web crawlers to crawl wireless network user data. In order to ensure the timeliness of user data, only user data within half a year is extracted. According to the user

Table 4. Experimental environment configuration

Name	Parameter
CPU	Intel Core(TM)i5 -4590@3.30 GHz
Memory	4.00 GB
Operating system	64 bit Windows 8 professional
Development language	Python
Programming environment	Eclipse4.2
Compiler	JetBrains PyCharm Community Edition

emotional space construction steps in Sect. 5, user emotional characteristics are gradually established, three representative users are selected for detailed analysis, and how to analyze the emotional tendency of network users based on emotional space and user emotional characteristics. Here, X, D, and J are used as user nickname codes. Count the emotional frequency of three users. Emotional frequency refers to the total number of wireless networks sent by users in a single day, reflecting the frequency with which users express emotions. Figure 5 is a statistical comparison chart of the frequency trend of three users' emotions.

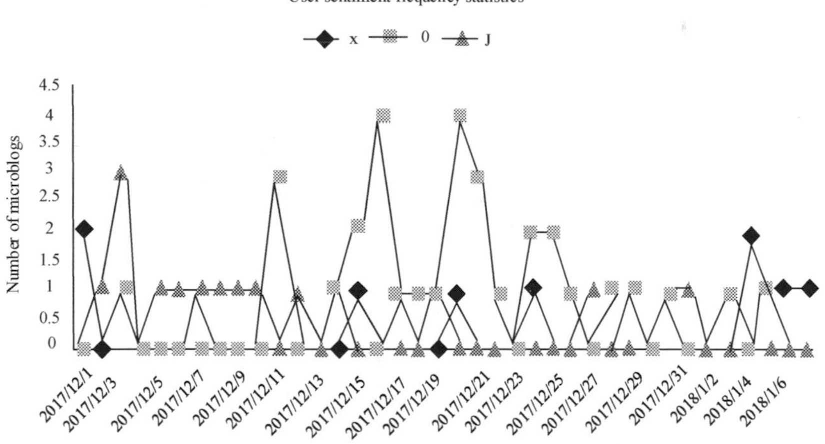

Fig. 5. Analysis of experimental results

Emotion density is the average daily emotion frequency value of users in a period of time, and the corresponding statistics of user emotion density are shown in the Table 5.

It can be seen from the table that the emotion density of user J is significantly higher than that of users X and D in the above time period, which indicates that user J is more inclined to express emotion in this time period. The experimental results of each eigenvalue based on implicit emotion space are discussed. It can be seen from the above table that the EI value of user x is much larger than the values of user D and J.

Table 5. Statistical table of emotional density of users

	X	J	D
Emotional density ω	0.447368421	0.868421053	0.421052632

according to the corresponding relationship between user characteristics and implicit emotion characteristics, it shows that the user has very strong emotion transitivity. By comparing their personal data, it is found that the number of fans of the user is 97,368,813, and the number of concerned users is 681. It really has a strong public opinion appeal and influence on the wireless network platform, and should be the key monitoring object of public opinion transmission guidance and supervision. Its D value is relatively large, which indicates that it has strong controllability, and the probability of needing external force interference is small, which indicates that the user's D emotion fluctuates greatly, the stability of emotion is low, and it may need to use external force to interfere with its emotional state when necessary. It shows that its emotional transitivity is relatively small, and the number of fans is only 136 through personal data check, which has very little influence on the wireless network platform and conforms to the quantitative judgment rules of transitivity. User J has a relatively large EI value. By comparing personal data, the number of fans is 1,492,514, and the number of concerned users is 1,277. It also has strong transitivity on the wireless network platform.

Based on the above experimental results, we test the accuracy of different methods. Methods of literature [2] and literature [3] were selected as experimental control group. The comparison results are shown in Fig. 6.

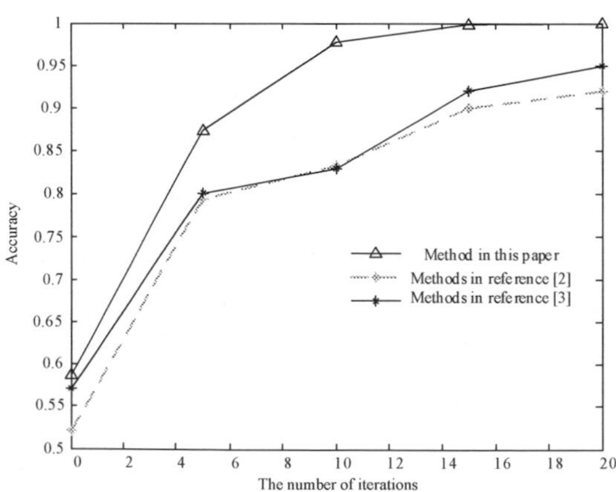

Fig. 6. Comparison of data mining accuracy

Analysis of Fig. 6 shows that with the increasing number of iterations, the accuracy of the user sentiment fluctuation analysis method proposed in this paper continues to improve, and the growth rate is fast. When the number of iterations is 4–12, the mining accuracy of literature [2] and literature [3] grows slowly, and the convergence is poor and the accuracy is low. Compared with the traditional method, the analysis of this method is more accurate and more applicable.

4 Conclusion and Prospect

Emotional expression is an innate instinct of human beings, which is the basis of human survival, mutual communication and social life. In recent years, the prediction based on Internet user sentiment analysis is an important field of sentiment analysis application research. At present, the typical prediction methods based on sentiment analysis can be summarized as the prediction methods based on sentiment analysis results and the prediction methods based on sentiment analysis results. In the process of prediction, the choice of emotion analysis source, the determination of prediction time advance and the statistical processing method of emotion words determine the real effect of prediction.

However, different social networks have different characteristics, for example, in anonymous social networks, anonymity will have an impact on emotional expression, the emotional expression of social networks used at different ages will have its own unique characteristics, these characteristics on the emotional expression of the impact can also be used as the content of the future study. In addition, the representation of user's emotion in different fields, the timely, accurate and comprehensive acquisition of corpus, and the correct analysis and statistics of user's emotion are the contents to be studied in the future.

Fund Projects. This work was supported by Jiangxi Education Department GJJ190316.

References

1. Faruolo, G., Santopietro, L., Saganeiti, L., Pilogallo, A., Scorza, F., Murgante, B.: The design of an urban Atlas to spread information concerning the growth of anthropic settlements in Basilicata Region. In: Gervasi, O., et al. (eds.) ICCSA 2020. LNCS, vol. 12255, pp. 214–225. Springer, Cham (2020). https://doi.org/10.1007/978-3-030-58820-5_17
2. Wang, J.L., Luo, W.L.: Prediction of user information behavior intention in mobile terminal message push process. Comput. Simul. **36**(03), 440–443 (2019)
3. Huang, K., Yuan, X., Li, L., et al.: Research on users' negative emotion and related influence factors during collaborative information searching based on affective load theory. Doc. Inf. Knowl. **01**, 42–52 (2020)
4. Liu, S., Lu, M., Li, H., et al.: Prediction of gene expression patterns with generalized linear regression model. Front. Genet. **10**, 120 (2019)
5. Xu, X., Lin, J., Xiao, Y., et al.: An approach to generating the sequence of part variant design based on information transfer utility. Assembly Autom. **39**(1), 186–199 (2019)
6. Alegre Sepúlveda, T., Norambuena, B.K.: Twitter sentiment analysis for the estimation of voting intention in the 2017 Chilean elections. Intell. Data Anal. **24**(5), 1141–1160 (2020)
7. Bibi, M., Aziz, W., Almaraashi, M., et al.: A Cooperative binary-clustering framework based on majority voting for Twitter sentiment analysis. IEEE Access **8**(11), 68580–68592 (2020)
8. Fu, W., Liu, S., Srivastava, G.: Optimization of big data scheduling in social networks. Entropy **21**(9), 902 (2019)

9. Liu, S., Bai, W., Zeng, N., et al.: A fast fractal based compression for MRI images. IEEE Access **7**, 62412–62420 (2019)
10. Wu, T., Weld, D.S., Heer, J.: Local decision pitfalls in interactive machine learning: an investigation into feature selection in sentiment analysis. ACM Trans. Comput.-Hum. Interact. **26**(4), 1–27 (2019)

Variable Bandwidth Receiving Method of Civil Aircraft Radar Signal Based on FPGA Technology

Ming-fei Qu(✉) and Dong-bao Ma

College of Mechatronic Engineering, Beijing Polytechnic, Beijing 100176, China

Absrtact. Traditional variable bandwidth receiving method has the problem of digital channel division confusion, which leads to the low anti-interference performance of the receiving method. Therefore, this paper designs a variable bandwidth receiving method of civil aircraft radar signal based on FPGA technology. After measuring the distance between two carrier frequency signals, the frequency modulated civil aircraft radar signal is continuous wave. Then, according to the signal bandwidth, the radar scattering surface characteristics of the radar signal bandwidth are obtained. Then, the FPGA technology is used to divide the digital channel, determine the number of effective signals, detect the shape of interference echo, and set up the variable broadband receiving module, so as to realize the design of civil aircraft radar signal variable bandwidth receiving method. The experimental results show that the anti-jamming degree of the proposed method is 9.384% ~ 17.426% higher than that of the two existing receiving methods under different jamming intensities, which proves that the proposed method is more suitable for civil aircraft radar signal receiving field under different jamming intensities.

Keywords: FPGA technology · Civil aircraft · Radar signal · Variable bandwidth

1 Introduction

With the continuous development of electronic information technology, civil aircraft need to adapt to more complex electromagnetic environment. In the complex electromagnetic environment, the radar signal receiving equipment of civil aircraft often receives multiple radar signals with different bandwidth in its monitoring bandwidth. In order to separate radar signals effectively, it is necessary to have multi-channel receiving ability. At the same time, due to the uncertainty of the bandwidth and position of the radar signal, civil aircraft also needs to have the ability to realize the variable bandwidth receiving technology.

In the early days, the radar signal receiving equipment of civil aircraft is to receive the radar signal in the way of single channel receiving, or on the basis of single channel receiver, they are used in parallel [1]. However, such radar signal receiving equipment of civil aircraft often needs to constantly change the local oscillator of the equipment in the search process because of the uncertainty of channel and the jump of radar signal

S. Liu and X. Ma (Eds.): ADHIP 2021, LNICST 416, pp. 155–166, 2022.
https://doi.org/10.1007/978-3-030-94551-0_13

when detecting and collecting radar signals. However, due to the limited search speed, it is easy to cause the omission of radar signal, which leads to the failure of the search equipment to intercept the complete radar signal. This indicates that the single channel receiver has been difficult to adapt to the needs of modern development through simple modification of architecture and improvement of details [2].

The radar signal receiving equipment of civil aircraft is mostly multi-channel receiver, which can achieve full probability radar signal interception. Their structures are similar to each other. Most of the front ends use the classic superheterodyne structure for reference, but the specific radar signal receiving and processing flow is different. Reconnaissance antenna is used to detect and receive radar signals within a certain frequency band.

FPGA (Field Programmable Gate Array) technology is a product of further development on the basis of PAL, GAL and other Programmable devices. As a semi-custom circuit in ASIC field, it not only solves the shortcoming of custom circuit, but also overcomes the shortcoming of limited gate number of original programmable device. FPGA technology has a programmable delay digital unit, which is widely used in communication system and various electronic equipment. Therefore, combined with the above research background, this paper designs a variable bandwidth receiving method for civil aircraft radar signal based on FPGA technology. The design idea of this method is as follows:

(1) After measuring the distance between two carrier frequency signals, frequency modulated continuous wave of radar signal of civil aircraft.
(2) According to the signal bandwidth, the radar scattering surface features of the radar signal bandwidth are obtained. Then, FPGA technology is used to divide digital channels to determine the number of effective signals and detect the shape of jamming echoes.
(3) Set up the variable broadband receiving module, so as to realize the variable bandwidth receiving of civil aircraft radar signals.

2 Variable Bandwidth Receiving Method of Civil Aircraft Radar Signal Based on FPGA Technology

2.1 Continuous Wave of FM Civil Aircraft Radar Signal

The simple rectangular pulse radar transmits a single frequency sine wave signal during pulse modulation. The signal frequency of pulse compression radar increases linearly during pulse modulation. The carrier frequency of frequency continuous wave radar system increases or decreases in the whole signal period. Pulse compression radar system has the advantages of high ranging accuracy. Its range resolution is directly proportional to the bandwidth of the transmitted signal. The larger the bandwidth of the transmitted signal, the higher the range resolution. FMCW is a radar system widely used in anti-collision radar [3]. Continuous wave can be divided into constant frequency continuous wave, frequency shift keying, phase shift keying and linear frequency modulation continuous wave. Among them, the frequency of the transmitted signal in the constant frequency continuous wave mode is single, and only the Doppler frequency of the target can be obtained after mixing. Therefore, it can only measure the velocity, but has no ranging ability, so it can not be used in the anti-collision radar. Compared with the above types of radar, LFMCW has the advantages of no range blind zone, wide signal bandwidth, simple transceiver system and high sensitivity, so it is most widely used [4, 5]. De slope is an effective way to process the frequency domain continuous wave of FM civil aircraft radar signal. The radar mixes the target echo signal with the local oscillator signal to get the beat signal. For the point target, the beat signal is a single frequency signal. Then, the baseband digital signal is processed by low-pass filtering and amplification of beat signal, and the obtained digital signal is processed by Fourier transform for spectrum analysis. The range information of target relative to radar can be extracted from the frequency domain information. When the radial velocity is 0, then:

$$L = \frac{2AQ}{PJ} \tag{1}$$

In formula (1), L is the frequency domain of radar signal, Q is the relative distance between radar and target, A is the baseband signal, P is the phase shift range, and J is the peak value of radar signal. Because the relative speed of radar and target is very small relative to the speed of light, and the multi cycle time is also relatively short, the change of radar received signal strength caused by the change of relative distance between radar and target during the cycle can be ignored, that is, the amplitude change of spectrum peak value of multi cycle echo beat signal can be ignored. For the chirp signal, only the peak value of the chirp signal changes with the time, that is to say, the variation of the echo phase of the chirp signal is ignored The larger the degree, the larger the envelope frequency. Its working principle is shown in Fig. 1.

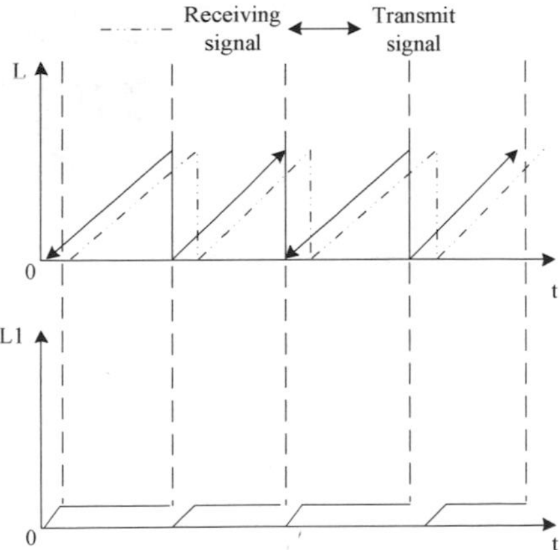

Fig. 1. Schematic diagram of frequency modulation

It can be seen from Fig. 1 that the continuous wave in the frequency domain of the signal of civil aircraft is related to its peak within a fixed distance, while the maximum peak is related to the motion state of the target within the receiving distance. In modern civil aircraft, the method of separating stationary target and moving target is often used to suppress clutter. When the aircraft is in flight, there is relative motion between the aircraft and the obstacles it faces. Therefore, in the process of frequency modulation, the error caused by relative motion should be solved first [6].

The clutter that radar may encounter when working includes fixed clutter and moving clutter. The Doppler frequency of fixed clutter and slow moving clutter is almost zero, the amplitude and phase of the signal after phase detection will not change with time or change slowly with time, while the amplitude and phase of the moving echo after detection will change greatly with time. In this case, the same range unit of the adjacent repetition period can be subtracted, and the fixed target can be completely cancelled and the slow moving echo can be eliminated The moving clutter is also attenuated to a great extent, and only the moving echo is retained, thus completing the frequency domain continuous wave working steps of FM civil aircraft radar signal.

2.2 Feature Extraction of Variable Wideband Radar Scattering Surface

It is very important to obtain radar scattering surface characteristics of radar signal bandwidth in the whole process of receiving radar signal. Generally speaking, in order to be able to receive signals of different frequency bands and different bandwidths, current digital receivers all have multi-channel receiving capability. For the multi-channel receiver, the radar signal bandwidth change judgment module can be located either before the original sampling signal enters the multi-channel receiving module or after the processing of the multi-channel receiving module [7]. In order to judge the signal bandwidth before the receiving module, panoramic monitoring is needed for the input, and in order to judge the signal bandwidth after the receiving module, channel judgment is needed for each subchannel. But both of them are time-frequency detection. Short time Fourier transform, wavelet transform and other classical methods can be used in the spectrum feature extraction module to analyze the spectrum of the original sampling signal in real time. The specific process is shown in Fig. 2.

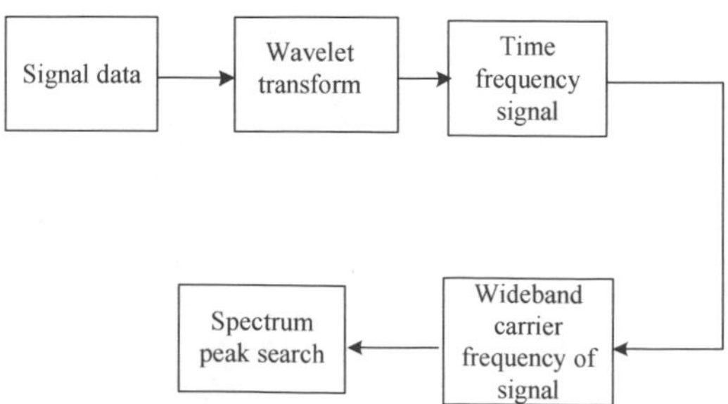

Fig. 2. Signal acquisition process

As can be seen from Fig. 2, after the broadband signal passes through the channelization module, the bandwidth of the broadband signal is greater than that of a single subchannel, and the signal spectrum is cut by multiple adjacent subchannels. Taking the wideband LFM signal as an example, the signal detection module of the left subchannel will detect the signal energy first, then the middle subchannel and the right subchannel in turn, and the signal occurrence time of each subchannel will be partially overlapped in the first place. But at this time, the parameters measured in each sub channel are incomplete or even wrong, so it is difficult to obtain the real parameters of broadband LFM. According to the theory of electromagnetic wave, the intensity of electromagnetic wave is determined by the intensity of electric field. Then the expression formula of RCS is as follows:

$$\delta = l \underset{H \to \infty}{im} \times 4\pi H^2 \left| \frac{W_1}{W_2} \right|^2 \tag{2}$$

In formula (2), δ is the area of radar cross section, H is the radar signal target with feature extraction, W_1 and W_2 are the intensity of incident electric field and scattering electric field respectively. The scattering modes of scattering sources include mirror echo, edge diffraction echo, traveling wave echo, creeping wave echo, tip diffraction echo, dihedral angle scattering, corner body scattering, cavity scattering, etc. [8].

RCS is calculated by theoretical calculation and practical measurement. The theoretical calculation includes geometric optical approximation, physical optical approximation and geometric diffraction theory. The scattering cross section of radar is closely related to the shape, size, structure, material and polarization mode of the radar. At the same time, it is also the fluctuation function of the frequency and angle of incidence of the incident electromagnetic wave. When other conditions are constant, the larger the target size is, the larger the radar cross-section is. For a certain radar frequency and fixed angle of view, the radar cross-section of the target is determined by the polarization mode. The time delay, Doppler frequency shift and amplitude factor attenuation in radar echo signal are all related to the distance and motion state of the target relative to the radar. Based on the above description, the acquisition step of the scattering surface feature of variable wideband radar is completed.

2.3 Using FPGA Technology to Divide Digital Channel

FPGA technology has the advantages of high integration, strong logic implementation ability, good design flexibility, short development cycle and fast working speed, which is widely used in the field of radar signal processing. Using FPGA technology to divide the digital channel is one of the important links in the design of civil aircraft radar signal variable bandwidth receiving method.

Because each subchannel filter has a certain center frequency and bandwidth, the channel decision for each subchannel can make a rough judgment on the signal frequency band and bandwidth [9]. Therefore, the channel decision is equivalent to depicting a discrete time-frequency diagram. It focuses on when signals appear in some subchannels. The common channel decision methods are energy detection and autocorrelation detection. Energy detection belongs to the non coherent detection of signal, which can be carried out in time domain or frequency domain. Time domain detection is to accumulate the time domain signal, and then compare with the time domain detection threshold to determine whether the signal exists.

Frequency domain detection is to convert the time domain input to the frequency domain, then carry out some accumulation, and then compare with the frequency domain threshold to determine whether the signal exists. Frequency domain detection accuracy is high, but the algorithm is complex and resource consumption is high. Time domain detection algorithm is simple, but the accuracy is not high. Both have their own advantages and can be applied in different occasions. Frequency domain detection is to detect the signal from the spectrum, so as to determine whether there is a signal in the subchannel. The length of external input data is unknown, so the length of output data of subchannel, that is, the length of input data detected in frequency domain, is also unknown. In the actual detection, the input data needs to be windowed to fix the length of the input data points. According to the different ways of data windowing, the frequency domain detection method can be roughly divided into three cases, as shown in Fig. 3.

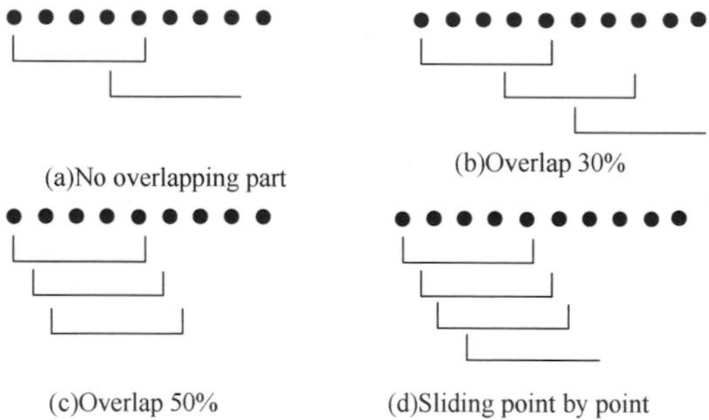

Fig. 3. Schematic diagram of data point sliding

As can be seen from Fig. 3, different overlapping methods will present different detection results after windowing detection. The higher the overlap degree of the sliding window, the greater the amount of computation, but too little overlap of the sliding window will lead to the loss of signal information.

In practical application, the balance between calculation and detection performance should be considered comprehensively. On the premise of ensuring reliable detection, the method with appropriate calculation should be selected. Finally, the output is compared with the set threshold to determine whether there are valid signals and the number of valid signals. The signal received by radar receiving equipment is non cooperative signal, and the key information is unknown. In the design of digital channelized receiver, the blindness of channel division occurs, which leads to the problem of cross channel when reconnaissance receives non cooperative signal [10].

Single stage Digital Channelization structure, usually the sub channels are evenly divided, only providing a single sub channel bandwidth. If the bandwidth is too wide, the signal is channelized in the frequency domain and distributed in multiple adjacent subchannels, resulting in distortion of the output signal. When the channel is divided into wide channels, the SNR gain of narrow channels is small, which makes the subsequent signal processing difficult. Autocorrelation detection method is also a practical channel detection decision method. The main process of the detection method is to first perform some autocorrelation operations on the received data, then accumulate the autocorrelation results, and finally compare the cumulative value with the threshold. From the detection process of autocorrelation method, we can see that it has the characteristics of time-domain energy detection method for accumulation, and frequency-domain energy detection method for some kind of operation. Considering that the received signal is a complex signal, it does not affect the actual effect:

$$Y = U^{2\pi Ej} + m \tag{3}$$

In the formula (3), Y is the complex signal type of civil aircraft, U is the signal amplitude, E is the carrier frequency, j is the sampling period, and E is the zero mean additive white Gaussian noise.

To sum up, it can be seen that the theoretical basis of this method is that the signal is not correlated with noise. The autocorrelation method has the advantages of high efficiency and fast detection of signal in time domain energy detection method, and also has the advantages of detection of signal in frequency domain energy detection method in the case of low signal-to-noise ratio. In engineering implementation, digital filter is not an ideal filter, there is a transition band between passband and stopband, and there is excessive band overlap between channelized channels. For multiple signals arriving at the same time, they overlap in time domain and do not overlap in frequency domain. Channel division is to use this feature to separate signals.

2.4 Setting Variable Broadband Receiving Module

Setting up the variable broadband receiver module is one of the important links in designing the bandwidth receiving method. The single-channel monopulse tracking receiver is an important part of the radar signal processing system, which is mainly used to receive and track the radar signal and other signals, so as to realize the automatic tracking of the target.

When radar antenna is receiving signals, its target does not always appear in the center of radar antenna beam. In general, the target will enter the antenna beam at 3 dB point, and then leave the antenna beam from 3 dB point. In this way, the target echo is double-modulated in the directivity diagram, and its expression formula is:

$$k = \frac{S_r \times \sigma_{0.5}}{\Omega} \tag{4}$$

In the formula (4), k is the number of received pulses of radar signal, S_r is the radar pulse frequency, $\sigma_{0.5}$ is the half power point width of radar beam, and Ω is the rotation speed of radar antenna. The radar with the same frequency band and the same model will cause synchronous jamming while the radar with the same frequency band and different model will cause asynchronous jamming. Through the observation of radar display, we can clearly see the different effects of two kinds of same frequency jamming on Radar: when two radars have the same or integral multiple pulse repetition period, the jamming echo presents concentric circle shape, which is called synchronous jamming.

If the pulse repetition periods of two radars are different and not integral multiple, the jamming echo presents the shape of rotating petals, which is called asynchronous jamming. When the channelization module processes different signals, according to the modulation type, carrier frequency, bandwidth and other parameters of the signal, the signal will appear different distortion in the cross channel frequency domain.

The radar antenna emits electromagnetic pulse or continuous wave to the aircraft or target in the air. When the signal propagates to the aircraft, the energy of the signal beam expands to the air. The signal strength of any point in the signal beam can be expressed by energy density. When the signal irradiates and passes through the aircraft surface, part of the incident energy is absorbed as heat energy, the other part of the energy completely passes through the aircraft parts, and the rest is re radiated in all directions by different conductive surfaces on the aircraft. The reradiation electromagnetic signal is generated by the surface current oscillation generated by the electromagnetic field oscillation of the incident radar signal that irradiates the target, which is usually called the scattered or reflected signal. The setting step of the variable broadband receiving module is completed.

3 Experimental Study

3.1 Experimental Preparation

In order to verify the stability of the civil aircraft radar signal variable bandwidth receiving method based on FPGA technology, the experimental comparison is carried out, and the experimental environment is set up as shown in Table 1.

Table 1. Experimental environment

Experimental environment	Application server	Database server	Client
Hardware parameters	Windows Server 8 × 2core 2.1 GHz Proc CPU 16G	IBM eServer P570 16× 2.1 GHz Proc CPU 48G	CPU: P41G Memory:2G
Software parameters	Apache-Tomcat-6.0 JDK 1.7.0 PostgreSQL9.0	WebSphere OS:AIX5.3	Windows 2007 IE7.0
Network environment	VPN		

Using the above experimental configuration, the radar signal noise jamming type index is obtained, and the expression formula is as follows:

$$Z = \frac{2R \times V}{R + V} \tag{5}$$

Where Z is the reception rate of the interference profile, V is the total number of interference profiles, and P is the number of real interference profiles.

In order to enhance the contrast between the experimental results, the traditional deep learning-based radar signal receiving method for civil aircraft and the narrow-band IoT based radar signal receiving method for civil aircraft were compared, and the performance verification was completed together with the proposed method.

3.2 Experimental Result

Based on the above attack type index, two existing receiving methods are selected to test the anti-interference performance with the designed receiving method. According to the same experimental parameters and under different interference intensity conditions, the experimental results are shown in Fig. 4.

As can be seen from Fig. 4, although the degree of interference is constantly changing, the anti-interference degree of the proposed method is higher than that of the traditional method, which proves that the proposed method has higher stability and is more suitable for application in the field of civil aircraft radar signal reception.

To sum up, the variable bandwidth receiving method of civil aircraft radar signal designed in this paper based on FPGA technology has a higher degree of anti-jamming, and can efficiently accept civil aircraft radar signal under different jamming intensity.

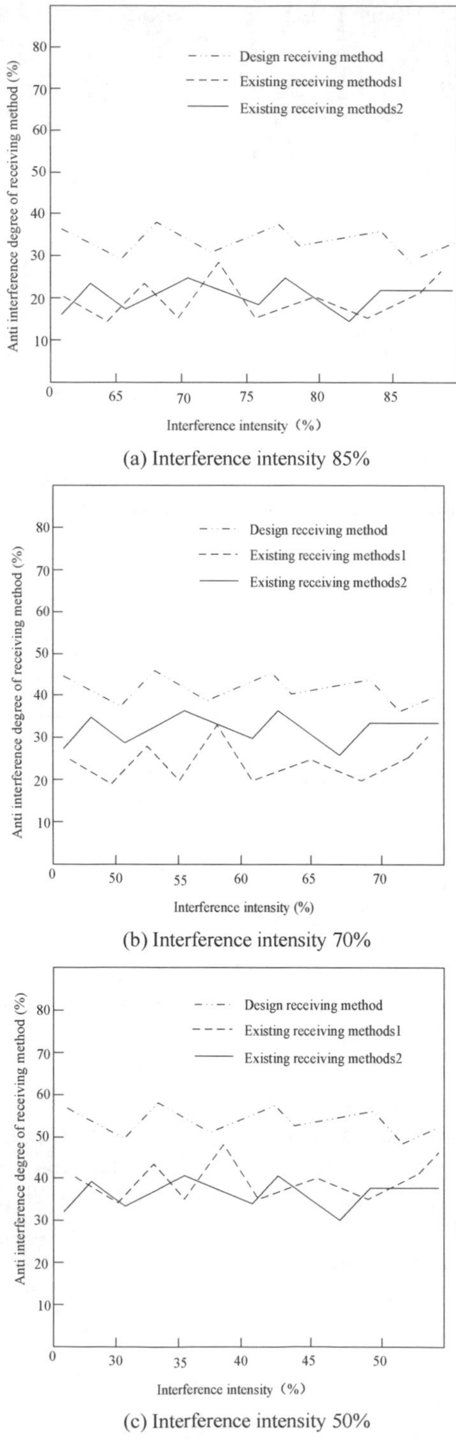

(a) Interference intensity 85%

(b) Interference intensity 70%

(c) Interference intensity 50%

Fig. 4. Test results of anti-interference degree

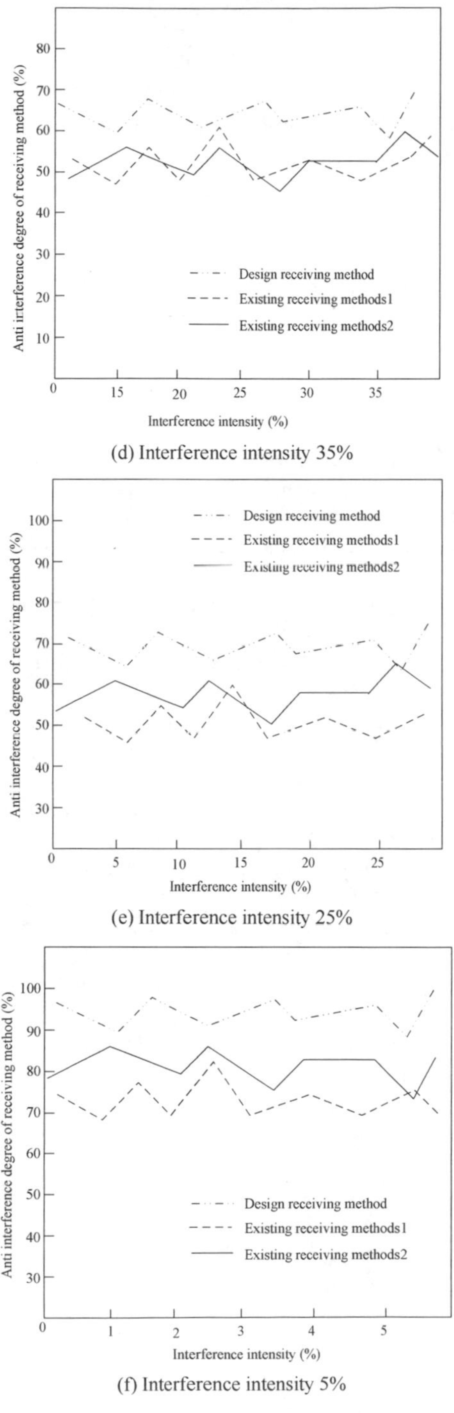

(d) Interference intensity 35%

(e) Interference intensity 25%

(f) Interference intensity 5%

Fig. 4. continued

4 Conclusion

This paper designs a variable bandwidth receiving method of civil aircraft radar signal based on FPGA technology. The experimental results show that it has better anti-jamming performance and can receive radar signal more accurately in the process of aircraft navigation. At the same time, it enriches the research content in the field of civil aircraft radar signal reception.

Fund Projects. Key topics of Beijing Polytechnic

References

1. Zhang, N., Gu, W., Li, P.: A direct way of generating wideband radar signals based on polyphase decomposition. Mod. Radar **42**(1), 82–85 (2020)
2. Wang, L., Zhang, L., Gu, P., et al.: Real-time implementation of frequency-modulated continuous-wave synthetic aperture radar imaging based on FPGA. J. Detect. Control **41**(2), 86–91 (2019)
3. Cao, H., Dong, Y., Li, Y.: Fast realization of multi-beam coherent integration based on FPGA. Radar Sci. Technol. **17**(4), 449–454, 460 (2019)
4. Liu, M., Ying, L., Wang, X., et al.: Wideband digital array radar beamforming based on RF direct sampling. Radar Sci. Technol. **18**(4), 394–398 (2020)
5. Tian, E., Li, Z.: Research on position correction method of picking robot-based on wireless sensor network and ultra wide band FM method. J. Agric. Mechanization Res. **41**(2), 216–219, 224 (2019)
6. Cao, J., Lir, U., Yang, J., et al.: Dual-band LFM-CW radar scheme based on photonic stretch processing. J. Radars **8**(2), 189–196 (2019)
7. Pan, W., Zhang, Q., Zhang, Q., et al.: Identification method of aircraft wake vortex based on Doppler lidar. Laser Technol. **43**(2), 233–237 (2019)
8. Chen, T., Cai, X., Pan, D.: Design of wideband digital receiver based on pulse compression. J. Tianjin Univ. **52**(4), 441–448 (2019)
9. Liu, S., Liu, G., Zhou, H.: A robust parallel object tracking method for illumination variations. Mobile Netw. Appl. **4**(1), 5–17 (2019). https://doi.org/10.1007/s11036-018-1134-8
10. Wang, K., Qian, J., Zhang, J., et al.: Multi-radar signal fusion in range-doppler domain. Mod. Radar **41**(11), 35–38 (2019)

Recognition of Aerobics Movement Posture Based on Multisensor Movement Monitoring

Ying Liu[1] and Zhong-xing Huang[2(✉)]

[1] Department of Public Basic Courses, Wuhan Institute of Design and Sciences, Wuhan 430205, China
[2] Guangzhou Metro Design and Reserch Institute Co., Ltd., Guangzhou 510000, China

Abstract. In the traditional attitude recognition methods, the final recognition rate is low because of the inadequate processing of the motion attitude data. Therefore, based on the multi-sensor mobile monitoring of aerobics movement posture recognition. In the design process of aerobics movement posture recognition method, first of all, based on multi-sensor movement monitoring, collected aerobics movement posture data. Then in order to improve the recognition rate of aerobics posture, the collected data is preprocessed. Will process the good data, according to the time frequency characteristic complete the data characteristic extraction. Comparing the extracted features with the multi-sensor moving monitoring images, through the multi-level attitude recognition algorithm, the movement attitude recognition of aerobics is finally realized. Experimental results show that the proposed method has a higher recognition rate than the traditional method. Even under the influence of changing factors such as penalty factors and kernel parameters, the proposed method is still predominant.

Keywords: Sensor · Mobile monitoring · Attitude recognition · Feature extraction

1 Introduction

The technology of human motion recognition has been developed for many years. It is a kind of pattern recognition method to judge the state of human motion by analyzing the related information reflecting human motion behavior [1]. This technology can provide users with motion state information, so it has a wide range of applications in sports health, sensory games, user social behavior analysis, safety monitoring, patient rehabilitation training, indoor positioning and navigation, personal feature recognition and other fields [2]. The change of living habits and the increase of life stress lead to a variety of diseases troubling people, making people begin to realize the importance of health. A variety of sports should be out of time [3], aerobics has become a popular sport of the masses. Aerobics combines many dance elements, gymnastics, dance, music, fitness, entertainment in one, is conducive to reducing the accumulation of hip and abdominal fat, while the movement of coordination and flexibility improvement [4]. Aerobics to play a role, need

S. Liu and X. Ma (Eds.): ADHIP 2021, LNICST 416, pp. 167–178, 2022.
https://doi.org/10.1007/978-3-030-94551-0_14

to complete the movement posture control. In the literature [5], it is proposed to install sensors on human body, fuse information and mine association according to the oscillation amplitude of sensors, and obtain the data of motion posture. The data is monitored by the method of sensor quantification tracking and recognition, and the data is fuzzy adaptive fused by feature clustering, thus realizing the monitoring of motion data. In the literature [6], the recognition of human motion based on data fusion of acceleration sensor and spiral instrument is put forward, which mainly refers to the spiral instrument outputting human motion information, selecting appropriate algorithm to repair and fuse the output information, and realizing accurate measurement of human motion posture.

Therefore, the movement of multi-sensor monitoring as a support to achieve aerobics posture recognition. After the rapid development of sensor technology, it is gradually developing towards miniaturization, digitalization, intelligentization and multifunction, which makes the recognition technology of human daily motion posture based on sensor signal develop rapidly.

In this paper, multi-sensor movement monitoring is applied to aerobics posture recognition. Multi-sensors are used to collect the movement posture, preprocess the collected data, remove the interference of redundant data, extract the data features according to the time-frequency characteristics, and use multi-level posture recognition algorithm to complete the movement posture recognition of aerobics. Experiments show that the designed method has a good recognition rate. It can provide some support for the handling of aerobics posture.

2 Design of Exercise Posture Recognition Method Based on Multisensor Mobile Monitoring

Aerobics movement posture recognition method design, based on multi-sensor movement monitoring. First of all, the typical aerobics movement posture data collection. In order to ensure better recognition effect, the collected data are preprocessed. Then, the data feature is extracted, and a multi-level attitude recognition algorithm is constructed to realize the recognition of the motion attitude.

2.1 Multi-sensor Based Data Acquisition for Aerobics Posture

Because this paper is aimed at a given time series of motion data, stable and accurate pattern recognition is carried out. Mainly based on the sensor movement monitoring, the aerobics movement posture recognition [7]. Taking the posture of human body in aerobics as the key target, a small data acquisition terminal is designed to effectively collect and store the movement information of human body. Multi-sensor mobile monitoring of motion and attitude data acquisition, involving the sensor unit [8], control unit, storage unit and power management unit. The data collection process is shown in Fig. 1.

The sensor module adopts MPU9250 nine-axis sensor chip integrated with many types of sensors, which can be used to extract the acceleration, angular velocity, magnetic field intensity and other related vector information [9]. The MPU9250 chip is composed of a moving attitude module and a magnetic module. The MEMS inertial measurement unit and the MPU6050 six-axis sensor chip can obtain the acceleration, angular velocity

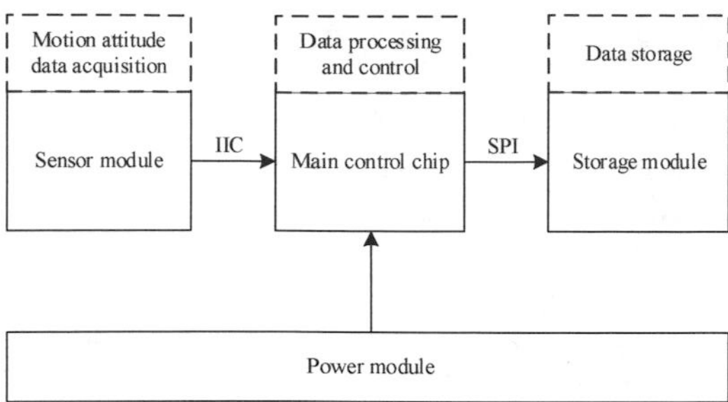

Fig. 1. Motion attitude data acquisition flow

and temperature of the target. The magnetic field intensity and other vector information of the target can be obtained by the AK8963 three-axis magnetometer. According to the motion characteristics of human body, the sample rate of the sensor is 25 Hz, and the sensor data can be read by IIC. So the communication between MPU9250 chip and MCU STM32F103RBT6 is realized by IIC protocol [10]. In the circuit design, the pull-up resistor is used to control the signal line, and the signal line is driven in turn by the IC interface of the main control chip and MPU6050, as shown in Fig. 2.

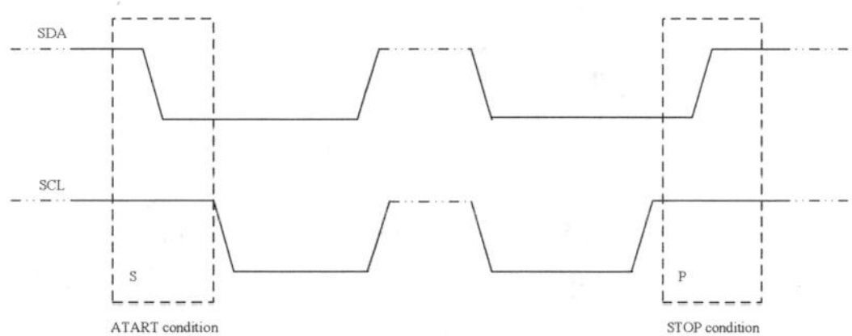

Fig. 2. IIC communication timing block diagram

In the main control chip plate, because the control center needs to carry on the reorganization to the MPU6050 chip data, coordinates between the module the data transmission [11], completes the data on chip storage. For the corresponding requirements of portable devices, the control center needs to have the following characteristics:

(1) Low power consumption.
(2) Fast response.
(3) Stable functions.
(4) Having sufficient communication interfaces.

Based on the above characteristics, this research chooses STM32 chip as the total on-chip control chip. Finally, the storage plate, using MICRO SD CARD as storage media. The layered structure of the medium has the features of low cost, SPI Slave UART, industry-mature interface, compatibility with most devices, and chip solidification of FAT 12/FAT 16/FAT32 file system. Through the combination of the above plates, the joint role of multi-sensor movement monitoring of aerobics movement data acquisition.

2.2 Sample Data Preprocessing

There are always some problems in the collection of motion attitude data by multi-sensor mobile monitoring. The first is that the collected data can not avoid the existence of noise, and prominent noise will affect the recognition algorithm [12]. The second point is due to the difference in the work, the data collected by the data acquisition terminal will have a certain zero deviation error. The third point is that the stability of gravitational acceleration will disturb the posture information of human body. In this study, low-pass filter is used to remove useless noise and extract useful information. The Butterworth filter is used in this paper. The system functions are shown in Fig. 3.

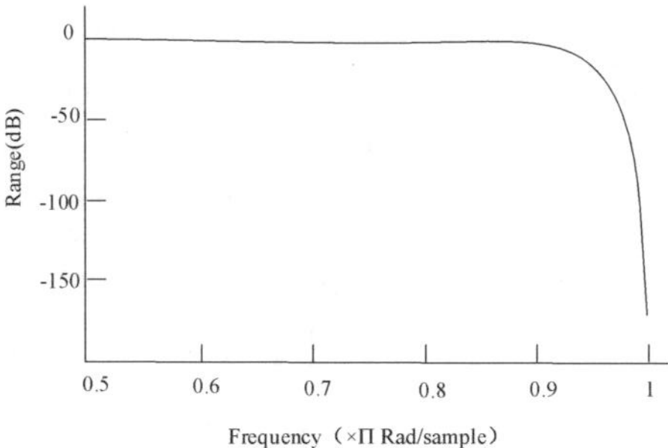

Fig. 3. Filter system functions

According to statistics, the filter has a certain degree of inhibition of interference noise, when the data itself is very stable, its order of magnitude in 10^{-2}–10^{-3} g, and the general movement of the human body in 10^{-1}–100 g change, so the impact of noise is not very large. In addition, the data collected by the terminal should be calibrated. When the objective value of the sensor's data is zero, the actual value of the sensor's data is zero [13]. So it is necessary to calculate the bias error by experiment and compensate the measured signal. In theory, if you put the mpu water at rest, the x-axis, the y-axis reading will be zero, and the z axis will be 1 g or –l g. But this is not the case, each chip due to manufacturing problems, the actual output value of the theoretical value must be deviated, so we need to calibrate the chip.

The calibration formula is:

$$G_x = K \times ADC_x + offset \tag{1}$$

In the formula, ADCX is the actual output of the sensor, GX is the actual sensing data, offset means that the acceleration of 0g is the sensor's actual sensing data, that is, zero bias error. K is the scaling factor. The final step is to remove the gravitational acceleration interference. The gravity acceleration is always included in the human posture data. For the static state, the gravity acceleration is always downward, which makes the three axes of the sensor coordinate system produce stable components. In addition to the gravity acceleration interference filter also used to aerobics posture in the leg, for example, can be filtered before and after the contrast of the time domain, as shown in Fig. 4.

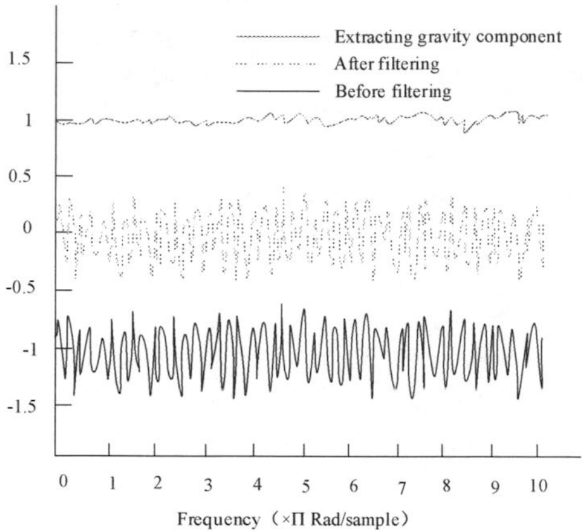

Fig. 4. Comparison of time domain before and after filter of swinging sample

As can be seen from Fig. 4, the motion signal after filtering out the gravity component changes around the zero point, and has a strong regularity, which is convenient for subsequent attitude recognition.

2.3 Feature Extraction of Aerobics Posture Data

It is difficult to get the direct basis of learning decision directly from the original data. Therefore, it is necessary to extract the typical features of various kinds of posture time series to serve the construction and verification of posture recognition algorithms. Before the feature extraction, we need to integrate the initial information to obtain further time series. The acceleration and angular velocity of human body reflect the intensity of human motion to a great extent. Another characterization of human motion is the rate of change of each axis data. It is therefore possible to generate a sequence of differences for each axis. Based on the difference sequence, different motion states can be obtained. In the research of human posture recognition algorithm, its features are generally time-frequency features. The statistical features used in this study include mean value, standard deviation, median absolute deviation, quartile distance, maximum value, minimum value, square and mean value, AR model coefficients with Berg order of 4, correlation of each axis series and information entropy. The formula for calculating mean value is as follows:

$$mean\_accel\_a\_x = \sum\nolimits_n accel\_a\_x_n \tag{2}$$

These formulas represent the total size of the time series. The amplitude of jitter of a signal represented by the standard deviation std. The formula is as follows:

$$std\_accel\_a\_x = \sqrt{\sum\nolimits_n (accel\_a\_x_n - mean\_accel\_x\_a)^2/N} \tag{3}$$

The two formulas show that the range of motion and intensity of the different postures of the human body are different. Based on this, the salient characteristics of signal frequency can be found. Due to the frequency domain waveform of the moving state, the peak spikes regularly appear on the whole spectrum line. Based on this attitude frequency domain information, including peak frequency point, sub-peak frequency point, peak rising along 3 dB frequency point, descending along 3 dB frequency point, number of outburst frequency point, weighted average frequency point. Peak frequency points represent the most prominent frequency information of attitude. The sub-peak frequency features characterize the attitude sub-salience in the frequency domain, and the peak bandwidth is described along the 3 dB frequency points and the 3 dB frequency points. The number of peak frequency points indicates the frequency richness of human body. The weighted average frequency points represent the average frequency points in the frequency domain. The formula is as follows:

$$\overline{f} = \frac{\sum_{i=1}^{N} \omega_i \times h_i}{\sum_{i=1}^{N} h_i} \tag{4}$$

Under various motion postures, the weighted average frequency maps were compared. According to the contrast graph, it can be seen that this feature can be better classified into aerobics posture categories. To sum up, the peak frequency point, weighted average frequency point and the number of outburst frequency points can all be used as effective data features. These features are numbered in practical applications to facilitate the subsequent expression and application.

2.4 Build a Multi-level Attitude Recognition Algorithm

Aerobics movement posture mainly includes many posture movements, but each movement has its own unique. Therefore, by roughly classifying the movement postures of aerobics, we can narrow the range of gesture recognition, improve the efficiency of the algorithm, and finally achieve accurate recognition of movement postures. Aerobics posture can be seen as a series of coarse classification of movement in space and time, the movement posture in this paper according to the movement intensity, frequency coefficient rough classification of the movement posture. Based on the sports characteristics of aerobics, the paper chooses “kick”, “swing”, “side waist extension”, “high lift leg”, “left and right level lift” five sports posture. Among them, “kick”, “swing leg”, “high lift leg” can be seen as dynamic, “side waist extension”, “left and right level lift” two actions relatively static, called static.

Firstly, the attitude is classified according to the motion intensity. In this paper, the standard deviation of the attitude angles is used to analyze the motion intensity. The standard deviation of the movement attitude angle expresses the change of the movement attitude angle of calisthenics in a certain time, which reflects the movement intensity of the movement attitude; the standard deviation of the combined attitude angle reflects the intensity of the movement attitude angle, which causes the error in order to prevent the special circumstance from separately calculating one angle of the attitude angle; meanwhile, the movement shows the change of the three angles on the attitude angle data, so the value of the combined attitude angle is introduced, and the standard deviation is defined as follows:

$$sum\sigma = \sigma_1 + \sigma_2 + \sigma_3 \tag{5}$$

The σ1, σ2, σ3 in the formula respectively represent the standard deviation of yaw angle, pitch angle and roll angle in the motion attitude angle. According to the experimental analysis, the standard deviation of the combined attitude angle in the rough classification based on motion intensity is set to 1.5°. When the standard value of the combined attitude angle is greater than this value, it is marked as dynamic and less than this value, it is recorded as static.

Secondly, based on the rough classification of pose in frequency domain, it is found that the normal stride frequency of aerobics is not more than 2.5 Hz, and it is periodic. Because the characteristics of aerobics movement in time domain are not obvious, it needs to be analyzed from time domain to frequency domain. When aerobics movements are carried out, the posture recognizer worn on the human body will swing up and down with the movements. Based on the signal of Z-axis acceleration sensor, because the data of Z-axis is affected by gravity field, the Z-axis is filtered at high speed, and the effect of gravity g is filtered out. And the range of exercise energy concentrates on 0–10 Hz, which satisfies the frequency distribution of activity, and the corresponding frequency at the maximum amplitude is 3.972 Hz, which accords with the normal exercise frequency of aerobics. Therefore, the specific action is dynamic.

Finally is the transition movement classification, some movements in the aerobics movement, said separately from the standing posture to the lying posture and the lying posture to the standing posture movement process, because it expresses the movement process, therefore calls it as the transition movement. Compared with other actions, the duration of these two actions is shorter. According to observation and analysis, the duration of transitional actions is usually 3–8 s, and the time is not fixed. In this paper, an adaptive sliding window method is used to segment the data, which can not meet the needs of recognition of transitional actions. Therefore, an adaptive sliding window method is used to segment the data. According to the characteristics of the transition movement, this paper uses the peak-peak and its slope to judge the transition movement of calisthenics. When the peak-peak pitch angle is greater than 34° but less than 50°, whether the peak peak value slope is less than –1.35° is judged. If the slope satisfies the peak value, it is judged to be a suspicious standing movement. Then calculate whether the sum of the standard deviation vectors of the attitude angle in the two time windows is less than 10°, and if it is satisfied, it can be judged to be standing movement. When the peak pitch angle is greater than 50° and its slope is greater than 4.5°, it is judged to be a suspected lying movement. Then calculate whether the sum of the standard deviation vectors of the attitude angle in the two time windows is less than 10°, and if it is satisfied, it is judged to be a lying movement. The algorithm flowchart for the transition action is shown in Fig. 5.

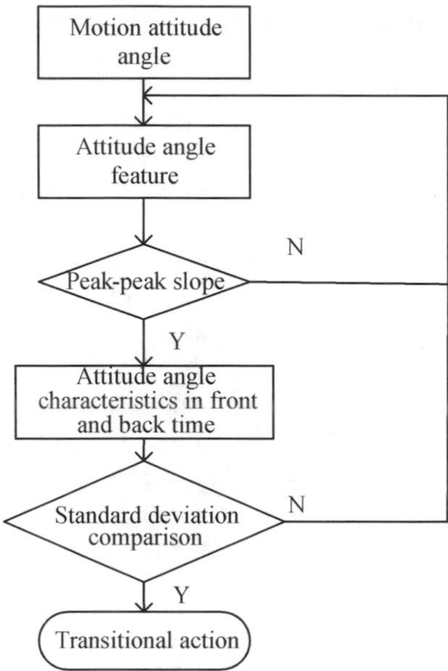

Fig. 5. Algorithm flow for transition action

The attitude recognition algorithm used in this paper is a decision tree method, which can effectively reduce the classification complexity and improve the recognition efficiency by transforming complex multi-classification problems into multi-level binary classification problems. Aerobics movement posture recognition algorithm flowchart is shown in Fig. 6. The attitude data of calisthenics are collected by the attitude recognizer, and the attitude angle is calculated, and the feature is extracted, including mean, standard deviation, frequency domain transform and peak slope. When the standard deviation of the combined attitude angle is higher than 1.5°, the movement is classified as dynamic, otherwise as static. Then the attitude is transformed and the frequency at the maximum

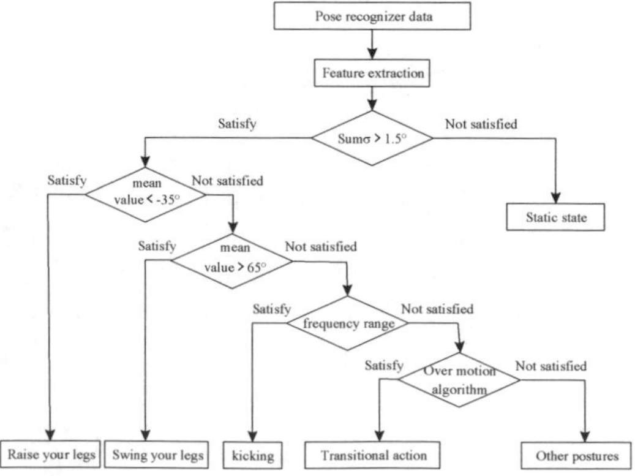

Fig. 6. Flow chart of attitude recognition algorithm

amplitude of the frequency is calculated to classify the motion attitude according to the frequency range.

3 Experiment

3.1 Experimental Environment and Methods

In order to verify the good effect of the proposed method in practice, the experiment is carried out. In order to make the experiment more scientific, this paper chooses two traditional methods as the control group, records and compares the results of the three methods, and analyzes the performance of different methods. The experiment is divided into two parts: training and testing. First, the classification model is trained. In order to intuitively feel the experimental results, this paper selects five kinds of aerobics postures, namely, "kick", "leg", "side waist extension", "high lift leg", "left and right level lift" and posture data, each aerobics movement posture lasts 1 min, according to the sampling rate of 50 Hz, while in most cases, the stability of the human body is about 2 s, the sampling time is set to 1 s, the repetition is 50%. In this experiment, 6 volunteers were selected to complete the whole demonstration of aerobics independently. From the demonstration video, the clear data of 2 h monitoring video were randomly extracted, and a total of 10800 sample window sets were collected. In this paper, the bias error of the sample is calculated, and the static data of the terminal is put horizontally, then the noise is filtered by the filter, and the error is compensated. Complete the de-interference processing, in the time window interception, extraction of sample frames. Using MATLAB software to verify the algorithm, and then use a clear picture of the monitoring video and experimental results were compared.

3.2 Comparison Results of Posture Recognition in Aerobics

After the sample training, three methods are used to recognize the posture. For aerobics posture recognition, the results are shown in Table 1.

Table 1. Comparison of attitude recognition rates of three methods

Movement posture	Methods in the paper (%)	Traditional methods 1(%)	Traditional methods 2 (%)
Kick	92.95	75.42	80.69
Swinging legs	89.91	78.53	86.37
Lateral lumbar extension	90.36	76.09	84.13
Raise your legs	95.27	82.34	84.00
Horizontal lift	96.33	85.29	79.13

According to the table, it can be found that the recognition rate of the proposed method is 96.33%, while the recognition rate of the two traditional methods is 85.29%

and 86.37%. The recognition rate of aerobics posture by this design method is much higher than other methods, which shows that this design method preprocesses the sensor data, can remove redundant data, and provides a good support for the subsequent accurate recognition of sports posture.In addition, the main factors affecting the motion attitude recognition are input feature selection, kernel function selection, penalty factor and kernel function parameters. Taking kicking posture as an example, starting with penalty factor and kernel function parameters, the variable parameters of recognition rate are obtained, as shown in Fig. 7, Fig. 8.

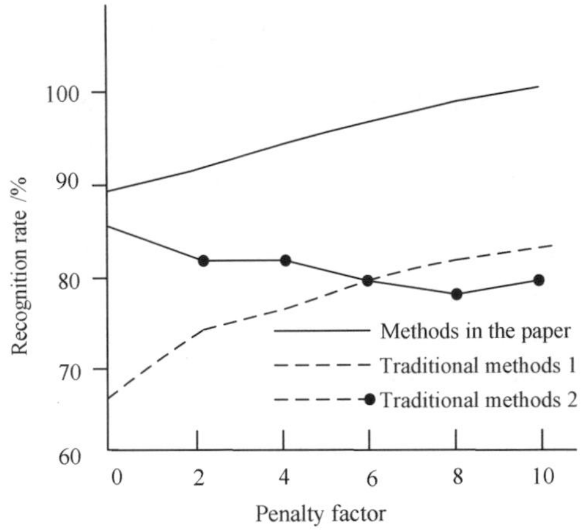

Fig. 7. Position recognition rate with penalty factor curve

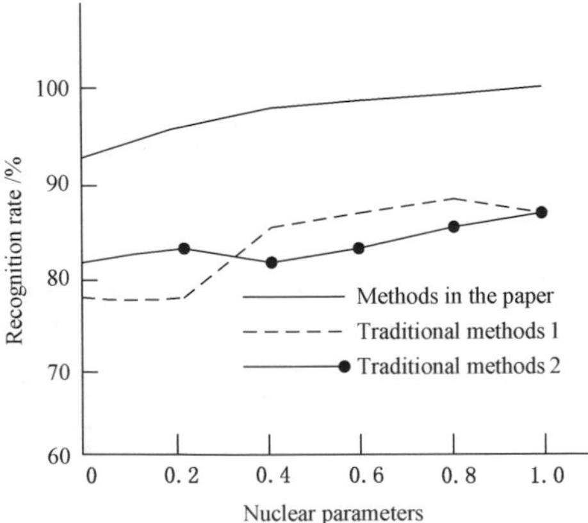

Fig. 8. Curve of gesture recognition rate with nuclear parameters

According to Fig. 7, when Gaussian kernel function is used to recognize posture, the kernel parameter is fixed and the penalty factor is changed. The recognition rate of the three methods is changed as Fig. 7 shows. The recognition rate of the method increases with the increase of the penalty factor, and reaches 100% when the penalty factor is 10. Although the recognition rate of traditional method 1 is also increasing, it is increasing slowly. Traditional method 2 position recognition rate is in a state of continuous decline. It can be found from Fig. 8 that when the fixed penalty factor is 10, the recognition rate of the proposed method is 100% when the kernel parameter is changed and increased to about 0.4, and then stably maintained. The recognition rate of the two traditional methods fluctuates continuously, and the recognition rate reaches 85% at most. To sum up, regardless of the influence factors, the proposed method is always better than the traditional method in the practical application, and can effectively improve the recognition rate. This is because this method can extract the characteristics of aerobics posture according to the time-frequency characteristics, and improve the recognition accuracy of multi-level posture when processing the data of aerobics posture.

4 Concluding Remarks

Based on the multi-sensor mobile monitoring, this paper designs the method of aerobics movement posture recognition. Through the acquisition of motion attitude data, data preprocessing, feature extraction, attitude recognition algorithm construction, the final realization of posture recognition. Through the research of this paper, the recognition rate of human motion posture is improved and the development of human-computer interaction is promoted.

Fund Projects. Guiding project of science and technology research plan of Hubei Provincial Department of Education (B2017373)

References

1. Zhang, H., Zhang, T.: Parallel processing method of inertial aerobics multisensor data fusion. Math. Probl. Eng. **2021**(4), 1–11 (2021)
2. Xie, Z., Lu, Y., Chen, X.: A multi-sensor integrated smart tool holder for cutting process monitoring. Int. J. Adv. Manuf. Technol. **110**(2), 1–12 (2020)
3. Ferrer-Cid, P., Barcelo-Ordinas, J.M., Garcia-Vidal, J., et al.: Multisensor data fusion calibration in IoT air pollution platforms. IEEE Internet Things J. **7**(4), 3124–3132 (2020)
4. Shan, P., Lv, H., Yu, L., et al.: A multisensor data fusion method for ball screw fault diagnosis based on convolutional neural network with selected channels. IEEE Sens. J. **20**(14), 7896–7905 (2020)
5. Xiao, F.: A new divergence measure for belief functions in D−S evidence theory for multisensor data fusion. Inf. Sci. **514**, 462–483 (2020)
6. Sowah, R.A., Apeadu, K., Gatsi, F., et al.: Hardware module design and software implementation of multisensor fire detection and notification system using fuzzy logic and convolutional neural networks (CNNs). J. Eng. **2020**, 1–16 (2020)
7. Liu, S., Liu, D., Srivastava, G., Połap, D., Woźniak, M.: Overview and methods of correlation filter algorithms in object tracking. Complex Intell. Syst. **7**(4), 1895–1917 (2020). https://doi.org/10.1007/s40747-020-00161-4
8. Fan, X., Guo, Y., Ju, Y., et al.: Multisensor fusion method based on the belief entropy and DS evidence theory. J. Sens. **2020**(10), 1–16 (2020)
9. Voitechovi, E., Pauliukaite, R.: Multisensor systems and arrays for medical applications employing naturally-occurring compounds and materials. Sensors **20**(3551), 1–38 (2020)

10. Fu, W., Liu, S., Srivastava, G.: Optimization of big data scheduling in social networks. Entropy **21**(9), 902 (2019)
11. Ni, P., Zhang, B., Song, Y., et al.: Multisensor distributed dynamic programming method for collaborative warning and tracking. Math. Probl. Eng. **2020**(2020), 1–19 (1955)
12. Xu, Y., Liu, J., Zhai, Y., et al.: Weakly supervised facial expression recognition via transferred DAL-CNN and active incremental learning. Soft. Comput. **24**(8), 5971–5985 (2020)
13. Liu, S., Bai, W., Zeng, N., et al.: A fast fractal based compression for MRI images. IEEE Access **7**, 62412–62420 (2019)

A Real-Time Detection Algorithm for Abnormal Users in Multi Relationship Social Networks Based on Deep Neural Network

Ai-ping Zhang[1] and Ying Chen[2(✉)]

[1] School of Software, East China JiaoTong University, Nanchang 330013, China
[2] College of Art, Xinyu University, Xinyu 338000, China

Abstract. In view of the imbalance of abnormal data in social networks, the bandwidth value of detection algorithm is high. To solve this problem, a real-time anomaly detection algorithm based on deep neural network is designed. A multi relationship social gathering model was set up, and random forest was used to process tag data. The deep neural network is used to create a set of suspicious network abnormal nodes, and the time-varying component of abnormal data is set. The wavelet transform square integral is used to deal with the abnormal data acquisition process, and the real-time detection algorithm is finally constructed. Prepare the environment parameters required by the algorithm, build the algorithm running environment, prepare two kinds of traditional detection algorithm and design detection algorithm for experiments, the results show that the designed detection algorithm has the largest bandwidth and the best performance.

Keywords: Deep neural network · Multi relationship social network · Abnormal user · Real time detection

1 Introduction

With the popularity of social networks, the security of users' social networks is becoming more and more important. The huge number of users of social network platform has attracted the attention of many attackers for profit. Attackers can create a large number of false accounts and embezzle normal accounts to send users false advertising, phishing, pornography, fraud and other bad information. This kind of malicious behavior seriously affects the user's online experience and the user's information property security [1]. Because social network platform provides users with many social functions, users can establish friend relationship, and the information released is public, which leads to the bad information in social network is more threatening than traditional spam information. As far as Sina Weibo social platform is concerned, attackers gain benefits through malicious likes, concerns, comments and sending malicious links.

Some bad users gain benefits by providing malicious access, attention and other abnormal microblog services. Nowadays, shopping websites can't directly search the

S. Liu and X. Ma (Eds.): ADHIP 2021, LNICST 416, pp. 179–190, 2022.
https://doi.org/10.1007/978-3-030-94551-0_15

shops that provide this service, but they can still purchase this service through the underground market, which seriously affects the social relationship of users and the credit system of the website [2]. Abnormal users generally do not engage in social interaction, but will send a large number of friend requests, and publish spam content in some popular microblogs to attract the attention of normal users. For the current social network abnormal account detection technology, the attacker will constantly update the attack method, by hiding identity and other means to avoid being detected. This phenomenon brings great challenges to the detection technology of abnormal users in social networks. The academia and industry need to constantly improve the detection technology of abnormal users to deal with different attack modes of abnormal accounts. Therefore, the abnormal user detection technology of social network needs to be innovated with the change of attacker attack mode, which has been a technical problem in the field of information security.

2 Real Time Detection Algorithm for Abnormal Users in Multi Relational Social Networks Based on Deep Neural Network

2.1 Setting up a Multi Relationship Social Gathering Model

In the social network environment, the number of abnormal users is less than that of normal users, so in the process of data collection, the collection of abnormal users is relatively troublesome. When setting the multi relationship social gathering model, we first mark the abnormal users in the social network and collect the data from the abnormal users. Compared with the marked normal users, the number of abnormal users is less [3]. Therefore, in the process of data modeling, we should fully consider the imbalance of data. In order to eliminate the imbalance of this part, the random forest is used to process the labeled data, and the labeling process can be expressed as the following.

$$C_R = \frac{O_c}{O_t} \tag{1}$$

In formula (1), C_R denotes tag parameters, O_c denotes exception data sets, and O_t denotes all social network data sets. After label processing, a classifier trained by a deep neural network is used to weight the labeled data set into a relation collection model, which can be expressed as a relation collection model.

$$V = \frac{\sum_{a=1}^{T} T_a}{C_R} \tag{2}$$

In formula (2), V is the relation parameter, T_a is the data set after labeling, and the meaning of other parameters remains unchanged. Using the model constructed above, the collected data sets are continuously generalized. In order to improve the fitting speed of multiple relationships, the trees dealing with neural network are independent from each other [4]. Based on cross validation, the abnormal user detection model is constructed. Random forest algorithm determines the category of the final sample by

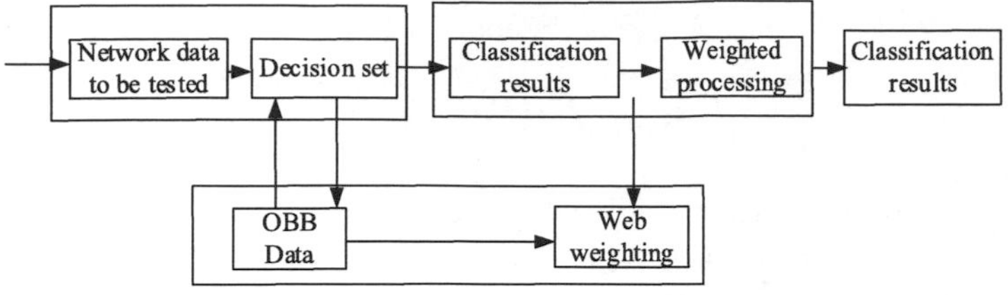

Fig. 1. Processing flow of acquisition model

simply voting the classification results of multiple trees. After verification, the structure of the acquisition model is shown in Fig. 1.

In the process flow shown in Fig. 1, the multi relationship data collected by the model is summarized. The collected data correspond to different relationship types, and are sorted into different data sets, then the deep neural network is used to create the suspicious network abnormal node set.

2.2 Using Deep Neural Network to Create Abnormal Node Set of Suspicious Network

Using the data set obtained from the above collation, because the overall structure of most real social media is gradually evolving, there will be no significant structural change in a short time [5]. This means that successive snapshots can be very similar to each other. Therefore, we can only focus on some important components that change over time, so as to find out the users who may be abnormal. This kind of component is called time-varying component, and the meaning and time parameters of this part of variables are set, as shown in Table 1.

Table 1. Time-varying components and time parameters set

Serial number	Time varying component	Time changing process
1	Node	New node insertion
2		Old node deletion
3	Edge	New edge generation
4		Old edge delete
5	Weight	Weight increase
6		Weight reduction

Through Table 1, we can see that the time-varying components include nodes, edges and weights, which are important indicators reflecting the structural changes of the graph. The exception usually occurs in the place where the structure changes, so the collection of suspicious abnormal nodes can be constructed by analyzing the range of nodes affected by time-varying components. Through the above description, the suspicious abnormal node set is defined as formula (3) quantity relationship.

$$S(n) = \frac{E(v) \cup E(v^-)}{u} \tag{3}$$

In formula (3), $S(n)$ is the set of suspicious nodes, $E(v)$ is the number of new suspicious nodes, $E(v^-)$ is the number of new edge endpoints, and u is the network node parameters. From the connectivity of the structure, we can use them to build a local substructure which has a close relationship with the node [5]. But in real social media, the close relationship between users usually follows the network node structure as shown in Fig. 2.

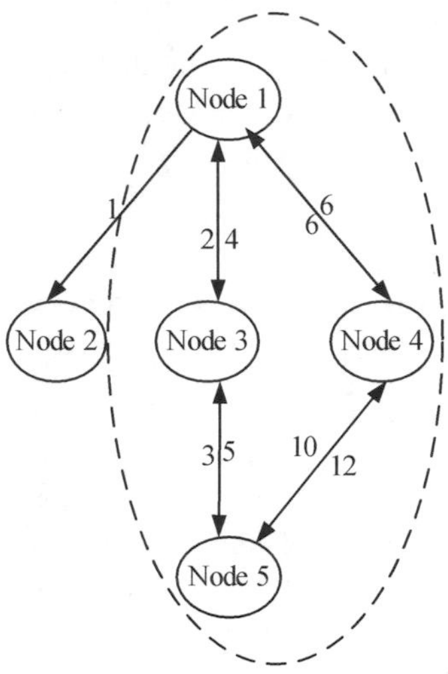

Fig. 2. Structure of social tight network

From the network structure shown in Fig. 2, it can be seen that the self network of nodes only focuses on structural connectivity. The former completely ignores the weight information of edges, while the latter ignores the risk of reliability reduction after continuous transmission [6]. In order to control the risk of data transmission, a weight threshold is set, which can be expressed as.

$$d = \frac{1}{k} \sum_{i=1}^{k} T_i \tag{4}$$

In formula (4), k is the number of network nodes and T_i is the transfer function. According to the calculation formula, only the nodes in the super self network of the node to be processed need to be calculated the intimacy with the node to be processed when constructing the core neighborhood of the node more accurately. Moreover, the maximum size of the core network is the neighbor within two hops, and the core network considers both structural connectivity and intimacy transfer [7]. To process the passed nodes as a collection is to create a collection of suspicious network abnormal nodes. Using this node set, a real-time detection algorithm is constructed.

2.3 Complete the Construction of Detection Algorithm

Using the node and data set [8] obtained from the above processing, the calculation formula (2) and calculation formula (3) of wavelet transform processing are defined, and the square integrable function $\varphi(t)$ of the two formulas satisfies the conditional wavelet mother condition. $\int_{-\infty}^{+\infty} \left|\hat{\varphi}(\omega)\right|^2 |\omega|^{-1} d\omega < +\infty$, where ω is the wavelet coefficient [12]. Set formula (2) as letter a and formula (3) matrix as letter b. set a and B to meet formula (5) at the same time.

$$\varphi_{a,b} = \frac{1}{\sqrt{|a|}} \varphi[\frac{t-b}{a}] \tag{5}$$

In formula (5), t represents the pulse coefficient between two formulas a and B. So the final abnormal user real-time detection expression (6) is obtained.

$$f(t) = \frac{1}{C} W_f(a,b) \varphi(\frac{t-b}{a}) \tag{6}$$

In formula (6), C is a constant between 0.5 and 0.7, W_f is the band coefficient and φ is the covered signal band. In order to maintain the bandwidth value of the detection algorithm in real-time detection [9]. Set the processing steps of the detection algorithm, and the set processing steps are shown in Fig. 3.

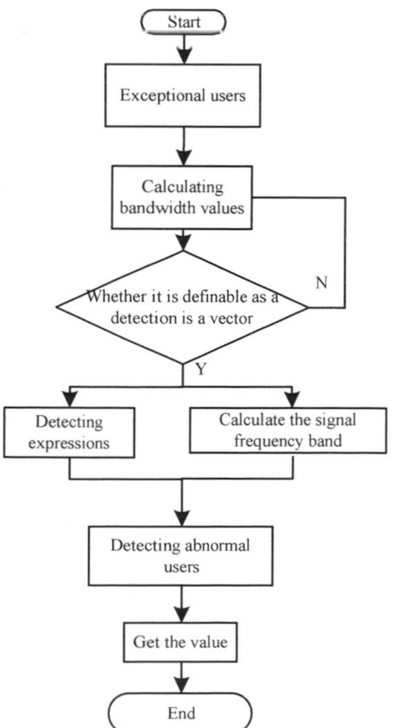

Fig. 3. Steps of electronic music type detection algorithm

In the calculation process of the algorithm shown in Fig. 3, use the formula (4) to calculate the network abnormal data in a variety of data transmission environments [10–12]. Control $f(t)$ to keep the minimum in the detection process to realize the real-time detection of abnormal users in multi relationship social network. The real-time anomaly detection algorithm for social network users studied in this paper, the basic knowledge of the multi-relational social model constructed in this paper, uses the random forest algorithm to weight the labeled data. Use wavelet transform to detect abnormal user data in real time.

3 Simulation Experiment

3.1 Experimental Data Set Preparation

The experimental preparation is shown in Table 2, and the processor parameters are shown in Table 2.

Table 2. Computer parameters prepared

Serial number	Environment	Explain
1	Computer model	Dell XPS 8910
2	Hardware environment	CPU: Intel i76700 Memory: 8G
3	Operating system	Windows 8.1
4	Development language and environment	J2EE, Struct2, Spring,
5	Web server	Tomcat
6	Database	MySQL 5.6
7	Language technology tools	Lucene3.50 ICTCLAS2013

Under the above computer parameters, the social network in the regional network environment is taken as the processing object. Randomly select a wireless sensor network as the communication information network of the experiment, take the sensor node data as the collection object, collect the experimental sample data. The sample data obtained are shown in Table 3.

Table 3. Collected data of abnormal network users

Serial number	Data group name	Number of sample data
1	Data group 1	200
2	Data group 2	400
3	Data group 3	600
4	Data group 4	800
5	Data group 5	1000
6	Data group 6	1200
7	Data group 7	1400
8	Data group 8	1600
9	Data group 9	1800
10	Data group 10	2000
11	Data group 11	2200
12	Data group 12	2400
13	Data group 13	2600
14	Data group 14	2800
15	Data group 15	3000
16	Data group 16	3200
17	Data group 17	3400
18	Data group 18	3600
19	Data group 19	3800
20	Data group 20	4000

Taking the sample data collected in Table 3 as the exception data processing object, the database structure of social network exception users is built. The database structure is shown in Fig. 4.

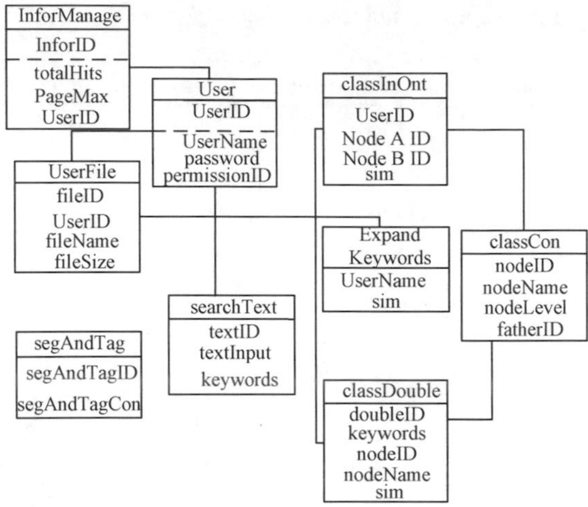

Fig. 4. Structure of exception database constructed

In the above abnormal database structure, the traditional detection algorithm 1, traditional detection algorithm 2 and the detection algorithm designed in this paper are used to experiment. The performance of the three detection algorithms is compared.

3.2 Results and Analysis

Based on the above experimental preparation, the error detection rate of three detection algorithms is defined as the evaluation result of the detection algorithm. The formula (7) can be expressed as.

$$F = \frac{FN}{FN + FP} \times 100\% \tag{7}$$

In formula (7), F is the false detection rate, FN is the number of abnormal data detected as normal, and FP is the number of correct data detected. Under the control of the above numerical relationship, the false detection rate results of the three detection algorithms are shown in Table 4.

Table 4. Results of false detection rate of three detection algorithms

Dataset name	False detection rate results/%		
	Traditional detection algorithm 1	Traditional detection algorithm 2	Design detection algorithm
Data group 1	29.1	15.7	8.5
Data group 2	22.4	15.7	8.7

(*continued*)

Table 4. (*continued*)

Dataset name	False detection rate results/%		
	Traditional detection algorithm 1	Traditional detection algorithm 2	Design detection algorithm
Data group 3	20.9	16.8	8.1
Data group 4	25.5	17.5	8.2
Data group 5	21.4	17.9	8.5
Data group 6	26.4	17.9	8.9
Data group 7	24.7	17.9	8.4
Data group 8	20.5	15.9	8.7
Data group 9	28.4	17.4	8.2
Data group 10	25.9	17.3	8.7
Data group 11	20.1	16.5	8.3
Data group 12	26.5	16.1	8.1
Data group 13	22.6	15.6	8.1
Data group 14	20.6	16.8	8.5
Data group 15	29.5	17.5	8.2
Data group 16	27.8	16.5	8.3
Data group 17	23.9	15.9	8.5
Data group 18	29.7	15.5	8.5
Data group 19	22.2	16.7	8.2
Data group 20	25.8	15.5	8.9

The three detection algorithms are controlled to process the data set prepared for the experiment. According to the calculation formula of the false detection rate defined above, the false detection rate value of the traditional detection algorithm 1 is about 25%, and the false detection rate value is the largest. The false detection rate of the traditional detection algorithm 2 is about 16%, and the actual false detection rate of the algorithm is large. Compared with the two traditional detection algorithms, the designed detection algorithm has the lowest false detection rate and the best detection effect.

Keep the above experimental environment unchanged, corresponding to the prepared experimental data set, repeat 100 iterations to process the data set processed in the preparation stage. Taking the average time generated by the iterative detection data set as the final experimental result, the detection time of the three detection algorithms is measured and calculated. The detection results are shown in Table 5.

Table. 5 Detection time results of three detection algorithms

Dataset name	Test time/s		
	Traditional detection algorithm 1	Traditional detection algorithm 2	Design detection algorithm
Data group 1	11.5	8.5	6.1
Data group 2	11.4	9.2	5.8
Data group 3	14.6	8.1	5.3
Data group 4	11.8	9.4	5.2
Data group 5	13.2	8.1	5.4
Data group 6	14.8	8.6	5.3
Data group 7	13.8	9.4	5.2
Data group 8	12.9	9.9	5.3
Data group 9	12.6	8.8	6.2
Data group 10	13.7	9.7	5.1
Data group 11	14.5	8.9	6.4
Data group 12	14.8	9.9	6.2
Data group 13	13.3	8.8	6.3
Data group 14	11.8	9.7	5.5
Data group 15	13.7	8.1	5.2
Data group 16	10.4	9.8	5.8
Data group 17	11.9	9.4	5.7
Data group 18	13.6	9.7	6.9
Data group 19	11.2	9.8	6.3
Data group 20	12.5	9.2	6.4

According to the statistical calculation results, the average detection time of traditional detection algorithm 1 is about 12.9 s, and the actual detection time is the longest. The average detection time of traditional detection algorithm 2 is 9.15 s, which takes a long time. The average detection time of the designed algorithm is about 5.78 s. Compared with the two traditional detection algorithms, the detection time of the designed algorithm is the shortest.

In the above experimental environment, after selecting the test data set prepared for the experiment, set the operation name of the multi relationship social network and the corresponding data size, as shown in Table 6.

Under the operation set in Table 6, the three detection algorithms are controlled to process the above operation 10 times. Take the operation serial number 1 ~ 10 as a processing process, measure the time required for a processing process. Calculate the

Table 6. Operation name of multi relationship social network and corresponding data volume

Operation number	Operation name	Data volume size/MB
1	File write operation	15.7
2	Relation call	42.6
3	Process creation	21.7
4	Process hiding	39.2
5	File hiding	40.9
6	Invalidation attack	21.7
7	Resource consumption	45.5
8	Kernel boot	42.8
9	Module compilation	38.2
10	End of process	46.7

measurement bandwidth of different detection algorithms, and the bandwidth results arc shown in Fig. 5.

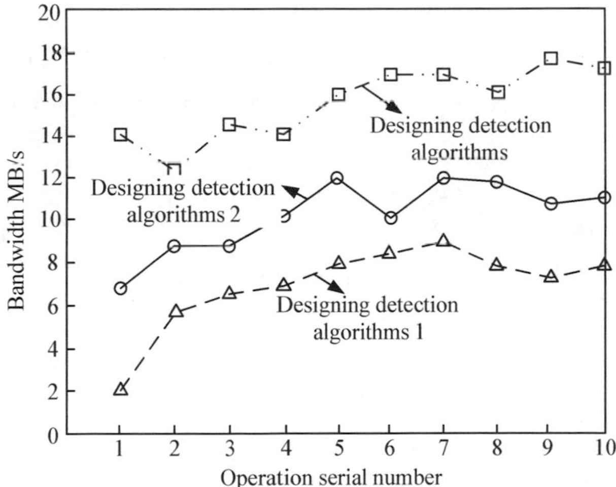

Fig. 5. Bandwidth results of three detection algorithms

According to the experimental results shown in Fig. 5, under the control of the three detection algorithms, corresponding to different operations, the average detection bandwidth generated by the traditional detection algorithm 1 is about 7 MB/s, the processing bandwidth is small, and the detection process is slow. The average detection bandwidth generated by the traditional detection algorithm 2 is about 10 MB/s, and the bandwidth generated by the detection process is larger. The detection bandwidth of the designed detection algorithm is about 16 MB/s. compared with the two traditional detection algorithms, this detection algorithm has the largest detection bandwidth and can process the most information.

4 Conclusion

In today's information explosion era of big data, while social networks provide people with convenient and diversified services, there are also some hidden information security risks. In order to improve the detection efficiency of abnormal network users, this paper uses deep neural networks to develop a personalized crawling strategy to crawl abnormal social user data, deeply analyze the difference between normal users and abnormal users in user information and behavior characteristics, and extracts that can distinguish between normal users and abnormal users. The important characteristics of the user to realize the real-time detection process. Experiments show that the average detection time of this method is about 5.78 s, and the detection band frame is about 16 MB/s, which has good performance. In the future, we will study more anomaly detection conditions corresponding to different social network scenarios, and consider in-depth study of the design of multiple social network user interaction models as the focus of subsequent work.

References

1. Chen, S., Zhu, G.-S., Qi, X.-Y., et al.: Custom user anomaly behavior detection based on deep neural network. Comput. Sci. **46**(S2), 442–445, 472 (2019)
2. Yin, J., Peng, Y., Lu, Y., et al.: Research on user abnormal behavior prediction of enterprise information system based on deep neural networks. J. Manage. Sci. **33**(01), 30–45 (2020)
3. Li, H., Zhu, M.: A small object detection algorithm based on deep convolutional neural network. Comput. Eng. Sci. **42**(04), 649–657 (2020)
4. Liu, S., Liu, G., Zhou, H.: A robust parallel object tracking method for illumination variations. Mob. Netw. Appl. **24**(1), 5–17 (2019)
5. Zhang, Y., Xu, T., Feng, D., et al.: Research on faster RCNN object detection based on hard example mining. J. Electron. Inform. Technol. **41**(06), 1496–1502 (2019)
6. Yang, X.-X., Li, H.-B., Hu, G.: An abnormal behavior detection algorithm based on imbalanced deep forest. J. China Acad. Electron. Inform. Technol. **14**(09), 935–942 (2019)
7. Xu, H.-J., Zhang, H., He, W.: A cloud user anomaly detection method based on mouse behavior. J. Harbin Univ. Sci. Technol. **24**(04), 127–132 (2019)
8. Li, X., Han, X., Wang, Z., et al.: A deep heterogeneous network embedding algorithm based on twin neural network. Telecommun. Eng. **60**(11), 1271–1277 (2020)
9. Shen, X., Shen, Z., Huang, Y., et al.: Deep convolutional neural network for parking space occupancy detection based on non-local operation. J. Electron. Inform. Technol. **42**(09), 2269–2276 (2020)
10. Li, J., Yun, X., Li, S., et al.: HTTP malicious traffic detection method based on hybrid structure deep neural network. J. Commun. **40**(01), 24–33 (2019)
11. Liu, S., He, T., Dai, J.: A survey of CRF algorithm based knowledge extraction of elementary mathematics in Chinese. Mob. Netw. Appl. **26**(5), 1891–1903 (2021). https://doi.org/10.1007/s11036-020-01725-x
12. Liu, S., Fu, W., He, L., et al.: Distribution of primary additional errors in fractal encoding method. Multimedia Tools Appl. **76**(4), 5787–5802 (2017). https://doi.org/10.1007/s11042-014-2408-1

Research on Personalized Recommendation Algorithm Based on Mobile Social Network Data

Xiang Yan[1] and Wen-hua Deng[2(✉)]

[1] Information Center of Pu'er University, Puer 665000, China
[2] Wuhan Railway Vocational College of Technology, Wuhan 430205, China

Abstract. Due to the lack of mining of hidden data in traditional personalized recommendation algorithms, the algorithm is interfered by the mobile social network environment, and it is difficult to accurately recommend targeted data for users. Therefore, research on personalized recommendation algorithms based on mobile social network data. By dividing mobile social network user categories, user information is obtained; based on mobile social network data, user demand characteristics are extracted; potential association rules between users and service needs are mined to build personalized recommendation algorithms. The experimental results show that compared with the traditional recommendation algorithm, the research algorithm has stronger perception and recognition ability, and it can recommend more matching information for users according to different user needs when facing different network environments.

Keywords: Mobile social network data · Personalization · Recommendation algorithm · Association rules

1 Introduction

At present, with the rapid development of Internet and computer technology, the fields of industry, commerce, agriculture, aerospace, environmental measurement, education and teaching are inseparable from the use of the network, so a large number of network users bring great work to the network Pressure, in order to improve the efficiency of the network, research a new personalized recommendation algorithm, for users of different ages, different work fields and different living environments, recommend network information that meets their respective needs [1].

Literature [2] proposed a hybrid algorithm for personalized recommendation for interactive experience devices. The algorithm sequentially uses collaborative filtering and content-based recommendation methods for recommendation. In the initial recommendation, the latent Dirichlet allocation (LDA) topic model is used to reduce the dimensionality of high-dimensional user behavior data, and the user writing topic matrix is established to reduce the recommendation inaccuracy caused by the high sparse data in the collaborative filtering algorithm. The user interest list is obtained by calculating the similarity between users. Then, on the basis of the preliminary recommendation results,

S. Liu and X. Ma (Eds.): ADHIP 2021, LNICST 416, pp. 191–203, 2022.
https://doi.org/10.1007/978-3-030-94551-0_16

the VGG16 model is used to extract the feature vector of the calligraphy image, and the similarity between the user's calligraphy words and the primary recommendation calligraphy words is calculated, so as to obtain the final recommendation result. The recommendation accuracy of this algorithm is high, but the efficiency is poor. Literature [3] proposes a personalized recommendation system for academic libraries based on a hybrid recommendation algorithm. First, the paper studies the application of collaborative filtering and content-based recommendation algorithms in college book recommendation, involving reader classification, user-item scoring matrix establishment, vector space model construction, and similarity calculation. user. And considering the characteristics of college books and readers, the user-item rating matrix is improved, and clustering is used to alleviate the problem of data sparseness. This method can quickly realize personalized recommendation, but the recommendation matching degree is low.

This research combines the problems of a variety of traditional algorithms and introduces mobile social network data. This type of data has the characteristics of many types, large amounts, and wide coverage. These data provide the most basic raw data for the recommendation work [4]. This research combines the characteristics of mobile social network data to establish a new personalized recommendation algorithm. Obtain user information by classifying mobile social network users; extract user demand characteristics based on mobile social network data; mine potential association rules between users and service demands, and build personalized recommendation algorithms. Compared with the traditional algorithm, the research algorithm has stronger perception and recognition ability, and can recommend more matching information for users according to different user needs.

2 Personalized Recommendation Algorithm Based on Mobile Social Network Data

2.1 Classification of Mobile Social Network Users

According to the browsing trajectory of the mobile social network user when browsing the web page information, the user attribute preference is calculated, and the score is used as the basis for user classification. Set user tags, including: finance, technology, digital, social, transportation, weather, news, law, brand, food, insurance, etc. Using mathematical algorithms, calculate the logic and similar preferences that exist in the user's browsing trajectory to form a classification definition. Assuming that there is m user preference A, the calculation result of its mathematical algorithm is:

$$Y_i \sim Z\left(\alpha_i, \beta^2\right), i = 1, 2, \ldots, m \tag{1}$$

In the formula: Y_i represents the sample data set; α_i represents the label offset threshold affected by the change of preference A; β represents the preference difference. The problem of saliency is transformed into a question of whether preference A is in the Z space, and whether it affects web browsing selection behavior, that is, whether $G_0 : \alpha_1 = \alpha_2 = \cdots = \alpha_m$ is valid or not. The following equations are given, and the various parameters are the indicators required for verification.

$$\begin{cases} \overline{y} = \frac{1}{n}\sum_{i=1}^{m}\sum_{j=1}^{n_i} y_{ij} & S_1^2 = \sum_{i=1}^{m}\sum_{j=1}^{n_i}\left(y_{ij}-\overline{y}\right)^2 \\ S_2^2 = \sum_{i=1}^{m}\sum_{j=1}^{n_i}\left(y_{ij}-\overline{y}_i\right)^2 & S_3^2 = \sum_{i=1}^{w}\sum_{j=1}^{n_i}(y_i-\overline{y})^2 \end{cases} \tag{2}$$

In the above formula, n represents the total number of results; $\overline{y}$ represents the total mean; S_1^2 represents the sum of squares of the total variance; S_2^2 represents the sum of squares within the group; S_3^2 represents the sum of squares between the groups [5]. According to the above indicators, G_0 rejection domains are obtained:

$$W = \left\{\frac{(n-m)S_3^2}{(m-1)S_2^2} > G(m-1, n-m)\right\} \tag{3}$$

The test results obtained can be divided into four cases, namely highly significant impact, significant impact, certain impact and no significant impact. According to the result, the influence degree of the mobile social network data selection under the change of user preference A is obtained, and a user classification data table is established, as shown in Table 1.

Table 1. Classification data table of mobile social users

High-end business crowd	White-collar user population	Campus user population	Rural user population	Other user groups
Financial management	Shopping	Home	Medical treatment	Communication
Country	Film and television	Social	Real estate	The internet
Finance	Weibo	Digital	Cell phone	Keep in good health
Fashion	Game	Real estate	Technology	Music
Read	Diet	Apparel	Education	News

According to the above decomposition results, a detailed user classification strategy is formulated, and the improved svm design classification model is used to classify mobile social network users. In the case where the user's non-linearity is separable, assuming that the selection vectors of the two users are a and b, the non-linear function f of the improved svm is used to map the user selection vector into the feature space T. Then the Euclidean of the two vectors The distance is:

$$d^T(a,b) = \sqrt{H(*) - 2H(a,b) + H(a,b)} \tag{4}$$

In the formula: $H(*)$ represents the kernel function. Then the center vector C of the feature space is:

$$C_f = \frac{1}{n}\sum_{i=1}^{n} f(a_i) \tag{5}$$

Calculate the class center according to the above formula, and then calculate the distance between the two class centers, the formula is:

$$D = \left|C^+ - C^-\right| \tag{6}$$

In the formula: C^+ represents the center of the positive class; C^- represents the center of the negative class [6]. Calculate the distance between the two types of samples and other user sample information. When the distance is less than the calculation result of formula (6), the sample is taken as a valid candidate support vector, that is, there is:

$$D' = |a_i - C| \tag{7}$$

Figure 1 below is a schematic diagram of the sample as a valid candidate support vector when the reservation satisfies $D' < D$.

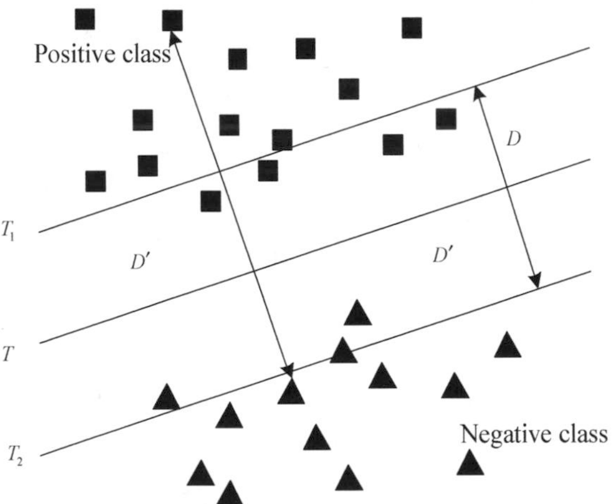

Fig. 1. Pre-select valid candidate support vectors

According to the above schematic diagram, the mobile social network users in the feature space are divided according to D' and D, so that users with the same selection preference are grouped into a data set, and the classification results shown in Table 1 above are obtained.

The needs of mobile social network users are diverse, so the need elements between different users are classified fuzzy in order to find out the correlation characteristics between needs. Collect raw data for the types of users who use personalized recommendation services. The source of the data is mobile social network data, and the data should be collected according to factors such as different age groups, different genders, and different occupations to ensure the universality and breadth of the data. Assuming that user needs are represented by q, there is user demand set $Q = \{q_1, q_2, \ldots, q_n\}$, where q_n represents n different needs of the user; at the same time, there are w characteristic indicators under the influence of different users for each demand, that is, there is $q_c = \{q_{c1}, q_{c2}, \ldots, q_{cn}\}$, so the user is obtained The original data matrix of the demand [7]. According to the original mobile social network data actually collected, the translation standard deviation transformation process is implemented, and the calculated result is:

$$q'_{ci} = \left(q_{ci} - \overline{q}_i\right)/D'B_i \tag{8}$$

In the formula: q'_{ci} represents the standard deviation value of the demand data after translational change; q_{ci} represents the original value of the i original data under the influence of the characteristic index c; $\overline{q}_i$ represents the average result of the i original data; B_i represents the transformation matrix; of which $c = \{1, 2, \ldots, n\}$, $i = \{1, 2, \ldots, m\}$. After the data transformation of the above formula, the mean value of each demand variable is 0, and the standard deviation is 1, eliminating the influence of dimensions. It is known that there is a demand set Q, from which two demand quantities are randomly selected, denoted as q_a and q_b respectively, and the degree of similarity between the two is calculated as $s_{ab} = f(q_a, q_b)$, the calculation expression is:

$$s_{ab} = \sum_{i=1}^{w} q_{ai} \cdot q_{bi} / \sqrt{\sum_{i=1}^{w} q_{ai}^2} \sqrt{\sum_{i=1}^{w} q_{bi}^2} \tag{9}$$

Use the above formula to find the degree of similarity between the demand data, and realize the fuzzy clustering of user demand through the optimal threshold method:

$$q_i^{(\mu)} = \frac{1}{s_{ab}} \sum_{k=1}^{n_\mu} q_{ki}^{(\mu)} \tag{10}$$

In the formula: $q_i^{(\mu)}$ represents the clustering result of i sample data under the influence of the μ types of demand; $q_{ki}^{(\mu)}$ represents the dynamic change constant of the demand data under the action of the change factor k [8]. According to the above calculation steps, fuzzy clustering of requirements is realized. The requirements after clustering have different characteristic quantities, so on the basis of demand correlation, by quantitatively analyzing the correlation between demand data, judging the hidden correlation between demands, so as to determine the degree of dispersion of demand, so as to extract different demands Similar characteristics of type data. Assuming that each cluster subset is represented by a feature vector λ, the category attributes of different data are calculated for n different user needs. When real number type demand data exists, $\lambda_u = [\lambda_{u1}, \lambda_{u2}, \ldots, \lambda_{un}]$, $\lambda_v = [\lambda_{v1}, \lambda_{v2}, \ldots, \lambda_{vn}]$. Then the distance components between different vector subsets are:

$$l_{uv} = \frac{|\lambda_{ui} - \lambda_{vi}|}{\max|\lambda_{ui} - \lambda_{vi}|} \tag{11}$$

In the formula: l_{uv} represents the distance between the two subsets whose subset vector is u, v; λ_{ui}, λ_{vi}. represents the value under the influence of attribute 4. At this time, in order to meet the universality of the data, it is necessary to extract the sample features uniformly. When extracting data features, it is necessary to ensure that the data features are evenly distributed in different spaces, and for both strong feature data and weak feature data, extraction must not be ignored. Then the extraction result of the demand feature quantity is:

$$q' = \sum_{k=1}^{n} \lambda_k / l_{uv} \tag{12}$$

In the formula: q' represents the user demand feature; λ_k represents the vector feature value under the control of feature intensity k. According to the above formula, the user needs feature extraction is completed.

2.2 Mining the Association Rules Between Users and Service Needs

Combining the above classification results and demand feature extraction results, mining association rules between users and service needs. First, describe the entire sample document abstractly, and represent the text in a form that can be quickly processed by the computer. Therefore, all user information is represented by a set Z, and there is $Z=\{Z_1, Z_2, \ldots, Z_n\}$, of which $Z_i \in Z_n$ indicates that the i user information is included in the transaction set with n user information. The document is abstracted into a collection of things $X = \{X_1, X_2, \ldots, X_n\}$, where X_n indicates that the collection consists of n documents. The keywords are abstracted into item sets e, and $e = \{e_1, e_2, \ldots, e_n\}$ and e_n represent n item sets in the document, which are the keywords of user demand information. The service demand problem is regarded as an abstract collection, and this collection is scanned multiple times and data is mined [9]. Assume that the structure list I is a set of service demand keywords, and $I = \{I_1, I_2, \ldots, I_n\}$ is used to represent each demand information, that is, a sample document. When mining the association rules between requirements and basic information, all feature keywords are collected. Assuming that there are i sample documents in a certain category, and each document has j feature words, then all keywords are collected, totaling $i \times j$ A. But there will be some repetitive keywords, so before mining association rules, remove these repetitive data to get a set of all keywords that are not repetitive [10–12]. The process is shown in Fig. 2 below.

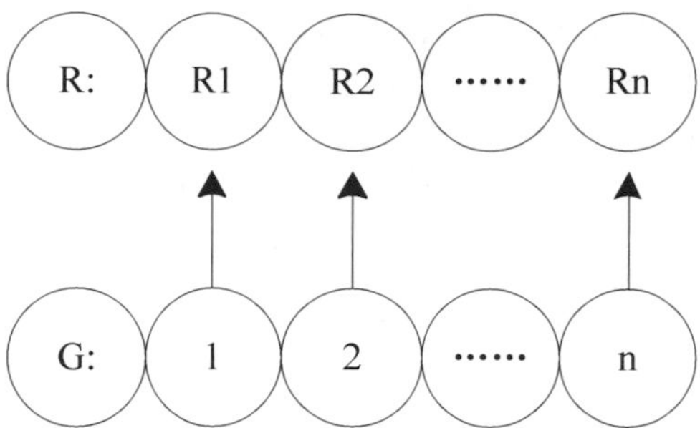

Fig. 2. Schematic diagram of deduplication

The 1 in Fig. 2 represents the marking of all keywords. In actual mining, category tags should be added, so the category occupancy should be set to 0 occupancy. The mining restrictions are: Rule: keyword 1, keyword 2,..., keyword n → category name. According to the association rules between users and service needs obtained by the above mining, a personalized recommendation algorithm is designed.

2.3 Building a Personalized Recommendation Algorithm

According to the above association rules, a personalized recommendation algorithm based on the user's temporal characteristics is constructed. The algorithm is mainly composed of two parts, which are perceptual data compression and perceptual feature recognition and recommendation. The construction plan of the algorithm is: the algorithm needs to meet the preprocessing network flow, sort it according to the timestamp of the data packet, and construct a time series set. Then the algorithm performs perceptual compression on the data to obtain a significantly robust private data set. The new data set is identified and recommended, and finally a traceability summary is generated according to the hash principle. According to the above construction scheme, the perception data is compressed first. Think of the network flow as a set of data packets with the same 5-tuple, and then divide the network flow with fixed-length time slots, so that the network flow can start from a random point in time to form a sequence set of several time slots, which means The formula is $T = \{t_1, t_2, \ldots, t_n\}$, where T represents the data packet contained in a certain time slot. In order to effectively distinguish each user demand network flow, by extracting burst period data, the unique perceptual characteristics of the network flow are retained [13, 14]. In the time series set divided by time slot, the bursty traffic period represents the time slot corresponding to the peak flow rate, so it is regarded as the new time series data set T'. The result is as follows:

$$T' = Dcomp(T) \tag{13}$$

When extracting the time slot according to the above calculation formula and calculating the flow rate of the time slot at the same time, the default time slot T contains m data packets, and the flow rate rate corresponding to the time slot is:

$$\varphi_i = \frac{L_{i,1} + L_{i,2} + \cdots + L_{i,j} + \cdots + L_{i,m}}{D} \tag{14}$$

In the formula: D represents the length of time slot i; $L_{i,j}$ represents the length of the j data packet in time slot i. The time slot corresponding to φ_i is selected from the set T, and a brand new time slot set $T' = \{t'_1, t'_2, \ldots, t'_n\}$ is obtained. In order to ensure the robustness of the data set, information entropy is used to further process the set T':

$$T'' = Hcomep(T') \tag{15}$$

Through the above calculation process, the algorithm realizes the perceptual data compression of the mobile social network data. Then use the traceability function to perceive feature extraction and recommendation. According to the research results, it can be known that the data rate reflects the burst characteristics of network traffic, so φ_i feature is extracted, and its recommended encoding is set as a perceptible hash sequence to obtain a set containing γ time slots. Divide each time slot into smaller time slices to reflect the detailed characteristics of the network flow, and the new time slice matrix after the original time slot set is mapped is:

$$Matrix_\theta = \begin{bmatrix} \theta_{1,1} & \cdots & \theta_{1,\gamma} \\ \vdots & \ddots & \vdots \\ \theta_{\gamma,1} & \cdots & \theta_{\gamma,\gamma} \end{bmatrix} \tag{16}$$

In the formula: θ is the length of the time slice, which means that there are several data packets in each time slice. According to the calculation result of formula (16), the characteristic matrix is obtained:

$$Matrix_\varphi = \begin{bmatrix} \varphi_{1,1} & \cdots & \varphi_{1,\gamma} \\ \vdots & \ddots & \vdots \\ \theta_{\gamma,1} & \cdots & \theta_{\gamma,\gamma} \end{bmatrix} \tag{17}$$

It is known that θ and φ are corresponding. Calculate the average data rate, represented by $\overline{\varphi}$ [15]. Perform an average hash on each column of the data, then compare the value of each data with the average of the set, and encode:

$$K(i) = \begin{cases} 0 \ if \ \varphi_{i,j} \le \overline{\varphi}_i \\ 1 \ if \ \varphi_{i,j} > \overline{\varphi}_i \end{cases} \tag{18}$$

When the above calculation result is less than the average value, the code is 0, otherwise it is 1. So far, the construction of a personalized recommendation algorithm based on mobile social network data has been completed.

3 Experimental Research

Apply the designed algorithm and the traditional algorithm to the same system separately, set up a simulated experiment environment, use efficient blockchain users as the test object, map the network flow into summary information, and send it to the test system. From the perspectives of perception recognition and demand recommendation, the performance of different recommendation algorithms is analyzed. In order to facilitate the description of the experimental test results, the algorithm of this study is used as the experimental group, and the traditional algorithm is used as the control group, and a two-stage experimental test task is carried out.

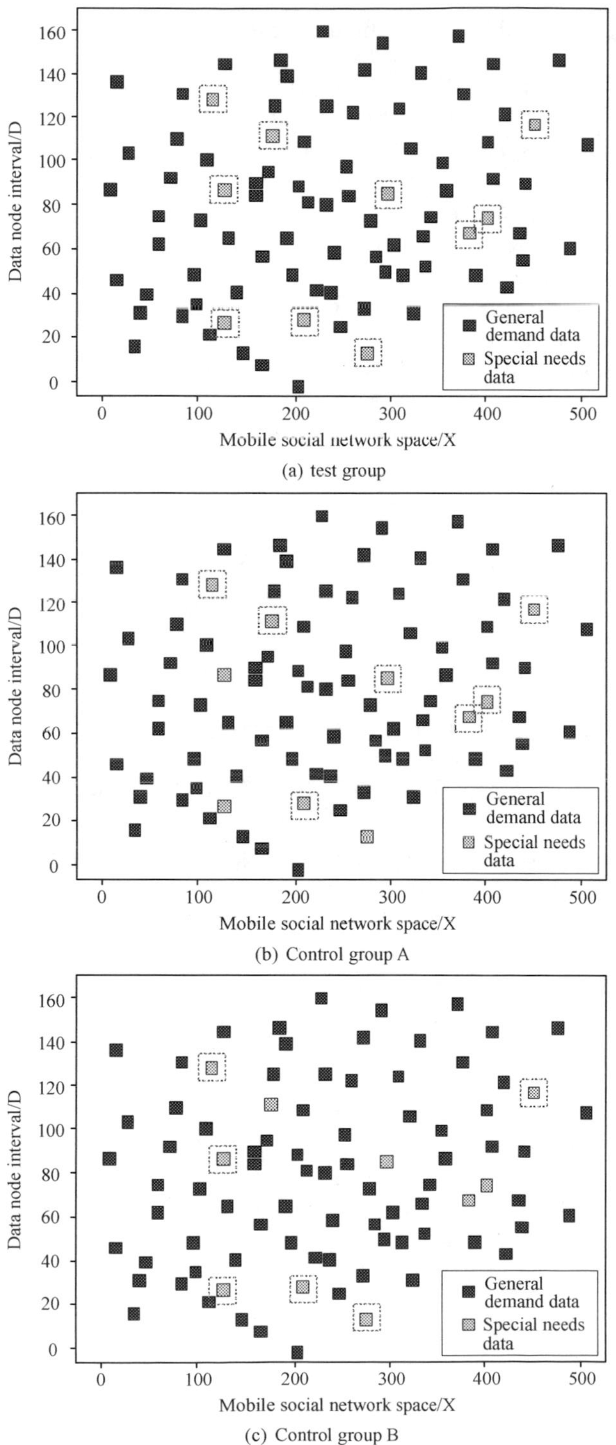

(a) test group

(b) Control group A

(c) Control group B

Fig. 3. Algorithm perception and recognition performance test

3.1 Perceptual Recognition Performance Test

Design two sets of experiments to test whether the two algorithms can identify the different needs of a large number of users in mobile social networks. The experiment set up a large number of conventional demand data and a type of special demand data. The following Fig. 3 shows the test results of the perception and recognition performance of different recommendation algorithms.

According to the test results in the above figure, it can be seen that under the same test conditions, the experimental group algorithm found all the special demand nodes among the massive network nodes; while the two control group algorithms only found 10 special demand nodes respectively. 7 and 6 of the nodes. The perceptual efficiency of calculating traditional algorithms is 70% and 60%, which are far lower than the perceptual recognition effect of the algorithm in the article. It can be seen that the recommendation algorithm of this research has a better perception and recognition effect when facing special demand nodes hidden in massive data. The method in this paper performs perceptual data compression on mobile social network data. Then use the traceability function to perceive feature extraction and recommendation. The data rate reflects the burst characteristics of network traffic, thereby extracting the characteristics, and realizing the personalized recommendation corresponding to the characteristics.

3.2 Recommended Effect Test

Set up a multi-stream mixed model, and perform multi-stream recommendation tests on the experimental group and the control group. The model contains multiple network flows, and the demand nodes among them increase with the increase of projects. The experiment sets up four different addition schemes, establishes four different test environments, compares the three groups of different algorithms when facing different user needs, and their ability to recommend information according to the needs. The test results are shown in Fig. 4 below.

According to the test results shown in Fig. 4, it can be seen that in the face of jitter environment, packet loss environment, packet injection environment, and multiple interference environment, the experimental group algorithm has better recommendation ability. The algorithm recommendation ability of the other two control groups is significantly weaker than the experimental group due to the influence of different test environments. It can be seen that the algorithm constructed this time can be used to solve network user service work.

Combined with the above experimental results, it can be seen that the personalized recommendation algorithm based on mobile social network data proposed in this paper has more accurate data perception and recognition effects, and can achieve more accurate personalized recommendations.

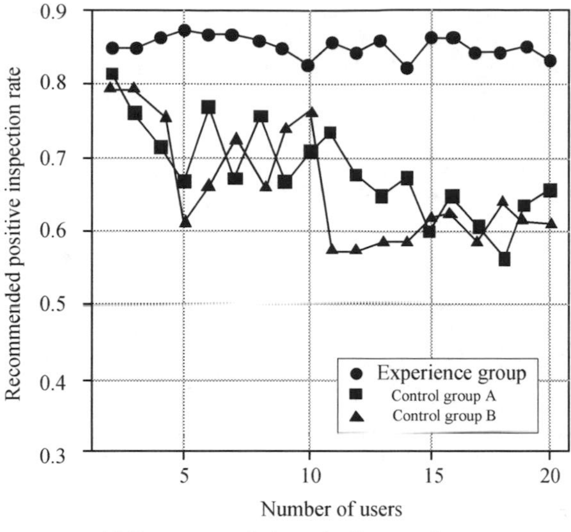

(a) Recommended test in jitter environment

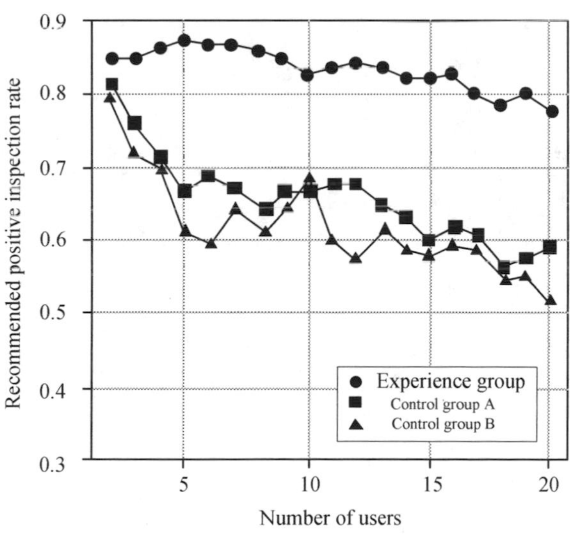

(b) Recommended test in packet loss environment

Fig. 4. Multi-stream hybrid traceability performance test of the algorithm

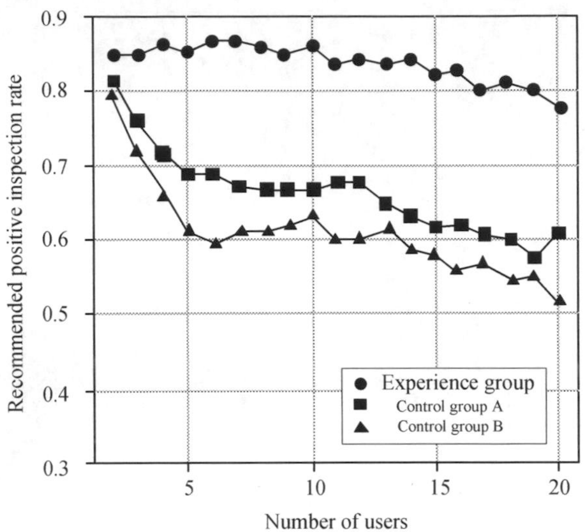

(c) Recommended testing in package injection environment

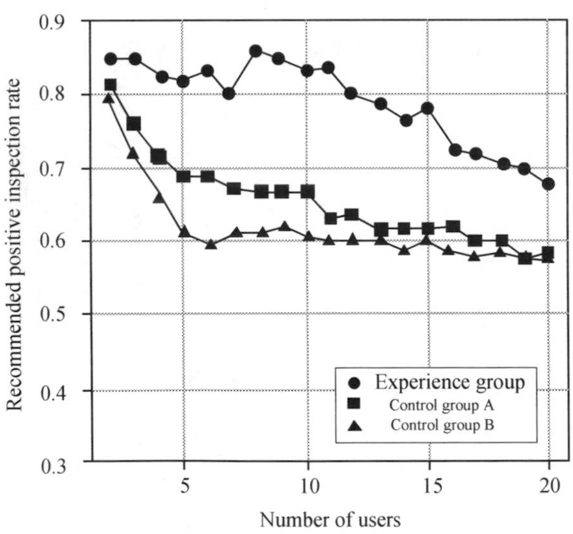

(d) Recommended test under multiple interference

Fig. 4. continued

4 Concluding Remarks

This research optimizes the personalized recommendation algorithm based on traditional algorithms and combined with mobile social network data, and has achieved relatively satisfactory research results. However, considering the design process of the entire algorithm, there are a large number of calculation formulas in this design, so calculation errors are prone to occur during calculations. At the same time, the calculation efficiency of this algorithm may be weaker than other algorithms. Therefore, future research should focus on calculation errors and efficiency. To provide users with more reliable network technology.

References

1. Shi, Y., Zeng, Z.: Temporal segment networks based on feature propagation for action recognition. J. Comput.-Aided Des. Comput. Graph. **32**(04), 582–589 (2020)
2. Chen, S., Huang, L., Lei, Z., et al.: Research on personalized recommendation hybrid algorithm for interactive experience equipment. Comput. Intell. **36**(3), 1348–1373 (2020)
3. Yt, A., Bing, Z.A., Yw, A., et al.: College library personalized recommendation system based on hybrid recommendation algorithm. Procedia CIRP **83**, 490–494 (2019)
4. Li, J., Hong, T., Wang, X., et al.: A data dissemination mechanism based on group structure in opportunistic mobile social networks. J. Comput. Res. Dev. **56**(11), 2494–2505 (2019)
5. Cao, H., Ma, Z., Yang, X., et al.: Pilot allocation based on users' location and user classification in massive MIMO system. Telecommun. Sci. **35**(10), 92–99 (2019)
6. Xie, G.-R., Zheng, H., Lin, W., et al.: Power outage sensitive customers classification based on improved random forest algorithm. Comput. Syst. Appl. **28**(03), 104–110 (2019)
7. Liu, C., Huang, J., Zhou, T.: Hierarchical characteristics and driving force of user demand for paid digital reading based on grounded theory. Modern Inf. **39**(11), 80–89 (2019)
8. Sun, Y., Yang, G., He, L.: Prediction of advertising conversion rate based on store characteristics and user needs. J. Qingdao Univ. (Eng. Technol. Edition) **34**(03), 16–20 (2019)
9. Wang, X., Zhu, J., Meng, X., et al.: A model of safety monitoring and early warning for coal mine based on ontology and association rules. Mining Saf. Environ. Prot. **46**(03), 27–31 (2019)
10. Li, P., Wang, Y.: Simulation study on the periodic association of book circulation data association rules. Comput. Simul. **37**(05), 354–357+452 (2020)
11. Zhao, C., Zhang, K., Liang, J.: Asymmetric recommendation algorithm in heterogeneous information network. J. Front. Comput. Sci. Technol. **14**(06), 939–946 (2020)
12. Zhang, Q., Wang, B., Cui, N., et al.: Attention-based regularized matrix factorization for recommendation. J. Softw. **31**(03), 778–793 (2020)
13. Liu, S., Liu, G., Zhou, H.: A robust parallel object tracking method for illumination variations. Mob. Networks Appl. **24**(1), 5–17 (2018). https://doi.org/10.1007/s11036-018-1134-8
14. Liu, S., Bai, W., Liu, G., et al.: Parallel fractal compression method for big video data. Complexity **2018**, 2016976 (2018)
15. Liu, S., He, T., Dai, J.: A survey of CRF algorithm based knowledge extraction of elementary mathematics in Chinese. Mob. Networks Appl. (2021). https://doi.org/10.1007/s11036-020-01725-x

Building Construction Information Real-Time Sharing Method Based on 3D Scanning Technology and Social Network Analysis

Qiang Li(✉) and Xiang-yun Tuo

School of Human Settlements and Civil Engineering, Xi'an Eurasia University, Xi'an 710065, China

Abstract. In order to reasonably improve the management effect of the construction personnel, and reduce the construction costs, arrange the progress of the construction project scientifically and reasonably to ensure the overall construction quality, a real-time sharing method of building construction information is proposed, which combines 3D scanning, augmented reality, intelligent wearable equipment, Internet of Things, regional network communication and RFID.

Keywords: 3D scanning · Social network · Construction · Information sharing

1 Introduction

With the rapid development of the urban construction industry, the construction process and completion safety management problems are gradually highlighted. In particular, the construction problems encountered are complex and diverse, such as real-time sharing of construction information collection, construction progress coordination, construction personnel behavior management, etc. In order to solve the above construction problems, improve the construction management effect, reduce the construction costs, and ensure the smooth progress of the construction, use the scientific and effective 3D scanning technology and network information processing technology in the real-time sharing management of construction information. The scheduling of construction projects is scientific and appropriate, that can ensure the whole construction quality, and the use of information collection and analysis can promote the optimization of resource allocation, which can effectively reduce the waste of resources in each construction link. The application of 3D scanning technology has a great promoting effect on construction management [1]. Three-dimensional scanning is a kind of technology that can fully display the computer simulation environment and make the user feel immersive. Three-dimensional scanning technology builds on augmented reality technology by adding a virtual and real-world nexus that allows workers to more realistically perceive the real world [2]. Therefore, the application of 3D scanning technology to building construction will not only greatly improve the safety and reliability of the construction phase, but also lay a solid foundation for the follow-up construction industry.

S. Liu and X. Ma (Eds.): ADHIP 2021, LNICST 416, pp. 204–216, 2022.
https://doi.org/10.1007/978-3-030-94551-0_17

In the process of real-time sharing and processing of construction project information, the construction party is the main body of the project construction, and an information-based safety management system for construction shall be established in light of the safety management problems of the construction party in the construction phase, and three closed-loop information sharing management systems, including the collection system of construction information in the intelligent era, the pre-prevention safety management system based on three-dimensional scanning technology and the safety management system for construction process based on three-dimensional scanning technology, shall be formed [3]. It is used to standardize and guide the actual construction process, ensure the safety performance coefficient of the construction process, and achieve the goal of guaranteeing the intrinsic safety of the construction system.

2 Method of Real-Time Sharing of Construction Information

2.1 Building Construction Information Processing Algorithm Based on 3D Scanning Technology

The building information model based on 3D scanning technology can integrate all kinds of effective information, realize collaborative operation and management of the whole life cycle of building, effectively grasp the information of each stage and ensure that this information can be shared in the whole stage. Combining the latest research results, a 3D scanning system based on machine vision, we establish the information transmission path to guide the final data processing method. The photo-type three-dimensional scanning system includes a photo-type 3D scanner, optical projector, CDD camera, stability bracket, 3D point cloud data processing methods, power supply and other accessories. CCtrlDevice, ClmageProcess, CRenderOpenGL, CDataManager four software module systems are embedded in the 3 D scanner and CCD and internal control calibration systems are integrated [4]. Based on the principle of laser ranging, the three-dimensional coordinate value in space is measured instantaneously by a high-speed laser transmitter [5].

In the process of building construction information real-time sharing, the camera coordinate system is used as the coordinate system (x, y, z) of 3D sensor, and the imaging mode of 2D camera can be seen:

$$\begin{cases} X_{ax} = \frac{(x_m - x_0 + \Delta x)}{c} Z_{aw} \\ Y_{an} = \frac{(y_m - y_n + \Delta y)}{c} Z_{av} \end{cases} \tag{1}$$

Where: (x_m, y_m, z_m) is the three-dimensional coordinate of the measured point m in the camera coordinate system, (x_m, y_m) is the image coordinate of the imaging point of the measured point on the camera; C is the effective focal length of the camera, (x_0, y_0) is the center of the image plane, and $(\Delta x, \Delta y)$ is the comprehensive distortion, which can be expressed by the following formula:

$$\begin{cases} \Delta x = x_c r^2 k_1 + x_c r^4 k_2 + x_c r^6 k_3 + \left(2x_c^2 + r^2\right)p_1 + 2p_2 x_c y_c + b_1 x_c + b_2 y_c \\ \Delta y = y_c r^2 k_1 + y_c r^4 k_2 + y_c r^6 k_3 + 2p_1 x_c y_c + \left(2y_c^2 + r^2\right)p_2 \end{cases} \tag{2}$$

Where:

$$\begin{cases} x_c = \Delta x x_m - x_0 \\ y_c = \Delta y y_m - y_0 \end{cases} \tag{3}$$

$$r = \sqrt{Y_{an} x_c^2 + X_{ax} y_c^2} \tag{4}$$

Among them, $c, x_0, y_0, k_1, k_2, k_3, p_1, p_2, b_1, b_2$ is the model parameter of camera coordinate system, which can be obtained by camera calibration technology [6]. Taking the plane structured light as an example, the plane of structured light has a definite mathematical description in the reference coordinate system. The specific form is determined by the mode of structured light

$$Z_r = f(X_r, Y_r) \tag{5}$$

The transformation relationship between reference coordinate systems can be described by rotation matrix R and translation matrix T. As shown in the formula, R and T can be obtained by sensor calibration technology:

$$\begin{pmatrix} X_c \\ Y_c \\ Z_c \end{pmatrix} = Z_r R \begin{pmatrix} X_r \\ Y_r \\ Z_r \end{pmatrix} + T \tag{6}$$

$$R = \begin{pmatrix} r_{11} & r_{12} & r_{13} \\ r_{21} & r_{22} & r_{23} \\ r_{31} & r_{32} & r_{33} \end{pmatrix} \tag{7}$$

$$T = \begin{bmatrix} T_1 \\ T_2 \\ T_3 \end{bmatrix} \tag{8}$$

The equation of structured light plane in 3D sensor camera coordinate system can be solved by the above formula

$$Z_c = Tf(X_c, Y_c) \tag{9}$$

The mathematical model of structured light vision sensor is obtained

$$\begin{cases} X_c = \frac{(x - x_0 + \Delta x)}{c} Z_c \\ Y_c = \frac{(y - x_0 + \Delta y)}{c} Z_c \\ Z_c = f(X_c, Y_c) \end{cases} \tag{10}$$

By introducing the structured light plane, the relationship between the light plane and the camera coordinate system is obtained by using the pre calibration technique. As a supplementary constraint, the ambiguity of the inverse mapping from the two-dimensional image space to the three-dimensional space is eliminated [7]. Further initialize the information, and then check whether it is in a sleep state under the sharing method. If not, send the information to the management module. After the successful transmission, they

Table 1. Construction initialization information

Initialization information	Information acquisition		Information management	
	a	b	a	b
1000	27.8	24.7	25.2	23.7
10000	27.6	25.4	170.9	28.0
100000	29.1	237.3	145.5	25.4

receive the information and process it. The data is analyzed as shown in the table (Table 1).

Three dimensional scanning is convenient to detect the contour data of the working environment. All kinds of monitoring devices arranged in the construction workplace can be transmitted to the technicians through the network, and then the three-dimensional modeling can be integrated into the BIM method, and the risk identification can be carried out in advance on the BIM method, so as to reduce the information asymmetry of all parties in the initial stage of the system construction, which is convenient for better management In order to avoid the construction safety problems that may occur in the construction process, we should coordinate all parties effectively. Based on this, the construction information processing steps are optimized, as shown in the following Fig. 1:

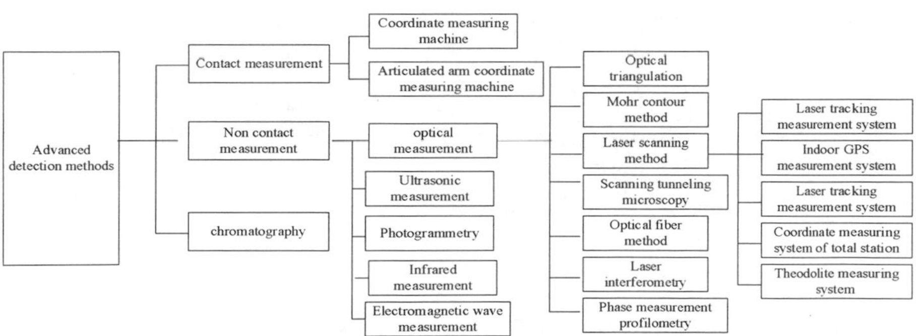

Fig. 1. Construction information processing steps

As shown in the figure, the most advanced heterodyne multi frequency phase-shifting three-dimensional optical measurement technology is used for single measurement format, measurement accuracy, measurement speed and other performance. Compared with the traditional gray code plus phase-shifting method, it has reached the world's most advanced level, with higher measurement accuracy, larger single measurement range, strong anti-interference ability, and little impact on the measured data, so as to scan and measure objects from several millimeters to tens of meters Volume information [8, 9].

The products represented by surface scanning equipment are especially suitable for the modeling of large plane, super large surface and complex surface parts.

2.2 Building Construction Information Sharing Model Based on Social Network

Social network related equipment can be divided into three categories according to its measurement methods. Compared with the traditional risk analysis mode, this risk identification method will be more scientific, reasonable and effective. Based on such practical problems, we should establish an information collection and modeling system in the early stage of social network construction, which is convenient for the follow-up work [10]. The effective construction information collected is integrated into the social network BIM integrated information management method, which is convenient for the later engineering modeling and real-time information sharing in the whole process. The whole information collection system is shown in Fig. 2

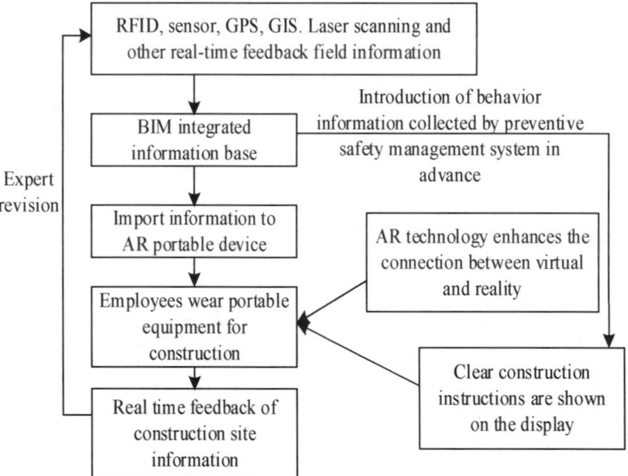

Fig. 2. Information collection system at the initial stage of construction

In the early stage of construction, fine information collection and three-dimensional modeling will greatly reduce the sudden accidents that may occur in the whole life stage. Integrating all kinds of building information, simulating the corresponding construction environment, and relying on BIM information method for all aspects and multi-level management will improve the safety and reliability in the actual construction process. BIM can store all geometric information and engineering technical information in engineering design scheme [11, 12]. On the basis of building construction information collection in the early stage, it is applied to BIM method and combined with digital network technology for 3D modeling. In the process of modeling, the design planning problems are discussed and analyzed to find the optimal solution. So as to better adapt to the technical development of the current digital era, and constantly introduce new science and technology to enhance the operation safety of the current facilities. Based on this, the informatization safety management system before construction is optimized, and the specific structure is shown in Fig. 3.

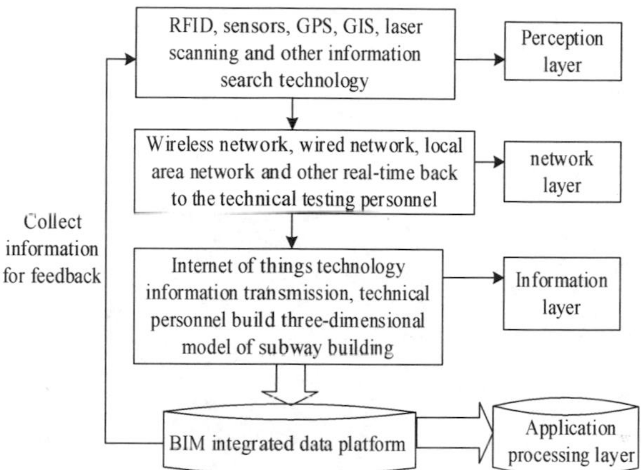

Fig. 3. Information sharing prevention safety management system

In the early stage of construction, the information collection system has been able to ensure technicians to establish a relatively complete project model based on BIM method, so as to find the design conflicts in the process of information integration modeling. The complete BIM model is connected with 3D scanning technology, the project model is imported into the virtual technology, and then the model is optimized. The 3D technology is used to integrate the virtual reality environment and BIM model seamlessly, and all the information collected is verified and corrected before the actual construction.

2.3 Realization of Real-Time Sharing of Construction Information

In the established three-dimensional model, VR technology is used to import. The main construction personnel wear a head mounted display and immerse themselves in the virtual construction environment for relevant construction operations. The safety management personnel will track and record their behavior, and finally feed back to the BIM information integration method to form a dynamic and real-time updated safety behavior log. The main personnel of intelligent construction method can also continuously improve their own reliability through the application of scientific information technology. The emergence of 3D scanning technology and its application equipment HMD makes virtual construction possible. The significance of building construction is great. The objective risk factors can be identified before the actual construction, and then the corresponding preventive measures can be formulated The implementation of man who will present these steps to prevent potential hazards and accidents on HMD equipment, increase the safety and reliability of construction, and then reduce the risk probability of people in the whole safety system, so as to achieve the intrinsic safety of system construction. Based on this, the construction information security management model is constructed to avoid all kinds of emergencies and improve the intrinsic safety of construction. The specific construction information security management model is shown in Fig. 4.

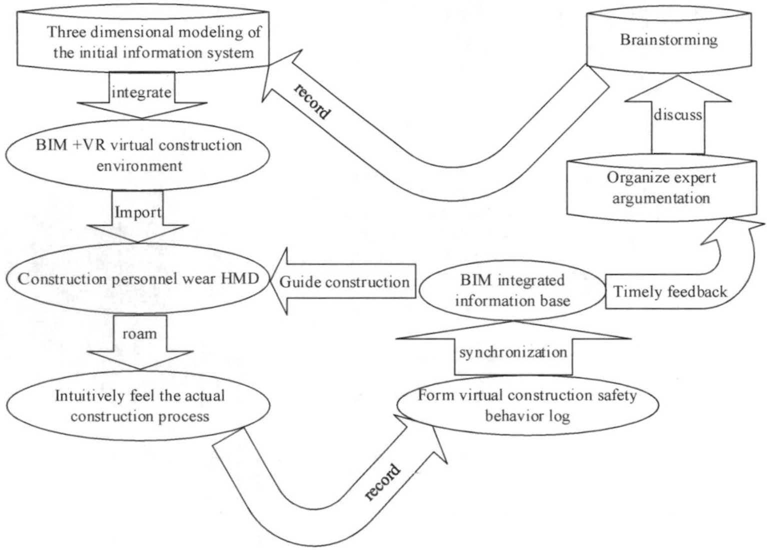

Fig. 4. Construction information security management model

According to the BIM target, the specific plan of building construction is planned. The implementation of real-time information sharing technology will change the production process of building construction. Through the establishment of real-time information sharing scheme, it can realize the smooth migration, sharing and checking of information, such as document management, query prevention authority, folder maintenance and notification. According to the technical ability of building construction, the technical framework is established by analyzing BIM tools, hardware and software requirements. For employees, including training system, number of trainees, time, expert selection, etc.; BIM application organization structure design. In the social network, the browser server architecture is adopted, which is based on the secondary development of scalable data collector. With network data as the technical support, the design and development of building construction information sharing method under the background of big data is realized. As shown in the figure is the overall architecture of the construction information sharing method (Fig. 5).

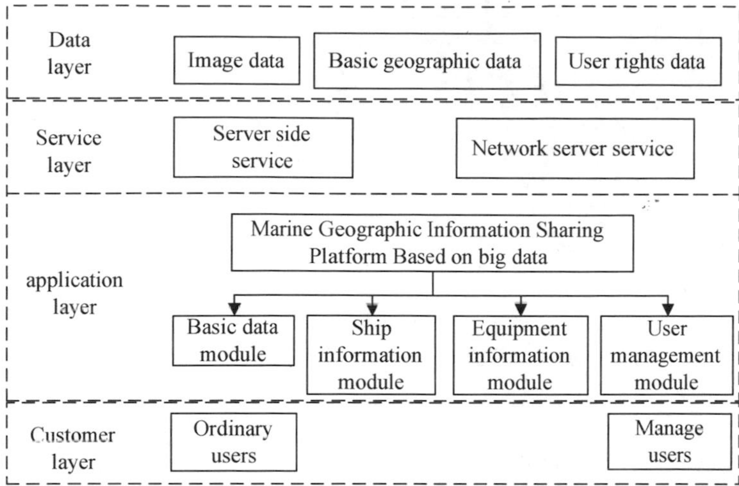

Fig. 5. Overall framework of construction information sharing method

The organizational structure includes the new department and new post specially responsible for managing BIM Technology and services, the division of work interface and the responsibilities of relevant parties. The final foothold of project BIM implementation and construction BIM Technology implementation is the construction project. At the project level, the implementation content of BIM includes project BIM implementation preparation, construction information model creation, model application, BIM results output; project BIM implementation preparation. BIM implementation preparation includes setting BIM implementation objectives, BIM implementation plan, BIM team and BIM facility environment. Construction information model creation. Based on the CAD drawings or BIM design model provided by the design unit, the construction model is constructed, and specific information is added according to the types of BIM application. Taking the BIM application of construction organization simulation as an example, the construction information model should include the contents of construction drawing, construction organization design, construction schedule, resource organization and plane layout; project BIM application. The BIM model is used in the management of construction process, such as schedule management, cost management, multi-party collaboration, etc. BIM output. Through the BIM application, the targeted solutions are obtained. Taking the BIM application of concrete prefabricated components as an example, through the BIM implementation in the above stages, the resource allocation plan, process instruction, processing drawing, report list, etc. Required for component production can be formed. Based on this, the architecture construction information management framework is optimized, as follows (Fig. 6):

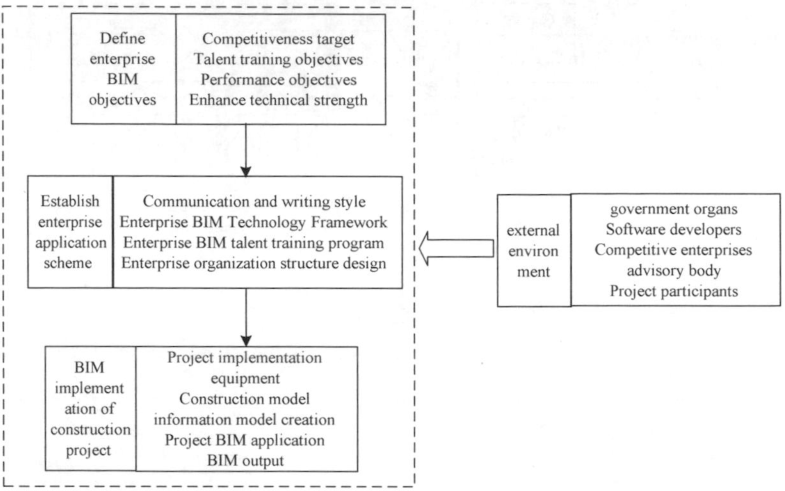

Fig. 6. Construction information management framework

The purpose of information sharing development of BIM Technology in construction building is to preliminarily determine the framework of success factors and define the scope for subsequent analysis of key success factors. The following figure describes the development steps of information sharing list processing (Fig. 7).

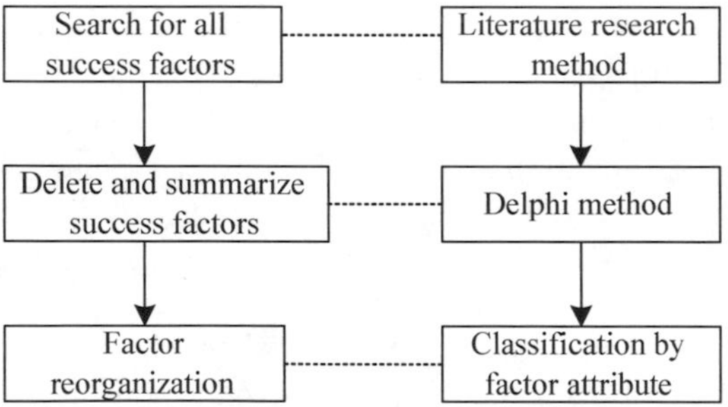

Fig. 7. Processing steps of information sharing list

According to the application layer of the overall framework of the above building construction information sharing method, the overall functions of the development and construction of the sharing method under the background of big data are designed, as shown in the Fig. 8.

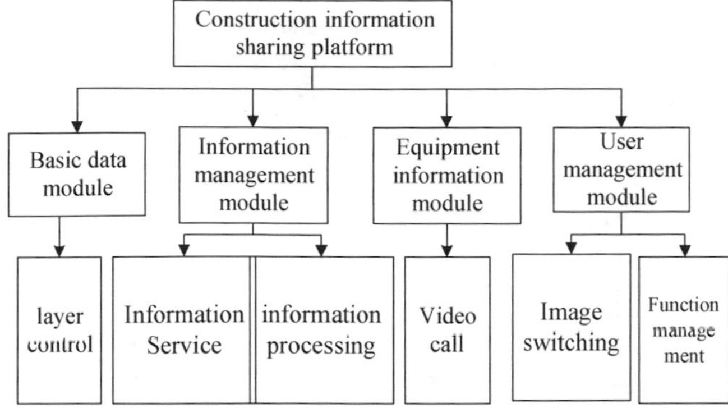

Fig. 8. Overall function design of information sharing model

It can be seen from the figure that the functional part of the building construction information sharing method based on 3D scanning technology and social network analysis is composed of basic data, building information, information and user management modules in the overall framework, in which the basic data module can realize image hierarchical control, the building information module can realize information query and processing, and the information module can realize video adjustment The user management module can realize image switching and function management. BIM direct cost control should be paid attention to in construction. Excessive purchase of BIM related software and hardware facilities will not only confuse the application objectives of BIM, but also cause the risk of construction operation. At the same time, due to the uncertainty in the process of BIM implementation, such as the change of design and business content, the potential cost of BIM Technology should bc controlled to improve the economic benefits of BIM.

3 Analysis of Experimental Results

In view of the practical application of the real-time sharing method of building construction information based on 3D scanning technology and social network analysis, the design is verified by experiments. The experimental parameters were optimized as shown in Table 2.

Table 2. Parameter setting of experimental data acquisition

Unit number	Acquisition node	Parameter setting
1	A1	60
2	A2	75
3	A3	55
4	A4	80

The traditional method and the information sharing method in this paper are compared. Under the noise interference, the construction information sharing situation is compared and analyzed. Set the information sharing time as 20 s, select 8 building construction information as the experimental object, respectively view the two methods information sharing situation, and the comparison results are shown in the Fig. 9.

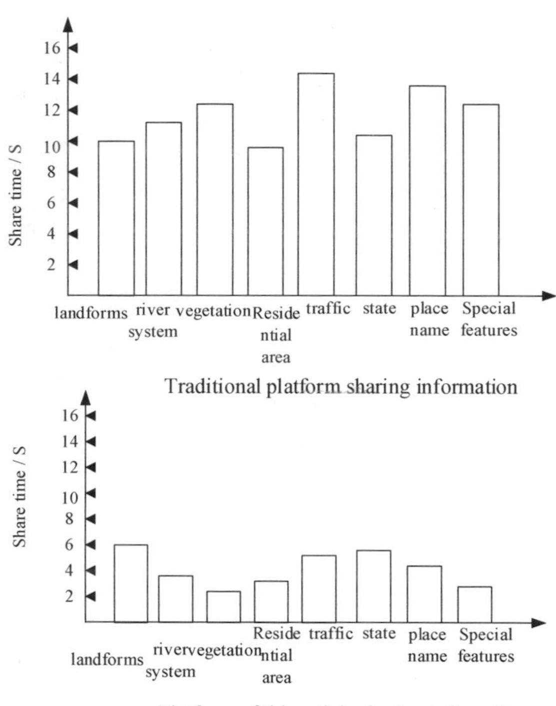

Fig. 9. Time consumption of information sharing between the two methods

It can be seen from the Fig. 9 that the information sharing time of the traditional method is longer than that of the data sharing method in this paper. Therefore, the real-time information sharing method of the three-dimensional scanning technology combined with social network analysis proposed in this paper has significantly shorter time-consume and lower time-consume in the practical application process, which fully meets the research requirements.

Furthermore, the two methods share the differences in time may be due to noise interference. Therefore the sharing effect of the two methods is compared under noise interference, and the results are shown in Fig. 10.

It can be seen from the Fig. 10 that the traditional method is more disturbed by noise and has a longer time, building construction information real-time sharing method based on 3D scanning technology and social network analysis is less disturbed by noise, and the performance of sharing information has remained stable. Therefore, the design method is better.

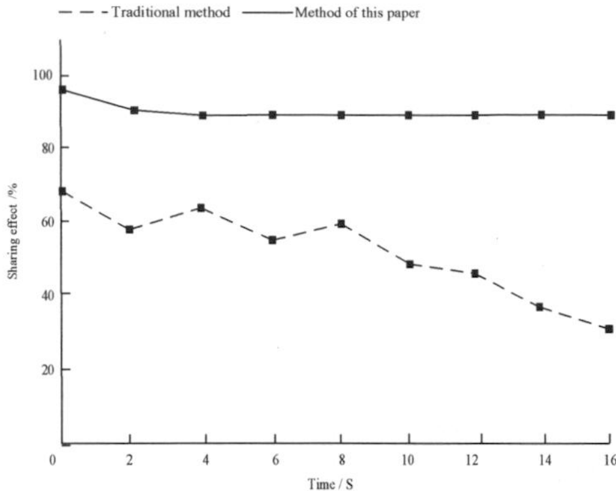

Fig. 10. Comparison of sharing effect of two methods under noise interference

4 Conclusion and Outlook

This paper briefly discusses the significance of information management for construction projects and the strategy of realizing information management. At present, society has entered information society, and it has become inevitable for all walks of life to be included in information management. As a basic industry of development, construction project shoulder a heavy list of responsibilities. In order to effectively improve the efficiency and safety of its work, it will be built It is an inevitable choice to include the informatization of construction project management, which can lay a solid foundation for the long-term development of China's construction projects. With the rapid development of information technology, the construction process is inseparable from the application of information technology. Construction management is closely related to construction quality and construction cost control. Computer network technology is also widely used in construction management, which effectively improves the level of construction management. Therefore, the analysis of information technology application in construction management plays a great role in promoting the development of the domestic construction industry. This paper deeply analyzes the application of computer in construction project management, and probes into the application status of information technology in construction management. Based on this, this paper puts forward the application strategy of information technology in construction management, in order to improve the application level of information technology in domestic construction management. With the continuous development of the economy and society, project construction is more and more strict, with higher requirements for information management technology. The role of real-time sharing of construction information is increasingly important in the construction industry and information technology application. In the future, information sharing methods with higher sharing accuracy and shorter sharing time will remain the main focus of research in this field, and ultimately promote the development of information sharing technology to a higher level.

Fund Projects. The Research on Emerging Engineering Education and Practice Project of Shaanxi province (202075).

References

1. Brna, M., Cingel, M.: Comparison of pavement surface roughness characteristics of different wearing courses evaluated using 3D scanning and pendulum. MATEC Web Conf. **313**(10), 13 (2020)
2. Emilio, A., Ramírez, P.A.U., Durazno, G.A., et al.: Positioning assessment of surgical cutting guides for osteosarcoma resection utilizing 3D scanning technology. Procedia CIRP **89**(12), 176–181 (2020)
3. Zhu, F., Zheng, S., Wang, X., et al.: Real-Time efficient relocation algorithm based on depth map for small-range textureless 3D scanning. Sensors **19**(18), 3855–3859 (2019)
4. Zhu, M.: Research on camera-based 3D scanning system based on machine vision. China Comput. Commun. **32**(2), 44–45, 48 (2020)
5. Ling, M., Lu, X., Wang, G., et al.: Analytical modeling the multi-core shared cache behavior with considerations of data-sharing and coherence. IEEE Access **PP**(99), 1 (2021)
6. Epiphaniou, G., Pillai, P., Bottarelli, M., et al.: Electronic regulation of data sharing and processing using smart ledger technologies for supply-chain security. IEEE Trans. Eng. Manage. **PP**(99), 1–15 (2020)
7. Fu, W., Liu, S., Srivastava, G.: Optimization of big data scheduling in social networks. Entropy **21**(9), 902 (2019)
8. Liu, S., Li, Z., Zhang, Y., et al.: Introduction of key problems in long-distance learning and training. Mob. Networks Appl. **24**(1), 1–4 (2019)
9. Liu, S., Bai, W., Zeng, N., et al.: A fast fractal based compression for MRI images. IEEE Access **7**, 62412–62420 (2019)
10. Ko, B., Liu, K., Son, S.H., et al.: RSU-Assisted adaptive scheduling for vehicle-to-vehicle data sharing in bidirectional road scenarios. IEEE Trans. Intell. Transp. Syst. **PP**(99), 1–13 (2020)
11. Fujiwara, T., Mitsuya, Y., Fushie, T., et al.: Demonstration of soft X-ray 3D scanning and modeling with a glass gas electron multiplier. J. Instrum. **14**(11), P11022–P11022 (2019)
12. Xu, Z.X., Chen, G.S.: Modeling and simulation of dynamic focusing control system for 3d scanning galvanometer. Model. Simul. **10**(2), 8 (2021)

A Mining Algorithm for Relevance of Business Administration Based on Complex Social Information Network

Zhao-xi Chen[1](✉) and Wen Zhang[2]

[1] Nanchang JiaoTong Institute, Nanchang 330100, China
[2] School of Sciences, East China University of Technology, Nanchang 330013, China

Abstract. With the rapid development of economy, the reform of business administration system in our country has changed the contents of business administration. Based on this, this paper analyzes the present situation of business administration in our country under the new situation, and advances an algorithm for mining the relevance of business administration based on complex social information network. Through collecting the characteristic behavior of business administration and integrating the structure of complex social information network, this paper analyzes the correlation parameters of business administration behavior. By using the basic idea of association rules mining and extending the traditional methods of co-word analysis and text clustering, a mining model of association rules based on topic keywords and abstract salient words is constructed, and combined with the econometric analysis of related business management behaviors in social information network and Web Science database, the steps of mining association of business management behaviors are optimized. Finally, the experiment proves that the algorithm of mining association of business management behaviors based on complex social information network is highly effective in the practical application and fully meets the research requirements.

Keywords: Social information network · Wireless network · Business administration · Relevance mining

1 Introduction

In the current environment of economic development, the management and law enforcement activities of industrial and commercial administration departments play a pivotal role in its economic operation and development. But many problems in its development can not be ignored. First of all, the business administration department lacks professional management personnel and is unable to perform management functions effectively [1]. Secondly, since China's reform and opening up, the pace of economic development is relatively rapid, industrial and commercial departments for its management, the lack of a sound management mechanism. In the concrete management behavior, because the system is imperfect, the local protectionism is widespread, causes its law enforcement dynamics to be weaker. In management, the lack of effective supervision of its

S. Liu and X. Ma (Eds.): ADHIP 2021, LNICST 416, pp. 217–228, 2022.
https://doi.org/10.1007/978-3-030-94551-0_18

internal control management [2]. Therefore, the industrial and commercial administration departments should pay more attention to the existing problems and take effective countermeasures in time. Under the current economic development environment, the business administration department of our country is not only facing the opportunity of development, but also facing the huge challenge. Therefore, it needs the business administration department, according to its concrete development present situation, to carry on the discussion of the solution, and proposes the optimization of the mining algorithm of the relevance of the business administration behavior based on the complex social information network, in order to carry on the effective consummation to the management method, thus promotes the comprehensive and stable development of the market economy of our country [3]. To solve the above problems, in order to promote the management effect realization, promote our country economy's long-run, the health, the stable development. Therefore, a mining algorithm of enterprise management relevance based on complex social information network is proposed. According to the characteristics of enterprise management behavior, H-P filter is used to analyze the trend of enterprise management behavior characteristics and estimate the characteristic components, so as to analyze the related parameters of enterprise management behavior. The association rule mining algorithm is used to statistically analyze the distribution of high-frequency keywords and the probability distribution of unexpected events in the process of enterprise management behavior, and the association rule mining model is constructed with subject keywords and abstract salient words. Combining social information network and database, the operation process of association mining enterprise management behavior is optimized.

2 Algorithm for Mining Association of Business Administration Behaviors

2.1 Feature Mining Algorithm for Business Administration

In order to ensure the effectiveness of the relational mining algorithm for business administration, firstly, the characteristics of business administration are mined, and the association relationship between business administration related parties is calculated based on the characteristics of business administration. Due to the strong concealment of the relativity of business administration characteristics of business administration, in the process of mining the characteristics of business administration, it is necessary to make a comprehensive analysis on the processes of surplus manipulation such as purchase and sale of commodities and transfer of assets [4]. Profit achieved through earnings manipulation in order to avoid losses and achieve listing purposes [4]. This paper studies and analyzes the relevant characteristics of the behavior characteristics of business administration. In this paper, H-P filter is used to decompose the trend of typical business management behavior characteristics, so as to explore the relevance of business management behavior characteristics. The principle of H-P filtering is to treat the trend value of the feature sample as the potential level, and then estimate the feature component by minimizing the deviation between the true sequence and the trend value of the sample.

$$\min A_{x_e,t=1,2,\ldots,T} = \left\{ \sum_{i=1}^{T} (Y_i - X_t)^2 + \lambda\left[(X_{t+1} - X_i) - (X_t - X_{t-1})\right]^2 \right\} \quad (1)$$

Among them, y is the growth rate sequence of the characteristics of business administration, and x is the trend component of the growth rate, which is used to describe the long-term trend of the economic growth rate of a country over time, and is mainly determined by the basic characteristics of the economic development of that country. Parameter λ is the penalty factor of the fluctuation of the trend component, and the difference of t is the deviation of the variation of the trend component of the adjacent two periods. Command C = y − x, representing the feature components, if C and 2x are subject to independent congruent distributions, when $L =$ var(C)/var(2x) filtering can achieve a minimum. The trend component may be expressed as follows:

$$X_i = \left[1 + \lambda\left(1 - L^2\right)^2\left(1 - L^{-1}\right)^2\right] Y_t \min A_{x_e,t=1,2,\ldots,T} \quad (2)$$

$$C_2 = \frac{\lambda(X_i - L^2)^2(1 - L^{-1})^2}{1 + \lambda(1 - L^2)^2(1 - L^{-1})^2} Y_t \min A_{x_e,t=1,2,\ldots,T} \quad (3)$$

Based on the abovementioned algorithms, further advance the arrival of the modern economic era with business administration and features and categories as units, collect data, communicate information, share knowledge, transform knowledge into wisdom, realize knowledge innovation and business process reengineering of business administration, transform traditional business administration marketing strategies and production processes with knowledge management and knowledge commerce, build business administration into a learning, innovative and intelligent business administration, and enhance the core competitiveness of business administration [5]. According to the understanding of industrial and commercial administration, the characteristics of information behavior management are mainly reflected in the four-level model, which is shown as follows (Fig. 1):

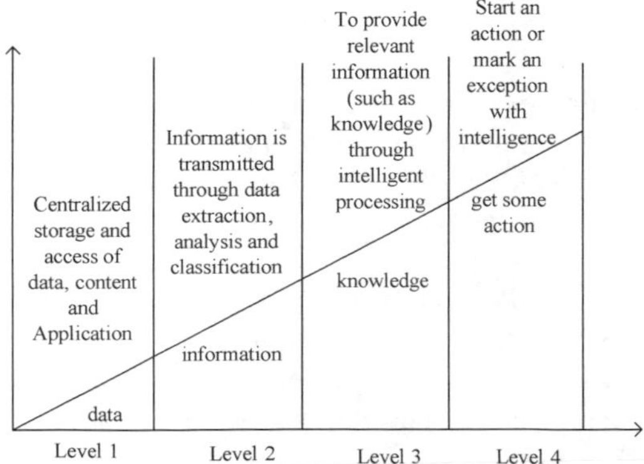

Fig. 1. Business administration information transmission maturity type

Facing the complex and changeable competitive environment, many business management can only make routine decisions, it is difficult to timely understand the abnormal situation and potential behavior management characteristics. At this point, business administration requires the use of appropriate intelligent software and social networks to assist policymakers in freeing them from duplicated efforts [6]. And set up customized rules, application scenarios, business needs and other ways to achieve the timely discovery of effective information, and automatically take necessary measures [7]. In the process of data feature mining, the data features are similar to the maturity model of business administration information transmission. The process of feature mining for business administration behavior management is a complete process composed of data collection, integration, analysis, application and decision-making, and several links. The main procedures and basic characteristics of competitive intelligence work of business administration are summarized, and different mining process models for the relevance of competitive intelligence features are constructed. The specific mining steps are shown in the figure below (Fig. 2).

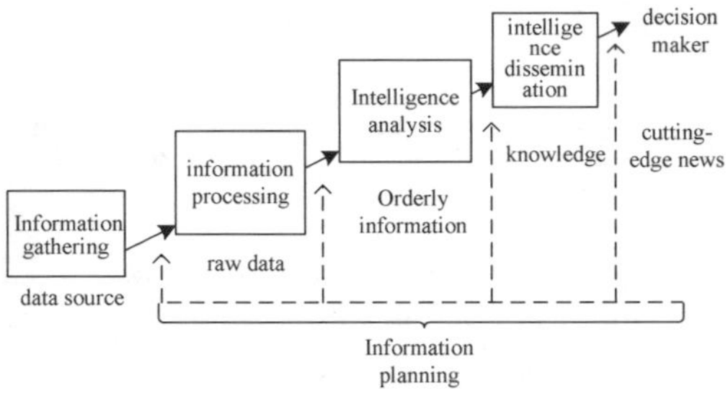

Fig. 2. Business administration behavior feature processing flow

It can be seen from the contrast graph that there is a main line of data-information-knowledge-wisdom in both the information transmission maturity model and the competitive intelligence processing process of business management to achieve the effect of business management.

2.2 Feature Association Algorithm for Business Administration

Based on the current status of research on the feature association of business administration behaviors, optimize the feature association algorithm of business administration behaviors based on the structure of social information network, and conduct statistical analysis on the distribution of high-frequency keywords and the probability distribution of emergent words in the process of business administration behaviors by combining statistical principles [8]. In the process of managing and mining the behavioral characteristics of business administration, it is necessary to comprehensively evaluate the change process of relevant data and information respectively, and realize the process of understanding from low level to high level. The higher the level is, the extension, depth, meaning, conceptualization and value are explored. The specific steps for mining behavioral characteristics of business administration without business administration are shown in the following figure (Fig. 3).

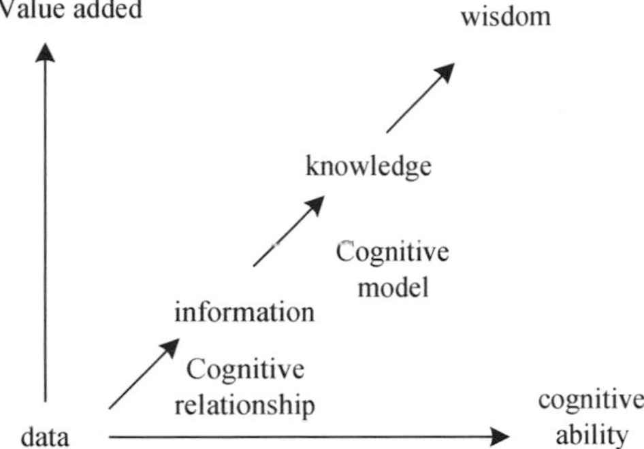

Fig. 3. Steps for mining characteristics of business administration behavior

The system framework of intelligent analysis of competitive intelligence of business administration based on data mining is to develop and implement intelligent activities in the environment of data mining, such as semantic organization of competitive intelligence of business administration, design of intelligent analysis strategy and method, visualization of intelligent analysis results, etc., in order to realize the intelligent processing of data information, including the research of supporting theory and method, intelligent analysis strategy and method, and display of intelligent analysis results, such as graph (Fig. 4).

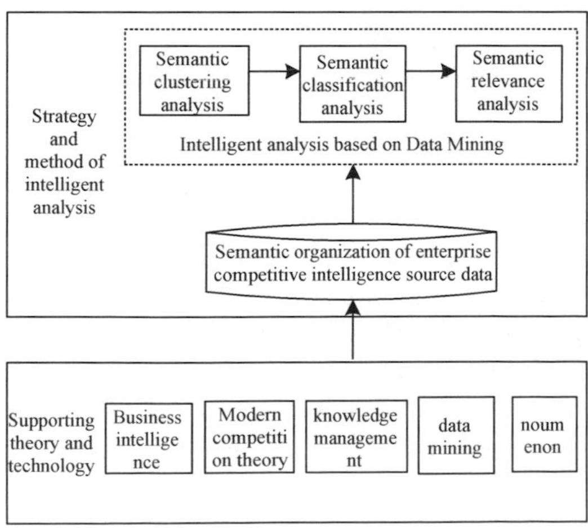

Fig. 4. Information intelligent analysis model based on data mining

In the process of management mining, each indicator (variable) has its own dimension and unit of order of magnitude, and the conversion rules between indicators are different.

In order to compare the data of different orders of magnitude and different dimensions, it is necessary to convert the data accordingly. In general, the central transformation is used, that is, the average value of each training set variable is calculated, and then the mean value of the training set variable is subtracted from the original data [9]. The result of the transformation makes the data mining of each column not only a simple retrieval, query and call oriented to a specific database, but also a micro-, meso-and even macro-statistics, analysis, synthesis and inference of these data, so as to guide the solution of practical problems, and try to discover the relevance between events, even use the existing data to predict future activities. Each indicator (variable) has its own dimension and unit of order of magnitude, and the conversion rules between indicators are not the same. In order to be able to compare the data of different orders of magnitude and different dimensions, it is necessary to convert the data accordingly. In general, the central transformation is used, that is, the average value of each training set variable is calculated, and then the mean value of the training set variable is subtracted from the original data [10].The result is that the sum of the columns is 0, the sum of the squares of the columns is $n - 1$ times the variance of the column, and the cross product of any two different columns is $n - 1$ times the covariance of the two columns. Let U be the total number of objects to be classified. Build the similarity coefficient on U between r and r_{ij} for the similarity between i and j. When U is a finite set, R is a matrix called the similarity coefficient matrix. There are eight common methods for building similarity matrices for data sets.

$$r_{ij} = \frac{C_2 \sum_{k=1}^{m} (x_{ik} - \overline{x}_i)(x_{jk} - \overline{x}_j)}{\sqrt{\sum_{k=1}^{m} (x_{ik} - \overline{x}_i)^2} \sqrt{\sum_{k=1}^{m} (x_{jk} - \overline{x}_j)^2}}, (i, j \leq n) \tag{4}$$

Absolute index method [11]:

$$r_{ij} = e^{-\sum_{k-1}^{m} 1x_{ik} - x'_{jk}}, (i, j \leq n) \tag{5}$$

The similarity relation r_{ij} established by the above method only satisfies the reflectivity and symmetry, but not the transitivity, so it is not a fuzzy equivalence relation. Therefore, it is necessary to use the transitive closure method to find R fuzzy equivalent matrices. That is, r_{ij} from the transformation into R, and then from the $R-$, so continue, until a step $R^{2k} = R^k = R^*$. At this point R satisfies the transitivity, that is, for the fuzzy equivalence matrix of R, the semantic proximity range of a concept c is defined as all its direct subconcepts and direct hyperconcepts, which are expressed as follows:

$$V(c) = r_{ij}\{R^{2k} b \in C \big| \; c\pi b \quad \text{or} \quad b\pi c\} \tag{6}$$

Second, get the set of keywords in all concepts within the semantic proximity of concept c that can express these concepts:

$$U(c) = \cup \text{Ref}^{-1}_{b \in V(c)}(b) \tag{7}$$

Disambiguation function is used to obtain the most appropriate concept for the meaning of keywords t based on d context [13]. The meaning of disambiguation function $dis(d, t)$ is as follows: the concept corresponding to keyword t should be the first concept corresponding to those keywords which obtain the largest weight value in text d in keyword $U(c)$ set. The formula for calculating the disambiguation function is:

$$\mathrm{dis}(d, t) = \mathrm{first}\{c \in \mathrm{Re}f_C(t)| \quad c\, maximizes\, w(d, U(c))\} \tag{8}$$

Calculate the weights of Concept c in Text d using the following formula:

$$w(d, c) = \mathrm{dis}(d, t)w(d, \{t \in T| \quad dis(d, t) = c\}) \tag{9}$$

Through data selection, preprocessing, transformation, pattern extraction, knowledge evaluation and process optimization, we use discriminant analysis, clustering analysis and exploratory analysis to discover and acquire knowledge. The basic process of knowledge discovery is a multi-step process, including data selection, preprocessing, data transformation, data mining and result interpretation and evaluation, as shown in the figure (Fig. 5).

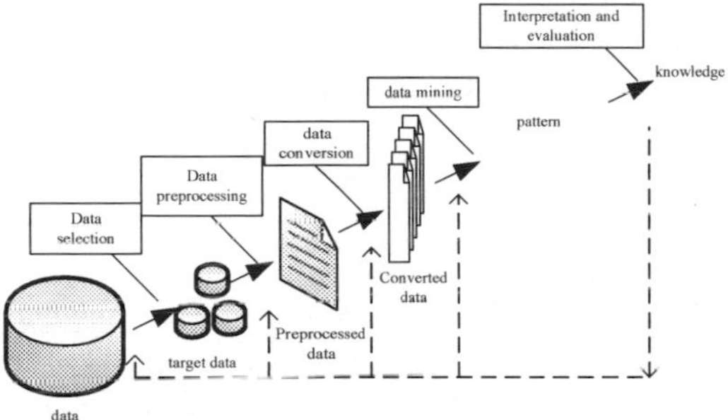

Fig. 5. Analysis of the characteristics of industrial and commercial administration

The application of complex social network technology in competitive intelligence analysis of business administration makes the analysis system of competitive intelligence of business administration automatically analyze the data and information in database, data warehouse and interconnected network, from which the potential comprehensive intelligence knowledge can be mined and the automation and intelligence of competitive intelligence analysis and knowledge discovery can be realized.

2.3 Realization of Business Administration Behavior Relevance Mining

As long as we follow the market rules, standardize the behavior of business administration, and enhance the openness and transparency of the relevance of business administration characteristics. At present, the disclosure of the relevance of business management characteristics in most companies' financial reports is incomplete and untrue, which reduces the efficiency and reliability of the use of accounting data. Below is a brief introduction to the key factors in mining the relevance of business administration behavior, as shown in the table below (Table 1).

Table 1. Factors affecting the mining of relevance of business administration behaviors

Name	Meaning
Motivation purpose	The effect of equity incentive plan
Incentive object	The object of equity grant is the beneficiary of equity
Incentive mode	The way to grant equity to the incentive object, that is, how the incentive object benefits
Stock source	It mainly refers to the source of the stock at the time of exercise
Incentive level	Number of shares granted to incentive objects
Exercise price	The price of the company's stock when the incentive object exercises the right
Source of funds	The source of funds used by the incentive object to exercise or purchase shares
Effective period of incentive	Time range of equity incentive implementation process
Exercise arrangement	The setting of waiting period (sales restriction period), unlocking period and exercise period
Performance evaluation index	The criteria to judge whether the incentive object can exercise the right include performance indicators and non performance indicators

Association rules are simple statements about the probability of certain events occurring together in a database. Mining association rules can reveal the dependencies among specific objects, and then infer the information of other objects from the information of one object. For itemset I, transaction dataset D, each transaction T_i, where $T_i \subset I$.

If $A \subset D, B \subseteq D$, and $A \subseteq I, B \subseteq I, A \cap B = \phi$, then the expression such as $A \Rightarrow B$ can be called an association rule. The process of mining association rules is divided into two stages: finding all frequent itemsets from the original data set. Generate association rules from frequent itemsets. Item set I is a collection of data items, namely $I = \{i_1, i_2, \cdots, i_n\}$. D is the set of transaction transactions, that is, $D = (T_1, T_2, \cdots, T_n\}$, T_i represents the ith transaction, with each transaction corresponding to a transaction identifier *TID*. Support S is the ratio of the number of transactions containing both A and B in transaction set D to the total number of transactions, which can be expressed as:

$$S(A \Rightarrow B) = \frac{||(T \in D|\ A \cup B \subseteq T)}{w(d, c) * ||D||} \tag{10}$$

Confidence C is the ratio of the number of transactions containing both A and B in transaction set D to the number of transactions containing only A, which can be expressed as:

$$C(A \Rightarrow B) = \frac{||\{T \in D|\ A \cup B \subseteq T\}}{||\{T \in D|\ A \subseteq T\}} \tag{11}$$

All stochastic quantities are the process of mining business management behaviors that change within a specific scope and at a specific time. For the processing of business administration behavior, it is not to analyze and find its statistical law and probability distribution, but to analyze and find the law from the original data without laws, that is, after processing the original data in a certain way, make it become a relatively regular time series data, and then establish a model. Prediction mainly refers to the use of $GM(1,1)$ model to predict the amount of time series data. For example, population forecast, labor force forecast, product output forecast, is the use of statistical data over the years, the future development of its forecast. Here are the basic steps to build a $GM(1,1)$ model:

Do first order cumulative generation for the following series:

$$\left\{x^{(0)}(k)\right\} = \left\{x^{(0)}(1), x^{(0)}(2), \cdots, x^{(0)}(N)\right\} \tag{12}$$

Among them, k represents the time of data sequence, (1) represents a cumulative sequence. Make a first order cumulative generation of it:

$$\left\{x^{(1)}(k)\right\} = \left\{x^{(1)}(1), x^{(1)}(2), \cdots, x^{(1)}(N)\right\} \tag{13}$$

$$x^{(1)}(k) = \sum_{i=1}^{k} x^{(1)}(i), k = 1, 2, \cdots, N \tag{14}$$

Construct the cumulative matrix and the constant term vector, the first order mean of x to generate:

$$x = (x(2), x(3), \cdots, x(n)) \tag{15}$$

Of which:

$$x(k) = \frac{1}{2}\left(x^{(1)}(k) + x^{(1)}(k-1)\right), k = 2, 3, \cdots, n \tag{16}$$

Then:

$$B = \begin{pmatrix} -x(2) & 1 \\ -x(3) & 1 \\ \vdots & \vdots \\ -x(n) & 1 \end{pmatrix}, \quad Y_N = \begin{bmatrix} x^{(0)}(2) \\ x^{(0)}(3) \\ \vdots \\ x^{(0)}(n) \end{bmatrix} \tag{17}$$

Using the least square method to solve the grey parameter to construct the white differential equation:

$$\frac{dx^{(1)}}{dt} + ax^{(1)} = \mu \tag{18}$$

Solve parameters as follows:

$$\hat{a} = \begin{pmatrix} a \\ \mu \end{pmatrix} = \left(B^T B\right)^{-1} B^T Y_N \tag{19}$$

Based on the above algorithms, effective mining of the relevance of formula management behaviors is realized, and detailed provisions are made for the disclosure matters, disclosure methods and disclosure format of the relevance of business management characteristics according to the calculation results.

3 Analysis of Experimental Results

In order to verify the practical application effect of the association mining algorithm based on complex social information network, the empirical analysis is made according to arbitrage pricing theory. To ensure the research effect, the equipment and relevant parameters used in the experiment are standardized, which is shown in the following table (Table 2).

Table 2. Experimental parameter settings

Parameter	Measurement
Video editor	2
Media editing software	2
Particle editing software	1
Vehicle identification software	1

Based on the theoretical basis of the previous experiment, hypothesis 1 tests the earnings management of the company and the earnings management of the control group by matching samples T test, and carries on multiple linear regression analysis, and validates hypothesis 2 to hypothesis 5 according to the results of the analysis. According to this parameter, the experimental analysis is carried out. Based on the calculation of SPSS software, the characteristics of behavioural correlation can be calculated and expressed in terms of a1.a2.a3. In order to ensure the accuracy of the results, all the experiments were repeated 20 times, and the results were taken as the average. Because the calculation process is complicated, the calculation results of the correlation shall be directly displayed instead of being described here, specifically as follows (Table 3):

Table 3. Row as correlation feature calculation result

Result	Nonstandard coefficient		Standard coefficient	T	Sig.
	B	Standard error	Beta value		
Variable	−0.001	0.010	–	−0.112	0.027
X1	4982579.439	4582519.253	0.048	1.087	0.001
X2	0.044	0.016	0.125	2.791	0.000
X3	−0.065	0.018	−0.168	−3.714	0.000

Furthermore, the correlation coefficient between fiscal expenditure and output is only 0.41, which is a strong cis-character. But for the related variables of monetary policy, it has a weak correlation with output volatility, which shows that the impact of monetary regulation on output is more moderate. Further comparison of China's economic characteristics in the process of fluctuations, the primary industry, the secondary industry and the tertiary industry in accordance with their own characteristics of the development of business management behavior characteristics of correlation mining results for comparative analysis, to judge the state characteristics of the entire feature fluctuations and mining results, the specific results are as follows. Tools Options Options Page (Figs. 6 and 7).

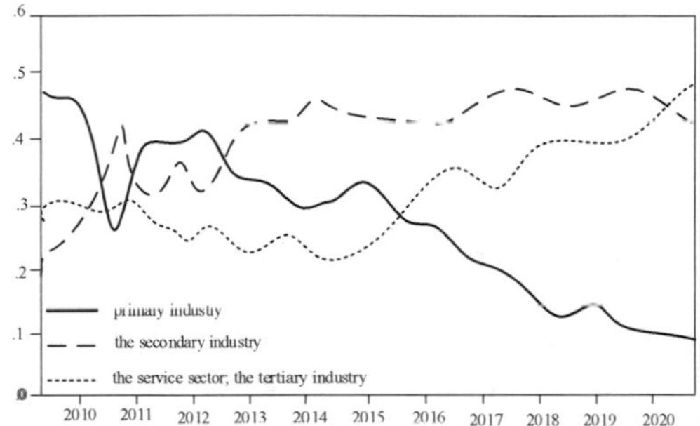

Fig. 6. Detection results of dynamic relational mining of business administration behavior under traditional methods

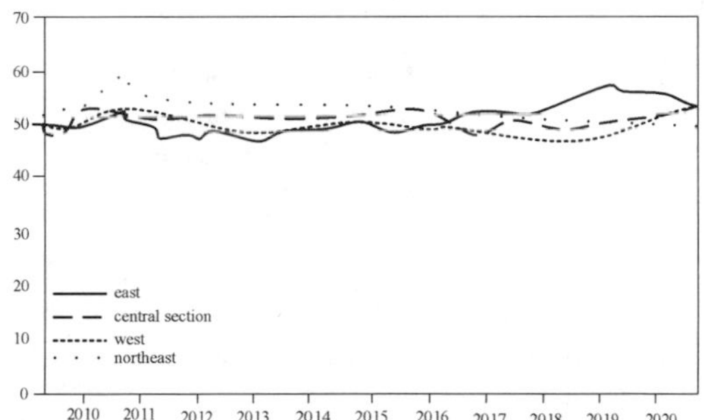

Fig. 7. Dynamic mining detection results of business administration behavior association under the method presented

Compared with the traditional methods, the algorithm based on complex social information network has higher accuracy and stability in the practical application, which fully meets the research requirements.

4 Closing Remarks

In the current rapid economic development situation, there are many problems in the business management activities of our country. Therefore, it is necessary for the managers of this department to put forward an algorithm for mining the relevance of business management behaviors based on complex social information networks, so as to update their management concepts and means in a timely manner, so as to make their specific management behaviors better adapt to the needs of the current economic development.By collecting the characteristics of enterprise management behavior, the trend is analyzed, and the characteristic components are estimated accordingly. The association rule mining algorithm is used to statistically analyze the distribution of high-frequency keywords and the probability distribution of unexpected events in the process of enterprise management behavior, and the association rule mining model is constructed in combination with social information network. However, the method designed in this paper still has some shortcomings, which need to be further adjusted in practical application to make the method more applicable.

References

1. Eyel, C.Ş, Durmaz, İB.V.: The effect of emotional capital on individual innovativeness: a research on Bahcesehir University business administration undergraduate students. Procedia Comput. Sci. **158**, 680–687 (2019)
2. Salim, A., George, J.: General well being and vitamin D practices: a study among female management students in Saudi Arabia. J. Dyn. Control Syst. **11**(1), 3154–3163 (2019)
3. Gao, Y.: Educational resource information sharing algorithm based on big data association mining and quasi-linear regression analysis. Int. J. Contin. Eng. Educ. Life-long Learn. **29**(4), 336–348 (2019)
4. Fu, W., Liu, S., Srivastava, G.: Optimization of big data scheduling in social networks. Entropy **21**(9), 902 (2019)
5. Mou, N., Wang, H., Zhang, H., Xin, F.: Association rule mining method based on the similarity metric of tuple-relation in indoor environment. IEEE Access **8**, 52041–52051 (2020)
6. Czibula, G., Mihai, A., Crivei, L.M.: S PRAR: a novel relational association rule mining classification model applied for academic performance prediction. Procedia Comput. Sci. **159**(10), 20–29 (2019)
7. Liu, S., Lu, M., Li, H., et al.: Prediction of gene expression patterns with generalized linear regression model. Front. Genet. **10**, 120 (2019)
8. Pang, H., Wang, B.: Privacy-preserving association rule mining using homomorphic encryption in a multikey environment. IEEE Syst. J. **15**, 3131–3141 (2021)
9. Khedr, A.M., Aghbari, Z.A., Ali, A.A., et al.: An efficient association rule mining from distributed medical databases for predicting heart diseases. IEEE Access **9**(10), 15320–15333 (2021)
10. Liu, S., Liu, D., Srivastava, G., Połap, D., Woźniak, M.: Overview and methods of correlation filter algorithms in object tracking. Complex Intell. Syst. **7**(4), 1895–1917 (2020)
11. Zhang, C., Xue, X., Zhao, Y., Zhang, X., Li, T.: An improved association rule mining-based method for revealing operational problems of building heating, ventilation and air conditioning (HVAC) systems. Appl. Energy **253**, 113492 (2019). https://doi.org/10.1016/j.apenergy.2019.113492
12. Nomura, K., Shiraishi, Y., Mohri, M., Morii, M.: Secure association rule mining on vertically partitioned data using private-set intersection. IEEE Access **8**, 144458–144467 (2020)
13. Apiletti, D., Pastor, E.: Correlating espresso quality with coffee-machine parameters by means of association rule mining. Electronics **9**(1), 100 (2020)

Classification Method of Regional Differentiation Characteristics of Enterprise Management

Zhao-xi Chen(✉)

Nanchang JiaoTong Institute, Nanchang 330100, China
ahga84@tom.com

Abstract. In the face of the risks and challenges brought by the downward pressure and regional differentiation, it is necessary to improve the enterprise management system, promote the reform of the financial system, create institutional conditions, and enhance the endurance of local finance and economy in order to prevent and resolve major risks. This paper puts forward the classification method of regional differentiation characteristics of enterprise management. Starting from the industrial and commercial management system, this paper further analyzes the characteristics of the rising trend of regional differentiation and the deep-seated causes of risk classification, and puts forward the general idea of preventing and resolving risks. This paper puts forward the corresponding countermeasures and suggestions from the aspects of improving the industrial and commercial management system, promoting the integration of fairness and efficiency, adjusting the direction of industrial and commercial management, and accelerating the reform of financial system.

Keywords: Social information background · Business administration · Regional differentiation · Feature classification

1 Introduction

Enterprise management is the process of obtaining and using resources in order to achieve the goal of market economy organization and development. Its purpose is to ensure that the enterprise economy and industrial structure can effectively adapt to the changes of the external environment, and measure the employees' work behavior according to a series of business objectives, so as to coordinate and correct the differences between the two, so as to accurately analyze and guide the regional differentiation of business administration, and understand and grasp the behavior characteristics of the controlled objects [1]. Differentiating different objects for management is a fundamental issue related to the effectiveness of management control.

Based on the background of the information society, explore the determinants and consequences of the regional differentiation of enterprise management. Firstly, the differentiation of different regions in enterprise management is investigated. On this basis,

S. Liu and X. Ma (Eds.): ADHIP 2021, LNICST 416, pp. 229–241, 2022.
https://doi.org/10.1007/978-3-030-94551-0_19

the evaluation algorithm of regional differentiation situation characteristics of enterprise management is optimized to realize the classification of regional differentiation characteristics. By investigating the characteristics of the current regional differentiation of enterprise management, using values, attitudes, and motivations as the behavioral characteristics of the division variables, empirical research on the behavioral characteristics of different groups of enterprise management in enterprise enterprise management control [2]. Through cluster analysis, the business administration situation characteristics are divided into collectivism tendency type, rule compliance type, self value realization type and environmental cognition type. The research shows that each category presents the statistical characteristics of differentiation. Furthermore, the relationship between different types of management performance and the control links such as the moderation of target difficulty, the timeliness of information feedback and the fairness of performance appraisal is investigated.

2 Classification of Regional Differentiation of Business Administration

2.1 Investigation on Regional Differentiation of Business Administration

Enterprise management has played an important role in promoting the equalization of regional basic public services. A survey of the changes in the regional differentiation of enterprise management in recent years found that China's enterprise management regional differentiation has new characteristics: From the level of economic development, economic growth and other quantitative indicators, the differentiation of business administration among different regions converges, but the differentiation characteristics within different regions are significant, and the differentiation characteristics are prominent [3]. From the perspective of investment efficiency of business administration, number of patents, R & D investment of industrial enterprises and other quality indicators, the regional differentiation trend has obvious characteristics of development quality differentiation, and the advantages of development quality indicators in developed regions are obvious. From the perspective of regional differentiation situation indicators of business administration, the regional differentiation of socialization indicators is obvious. Regional differentiation will eventually be passed on to financial indicators. The regional differentiation of business administration has brought income changes, and the differentiation has affected the differentiation of some key areas of business administration. The difference of financial self-sufficiency rate is significant, and the contradiction between revenue and expenditure at the grass-roots level is prominent [4]. In order to cope with the imbalance of regional development, business administration has achieved certain results in reducing the financial differences between regions and promoting the equalization of basic public services. The combination of economic downturn and tax reduction and fee reduction leads to the decline of local fiscal revenue growth, which makes the local dependence on business administration continue to rise. The overall economic downturn makes the developed regions, as the main source of business administration, also face the trend of growth decline. Therefore, the sustainability of business administration growth will face certain challenges [5]. Based on this, this paper investigates the variation trend of regional increment of business administration. The specific results are as Fig. 1:

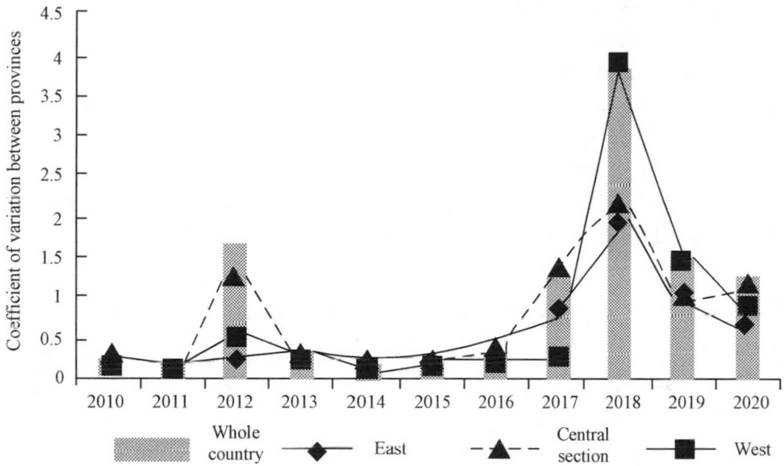

Fig. 1. Variation trend of regional increment of business administration

In the business administration industry, good financial operation is the basis of the orderly and healthy operation of business administration, and the differentiation of business administration will be reflected through financial differentiation to a certain extent. It is found that there are great differences in the standardization of the use and management of business administration funds in different places, which reflects the regional financial differentiation and tight balance pressure. It is found that the gap of per capita general public budget expenditure among provinces within the region shows different trends. In recent years, the gap of business administration situation has expanded, while

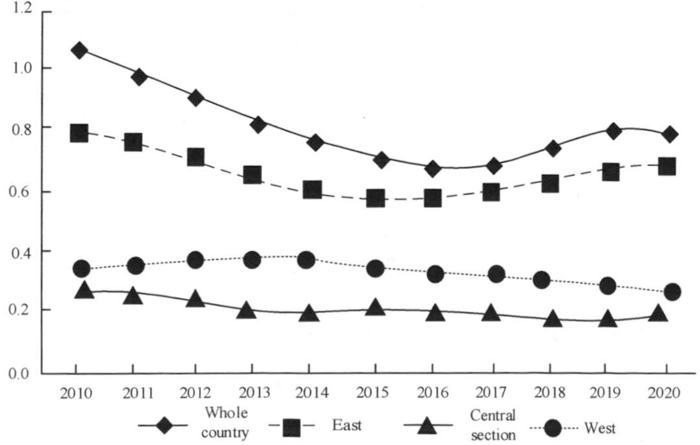

Fig. 2. Changes of business administration situation characteristics

the gap in the central region has decreased steadily. Based on this, this paper further draws the changes of business administration situation characteristics, as shown in the Fig. 2:

As shown in the figure, the regional differentiation in operation has expanded from the traditional gap between the East, the West and the east to the gap between enterprises, and the differentiation characteristics have changed from quantity to quality, reflecting that China's economic development is still in the period of climbing over the threshold and needs to cope with many unpredictable new challenges. From the perspective of development process, local governments are actively working hard to change the mode and adjust the structure, and strive to maintain the stable operation of the economy. The resilience of economic development is enhanced, and the risks are generally controllable.

2.2 Optimization of Evaluation Algorithm for Regional Differentiation Situation Characteristics of Business Administration

In order to measure the performance of enterprise management, it is a sub-item of the overall evaluation that combines the eight-dimensional evaluation method to perform evaluation calculation. The correlation coefficient between the first 8 sub-items and the last comprehensive evaluation sub-item is 0.714. Therefore, the total average of the first eight sub items is used to express the performance of Business Administration [6]. Klirst scale was used to measure the suitability of target difficulty and the timeliness of information feedback. The fairness of performance appraisal is measured by Hofstede's scale. This paper uses multiple linear regression to analyze the management performance of different types of business administration in management control. Set "management performance" as the dependent variable y, and set the three control links as explanatory variables X_1, X_2, X_3, which represent "target difficulty suitability", "timeliness of information feedback" and "fairness of performance appraisal". The multivariate linear model is:

$$Y = a_1X_1 + a_2X_2 + a_3X_3 + b \tag{1}$$

Where a_1, a_2 and a_3 are coefficients and b is constant. The utility function is assumed to be Coriolis function and CES function:

$$\mathrm{U} = \mathrm{C}_{\mathrm{m}}^{\mu}\mathrm{C}_{\mathrm{A}}^{1-\mu}, \quad \mathrm{C_M} = \left[\int_{\mathrm{i}}^{\mathrm{n+n^*}} \mathrm{c_i^p d_i}\right]^{1/\rho}; \quad 0 < \mu < 1, 0 < \mathrm{p} < 1 \tag{2}$$

Through the budget constraint, we can get the following conclusions:

$$\mathrm{P_A C_A} + \int_{\mathrm{i}=0}^{\mathrm{n+n^*}} \mathrm{P_i c_i d_i} = \mathrm{E} \tag{3}$$

The optimal constraint conditions are further derived:

$$C_{\mathrm{A}} = (1-\mu)\mathrm{E}/\mathrm{P_A}; \quad C_{\mathrm{j}} = \mu \mathrm{E}_{\mathrm{P_j}}^{-\sigma}/\mathrm{P}_{\mathrm{M}}^{1-\sigma}\mathrm{E}_{\mathrm{P_j}}^{-\sigma} \tag{4}$$

If the profit function of the product manufacturer is:

$$\pi = px - (r + a_M wL_x) \tag{5}$$

From this, we can get the share of regional expenditure:

$$S_E = (1 - b)S_L + bS_k \tag{6}$$

$$S_L = \frac{L}{r, w} \tag{7}$$

$$S_k = \frac{K}{K^w} \tag{8}$$

Furthermore, the important long-term equilibrium inter regional factor flow equation is obtained as follows:

$$\mathrm{s_n} = \left(\mathrm{r} - \mathrm{r}^*\right)\mathrm{S_n}(1 - \mathrm{S_n}) \tag{9}$$

Based on the research results of regional differences in business administration, the Gini coefficient formula reflecting the development history and current situation of regional business administration is as follows:

$$\mathrm{GiNi} = \frac{2}{\mathrm{n^2 u_y}} \sum_{\mathrm{i}} \mathrm{iy_i} - \frac{\mathrm{n}+1}{\mathrm{n}} \tag{10}$$

The results of multiple linear regression analysis of this data are shown in the Table 1.

Table 1. Regression results of target difficulty, information feedback and fairness of performance appraisal on management performance

Independent variable	Category 1 (environmental cognitive)	Category 2 (collectivist tendency)	Category 3 (rule compliant)	Category 4 (self value realization)
Constant	1.485** (2.168)	2.774*** (5.357)	1.119 (1.595)	2.417*** (4.331)
X_1	−0.201 (−0.183)	0.039 (0.147)	−0.079 (−0.549)	0.051 (0.481)
X_2	0.302** (2.000)	0.147 (1.337)	0.273* (1.951)	0.202** (2.018)
X_3	0.550*** (3.403)	0.183 (1.493)	0.518*** (3.678)	0.233* (1.722)
R2	0.425	0.115	0.334	0.191
F test	8.114***	3.611**	6.675***	4.013***

Business administration includes people's social perception, personality, attitude, values, needs, motivation and other issues. Human behavior is determined by the interaction between internal needs and surrounding environment. Behavior is a function and result of environmental interaction. From the perspective of management, the purpose of psychology is to induce the behavior needed by the organization through the study of psychology [7]. At present, there are great differences in the management status, management level and management performance of Chinese enterprises, and the needs, psychology and behavior of employees are also very different. To grasp the behavior characteristics, we must consider the variables closely related to the control activities.

2.3 The Realization of the Classification of the Characteristics of Regional Differentiation in Business Administration

The classification of regional differentiation situation characteristics of business administration can compare the nature of individual things, and classify Business Administration with similar nature or greater similarity into one category, with great differences among different categories of Business Administration [8]. After defining the factor composition and connotation of management behavior characteristics, this paper classifies business administration by cluster analysis, taking behavior characteristics factors as classification variables. There are many clustering methods, the most commonly used is hierarchical clustering method, which can be used to complete the system clustering analysis. The hierarchical clustering method is used to determine the optimal number of groups. The extracted factors are taken as hierarchical variables, and the most similar objects are combined. The samples are grouped by successive aggregation. Finally, the K-means clustering method is used to merge the sample values into various types [9–11]. The specific enterprise management regional differentiation situation motivation factor component matrix is shown in the following Table 2:

Table 2. Motivation factor matrix of regional differentiation of Business Administration

	Factor 1	Factor 2
Completed the targets to avoid being criticized or fined	0.899	
Finish all the targets in order to get a reward	0.812	
Complete the indicators in order to show my ability		0.809
Complete the indicators in order to work in the future		0.864

In hierarchical clustering method, the optimal number of clusters is generally determined according to the variation characteristics of clustering coefficient. If the increment of clustering coefficient rises suddenly, the optimal number of clusters can be obtained. Hierarchical clustering is carried out by the inter group method, and the change of agglomeration coefficient is shown in table [10]. According to the table, when the samples are clustered into 4 groups, the increment of distance is the largest, so it is more appropriate to divide the samples into 4 groups. The changes of hierarchical clustering coefficients are shown in the following Table 3:

Table 3. Change table of clustering coefficient

Number of clusters	Coefficient	Variation of coefficient	Change increment of coefficient
1	234.010	0	0
2	293.352	59.352	20.23%
3	299.616	6.264	9.87%
4	332.433	32.817	49.7%
5	354.464	22.031	6.22%
6	383.592	29.128	7.59%
7	399.742	16.15	4.04%

Through the classification and analysis of the characteristics of regional differentiation of business administration, the characteristics of business administration can be divided into five categories, and the basic statistical characteristics of different types of business administration are compared. For "target difficulty" and "information feedback", the scale is used to measure; For "fairness of performance appraisal", the scale is used to measure. Using multiple linear regression to analyze the management performance of different groups in the management control, the reform under the current industrial and commercial management structure can only adjust the regional industrial and commercial management on the basis of the stock. The establishment of classified and graded Business Administration of common financial powers can be based on the division of central and local common financial powers and expenditure responsibilities in the field of basic public services, as well as the actual situation of the cost and implementation of basic public services. The incremental adjustment will help to break the current interest pattern of various regions in business administration, better solve the problems existing in the current business administration, and further promote the fairness of business administration and the equalization of basic public services among regions. Taking full account of the unbalanced development of business administration and the great differences in the cost and financial resources of basic public services in different parts of China, the central government should bear different expenditure responsibilities, reflecting the preference to difficult areas. At the same time, the national basic standards for classification and classification are also set. In the future, combined with the practice of the reform of the division of central and local common financial power and expenditure responsibility in the field of basic public services, it is necessary to further standardize the sharing mode of common financial power and expenditure responsibility in the field of basic public services, and reasonably formulate the national basic standards for the guarantee of basic public services.

3 Analysis of Experimental Results

Take the Cronbachα coefficient as the reliability test index. According to data analysis, the reliability test results show that the Cronbachα coefficient of each variable is above 0.78. The mature scale 1 in psychology and management control research is used to set the measurement items of each indicator. Therefore, the questionnaire used has considerable content validity. SPSS statistical analysis software was used to analyze the behavior characteristic variables, and the factors of cluster analysis were determined by maximum variance orthogonal rotation. The kmo values of all factors are greater than 0.6 except that some factors are slightly less than 0.6, which indicates that the choice of variables is more in line with the requirements. The extracted factors are highly consistent with the contents of the corresponding questionnaire items, indicating that the questionnaire has high structural validity (Table 4).

After analysis, three factors with eigenvalues greater than 1 were extracted, and the cumulative degree of interpretation reached 83.157%. Factor 1 is composed of two elements, which involve the priority of collective goals over individual goals and the tendency of cooperation, which is called collectivism. Factor 2 is composed of two parts,

Table 4. Factor analysis results of business administration behavior characteristics

Variable	Subvariable	Survey items	Factor 1	Reliability /%	Cumulative Explanation /%
Cultural values	Collectivism	Members of the work team should be aware that sometimes they have to sacrifice for the overall interests of the work team	0.925	0.531	42.494
	Individualism	If I have a choice, I will choose to finish the work independently, not those who have to cooperate with colleagues			62.610
	Power distance	The best way to avoid mistakes is to follow your boss's instructions			83.157

(*continued*)

Table 4. (*continued*)

Variable	Subvariable	Survey items	Factor 1	Reliability /%	Cumulative Explanation /%
		I am not interested in the work can immediately do			
Attitude	Sense of organizational justice	Compared with colleagues in the same job or position, my salary is reasonable and fair	0.848	0.803	24.345

which involve the promotion of independent work and the pursuit of personal work efficiency, which is called individualism. Factor 3 consists of two terms, involving right distance, which is called right distance. A total of 22 measurement items were designed, involving cognitive factors, emotional factors and intention factors of management control. After factor analysis, three factors with eigenvalues greater than 1 were extracted, and the cumulative degree of explanation reached 69.027%. Factor 1 is composed of four items, mainly involving distributive justice and procedural justice, which is called organizational justice; Factor 2 is composed of one item, mainly involving the response to control measures, which is called compliance. Factor 3 is composed of four items, which involves the cognitive factor of management control, namely environmental cognition. Measured from two perspectives: Utilitarian needs and expressive needs. Through the analysis, all four items are reserved. Two factors with eigenvalues greater than 1 are extracted, and the cumulative interpretation rate is 75.729%. In accordance with the theoretical analysis, these two factors are called utilitarianism demand and expressionism demand respectively. By cluster analysis, the samples were divided into four categories: 73 cases (18.5%) in the first category, 127 cases (32%) in the second category, 87 cases (22%) in the third category and 109 cases (27.5%) in the fourth category. The results are shown in the Table 5.

Table 5. Cluster analysis results

Category	Frequency	Proportion /%	Effective ratio /%	Cumulative percentage /%
Category 1	73	18.5	18.5	18.0
Category 2	127	32.0	32.0	50.5

(*continued*)

Table 5. (*continued*)

Category	Frequency	Proportion /%	Effective ratio /%	Cumulative percentage /%
Category 3	87	22.0	22.0	72.5
Category 4	109	27.5	27.5	100.0
Total	396	100.0	100.0	

After confirming that the characteristics of business administration situation are divided into four categories, the following coefficient table is obtained. This table shows the coefficient relationship of each group of eight features. The smaller the coefficient,

Table 6. Nomenclature of business administration categories

Category	Category 1	Category 2	Category 3	Category 4
Collectivism	−0.848 68	−0.462 11	−0.135 46	0.198 48
Individualism	0.561 64	0.619 89	0.218 89	−0.449 19
Power distance	−0.768 32	−0.543 61	−0.386 74	0.137 63
Utilitarian demand	0.286 75	0.729 85	0.218 02	−0.867 76
Expressionism demand	−1.213 55	0.369 31	0.023 59	−0.367 77
Compliance	−1.139 37	−0.353 01	−0.192 53	0.209 74
Sense of organizational justice	−0.762 47	−0.129 36	0.677 69	0.121 31
Environmental cognition	−1.273 20	0.720 74	−0.282 80	0.428 65
Practical significance of category	The work attitude will be influenced by the environment and others, and has certain expressive needs and collectivism tendency	Collectivism tends to be strong, with a certain degree of conformity, a strong perception of the environment, and a weak utilitarian demand	They have a strong sense of obedience, strong compliance with rules, easy to be influenced by others, and low self orientation	Pursuit of utilitarianism, strong self-directed and performance
Category naming	Environmental cognitive	Collectivist tendency	Rule compliant	Self value realization

the higher the correlation. 0 means neutral. According to the actual meaning of each category, the category is named as Table 6:

The first type of enterprise management work attitude will be affected by the environment and others, with a certain performance needs and collectivism. Among the attitude options, the ones with high consistency were as follows: "Through continuous publicity, training and education, I gradually realize the importance of abiding by the rules and regulations", "I abide by the rules and regulations of the unit, because everyone can abide by them", "the serious behavior of leaders will affect my behavior", which is easily affected by the environment and others, so it is called environmental cognition. The second type of business administration has higher collectivism tendency, certain conformity, lower self orientation, stronger environmental perception and weaker utilitarian demand. This kind of business administration accounts for the largest proportion of the respondents, accounting for about 30% of the total number, which is called collectivism tendency type. The third kind of managers obey the leadership, organization or group, their own working environment and working principles. They are easily influenced by others and have a certain tendency of collectivism. They agreed that "the best way to avoid mistakes is to follow the boss's instructions" and "I can do the work I'm not interested in immediately". So they are called rule obedience. This may be related to the pursuit of the golden mean and the cultural consciousness of "face" and "relationship", which makes these groups show a high degree of obedience. The fourth type of business administration pursues utilitarianism, is self-directed and expressive, and is not easy to be influenced by others. Therefore, they are called self-worth actualization type. Compared with the other three groups, this group has the characteristics of younger age, higher education level and higher management level.

At present, regional differentiation presents new characteristics. Economic growth in the north and West is not as strong as in the South and East. In fact, regional differentiation is the differentiation of regional efficiency, while business administration embodies regional equity. Although the industrial and commercial administration policy has been increasing to promote fairness, it has fallen into the "paradox" of regional fairness and efficiency. The deep-seated reason is that there is no virtuous circle between the industrial and commercial management funds and the aftereffect of local economic development. The relatively low performance of business administration funds reflects that the central and local financial systems have failed to achieve "incentive compatibility", which leads to the adverse selection of local governments and the increase of dependence on the central business administration. With the aggravation of regional differences, the aggravation of inequality between regions and the expansion of the gap, the demand for business administration will be increasing until it is unsustainable. This will eventually lead to the increase of the uncertainty of the development of the whole country, the decline of the overall efficiency, and then turn into a public risk or even crisis. Therefore, we need to change our way of thinking from certainty to uncertainty, inject risk rationality into the design of financial system and enterprise management system, break the "risk cauldron" under the "one body, two wings" reform framework, and design enterprise management system according to the principle of risk decision-making, risk sharing and risk matching. In essence, it is to realize the effective development of regional equity and efficiency.

4 Conclusion

Industrial and commercial management is an important way of economic management, but also on the new economic background of the market economy mode of operation of a supplement. With the continuous development of management control system, many researchers pay attention to the research of enterprise management behavior in the organization. There is less research on the characteristics of behavior, especially in the aspect of behavior classification. Behavior research in management control has become a prospective research direction in the field of management control. To understand and grasp the characteristics of human behavior in management control is the basis of effective behavior management. Taking values, attitude and motivation as behavioral variables, this paper makes an empirical study on the behavioral characteristics of business administration in enterprise management control. The research shows that the behavior of business administration in China's enterprise management control can be divided into different types, and each type of personnel shows different statistical characteristics and behavioral characteristics.

At present, the economic background has begun to develop into a new normal of the economy. How to effectively exercise the functions of business administration is also worth considering. This is also a very important issue in the current business administration. In the future, the research will focus on the functions of business administration in order to provide a positive reference for business administration.

References

1. Cafer Şafak, E., İsmet Burçak, V.D.: The effect of emotional capital on individual innovativeness: a research on Bahcesehir University business administration undergraduate students. Procedia Comput. Sci. **158**(10), 680–687 (2019)
2. Salim, A., George, J.: General well being and vitamin D practices: a study among female management students in Saudi Arabia. J. Dyn. Control Syst. **11**(1), 3154–3163 (2019)
3. Ye, X., Zhang, Z., Xu, C.Y., et al.: Attribution analysis on regional differentiation of water resources variation in the Yangtze River basin under the context of global warming. Water **12**(6), 1809 (2020)
4. Philip, S.Y., Kew, S.F., Wiel, K.V.D., et al.: Regional differentiation in climate change induced drought trends in The Netherlands. Environ. Res. Lett. **15**(9), 094081 (2020)
5. Chen, F., Fu, B., Xia, J., et al.: Major advances in studies of the physical geography and living environment of China during the past 70 years and future prospects. Sci. Chin. Earth Sci. **62**(11), 1665–1701 (2019)
6. Schilling, K.E., Streeter, M.T., Seeman, A., et al.: Total phosphorus export from Iowa agricultural watersheds: quantifying the scope and scale of a regional condition. J. Hydrol. **581**(5), 124397 (2019)
7. Jiang, X., Zheng, P., Cao, L., et al.: Effects of long-term floodplain disconnection on multiple facets of lake fish biodiversity: decline of alpha diversity leads to a regional differentiation through time. Sci. Total Environ. **763**(10), 144177 (2020)
8. He, Y., Wang, Y., Chen, X.: Spatial patterns and regional differences of inequality in water resources exploitation in China. J. Clean. Prod. **227**(1), 835–848 (2019)
9. Liu, S., Bai, W., Zeng, N., et al.: A fast fractal based compression for MRI images. IEEE Access **7**, 62412–62420 (2019)

10. Jiang, X., Ding, C., Brosse, S., et al.: Local rise of phylogenetic diversity due to invasions and extirpations leads to a regional phylogenetic homogenization of fish fauna from Chinese isolated plateau lakes. Ecol. Ind. **101**(6), 388–398 (2019)
11. Villanueva-Rey, P., Vazquez-Rowe, I., Quinteiro, P., et al.: Regionalizing eco-toxicity characterization factors for copper soil emissions considering edaphic information for Northern Spain and Portuguese vineyards. Sci. Total Environ. **686**(10), 986–994 (2019)

Research on Price Stickiness of Consumer Goods Based on Real-Time Social Information Flow

Hai-ying Chen[1] and Liang Xu[2](✉)

[1] Department of Public Courses,
Tianmen Vocational College, Tianmen 431700, China
[2] Department of Public Basic Courses,
Wuhan Institute of Design and Sciences, Wuhan 430205, China

Abstract. Due to the lack of consumer goods characteristics in the construction of the evaluation index of consumer goods price stickiness level, the loss aversion rate of consumer goods decreases. Therefore, based on the real-time social information flow of consumer goods price stickiness research. Through the real-time social information flow to adjust the price stickiness of consumer goods, through the study of the existence of price stickiness, we analyze the price stickiness data of consumer goods. Based on the characteristics of consumer goods, the index of product price stickiness level is established. The price adjustment frequency in the data period is calculated to determine the price stickiness of consumer goods. The range and absolute value of commodity prices are determined to adjust the price stickiness tendency of consumer goods. The experimental results show that the price stickiness is negatively correlated with the price adjustment cycle, and the effect of price stickiness adjustment based on real-time social information flow is better.

Keywords: Real-time social information flow · Consumer goods · Price stickiness

1 Introduction

Manufacturers can achieve equilibrium in the product market through their flexible pricing strategies [1]. However, in the real market economy, the adjustment of product price is often slow, which makes the price of the manufacturer can not keep up with the rhythm of the market environment, so that the equilibrium state can not be realized immediately. In order to explain this phenomenon, researchers begin to add price rigidity theory into Keynesian macroscopical theory, but there are some defects in the microcosmic explanation. Neo-Keynesian economists began to put forward relative wage theory, menu cost theory and approximation theory, and applied these theories to analyze the causes of wage stickiness, and emphasized the ability of Keynesian theory to analyze price rigidity from a microscopic perspective. Neoclassicism and Neo- Keynesianism have many differences in the study of the impact of price on the overall welfare level of a society. Reference [2] proposes to interpret the complexity of the competition in the consumer

S. Liu and X. Ma (Eds.): ADHIP 2021, LNICST 416, pp. 242–253, 2022.
https://doi.org/10.1007/978-3-030-94551-0_20

goods market – a complex Hilbert PCA analysis. Today's consumer goods market is developing rapidly, and the number of information media and competitive products is increasing significantly. In this environment, the quantitative grasp of heterogeneous interaction between enterprises and customers has aroused the interest of management scientists and economists, which requires the analysis of very high dimensional data. The existing quantitative research methods can not deal with such data without any reliable prior knowledge or strong assumptions. In addition, a complex Hilbert principal component analysis (chpca) method is proposed, and a synchronization network is constructed by Hodge decomposition. Significant motion with time lead/delay can be extracted from the data, and Hodge decomposition is helpful to identify the time structure of correlation. This method is applied to the Japanese beer market data, and reveals the changes of related variables in the selection process of consumers in a variety of products. In addition, significant customer heterogeneity is found by calculating the coordinates of each customer in the space from chpca results. Reference [3] proposed the characteristics and environmental value proposition of fast moving consumer goods reuse model. Fast moving consumer goods (FMCG) refers to the products that are often purchased and consumed in order to meet the continuous needs of consumers. In the online economy, fast moving consumer goods are usually disposable products. Product design limitations, inadequate collection systems and inefficient recycling processes prevent high recovery rates. The purpose of the study is to develop a comprehensive feature reuse model and evaluate its potential environmental value. 92 kinds of reusable products are selected for analysis to determine the reusable system elements. The analysis identified a framework of five reuse models and evaluated them to establish their environmental value compared with traditional disposable and disposable FMCG. Currently, consumers can access five reuse models and choose between exclusive and sequential reuse behaviors. When combined with recycling, recycling mode can make FMCG more effective consumption. Providing the infrastructure needed to support reuse and recycling is the key to successful and sustainable deployment of reuse models. However, according to the theoretical research of the literature, the similarities between them lie in the fact that they both believe that there is correlation between price changes and social welfare level, and that price adjustment plays a vital role in the change of welfare level. Based on the real-time social information flow, this paper proposes a research on the price stickiness of consumer goods. The frequency of price adjustments during the data period is calculated to determine the price stickiness of consumer goods. The range and absolute value of commodity prices determine the adjustment trend of the price stickiness of consumer goods. The proposed price stickiness of consumer goods has a good adjustment effect, which is not only the basis to solve the real macroeconomic problems, but also the theoretical basis for the government to deal with the economic shocks.

2 Price Stickiness Adjustment of Consumer Goods Based on Real-Time Social Information Flow

2.1 Analysis of Consumer Price Stickiness Data

Before studying the price stickiness data of consumer goods, it is necessary to verify the existence of price stickiness. According to the classification of different goods, the prices of some goods change more frequently, and manufacturers of such goods often change their prices, while the prices of some goods change more slowly, or even remain unchanged for a period of time [4]. These goods because of their own special attributes or macroeconomic environment and other reasons, leading to manufacturers will deliberately manage the price of such goods, control its price change rate. At this point, the product price becomes sticky, as shown in Fig. 1.

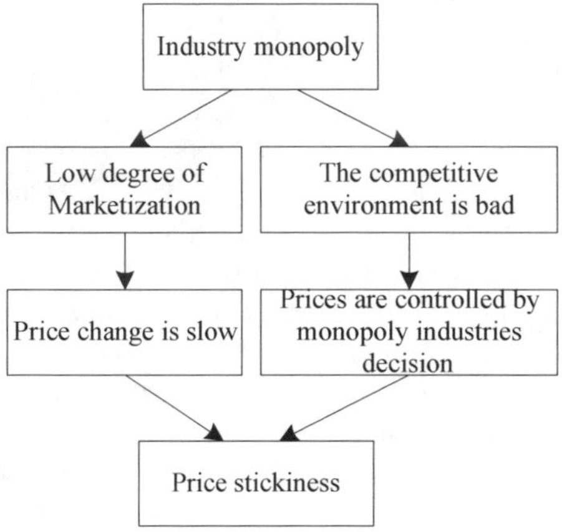

Fig. 1. Generation of price viscosity

Product price stickiness analysis based on data provides more microscopic data in line with the actual macro-economic model, and changes in product prices may occur frequently within a month. Therefore, price behaviors such as sales promotion that occur within a month cannot be taken into account, and when the time dimension of a month is taken as the time dimension, manufacturers are also likely to stop selling or stop production, which results in a relatively large measurement error [5, 6]. On the other hand, the time and money required to analyse such data is significant, which limits the analysis of such data [7]. Next may choose the supermarket to sell the data, the supermarket when sells can leave information and so on commodity name, price and quantity. The cost of purchasing the data is relatively high by paying a certain fee to the supermarket. This price data is accurate and in line with the current market price. However, the data sample only covers some products such as daily necessities, food products, tobacco and alcohol products, etc., but industrial consumer goods are not included. The data are all from a certain supermarket when analyzing the price data. Compared with other data analysis methods, the type of sample commodities is not representative, and cannot accurately reflect the trend of changes in manufacturer prices in the whole market environment. On the other hand, because the data of supermarket is usually weekly, there will be a big time error, and the supermarket will usually carry out many promotions at festivals or celebrations, showing that the price of a certain commodity has dropped sharply for a period of time, while the price of the commodity in the whole market may remain the same as the original price, which will result in a big measurement error. Finally, with the development of Internet sales platform and computer technology, in recent years, scholars gradually choose network data to analyze web page price data automatically by setting up crawler software program. This kind of data usually enters Taobao, Jingdong and other online sales platforms at a certain time to obtain web page price data. Compared with the former two kinds of data, the cost of these data is lower, and the daily regular analysis data can observe the frequency of price change in a shorter period, reducing the late-stage measurement error. Second, the data contains more detailed information, including but not limited to the name, attributes, prices, etc., the most important is that each commodity has a fixed number, which facilitates the price tracking of goods. On the other hand, data samples are also incomplete, and for a small number of industrial consumer goods the price data cannot be accessed on the Web because online sales are not yet widespread [8]. With the development of E-commerce in recent years, the variety of goods sold on the Internet begins to increase greatly, which makes up for the lack of data sample breadth, but also makes the analysis of data more dependent on e-commerce sales. Data analysis needs to master a certain knowledge of computer programs, so that the late analysis work can be carried out smoothly, and data analysis and collection work in batches, unlimited time breadth limit. Thus completes to the resident expense price stickiness data analysis.

2.2 Establishing Evaluation Indexes for Price Stickiness Level of Products

By analyzing the price stickiness data of consumer goods, we get the evidence of establishing the evaluation index of price stickiness level. Evaluation indicators are constructed on the basis of data and are more representative in terms of practice testing [9]. The real time social information flow is used to calculate the price stickiness of various consumer goods in our country, and the price characteristics of consumer goods are discussed. The calculation steps are as follows: First, calculate the price adjustment frequency F_i for each product i sample time period, namely:

$$F_i = X/N \tag{1}$$

In the formula, X represents the times of price adjustment and N represents the specific sample period. Using the above expressions, we can get the price adjustment frequency of each product in the sample period, which is also an important index for the evaluation of product price viscosity. Furthermore, the median frequency of price adjustment of the products under the classification of each industry, including 19 secondary classification products classified according to the Resident Commodity Classification Standards of the National Bureau of Statistics, was calculated. Finally, the weighted median of price adjustment frequency for each category of products is calculated, and the result is used to represent the total price frequency of the product market. The evaluation index of price stickiness level can be constructed. Through the calculation of the product price adjustment frequency, the price viscosity measure is affected by the extreme abnormal value, which leads to the calculation error. The real-time social information flow is used to calculate the price stickiness. Firstly, the price adjustment frequency of each product is calculated, and secondly, the average of the price adjustment frequency of the products in the lower classification is calculated. Finally, the weighted average number of each product is calculated, and the total market price frequency is represented by the result. At the same time, the method is used to calculate the price change cycle in order to serve as the product price viscosity change cycle. The specific formula is as follows:

$$D = -1/\ln(1 - F_i) \tag{2}$$

In the formula, D is the price change cycle, F_i is the price adjustment frequency of the product. It can be seen from the above expression that when F_i is larger, the price change cycle D will be smaller, and the price stickiness level of the product will be weaker. There is a negative correlation between fi and D. In the monopolistic competition market, it is assumed that the product demand of the manufacturer in the t period is:

$$Y_{it} = D(P_{it}/p_t)^{-\theta} \tag{3}$$

In the formula, θ stands for product demand elasticity; P_{it} stands for product price. p_t is the average price of the product cycle, and the total output level of the manufacturer is:

$$Y'_{it} = Y_{it}[\int D^{(\theta-1)/\theta} di]^{\theta/(\theta-1)} \tag{4}$$

In the formula, di is the elasticity coefficient of the product, and the price level is:

$$pt' = [Y'_{it} \int D^{(1-\theta)} di]^{1/(1-\theta)} \tag{5}$$

The firm's actual marginal cost depends on two factors: productivity shock X_{it} and overall output gap MC_{it}. In this case, the actual marginal cost can be expressed as follows:

$$MC_{it} = \delta X_{it}(Y'_{it}/pt')^2 \tag{6}$$

In the formula, X_{it} represents the natural output rate under total factor information, and Y'_{it} represents the actual rigidity. When Y'_{it} is small enough, the output gap will be relatively small. The total demand is determined by the quantity equation, the total demand for nominal money M_t is expressed as follows:

$$M_t = P_t Y_{it} \tag{7}$$

To sum up, the profit function of an enterprise is:

$$\Delta M_t = \mu(1 - \rho) + \rho \Delta M_{t-1} + \varepsilon_t - t^2 \tag{8}$$

In this case, μ is the exogenous variable, ρ is the rigidity factor, ΔM_{t-1} is the profit amount of the last cycle, ε_t is the price coefficient of the cycle product, t is the generation time of the cycle profit, In the steady state, with the increase of productivity. However, according to the above analysis, the calculation results under the real-time social information flow will be affected by extreme outliers, so when the update probability is t, the company does not update the information of new macroeconomic conditions. In the case of 2 = t, the company always updates new information in time, so the model can be simplified to full information model. At the same time, changing prices from previous levels can be costly for a manufacturer. Specifically, if the manufacturer realizes the price change in the cycle, it must pay the cost, which makes the pricing decision of the manufacturer depend on the government.

2.3 Estimation of Consumer Price Stickiness Based on Real-Time Social Information Flow

This paper uses real-time social information flow to calculate the price stickiness of all kinds of consumer goods in China, and discusses the price characteristics of consumer goods in China through the calculation results. First, calculate the price adjustment frequency F_i of each product i sample period

$$F_i = X/N \tag{9}$$

In the formula, X represents the number of product price adjustment, and N represents the specific sample period. Using the above expression, the price adjustment frequency of each product in the sample period can be obtained, which is also an important index for evaluating the price stickiness of products. In addition, we continue to work out the median price adjustment frequency of each industry's sub level classified products, including 19 secondary classified products according to the National Bureau of statistics resident commodity classification standard. Finally, the weighted median of the price adjustment frequency of each large category of products is calculated, and the calculation result is used to represent the overall price frequency of the product market [10, 11].

2.4 Adjusting the Stickiness of Consumer Goods

According to the stickiness of the price of consumer goods, the price propensity shall be adjusted, that is, the direction of price change shall be adjusted, and the commodity prices may be divided into upward and downward tendencies according to the price risc and downward tendencies of the commodity prices as a whole [12]. For the calculation of commodity price propensity under the social information flow, the average price adjustment range of each commodity is obtained by first calculating the sum of each commodity price adjustment range and its absolute value in the sample period, and then dividing the sum of ranges by the total price adjustment frequency of the commodity in the sample period. Then, the prices of the secondary commodities are summed and averaged. Finally, the median adjustment range of the secondary commodities is used to replace the adjustment range of the primary commodities. Describe the ratio of the times of commodity price adjustment and reduction to the total times of commodity price change. According to the formula, the average price adjustment propensity of every second class is obtained, and the median value is obtained as the price propensity of the first class. Suppose that the price change index is I_{it}, I_{it}^+ represents the rising frequency of commodity price, I_{it}^- represents the falling frequency of commodity price, and the calculation formula of price adjustment tendency is as follows:

$$\begin{aligned} g_s^- &= \sum_t w_{it} I_{it}^- / \sum_t w_{it}(I_{it}^+ + I_{it}^-) \\ g_s^+ &= \sum_t w_{it} I_{it}^+ / \sum_t w_{it}(I_{it}^+ + I_{it}^-) \end{aligned} \tag{10}$$

In the formula, g_s^+ represents the rising tendency of price and g_s^- represents the declining tendency of price. w_{it} is the fluctuation value of cyclical commodity price. The calculation results of commodity price variation range under each classification are shown in Table 1:

Table 1. Sticky periodic changes of price of consumer goods

Category descend	Increase (%)	Descend (%)	Category	Increase (%)	Descend (%)
Grain	52.83	47.17	Cloth	61.03	38.07
Grease	73.93	26.06	Tobacco	54.92	45.08
Meat fowl	51.54	48.46	Liquor	65.23	34.77
Egg	41.82	58.18	Durable goods	65.67	34.33
Aquatic products	43.48	56.62	Consumable goods	66.10	33.90
Fresh vegetables	56.63	43.37	Service	53.24	46.67
Medical treatment	51.28	26.34	Transportation	73.66	70.63
Education	92.86	7.14	Communications	80.06	25.76
Others	68.62	25.76	Hydropower	74.24	16.67

From Table 1, it can see that there is a significant correlation between the price change under the social information flow and the macroeconomic data, which shows that the pricing behavior of firms will be affected by the social information flow, that is, the state-related pricing model exists under the special economic conditions of our country. On the one hand, there is a positive correlation between the change of price and each variable, which indicates that the inflation rate will have a positive effect on the change of the price of consumer goods to some extent. On the other hand, there is a negative correlation with the price change, that is, manufacturers reduce their own price adjustment speed when the price rises, thus increasing market price stickiness. When implementing the tightening policy, with the increase of the reserve requirement of the bank, the manufacturers slow down the adjustment of the price of their own products, and the stickiness of the product market increases. On the contrary, when the expansionary policy is implemented, the reserve that banks must pay will be reduced. Generally speaking, the prices of the three kinds of consumer goods will increase more than decrease, and the price of the three kinds of consumer goods will increase more than 50%, which indicates that the prices of the products in our country will increase. In the classification of the next level, the products of different attributes and categories show obvious differences. Under the classification of food commodities, the highest price rise tendency belongs to the commodities of grain, and its price tendency is 73.93%. Under the classification of service goods, the highest price increase tendency is education, and the price tendency is as high as 92.86%, but the price increase tendency of communication goods is only 29.37%. Under the classification of industrial consumer goods, the price increase tendency of daily consumer goods is the highest, except for clothing and durable consumer goods, the price increase tendency of all other industrial consumer goods is higher than 60%. So far, the price stickiness of consumer goods based on real-time social information flow is adjusted.

3 Experimental Analysis

Through the contrast experiment, we compare the stickiness adjustment of consumer goods price based on real-time social information flow with the loss aversion rate of traditional methods 1 and 2.

For the study on the loss aversion rate of consumer goods, the selection of the data set is the same as the data used for the price stickiness measurement of consumer goods, and the monitoring data on commodity prices of residents in large and medium -sized cities nationwide issued by the Price Monitoring Center of the National Development and Reform Commission are used. The time reference dimension of the data set is measured by month, and a total of nine types of consumer goods are included, and the classification of these nine types of consumer goods is conducted by the method of the National Bureau of Statistics for the classification of consumer goods. According to the secondary classification standards, there are six types of food commodities and two types of service commodities, and the price data of food and service commodities are the average prices of commodities sold in the market, while the price of other categories of commodities is the representative data based on the expenses actually paid by consumers for enjoying the services. A regression model shall be established to conduct regression analysis on the factors affecting price changes. Suppose that the time of examining the micro data such as the frequency of price change is t, the correlation factor of inflation rate is C, the dummy variable of time factor in the equation is ε_t, β_i is the frequency of price change, and A_i is the amount of spillover of consumer goods, taking this as a representative, the paper studies the time-varying law of price change and calculates the loss aversion rate g_t of consumer goods as follows:

$$g_t = C + \sum\nolimits_{i=1}^{12} \beta_i A_i + \varepsilon_t \tag{11}$$

Secondly, the equation will make a distinction between the current period and the lag period. Finally, the conclusion is drawn by analyzing the regression coefficient of the loss aversion rate of consumer goods, and the correlation between price changes and various factors and the degree of correlation are discussed.

The calculation results of price adjustment frequency and period of three major categories of consumer goods and their sub categories are shown in Table 2.

According to the data in Table 2, we can get the index of stickiness level of product price through each period without frictional economy, so that firms can set product price and maximize profit. While obtaining and processing effective information, the manufacturer has to pay a certain commodity cost, that is, the cost of setting up the floating index of the product, which involves the cost of time and manpower. Infrequent information updates are combined with state-dependent pricing to update indicators with probability. Through the establishment of price stickiness level of consumer goods value indicators to complete the summary of real-time social information flow, thus completing the price stickiness estimation.

Table 2. Evaluation index of product price stickiness

Classification population	Weight	Frequency (%)	Cycle (month)
On the whole	100	32	2.6
Grain	37.3	88	0.5
Grease	4.0	32	2.6
Meat fowl	3.2	88	0.5
Egg	7.1	88	0.5
Aquatic products	0.8	92	0.4
Fresh vegetables	6.3	86	0.4
Fresh fruit	12.0	92	0.5
Tobacco	3.9	92	0.45
Liquor	27.0	90	0.5
Durable consumer goods	2.4	88	1.4
Daily consumables	4.8	92	0.7
Medical care	3.9	76	0.5
Transportation	11.1	88	0.4
Communications	4.8	94	5.2
Education	35.7	16	2.8
Rental and property services	7.9	30	7.8
Hydroelectric fuel service	6.3	12	11.9
Others	1.6	8	1.7

By studying the loss aversion rate of consumer goods, the traditional methods 1 and 2 are compared with the research methods in this paper. The experimental results are shown in Table 3.

Table 3. Comparison of loss aversion rate of consumer goods

Types of consumer goods	Traditional method 1 Loss aversion rate (%)	Traditional method 2 Loss avoidance rate (%)	This paper studies the method loss aversion rate (%)
Grain	13.41	22.06	52.43
Grease	12.95	29.17	43.75
Meat fowl	17.35	14.29	50.29

(*continued*)

Table 3. *(continued)*

Types of consumer goods	Traditional method 1 Loss aversion rate (%)	Traditional method 2 Loss avoidance rate (%)	This paper studies the method loss aversion rate (%)
Egg	13.36	16.31	52.86
Aquatic products	17.80	19.54	56.07
Fresh vegetables	15.95	29.22	47.52
Medical treatment	23.54	17.35	48.12
Education	13.21	24.66	42.01
Others	14.41	20.13	56.68

According to the data in the table, the traditional method 1 in consumer oil category loss avoidance rate was 12.95%, the highest in the medical category loss avoidance rate, 23.54%. Compared with the traditional method 1, the traditional method 2 has a higher rate of loss aversion of various consumer goods, and the lowest rate of loss aversion is 14.29%. Based on the real-time social information flow, the loss avoidance rate of consumer goods is 42.01% and 56.68%, respectively.

The overall price adjustment frequency of consumer goods is calculated to measure the overall cycle length of consumer goods, and the price stickiness and price change cycle of different categories of products show strong differences. In the secondary classification under the same classification, different products also show different price adjustment frequency. But to judge whether the price change has stickiness or not, under the premise of no consistent standard, the more common method is to draw the conclusion of price stickiness through comparison. When using the real-time social information flow to measure the price stickiness, we can get the price stickiness change cycle of various consumer goods, as shown in Fig. 2.

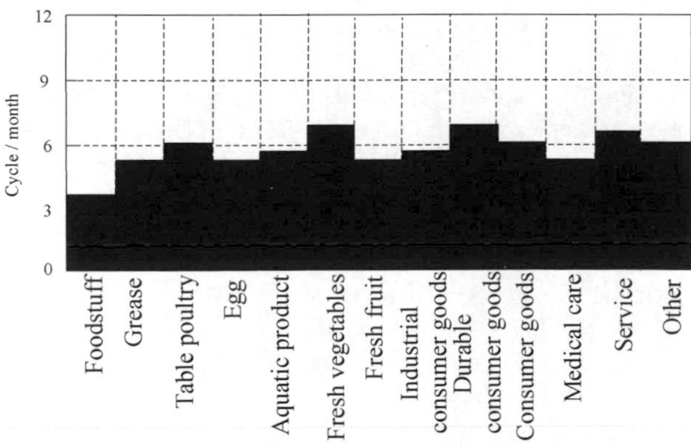

Fig. 2. Price stickiness cycle of consumer goods

From the cyclical results of the real-time social information flow in Fig. 2, we can see that the overall frequency of consumer goods price adjustment cycle implies that there is a monthly consumer goods price adjustment cycle. The result of the three-month price adjustment cycle for food products means that the number of monthly price changes in food accounted for 33.33%. A price adjustment period of 5.9 months for industrial consumables shall mean that the monthly price change of products accounts for 16.9% of the total price change of industrial consumables, among which the price stickiness of grain products is the lowest, and the adjustment period for daily consumables such as tobacco, alcohol, etc. is less than the overall price adjustment period for industrial consumables, and the price adjustment period for medical products is five months, which means that the monthly price change of products accounts for 20% of the total price change of medical services, compared with the price stickiness of services being the lowest, and the other prices are less than the overall price adjustment frequency for medical consumables. According to the cycle calculation results, the total cycle length of daily consumer goods is 6 months, which indicates that it takes 6 months for the consumer goods to realize a round of price change, and the price stickiness is negatively correlated with the price adjustment cycle. Therefore, the price stickiness of food and industrial consumer goods is shorter than that of service goods. By estimating the price stickiness of consumer goods, the adjustment tendency of price stickiness can be determined.

4 Conclusion

This paper summarizes the correlation between the general pricing mode and the state, and provides a reference for the formulation of monetary policy and the prediction of industry development trend. Price stickiness is negatively correlated with price adjustment cycle. The future discussion should gradually turn to empirical research based on market microscopic data. At the same time, we should pay more attention to the non-economic factors, such as cultural tradition, so as to fully and accurately understand the market operation mechanism of our country.

References

1. Gómez-Montoya, R.A., Cano, J.A., Campo, E.A., et al.: Improving cross-docking operations for consumer goods sector using metaheuristics. Bull. Electr. Eng. Inform. **10**(1), 524–532 (2021)
2. Mizuno, M., Aoyama, H., Fujiwara, Y.: Untangling the complexity of market competition in consumer goods—a complex Hilbert PCA analysis. PLoS ONE **16**(2), 24–31 (2021)
3. Muranko, A., Tassell, C., Laan, A.Z.V.D., et al.: Characterisation and environmental value proposition of reuse models for fast-moving consumer goods: reusable packaging and products. Sustainability **13**(5), 2609–2615 (2021)
4. Albizua, A., Bennett, E.M., Larocque, G., et al.: Social networks influence farming practices and agrarian sustainability. PLoS ONE **16**(1), 24–29 (2021)
5. Alassad, M., Hussain, M.N., Agarwal, N.: Comprehensive decomposition optimization method for locating key sets of commenters spreading conspiracy theory in complex social networks. CEJOR **25**(8), 1–28 (2021)

6. Ali, F., Ali, A., Imran, M., et al.: Traffic accident detection and condition analysis based on social networking data. Accid. Anal. Prev. **151**(9), 105–112 (2021)
7. Wu, Y., Shan, S.: Application of artificial intelligence to social governance capabilities under public health emergencies. Math. Prob. Eng. **2021**(2), 1–10 (2021)
8. Firdaniza, Ruchjana, B.N., Chaerani, D.: Information diffusion model using continuous time Markov chain on social media. J. Phys. Conf. Ser. **1722**(1), 12–19 (2021)
9. Sokalska, E.: Flaws and advantages of the polish local self-government in the 21st century: social consultations at the local level. Lex Localis **19**(1), 19–37 (2021)
10. Fu, W., Liu, S., Srivastava, G.: Optimization of big data scheduling in social networks. Entropy **21**(9), 902 (2019)
11. Liu, S., Bai, W., Zeng, N., et al.: A fast fractal based compression for MRI images. IEEE Access **7**(5), 62412–62420 (2019)
12. Liu, S., Li, Z., Zhang, Y., et al.: Introduction of key problems in long-distance learning and training[J]. Mobile Networks and Applications **24**(1), 1–4 (2019)

An Algorithm of Employment Resource Allocation for College Students Based on Social Network Mining

Mei-bin Qi(✉)

Admission and Employment Office, Dongying Vocational Institute, Dongying 257090, China
hbdsfisdf545@sohu.com

Abstract. Aiming at the current uneven distribution of employment resources and poor accuracy of college students, social network mining is applied to the design of college students' employment resource allocation algorithm, and a college student employment resource allocation algorithm based on social network mining is proposed. First, build an LTE (Long Term Evolution) system. LTE interference suppression is performed through inter-cell interference randomization technology, inter-cell interference cancellation technology, and inter-cell interference coordination technology. Construct a resource allocation model based on the constructed LTE system, and the constructed resource allocation model is a continuous decision model. In terms of resource allocation, virtual machines can be simulated and placed through the established energy consumption model and performance loss model, so the state value can be obtained through the Monte Carlo method. Based on social network mining, according to the constructed resource allocation model and the obtained state value, the employment resource allocation algorithm for college students is realized. Experiments verify that this method has short task scheduling time, better resource allocation accuracy and efficiency, and it can optimize the allocation of employment resources for college students to a certain extent.

Keywords: Allocation algorithm · College students · Employment resource · Social network mining

1 Introduction

Since the expansion of college enrollment, the number of college students graduating each year has been increasing, but the number of employers has not increased rapidly, resulting in an imbalance in market supply and demand, and the employment situation for college students is very severe. This problem has become one of the key issues of concern to the whole society [1]. In order to alleviate the employment difficulties of college students, the state has issued a series of related policies, such as encouraging graduates to work at the grassroots level in urban and rural areas, central and western regions, and small and medium-sized enterprises, encouraging graduates to enlist in

S. Liu and X. Ma (Eds.): ADHIP 2021, LNICST 416, pp. 254–265, 2022.
https://doi.org/10.1007/978-3-030-94551-0_21

the army, and encouraging graduates to start their own businesses. Self-employment of graduates has been the focus of work in recent years. However, these policies cannot completely solve the problem of employment difficulties for college students [2]. Since China has been a "relationship" society since ancient times, many college students will try their best to use family or personal social capital in order to find an ideal job. The phenomenon of using social capital for employment is very common in China. It can allow job seekers to obtain more comprehensive employment information, increase employment opportunities, and reduce employment pressure. At the same time, it can also cause unfairness and affect social order. It can be said that social capital will have a great impact on the employment rate and employment quality of college students [3].

However, too much social capital will also have a negative impact on college students and society as a whole. Therefore, the government must regulate the employment order, create a relatively fair employment environment, and allocate employment resources for college students. In this context, an algorithm for the allocation of employment resources for college students based on social network mining is proposed. At present, relevant scholars have done research on the allocation algorithm of college students' employment resources. Literature [4] proposes to give priority to the allocation of medical care and employment resources during the period of new coronary pneumonia. Designed an online survey on how much priority should be given to the elderly in healthcare-related scenarios and three related employment scenarios. The results show that benevolent age discrimination predicts that older people will have higher priority for access to healthcare (classification, COVID-19 vaccine, COVID-19 testing) and employment resources (reserving jobs, working from home), and more recognition of hostile age discrimination Significantly predicts a lower priority rating. Literature [5] proposes employment resources for patients with Parkinson's disease: resource review and needs assessment. Parkinson's disease patients (PwP) exit the labor market five years earlier than non-Parkinson's patients due to motor, cognitive, communication, and emotional symptoms. The reduction in employment will result in huge personal and social costs. The purpose of this research is to determine the advantages and disadvantages of employment resources, and to evaluate the needs of consumers and clinical stakeholders to improve job retention. The research uses a qualitative content analysis and quasi-needs assessment framework. Sixteen PwP and 10 clinician stakeholders participated in two rounds of stakeholder discussion groups.

Interference suppression by LTE is to use inter-cell interference randomization technology to construct a resource allocation model based on the constructed LTE system, and the constructed resource allocation model is a continuous decision model. In resource allocation, the virtual machine can be simulated and placed through the established energy consumption model and performance loss model, so that the state value can be obtained through the Monte Carlo method. On the basis of social network mining, according to the constructed resource allocation model and the obtained state value, the employment resource allocation algorithm for college students is realized.

2 The Algorithm Design of Employment Resource Allocation for College Students Based on Social Network Mining

2.1 Building an LTE System

The LTE system is different from the previous wireless communication system. It mainly supports packet switching services by establishing a seamless mobile IP connection between the user terminal and the packet data network. First, the LTE system is constructed [6]. The constructed LTE network consists of two parts: a core network and an access network. The core network is composed of many logical nodes, and the access network basically has only one node, which is an evolution type, and all network elements are connected to each other through interfaces. The main function of the core network is to establish relevant bearer and control user terminals. The main logical nodes of EPC (Electronic Product Code) mainly include the following three parts: PDN (Public Data Network) gateway, service gateway, mobility management entity, in addition to some registers such as home location register, etc. [7].

Among them, P-GW is responsible for user IP address allocation and QoS guarantee, and performs traffic-based charging according to PCRF rules. The S-GW is responsible for sending user IP data packets. The MME is responsible for handling the control of the signaling interaction between the UE and the core network.

E-UTRAN is composed of eNodeB networks, which are connected to each other through interface X2 and to MME through interface S1. The main functions of the access network include: wireless resource management, IP header compression, and security assurance. The main functions of radio resource management include: radio access control, radio bearer control, dynamic resource allocation, UE uplink and downlink scheduling, etc. Security is mainly reflected in the need to encrypt all data packets sent through the wireless interface. The access network is also responsible for the MME signaling and the establishment of the bearer path to the S-GW. LTE not only cancels the radio network controller node, but also adopts the all-IP network architecture, and at the same time improves the air interface physical layer technology, so that it can support wider system bandwidth and shorter transmission delay, and in terms of spectrum utilization Three to five times the previous standard. The system not only optimizes the traditional UMTS network architecture, but also supports both FDD and TDD duplex modes.

2.2 LTE Interference Suppression

In the LTE system, there are the following three technologies according to different interference processing methods: inter-cell interference randomization technology, inter-cell interference cancellation technology, and inter-cell interference coordination technology. These three technologies can suppress inter-cell interference [8].

Inter-cell interference randomization techniques include scrambling and interleaving. Scrambling is performed by using different pseudo-random codes for the signals of each cell. Interleaving is to use different interleaving patterns for channel interleaving to obtain interference whitening effects.

Inter-cell interference cancellation technology includes spatial interference suppression technology based on multi-antenna receiving end and interference cancellation technology based on interference reconstruction. The principles of these two methods are to demodulate or even decode the interference signal of the interfering cell to a certain extent, and then use the processing gain of the receiver to eliminate the interference signal component from the received signal.

Inter-cell interference coordination techniques include soft frequency reuse and partial power control. The principle of soft frequency reuse is to allow cell center users to freely use all frequency band resources, while cell edge users can only use a certain part of the frequency band resources.

2.3 Building a Resource Allocation Model

According to the constructed LTE system, the resource allocation model is constructed. The constructed resource allocation model is a continuous decision-making model. The goal of the model is to minimize the energy consumption and performance loss of cloud computing. The resource allocation model is constructed by the cloud computing simulation tool CloudSim. CloudSim is written in java language, so it can run normally on different systems. The simulation tool provides a power component, which includes the placement strategy class of the virtual machine, the energy consumption model class of the virtual machine, and the dynamic migration selection strategy class of the virtual machine. By inheriting or modifying existing classes, users can implement their own virtual machine placement strategies and simulate them on the platform. The constructed resource allocation model consists of the following parts: Energy consumption evaluation model: This model performs energy consumption evaluation after simulation placement based on the possible relationship between the energy consumption of cloud computing and the resource occupancy ratio, which makes the obtained solution close to the optimal solution. An indispensable model. Resource allocation model: This model is mainly used to model and constrain resources in cloud computing. The constraint conditions are generally that a virtual machine can only be placed on one physical machine, and the required memory of the placed virtual machine should not exceed the physical machine's Available memory, etc.

2.4 Evaluation of State Value

In resource allocation, the virtual machine can be simulated and placed through the established energy consumption model and performance loss model, so the state value can be obtained by Monte Carlo method.

In a standard Monte Carlo process, a lot of random simulations are run, in this case, start with the board position where you want to find the best move position. Every possible move from this starting state retains statistical information, and then returns the move with the best overall result. However, the disadvantage of this method is that for any given branch in the simulation, there may be many possible moves, but only one or two moves are good. If a move is randomly selected for each round, the simulation is unlikely to find the best path forward. Therefore, the UCB1 strategy is used to enhance the selection. The strategy of UCB1 is different from random simulation, but the branch child nodes are selected according to the current number of games and the total number of wins of the node each time [9].

The Monte Carlo tree search is divided into four stages in total:

(1) The first stage: selection, when you use statistical data to deal with each position you reach, it is like a multi-gambling machine problem. Then, the move used will be obtained by the UCB1 strategy instead of randomly selected, and applied to obtain the next position to be considered. Then make a selection until you reach a sub-location that is not recorded with statistical information. The specific operation of selection is shown in Fig. 1. The number on the left of the figure represents the number of times the node is finally won, the number on the right of the figure represents the total number of times the node is selected, and the left and right sides of the top number represent respectively The total number of matches won and the total number of matches [10].

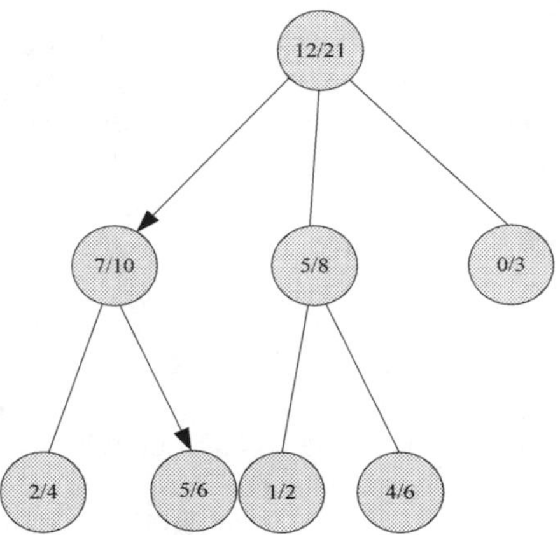

Fig. 1. Selection flow chart

(2) The second stage, expansion, occurs when the UCB1 strategy can no longer be used for selection [11]. Randomly select an unvisited sub-location, and add a new record node to the statistics tree. The specific operation of the extension is shown in Fig. 2. Compared with Fig. 1, the new node is explored, and the attribute of the new node is 0/0.

(3) The third stage, simulation. This is a typical Monte Carlo simulation. It is either purely random or some simple weighted heuristic algorithm. If it is a lightweight match, choose a random simulation. If it is a non-lightweight match, use the more computationally expensive one. Weighted heuristic algorithm for simulation. For games with fewer branches in each branch node, using random simulation can produce better results. Simulation is a complete process. From selection, exploration

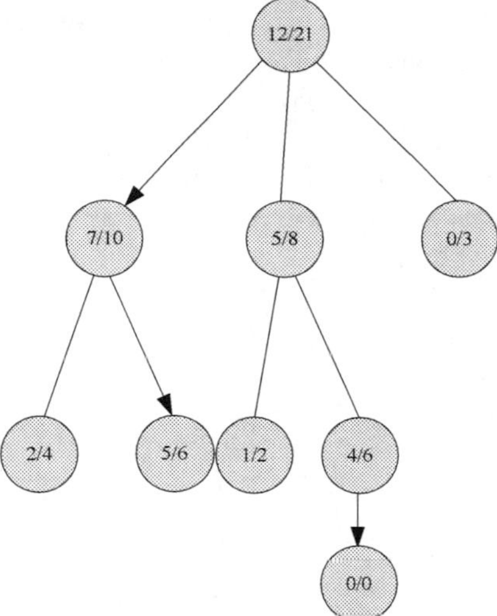

Fig. 2. Extended flowchart

and finally to the victory or defeat, the simulation operation is shown in Fig. 3. After the expansion, continue to select the child nodes, and finally get the game result 0 (negative) or 1 (win).

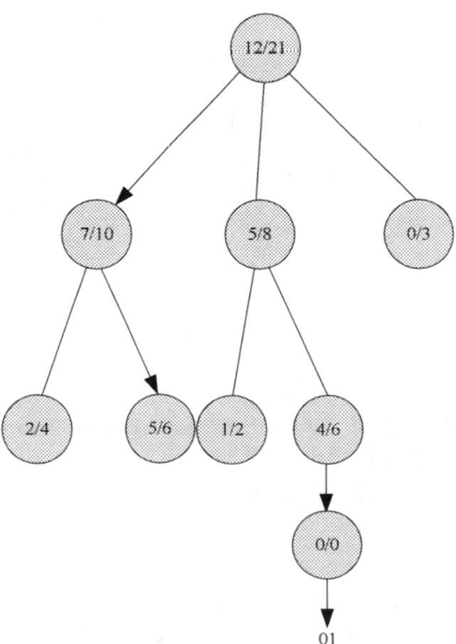

Fig. 3. Simulation flow chart

(4) Finally, the fourth stage is the update or back propagation stage. This happens at the end of a simulation. The number of visits to all locations visited during this simulation will increase, and if a player in that location wins the game, the number of wins will also increase. The update operation is shown in Fig. 4, which is based on the update operation after the final simulation result in Fig. 3 is victory. That is, the number of visits and the number of wins of all nodes traversed in the last simulation is increased by 1.

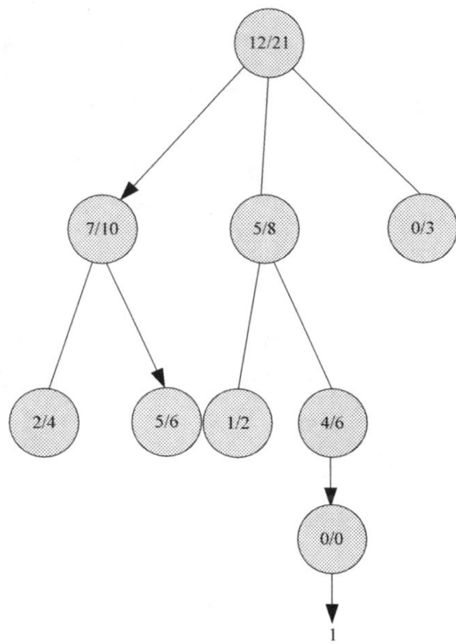

Fig. 4. Backpropagation flowchart

2.5 Allocation Algorithm of Employment Resources for College Students

Based on social network mining, according to the constructed resource allocation model and the obtained state value, the employment resource allocation algorithm for college students is implemented [12–14].

The basic description of the allocation algorithm of employment resources for college students is as follows:

(1) Enter the virtual machine list, physical machine list, and historical information list of college students' employment resources, and create a matrix Y of the advantage value of college students' employment resources.

(2) Based on social network mining, traverse the virtual machine list of college students' employment resources, and calculate the advantage value Yij for each virtual machine according to the state value and frequency matrix and the advantage frequency matrix [15].

(3) According to the advantage value Yij, the virtual machine placement of the employment resource agent of college students is carried out, and the placement strategy is as follows:

    When another agent does not place the placement, select the physical machine with the largest current placement times and meet the constraints between the virtual machine and the physical machine according to the placement matrix information for placement, and update the times matrix after placement. If another agent has already placed the selected virtual machine, according to the frequency matrix, select the physical machine with the frequency matrix value of 0, and then select the following from the selected physical machines: perform another based on the historical placement optimal information The agent's placement simulation predicts to obtain an energy consumption value and based on the historical placement information, the current agent is divided by the current virtual machine placement prediction to be placed (after the prediction, only the current virtual machine has not been placed) and an energy consumption value is obtained. Select the physical machine that can win the remaining part of the agent after the placement is completed and save it into a list.

(4) After traversing and selecting the physical machine of employment resources for college students, perform the following operations:

    Determine whether the list is empty. If it is empty, perform unfavorable expansion, that is, select the physical machine that satisfies the constraint condition and minimizes the additional energy consumption after placement from the physical machines with the value of 0 in the order matrix. If the list is not empty, traverse the virtual machines in the list, and find out the physical machine that minimizes energy consumption after placement.

(5) Return to the selected college student employment resource allocation physical machine.

The specific description of the allocation algorithm of college students' employment resources is shown in Fig. 5.

```
Input: VMS,PMS,discount factor
y,listl,list2,list3,C(S),U(S),Q(S),VMi
Output:Selected PM
create matrix Yu ,pm=null,listpm,count=O,double min=999
    for VMi in VMS do:
By C(S),U(S) use equal (5-4):
by ,u ,Q(S) use equal (5-9)
    end for
    get E2 by list2
    get E1 by listl
     for PMj in PMS do:
        e= getE(PMj,VMi)
        if E l+e<E2:
add PMj into listpm
              count++
endfor
     if count==0:
      pm=getSuit(PMS,VMi)
    else:
      for PMk in listpm do:
        if min>getE(PMk,VMi):
           min=getE(PMk,VMi)
           pm=PMk
endfor
  return pm
```

Fig. 5. Detailed description of the allocation algorithm of employment resources for college students

3 Experimental Test

3.1 Experimental Environment

The experiment used four types of virtual machines and two types of physical machines. The parameter configurations of the physical machine and virtual machine are shown in Table 1 and Table 2 respectively.

Table 1. Virtual machine parameter configuration

Type	VM1	VM2	VM3	VM4
MIPS	2500	2000	1000	500
PES	1	1	1	1
RAM/M	870	1740	1740	613
BW/Ms	100000	100000	100000	100000
SIZE/M	2500	2500	2500	2500

Table 2. Physical machine parameter configuration

Type	HPProLiant ML110 G4	HPProLiant ML110G5
MIPS	1860	2660
PES	2	2
RAM/M	4096	4096
BW/Ms	1000000	1000000
STORAGE	1000000	1000000

Build an experimental platform under the configuration of Table 1 and Table 2, set environment variables as required. The specific settings are shown in Table 3.

Table 3. Environment variable settings

Serial number	Path	Environment variable settings
1	ClassPath	D:\Program Files (x86)\Java\jdk1.6.0_12\lib;
2	Path	D:\Program Files (x86)\Java\jdk1.6.0_12\bin;

In the experiment, the computing power of each virtual machine ranges from 100-1000MIPS, and the task length ranges from 10000-100000MI.

The network topology built in the experiment is shown in Fig. 6.

3.2 Simulation Experiment Process

When starting the simulation, you first need to create a data center, then create virtual machines and cloud tasks in the data center, select appropriate scheduling strategies and set resource parameters, etc., and finally register resource information with the agent center so that users can use it The resources of the data center are simulated. In the simulation experiment, the task scheduling time data of the employment resource allocation algorithm based on social network mining is obtained as the experimental data.

3.3 Analysis of Experimental Results

The experimental data of the task scheduling time of the employment resource allocation algorithm based on social network mining is shown in Table 4.

According to the experimental data of task scheduling time in Table 4, the task scheduling time of the employment resource allocation algorithm for college students based on social network mining is shorter. This method uses LTE interference suppression technology to construct a resource allocation model, and the constructed resource allocation model is a continuous decision model. Through the established energy consumption model and performance loss model to simulate and place the virtual machine,

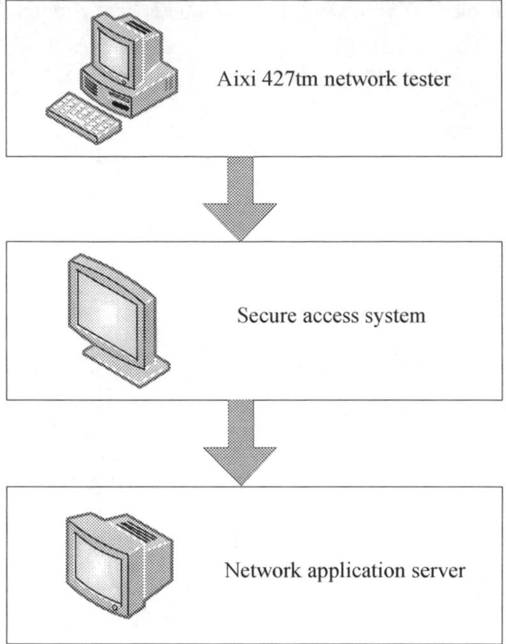

Fig. 6. Network topology built in the experiment

Table 4. Experimental data of task scheduling time

Serial number	Number of tasks	Task scheduling time(ms)
1	50	742.25
2	100	1183.16
3	150	2032.38
4	200	3206.24
5	300	4559.25
6	400	5324.57
7	500	5484.34

the state value can be obtained by the Monte Carlo method. It saves a lot of calculation steps and improves the efficiency of resource allocation.

4 Concluding Remarks

The employment resource allocation algorithm for college students based on social network mining realizes the improvement of task scheduling performance, effectively improves the accuracy and efficiency of college students' employment resource allocation, and has a certain significance for improving the employment prospects of college students.

In future research, the employment problems of college students of different majors should be classified, so as to improve the distribution of employment resources of college students in a more targeted manner.

References

1. Afif, M., Hassen, W.B., Tabbane, S.: A resource allocation algorithm for throughput maximization with fairness increase based on virtual PRB in MIMO-OFDMA systems. Wirel. Netw. **25**(3), 1083–1097 (2019)
2. Tian, Y.: Community structure division based on immune algorithm. Int. J. Performability Eng. **15**(4), 1103–1111 (2019)
3. Alvarado, S.E., Turley, R.N.L.: College-bound friends and college application choices: Heterogeneous effects for Latino and White students. Soc. Sci. Res. **41**(6), 1451–1468 (2012)
4. Apriceno, M.B., Lytle, A., Monahan, C., et al.: Prioritizing health care and employment resources during COVID-19: roles of benevolent and hostile ageism. Gerontologist **29**(12), 98–102 (2020)
5. Rafferty, M., Stoff, L., Palmentera, P., et al.: Employment resources for people with Parkinson's disease: a resource review and needs assessment. J. Occup. Rehabil. **1**, 1–10 (2020)
6. Yu, Y., Bu, X., Yang, K., et al.: Network function virtualization resource allocation based on joint benders decomposition and ADMM. IEEE Trans. Veh. Technol. **69**(2), 1706–1718 (2020)
7. Milkova, L., Crossman, C., Wiles, S., et al.: Engagement and skill development in biology students through analysis of art. Cbe Life Sci. Educ. **12**(4), 687–700 (2013)
8. Mohammadpoorasl, A., Ghahramanloo, A.A., Allahverdipour, H., et al.: Prevalence of Hookah Smoking in Relation to Religiosity and Familial Support in College Students of Tabriz, northwest of Iran. J. Res. Health Sci. **14**(4), 268–271 (2014)
9. Wang, L., Wen, C., Wu, L.: Target identity recognition method based on trusted information fusion. Int. J. Performability Eng. **15**(4), 1235–1246 (2019)
10. Bills, J.L., Vanhouten, J., Grundy, M.M., et al.: Validity of the medical college admission test for predicting MD–PhD student outcomes. Adv. Health Sci. Educ. Theory Pract. **21**(1), 33–49 (2016)
11. Witkow, M.R., Huynh, V., Fuligni, A.J.: Understanding differences in college persistence: a longitudinal examination of financial circumstances, family obligations, and discrimination in an ethnically diverse sample. Appl. Dev. **19**(1), 4–18 (2015)
12. Liu, S., Bai, W., Zeng, N., et al.: A fast fractal based compression for MRI images. IEEE Access **7**, 62412–62420 (2019)
13. Liu, S., Lu, M., Li, H., et al.: Prediction of gene expression patterns with generalized linear regression model. Front. Genet. **10**, 120 (2019)
14. Fu, W., Liu, S., Srivastava, G.: Optimization of big data scheduling in social networks. Entropy **21**(9), 902 (2019)

An Improved Mobilenetv3-Yolov5 Infrared Target Detection Algorithm Based on Attention Distillation

Ronglu Jin[1], Yidong Xu[1,2](✉), Wei Xue[1,2], Beiming Li[1], Yingwei Yang[1], and Wenjian Chen[2]

[1] College of Information and Communication Engineering, Harbin Engineering University, Harbin 150001, China
xuyidong@hrbeu.edu.cn
[2] College of Underwater Acoustic Engineering, Harbin Engineering University, Harbin 150001, China

Abstract. Most current target detection algorithms have problems such as high computational complexity and poor adaptability, which are difficult to meet actual application requirements and are deployed on mobile embedded platform devices. Towards solving the above problems, in this paper, we propose an improved Mobilenetv3-Yolov5 infrared target detection algorithm based on attention distillation. Firstly, the improved Mobilenetv3 and Recurrent FPN-PAN are suitable for reconstruction of the Yolov5 model. Secondly, Attention Distillation is employed to improve the accuracy in object detection. Further, that there doesn't exist public dataset in the field of security infrared target detection, this paper constructs an infrared image dataset (coming soon) that contains a variety of scenarios with pedestrians, motor vehicles, and non motorized targets. Experimental results of tests conducted on the above dataset show that compared with original Yolov5, the proposed algorithm has the following improvements: (1) 88.2% smaller with the model size as only 3.5 MB; (2) 2 times faster with 60.27; (3) 73.9% lower complexity with 4.7 GFLOPS; and (4) 0.8 accuracy higher with 92.1 overall mAP. The testing results on Ambarella platform show that our model achieves higher accuracy and faster running speed with 91.9 mAP and 32.48 FPS, which is 0.9 higher than Yolov5, 3.4 higher than Yolov3-SPP, 6.8 higher than Yolov3, 25 higher than Yolov2, 17.6 higher than FasterR-CNN, in terms of accuracy.

Keywords: Improved Mobilenetv3 · Recurrent FPN-PAN · Attention distillation

This research work was supported by International Science & Technology Cooperation Program of China (2014DFR10240), the Fundamental Research Funds for the Central Universities (3072021CF0802), the Key Laboratory of Advanced Marine Communication and Information Technology, Ministry of Industry and Information Technology (AMCIT2101-02), and Sino-Russian Cooperation Fund of Harbin Engineering University (2021HEUCRF006).

Supplementary Information The online version contains supplementary material available at https://doi.org/10.1007/978-3-030-94551-0_22.

S. Liu and X. Ma (Eds.): ADHIP 2021, LNICST 416, pp. 266–279, 2022.
https://doi.org/10.1007/978-3-030-94551-0_22

1 Introduction

With the rapid development of today's society, economy and technology, public safety issues are becoming increasingly complex and diversified. Therefore, it is of great significance to study intelligent security monitoring technology. In the field of security monitoring, it is often necessary to understand the scene content of a large number of images and videos. In recent years, due to the continuous improvement of image processing capabilities of convolutional neural networks in artificial intelligence technology, artificial intelligence technology has also been continuously developed in security equipment [1]. Among the hundreds of security companies in the world, most of the security products are based on visible light imaging, and the imaging effect is poor at night and under bad weather. Infrared thermal imaging uses a sensor to receive the heat radiation signal of an object, and then converts the signal into an image. Compared with visible light imaging, the infrared imaging system has the advantages of all weather work and penetration of clouds and fog, and it is very sensitive to temperature, and can highlight sensitive targets such as infrared people and cars. In order to make up for the shortcomings of poor imaging quality of visible light imaging in some scenes, infrared imaging equipment has gradually become an important part of security equipment [2]. Target detection technology is an important branch of security monitoring technology. However, most of the current researches on infrared target detection at home and abroad are mainly based on traditional methods, which cannot meet the actual application requirements in terms of detection effect and speed.

For infrared target detection, the existing research algorithms can be roughly divided into two categories, one is based on artificial feature extraction, and the other is based on deep learning.

The method based on artificial feature extraction mainly uses artificially designed features (such as SIFT [3], HOG [4], HAAR-like [5], LBP [6], etc.) to distinguish under the sliding window classifier. Ke et al. proposed PCA-SIFT [7], which reduces the feature dimension of processing and speeds up the calculation of the algorithm. Yu et al. added Harris corner operator in the process of extracting SIFT features [8]. However, because these methods mostly rely on artificially designed parameters, their scene adaptability and accuracy are poor, and they usually lack sufficient generalization ability for new images.

Compared with the method based on artificial feature extraction, the method based on deep learning uses a large number of parameters to fit various situations, and uses a multi-layer architecture to gradually abstract target information. It has significant advantages in detection accuracy and greatly improves the target. The generalization ability of the test. Methods based on deep learning can be divided into two categories according to the framework: two-stage detection algorithms and single-stage detection algorithms. Two-stage detection algorithms mainly include R-CNN [9], FastR-CNN [10] and FasterR-CNN [11]. Single-stage detection algorithms mainly include SSD [12] and YOLO [13]. These algorithms have achieved excellent results in the task of target detection in visible light images. Inspired by these algorithms, many researchers have tried to use deep learning algorithms in the field of infrared target detection. Du [14] et al. proposed a two-stage infrared target detection algorithm. They use convolutional neural network to extract infrared image features, and then use support vector machine (SVM)

[15] to complete target classification to achieve infrared target detection. Sommer [16] et al. used the Region Proposal Network (RPN) proposed in FasterR-CNN to generate a series of candidate regions, and then input these candidate regions into CNN for classification and determine whether they are real targets. The current two-stage detection network has a relatively complex model structure, slow running speed, and manual adjustment of parameters, which increases the difficulty of optimizing the overall model. The model structure of the single-stage detection network is relatively simple and has a faster detection speed, but its accuracy is usually inferior to that of the two stage detection network. In view of the poor accuracy of existing traditional methods and deep learning methods, and the poor real-time performance of the above methods, this paper uses Yolov5 as the network framework and proposes an improved Mobilenetv3-Yolov5 infrared target detection algorithm based on attention distillation. The method includes the following innovations: (1) In view of the large number of parameters of the Yolov5 model, the Yolov5 model is reconstructed using the improved Mobilenetv3 and Recurrent FPN-PAN (2) The attention distillation method is used to improve the detection accuracy of the model (3) This article created an infrared target detection data set (open source soon).

2 The Improved Mobilenetv3-Yolov5

In this section, we mainly introduce the model structure of the improved Mobilenetv3-Yolov5 and analyze its advantages.

2.1 Infrared Feature Analysis

Yolov5 is the most accurate model of the YOLO series network in recent years. It has made many improvements on the original Yolov3 [17] and Yolov4 [18] models, and introduced a variety of techniques for adjusting parameters. On the whole, the model greatly increases the width and depth of the network, improves the fitting and generalization capabilities of the model, and increases the accuracy of the target detection task. However, due to the introduction of a large number of convolution kernels in the backbone network to construct the residual structure, the front-end of the model is too large. And for target detection in visible light, most of the differences in feature information in the feature maps produced by convolutional neural networks come from different HSV color gamuts. These subtle differences in the same detection area can improve the performance of the Yolov5 model. Detection accuracy rate. For the target detection task of infrared images, most of the feature information learned by CNN is gray information, which does not have obvious color gamut discrimination like visible light detection, which results in a large number of repeated feature maps. These redundant feature maps have a limited effect on infrared detection and increase the parameters of the model, making the Yolov5 model run very slowly on some edge devices. Therefore, it is necessary to construct a lightweight backbone network for feature extraction.

2.2 The Improved Mobilenetv3

For infrared target detection, the Darknet structure in the Yolov5 model will generate many redundant feature maps, which will increase the burden of the detection algorithm. Therefore, in view of the unique characteristics of infrared images, the improved Mobilenetv3 structure is adopted as the backbone of the Yolov5 model in this paper. This can reduce the consumption of computing resources on the one hand, and on the other hand, it is more consistent with the characteristics of infrared images. An important part of Mobilenetv3 [19] is the depth separable convolution, and its core idea is to decompose the standard convolution operation into deepwise convolution and pointwise convolution, as shown in Fig. 1.

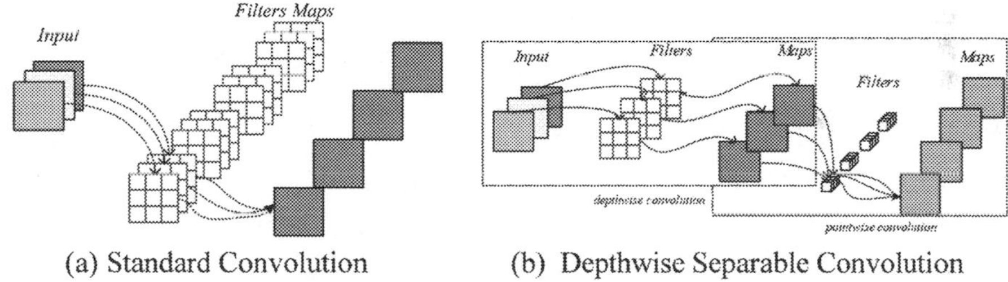

Fig. 1. Standard convolution and depthwise separable convolution.

Depth convolution is completely performed in a two-dimensional plane, a channel is convolved by only one convolution kernel, and the number of convolution kernels is the same as the number of channels in the previous layer. The number of feature maps after deepwise convolution is the same as the number of channels in the input layer, and the feature maps cannot be expanded. Moreover, this operation independently performs convolution operations on each channel of the input layer, and does not effectively use the feature information of different channels at the same spatial position. Therefore, pointwise convolution is needed to combine these feature maps to generate new feature maps. The pointwise convolution operation is very similar to the conventional convolution operation. The size of its convolution kernel is $1 \times 1 \times M$, and M is the number of channels in the upper layer. Therefore, the convolution operation here will weighted and combine the feature maps of the previous step in the depth direction to generate a new feature map. There are several output feature maps for several convolution kernels. According to the characteristics of depth separable convolution, the overall structure of the improved Mobilenetv3 is shown in Fig. 2.

As shown in Fig. 2, the bneck module is composed of depth separable convolutions, with depth convolution kernels of different sizes, used to complete down-sampling or feature extraction. In order to strengthen the feature extraction capability of Mobilenetv3, this paper introduces the SE [20] module between multiple bneck modules to enhance the correlation between channels, and builds a residual module to improve the feature aggregation capability of the backbone network. In this paper, the improved Mobilenetv3 is used to replace the Darknet structure in the Yolov5 model, so that the overall network structure has the capability of multi-scale detection while maintaining a moderate depth, making the model more suitable for target detection in infrared images.

3 Recurrent FPN-PAN

In this section, the problem of multi-scale feature fusion is analyzed first, and then the Recurrent FPN-PAN idea used in this article: effective two-way cross-scale connection and weighted feature fusion.

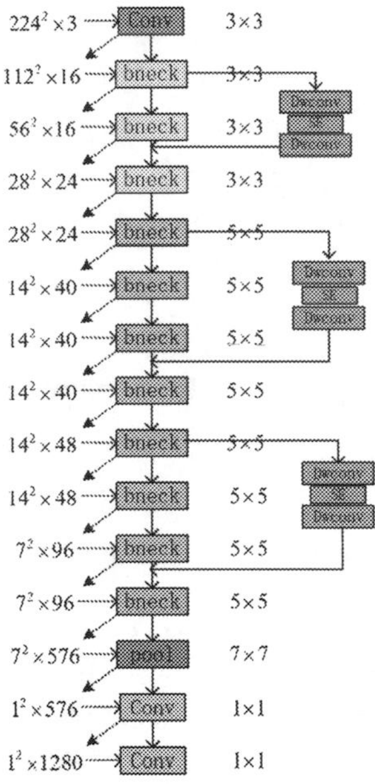

Fig. 2. The overall structure of the improved Mobilenetv3.

3.1 Multi-scale Feature Fusion Problem

Convolutional networks usually adopt a single-line structure from top to bottom. For large objects, its semantic information will appear in a deeper feature map; for small objects, its semantic information will appear in a shallower feature map. As the network deepens, its detailed information may completely disappear. In order to enhance semantics, a better method is to fuse the large-scale features of the shallow network with the small-scale features of the deep network, and obtain high-resolution, strong semantic features by gathering features of different resolutions. Given a list of multi-scale features: $\overrightarrow{T^{input}} = (T_{l_1}^{input}, T_{l_2}^{input}, \ldots)$, Among them, $T_{l_i}^{input}$ represents the features at the level l_i. The purpose of multi-scale feature fusion is to find a conversion method f that can effectively aggregate different feature information: $\overrightarrow{T^{output}} = f(\overrightarrow{T^{input}})$.

3.2 FPN and PAN

The feature fusion method of FPN [21] is used in the Yolov3 network structure. For the feature pyramid network of the FPN structure, the size of the feature map is of different scales, and different features can be fused, but due to the different size of the feature map, the features at the bottom of the feature pyramid cannot be merged with the features at the top. In Yolov4 and Yolov5, after connecting the PAN [22] structure to the FPN structure, it can play the role of path aggregation, making it easier for the bottom fine features to be transferred to the upper network and be integrated on the same scale. In this combination of operations, the FPN layer conveys strong semantic features from the top to the bottom, while the feature pyramid conveys strong positioning features from the bottom to the top, which expands the characterization content of the features and improves the detection performance. However, because such a combination method still does not get rid of the limitation of one-way flow of characteristic information, it is necessary to find a way to make the fusion of characteristic information more sufficient.

3.3 Recurrent FPN-PAN

The Recurrent FPN-PAN proposed in this paper is improved on the basis of the original FPN-PAN network structure, and a feedback connection is added between FPN and PAN, as shown in Fig. 3.

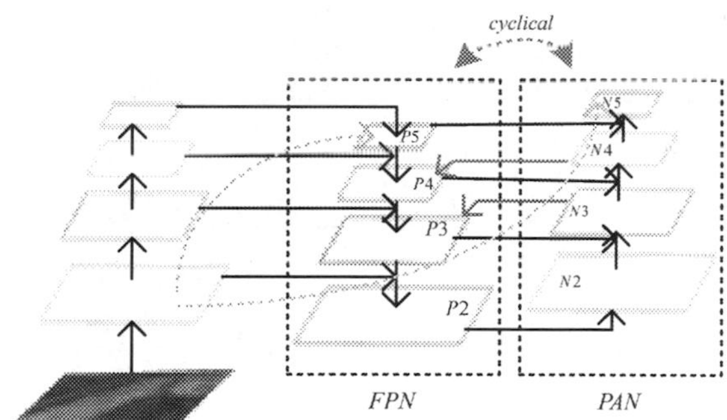

Fig. 3. Recurrent FPN-PAN structure.

As can be seen from Fig. 3, this feedback link forms a loop between FPN and PAN network, and splices the feature information originally in the PAN network into FPN. The Recurrent FPN-PAN network structure has not only positive feature information flow, but also reverse feature information feedback. More importantly, this feedback only needs to use channels with the same resolution features for splicing and does not introduce other modules, so It does not change the parameter amount and model size of the entire network, and greatly improves the detection capability of the model.

According to the above analysis, compared with the original Darknet backbone structure, the improved Mobilenetv3 structure greatly reduces the parameters of the overall network. The Recurrent FPN-PAN structure makes the feature information fusion more sufficient. This article uses the improved Mobilenetv3 and Recurrent FPNPAN to reconstructed the Yolov5 model in order to make the overall network structure capable of multi-scale detection while maintaining a moderate depth, making the model more suitable for target detection in infrared images.

4 Attention Distillation

Most of the research work on the combination of deep learning and visual attention mechanism focuses on the use of masks to form attention mechanisms. The principle of the mask is to use another layer of new weights to identify the key features in the image data. Through learning and training, the deep neural network can learn the areas that need to be paid attention to in each new image and form attention. This kind of attention is deterministic attention, which can be directly generated through the network after learning is completed, and more importantly, it is differentiable. The attention that can be differentiated can be calculated through the neural network to calculate the gradient and forward propagation and backward feedback to learn the weight of attention.

We use $D \in \mathbb{R}^{C,H,W}$ to represent the characteristics of the backbone network in the target detection model, where, *C*, *H*, *W* are respectively represent the number of channels, height and width. Then, the generation of the spatial attention feature map and the channel attention feature map is equivalent to finding the mapping function: $f^s : \mathbb{R}^{C,H,W} \to \mathbb{R}^{H,W}$ and $f^c : \mathbb{R}^{C,H,W} \to \mathbb{R}^{C}$, respectively. Since the absolute value of the feature pixel has its importance, we build a mapping by summing the absolute values in the channel dimension f^s, and building the mapping by summing the absolute values in the height and width dimensions f^c. The formula is as follows:

$$f^s(D) = \frac{1}{C}\sum_{C}^{k=1} |A_{k,\ldots}| \tag{1}$$

$$f^c(D) = \frac{1}{HW}\sum_{H}^{i=1}\sum_{W}^{j=1} |D., i, j| \tag{2}$$

Among them, *i*, *j*, *k* are represent the dimensions of the middle height, width, and channel, respectively. The spatial attention mask M^s and channel attention mask M^c are produced in the teacher model and the student model, the formula is as follows:

$$M^s = HW \cdot softmax((f^s(D^S) + f^s(D^\tau))/T) \tag{3}$$

$$M^c = C \cdot softmax((f^c(D^S) + f^c(D^\tau))/T) \tag{4}$$

Where *S* and $\mathcal{T}$ represent the student model and the teacher model respectively, and *T* are a hyperparameter in *softmax*, which are used to adjust the distribution of pixels in the attention mask. We chose the super-large model Scaled-Yolov4 [23] as the teacher model, and chose the improved Mobilenetv3-Yolov5 model as the student model. The distillation framework is shown in Fig. 4.

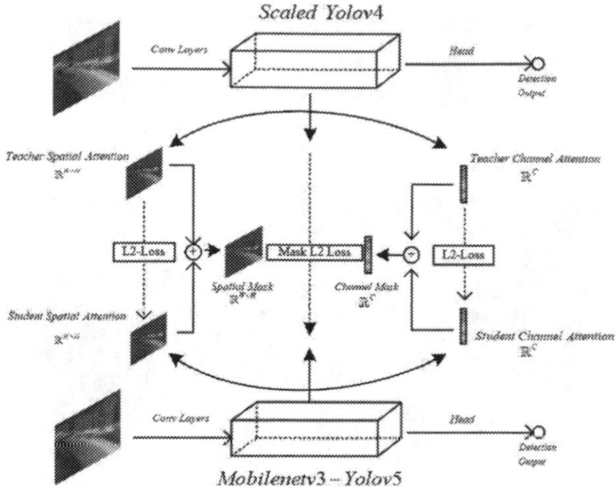

Fig. 4. Attention feature distillation framework.

The distillation loss L_{DTM} is composed of two parts: attention shift loss L_{DT} and attention mask loss L_{DM}.

L_{DT} is used to motivate the student model to imitate the space and channel attention of the teacher model, which can be expressed as:

$$L_{DT} = L_2(f^s(D^S), f^s(D^\tau)) + L_2(f^c(D^S), f^c(D^\tau)) \tag{5}$$

L_{DM} uses mask loss L_2 to encourage the student model to learn the characteristics of the teacher model, which can be expressed as:

$$L_{DM} = \left(\sum_{k=1}^{C}\sum_{i=1}^{H}\sum_{j=1}^{W}\left(D^{\tau}_{k.i.j} - D^{S}_{k,i,j}\right)^2 \cdot M^{S}_{i,j} \cdot M^{c}_{k}\right)^{\frac{1}{2}} \tag{6}$$

The advantage of this method is to let the small model learn the intermediate process and methods of solving the problem, so that it can learn more knowledge.

5 Experiments

In this section, first introduce the infrared target data set and evaluation system used in the experiment, and then design a series of experiments to evaluate the detection performance of the improved Mobilenetv3-Yolov5 infrared target detection algorithm based on attention distillation proposed in this paper.

5.1 Introduction to the Data Set and Experimental Platform

The data set used in this experiment is an infrared target data set in an outdoor scene. The data set contains a total of 10,000 images, more than 10 different scenes and more than 50,000 targets. Each target corresponds to a marked location, and each image corresponds to a marked file. The data set covers a variety of scenarios, including simple short-distance large target backgrounds and complex overlapping long-distance medium and small target backgrounds, as shown in Fig. 5.

Fig. 5. Infrared data images in multiple scenarios.

The research of infrared target recognition in this paper is the recognition of three categories of pedestrians, motor vehicles and non-motor vehicles. The amount of data in the prepared data set can fully meet the needs of recognition model training. The algorithm proposed in this paper is developed using Python3.8, and Pytorch is selected as the deep learning framework. The hardware platform configuration used in the infrared image target detection experiment is shown in Table 1.

5.2 Experimental Setup

In order to ensure the scientificity of the experiment, this paper divides all infrared data sets into training sets according to the principle of the same scene distribution, 10% of the data sets as the validation set, and the remaining 10% as the test set. In the infrared target detection experiment, the default hyperparameters are as follows: the total training rounds is 200 rounds. Using the stochastic gradient descent strategy, the initial learning rate is set to 0.01, momentum and weight decay are set to 0.937 and 0.0005, respectively. All models use a single GPU to perform multi-scale training with a batch size of 32.

Through sampling different training rounds in the process of model training, the change trend of model indicators is obtained. After 200 rounds of training, the accuracy rate, recall rate and average detection accuracy all tend to stabilize in the process of increasing. The evaluation indicators of the experiment include average detection

Table 1. Experimental platform parameter configuration.

Component	Model parameter
Operating system	Ubuntu18.04
CPU	Intel(R) Xeon(R) Gold 5218R CPU
CPU cores	20
GPU	Quadro RTX 5000
Video memory	48G
Embedded hardware platform	Ambarella CV2x

accuracy (mAP@0.5), model size, model reasoning speed (FPS), and model computing power (GFLOPS). First, the performance difference between the improved Mobilenetv3-Yolov5 model and the Yolov5 model in this article is tested. All comparison algorithms are tested using the infrared data set distribution given in the previous article. To ensure fairness, this experiment uses the same Training parameters, and do not use pre-training weights. The model parameter indexes of the improved Mobilenetv3Yolov5 and Yolov5 are shown in Table 2.

Table 2. The comparison of various indicators of the improved Mobilenetv3-Yolov5 and Yolov5 models.

Indicators	Improved-Mobilenetv3Yolov5	Improved-Mobilenetv3Yolov5 (distillation)	Yolov5
mAP	89.4	92.1	91.3
Model size	3.5 M	3.5 M	29.7 M
FPS	60.27	60.27	31.35
GFLOPS	4.7	4.7	18.0

From the experimental data in Table 2, it can be seen that the improved-Mobilenetv3-Yolov5 model and the Yolov5 model before the attention distillation have a recognition accuracy difference of only 1.9, which shows that for the infrared target detection task, the Yolov5 network model is in the process of detection. A large number of similar or even repeated infrared feature maps are generated, so that the model has been learning the same feature content repeatedly. These redundant feature maps are not very helpful to the overall performance of the model. Using Improved-Mobilenetv3 and Recurrent FPN-PAN modules to improve the Yolov5 model can also achieve better detection accuracy. Table 2 also shows that our attention distillation is a very effective method, which can reduce various errors, improve the positioning and classification capabilities of the network, and will not bring about an increase in model parameters and a decrease in speed at all. Compared with the Yolov5 model, the model size of Improved-Mobilenetv3-Yolov5 is reduced by 88.2%, the detection speed (FPS) is increased by 92.9%, and the model computing power GFLOPS is reduced by 73.9%.

In addition, we also train the two-stage detection model Faster-RCNN and the singlestage detection model Yolov2, Yolov3, and Yolov3-SPP, which have good current performance capabilities, on our compiled infrared data set, and compare the training results with ours. The designed final distillation models are compared, and the comparison results are shown in Table 3.

Table 3. Various indicators of the detection results of various models on the infrared data set (GPU).

Indicators	Faster-RCNN	Yolov2	Yolov3	Yolov3-SPP	Yolov5	Ours
mAP	74.7	67.1	85.3	88.9	91.3	92.1
Model size	135.1 M	9.4 M	98.5 M	117.6 M	29.7 M	3.5 M
FPS	9.46	244.35	30.27	17.42	31.35	60.27
GFLOPS	188.5	62.9	65.8	141.4	18.0	4.7

It can be seen from Table 3 that the model we designed greatly reduces the model size and model calculation power while improving the detection accuracy, and the model's inference speed (FPS) has also reached a relatively excellent level.

5.3 Embedded Platform Experiment

This article uses the Ambarella intelligent vision heterogeneous acceleration platform equipped with CV2x chip to complete the model inference process. The platform includes multiple hardware processing units such as CPU, DSP, CVFlow VP, and SDK development environment running on these hardware, as well as supporting tool chain development. surroundings. Among them, CVFlow VP is the abbreviation of CVFlow Vector Processor. It is a hardware unit in the Ambarella development board that is specifically designed to accelerate processing of neural networks, especially deep learning convolutional neural networks. It supports most of the existing public networks, such as Alexnet [24], VGG16 [25], Googlenet [26], Resnet18, Resnet50 and other classification networks, FasterR-CNN, Yolo, SSD, RFCN [27] and other detection networks, and SegNet [28], FCN [29] and other scene segmentation networks. In order to verify the effectiveness of the model designed in this paper on the platform, a variety of commonly used models that can be transplanted into the platform are selected for comparison. The experiment still uses the data set in 5.1 and uses the same distribution method. In the hardware platform, we use detection accuracy and model inference speed as performance evaluation indicators.

Table 4. The performance of each model in the Ambarella hardware platform.

Indicators	FasterRCNN	Yolov2	Yolov3	Yolov3-SPP	Yolov5	Ours
mAP	74.3	66.9	85.1	88.5	91.0	91.9
FPS	3.76	125.00	16.94	10.75	18.21	32.48

According to the experimental data in Table 4, the model designed in this paper has achieved good results on the Ambarella platform, and the detection accuracy of the model is higher than that of Yolov5, Yolov3-SPP, Yolov3, Yolov2, FasterR-CNN, which are commonly used in embedded platforms The model of this article, and the reasoning speed of this model is also at a relatively fast level among these commonly used models. For engineering applications, the generalization performance of the target detection model is also very important. In this paper, other infrared scenes that do not appear in the data set are also selected for detection. The visual effect of the detection is shown in Fig. 6. The test results found that the infrared target detection model in this paper has a strong generalization ability, and it has good detection results for large targets at short distances or for small and medium-sized targets at long distances. There are no errors in several types of scenes tested which the detection situation.

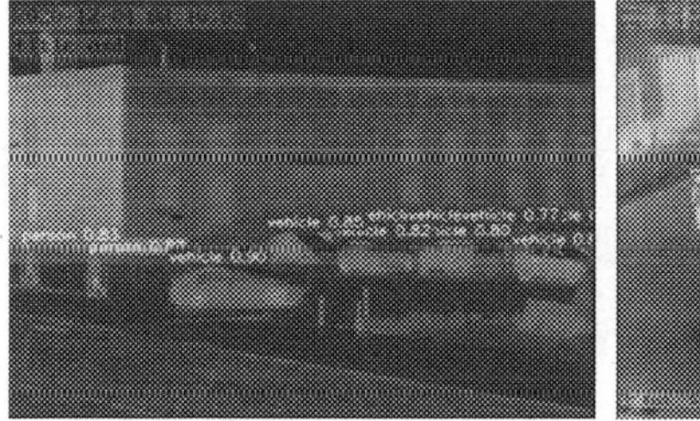

Fig. 6. The model detects visual effects.

6 Conclusion

We propose an improved Mobilenetv3-Yolov5 infrared target detection algorithm based on attention distillation. Based on the Yolov5 model structure, this algorithm firstly uses the improved Mobilenetv3 and Recurrent FPN-PAN to reform it. Secondly the attention distillation methods were used to improve the detection accuracy of the compressed model. This paper also constructs a data set containing a variety of reallife human, motor and non-motor objects. On the above data set test results show that the proposed algorithm using improved Mobilenetv3 and Recurrent FPN-PAN can cut the original size is reduced to 3.5 M, size decreased by 88.2% compared with the original model, detection speed (FPS) increased two times, (BFLOPS) model complexity was reduced by 73.9%, and by the method of attention distillation, the overall mAP value reached

92.1, 0.8 higher than Yolov5 model. On the Ambarella, the detection speed of the model can reach 32.48 frames and the average detection accuracy can reach 91.9, which is 0.9 higher than Yolov5, 3.4 higher than Yolov3-SPP, 6.8 higher than Yolov3, 25 higher than Yolov2 and 17.6 higher than FasterR-CNN.

References

1. Huang, S.L., Li, C.P., Chang, C.C., et al.: Real-time monitoring of the thermal effect for the redox flow battery by an infrared thermal imaging technology. Energies **13**, 6717 (2020)
2. Du, L., Zhang, W., Fu, H., et al.: An efficient privacy protection scheme for data security in video surveillance. J. Visual Commun. Image Represent. **59**, 347–362 (2019)
3. Lowe, D.G.: Distinctive image features from scale-invariant keypoints. Int. J. Comput. Vision **60**(2), 91–110 (2004)
4. Dalal, N., Triggs, B.: Histograms of oriented gradients for human detection. In: IEEE Computer Society Conference on Computer Vision and Pattern Recognition, pp. 886–893 (2005)
5. Lienhart, R., Maydt, J.: An extended set of Haar-like features for rapid object detection. In: International Conference on Image Processing, pp. 900–903 (2002)
6. Manheimer, E., White, A., Berman, B., et al.: Meta-analysis: acupuncture for low back pain. Ann. Intern. Med. **142**(8), 651–663 (2005)
7. Ke, Y., Sukthankar, R.: PCA-SIFT: a more distinctive representation for local image descriptors. In: Proceedings of the 2004 IEEE Computer Society Conference on Computer Vision and Pattern Recognition, 2004, CVPR 2004, vol. 2, pp. II–II. IEEE (2004)
8. Yu, L., Zhang, D., Holden, E.J.: A fast and fully automatic registration approach based on point features for multi-source remote-sensing images. Comput. Geosci. **34**(7), 838–848 (2008)
9. Girshick, R., Donahue, J., Darrell, T., et al.: Rich feature hierarchies for accurate object detection and semantic segmentation. In: Proceedings of the IEEE Conference on Computer Vision and Pattern Recognition, pp. 580–587 (2014)
10. Girshick, R.: Fast R-CNN. In: Proceedings of the IEEE International Conference on Computer Vision, pp. 1440–1448 (2015)
11. Ren, S., He, K., Girshick, R., et al.: Faster R-CNN: Towards Real-Time Object Detection with Region Proposal Networks. arXiv preprint, arXiv:1506.01497 (2015)
12. Liu, W., Anguelov, D., Erhan, D., et al.: SSD: single shot multibox detector. In: European Conference on Computer Vision, pp. 21–37. Springer, Cham (2016). https://doi.org/10.1007/978-3-319-46448-0_2
13. Redmon, J., Divvala, S., Girshick, R., et al.: You only look once: unified, real-time object detection. In: Proceedings of the IEEE Conference on Computer Vision and Pattern Recognition, pp. 779–788 (2016)
14. Du, L., Gao, C., Feng, Q., et al.: Small UAV detection in videos from a single moving camera. In: CCF Chinese Conference on Computer Vision, pp. 187–197. Springer, Singapore (2017)
15. Chen, P.H., Lin, C.J., Schölkopf, B.: A tutorial on ν-support vector machines. Appl. Stoch. Model. Bus. Ind. **21**(2), 111–136 (2005)
16. Sommer, L., Schumann, A., Müller, T., et al.: Flying object detection for automatic UAV recognition. In: 2017 14th IEEE International Conference on Advanced Video and Signal Based Surveillance (AVSS), pp. 1–6. IEEE (2017)
17. Farhadi, A., Redmon, J.: Yolov3: an incremental improvement. In: Computer Vision and Pattern Recognition (2018)
18. Bochkovskiy, A., Wang, C.Y., Liao, H.Y.M.: Yolov4: Optimal Speed and Accuracy of Object Detection. arXiv preprint, arXiv:2004.10934 (2020)

19. Howard, A., Sandler, M., Chu, G., et al.: Searching for mobilenetv3. In: Proceedings of the IEEE/CVF International Conference on Computer Vision, pp. 1314–1324 (2019)
20. Cheng, D., Meng, G., Cheng, G., et al.: SeNet: structured edge network for sea–land segmentation. IEEE Geosci. Remote Sens. Lett. **14**(2), 247–251 (2016)
21. Lin, T.Y., Dollár, P., Girshick, R., et al.: Feature pyramid networks for object detection. In: Proceedings of the IEEE Conference on Computer Vision and Pattern Recognition, pp. 21172125 (2017)
22. Liu, S., Qi, L., Qin, H., et al.: Path aggregation network for instance segmentation. In: Proceedings of the IEEE Conference on Computer Vision and Pattern Recognition, pp. 8759–8768 (2018)
23. Wang, C.Y., Bochkovskiy, A., Liao, H.Y.M.: Scaled-YOLOv4: Scaling Cross Stage Partial Network. arXiv preprint, arXiv:2011.08036 (2020)
24. Iandola, F.N., Han, S., Moskewicz, M.W., et al.: SqueezeNet: AlexNet-Level Accuracy with 50× Fewer Parameters and <0.5 MB Model Size. arXiv preprint, arXiv:1602.07360 (2016)
25. Simonyan, K., Zisserman, A.: Very Deep Convolutional Networks for Large-Scale Image Recognition. arXiv preprint, arXiv:1409.1556 (2014)
26. Szegedy, C., Liu, W., Jia, Y., et al.: Going deeper with convolutions. In: Proceedings of the IEEE Conference on Computer Vision and Pattern Recognition, pp. 1–9 (2015)
27. Dai, J., Li, Y., He, K., et al.: R-FCN: Object Detection via Region-Based Fully Convolutional Networks. arXiv preprint, arXiv:1605.06409 (2016)
28. Badrinarayanan, V., Kendall, A., Cipolla, R.: Segnet: a deep convolutional encoder-decoder architecture for image segmentation. IEEE Trans. Pattern Anal. Mach. Intell. **39**(12), 2481–2495 (2017)
29. Long, J., Shelhamer, E., Darrell, T.: Fully convolutional networks for semantic segmentation. In: Proceedings of the IEEE Conference on Computer Vision and Pattern Recognition, pp. 3431–3440 (2015)

A Novel Method of Combined Window Function Construction Filter Applied to F-OFDM System

Mingxin Liu[1(✉)], Wei Xue[1,3], Yidong Xu[1,3], Wenjian Chen[2], and Sergey B. Makarov[4]

[1] College of Information and Communication Engineering, Harbin Engineering University, Harbin 150001, China
{liumx,xuewei,xuyidong}@hrbeu.edu.cn

[2] College of Underwater Acoustic Engineering, Harbin Engineering University, Harbin 150001, China
chenwenjian@hrbeu.edu.cn

[3] Yantai Research Institute of Harbin Engineering University, Yantai 264000, China

[4] Institute of Physics, Nanotechnology and Telecommunications, Higher School of Applied Physics and Space Technologies, Peter the Great St. Petersburg Polytechnic University (SPbPU), Polytechnicheskaya, 29, 195251 St. Petersburg, Russia
makarov@cee.spbstu.ru

Abstract. The Orthogonal Frequency Division Multiplexing (OFDM) systems use rectangular pulse as baseband signal, and a main problem of the rectangular pulse is that its main lobe energy concentration is low and the out-of-band emissions (OOBE) is too high. The filtered orthogonal frequency division multiplexing (F-OFDM) system can make the F-OFDM system have a less energy radiation because of filtering in the time domain. Applying filtering operations can prevent signals from different communication scenarios from interfering with each other at the same time. In general, the filter used in the F-OFDM system is by applying a finite pulse to multiply the sinc function. In this study, a functional model is used, which can obtain a pulse that exhibits fast attenuation features in the frequency domain. The energy of this window function is mainly concentrated in the main pulse of the energy spectral density. This window function is combined with Hanning window and Blackman window to form a new window function form, truncating the sinc function, and constructing a filter. This filter has less energy leakage in F-OFDM system than filters constructed by other combined finite pulses to multiply the sinc function.

This research work was supported by International Science & Technology Cooperation Program of China (2014DFR10240) ,the Fundamental Research Funds for the Central Universities (3072021CF0802), the Key Laboratory of Advanced Marine Communication and Information Technology, Ministry of Industry and Information Technology (AMCIT2101-02), and Sino-Russian Cooperation Fund of Harbin Engineering University (2021HEUCRF006).

S. Liu and X. Ma (Eds.): ADHIP 2021, LNICST 416, pp. 280–291, 2022.
https://doi.org/10.1007/978-3-030-94551-0_23

Keywords: F-OFDM · Functional model · Combination window · Energy leakage

1 Introduction

For broadband channels, multi-carrier transmission schemes are used. Among them, OFDM system is the first choice [1,2]. OFDM system has many advantages, such as anti-multipath channel interference [3,4]. However, the problem of excessive energy leakage has always been the focus of research. Therefore, many emerging multi-carrier modulation systems have been derived. Among them, generalized frequency division multiplexing (GFDM) and filter bank multi-carrier (FBMC) schemes have been widely studied, and both of them also show excellent performance in terms of energy leakage [5–7]. However, the large amount of calculation of them and the inability to directly apply some mature algorithms of the OFDM system are the main problems. The F-OFDM system only adds a filtering operation, and the amount of calculation does not increase significantly. Most of the algorithms of the OFDM system can also be directly applied [8,9]. On the other hand, energy leakage can also be controlled in a small range through filtering [8]. First proposed the F-OFDM system, which used the Hanning pulse to multiply the sinc function to construct a filter, which effectively reduced the energy leakage. In [10], Blackman-Harris pulse is used in F-OFDM system. Compared with F-OFDM system based on Hamming pulse and Hanning pulse, this scheme effectively controls energy leakage. In [11], the traditional pulses were used to construct fiters, and the sinc window function is truncated to construct the filter. Compared with applying the window function alone, the combined window makes the energy leakage of the F-OFDM system lower. The window function method used in the above [8–11] is based on the commonly used window function truncated sinc function to construct the filter. In essence, the most important point of this method is the choice of window function, that is, different pulses will result in different energy leakage. In this study, a functional model is used, which can obtain a window function that exhibits fast attenuation characteristics in the frequency domain. The energy of this window function is mainly concentrated in the main part of the energy spectral density. Combine this pulse with Hanning window, Blackman window, etc. to form a new pulse form, multiplying the sinc function, and construct a filter. The results show that the new filters in the F-OFDM system compared with other combined window functions, the filter has less energy leakage.

The remaining sections of this study are arranged as follows:

The Sect. 2 models the F-OFDM system and the functional model with less energy radiation. Section 3 restricts the functional model and carries out numerical simulation. The window function obtained in Sect. 4 is combined with Hanning window and Blackman window to form new window function forms, truncating the sinc function, constructing the filters and applying them to F-OFDM system, and discussing the performance. Conclusions in Sect. 5.

2 System Model

2.1 F-OFDM System Model

The F-OFDM filters the entire full-band signal, which makes the implementation of the F-OFDM system not as complicated as FBMC. Here, only one embranchment part of the F-OFDM system is taken for a brief analysis. Figure 1 shows the structure of a embranchment of the F-OFDM system.

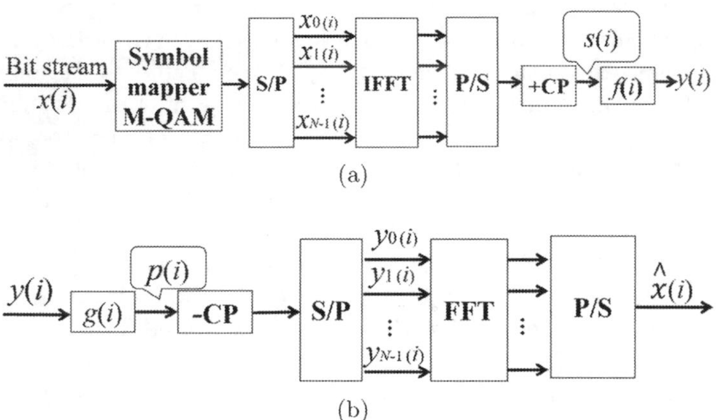

Fig. 1. A embranchment part of the F-OFDM system. (a) Transmitting end embranchment of the F-OFDM system. (b) Receiving end embranchment of the F-OFDM system.

In Fig. 1, the length of an F-OFDM signal is N, and the length of CP is N_{CP}. It can be seen that the length of the whole F-OFDM signal is $N_{F-OFDM} = N + N_{CP}$. Further, the baseband signal of F-OFDM is expressed as:

$$s(i) = \frac{1}{N}\sum_{k=0}^{N-1} x_k(i) \exp \frac{j2\pi k(i - N_{CP})}{N}, \tag{1}$$

Filtering $s(i)$, in fact, is $s(i)$ convolved with filter $f(i)$:

$$y(i) = \frac{1}{N}\sum_{l=-\infty}^{l=\infty} f(i-l) \sum_{k=0}^{N-1} x_k(i) \exp\{\frac{j2\pi k(i - N_{CP})}{N}\}. \tag{2}$$

The filter $f(i)$ is used on the transmission end, and the matched filter $g(i)$ can be used for equalization on the receiving end.

The traditional filter $f(i)$ is obtained by Hanning pulse multiplying the sinc function.

The expression of Hanning pulse is:

$$w_H(i) = 0.5\{1 - [\frac{\cos(2\pi i)}{N-1}]\}, -\frac{(N-1)}{2} \le i \le \frac{(N-1)}{2} \tag{3}$$

The form of the sinc function:

$$\sin c(i) = \frac{\sin(\pi BiT_s)}{\pi BiT_s}, -\frac{(N-1)}{2} \le i \le \frac{(N-1)}{2}, \tag{4}$$

where 1/B is the width of the main part of the sinc function, T_s is the sampling period, the length of the signal is N.

In general, the important design problem for F-OFDM system is how to design filter $f(i)$. By multiplying the sinc function with the Hanning pulse can obtain the filter, and the specific form is as follows:

$$f_H = w_H(i)\sin c(i) \tag{5}$$

At the receiving end, $g(i) = f(-i)$; when $f(i)$ is set as a real-valued even function, and $f(i) = f(-i)$, there, $g(i) = f(i)$. The receiving end signal $p(i)$ can be written:

$$p(i) = y(i) * g(i), \tag{6}$$

where the $*$ is the convolution symbol. After deleting the CP, serial-to-parallel conversion, fast Fourier transform (FFT) conversion, and parallel-to-serial conversion, the signal $\hat{x}(i)$ can be written as:

$$\hat{x}(i) - \frac{1}{N}\sum_{l=-\infty}^{l=\infty} p(i-l)\sum_{k=0}^{N-1} y_k(i)\exp\{\frac{j2\pi k(i-N_{CP})}{N}\} \tag{7}$$

2.2 Functional Model with Less Energy Leakage

The functional model with less energy leakage is established as [12]:

$$J = \frac{1}{2\pi}\int_{-\infty}^{\infty} g(\omega)|S(\omega)|^2 d\omega, \tag{8}$$

In (8), $g(\omega)$ is a function which restricts specific waveform and characteristics of the target pulse. The choice of $g(\omega)$ will determine the properties of the pulse $w(t)$. $S(\omega)$ is the Fourier transform of the pulse $w(t)$, and $|S(\omega)|^2$ is the energy spectral density of $w(t)$. First assume that the pulse $w(t)$ is an even function and the length is T, that is, $t \in [-T/2,T/2]$.

Specifically, $g(\omega)$ is expressed as $g(\omega) = \omega^{2n}$, n=1,2,. . . . . ., n is the variate. And $g(\omega)$ is symmetric about the ordinate and has a fast rising trend. Therefore, in order for J to satisfy the convergence property, $|S(\omega)|^2$ must have a rising speed faster than $g(\omega)$ The rate of decline. Therefore, it can be explained that $|S(\omega)|^2$ has the ability to obey the minimum out-of-band radiation criterion.

According to literature [12], formula (8) can be rewritten as:

$$J = (-1)^n \int_{-T/2}^{T/2} w(t)w^{(2n)}(t)dt, \tag{9}$$

where $w^{(2n)}(t)$ is the $2n$-order derivative of $w(t)$.

Therefore, to solve the pulse $w(t)$ is transformed into when its energy spectral density $|S(\omega)|^2$ obeys the minimum out-of-band energy radiation criterion, J obtains the minimum value, and the problem of the form $w(t)$ is solved.

3 Constraints and Numerical Simulation

3.1 Restrictions

In order to obtain a pulse with lower energy leakage, three constraints are added.

Restriction 1. The boundary constraints of the pulse:

$$w^{(n-1)}(\pm T/2) = \cdots = w'(\pm T/2) = w(\pm T/2) = 0, \tag{10}$$

Here, the value of the length T of the window function is a normalized value, that is, $T = 1$. Restriction 1 can make the less out-of-band radiation in the frequency domain.

Restriction 2. Energy restriction conditions of pulse:

For $w(t)$, according to the setting of its time domain length, its energy needs to be a certain value, that is, its energy cannot be infinite. The energy limit of a single symbol signal is expressed as:

$$E = \int_{-T/2}^{T/2} w^2(t)dt \tag{11}$$

here, the energy value of a pulse is taken as the normalized value.

Restriction 3. Energy spectrum constraints:

In order to further improve the efficiency of spectrum utilization, and then we add constraints in the frequency domain. The expressions are as:

$$|S(f_1)|^2 = X_1; |S(f_2)|^2 = X_2 \tag{12}$$

where f_1 is a point of the main lobe function, relative to the f_1, and the value of spectral density is X_1. In addition, f_2 is a point of the first side lobe, relative to the f_2, and the value of spectral density is X_2. The Lagrangian multiplier method can be used to control the values of these two points.

According to the above constraints, the Lagrangian function can be expressed as:

$$G = J + \alpha[\int_{-T/2}^{T/2} a^2(t) - E] + \beta[|S(f_1)|^2 - X_1] + \gamma[|S(f_2)|^2 - X_2] \tag{13}$$

where α, β, and γ are Lagrangian coefficients.

We can express $w(t)$ in the form of Fourier series.

$$w(t) = \frac{a_0}{2} + \sum_{k=1}^{m} a_k \cos(\frac{2\pi kt}{T}), t \in [-\frac{T}{2}, \frac{T}{2}], \tag{14}$$

$$a_0 = \frac{2}{T}\int_{-T/2}^{T/2} w(t)dt; \; a_k = \frac{2}{T}\int_{-T/2}^{T/2} w(t)\cos(\frac{2\pi kt}{T})dt \tag{15}$$

Applying Variations theory to solve the pulse $w(t)$, the Lagrange function needs to satisfy the following equations:

$$\frac{\partial G}{\partial a_k} = 0, k = 1, .., m; \frac{\partial G}{\partial \alpha} = 0; \frac{\partial G}{\partial \beta} = 0; \frac{\partial G}{\partial \gamma} = 0. \tag{16}$$

3.2 Numerical Simulation

As a demonstration, this article chooses the rising index n = 2, 4, and 6, and the maximum number of Fourier series is a_6. The following window function expressions are obtained through numerical simulation:

$$\begin{aligned} w_{2,6}(i) &= 1.6613/2 + 0.7866\cos(\frac{2\pi i T_s}{N-1}) - 0.0340\cos(\frac{4\pi i T_s}{N-1}) \\ &+ 0.0066\cos(\frac{6\pi i T_s}{N-1}) - 0.0021\cos(\frac{8\pi i T_s}{N-1}) \\ &+ 0.0009\cos(\frac{10\pi i T_s}{N-1}) - 0.0004\cos(\frac{12\pi i T_s}{N-1}) \end{aligned} \tag{17}$$

$$\begin{aligned} w_{4,6}(i) &= 1.4780/2 + 0.9362\cos(\frac{2\pi i T_s}{N-1}) + 0.1763\cos(\frac{4\pi i T_s}{N-1}) \\ &- 0.0168\cos(\frac{6\pi i T_s}{N-1}) + 0.0031\cos(\frac{8\pi i T_s}{N-1}) \\ &- 0.0008\cos(\frac{10\pi i T_s}{N-1}) + 0.0003\cos(\frac{12\pi i T_s}{N-1}) \end{aligned} \tag{18}$$

$$\begin{aligned} w_{6,6}(i) &= 1.3579/2 + 0.9982\cos(\frac{2\pi i T_s}{N-1}) + 0.3386\cos(\frac{4\pi i T_s}{N-1}) \\ &+ 0.0299\cos(\frac{6\pi i T_s}{N-1}) - 0.0046\cos(\frac{8\pi i T_s}{N-1}) \\ &+ 0.0009\cos(\frac{10\pi i T_s}{N-1}) - 0.0002\cos(\frac{12\pi i T_s}{N-1}) \end{aligned} \tag{19}$$

Blackman pulse and Kaiser pulse are respectively expressed as:

$$w_B(i) = \{0.42 - 0.5\cos(\frac{2\pi i}{N-1}) + 0.8\cos(\frac{4\pi i}{N-1})\} \tag{20}$$

where $t = iT_s$ is used for sampling, the window length is N. Among them, T_s is the sampling period.

4 Performance Analysis of F-OFDM System Under Different Filters

4.1 Combination of Optimized Pulse and Traditional Pulse

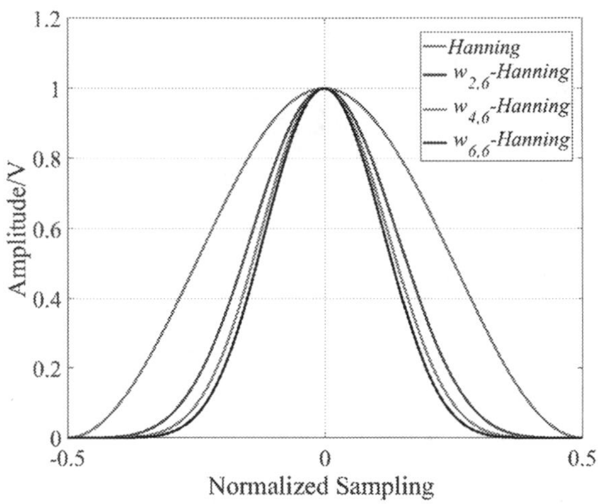

Fig. 2. The plots of different pulses under the normalized sampling in time domain.

The combined forms of the optimized pulses and Hanning pulse (Fig. 2):

$$\begin{aligned} w_{2,H}(i) &= w_{2,6}(i)w_H(i), \\ w_{4,H}(i) &= w_{4,6}(i)w_H(i), \\ w_{6,H}(i) &= w_{6,6}(i)w_H(i). \end{aligned} \tag{21}$$

The combined forms of the optimized pulses and Blackman pulse:

$$\begin{aligned} w_{2,B}(i) &= w_{2,6}(i)w_B(i), \\ w_{4,B}(i) &= w_{4,6}(i)w_B(i), \\ w_{6,B}(i) &= w_{6,6}(i)w_B(i). \end{aligned} \tag{22}$$

The feature of the pulse is that as the variate n increases, the peak is sharper (three optimized windows are all sharper than the Hanning pulse and the Blackman pulse).

4.2 Frequency Domain Performance of Filter

Use the finite pulse to multiply the sinc function to construct the filter as follows, where $-N/2 \leq i \leq N/2$, N is odd (Fig. 3).

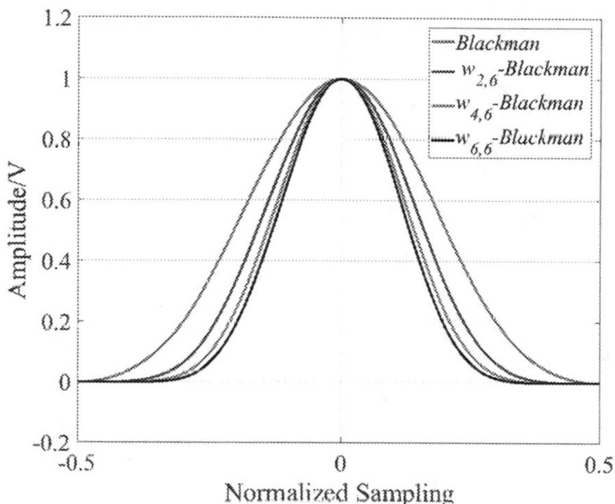

Fig. 3. Normalized PSD diagrams under different filters.

$$\begin{aligned} f_{2,6-Hanning}(i) &= w_{2,H}(i) \sin c(i), \\ f_{4,6-Hanning}(i) &= w_{4,H}(i) \sin c(i), \\ f_{6,6-Hanning}(i) &= w_{6,H}(i) \sin c(i), \\ f_{2,6-Blackman}(i) &= w_{2,B}(i) \sin c(i), \\ f_{4,6-Blackman}(i) &= w_{4,B}(i) \sin c(i), \\ f_{6,6-Blackman}(i) &= w_{6,B}(i) \sin c(i). \end{aligned} \tag{23}$$

The designed filter can solve the problem of high out-of-band radiation of the system. See the following figures for frequency normalization and amplitude normalized spectral density plots of the filters $f_{2,6-Hanning}(i)$, $f_{4,6-Hanning}(i)$, $f_{6,6-Hanning}(i)$, $f_{2,6-Blackman}(i)$, $f_{4,6-Blackman}(i)$, $f_{6,6-Blackman}(i)$.

For the parameter setting, according to the reference [11], the length of filter usually selected is $T_{filter} \leq 1/2T_{F-OFDM}$, where T_{filter} is the length of filter, and T_{filter} is the length of signal. Therefore, we choose the signal length to be 512, that is, the FFT/IFFT length is 512. The length of the filter selected in this study is 257, and the mapping method is 16QAM.

The normalized power spectral density (PSD) of the sinc-Hanning and the sinc-Blackman of the same length shown in Fig. 4 and Fig. 5.

Through Fig. 4 and Fig. 5, the normalized PSD of the filter constructed by the Hanning pulse and Blackman pulse converges to -162.7 dB and -171.6 dB, respectively; and the filter constructed by the combination of the optimized pulse and the Hanning pulse is normalized. The PSD of a single unit converges below

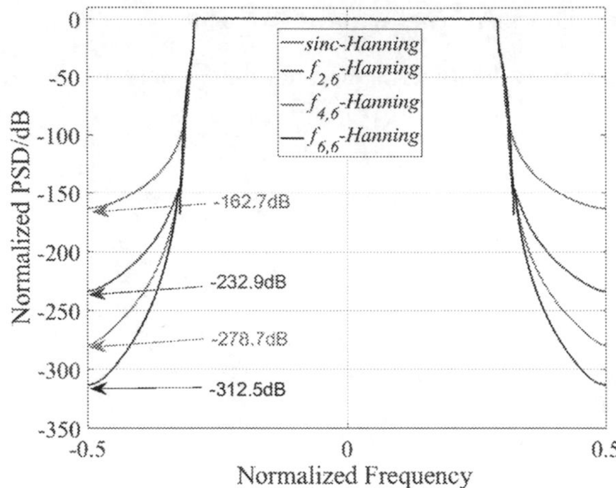

Fig. 4. Normalized PSDs under the different $f_{n,m-Hanning}(i)$ filter configurations.

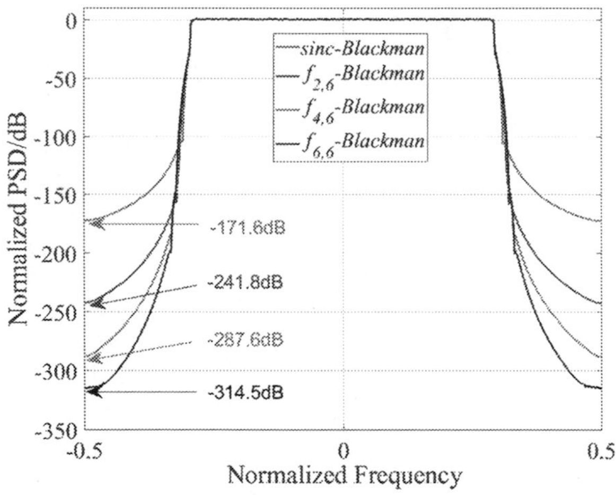

Fig. 5. Normalized PSDs under the different $f_{n,m-Blackman}(i)$ filter configurations.

−200 dB, and its out-of-band radiation is lower than that of a filter constructed with a single window function.

Choose the length of signal as 1024, that is, the FFT/IFFT length is 1024. The length of the filter selected in this study is 513, and the mapping method is 16QAM.

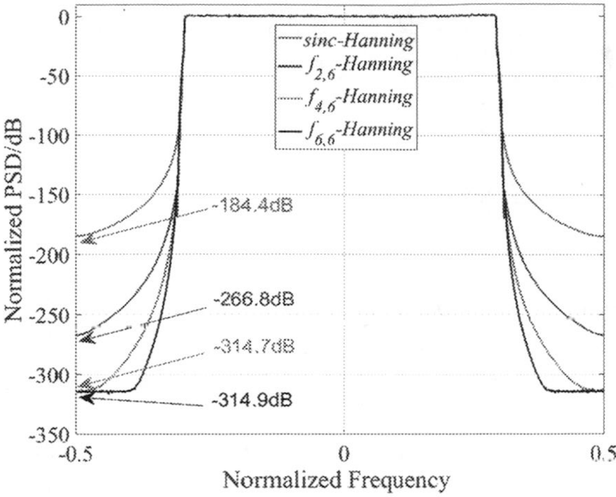

Fig. 6. Normalized PSDs under the different $f_{n,m-Hanning}(i)$ filter configurations.

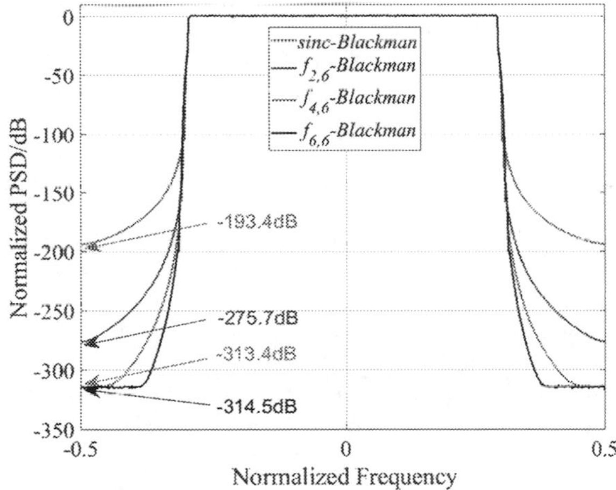

Fig. 7. Normalized PSDs of the F-OFDM system with different $f_{n,m-Blackman}(i)$ filter configurations.

Through Fig. 6 and Fig. 7, the normalized PSD of the filter constructed by the Hanning pulse and Blackman pulse converges to −184.4 dB and −193.4 dB, respectively; and the filter constructed by the combination of the optimized pulse and the Hanning pulse is normalized in the F-OFDM system. The PSD of a single unit converges below −250 dB, and its out-of-band radiation is lower than that of a filter constructed with a single window function.

At the same time, it can be seen that the filter of the window structure combined with the Blackman pulse has a better out-of-band radiation capability than the Hanning pulse. And as the number of sub-carriers increases, the length of the filter can also increase, so the out-of-band radiation of the system will also be significantly reduced.

5 Conclusions

In this study, through the basic model of the F-OFDM system, the key technology is how to design filter. The filter is usually constructed by multiplying the known impulse function by the sinc function. The key to the design of the window function lies in its fast-decaying side-lobes features. We propose a functional model which has the less energy leakage criterion to obtain a pulse with features of the above requirements. The pulse is expressed as an even function in the form of a Fourier series. Changing the variate n in the functional model can obtain optimized pulses with different decay speeds. The optimized pulse and the traditional pulses are combined to form a new pulse, and the filter is constructed by multiplying the sinc function. In the time domain, the feature of the combination pulse is that as the variate n increases, its peak is steeper and the edges are smoother, and the optimized pulses have steeper peaks than the Hanning pulse and Blackman pulse. The combination pulse and the sinc function forming filter are applied to the F-OFDM system, and they both show the characteristics of less energy leakage. Therefore, it can be said that it is more inclined to use combination window to construct filters.

References

1. Weinstein, S.B.: The history of orthogonal frequency-division multiplexing. IEEE Commun. Mag. **47**, 26–35 (2009)
2. Mesleh, R.Y., Haas, H., Sinanovic, S., Ahn, C.W., Yun, S.: Spatial modulation. IEEE Trans. Veh. Technol. **57**, 2228–2241 (2008)
3. Brandes, S., Cosovic, I., Schnell, M.: Reduction of out-of-band radiation in OFDM systems by insertion of cancellation carriers. IEEE Commun. Lett. **10**, 420–422 (2006)
4. Cosovic, I., Brandes, S., Schnell, M.: Subcarrier weighting: a method for sidelobe suppression in OFDM systems. IEEE Commun. Lett. **10**, 444–446 (2006)
5. Farhang-Boroujeny, B.: OFDM versus filter bank multicarrier. IEEE Signal Process. Mag. **283**, 92–112 (2011)
6. Nissel, R., Schwarz, S., Rupp, M.: Filter bank multicarrier modulation schemes for future mobile communications. IEEE J. Sel. Areas Commun. **358**, 1768–1782 (2017)
7. Michailow, N., et al.: Generalized frequency division multiplexing for 5th generation cellular networks. IEEE Trans. Commun. **629**, 3045–3061 (2014)
8. Abdoli, J., Jia, M., Ma, J.: Filtered OFDM: a new waveform for future wireless systems. In: Proceedings of the IEEE International Workshop Signal Process. Advances Wireless Communication, pp. 66–70. IEEE, Stockholm, Sweden (2015). https://doi.org/10.1109/SPAWC.2015.7227001
9. Zhang, X., Jia, M., Chen, L., Ma, J., Qiu, J.: Filtered-OFDM - enabler for flexible waveform in the 5th generation cellular networks. In: Proceedings of the IEEE Global Communications Conference. IEEE, San Diego (2015). https://doi.org/10.1109/GLOCOM.2015.7417854

10. Yang, L.Z., Xu, Y.H.: Filtered-OFDM system performance research based on Nuttall's Blackman-Harris window. In: 2017 IEEE 17th International Conference on Communication Technology (ICCT). IEEE, Chengdu (2017)
11. Taher, M., Radhi, H., Jameil, A.: Enhanced F-OFDM candidate for 5G applications. J. Ambient Intell. Hum. Compt. **121**, 635–652 (2020). https://doi.org/10.1109/ICCT.2017.8359724
12. Xu, Y., Xue, W., Shang, W.: A pan-function model for the utilization of bandwidth improvement and PAPR reduction. Math. Probl. Eng. **2014**, 1–9 (2014)

Passive Electromagnetic Field Positioning Method Based on BP Neural Network in Underwater 3-D Space

Chaoyi Wang[1], Yidong Xu[1,3](✉), Junwei Qi[1], Wenjing Shang[1], Mingxin Liu[1], and Wenjian Chen[2]

[1] College of Information and Communication Engineering, Harbin Engineering University, Harbin 150001, China
xuyidong@hrbeu.edu.cn

[2] College of Underwater Acoustic Engineering, Harbin Engineering University, Harbin 150001, China

[3] Yantai Research Institute of Harbin Engineering University, Yantai 264000, China

Abstract. This paper studies the positioning method of combining the passive electric field positioning and passive magnetic field positioning under three-dimensional (3-D) water. This technology can be applied to underwater submarine positioning, underwater leakage power supply positioning, underwater rescue and other occasions. We collect the samples data by electromagnetic sensors array, and recover the location of the targets. After the data preprocessing process includes data normalization, de-redundancy process, and generalization process, we use Back Propagation (BP) neural networks to build electric field and magnetic field distribution models of electric dipole source. Finally, we enter the test data to obtain the target position in the well-trained positioning model. We take the Euclidean distance between the ideal position and the model output target position as an absolute error. The results show that this method can effectively improve the accuracy of underwater target positioning and anti-interference ability of the training model, and the nonlinear function model trained by the neural network can be applied to the complex and changeable underwater environment.

Keywords: BP neural networks · Passive positioning · Anti-interference

---

This research work was supported by International Science & Technology Cooperation Program of China (2014DFR10240), the Fundamental Research Funds for the Central Universities (3072021CF0802), the Key Laboratory of Advanced Marine Communication and Information Technology, Ministry of Industry and Information Technology (AMCIT2101-02), and Sino-Russian Cooperation Fund of Harbin Engineering University (2021HEUCRF006).

S. Liu and X. Ma (Eds.): ADHIP 2021, LNICST 416, pp. 292–304, 2022.
https://doi.org/10.1007/978-3-030-94551-0_24

1 Introduction

With the development and enrichment of underwater electric and magnetic sensor technology, the acquisition accuracy and acquisition speed of the sensors have been greatly improved, and the conditions of data acquisition device can gradually meet the requirements of theoretical experiments. Research on passive positioning of underwater targets has also attracted more and more attention from scientific researchers [1]. This technology is widely used in underwater rescue, underwater exploration and maintenance of flooded urban facilities. In practical applications, small data fluctuations may introduce huge errors, especially in the complex and changeable underwater environment. The adaptive weakening of underwater interference and nonlinear multi-parameter inversion methods are the main difficulties in this research direction.

Passive electric field positioning technology set electric dipole sources as targets that emits an electromagnetic field in an infinite homogeneous medium [2,3]. We studies the nonlinear mapping relationship between the position of the electric dipole source and electromagnetic field that it emits. Magnetic field data positioning has stronger anti-interference ability, but the magnetic field data is particularly weak and requires higher accuracy of the magnetic sensor [4–6]. Therefore, there is less research on passive magnetic field positioning technology.

We mainly study the method of combining passive electric field positioning and passive magnetic field positioning based on BP neural network. BP artificial neural network is suitable for training nonlinear mapping models. Our innovation lies in the use of electromagnetic field data for positioning of targets, and the use of neural networks to deal with nonlinear inversion problems. It uses forward model for data preprocessing, automatically configures weighting factors, and combines electric field data and magnetic field data to co-locate targets. Judging the weighting factors of training samples by interference strength, on one hand, can improve the adaptive adjustment ability of training parameters, and then improve the system's anti-interference ability in the complex and changeable underwater environment. On another hand, it can remove useless parameter information, reduce the burden of neural network training, and greatly improve training efficiency. Therefore, positioning models trained by BP neural network obtain anti-jamming capabilities and rapid positioning capabilities based on high accuracy.

2 Forward Model Derived from Theory

In an infinite uniform conductive medium, there is a nonlinear mapping relationship between the position $\mathbf{r}$ of the electric dipole source and the electromagnetic field data it emits. The electric potential difference U distribution model and the magnetic field intensity $\mathbf{H}$ distribution model derived from $\mathbf{r}$ are defined as forward models [7]. On the contrary, the process of obtaining $\mathbf{r}$ from U and $\mathbf{H}$ is defined as the inversion process, that is, the positioning process. The forward model provides reference samples for building a neural network model in the inversion process. It can extract effective information, remove redundant samples, and reduce computing space.

As shown in Fig. 1, in an infinite uniform conductive medium, we set the electric dipole source at the origin of the spherical coordinate system, the electric dipole moment is $\mathbf{p}$, the electric field intensity $\mathbf{E}$ at point Q is $(E_r, E_\varphi, E_\theta)$, and the magnetic field intensity $\mathbf{H}$ at point Q is $(H_r, H_\varphi, H_\theta)$. Suppose the conductive medium is fresh water, the electric conductivity is σ, and the permittivity is ε. The electric dipole source radiates a DC signal, and the angular frequency of the signal is $\omega = 0$.

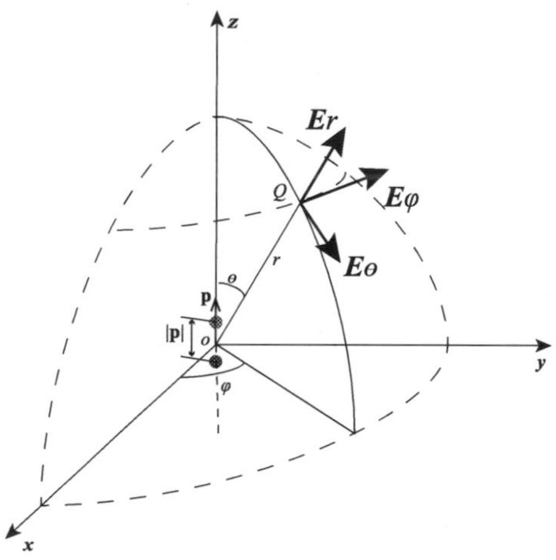

Fig. 1. Electric field distribution diagram of electric dipole source.

When $\sigma >> \omega\varepsilon$, under the spherical coordinate system, the radiant electromagnetic field of the DC electric dipole source is satisfied with the equation [8]:

$$E_r = \frac{|\mathbf{p}| \cos\theta}{2\pi\sigma r^3} \tag{1}$$

$$E_\theta = \frac{|\mathbf{p}| \sin\theta}{2\pi\sigma r^3} \tag{2}$$

$$H_\varphi = \frac{|\mathbf{p}| \sin\theta}{4\pi r^2} \tag{3}$$

Where $\mathbf{r} = (x, y, z)$, $r = |\mathbf{r}|$. Multiply $(E_r, E_\varphi, E_\theta)$ and $(H_r, H_\varphi, H_\theta)$ by the transition matrix. The transition matrix transitions from the spherical coordinate system to the rectangular coordinate system. Transform $(E_r, E_\varphi, E_\theta)$ and $(H_r, H_\varphi, H_\theta)$ into the expression of rectangular coordinate system:

$$E_x = \frac{3\,|\mathbf{p}|\, zx}{4\pi\sigma r^5} \tag{4}$$

$$E_y = \frac{3\,|\mathbf{p}|\, zy}{4\pi\sigma r^5} \tag{5}$$

$$E_z = \frac{3\,|\mathbf{p}|\,z^2}{4\pi\sigma r^5} - \frac{\mathbf{p}}{4\pi\sigma r^3\,|\mathbf{p}|} \tag{6}$$

$$H_x = \frac{-\,|\mathbf{p}|\,y}{4\pi r^3} \tag{7}$$

$$H_y = \frac{|\mathbf{p}|\,x}{4\pi r^3} \tag{8}$$

E can be deduced as:

$$\mathbf{E} = \frac{3(\mathbf{p} \bullet \mathbf{r})\mathbf{r} - r^2\mathbf{p}}{4\pi\sigma r^5} \tag{9}$$

In the actual measurement process, the data measured by the electric field sensor is the electric potential difference U. We need to transform **E** into U. Therefore, in an infinite uniform underwater environment, the forward model from the position information (**r** and **p**) to the electromagnetic field data (U and **H**) is:

$$U = \frac{\mathbf{p} \bullet \mathbf{r}}{4\pi\sigma r^3} \tag{10}$$

$$\mathbf{H} = (H_x, H_y, 0) \tag{11}$$

3 Data Preprocessing

3.1 Acquisition of Training Samples

We place 24 receiving array elements in an underwater 3-D space with a range of 100 m × 100 m × 30 m to form a sensor array [9]. Set the reference electrode of the electric sensors at the center of the array. Each array element is equipped with an electric sensor and a magnetic sensor to receive U and **H** respectively. In addition, we have arranged media that produce both electric and magnetic field disturbances as the sources of interference underwater. Because the conductivity and magnetic permeability of the media are different in strength, the disturbances caused by the media are biased magnetic field or biased electric field.

In an underwater environment where the signal-to-noise ratio (SNR) is 40 dB, after collecting N samples and normalizing them, the electric field training set $\mathbf{E}_\mathrm{N}$ and the magnetic field training set $\mathbf{M}_\mathrm{N}$ can be obtained. Where $\mathbf{E}_\mathrm{N}$ contains sample set $\mathbf{R}_{3\times\mathrm{N}}$ (Electric dipole position), $\mathbf{P}_{3\times\mathrm{N}}$ (electric dipole moment) and label set $\mathbf{U}_{24\times\mathrm{N}}$, $\mathbf{M}_\mathrm{N}$ contains sample set $\mathbf{R}_{3\times\mathrm{N}}$, $\mathbf{P}_{3\times\mathrm{N}}$ and label set $\mathbf{H}_{24\times\mathrm{N}}$.

3.2 De-redundancy and Generalization of Training Set

De-redundancy. The de-redundancy process can remove samples with large data deviations caused by the interferences in the measured area. We substitute $\mathbf{R}_{3\times\mathrm{N}}$ and $\mathbf{P}_{3\times\mathrm{N}}$ into the forward model of formulas (10), (11) to obtain the theoretical electromagnetic field data sets ($\mathbf{U}^\mathrm{o}{}_{24\times\mathrm{N}}$ and $\mathbf{H}^\mathrm{o}{}_{24\times\mathrm{N}}$). Next, we find the Euclidean distance $\mathbf{D}_\mathrm{N}$ of the column vectors between the theoretical set $\mathbf{U}^\mathrm{o}{}_{24\times\mathrm{N}}$ and the measurement set $\mathbf{U}_{24\times\mathrm{N}}$, and calculate the average value $\bar{D}$ of $\mathbf{D}_\mathrm{N}$. In this experiment, we removed the samples whose Euclidean distance is greater than $\bar{D}$ to obtain the new training set $\mathbf{E}_\mathrm{S}$ after de-redundancy. In the same way, after removing the magnetic samples with large interference, the new training set $\mathbf{M}_\mathrm{T}$ of the passive magnetic field positioning model is obtained.

Generalization. The process of de-redundancy may lead to non-uniform distribution of samples in the positioning area, which is manifested in the centralized distribution of data or missing data in individual areas. We solve such problems through a generalization process. The steps of generalization processing are as follows:

We cut the positioning area ($100\,\mathrm{m} \times 100\,\mathrm{m} \times 30\,\mathrm{m}$) into 25 sub-areas ($20\,\mathrm{m} \times 20\,\mathrm{m} \times 30\,\mathrm{m}$) through refinement processing. The number of samples in the i-th sub-area is defined as M_i, the maximum M_i in the 25 sub-areas is taken as $\mathrm{M}_{\mathrm{max}}$, and the minimum M_i is taken as $\mathrm{M}_{\mathrm{min}}$. We take the decision threshold as $a = (\mathrm{M}_{\mathrm{max}} + \mathrm{M}_{\mathrm{min}})/2$, and set the step size to $b = (\mathrm{M}_{\mathrm{max}} - \mathrm{M}_{\mathrm{min}})/2$.

The algorithm is used to traverse the samples in each sub-area. When judging $\mathrm{M}_i > a$, the training set samples in the i-th sub-area are sorted in ascending order, according to the Euclidean distance between the theoretical data and the measured data. The first b samples in this sub-area are removed to update the training set in this sub-area. When $\mathrm{M}_i > b$, keep the training set of this sub-area unchanged.

After generalization, the result shows that $\mathbf{E}_{\mathrm{S}}$ is updated to $\mathbf{E}_{\mathrm{K}}$, and the number of samples is K. $\mathbf{E}_{\mathrm{K}}$ includes sample set $\mathbf{U}_{24\times\mathrm{K}}$, label set $\mathbf{R}_{3\times\mathrm{K}}$, $\mathbf{P}_{3\times\mathrm{K}}$. We update $\mathbf{M}_{\mathrm{T}}$ to $\mathbf{M}_{\mathrm{L}}$ through the same steps, and the number of samples is L. $\mathbf{M}_{\mathrm{L}}$ is divided into sample set $\mathbf{H}_{24\times\mathrm{L}}$ and label set $\mathbf{R}_{3\times\mathrm{L}}$, $\mathbf{P}_{3\times\mathrm{L}}$ for training BP neural networks.

3.3 Weight Factors

The weight factors represent the ratio of the effective training data to the total training data in a certain sub-area. In the data sets $\mathbf{E}_{\mathrm{K}}$ and $\mathbf{M}_{\mathrm{L}}$, the number of updated samples in the i-th sub-area is K_i and L_i. We define the weight factor of $\mathbf{E}_{\mathrm{K}}$ in the i-th sub-area as $\alpha_i = \mathrm{K}_i/(\mathrm{K}_i + \mathrm{L}_i)$, and the weight factor of $\mathbf{M}_{\mathrm{L}}$ in the i-th sub-area as $\beta_i = \mathrm{K}_i/(\mathrm{K}_i + \mathrm{L}_i)$. The conductivity and magnetic permeability of the interference source are different in strength. The weight factors can judge the credibility of the electric field model [10,11] and the magnetic field model by proportion of the effective samples in the interference area, so as to allocate the proportion of the outputs from the electric field model and the magnetic field model, improve the accuracy and anti-interference ability of the positioning model.

4 Inversion Model Trained by BP Neural Network

Research shows that the training accuracy of BP neural network is not greatly affected by the number of layers. As long as the number of neurons in the hidden layer is large enough, a single hidden layer can also approximate a nonlinear function model with finite discontinuities in arbitrary accuracy. Therefore, we choose a single hidden layer BP neural network to train the passive electric field model and passive magnetic field positioning model.

4.1 Passive Electric Field Positioning Model (E-model) and Passive Magnetic Field Positioning Model (H-model)

The BP neural network for training E-model and H-model has three layers, namely the input layer, the hidden layer and the output layer (Fig. 2).

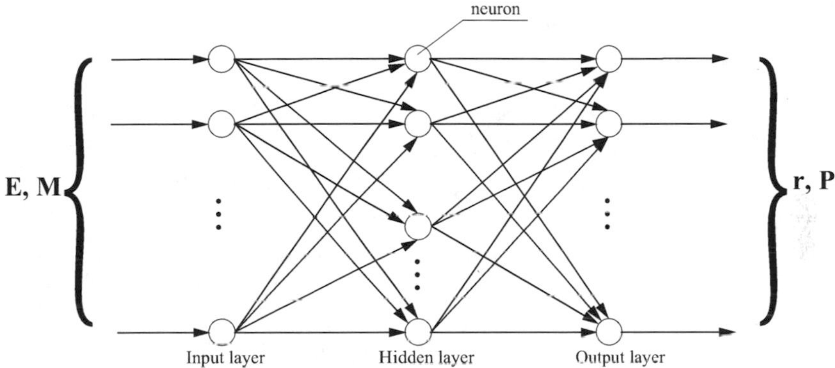

Fig. 2. The structure of BP neural network.

For E-model, characteristic parameters' number of the input layer is 24, and the number of samples is K. The number of neurons in the hidden layer is set to 240. characteristic parameters' number of the output layer is 6, and the number of samples is K. Since the data range of the feature parameter contains negative numbers, we choose the tansig function as the transfer function of the hidden layer. We select the adaptive gradient descent method (Levenberg-Marquardt algorithm) as the training method. The algorithm is characterized by a large memory requirement, a faster convergence speed. It is suitable for a network structure with many neurons and few layers. The loss function chooses the Mean Square Error function (MSE), because MSE is more suitable for neural networks for nonlinear fitting.

As for H-model, there are 24 characteristic parameters and L samples in the input layer. From formulas (7) and (8), $\mathbf{P}_{3\times\mathrm{L}}$ only affects the linear relationship during training, and what we invert is the non-linear mapping relationship. After normalization, $\mathbf{P}_{3\times\mathrm{L}}$ becomes invalid parameters. Due to the reduction of the valid characteristic parameters of the output layer and the reduction of the complexity of the network, the number of neurons in the hidden layer is taken as 200. The training function is the Levenberg-Marquardt algorithm, the hidden layer transfer function is the tansig function, and the loss function is MSE.

Figure 3 shows the convergence process of the E-model and H-model, the abscissa is the number of iterations, and the ordinate is MSE value between the positioning of the models and the label set. We observe that models are approaching the ideal models from the tenth iteration, and the convergence speed of the neural network fitting process is very rapid. In the positioning area with

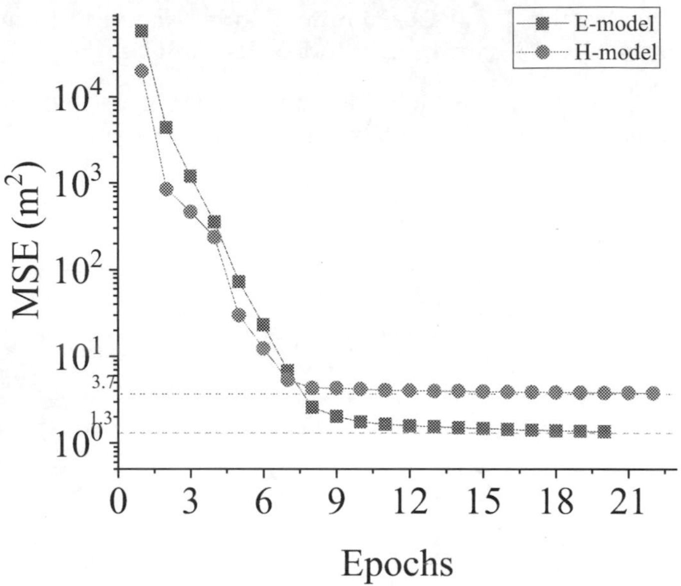

Fig. 3. Convergence processes of E-model and H-model.

a volume of 3000 m^3, MSE eventually tends to 1.3 and 3.7. All of these indicate that models trained by BP neural network has nicer precision and robustness.

4.2 Positioning Method Combining E-Model and H-Model

We measure the electric field data and magnetic field data generated by the electric dipole source at $\mathbf{r}_{test}$ [12], and perform normalization processing to obtain $\mathbf{U}_{test}$ and $\mathbf{H}_{test}$. Input $\mathbf{U}_{test}$ and $\mathbf{H}_{test}$ into E-model and H-model to find the positioning ($\mathbf{r}_{\mathrm{E}}$ and $\mathbf{r}_{\mathrm{H}}$).

Configure the weight factors vectors (α and β) of the E-model and the H-model based on the strength of the interference intensity of $\mathbf{E}_{\mathrm{K}}$ and $\mathbf{M}_{\mathrm{L}}$ in all sub-areas. The final positioning output of the positioning model(E-H-model) that combines the E-model and the H-model is $\mathbf{r}_{\mathrm{EH}} = \alpha \bullet \mathbf{r}_{\mathrm{E}} + \beta \bullet \mathbf{r}_{\mathrm{H}}$.

Figure 4 shows a flowchart of the E-H-model positioning method.

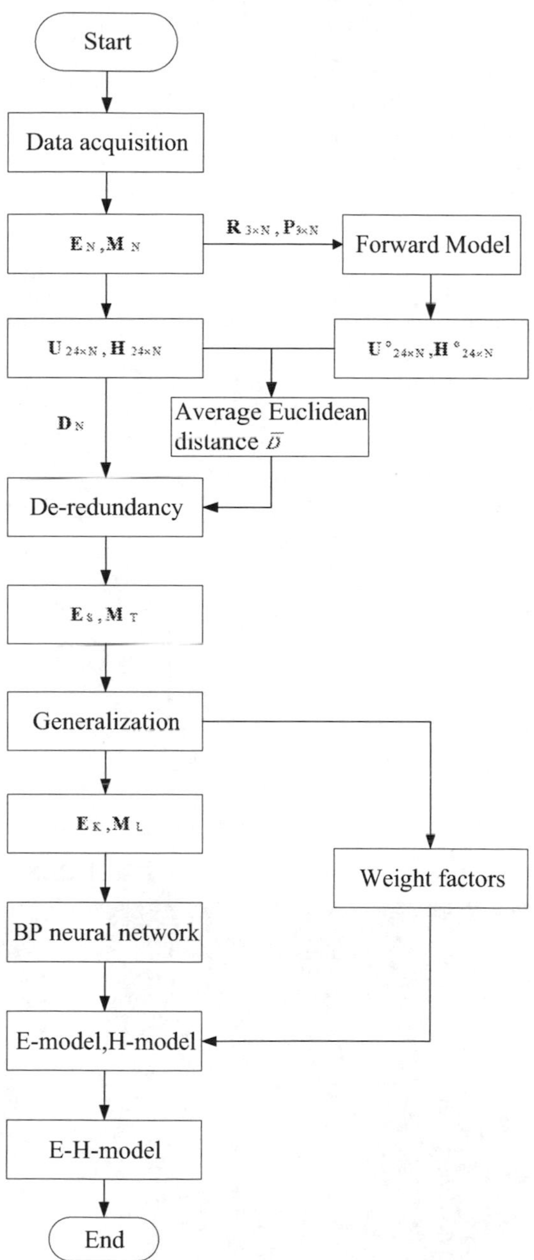

Fig. 4. Overall flow of the E-H-model positioning method.

5 Simulation

We use the three inversion models to positioning scattered targets distributed in the plane (z = 15), and visualize the positioning results. As shown in the Fig. 5, we set the a interference with strong magnetic disturbance and weak electrical disturbance at the coordinates (30, 10, 15). Another interference source with strong electric disturbance and weak magnetic disturbance is placed at the coordinate (−30, 10, 15).

E-model, H-model and E-H-model respectively locate the plane (z = 15) in the underwater environment of Fig. 5, and the positioning results are as Fig. 6, Fig. 7 and Fig. 8.

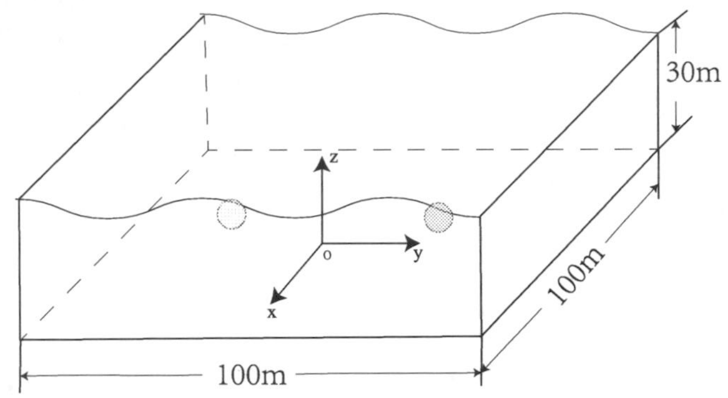

Fig. 5. Distribution of interference sources.

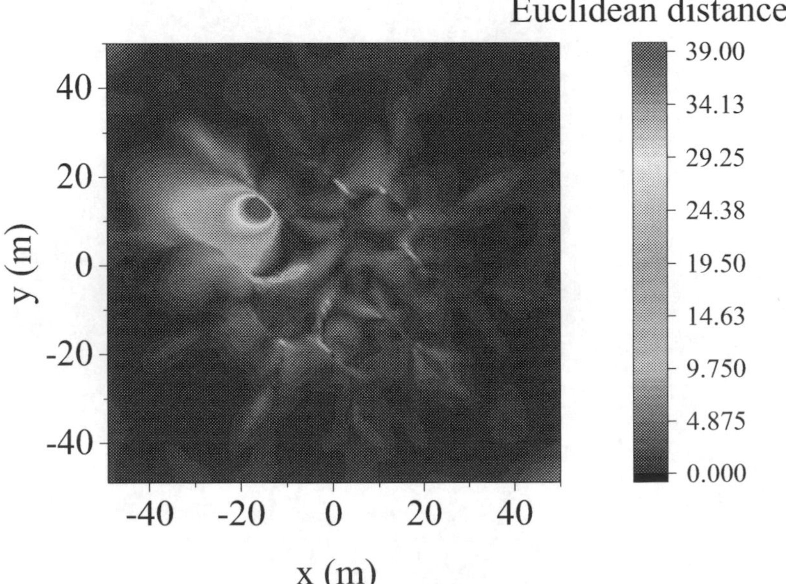

Fig. 6. E-model positioning.

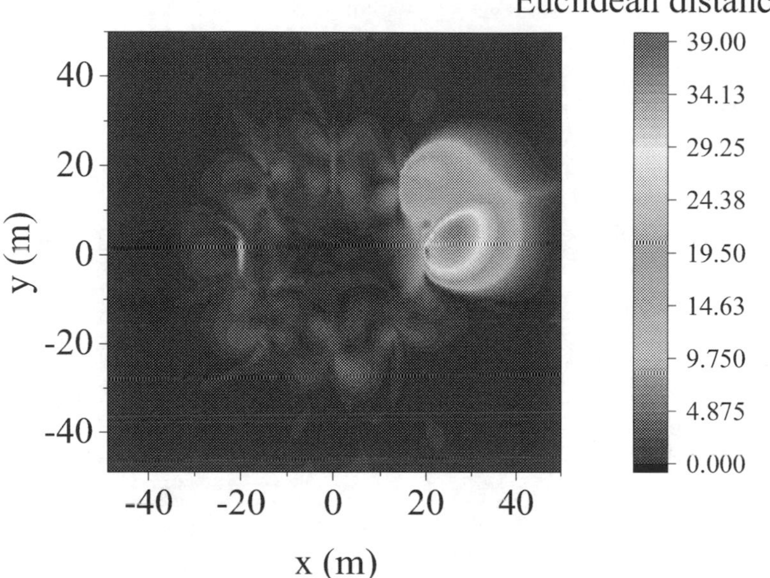

Fig. 7. H-model positioning.

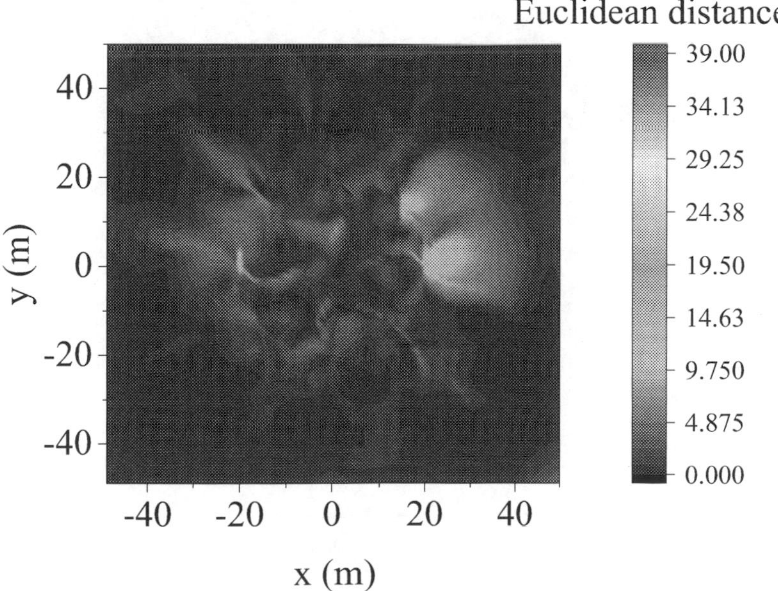

Fig. 8. E-H-model positioning.

In Fig. 6, Fig. 7 and Fig. 8, The axis of abscissa and ordinate indicates the position of the targets in the z = 15 plane. The color map bar represents Euclidean distance (absolute error) between model positioning and target's location.

The positioning error of E-model and H-model is kept within 1 m in the sub-areas without interference. Which indicates that the positioning model trained by the BP neural network has high positioning accuracy. E-H-model reduces the positioning error of E-model in the sub-areas ($x \in (-10, -40)$, $y \in (0, 20)$), reduces the positioning error of H-model in the sub-areas ($x \in (10, 40)$, $y \in (-10, 20)$), and reduces the influence of interference in all sub-areas. These results show that the E-H-model maintains high-precision positioning ability in the sub-areas with greater interference, and improves the anti-interference ability.

We fix the abscissa, and find the Root Mean Square Error (RMSE) of the Euclidean distance values of the column coordinates. The RMSE curves of the E-H-model, E-model, and H-model is drawn in Fig. 9.

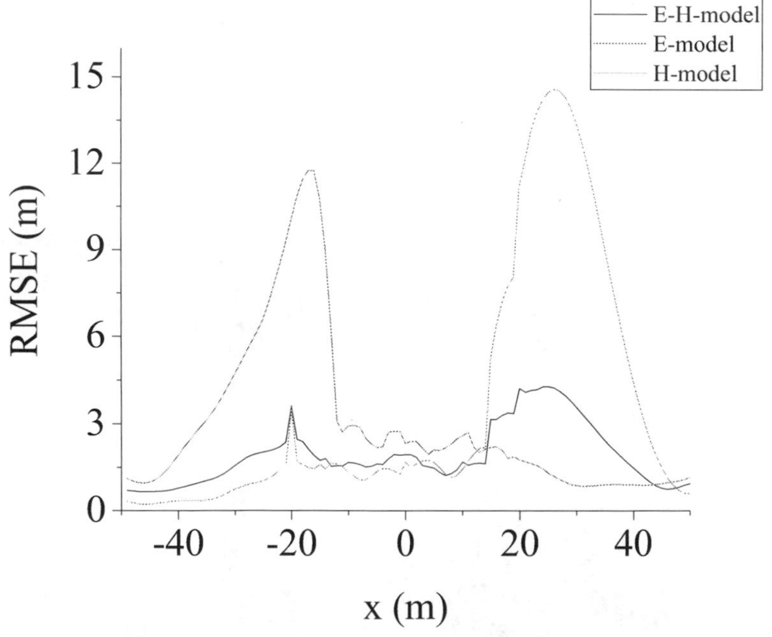

Fig. 9. Comparison of RMSE.

The larger the RMSE value, the larger the positioning error of the abscissa [13,14]. Obviously, the E-model curve is affected by electrical interference ($x = -30$), the H-model curve is affected by magnetic interference ($x = 30$), while the E-H-model curve minimizes these two interferences.

6 Conclusion

This paper applies BP neural network to train E-model and H-model, and proposes a passive electromagnetic field positioning method in underwater 3-D space. This method uses E-model and H-model to ensure positioning accuracy, and adaptively reduces the training samples with larger interference through weight factors. Experiments show that this method overcomes the problem that the E-model and H-model are greatly affected by interference, and reduces the error caused by underwater interference. This method can quickly fit underwater scenes and combine theoretical models with real scenes. It means that passive electromagnetic field positioning methods may no longer be limited to theoretical models. Applying this technology to real scenes is the main directions for deep exploration and improvement.

References

1. Jin, B., Xu, X., Zhu, Y., et al.: Single-source aided semi-autonomous passive location for correcting the position of an underwater vehicle. IEEE Sens. J. **19**(9), 3267–3275 (2019)
2. Wang, J., Li, B.: Electromagnetic fields generated above a shallow sea by a submerged horizontal electric dipole. IEEE Trans. Antennas Propag. **65**(5), 2707–2712 (2017)
3. Panyukov, A.V., Kandelousi, M.S.: Inverse problem for an electrical dipole and the lightning location passive monitoring system. In: Electric Field, pp. 283–300 (2018)
4. Pasku, V., De Angelis, A., De Angelis, G., et al.: Magnetic field-based positioning systems. IEEE Commun. Surv. Tutor. **19**(3), 2003–2017 (2017)
5. Pasku, V., De Angelis, A., Dionigi, M., et al.: A positioning system based on low-frequency magnetic fields. IEEE Trans. Ind. Electron. **63**(4), 2457–2468 (2015)
6. Shoaib, H.S.: Electromagnetic field analysis of a vertical magnetic dipole antenna in planar multi-layered media. IETE Tech. Rev. **38**, 1–8 (2020)
7. Fu, B., Kirchbuchner, F., von Wilmsdorff, J., Grosse-Puppendahl, T., Braun, A., Kuijper, A.: Performing indoor localization with electric potential sensing. J. Ambient Intell. Humaniz. Comput. **10**(2), 731–746 (2018). https://doi.org/10.1007/s12652-018-0879-z
8. Kraichman, M.B.: Handbook of Electromagnetic Propagation in Conducting Media, 1st edn. Headquarters Naval Material Command. University of Illinois at Urbana-Champaign (1970)
9. Zhang, M., Cai, W.: Multi-AUV aided cooperative 3D-localization for underwater sensor networks. Recent Adv. Electr. Electron. Eng. **13**(1), 80–90 (2020)
10. Jiang, J., Zhang, J., Zhang, L., et al.: Passive location resource scheduling based on an improved genetic algorithm. Sensors **18**(7), 2093 (2018)
11. Fang, G., Chen, H., Xie, L.: Precision-improved passive localization method in the underwater multipath environment. In: 2019 IEEE International Conference on Signal Processing. Communications and Computing (ICSPCC), pp. 1–6. IEEE, Dalian (2019)
12. Gao, L., Selleri, S., Battistelli, G., et al.: Passive target detection and tracking from electromagnetic field measurements. Int. J. RF Microw. Comput. Aided Eng. **30**(9), e22321 (2020)

13. Constable, S., Kowalczyk, P., Bloomer, S.: Measuring marine self-potential using an autonomous underwater vehicle. Geophys. J. Int. **215**(1), 49–60 (2018)
14. He, S., Liu, Y., Xiang, J.: A low cost visual positioning system for small scale tracking experiments on underwater vehicles. In: 2020 IEEE 29th International Symposium on Industrial Electronics, pp. 1370–1375. IEEE, Delft (2020)

Optimization of Cross-border E-commerce Marketing Strategy During the Pandemic

Yunfeng Wang[1] and Hongjie Fan[2(✉)]

[1] School of Management, Zhejiang University, 866 Yuhangtang Rd, Hangzhou 310058, China
[2] School of Information Management for Law, China University of Political Science and Law, 27 Fuxue Rd, Beijing 102249, China
hjfan@cupl.edu.cn

Abstract. Cross-border e-commerce, as a new business model, has gradually replaced the traditional trade model. However, during the COVID-19 pandemic, many companies encountered bottlenecks, mainly because products could not adapt to the rapid changes in consumption and products were out of stock due to weak logistics system. Through our observation of the current marketing situation and problems, we analyze that sellers need to enhance their adaptability and risk resistance in the pandemic environment, especially in the face of the new trends in cross-border e-commerce. Based on this, we propose a strategic optimization plan, including product optimization, sales channel optimization, brand optimization and logistics optimization. To ensure the implementation of the strategy, we designed a guaranteed roadmap, including user-demand oriented product development management, improvement of human resource management structure, optimization of operation structure and logistics supply chain enhancement. Finally, we hope this solution to be effective in a pandemic situation.

Keywords: Cross-border e-commerce · Marketing strategy · Optimization · Logistics supply chain · COVID-19 pandemic

1 Introduction

E-commerce is a representation of the digital transformation that continues to shape the behavior and practices of individuals and societies due to its advantages over traditional trade models, [1]. Over the past two decades, cross-border e-commerce in China has experienced continuous innovation and development from the nascent stage to the mature stage, from business-to-business (B2B) model to business-to-consumer (B2C) model, and has become the new engine of China's foreign trade exports. Driven by the cost and information advantages of cross-border e-commerce, the global market is accelerating [3].

Corresponding Author: Hongjie Fan.

S. Liu and X. Ma (Eds.): ADHIP 2021, LNICST 416, pp. 305–314, 2022.
https://doi.org/10.1007/978-3-030-94551-0_25

China's cross-border e-commerce exports started in 2003 and developed after 2013. As shown in Fig. 1, China's cross-border e-commerce exports increased from RMB 2.7 trillion in 2015 to RMB 6.7 trillion in 2020, with an average annual growth rate of 20%, making it another new market.

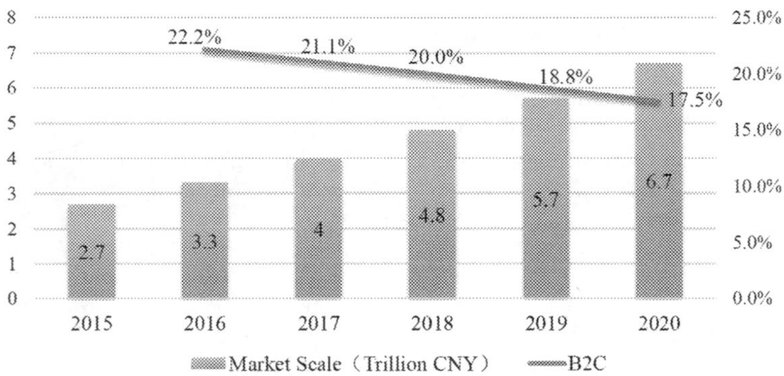

Fig. 1. China's cross-border E-commerce export market transaction scale

As shown in the Fig. 2, the share of cross-border e-commerce retail in the overall has increased from 5.2% in 2013 to 19.5% in 2019. It is estimated that by 2020, the global cross-border e-commerce retail market will reach nearly RMB 7 trillion, accounting for more than 30% of total transactions[1]. However, COVID-19 has brought serious impact and influence to the global economy, society and life, and the escalation of the pandemic has hindered economic growth. However, e-commerce, especially cross-border e-commerce, still has great potential for development. Therefore, it is valuable to analyze the impact of COVID-19 on Chinese cross-border e-commerce enterprises and explore countermeasures.

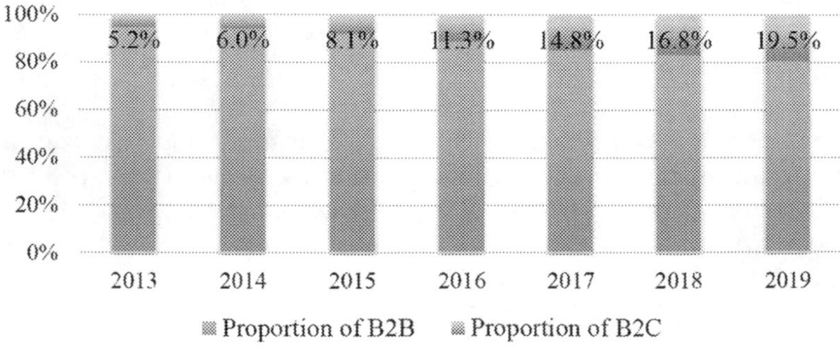

Fig. 2. B2B and B2C structure change of cross-border E-commerce

[1] http://www.jjckb.cn/2021-01/14/c_139666998.htm.

2 Analysis on the Current Situation and Problems of Marketing

Since 2015, Chinese companies have joined the market competition for B2C exports. Nevertheless, there are still some problems, such as international trade barriers [5]. Many enterprises have encountered bottlenecks in their development, mainly in the form of increasingly fierce price competition and decreasing profits. The homogenization and low quality of goods lead to fierce price competition and rising marketing costs, resulting in a rapid decline in profitability. In addition, cross-border logistics management and the difficulty of managing the logistics operations of overseas suppliers have led to high logistic costs and poor inventory planning [6,7], which seriously hinders the development of the business. In addition, due to the differences in culture and user habits, enterprises do not conduct in-depth research on the needs of the target market, resulting in products not being recognized by the market. In addition, some enterprises lack brand awareness and innovation spirit.

In North America and Western Europe, where e-commerce is highly developed, consumers are not only concerned about the price and quality of goods, but also about the brand connotation attached to these products. Although most enterprises engaged in cross-border e-commerce have registered their trademarks overseas, the proportion of investment in R&D is relatively low. In addition, some companies do not carefully analyze their target consumers and design products with brand connotations, but simply imitate foreign brands and competitive products.

In addition, the impact of COVID-19 is unprecedented. As the problems encountered in global economic and trade activities, the multinational operations of enterprises are mainly influenced by the following factors.

1. Overloaded Supply Chain Leads to Out of Stock

The global supply chain has been paralyzed by the epidemic. Due to the success of China's epidemic prevention efforts, China has become the only source of supply chain in the world. As a result, delivery times were repeatedly extended, a large number of orders could not be delivered on time, and sales shortages were a serious problem.

2. The Blocked Logistics Supply Causes Cost Rises Sharply

During the worst outbreak of the epidemic, there were significant delays and lost packages in the delivery of international postal parcels, resulting in many users' orders being cancelled. Sellers not only lost their orders, but also had to bear the cost of their products and high mailing costs. For example, shipping containers from Ningbo to Los Angeles increased nearly five-fold from $2,000 at the beginning of the year to nearly $10,000 in the fourth quarter of 2020.

3. The Capital Chain is Almost Broken

On the one hand, the supply chain has no goods to deliver and some goods are difficult to transport, leading to hoarding of goods in China. In addition, logistics costs and the devaluation of the U.S. dollar have led to a general decline in sales profits. On the other hand, many sellers could not hold on because of the break in capital.

3 New Trend of Cross-border E-commerce

1. Rapid Growth of the Global E-commerce Market

The epidemic has caused a number of problems, such as large numbers of people living in isolation and the closure of brick-and-mortar stores. This has led to a significant increase in the scale of online consumption and an acceleration in global e-commerce penetration. The formation of online consumption habits has become an irreversible trend, and cross-border e-commerce has ushered in new growth drivers.

Due to the influx of sellers into cross-border e-commerce, brand building and service upgrading to create quality products with higher added value has become a major trend.

2. Diversified Choice of Sale Channels

COVID-19 has further accelerated the pace of sales channel diversification. Cross-border e-commerce companies are no longer limited to a single channel. More and more companies are paying attention to vertical category platforms and independent online stores, such as shopify stores. In particular, independent online stores have outstanding advantages in terms of autonomy and growth, becoming an important channel for enterprises to expand into new markets.

3. Cross-border E-commerce Sellers are Polarized

While dominant companies accelerated their growth, weak sellers faced elimination. Some quality sellers saw a surge in sales and exhibited a number of characteristics. For example, they have high-quality supply and logistics chains. In addition, they have strong financial strength or stable cash flow to provide sufficient financial support for the implementation of the strategy.

On the contrary, some companies lack core competencies, such as reduced orders, declining sales, or even broken capital chains. These companies generally have poor organizational structure and sloppy management. In addition, they only have disorganized or single product lines and lack deep cooperative supply chain support, resulting in weak marketing ability and fierce price competition.

4. Changed Consumer Demand Preferences

Many companies in Europe and the United States shut down production and began to cut salaries and lay off employees. This led to a sharp rise in unemployment and a decline in the income of the general public, resulting in a lack of spending power at a later stage. Under the more serious situation of the epidemic, in addition to the rapid rise in demand for epidemic prevention materials, the demand for home office supplies, indoor fitness equipment and other goods around the “home economy” also rose sharply.

5. Insufficient Subsequent Consumption Power

The epidemic has reduced the trade activities and stagnated the economic development. Many countries have suspended work and production, and companies have begun to reduce wages and layoffs to protect themselves. Which have led to a sharp rise in the unemployment rate, and the decline in ordinary people's income will definite led to insufficient consumption capacity in the later period.

4 Optimization of Cross-border E-commerce Marketing Strategy During the Pandemic

4.1 Product Strategy Optimization

1. Product Diversification

We need to segment the differences in the needs of the population and collect the pain points of demand, so as to carry out the deep development and extension of the category. For example, through data analysis, we found that the proportion of women buying hardware tools in the U.S. is as high as 30%, so we need to deeply understand the characteristics and preferences of female tool users and conduct accurate mining. In addition, in order to fill the gaps in the product area, we need to cooperate deeply with high-quality suppliers for product complementation.

2. Product Differentiation

Taking saber saw blades as an example, they are favored by users in developed countries and are in great demand due to their safe, convenient and efficient cutting performance. However, there are still significant differences from region to region. For example, professionals in Japan and South Korea manipulate saber saw blades for indoor and outdoor cutting and equipment maintenance. In contrast, the United States has a high percentage of home DIY users, in addition to professionals. As you can see, products must be directionally targeted, and one cannot expect to develop one type for all users.

3. Brand Differentiation

A multi-brand strategy can enhance risk resistance. For example, the company's existing brand X is the core brand of the company, and all core products are aimed at high-end users in Europe and the US, which leads to relatively high prices. In this case, a new brand could fill a certain gap with the existing brand in terms of price category and target the DIY-oriented mid-range users. It can differentiate itself from the existing brand X, allowing different product types to operate under different brands, thus increasing consumer recognition and trust in the brand's professionalism.

4. Product Development and improvement

We need to strengthen the specificity of product development and the training of compound talents, thus improving the quality of the R&D team. Only in this way can we provide enough sales growth points for the future development of the company and form a systematic product competitiveness.

4.2 Marketing Channel Strategy Optimization

The traditional channel such as Walmart and HomeDepot have set up their online platform. To counter the effects of the COVID-19, these supermarkets have increase their investment for the online platform and lowered their entry requirements. With access to large supermarkets such as HomeDepot, the huge number of user visits will inevitably expand the company's business rapidly.

Although developed countries in Europe and the US have huge market size and huge consumer demand, the market competition is relatively fierce. Emerging markets such as Russia, Brazil and Southeast Asia have become the fastest growing B2C export markets. In this case, we can use B2B platform and Internet technology to integrate traditional B2B trade and develop the market with B2B2C (business-to-business, business-to-customer) business model.

In addition, China is the world's largest e-commerce consumer market, and the incremental space for e-commerce platforms is huge. In this case, we should study user groups in depth, grasp consumption trends, and design products that meet the needs of the domestic market.

4.3 Brand Strategy Optimization

When the product meets or even exceeds consumers' expectations, the good word of mouth formed can better promote the spread of value. At the same time, we need to provide professional pre-sales, sales and after-sales services to give consumers the best shopping experience and increase the product repurchase rate. In addition, we can visually show the advantages by shooting videos and comparing competing products to win consumers' trust.

5 Guarantee for the Implementation of Cross-border E-commerce Marketing Strategy

Marketing has an important impact on the enterprise whether can achieve the expect [8]. To ensure the success of various marketing strategies, managers need to establish a correct marketing perception and keep up with the times.

5.1 Establish the Management System of Product Development Oriented by User's Demand

1. Establish demand analysis and product selection

All cross-border e-commerce data plays an important role [9], and can be collected through technical means. Through a large amount of feedback, companies can easily understand the concerns of consumers and accurately locate their needs. Based on this, we can make full use of big data for user profiling and competitive analysis to achieve a shift from experience-based product development to product selection based on big data analysis. We believe that this is effective in improving the success of a company's product strategy.

2. Standardized project implementation process

The project implementation process should be standardized to ensure an optimal balance between product quality, project speed and total product development cost. By following these guidelines, we can smoothly implement the company's new product initiatives.

3. Establish consumption feedback mechanism

E-commerce makes it possible to provide timely feedback on consumer information. For consumer feedback on products, we need to track, analyze and evaluate them regularly. In addition, we need to use the product search volume, click rate and product conversion rate provided by big data to analyze consumers' product preferences and summarize effective product upgrade plans from consumers' feedback. Then product iteration and optimization will be continuously carried out.

5.2 Improve Human Resource Management Structure

With the development and depth of artificial intelligence, digitalization is gradually extending to the direction of intelligence. According to the needs of the company's overall development strategy, we need to increase recruitment efforts, optimize the talent structure and improve talent selection. At the same time, we need to improve the execution of employees and enhance their sense of belonging so that they will voluntarily devote themselves to work in line with the vision.

5.3 Optimize Operation Structure

The company's refined operations can effectively reduce product costs, logistics costs and marketing and promotion costs. We suggest using big data to improve the company's user demand analysis and market research capability, and complete user portraits of target customers from a series of data such as consumers' age, income, gender, and purchasing behavior. In addition, we can delve into the company's business rules through data statistics, and then adjust the product strategy and marketing strategy at any time.

5.4 Strengthen the Management of Logistics Supply Chain

The supply chain includes production and transportation, and finally the product is delivered to the consumer through the logistics system. Giuffrida et al. [10] collated 32 papers published in international journals or conferences from 2002 to 2016, and summarized the scientific research results in the field of logistics in China. Golan et al. [11] conducted a comprehensive literature review on supply chain resilience in the context of the COVID-19 pandemic.

While COVID-19 is impacting the global supply chain, we should pay close attention to and take seriously the diversification of the supply chain, including supplier diversification and logistics diversification. In addition, the use of third-party mature overseas warehouse centers as a transit point for export sales can effectively prevent economic losses caused by out-of-stock. For the long-term development of cross-border e-commerce, overseas warehouses are inevitable [12] although overseas warehouses require a certain amount of investment under the premise of relevant requirements for sellers, and sellers must be able to predict sales trends more accurately and arrange production and inventory in advance.

In order to ensure the smooth implementation of the company's marketing strategy, increasing overseas warehouses and forming a global logistics and warehousing layout will be an important guarantee for the efficient implementation of the company's marketing strategy. As shown in the Fig. 3, the addition of overseas warehouses will greatly improve the flexibility of the Company's logistics and warehousing and reduce the risk of product out-of-stock due to delays in logistics and distribution. According to the company's existing business distribution, it can give priority to choosing suitable third-party overseas warehouse service providers and prioritize the layout of the U.S. overseas warehouse.

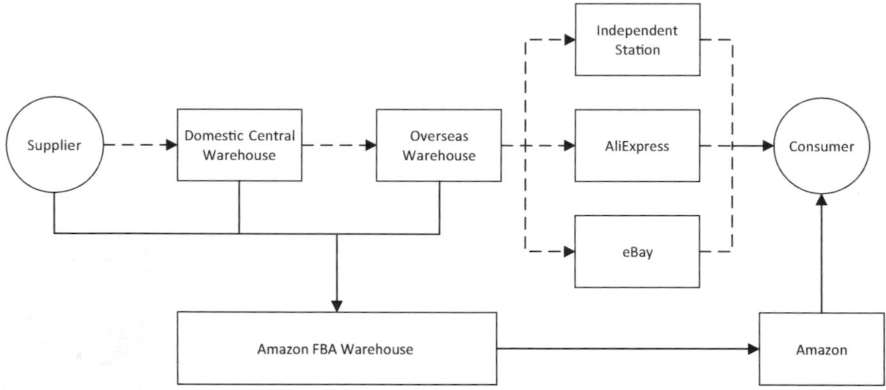

Fig. 3. Company logistics warehousing optimization

In addition, information technology can effectively help companies improve operational efficiency and better handle issues related to various aspects of supply chain operation, especially in the case of expanding category size. The use of information systems instead of manual forecasting to improve logistics efficiency has become more urgent. We need to integrate the resources and logistics information of each link in the whole supply chain, analyze the trend of product demand and warehouse warning based on historical sales data, so as to improve the efficiency of supply chain management. At the same time, we need to improve the depth of cooperation with high-quality foreign suppliers to complement our independent research and development capabilities. In addition, effective consumer demand feedback and market trends can help us determine the direction of our products.

5.5 Optimize Capital Management

In the development process, entrepreneurs should pay attention to the management of capital risk. The centralized management of capital can effectively reduce the capital risk of the enterprise [13]. The centralized management of capital in enterprises is conducive to improving the efficiency of capital use in enterprises and maximizing benefits in the true sense [14]. Andrew et al. [15] firstly combined supply chain management with overall capital management. Hofmann et al. [16] analyzed supply chain-based logistics and capital flows and compared them with the traditional single-item perspective of working capital management, thus shifting the optimization of working capital management to supply chain-based improvements.

We need to comprehensively analyze the inventory and capital flow of products from the market side to the supplier side so as to regularly sort out the total inventory of products and the capital tied up in each product. At the same time, we need to reduce or eliminate products that affect the capital turnover rate, such as controlling unreasonable capital occupation, in order to relieve capital pressure. At present, unsecured and unsecured fast financing method is the primary choice for working capital financing. Because the data of the e-commerce industry from logistics, commodity flow to transaction flow are convenient to assess the credit risk of an e-commerce enterprise. In addition, the scale of venture capital has grown rapidly in recent years. The cross-border e-commerce industry is characterized by fast liquidity and rapid growth, which makes the investment scale of the venture capital market remain large. In addition, banks are actively providing loans to enterprises engaged in cross-border e-commerce. They can provide good credit lines for small and medium-sized enterprises.

6 Conclusion

Cross-border e-commerce has gradually replaced traditional trade models and has shaped individual and societal behaviors and practices due to its advantages. Based on the analysis of the current situation and competitive environment of cross-border suppliers in the hardware tools industry under the influence of COVID-19, we believe that the cross-border supplier industry is in a period of accelerated growth, and the hardware tools category has great potential in the post-epidemic era.

We first study the current situation of the development of the cross-border e-commerce industry under the influence of the epidemic, including overloaded supply chain, obstructed logistics supply and tight capital chain, as well as the main issues and trends in the post-epidemic era. We analyze the need for cross-border e-commerce companies to enhance their adaptability and risk-resilience in the pandemic.

Based on this, we propose strategic optimization solutions, including product optimization, sales channel optimization and brand optimization. In this regard, to ensure the success of various marketing strategies, we need to establish the right marketing concepts and keep up with the times. To ensure the implementation of the strategy, we design and implement a roadmap. Through the research in this paper, we hope to provide reference for other cross-border e-commerce enterprises to optimize their marketing strategies, and to provide reference for traditional foreign trade export enterprises planning to enter the cross-border e-commerce field.

References

1. Nuray, T.: The impact of e-commerce on international trade and employment. J. Procedia Soc. Behav. Sci. **24**(1), 745–753 (2011)
2. Gomez, E., Martens, B., Turlea, G.: The drivers and impediments for cross-border e-commerce in the EU. J. Inf. Econ. Policy. **28**, 83–96 (2014)
3. Li, Y.M.: Analysis on E-commerce crisis in China in the international trade environment. J. Mod. Econ. **1**, 71–76 (2016)
4. Zhang, X.H.: Impact of COVID-19 on small and micro businesses in China and countermeasures. J. China Bus. Market **3**, 24–32 (2020)
5. Guy, H.G., Coral, R.S.: Designing e-commerce cross-border distribution networks for small and medium-size enterprises incorporating Canadian and US trade incentive programs. J. Res. Transp. Bus. Manage. **16**, 84–94 (2018)
6. Wang, H.: Research on the construction of international logistics supply chain management model from the perspective of cross-border ecommerce. In: Reform Strategy (in Chinese) (2017)
7. Wang, J., Zheng, H.: Research on integration of customs clearance in "the Belt and Road" region based on perspective of cross-border e-commerce. J. Beijing Technol. Bus. Univ. (2018) (in Chinese)
8. Chung, H.: International marketing standardization analysis: a cross-national investigation. J. Asia Pac. J. Mark. **19**, 140–145 (2007)
9. Zhang, H.: Study on cross-border e-commerce logistics optimization platform based on big data. J. Revista de la Facultad de Ingenieria **32**(4), 329–335 (2017)
10. Giuffrida, M., Mangiaracina, R., Perego, A. et al.: Cross-border B2C e-commerce to greater China and the role of logistics: a literature review. J. Int. J. Phys. Distrib. Logist. Manage. **47**(6) (2017)
11. Golan, M.S., Jernegan, L.H., Linkov, I.: Trends and applications of resilience analytics in supply chain modeling: systematic literature review in the context of the COVID-19 pandemic. J. Environ. Syst. Decis. **40**, 222–243 (2020). https://doi.org/10.1007/s10669-020-09777-w
12. Pan, H.L.: Research on international logistics supply chain management mode from the perspective of cross-border e-commerce. In: Xu, Z., Choo, K.-K.R., Dehghantanha, A., Parizi, R., Hammoudeh, M. (eds.) CSIA 2019. AISC, vol. 928, pp. 737–744. Springer, Cham (2020). https://doi.org/10.1007/978-3-030-15235-2_101
13. James, W., De, L.: Enterprise risk management: practical implementation ideas. In: Protiviti, pp. 1–7 (2005)
14. Robert, K.: Working capital management and shareholder's wealth. J. Rev. Finance **1**, 200–210 (2013)
15. Andrew, A., Roger, F., George, N., et al.: The changing role of builders merchants in the construction supply chain. J. Constr. Manage. Econ. **16**, 351–361 (1998)
16. Hofmann, E., Kotzab, H.: A supply chain-oriented approach of working capital management. J. Bus. Logist. **31**(2), 305–330 (2010)

Deep Residual Network with Transfer Learning for High Spatial Resolution Remote Sensing Scenes Classification

Ziteng Wang, Wenmei Li, and Guan Gui(✉)

College of Telecommunications and Information Engineering,
Nanjing University of Posts and Telecommunications, Nanjing 210003, China
{1019010409,liwm,guiguan}@njupt.edu.cn

Abstract. Deep residual network (DRN) is considered a promising image classification method for high spatial resolution remote sensing (HSRRS) images due to its great feature extraction capabilities. The classification performance of the DRN is greatly relies on the size of training samples. However, the sample size of HSRRS images is relatively small due to the high acquisition cost. Blindly increasing the sample size would requires huge computing resources and image annotation cost, but would not necessarily improve the classification performance of DRN. In this paper, a transfer learning-aided DRN method (TL-DRN) is proposed for a few shot learning to address the performance challenges associated with HSRRS with relatively small sample size and explore the impact of sample size on classification performance. In the experiment, the weights (shared knowledge) obtained by training the ImageNet datasets on the DRN model are transfered to the TL-DRN model. Experiments with ten different small-scale training sample sizes are conducted. Experimental results show that when the total training sample size is increased from 10 to 100, the classification performance of the TL-DRN model tends to be stable and the mean accuracy of its testing set has stabilized at around 94%. TL-DRN shows a superiority of up to 16% over DRN, in terms of classification accuracy.

Keywords: Convolutional neural network · Images classification · High spatial resolution remote sensing · Transfer learning

1 Introduction

With the enhancement of remote sensing satellite images resolution, we can obtain more information to conduct more impactful research related to land planning, disaster prevention and so on. Classification is one of the most important tasks in HSRRS image processing. HSRRS images have rich spatial, shapes, textures and colors features, which provide good basis for the classification [1].

Deep learning (DL)-based image classification methods usually adopt convolutional neural networks (CNNs) to automatically extract image feature, which

S. Liu and X. Ma (Eds.): ADHIP 2021, LNICST 416, pp. 315–325, 2022.
https://doi.org/10.1007/978-3-030-94551-0_26

gets rid of the complex artificial feature design [2–4]. Generally, the deeper the neural network, the better the classification performance [5]. On account of its excellent feature extraction capabilities, CNNs have been widely applied in the tasks of signal modulation recognition [6,7], voice signal processing and other fields. In recent years, CNNs have developed rapidly, and more and more image classification models based on CNNs achieved excellent results like VGG [8], ResNet [9], InceptionV3 [10] and so on. However, these classification models mainly rely on a great quantity of training samples to obtain high accuracy. Otherwise overfitting may occur [11], which means that trained models have the prefect performance on training samples, while perform badly on the independent test samples. Due to the difficulty of acquiring HSRRS images, the dataset of HSRRS is small. In recent years, a great deal of research have shown that transfer learning could greatly alleviate the phenomenon of over-fitting of the classification model with small sample size [12]. Y. Boualleg *et al.* proposed a CNN-DeepForest based on deep forest and CNNs transfer learning for HSRRS images classification [13]. Xue *et al.* proposed a MSDFF model based on multi-structure deep features fusion for HSRRS image classification [14]. However, there is still a gap in the drive to find optimum sample size for transfer learning to achieve the best classification performance.

In this paper, we propose a TL-DRN model for HSRRS image classification. The proposed model is used to train ten groups of datasets with different sample sizes to explore the influence of the sample size on the model training. The mean accuracy is used to evaluate the performance of the model. The main contributions of this paper can be summarized as below:

1. TL-DRN for HSRRS image classification with limited sample size is proposed. Experiments have proved that the TL-DRN model is more suitable for HSRRS image classification in the case of small sample size than the DRN model.
2. The impact of sample size on TL-DRN are studied. Experimental results confirm that the performance of TL-DRN tends to be stable when the sample size of training reaches six per category.

2 Theoretical Basis

In this section, CNNs, ResNet50 and transfer learning are introduced in detail.

2.1 CNNs

CNNs is a special artificial neural network. Its main feature is the ability to perform convolution operations. Therefore, CNNs is excellent in image classification, detection and segmentation [15].

The input of CNNs is often raw data such as images and audio. The structure of CNNs is a hierarchical model composed of convolutional layers, pooling layers, fully connected layers [16], and activation functions. The original input information undergoes layer-by-layer operation to extract feature information.

Then, those information is used for classification through the fully connected layer [17].

When designing a CNNs, the number of channels in the convolution layer should be equals to the input data. The number of convolution kernels should be the same as that of channels output from this layer. The convolution kernel generally has two attributes: stride (s) and padding (p). Output size of the feature layer after convolution can be obtained by the following calculation formula [18]:

$$n = \frac{N + 2p - f}{s} + 1 \tag{1}$$

where N, f and n represent the size of the input, the convolution kernel and the output respectively.

When the image passes through the convolutional part, low-level convolutional layers extract low level semantic features like texture and shape etc., while high level convolutional layers extract high level semantic features [19]. In general, high level semantic features are more convenient for image classification. Finally, the feature information output by the convolutional layer is mapped to the labeled sample space through the fully connected layer to complete the classification task.

2.2 ResNet50 and DRN

As the number of convolutional layers is increased, the high-level semantic features of the image can be better extracted. However, the deep network may have the problem of gradient disappearance or explosion, which hinders the convergence of the network, otherwise known as the degradation problem. To solve the issue of degradation, ResNet50 is proposed. The ResNet50 network is a stack of residual networks. The Fig. 1 shows the structure of the residual network. The principles of the residual network are as follows.

The residual network consists of one residual unit. First, the residual unit can be written as

$$y_l = h(x_l) + F(x_l, W_l) \tag{2}$$

$$x_{l+1} = f(y_l) \tag{3}$$

where x_l and x_{l+1} respectively represent the input and output of the l-th residual unit [20], and every residual unit includes a multi-layer structure generally. F is the residual function, indicating the learned residual. In addition, $h(x_l) = x_l$ indicates the identity mapping and f represents the rectified linear unit (relu) activation function which is expressed in Eq. (4). The way of adding a highway between the output and input of the network allows us to easily solve the problem of gradient dispersion and network performance degradation.

$$f(x) = \max(x, 0) \tag{4}$$

It should be note that the convolution part of the ResNet50 network is called as deep residual network (DRN) for the convenience of writing.

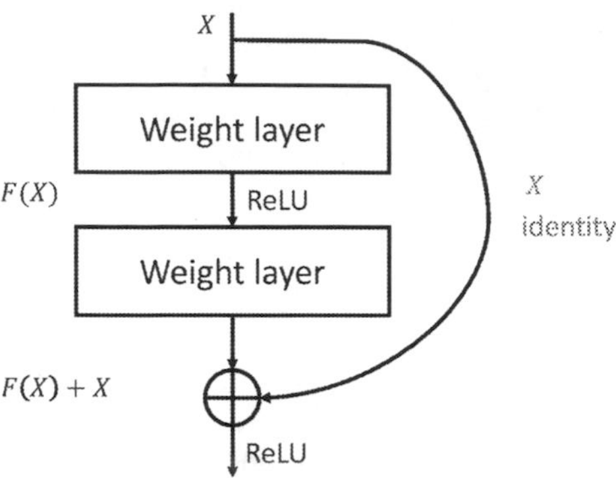

Fig. 1. The architecture of the residual block.

2.3 Transfer Learning

Transfer learning is a learning method for small sample training [21]. It applies the knowledge and experience learned in other tasks to the current task.

In transfer learning, domains (D) and tasks (T) are defined, domains are divided into source domains (D_s) and target domains (D_t), and tasks are divided into source tasks (T_s) and target tasks (T_t). The domain includes feature space and edge probability distribution. Given D_s, T_s, D_t, and T_t, transfer learning uses the knowledge learned from D_s and T_s to enhance the learning of the prediction function f for D_t, where $T = f(D)$, $T_s \neq T_t$ and $D_s \neq D_t$ [22].

In terms of image classification processing, some studies have found that no matter which image dataset is input into the CNNs, the features extracted from the low level convolutional layers are similar. A great deal of researches have proved that the features extracted by a dataset after CNNs are often applicable to another dataset [23]. Therefore, based on this feature, we conduct transfer learning on small samples.

3 The Proposed HSSRS Classification Based on DRN and Transfer Learning

In this section, the structure of TL-DRN network, the training method of TL-DRN, the objective function and evaluation indexs used in the experiments will be introduced in detail.

3.1 The Structure of TL-DRN

TL-DRN is a deep learning network model composed of DRN and transfer learning. First, we build the convolution part of the TL-DRN according to the structure of DRN to prepare for migration. After that, two sets of fully connected layers are added behind the DRN for classification. The framework of the model is presented in Fig. 2.

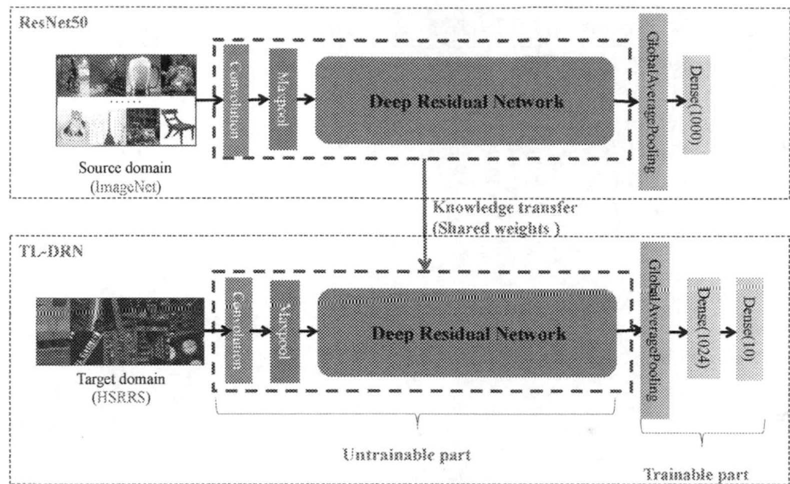

Fig. 2. The framework of ResNet50 and TL-DRN.

3.2 The Method of TL-DRN Training

The TL-DRN training method consists of three parts: ResNet50 training, network reconstruction and feature transfer, and TL-DRN training. The process is presented in Algorithm 1.

ResNet50 Training. ResNet50 is composed of CNNs and residual blocks, which can solve the gradient explosion problem caused by the increase of the network level. Figure 2 shows the model structure. ImageNet [24] dataset is used as the source domain to training ResNet50 and it can be expressed as:

$$D_S = \{(x_1, y_1), (x_2, y_2), \ldots, (x_n, y_n)\} \tag{5}$$

where $\boldsymbol{x}_i = [x_i^{(1)}, x_i^{(2)}, \ldots, x_i^{(k)}]^T$ $(i = 1, 2, \ldots\ldots, n)$, $x_i^{(j)}$ represents the j-th feature of the i-th input data of source domain, y_i represents the true label category of source domain. The model can be expressed as:

$$F_S = f_{ResNet50}(\theta_{DRN}, \theta_{fc}; x_i) \tag{6}$$

where F_S represents the output of the model. θ_{DRN} indicates the weight parameter obtained by training the deep residual convolution part and θ_{fc} indicates the weight parameter obtained from the fully connected layer training.

Network Reconstruction and Feature Transfer. TL-DRN is constructed according to Fig. 2. Since ResNet50 and TL-DRN have the same DRN structure, the weight of the convolution part obtained by ResNet50 training can be extracted and loaded into the convolution part of TL-DRN. The dataset used

Algorithm 1. The proposed TL-DRN method.

```
Input: 10 categories of HSRRS images;
Output: The TL-DRN;
  Training and testing the ResNet50 on the ImageNet dataset;
  Constructing TL-DRN network based on DRN network of ResNet50;
  for j-th layer in TL-DRN_layers do
    for i-th layer in DRN_layers do
      if j-th layer name == i-th layer name then
        Load the weights of layer i to layer j;
        Freeze the weights of layer j;
      else
        Pass;
      end if
    end for
  end for
  Training and testing the TL-DRN on the HSRRS images ;
  return TL-DRN.
```

in TL-DRN is ten-category images of HSRRS. In transfer learning, this dataset represents the target domain, so it can be written as:

$$D_T = \{(x_1, y_1), (x_2, y_2), \ldots, (x_n, y_n)\} \tag{7}$$

where $\boldsymbol{x}_i = [x_i^{(1)}, x_i^{(2)}, \ldots, x_i^{(k)}]^T$, $x_i^{(j)}$ represents the j-th feature of the i-th input data of target domain, y_i represents the true label category of target domain. The model can be expressed as:

$$F = f_{TL-DRN}(\theta_{DRN}, \theta_{newfc}; x_i) \tag{8}$$

where, the value of θ_{DRN} in Eq. (8) is the value of θ_{DRN} trained in Eq. (6). θ_{newfc} indicates the weight parameter obtained from the TL-DRN fully connected layer training.

TL-DRN Training. After network reconstruction and feature transfer, the TL-DRN model can learn the knowledge θ_{DRN} obtained by ResNet50 training. Therefore, we only need to train the parameters θ_{newfc} of the fully connection layer.

3.3 Loss Function

The multi-class categorical cross entropy loss function is used as the loss function and it can be written as:

$$L = -\sum_{i=1}^{N} y_i \log(F_i) \tag{9}$$

where y_i is the true label of i-th input and F_i is the result of the i-th output of the model. N is the number of categories. We aims to train the model to find a suitable set of θ to minimize the loss function.

3.4 Evaluation Index of Experimental Results

Firstly, line chart is used to show the training process, where the horizontal axis represents the epoch of training, the left vertical axis indicates the mean accuracy, and the right vertical axis indicates the loss value.

Secondly, mean accuracy (MA) judge the overall performance of the classification and its calculation formula can be written as [25]:

$$\mathrm{MA} = \frac{1}{N}\sum_{i=1}^{n} C_{ii} \tag{10}$$

where N is the total sample size of the testset, n is the total number of categories to be classified, C_{ii} is the number correctly classified for class i.

4 Experimental Results

In this section, we collect different datasets, conduct experiments according the Section III, and analyze the results of the experiments.

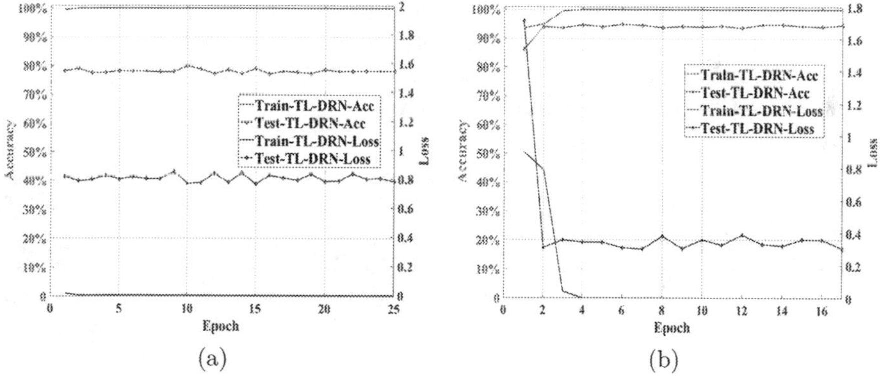

Fig. 3. Accuracy and loss line chart during the period of training and testing for training sample size is (a) one, (b) ten per category with TL-DRN.

4.1 Data Description

From the UCMerce LandUse [26] and RSI-CB128 datasets [27], ten-category HSRRS image samples are selected for classification. One image per category is randomly selected to form a training dataset. Afterwards, $\{2, 3, 4, \cdots, 10\}$ images are selected in the same way as in the above. These ten sample sets are used as small training samples to explore the influence of sample size on TL-DRN classification. Ten images per category are selected randomly to form a training dataset. After that, $\{20, 30, \cdots, 100\}$ images per category are selected in the same way as in the above. As a control group, these 10 sample sets with larger training samples are directly trained by ResNet50. The specific training sample size is shown in Table 1 and Table 2. In the experiment, all test sets used are the same, with one hundred samples in each category. It should be noted that the following operations are performed in the image preprocessing stage: (1) The input picture size is uniformly changed to the format of 224 × 224 × 3; (2) The data is effectively expanded by rotating and translating the image. Therefore, the data set has been expanded twice on the original basis.

4.2 Experiments Setting

There are three experiments in this part. The experiment one explores the influence of sample size on the transfer learning model. The experiment two and experiment three serve as control learning groups. First of all, for the first experiment, ten sets of training data with a small sample size were selected. The sample sizes of these ten training sets are $\{1, 2, 3, \cdots, 10\}$ per category. The TL-DRN model is built according to Fig. 2. Experiments with 10 sets of data are conducted according to the TL-DRN training method in Section III. Secondly, for the second experiment, the same data set as TL-DRN experiment is selected. The ResNet50 model is built according to Fig. 2 and used to train 10 sets of training samples separately. At last, for the third experiment, ten sets of training data with a larger sample size are selected. The sample sizes of these ten training sets are $\{10, 20, 30, \cdots, 100\}$ per category. After that, ResNet50 is used to train ten sets of training samples directly.

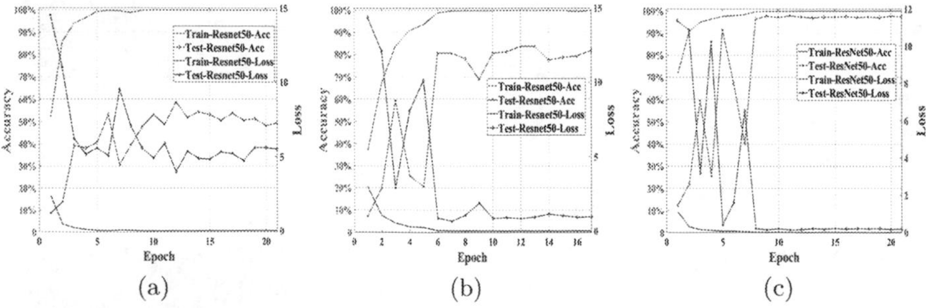

Fig. 4. Accuracy and loss line chart during the period of training and testing for training sample size is (a) one, (b) ten and (c) one hundred per category with ResNet50.

4.3 Experimental Results

First of all, the line graphs of the training process of the three experiments when the sample size is the smallest and the sample size is the largest are shown in Figs. 3 and 4. Each line graph shows the changes of the accuracy and loss of the training set and testing set as the number of training iterations increases. Comparing the training process diagram with the smallest sample size and the largest

Table 1. Testing set mean accuracy on TL-DRN and ResNet50.

TSSPC[a]	1	2	3	4	5	6	7	8	9	10
TMA/T[b] (%)	76.60	83.90	89.00	93.30	94.30	94.80	94.40	94.20	94.80	94.30
TMA/R[c] (%)	52.70	45.00	62.40	71.00	73.70	71.90	72.40	82.10	75.20	79.40

[a] Training sample size per category
[b] Test Mean Accuracy/TL-DRN
[c] Test Mean Accuracy/ResNet50

Table 2. Testing set mean accuracy on ResNet50.

TSSPC[a]	10	20	30	40	50	60	70	80	90	100
TMA[b] (%)	79.40	91.50	90.10	90.60	92.90	95.50	95.80	96.10	94.60	95.90

[a] Training sample size per category
[b] Test Mean Accuracy/ResNet50

sample size of each experiment, we can see that as the sample size increases, the overfitting problem of the model is better mitigated. For example, the MA of the training set and testing set of Fig. 3(a) differs by about 20% while differs by about 6% of Fig. 3(b). Moreover, for small samples, transfer learning can reduce the overfitting situation (This can be understood by comparing to the image of Fig. 3(b) and Fig. 4(b)).

For the first experiment, the test MA obtained by the test experiment in the case of each sample size are given in Table 1. It can be seen clearly that with the increase of the sample size, the classification performance of our proposed TL-DRN is also greatly improved. However, the classification performance of the test fluctuates little, by about 0.3% when the sample size reaches six per category. The MA of the testing set is around 94.4%.

For the second experiment, comparing the test results of TL-DRN and the test results of ResNet50 in the same sample size in Table 1, it can be found that transfer learning greatly improves the performance of classification with small samples size. At the highest level, TL-DRN improves accuracy by nearly 39% compared to ResNet50, and at the lowest level, it has a nearly 12% improvement.

For the third experiment, the test MA obtained by the test experiment in the case of each sample size are given in Table 2. Compared with the first experiment, in the case of ten times the training sample size of the first experiment, the classification performance of ResNet50 is slightly better than that of TL-DRN generally. At the highest level, ResNet50 improves accuracy by nearly 7.6% compared to TL-DRN, and at the lowest level, it has a nearly 2.7% drop.

5 Conclusion

In this paper, the influence of the sample size on the classification of TL-DRN model for ten-category of HSRRS images was investigated. When the sample size of ten types of HSRRS images reaches six per category, the classification performance of TL-DRN network tends to be stable. In addition, the classification effect of the TL-DRN model is far better than that of ResNet50 on training samples of the same magnitude. And when the training sample size of ResNet50 is 10 times that of TL-DRN, the classification effect of TL-DRN is only slightly lower than ResNet50. It was also shown that TL-DRN is a good candidate for classification of HSRRS images. However, when the sample size increases to a certain level, the continued increase of sample size has little effect on performance. We will continue to be committed to improving the classification performance of the model through other methods in the case of small samples.

References

1. Zhong, Y., Han, X., Zhang, L.: Multi-class geospatial object detection based on a position-sensitive balancing framework for high spatial resolution remote sensing imagery. ISPRS J. Photogramm. Remote Sens. **138**, 281–294 (2018)
2. Arredondo-Velzquez, M., Diaz-Carmona, J., Torres-Huitzil, C., Barranco-Gutirrez, A.I., Padilla-Medina, A., Prado-Olivarez, J.: A streaming accelerator of convolutional neural networks for resource-limited applications. IEICE Electron. Express **16**(23), 20190633 (2019)
3. Gui, G., Huang, H., Song, Y., Sari, H.: Deep learning for an effective nonorthogonal multiple access scheme. IEEE Trans. Veh. Technol. **67**(9), 8440–8450 (2018)
4. Wang, Y., Yang, J., Liu, M., Gui, G.: LightAMC: lightweight automatic modulation classification via deep learning and compressive sensing. IEEE Trans. Veh. Technol. **69**(3), 3491–3495 (2020)
5. Chen, Y., Jiang, H., Li, C., Jia, X., Ghamisi, P.: Deep feature extraction and classification of hyperspectral images based on convolutional neural networks. IEEE Trans. Geosci. Remote Sens. **54**(10), 6232–6251 (2016)
6. Wang, Y., Liu, M., Yang, J., Gui, G.: Data-driven deep learning for automatic modulation recognition in cognitive radios. IEEE Trans. Veh. Technol. **68**(4), 4074–4077 (2019)
7. Wang, Y., et al.: Automatic modulation classification for MIMO systems via deep learning and zero-forcing equalization. IEEE Trans. Veh. Technol. **69**(5), 5688–5692 (2020)
8. Simonyan, K., Zisserman, A.: Very deep convolutional networks for large-scale image recognition. arXiv preprint arXiv:1409.1556 (2014)
9. He, K., Zhang, X., Ren, S., Sun, J.: Deep residual learning for image recognition. In: Proceedings of the IEEE Conference on Computer Vision and Pattern Recognition, pp. 770–778 (2016)
10. Szegedy, C., Vanhoucke, V., Ioffe, S., Shlens, J., Wojna, Z.: Rethinking the inception architecture for computer vision. In: Proceedings of the IEEE Conference on Computer Vision and Pattern Recognition, pp. 2818–2826 (2016)
11. Liu, S., Deng, W.: Very deep convolutional neural network based image classification using small training sample size. In: 2015 3rd IAPR Asian Conference on Pattern Recognition (ACPR), pp. 730–734. IEEE (2015)
12. Torrey, L., Shavlik, J.: Transfer learning. In: Handbook of Research on Machine Learning Applications and Trends: Algorithms, Methods, and Techniques, pp. 242–264. IGI global (2010)
13. Boualleg, Y., Farah, M., Farah, I.R.: Remote sensing scene classification using convolutional features and deep forest classifier. IEEE Geosci. Remote Sens. Lett. **16**(12), 1944–1948 (2019)

14. Xue, W., Dai, X., Liu, L.: Remote sensing scene classification based on multi-structure deep features fusion. IEEE Access **8**, 28746–28755 (2020)
15. Krizhevsky, A., Sutskever, I., Hinton, G.E.: ImageNet classification with deep convolutional neural networks. Adv. Neural Inf. Process. Syst. **25**, 1097–1105 (2012)
16. Wu, H., Gu, X.: Max-pooling dropout for regularization of convolutional neural networks. In: Arik, S., Huang, T., Lai, W.K., Liu, Q. (eds.) ICONIP 2015, Part I. LNCS, vol. 9489, pp. 46–54. Springer, Cham (2015). https://doi.org/10.1007/978-3-319-26532-2_6
17. Wiatowski, T., Bölcskei, H.: A mathematical theory of deep convolutional neural networks for feature extraction. IEEE Trans. Inf. Theory **64**(3), 1845–1866 (2017)
18. LeCun, Y., Bottou, L., Bengio, Y., Haffner, P.: Gradient-based learning applied to document recognition. Proc. IEEE **86**(11), 2278–2324 (1998)
19. Long, M., Zhu, H., Wang, J., Jordan, M.I.: Deep transfer learning with joint adaptation networks. In: International Conference on Machine Learning, pp. 2208–2217. PMLR (2017)
20. Zagoruyko, S., Komodakis, N.: Wide residual networks. arXiv preprint arXiv:1605.07146 (2016)
21. Pan, S.J., Yang, Q.: A survey on transfer learning. IEEE Trans. Knowl. Data Eng. **22**(10), 1345–1359 (2009)
22. Li, W., et al.: Classification of high-spatial-resolution remote sensing scenes method using transfer learning and deep convolutional neural network. IEEE J. Sel. Top. Appl. Earth Obs. Remote Sens. **13**, 1986–1995 (2020)
23. Yosinski, J., Clune, J., Bengio, Y., Lipson, H.: How transferable are features in deep neural networks? Adv. Neural Inf. Process. Syst. **27**, 3320–3328 (2014)
24. Marmanis, D., Datcu, M., Esch, T., Stilla, U.: Deep learning earth observation classification using ImageNet pretrained networks. IEEE Geosci. Remote Sens. Lett. **13**(1), 105–109 (2016)
25. Luo, W., Yang, W., Yu, X., Wang, Y., Tan, K.: Lightweight convolutional neural network for high-spatial-resolution remote sensing scenes classification. In: 2020 International Conference on Wireless Communications and Signal Processing (WCSP), pp. 104–108 (2020)
26. Yang, Y., Newsam, S.: Bag-of-visual-words and spatial extensions for land-use classification. In: Proceedings of the 18th SIGSPATIAL International Conference on Advances in Geographic Information Systems, pp. 270–279 (2010)
27. Li, H., et al.: RSI-CB: A large scale remote sensing image classification benchmark via crowdsource data. arXiv preprint arXiv:1705.10450 (2017)

Research on E-commerce Logistics Transportation Route Planning Method Based on Recurrent Neural Network

Xian-Bin Xie[1] and Chi-ping Li[2(✉)]

[1] School of Economics and Management, Hunan Software Vocational College, Xiangtan 411100, China
[2] Guangzhou Maritime University, Guangzhou 510725, China

Abstract. Aiming at the problem that the traditional transportation route planning method can adaptively allocate a small number of transportation points, which leads to the long time required for the final route planning, a recurrent neural network-based e-commerce logistics transportation route planning method is studied. Use recursive neural network to extract e-commerce logistics characteristics, calculate the adaptive probability parameters shown by individual transportation individuals, combine gene blocks to control adaptively processed transportation points, set transportation route neighborhood search mechanisms, and build transportation route cost constraint numerical relationships. Combining the congestion parameters generated when the vehicle is running, set the route planning plan, and finally complete the research on the transportation route planning method. After preparing the e-commerce logistics transportation data, simulate the planning of the e-commerce logistics transportation route, apply two traditional transportation route planning methods and the designed route planning method to experiment, and the results show: the planning time required for the designed route planning method The shortest.

Keywords: Recurrent neural network · E-commerce logistics transportation route · Planning method · Transportation point

1 Introduction

With the rapid development of cross-border e-commerce, cross-border e-commerce logistics has shown a relatively lagging situation. Data shows that 80% of the current international express delivery market is dominated by the "four major" international express companies, namely DHL, UPS, FEDEX and TNT. Some domestic express companies engaged in international express business still occupy a relatively small market share. In this context, domestic cross-border electronics suppliers generally cost high freight costs to deliver goods to overseas consumers [1]. In addition, the delivery cycle of these international express goods is generally too long, the overseas warehouses are seriously underweight, the return and exchange procedures are complicated and the cost

S. Liu and X. Ma (Eds.): ADHIP 2021, LNICST 416, pp. 326–343, 2022.
https://doi.org/10.1007/978-3-030-94551-0_27

is high. The lack of cooperation among cross-border e-commerce companies in our country, the loose body of supply chain logistics, and the unreasonable structure are another prominent problem.

Most cross-border e-commerce companies establish cooperation with cross-border logistics companies independently, without considering joint navigation to reduce the total cost of international express transportation. According to reports, since international express mails often need to meet a certain amount before they start shipping, if small items want to be shipped to customers quickly, they must pay a higher fee separately, which makes cross-border e-commerce companies "timeliness and cost" The contradiction between the two is in a dilemma. The other is the serious lack of after-sales service capabilities. According to relevant research released by the China E-commerce Research Center, returns are almost impossible to achieve in cross-border e-commerce. One is because the return of goods needs to go through various procedures and the process is very complicated, and the other is because of returns. The cost of logistics payment almost exceeds the price of goods, which leads to the phenomenon that many returned goods are directly destroyed in the bonded warehouse, and merchants send new products to customers [2]. Therefore, the lag of cross-border e-commerce logistics is not only reflected in the backward logistics infrastructure, but also in the lack of awareness of logistics cooperation and environmental constraints.

From the practice of the pilot, it can be seen that promoting the integration and coordinated development of logistics resources is an inevitable trend to promote the development of cross-border e-commerce. The strength of logistics capabilities determines the prosperity of cross-border e-commerce in a region. In the practice of promoting the development of collaborative logistics, the policies implemented by each pilot can be reflected in these aspects. First, establish industry common standards to promote win-win cooperation between logistics companies [3]. Because in the environment of lack of cooperation, logistics companies pay more attention to their competitive position among the peers, and pay more attention to competition rather than cooperation. A development model oriented to pure competition will cause logistics companies to only care about their own profits, and not to pay attention to the benefits that environmental improvements can bring to individual companies.

Aiming at the problem that the number of allocated transportation points is small and the time required for route planning is too long, this paper designs an e-commerce logistics transportation route planning method based on recurrent neural network. The specific research ideas are as follows:

Firstly, the recursive neural network is used to extract the characteristics of e-commerce logistics and calculate the adaptive probability parameters of individual transportation,

Secondly, by combining gene blocks to control the adaptive transportation points, the neighborhood search mechanism of transportation path is set, and the numerical relationship of transportation route cost constraint is constructed;

Then, combined with the congestion parameters generated by vehicles, set the route planning scheme, and finally complete the research on the transportation route planning method.

Finally, the effectiveness of this method is verified by experiments.

2 Research on E-commerce Logistics Transportation Route Planning Method Based on Recurrent Neural Network

2.1 Use Recurrent Neural Network to Extract E-commerce Logistics Features

When using the recurrent neural network to extract the characteristics of e-commerce logistics, the memristive numerical model in the network is called to sort the characteristics of e-commerce logistics into feature parameters [3], and the numerical relationship can be expressed as:

$$k = \frac{\left(R_o - R_f\right)^2}{D^2} \tag{1}$$

Among them, R_o represents the recursive cycle data set, R_f represents the e-commerce logistics characteristic function, and D represents the calibration parameter. Corresponding to the sorted characteristic parameters, a characteristic control process is formed, and the numerical relationship can be expressed as:

$$M(t) = \begin{cases} q(t) & t \leq 1 \\ M(o) + k_i & 0 < k_i < c_1 \\ q(k) & k > 0 \end{cases} \tag{2}$$

Among them, $q(t)$ represents a characteristic control function, $M(o)$ represents a continuous function, k_i represents an internal characteristic parameter, c_1 represents a memory parameter, and $q(k)$ represents a characteristic function. Under the control of the characteristic parameters, sort out the numerical changes of the characteristic function in the numerical interval. The numerical changes are shown in the following figure (Fig. 1):

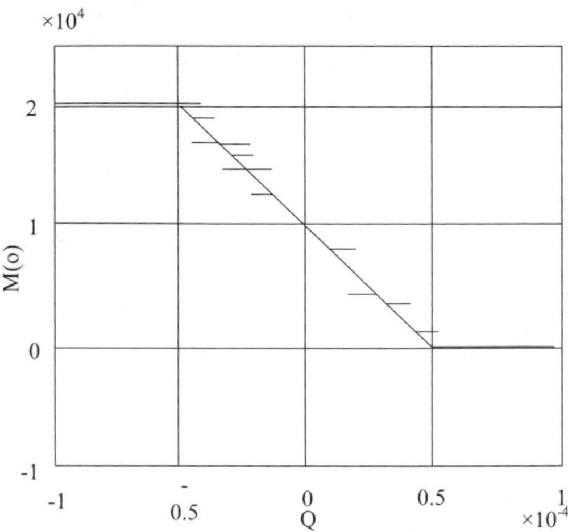

Fig. 1. Characteristic function changes

Under the change of the characteristic function shown in the figure above, combined with the characteristics of the numerical change, the intercept generated during the change is obtained. The numerical relationship can be expressed as:

$$b_0 = R - \sqrt{\frac{2(b - R_f)^2}{D}} \tag{3}$$

Among them, R represents the characteristic parameter, and b represents the calculated intercept parameter. Based on the above processing process, construct an intelligent value acquisition process, which can be expressed as:

$$u(e) = e\frac{\sum_{i=1}^{p} e(i)}{k-1} \tag{4}$$

Among them, $p_i = e^{\frac{t_1}{\sum f_i}}$ represents the control input parameter, $p_i = e^{\frac{t_1}{\sum f_i}}$ represents the numerical proportional function, and $p_i = e^{\frac{t_1}{\sum f_i}}$ represents the integral parameter of the e-commerce logistics movement path, and the meaning of the remaining parameters remains unchanged. Corresponding to the transportation paths of different control layers, combined with different types of e-commerce, the single neuron structure is set to a control process, which can be expressed as:

$$J = -k\frac{e}{q(i)} \tag{5}$$

Among them, $p_i = e^{\frac{t_1}{\sum f_i}}$ represents the neuron function in the control layer, and the meaning of the remaining parameters remains unchanged. Under the control of the above single neural structure, the transportation network is fixedly connected to the coordinate system, and the movement trajectory of the transportation logistics vehicle control point is processed into a homogeneous form, which can be expressed as:

$$\begin{cases} a = \left[a_x, a_y, a_z\right]^T \\ b = \left[b_x, b_y, b_z\right]^T \\ c = \left[c_x, c_y, c_z\right]^T \end{cases} \tag{6}$$

Among them, a, b, c represent the unit vector generated during the transportation process, and T represents the degree of freedom period. After sorting out the collected characteristic data, set up a neighborhood search mechanism for transportation routes.

2.2 Set Up a Nearby Search Mechanism for Transportation Routes

There are many configurable routes in the actual e-commerce transportation route. When setting the transportation route neighborhood search mechanism, adjust the neighborhood search probability according to its own fitness level, and obtain new ones through neighborhood search for individuals with low fitness. Individual, the adaptive probability shown by the transport individual can be expressed as:

$$p_i = e^{\frac{t_1}{\sum f_i}} \tag{7}$$

Among them, f_i represents the e-commerce transportation individual, t_1 represents the time required for the transportation process, and p_i represents the adaptive function probability. It can be seen from the above calculation formula that the adaptive probability depends on the fitness of the population and the fitness of the individual. If the population fitness remains unchanged, it will decrease as the individual fitness increases, that is, the smaller the individual fitness [4], the greater the probability of neighborhood search. On the contrary, the greater the fitness of the individual, the smaller the probability of neighborhood search, which is accomplished by gene inversion operations.

The above logistics distribution cost model is designed based on the problem of vehicle route planning. How to design a fast and effective intelligent optimization algorithm is the key to solving this problem. The basic Drosophila optimization algorithm is a fast optimization algorithm with high efficiency and few adjustable parameters, but its main disadvantages are that it is easy to fall into local search and cannot solve optimization problems with discrete variables. Here, under its basic framework, the idea of simulated annealing algorithm and genetic algorithm is introduced, and an optimization algorithm suitable for the above model is discussed. The algorithm flow chart is shown in the figure below (Fig. 2):

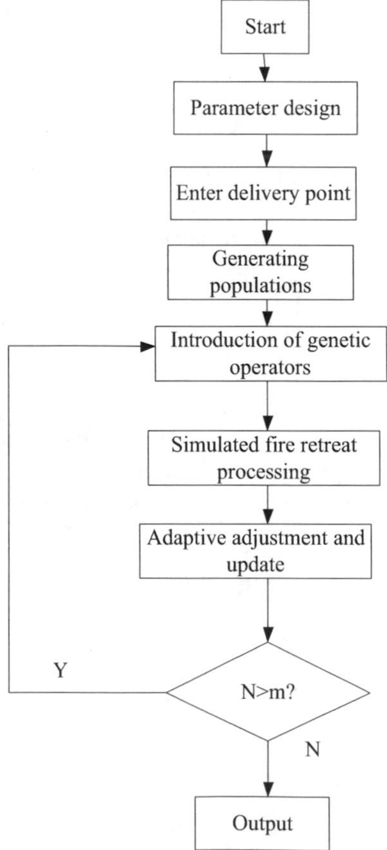

Fig. 2. Constructed neighborhood search algorithm

Under the neighborhood search algorithm shown in the figure above, the number of delivery vehicles and the number of vehicles are deployed, and the delivery task of each truck is abstracted into a set of gene blocks. The delivery plan is expressed as a combination of different gene blocks, and the delivery combination processing process can be expressed for:

$$m = \frac{l + r}{h} \tag{8}$$

Among them, l represents the length of the delivery route, r represents the radius of the similar path, and h represents the length in the gene block. Continuously adjust the distribution process corresponding to the node, adjust the distribution process, the adaptive processing process can be expressed as:

$$f(u) = \frac{\gamma \sum_{k=1}^{K} u_k}{v} \tag{9}$$

Among them, γ represents the delivery parameters, u_k represents the path planning function, and v represents the delivery speed. Under the corresponding distribution speed control, the trend of route planning changes, as shown in the following figure (Fig. 3):

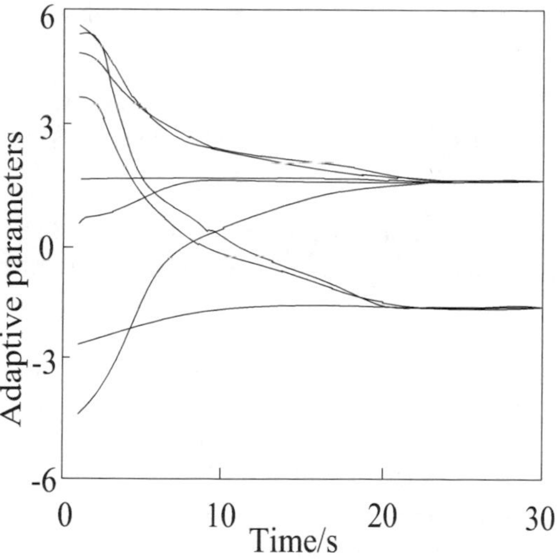

Fig. 3. Route planning trends

Under the route planning trend control shown in the figure above, the pheromone generated during the initial delivery trajectory correction process is used to select the planning direction, and the execution heuristic parameters of the lower layer can be expressed as

$$\rho(t) = \frac{\alpha}{\sum_{k} (\tau(t)\eta_k)} \tag{10}$$

Among them, α represents the number of pheromone produced, $\tau(t)$ represents the visibility function, η_k represents the transfer parameter, and k represents the number of paths. In the process of continuous execution, in order to enhance the accuracy of executing adaptive parameters and control the adjustment amount of pheromone, the processing process can be expressed as:

$$\Delta\tau = \frac{p^k}{d} \tag{11}$$

Among them, $\Delta\tau$ represents the adjustment amount, p^k represents the pheromone parameter, and d represents the calibration path length. After controlling the distance of processing pheromone, introduce a random direction, execute the above-mentioned structural formula, set the random probability constant, and fix the pheromone within a logistics adaptation range [5], and build an adaptive pheromone update process. Expressed as:

$$L\tau = \frac{1}{n}\sum_{i=1}^{n}|B_i|^2 \tag{12}$$

Among them, n is the number of environmental intervals in the logistics matching interval, and B_i represents the logistics structure function. Taking the search mechanism set above as the restriction condition, construct the numerical relationship of the transportation route cost constraint.

2.3 Constructing Numerical Relations of Cost Constraints for Transportation Routes

Under the control of the above-mentioned transportation path neighborhood search mechanism, whether it is to establish a new distribution center or transform the original equipment base into a logistics product distribution center, the enterprise needs to pay related on-site purchase fees, construction fees, construction facility rental fees, and building materials Expenses, etc., are construction costs. After the construction of the distribution center is completed, it will enter the operation stage [6]. In order to ensure the seamless connection between the upstream and downstream of the supply chain, the products can be efficiently transferred in the distribution center, and certain labor costs need to be paid, which is the operating cost of the distribution center. Due to the small difference in daily transaction volume, equipment purchase fees, water and electricity fees, and therefore operating costs have relatively small changes. Assuming that the construction and operating costs of each candidate center are fixed and known conditions, the construction and operating costs of the e-commerce transportation transfer station can be expressed as:

$$\begin{cases} C_r = \dfrac{\sum_{r\in G} Z_r}{F} \\ C_a = \dfrac{\sum_{r\in G} A_r}{F} \end{cases} \tag{13}$$

Among them, C_r represents the construction cost, C_a represents the operating cost, Z_r represents the project construction function, F represents the construction period, A_r represents the maintenance cost function, and G represents the distance parameter of the transit center. The transportation cost specifically refers to the vehicle cost incurred by fuel and vehicle maintenance during the delivery of the product to the customer. Transportation costs are closely related to distance traveled, vehicle speed and time. The influencing factors are the vehicle model, the unit distance cost of the vehicle, the driving speed, the distance from the supplier to the distribution center, the distance from the distribution center to each retailer [7], the distance between the retailer and the retailer, and the distance of the vehicle Travel time, etc. The transportation cost of fresh food cold chain logistics mainly includes two parts. One part is the transportation cost from the supplier to the distribution center, and the other part is the transportation cost from the distribution center to the retailer:

$$C_i = \frac{C_p + d_r}{Z_r} \tag{14}$$

Among them, C_p represents the cost incurred during driving, d_r represents the distribution function, and the meaning of the remaining parameters remains unchanged. In order to control the redundant time generated in the distribution process, using the hybrid time window to control the time generated during the distribution process and control the loss, the numerical relationship can be expressed as:

$$C_p = \frac{t \sum\limits_{r \in G} Z_r}{E_k} \tag{15}$$

Among them, E_k represents the redundant function generated by the e-commerce distribution process, and the meaning of the remaining parameters remains unchanged. The control cost constraint is divided into five processing levels, and the cost constraint level changes are shown in the following figure (Fig. 4):

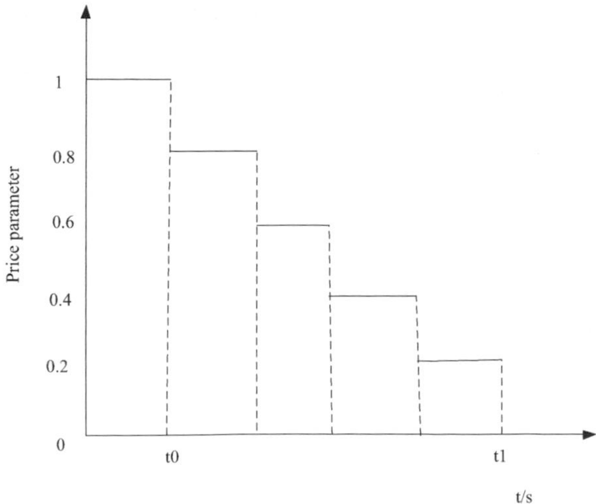

Fig. 4. Cost constraint level

Under the cost level control shown in the above figure, the two-level planning model is used to control costs, and both the upper and lower levels have unique target decision-making powers. It adheres to the principle of implementing planning and decision-making under the condition of maximizing the interests of all people, and discovering the best decision among the upper and lower goal decisions, so that its decision-making strategy achieves a systematic and complete effect [8]. In this way, the control power is in the hands of different levels of managers, so that the overall planning can be carried out more fully and comprehensively, and the value standards of all parties have been considered. Therefore, this method is closer to the reality. Under the control of the set value agreement value relationship, corresponding to the value size of different parameters, set the transportation route planning plan.

2.4 Realize Transportation Route Planning

Under the above-mentioned e-commerce transportation cost constraint numerical relationship, combined with the impact of urban road network traffic conditions, the distribution process is treated as a time-sensitive process [9], and the driving speed is time-varying. It is difficult to directly calculate the travel time of the vehicle on the distribution section, and it needs to span a time period to complete a certain section. Therefore, when constructing the transportation route plan, calculate the vehicle travel time under the time-varying road network condition to control the congestion speed of the vehicle. At this time, the vehicle is congested The duration can be expressed as:

$$t_h = \begin{cases} \frac{d_i}{v_c}, v_c \leq a \\ \frac{v_i - v_c}{v_f}, 0 < v_c \leq u \\ \frac{d_i}{v_f}, t_a \geq a + 1 \end{cases} \quad (16)$$

Among them, v_i represents the travel speed of the logistics vehicle when the traffic is not congested, v_f represents the average speed during half-way congestion, u represents the congestion coefficient, t_a represents the travel time, and v_c represents the travel speed when the entire journey is congested. Construct an adaptive motion process controlled by a single neuron, which can be expressed as:

$$\frac{\overline{M}}{V}\kappa = \overline{B}\tau \quad (17)$$

Among them, $\overline{M}$ represents the angular velocity of the moving vehicle, $\overline{B}$ represents the attitude function generated by the moving vehicle, and κ represents the formation parameter, and the meaning of the remaining parameters remains unchanged. Under different paths of logistics transportation, the corresponding formation parameter values are different. After the formation parameters are fuzzy processed, a set of output parameters are output, which can be expressed as:

$$\begin{cases} \tau_1 = \dfrac{\sum\limits_{k=1}^{n} \mu_{k-1}}{\kappa} \\ \tau_2 = \dfrac{\sum\limits_{k=1}^{n} \mu_{k+1}}{\kappa} \end{cases} \tag{18}$$

Among them, τ_1 and τ_2 represent the output parameters, μ_k represents the fuzzy output function, and the meaning of the remaining parameters remains unchanged. Processing the above output processing parameters are integrated into a data set, in order to arrange all output parameters into a reasonable output sequence of path nodes, and control the convergence of the movement speed function relationship of logistics vehicles, and control the movement speed to satisfy the following numerical relationship:

$$\lim_{t \to t+1} \left\| \frac{U_F}{U_t} \right\| = 0 \tag{19}$$

Among them, U_F represents the ideal driving process of the logistics vehicle, and U_t represents the actual convergence function produced by the logistics transportation vehicle in the time period. In order to control the logistics transportation vehicle as the best operating point, the driving path is treated as a controllable space, as shown in the following figure (Fig. 5):

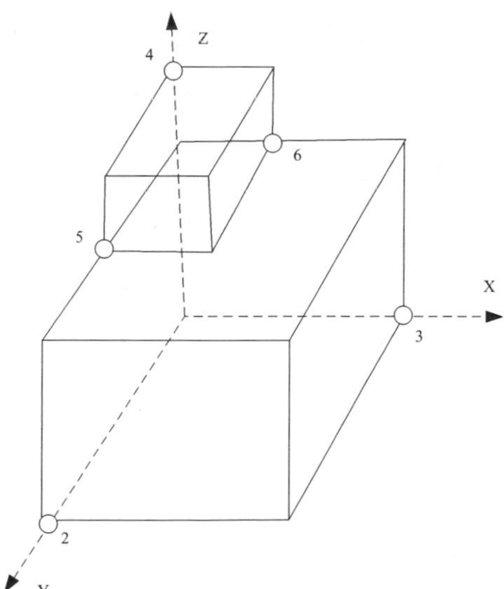

Fig. 5. Regulation of operating space

In the operating space shown in the figure above, when the transport vehicle travels to a point in the space, use MATLAB to compile a calculation example program to optimize the model and design algorithm for the established three-dimensional loading constraint of the fresh product recycling path Perform calculations to solve. In the process of calculation, the vehicle loading inspection algorithm is called for each pick-up route generated by the genetic taboo algorithm, and the loading inspection algorithm is used to check whether the fresh packaging of each route can be loaded successfully [10] to ensure that each cycle All the pickup routes can be loaded successfully. Within the calibrated value points, the circular transportation route formed is as shown in the figure below (Fig. 6):

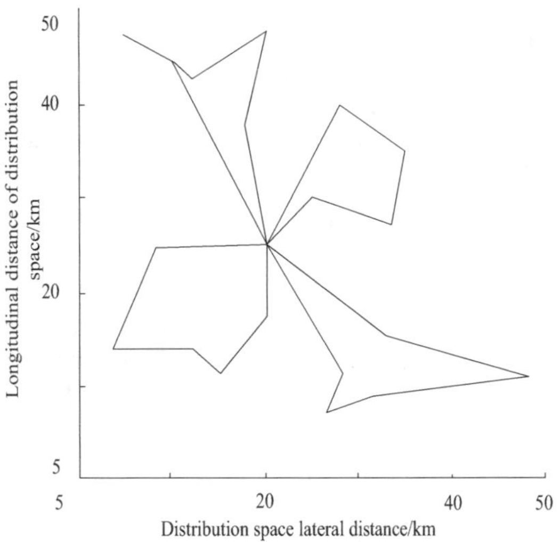

Fig. 6. Formed transportation route plan

Under the transportation route plan shown in the above figure, the existing operation center is used as the processing point, and after the traffic congestion route is reasonably avoided, the planning of the logistics transportation route is finally realized.

3 Simulation

3.1 Experiment Preparation

Since there is currently no standard case database for multi-logistics center joint delivery, and considering that there are many situations in the customer orders of e-commerce companies, the calculation examples 8201, 8202, 8203, 8204, 8205, Solomon's VRP standard test problem database are used. 8206, 8207, 8208, 8209, 8210, 8211, 8212, RC201, RC202, RC203, RC204, RC205, RC206, RC207 and RC208, use the customer coordinates, time window, demand, service time and other data of the above examples as experiments Basic data of the test case. The number of customers in each example is 100, divided into two different customer coordinate distributions. The customer coordinates of the R type example are random distribution, and the customer coordinates of the RC type example are mixed distribution. Each calibration example has 2 logistics centers, the coordinates are (33, 30), (67, 61).

According to the actual situation of urban distribution, the total service time of the logistics center is set to 960 min, and the earliest time for the vehicle to depart from the logistics center is 7:00, set to 0 time, corresponding to the READY TIME of the logistics center in the calculation example, and return to the logistics center The latest time is 23:00. Every 15 min is regarded as a time period, divided into 64 time periods, that is, 7:00 to 7:15 is the first time period, and so on. According to the urban traffic law, set 8:00 to 9:00 and 18:00 to 19:00 as the traffic jam time period, and the other time periods as the normal driving time period. The average vehicle speed is set to 60 km/h, and the vehicle speed during the traffic jam time period is set to 30 km/h. After the basic parameters are set, use Matlab R2016a to program and run on a computer with a 2.30 GHz processor and 4G memory. In order to verify the effectiveness of the multi-logistics center joint distribution model, a comparison experiment was conducted between the multi-logistics center joint distribution mode and the multi-logistics center independent distribution mode. Within the area, plan the distribution path of e-commerce logistics, and the planned distribution path structure is shown in the following figure (Fig. 7):

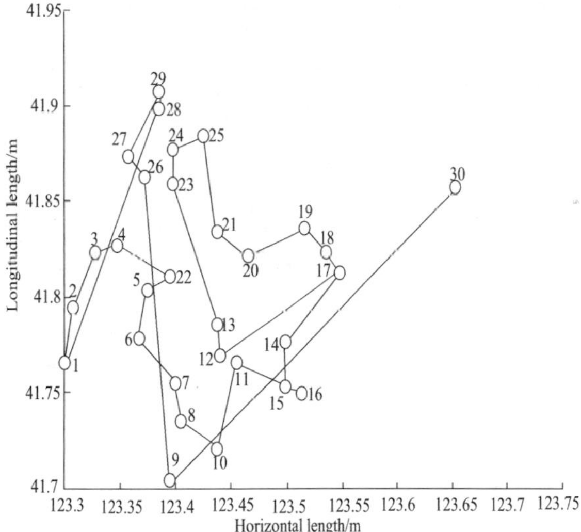

Fig. 7. Set delivery route

Under the distribution route set in the above figure, a distribution center finally determined in 10 alternative logistics centers is No. 37, the location is (123.45, 41.8), 4 distribution vehicles are used, and two traditional route planning methods are used. The designed route planning method is tested to compare the performance of the three route planning methods.

3.2 Results and Analysis

Based on the above experimental preparations, 20 examples in the test question library were randomly selected, and the e-commerce logistics timeliness was defined according to the parameters of the examples. The numerical relationship of the timeliness parameters can be expressed as:

$$PX = \frac{\left(\sum \theta(t_i)d_i\right)}{\sum d_i} \tag{20}$$

Among them, $\theta(t_i)$ represents the time cost consumed, and d_i represents the actual time of the vehicle's arrival. Under the control of the above-mentioned numerical relationship, the final time-dependent parameter results of the three route planning methods are shown in the following Table 1:

Table 1. Results of the time-dependent parameters of the three route planning methods

Study name	Aging parameter		
	Traditional route planning method 1	Traditional route planning method 2	Route planning method
8201	0.31	0.67	0.87
8202	0.33	0.57	0.85
8203	0.34	0.51	0.92
8204	0.36	0.69	0.86
8205	0.39	0.51	0.91
8206	0.37	0.61	0.88
8207	0.36	0.53	0.86
8208	0.33	0.55	0.88
8209	0.32	0.62	0.88
8210	0.33	0.67	0.88
8211	0.34	0.53	0.89
8212	0.38	0.52	0.91
RC201	0.38	0.64	0.91
RC202	0.33	0.54	0.92
RC203	0.32	0.58	0.89
RC204	0.35	0.56	0.89
RC205	0.32	0.59	0.85
RC206	0.39	0.52	0.93
RC207	0.39	0.66	0.88
RC208	0.34	0.58	0.94

According to the defined numerical relationship of timeliness, it can be seen that under the control of three route planning methods, the planned routes show timeliness of different numerical magnitudes. From the parameter results in the above table, it can be seen that the timeliness parameter of the traditional route planning method 1 is around 0.34, and the route planning by this route planning method has the worst timeliness. The traditional route planning method 2 has an average time-efficiency parameter of about 0.6, and this route planning method has better time-efficiency for planning routes. The time-efficiency parameter of the designed route planning method is about 0.89. Compared with the two traditional path planning methods, the time-efficiency parameter of the designed route planning method is the largest, and the time-efficiency of the planned transportation route is the strongest.

In the above experimental environment, point 1 is calibrated as the starting point of logistics transportation, and the three route planning methods are controlled. The planning points are 8, 16, 20, 13, 22, 27, 26, and point 30. In actual delivery, compare the three The route result of transportation route planning, the route result of route planning is shown in the figure below (Fig. 8):

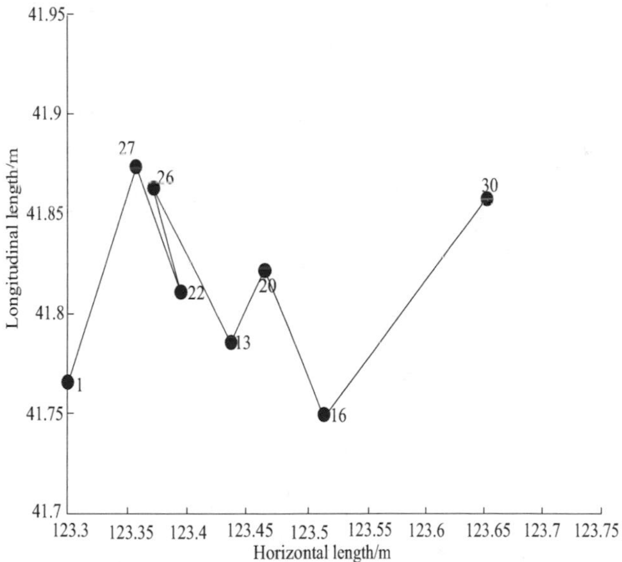

(a) Traditional route planning method 1 Route planning results

Fig. 8. Three route planning methods planning route results

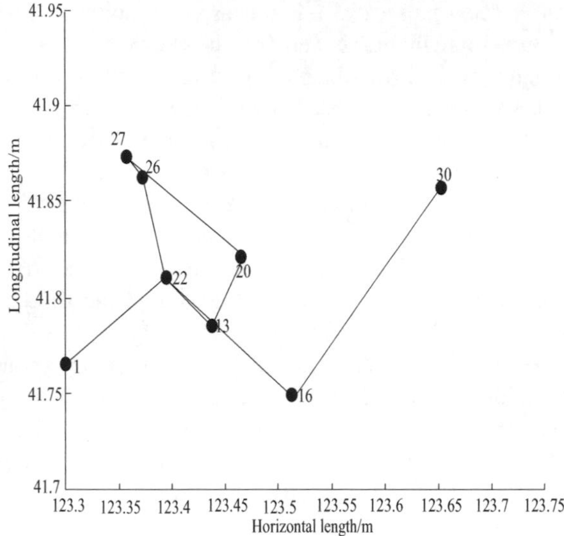

(b) Traditional route planning method 2 Route planning results

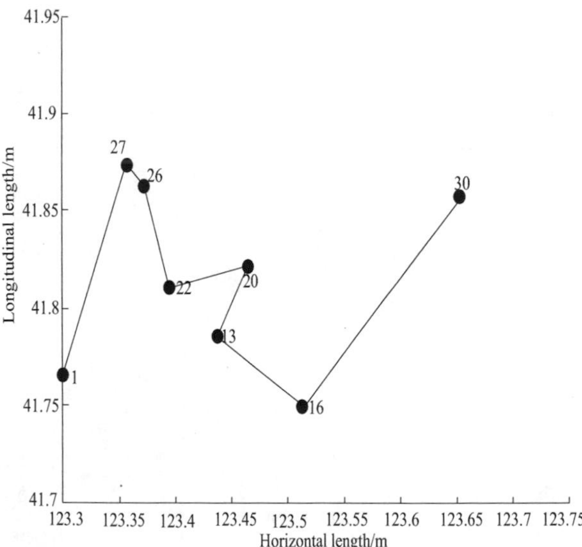

(c) Designed route planning method route planning results

Fig. 8. continued

Under the control of three kinds of route planning, the points in the route of the experimental preparation distribution plan are selected. According to the actual planning results, it can be known that the traditional route planning method 1 can obtain the longest route and the actual transportation cost is relatively large. The traditional route planning method 2 The obtained route is longer, and the actual cost is lower. The designed transportation route planning method can obtain the shortest route. Compared with the two traditional route planning methods, the designed route planning method has the shortest route planning.

Keeping the above experimental environment unchanged, corresponding to the experimental data sets prepared for the experiment, apply three route planning methods to process the e-commerce transportation data in the calculation example, take the host computer to start planning and processing the e-commerce logistics transportation route as the time statistics starting point, and display the final route structure as the time statistics cut-off point, and finally the planning time results required by the three route planning methods, The mean planning time of the three route planning methods under 50 iterative experiments is obtained, as shown in Table 2.

Table 2. Planning time required for the three route planning methods

Study name	Planning time/s		
	Traditional route planning method 1	Traditional route planning method 2	Route planning method
8201	9.1	6.4	4.6
8202	13.7	6.7	4.5
8203	10.8	6.8	4.3
8204	12.9	6.4	4.3
8205	12.3	7.1	4.4
8206	9.4	6.4	4.4
8207	11.2	6.6	4.5
8208	13.1	6.7	4.5
8209	12.7	6.7	4.5
8210	12.9	6.8	4.6
8211	11.5	6.9	4.7
8212	11.4	7.7	4.7
RC201	12.4	6.6	4.4
RC202	10.8	7.2	4.1
RC203	12.9	6.1	4.7

(*continued*)

Table 2. (*continued*)

Study name	Planning time/s		
	Traditional route planning method 1	Traditional route planning method 2	Route planning method
RC204	12.9	6.9	4.3
RC205	10.9	7.1	4.1
RC206	12.5	6.9	4.1
RC207	11.1	7.2	4.5
RC208	12.3	6.2	4.8

Under the time statistical period defined above, according to the planning time results obtained from the statistics in the above table, the average planning time of the traditional route planning method 1 is about 11.8 s, and the actual planning time required is the longest. The average planning time required by the traditional route planning method 2 is about 6.7 s, and the actual planning time required for the transportation route is relatively short. The average planning time required by the designed route planning method is about 4.4 s. Compared with the two traditional route planning methods, the designed route planning method requires the shortest planning time.

4 Concluding Remarks

The advantage of cross-border e-commerce is that it has stronger information aggregating and processing capabilities than traditional trade. The reduction of information asymmetry increases the mutual understanding of the two trades and reduces unnecessary doubts. At the same time, the cost and efficiency of information interaction are also Greatly increase. Use information advantages to reduce its costs and increase its transaction volume. Second is the effectiveness of information. Comprehensive information helps both parties to the transaction establish intuitive trust. However, after all, the differences in the field, product, and characteristics of the transaction cause huge differences in information. The buyer needs to be able to find it for his own field or product. Comprehensive and true information, and this information has a strong reference value. The availability of logistics data is a characteristic embodiment of the improvement of logistics service quality. For route planning, data is an intuitive tool for observing its past credit, experience, and achievements. Its value lies in reflecting the development of e-commerce logistics and transportation.

This paper will study how to obtain lower e-commerce logistics transportation cost in the future.

References

1. Wang, L., Wang, B.-q., Liu, J.-g., et al.: Study on malicious program detection based on recurrent neural network. Comput. Sci. **46**(07), 86–90 (2019)
2. Chen, H., Chen, J.: Recurrent neural networks based wireless network intrusion detection and classification model construction and optimization. J. Electron. Inf. Technol. **41**(06), 1427–1433 (2019)
3. Tan, F., Li, C., Xiao, H., et al.: A thermal error prediction method for CNC machine tool based on LSTM recurrent neural network. Chin. J. Sci. Instrum. **41**(09), 79–87 (2020)
4. Li, F., Xiang, W., Chen, Y., et al.: State degradation trend prediction based on double hidden layer quantum circuit recurrent unit neural network. J. Mech. Eng. **55**(06), 83–92 (2019)
5. Li, S., Wang, S.: Recurrent convolutional neural networks-based mobile robot localization algorithm. Comput. Eng. Appl. **55**(10), 240–243+249 (2019)
6. Liu, S., Liu, X., Wang, S., Muhammad, K.: Fuzzy-aided solution for out-of-view challenge in visual tracking under IoT-assisted complex environment. Neural Comput. Appl. **33**(4), 1055–1065 (2020). https://doi.org/10.1007/s00521-020-05021-3
7. Liao, L., Zhang, X.: Multi-depot low-carbon logistics vehicle routing considering customer satisfaction. Inf. Control **49**(04), 420–428 (2020)
8. Liu, S., Li, Z., Zhang, Y., Cheng, X.: Introduction of key problems in long-distance learning and training. Mobile Netw. Appl. **24**(1), 1–4 (2018). https://doi.org/10.1007/s11036-018-1136-6
9. Juntao, L.I., Mengmeng, L.U., Doulin, L.I., et al.: Research on the logistics path planning of fuzzy time window multi-objective cold chain. J. China Agric. Univ. **24**(12), 128–135 (2019)
10. Lu, S., Hu, Y., Qu, S.: Joint optimization of tow-trains dispatch and conflict-free route planning in mixed-model assembly lines. Procedia CIRP **97**(13), 253–259 (2021)

Has the Construction of the Guangdong-Hong Kong-Macao Greater Bay Area Promoted Outward FDI from Provinces Along the Route?—Empirical Analysis Based on the DID Method

Xin-ye Liu[1(✉)] and Li-jun Qiu[2]

[1] College of Economics and Management, China Agriculture University, Beijing 100083, China
iuxinye6565@yeah.net

[2] College of Mechanical and Electrical Engineering, Beijing Polytechnic, Beijing 100176, China

Abstract. The introduction of the Outline of the Guangdong-Hong Kong-Macao Greater Bay Area Development Plan signifies that the Greater Bay Area has become the "new fulcrum" of China's new round of opening up to the outside world, and the construction of it will have a direct and significant impact on the development of outward investment in the regions along the route. With this background, this paper empirically examines the policy effects of the construction on the development of outward foreign direct investment in the provinces along the route using provincial panel data from 2011 to 2017 using the double difference model (DID model). The empirical test results show that the Guangdong-Hong Kong-Macao Greater Bay Area construction initiative has significantly contributed to the growth and development of outward FDI from the provinces along the route, and is an important driving force for the growth of outward FDI. Regions along the GBA construction should strengthen outward investment cooperation, integrate outward investment resources and implement outward investment projects, and the state should increase corresponding support for going global in order to build a high-level open economy.

Keywords: The Guangdong-Hong Kong-Macao Greater Bay area · Outward foreign direct investment · DID · Effect assessment

1 Introduction

As an important step in the "One Belt, One Road" initiative, the concept of the Guangdong-Hong Kong-Macao Greater Bay Area has aroused much attention and discussion among academics, politicians, businessmen and ordinary people at home and abroad since its formal introduction in 2015. The Greater Bay Area has strong economic strength, a complete industrial system, obvious cluster advantages and strong economic complementarities, and is China's external financial hub. In the past 40 years, its local

S. Liu and X. Ma (Eds.): ADHIP 2021, LNICST 416, pp. 344–356, 2022.
https://doi.org/10.1007/978-3-030-94551-0_28

economic and trade cooperation has evolved through three stages, namely the "front shop and back factory", CEPA and the innovative cooperation in the Pilot Free Trade Zone. Substantial strategies, the Bay Area cooperation is also facing critical transformation and great challenges.

The construction of the GBA has brought tremendous development opportunities for OFDI. OFDI has integrated investment resources along the Greater Bay Area, promoted the interconnection of its financial markets and economic factors, facilitated the flow of goods and capital with the provinces along the area, and enhanced international competitiveness. China has always advocated the "going out" positioning of the GBA, and outward direct investment along the provinces is an inevitable choice for enterprises to "go out". With the continuous construction of the Greater Bay Area, outward investment from provinces along the area has developed rapidly. According to provincial-level panel data from 2011 to 2017, the amount of outward direct investment from 30 provinces and cities across China has been increasing during this period, especially in Guangdong, Shenzhen and other provinces along the Greater Bay Area, where the growth rate of outward direct investment has accelerated and the structure has gradually become more diversified. Therefore, it is important to empirically examine the impact of the construction of the GBA on the development level of outward foreign direct investment in China's trade in the provinces along the area, and the findings of the study are of reference significance for improving the outward investment of the provinces along the area, improving the quality of foreign economic cooperation, and building a higher level of open economy and going global. The specific research path of this article is as follows:

1. First analyze the current research status;
2. Secondly, the 2011–2017 provincial panel data used the double-difference model to empirically test the policy effect of construction on the development of foreign direct investment in provinces along the route;
3. Finally, an empirical analysis shows that the construction of the Guangdong-Hong Kong-Macao Greater Bay Area has made significant contributions to the growth and development of foreign direct investment in provinces along the route, and is an important driving force for the growth of foreign direct investment.

2 Review of the Literature

Under the new normal of economic development, OFDI development is an urgent requirement to promote structural reform on the supply side, make full use of international markets and resources to promote industrial restructuring, and accelerate the cultivation of new advantages in international economic cooperation and competition. After sorting through previous literature, it is found that the factors affecting OFDI development include economic factors, institutional policy factors and other factors. At the economic level, Li Yang et al. (2013), based on a panel data econometric model of Jiangsu province's investment in other countries or regions, concluded that the GDP, market size and natural resource endowment of the host country have a significant impact on OFDI in the province, and the greater the demand of the host country, the more enterprises are inclined to make direct investment in it. In addition, some scholars have also suggested other influencing factors such as government intervention, industrial structure, human capital, level of investment, and level of openness to the outside world. For the institutional policy level, Yang Dongxu et al. (2019) point out that domestic macroeconomic policy uncertainty plays a facilitating role in the development of China's OFDI, and economic policy uncertainty is significantly and positively related to the probability of firms' OFDI. Not only that, the institutional quality of the host country (Yao Fengge 2021), institutional factors of the home country (Li Shusheng 2012) and corporate system (Wu Xianming 2011) also affect OFDI performance by influencing market entry, resource allocation and exit mechanisms, and thus OFDI performance. At the political level, bilateral political-diplomatic relations (Yang Lianxing et al. 2019) and political risk (Yang Ming and Sun Yufeng 2021) can also promote or inhibit the efficiency and scale of OFDI to some extent.

As a policy that promotes the orderly and free flow of economic factors, efficient allocation of resources and deep market integration, the Greater Bay Area has brought about a huge impact on both China and countries along the route. For the domestic level, Li and Affirmation (2017) points out that the Guangdong-Hong Kong-Macao Greater Bay Area has boosted the overall economic level of the Pearl River Delta region, thus contributing to China's crossing of the middle-income trap, enhancing the status and functions of Hong Kong and Macao in the country's economic development and opening up to the outside world, and deepening the cooperation between the Mainland and Hong Kong and Macao. Qin Chenglin and Liu Liling et al. (2017) argue that the Greater Bay Area city cluster has a strong innovation drive and considerable innovation resources, contributing to the formation of a new mechanism for economic development with the interaction of innovation, reform and openness in China. For foreign countries along the route, Qiu Suan (2017) believes that the Greater Bay Area is expected to help developing countries along the route to establish operational standards, set up economic and trade offices, reduce duplication of locations along the route, save costs, and jointly carry out infrastructure investment and capacity expansion. Deng Zhixin (2017) outlines that it has led to the synergistic development of the surrounding regions, promoted economic and trade cooperation with ASEAN countries, and become a bridgehead of the "Belt and Road".

Outward FDI and the construction of the Guangdong-Hong Kong-Macao Greater Bay Area are both important initiatives to promote international financial development and financial market interoperability, and are key steps in the implementation of the country's "going out" strategy, but from the existing literature, few scholars have paid attention to the link between the two. Based on this, this paper will consider both the GBA policy and outward foreign direct investment, providing new empirical evidence with certain theoretical significance for understanding the impact of construction on outward foreign direct investment in China's provinces along the route; secondly, this paper adopts the propensity score matching-dual difference method to evaluate the impact of the GBA policy on outward foreign direct investment Through empirical testing, the paper analyses the new prospects for OFDI development in the context of the GBA policy in the provinces along the route, and provides a reference for OFDI work in the provinces along the route, which has certain practical significance.

3 Research Design

3.1 Model Setting: Double Difference Method

This paper considers the GBA policy as a 'quasi-natural experiment' conducted in the provinces along the route (Shuai Liu 2019). As the double difference method has been commonly used in recent years to assess policy effects, this paper proposes to adopt the double difference method to assess the impact of the Great Bay Area policy initiative on the development of OFDI in the provinces along the route. This paper divides China's 30 provinces and cities into two groups according to the status of the GBA policy (due to the incompleteness of the Tibetan data, the sample of provinces along this route is excluded from this paper), using du as the grouping dummy variable, with the treatment group being Guangdong Province, which is covered by the GBA policy, and the control group being the remaining 29 provinces and cities, whose du = 0 (Shuai Liu et al. 2021). The time dummy variable is dt, before the initiative dt = 0, and dt = 1 after the initiative; as the GBA policy was proposed at the end of 2015, this paper takes 2015 as the policy node. Based on the above analysis, the baseline regression model of the double difference model is as follows.

$$\mathrm{OFDI}_{it} = \beta_0 + \beta_1 \mathrm{du}_{it} + \beta_2 \mathrm{dt}_{it} + \beta_3 \mathrm{du}_{it} \times \mathrm{dt}_{it} + \beta_4 \mathrm{X}_{it} + \varepsilon_{it} \tag{1}$$

where the subscripts i and t represent the ith province and city and year t respectively, OFDI denotes the outward direct investment of each province, X denotes a series of control variables, which are random disturbance terms, and the product denotes the estimated amount of the effect of the Guangdong-Hong Kong-Macao Greater Bay Area initiative on the outward direct investment of the provinces along the route, which is also the variable of focus in this paper (Shuai Liu 2019).

Some coefficient-related terms in the DID model are shown in Table 1: the change effect of OFDI in the treatment group before and after the implementation of the Guangdong-Hong Kong-Macao Greater Bay Area policy is $\Delta \mathrm{ofdi}_t = \beta_2 + \beta_3$, the change effect of OFDI in the control group is $\Delta \mathrm{ofdi}_0 = \beta_2$, the change effect of OFDI in the treatment group minus the change effect of OFDI in the control group is the net policy effect of the GBA policy on the development of OFDI in the provinces along the route. The coefficient of the interaction term of these two dummy variables is β_3, which is the focus of this paper, and du is the dummy variable for the subgroup, dt is the dummy variable for the time and du * dt is the experimental group of the policy. If the coefficient is positive, it means that the policy has a positive impact on the development of OFDI in the provinces along the route in China; if negative, it means not.

Table 1. Table of interpretations of DID coefficient related terms

	Before the policy implemented ($dt = 0$)	After the policy implemented ($dt = 1$)	Difference
Processing group ($du = 1$)	$\beta_0 + \beta_1$	$\beta_0 + \beta_1 + \beta_2 + \beta_3$	$\Delta \text{ofdi}_\text{t} = \beta_2 + \beta_3$
Control group ($du = 0$)	β_0	$\beta_0 + \beta_2$	$\Delta \text{ofdi}_0 = \beta_2$
Difference	—	—	$\Delta\Delta \text{ofdi}_0 = \beta_3$

Data Description

This paper uses Chinese provincial data, including panel data from 2011–2017 for 30 provinces and cities across China (except Tibet) to measure the level of provincial OFDI development and the policy effects of the Greater Bay Area policy on OFDI development, with all data collated and calculated through the China Statistical Yearbook, the China Foreign Investment Statistics Bulletin, the National Bureau of Statistics and provincial statistical yearbooks.

Variable Selection

The explanatory variable is OFDI, i.e. outward foreign direct investment by province, sourced from the China Outward Investment Statistics Bulletin. As for the key explanatory variables, for the background that this paper considers the policy effects of the Greater Bay Area, the dummy variables for the subgroups du and the time dt at which the policy was proposed and the interaction term between the two du * dt are used as the core explanatory variables, with the coefficient of the interaction term being the main variable of interest in the study.

As for the control variables, a series of other factors affecting the development of provincial OFDI are selected as control variables, specifically GDP per capita, industrial structure, level of urbanization, financial development, level of openness to the outside world, technological progress and foreign direct investment, with the specific variables defined and interpreted in the following table (Table 2).

Table 2. Variable definition table

Variable name	Variable meaning	Variable calculation method
ofdi	Foreign direct investment	Outward foreign direct investment by province
du	Grouping variables	1 for the experimental group and 0 for the control group
dt	Policy variables	1 before policy implementation, 0 after policy implementation

(*continued*)

Table 2. (*continued*)

Variable name	Variable meaning	Variable calculation method
du × dt	Net policy effect	Product of sub-group variables, policy variables
pergdp	GDP per capita	Real value of GDP per capita by region
industry	Industrial structure	Value added of tertiary sector by region/regional GDP
urban	Level of urbanization	Number of urban population by region/total regional population
open	Level of external openness	Total imports and exports by region/regional GDP
fin	Financial Development	Deposit and loan balances of financial institutions by province/regional GDP
Tec	Technological advances	Technology market turnover per capita
FDI	Foreign Direct Investment	Foreign direct investment by province

4 Results of the Empirical Analysis

Results of Descriptive Statistics for Variables

The time dimension selected for this paper is 2011–2017, with data on GDP per capita, the number of urban population in each region, and the total amount of imports and exports in each region sourced from the China Statistical Yearbook, the China Foreign Investment Statistical Bulletin, the National Bureau of Statistics, and provincial statistical yearbooks in previous years, and descriptive statistics for the relevant variables are shown in Table 3.

As can be seen from the following descriptive statistics, there are large individual differences in OFDI, GDP per capita, industrial structure, urbanization level, opening-up level and financial development among the 30 provincial administrative regions in China from 2011–2017. The large standard deviation of OFDI in different provinces, cities or regions indicates that there are large fluctuations and strong heterogeneity in the sample data.

Table 3. Descriptive statistics table

Variables	Number of samples	Average	Standard deviation	Minimum	Maximum
Foreign direct investment	210	1049657.5	2163559.1	1304	18971365
Financial Development	210	3.076	1.15	1.518	8.131
GDP per capita	210	47377.481	21218.442	16413	113051.96

(*continued*)

Table 3. (*continued*)

Variables	Number of samples	Average	Standard deviation	Minimum	Maximum
Industrial structure	210	0.448	0.094	0.297	0.806
Level of urbanization	210	0.566	0.124	0.35	0.896
Level of external openness	210	0.279	0.319	0.017	1.548
Technological advances	210	4035093.8	4471476.3	102511.93	20617500
Foreign Direct Investment	210	1417.634	2192.541	28.292	17622.301

Parallel Trend Test

In order to visually examine whether there is an impact of the GBA policy on the development of OFDI in the provinces along the route, this paper intends to draw a control chart of the time trend of OFDI development in the experimental group and the control group to observe the change trend of OFDI development in the two groups of provinces. After calculating the mean value of OFDI development in the two groups, a chart of the change trend of the provinces along the route of the Greater Bay Area policy and the provinces not along the route is drawn as shown in the figure below.

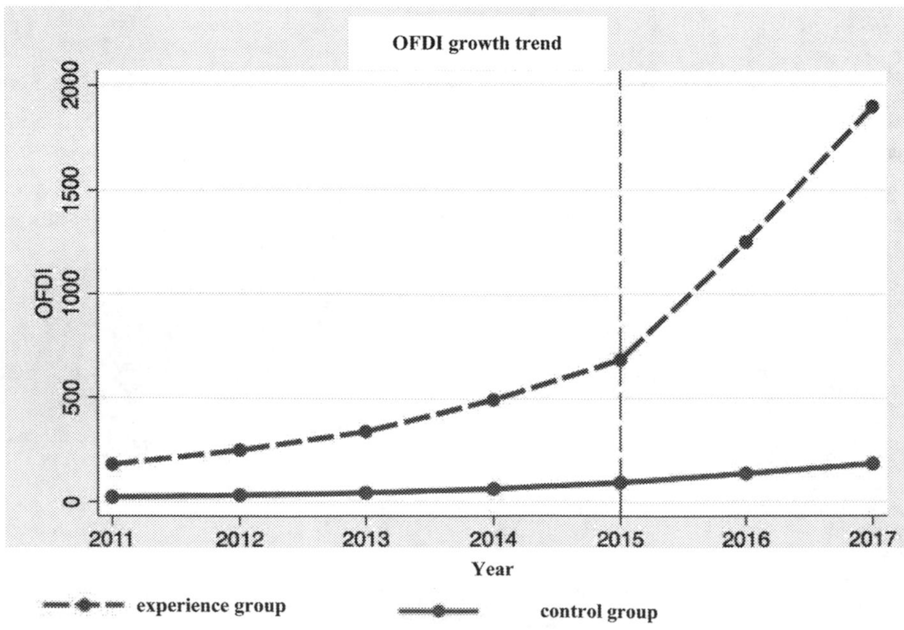

Fig. 1. OFDI growth trends

As can be seen from Fig. 1, before the policy shock, the OFDI growth trends of the experimental and control groups were close, while after the GBA policy was proposed, the OFDI of the experimental group rose significantly more than that of the control group, indicating a significant and positive effect of the policy, and the graph basically satisfies the assumption of parallel trends.

Further, this paper adopts a control time trend approach by doing the hypothetical treatment of participation in the Greater Bay Area for the first two years of 2015 and regressing it on the control variables. As can be seen from Table 4, the regression results for the two variables du * dt 2013 and du * dt 2014 are not significant, indicating that there is no significant difference in the change of OFDI between the experimental and control groups before the policy shock, thus it can be determined that the experimental and control groups satisfy the "parallel trend hypothesis" before the policy is proposed, indicating that this study can adopt a double difference model for regression testing.

Table 4. Parallel inspection

Variables	OFDI
du × dt	228.517* (0.082)
du * *dt* 2014	20.666 (0.773)
du * *dt* 2013	−19.839 (0.72)
GDP per capita	0 (0.121)
Industrial structure	−356.382 (0.288)
Level of urbanization	−3281.794** (0.048)
Level of external openness	45.829 (0.792)
Financial development	−7.038 (0.871)
Foreign direct investment	0.077 (0.000)
Provincial fixed effects	Yes
Year fixed effects	Yes
Constant term	1730.647 (0.049)
Sample size	210

Note: *, **, *** denote significant at the 10%, 5% and 1% levels respectively

Baseline Regression Results

The results of the double difference model regressions are presented in Table 4, where we focus on the regression coefficients, as they reflect the interaction between the provinces affected by the policy and the provinces not affected by the policy in terms of OFDI development. Column (1) of the table shows the results of the double-difference regression without the inclusion of control variables. We can see that the regression coefficient is positive and significant at the 1% level without the inclusion of control variables; the regression coefficient is still positive and significant at the 1% level after the inclusion of control variables such as GDP per capita, industrial structure, urbanization level and openness level. This indicates that the policy of the Greater Bay Area has a positive effect on the outward foreign direct investment of the provinces along the route, and the policy effect in pulling up the level of opening up of the provinces along the route has come to the fore (Table 5).

Table 5. Double difference baseline regression table

Variables	(1)	(2)	(3)	(4)
du × dt	960.8619*** (0.000)	416.185*** (0.004)	392.69*** (0.007)	208.157* (0.067)
GDP per capita		0*** (0.000)	0*** (0.005)	0*** (0.108)
Industrial structure		966.942* (0.053)	393.113 (0.274)	358.4717 (0.159)
Level of urbanization		−2272.623** (0.042)	−2080.775 (0.046)	−1137.378 (0.146)
Level of external openness			−297.176** (0.031)	−184.8778 (0.110)
Financial development			64.314 (0.224)	17.8405 (0.676)
Foreign direct investment				0.0856*** (0.000)
Provincial fixed effects	Yes	Yes	Yes	Yes
Time fixed effects	Yes	Yes	Yes	Yes
Constant term	91.239	563.58	643.827*	303.0527
Sample size	210	210	210	210

Note: *, **, *** denote significant at the 10%, 5% and 1% levels respectively

Excluding Other Events

In order to exclude the interference of other events, this paper selects the study period in which the policy is advanced by one year and two years for the temporal counterfactual test, resetting the temporal dummy variables year 2013 and year 2014, i.e. the policy is proposed in 2013 and 2014: where year 2013 indicates a value of 1 after 2013 and 0 otherwise. The results of the test are shown in the table below. in the two counterfactual scenarios where the policy is assumed to be one year earlier and two years earlier, the development of OFDI in the experimental group of provinces shows a significant negative correlation compared to the control group of provinces, whereby the factor before the policy was proposed can be ruled out as triggering the growth of OFDI The possibility that factors prior to the introduction of the policy triggered OFDI growth can be ruled out. Therefore, it can be argued that the proposed policy has had a policy effect on the development of OFDI in the provinces along the route.

In order to exclude the shock effect of other events on provincial OFDI development levels during the period when the policy was proposed, and because the construction of the Yangtze River Economic Belt was elevated to a national strategy in 2014, very close to the time when the policy was proposed, and thus to exclude the exogenous shock results, this paper considers excluding the eleven provinces along the Greater Bay Area in the Yangtze River Economic Belt (Shanghai, Jiangsu, Zhejiang, Anhui, Jiangxi, Hubei, Hunan, Chongqing, Sichuan, Yunnan, Guizhou). The excluded samples are then regressed on a double-difference model again. According to the test results, it is found that the regression results are still robust after proposing the eleven provinces in the Yangtze River Economic Zone, and the policy effect is still reflected as the policy promotes the level of OFDI development in the provinces along the zone at the 1% significance level (Table 6).

Table 6. Exclusion of other events checklist

Variables	Policy advanced by one year	Policy brought forward two years	Excluding the four provinces of the Yangtze River Economic Belt
du × dt	5757840 (0.000)	4802960 (0.000)	7556788 (0.000)
GDP per capita	181.4813 (0.002)	167.2728 (0.004)	43.58714 (0.469)
Industrial structure	3095485 (0.251)	2204586 (0.459)	3552498 (0.314)
Level of urbanization	−3.63e+07 (0.021)	−3.58e+07 (0.021)	−133940.9 (0.992)
Level of external openness	−1379256 (0.586)	−2575280 (0.458)	−5994806 (0.002)
Financial development	751470.9 (0.220)	583016.8 (0.344)	−474399.2 (0.394)
FE	Yes	Yes	Yes
IYEAR	Yes	Yes	Yes

(*continued*)

Table 6. (*continued*)

Variables	Policy advanced by one year	Policy brought forward two years	Excluding the four provinces of the Yangtze River Economic Belt
Constant term	9512935	1.10e+07	307428.8
Sample size	210	210	210

Note: *, **, *** denote significant at the 10%, 5% and 1% levels respectively

5 Conclusions and Insights

Conclusion

Based on the "Vision and Actions for Promoting the Construction of the Silk Road Economic Belt and the 21st Century Maritime Silk Road" jointly released by the National Development and Reform Commission, the Ministry of Foreign Affairs and the Ministry of Commerce in 2015, this paper formally specifies the concept of "the Guangdong-Hong Kong and Macao Greater Bay Area" as a quasi-natural experiment, and uses a panel of data from 30 provinces and cities across China from 2011 to 2017. This quasi-natural experiment uses a panel of 30 provinces and cities from 2011 to 2017 to explore the impact of the Greater Bay Area policy on the development of OFDI in the provinces along the route. After clarifying the basic research idea, this paper empirically examines the impact of the Greater Bay Area policy on the level of OFDI development in the provinces along the route using the double difference method (DID model). It is found that the policy has indeed effectively enhanced the level of OFDI development in the provinces along the route in China. And the robustness of the regression results is further verified by tests that exclude exogenous event shocks, such as advancing the timing of the policy and proposing eleven provinces in the Yangtze River Economic Belt, indicating that the promotion effect of the policy on OFDI in the provinces along the route is neither affected by pre-policy factors nor disturbed by other events.

Revelation

1. The Guangdong-Hong Kong-Macao Greater Bay Area policy has significantly promoted the development of outward foreign direct investment in the provinces and regions along the route, raising the level of opening-up of the provinces and regions along the route, and the Greater Bay Area is capable of acting as the "new pivot point" for the next round of opening-up. Under the influence of the policy, the level of outward FDI from the provinces and cities along the route has increased significantly, which is conducive to accelerating the adjustment and optimization of the industrial structure of the GBA, driving the overall level of China's industrial development, creating a favourable international environment, raising the overall level of opening-up, and providing experience and institutional framework for promoting the construction of a wider FTA and the next round of opening-up process.

2. China should continue to encourage and strengthen the construction of various forms of open zones, such as "bay areas" and "economic zones", combining administrative tools and economic instruments to lead the way by achieving a significant increase in the level of openness of local areas, so as to drive and radiate the opening up of a larger area. The opening up of the region to the outside world will be driven and radiated by the significant increase in the level of opening up of local areas. Provinces and cities should also respond positively to the policy call to co-ordinate the opening up and cooperation of city clusters, vigorously promote the integration of industrial and social development, and enhance the development synergy. For cities or regions with a relatively high level of openness to the outside world, they should give full play to the leading, organizing and radiating role of national growth poles, vigorously promote the flow of factors and market interconnection, and realize a higher level of openness to the outside world and a larger region.

3. Strengthen the construction of supporting services and measures in the areas of institutions, education and finance to create an excellent environment for the creation of a new round of distinctive external opening areas and to achieve positive interaction and circulation. China should continue to strengthen the construction of infrastructure platforms in the Greater Bay Area, formulate complementary fiscal policies to provide convenient conditions for the circulation of factors and mutually beneficial cooperation, and strengthen the training and introduction of innovative talents, so as to inject impetus for sustained urban innovation and regional economic development. At the same time, the Greater Bay Area and the provinces along its route should focus on building supporting facilities services and platforms that are in line with international standards, so as to provide the necessary support to create a good international trading environment and a high standard international trade platform.

References

Yang, L., Xin, Z., Mantian, X.: Economic resources, cultural institutions and location choice of outward foreign direct investment: an empirical study based on panel data in Jiangsu Province. Int. Trade Issues **04**, 148–157 (2013)

Yang, D., Yu, J.: The impact of investment facilitation on Chinese outward foreign direct investment in countries along the Belt and Road: theory and empirical evidence. Int. Econ. Trade Explor. **37**(03), 65–80 (2021)

Fengge, Y.: The impact of institutional distance and the establishment of overseas industrial parks on outward foreign direct investment: an example from countries along the Belt and Road. Investment Cooperation **2021**(02), 68–71+88 (2021)

Shusheng, L.: A literature study on home country institutional factors in Chinese enterprises' outward foreign direct investment. Reform Strategy **28**(12), 134–137 (2012)

Xianming, W.: Institutional environment and overseas investment entry mode of Chinese enterprises. Econ. Manage. **33**(04), 68–79 (2011)

Lianxing, Y., Chaohai, S., Yancheng, M.: A study on the impact of bilateral political relations on China's cultural trade - based on the binary marginal perspective of exports[J]. J. Nanjing Univ. Finance Econ. **05**, 96–108 (2019)

Yang, M., Sun, Y.F.: Research on the impact of political risk on China's outward foreign direct investment. Natl. Circ. Econ. **02**, 53–56 (2021)

Li, X., Affirmation, H.: Discussion on the development strategy and construction path of Guangdong, Hong Kong and Macao Greater Bay Area in the context of the new round of opening up to the outside world. Int. Econ. Trade Exploration **33**(09), 4–13 (2017)

Chenglin, Q., Liling, L., Wenhao, Q.: Strategic thinking on the development of Guangdong-Hong Kong-Macao Greater Bay Area city cluster. Reg. Econ. Rev. **05**, 113–118 (2017)

Suan, Q.: Dimensional analysis of the choice of the way forward for the development of the Guangdong-Hong Kong-Macao Greater Bay Area city cluster. Guangdong Soc. Sci. **04**, 15–20 (2017)

Zhixin, D.: Guangdong, Hong Kong, Macao and the Greater Bay Area: a new engine for the development of the Pearl River Delta. Guangdong Econ. **05**, 32–35 (2017)

Liu, S., Liu, D., Muhammad, K., Ding, W.: Effective template update mechanism in visual tracking with background clutter. Neurocomputing **458**, 615–625 (2020) online first. https://doi.org/10.1016/j.neucom.2019.12.143

Liu, S., Liu, X., Wang, S., Muhammad, K.: Fuzzy-aided solution for out-of-view challenge in visual tracking under IoT assisted complex environment. Neural Comput. Appl. **33**(4), 1055–1065 (2021)

Liu, S.: Introduction of key problems in long-distance learning and training. Mobile Netw. Appl. **24**(1), 1–4 (2019). https://doi.org/10.1007/s11036-018-1136-6

Key Technology of Offshore Small Buoy Deployment Based on Bayesian Network

Ji-ming Zhang[1,2,3](✉) and Xuan-qun Li[1,2,3]

[1] Institute of Oceanographic Instrumentation, Qilu University of Technology (Shandong Academy of Sciences), Qingdao 266000, China

[2] Shandong Provincial Key Laboratory of Ocean Environmental Monitoring Technology, Qingdao 266000, China

[3] National Engineering and Technological Research Center of Marine Monitoring Equipment, Qingdao 266000, China

Abstract. In view of the high probability of natural or man-made damage during deployment and operation of offshore small buoys, after summing up the research institute's years of experience and knowledge in the development, production and maintenance of offshore small buoys, a new method based on shellfish has been proposed. The key technology for the deployment of small buoys in the coastal waters of the Yes Networks, through the analysis of marine environmental characteristics and buoy release requirements, effectively guides the deployment and maintenance of small buoys. The popularization and application of this technology shows that this technology can improve the deployment and maintenance of small buoys. Maintain efficiency, work efficiency and operational safety.

Keywords: Bayesian network · Offshore buoy · Buoy deployment · Release demand

1 Introduction

The small multi-parameter marine environment monitoring buoy has the characteristics of small size, convenient transportation, and the ability to comprehensively measure various marine data. It is widely used in offshore waters and exerts the function of unmanned transmission of data and information [1]. But also because of its small size and light weight, its deployment location does not belong to the remote deep sea area, and the probability of various natural and man-made damages is greater. Therefore, in order to reduce the probability of damage to the small buoy, make it as long as possible. Extending, the various tasks before, during and after the deployment cannot be ignored. Since the deployment location, sea conditions, and requisitioned vessels of small buoys are different each time, we will not describe in detail the various links of the entire deployment of buoys, only the links that may affect the normal operation of the buoys after they enter the water. Put forward and provide corresponding solutions for reference [2]. At present, there are still differences in the operating environment of buoys at home

S. Liu and X. Ma (Eds.): ADHIP 2021, LNICST 416, pp. 357–376, 2022.
https://doi.org/10.1007/978-3-030-94551-0_29

and abroad. Foreign countries generally operate fully mechanized, using large ships to load several buoys, and after reaching the designated position, they are hoisted into the sea with a ship-mounted boom. At present, only a few locations in China can adopt this method. With sufficient funds and smooth coordination, vessels with ample aft deck, hoisting booms and winches can be requisitioned; if conditions do not permit, a vessel with a wide aft deck and lower ship can be used instead, and a ship with better sea conditions can be selected. Go to sea for deployment [3]. The following are the key links that should be paid attention to in the whole process of offshore small buoys.

2 Key Technology for the Deployment of Offshore Small Buoys

2.1 Placement of Small Offshore Buoys

Since the deployment location is offshore, it is often close to the land, and there are many ships in the deployment area, and collisions may occur regardless of day and night. Therefore, the surrounding sea area should be carefully inspected before deployment, and the latitude and longitude are roughly unchanged. Make small-scale adjustments. Pay attention to choosing the sea area away from the channel port, the sea area away from the fishermen's work area, and the sea area that is more prone to marine accidents, such as sea reef areas. The reef exposed at low tide will cause serious damage to the buoy body. For example, the oil terminal, because oil leakage often occurs in the sea near the oil terminal, the leaked oil directly has a serious adverse effect on the buoy body [4]. At present, there are roughly four ways for people to obtain marine information, namely: marine monitoring buoys, satellite remote sensing, ship sailing, and coastal marine stations. Marine buoys play a huge role in the observation of marine environment and hydrological information. How to accelerate the development of buoy technology is of great significance to the development and utilization of the ocean [5]. In the overall technology of marine buoys, the buoy body as the carrier of each monitoring instrument is related to whether it can provide the required working conditions for the equipment. Therefore, the design of the buoy body structure is very important in the overall design of the buoy [6]. As a floating structure on the sea, the offshore buoy structure combines the characteristics of offshore engineering structure and floating structure. As a small offshore buoy, compared with large offshore buoys, offshore buoys are more susceptible to damage from the complex marine environment. Therefore, the structural design and research of offshore buoys is the key and difficult point in the overall design of marine buoys [7]. These four observation methods have their own advantages and disadvantages, as shown in the table (Table 1):

Marine buoy monitoring is an advanced and practical marine monitoring method. This method is developed on the basis of traditional marine monitoring methods. It can realize unmanned, automatic, fixed-point, regular and continuous monitoring of various hydrological elements at sea. The monitoring of environmental elements is one of the main tools for studying various marine physical, chemical and biological phenomena [8]. It forms a three-dimensional marine monitoring system with satellites, marine survey ships, marine surveillance aircraft, and submersibles. Compared with the other three monitoring methods, marine monitoring buoys have the advantages of mobile and flexible layout, long-term continuous, all-weather fixed-point monitoring, and are used

Table 1. Detection methods and advantages and disadvantages

Observation approach	Inferiority	Advantage
Satellite remote sensing	The accuracy of data is not high, and the types of measurable sample parameters are few	Continuous data monitoring, fast data transmission and large work coverage
Coastal marine station	Only the data of coastal marine environment can be obtained	Data continuity, high accuracy
Ocean detection buoy	It is difficult to design, launch and maintain the ocean buoy	Multifunction, continuous detection, long life, low cost, strong viability
Ship sailing	The cost of monitoring is high, and it cannot be monitored continuously for a long time	It has high measurement accuracy and many kinds of measurable parameters

by more and more countries and regions in the development of marine energy resources, marine environmental forecasting, and maritime defense construction.

Small buoys need to be transferred from land to the ocean. Generally, a crane or crane is used. It is recommended that the boom should be longer than 5t. If it is towed at sea, it should be transferred to the sea at full tide as much as possible to avoid collisions under the sea. Reef: When towing, pay attention to keeping a certain distance between the buoy and the ship to reduce the possibility of collision [9]. The towing speed is within 5 knots to prevent the target body from tipping over. If it is hoisted to the deck, ample storage space and a stable base are required. Before launching into the water, a final inspection of each key part of the seal is necessary, which has a direct relationship with the life of the instrument inside the buoy. The connection work of the anchor system is generally completed at sea, and the installation personnel must be skilled. Pay attention to the placement and fixing of the anchor and the anchor chain during the work to avoid the anchor chain from knotting or falling into the water. The specific placement depends on the hull of the ship [10]. The derrick of the underwater equipment needs to reach the sea area where it is deployed, and then install it after anchoring and stabilization, so as to avoid scraping the fishing net or submerged reef during towing. When planning the buoy search submarine formation, it is usually necessary to make the two buoys have a certain overlap [11]. The maximum allowable error of the buoy array should meet the buoy's non-missing warning coverage condition, that is, there should be no coverage gap between two buoys in the formation, so that the submarine may pass through the gap between the buoys without being detected. As shown (Fig. 1).

In the figure, Δd is the deviation distance between the actual position of the buoys and the position of the planned lattice point when the two buoys' range circles are tangent. In order to avoid missing alarms, the error Δd when the two buoys' range circles are exactly tangent is defined as the maximum delivery error. Assuming that the buoy's operating distance is 3 km and the planned buoy spacing is 4.8 km, it can be known from the geometric relationship that the maximum delivery error of the buoy without missing alarm is 600 m. In the process of random placement, the normal distribution model is

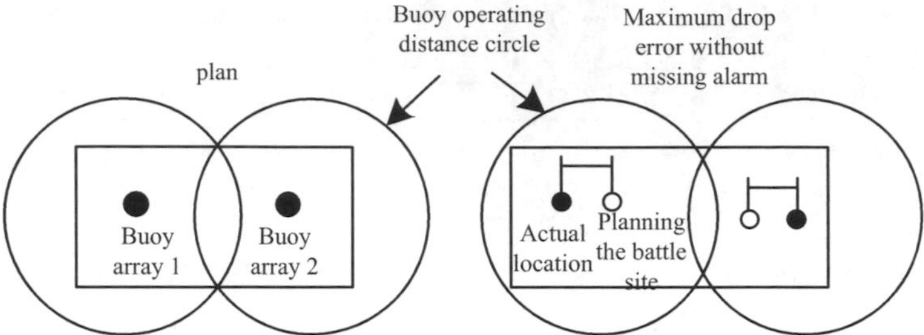

Fig. 1. Schematic of the maximum release error of the buoy formation

used to describe the error distribution, and the ±3σ range is required to be (−600 m, 600 m), that is, the standard deviation of the error distribution requires 3σ < 600 m and 1σ < 200 m.

After the deployment location is selected, the role of warning signs cannot be ignored. First of all, the sign should be eye-catching, and the installation location should be as close as possible to the upper part of the buoy, so that ships coming from any direction can notice and avoid in advance [12]. Secondly, the sign should be durable, not easy to deform or break, and the text and icons on it should not fade easily, and then lose its warning effect; the installation position should be stable enough to avoid falling. The anchor light is the only warning sign of the buoy at night. The brightness should meet the warning requirements, avoid other large-volume sensors, and the signal is stable. Once a fault occurs, it should be found as soon as possible and dealt with quickly.

The area around the sea buoy is open. In most cases, the signal strength is good, but there are also some areas where the signal strength is weak, or there are other signal interference, which will affect the success rate of shore station reception. Therefore, there is It is necessary to confirm whether the shore station receives the data smoothly, so that once the buoy has a problem, it can be observed at the first time, which is convenient for future maintenance [13].

The failure of buoys is often caused by man-made. The main reason is that the masses are curious about buoys and the awareness of public and civil facilities is not enough to protect [14]. Propaganda in this area needs to be popularized, and the people around need to know and understand the contribution of buoys to marine meteorology. The good work of buoys is helpful to their navigation planning and ship safety.

2.2 Analysis of the Errors in the Placement of Small Buoys in Offshore

The source of the buoy array launch error is the deviation from the course error caused by the automatic navigation deployment, which is described by the yaw distance, which can be called the lateral error (the direction of the route is the longitudinal direction); the second is the deviation from the target when the buoy is launched forward. The point error, described by the cast radiometer, can be called the longitudinal error; the third is the influence of wind speed and direction [15]. In order to analyze the lateral error, for the square buoy array, for a certain type of helicopter, two deployment routes are designed, as shown in the figure (Fig. 2):

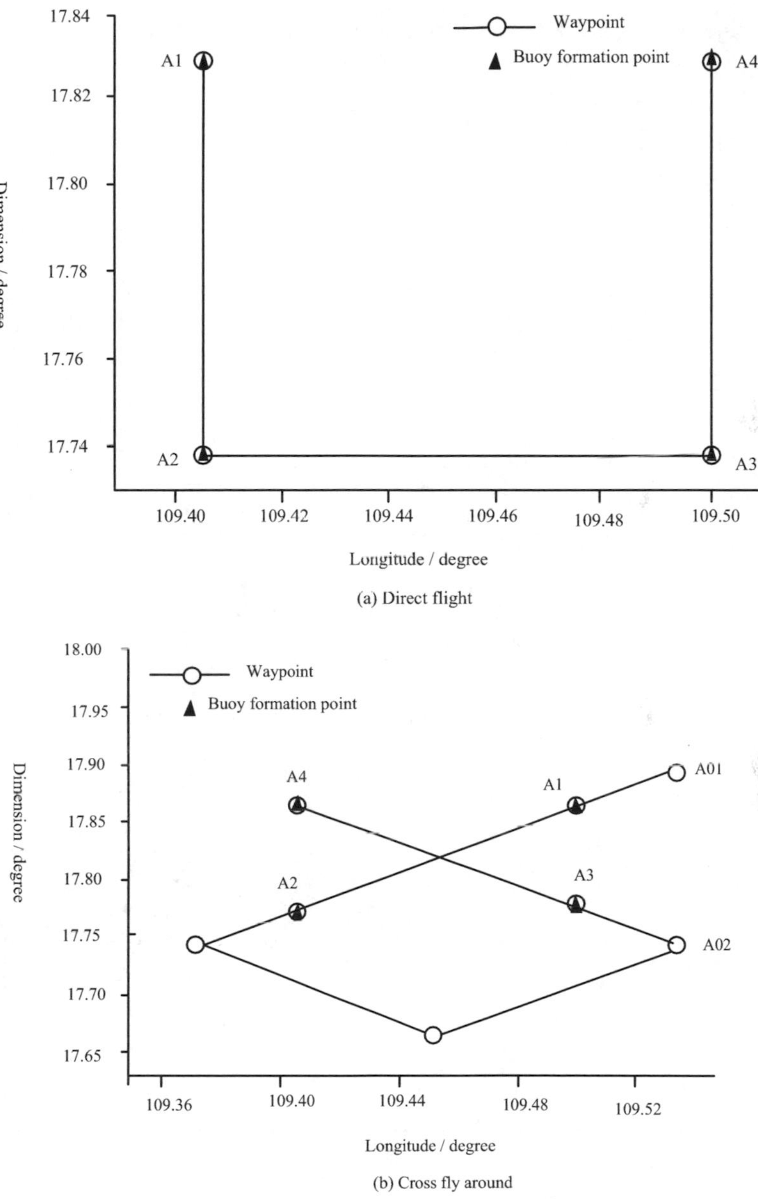

(a) Direct flight

(b) Cross fly around

Fig. 2. Arrangement curve of square array

The trajectory of direct deployment is shown in the figure (Fig. 3).

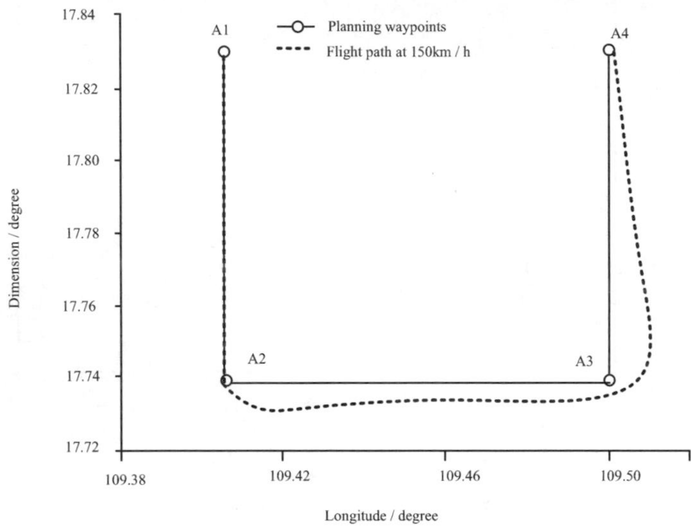

Fig. 3. Directly deploy trajectory

The yaw distance is used to describe the lateral error, and the yaw distance distribution is shown in the figure (Fig. 4).

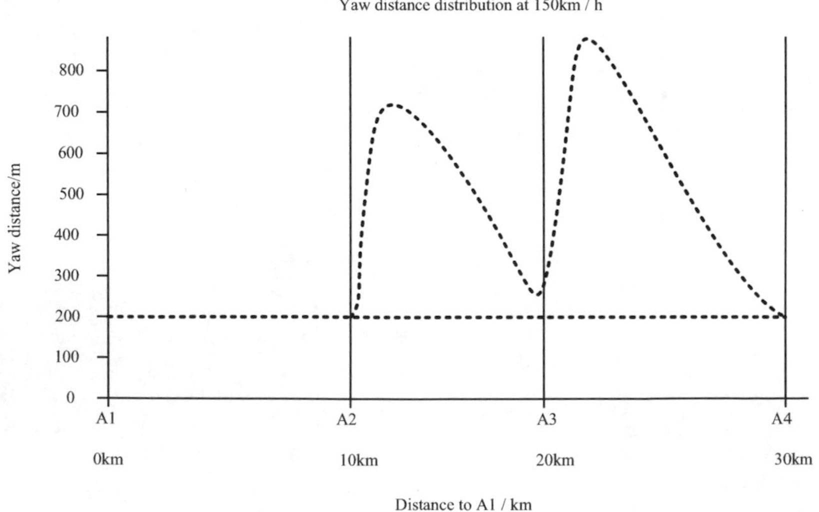

Fig. 4. Distribution of yaw distance for direct deployment

From the figure, at the turning point (A2, A3) of the square array, the yaw distance increases rapidly and then gradually converges. For a speed of 150 km/h, 10 km after the A2 turning point, the yaw distance converges to 241 m. Deploy 10 km after the A3 turning point, and the yaw distance will converge <200 m. The A3 yaw distance exceeds the maximum deployment error requirement of the formation. The deployment trajectory of the cross-wound fly is shown in the figure (Fig. 5).

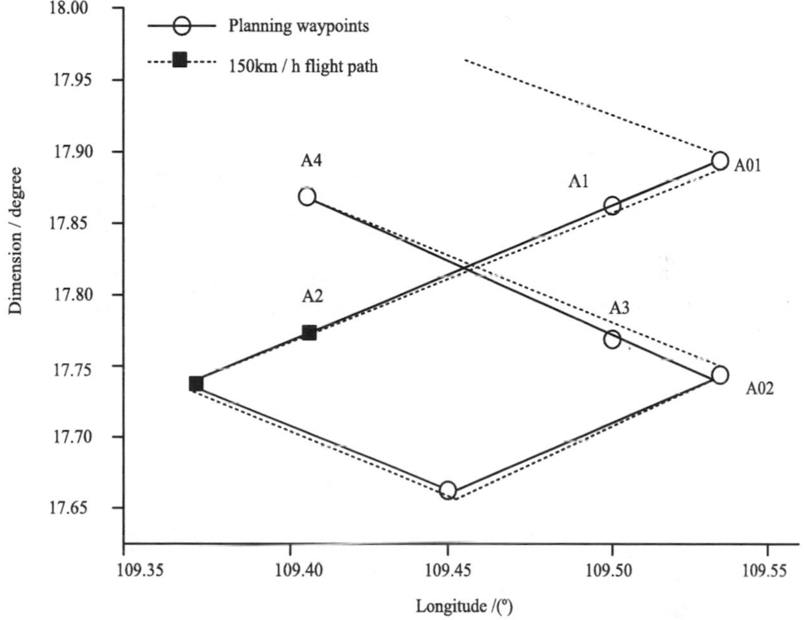

Fig. 5. Cross-wound deployment trajectory

Based on the foregoing analysis, in order to meet the requirements of the maximum launch error of the buoy formation, when designing the launch route, a route guidance point can be added before the formation route turning point, and the guidance point is more than 9 km away from the route turning point. The longitudinal error is the distance between the actual water entry point of the buoy and the planned array point when the aircraft is flying forward along the route. The longitudinal error can be controlled by calculating the buoy projection radiation table and launching according to the firing table [16]. When the influence of wind is not considered, after the buoy leaves the aircraft, it freely falls in the vertical direction, opens the umbrella and enters the water; moves at a constant speed in the horizontal direction, and opens the umbrella to slow down. The longitudinal error is the distance between the actual entry point of the buoy and the planned array point when the aircraft is flying forward along the route. The longitudinal error can be controlled by calculating the buoy emission meter and launching it according to the launch meter. When the wind influence is not considered, after the buoy leaves the aircraft, it will free fall in the vertical direction and open the umbrella into the water [17]. Perform a uniform movement in the horizontal direction and decelerate when the parachute is opened. Each time period is shown in the figure (Fig. 6):

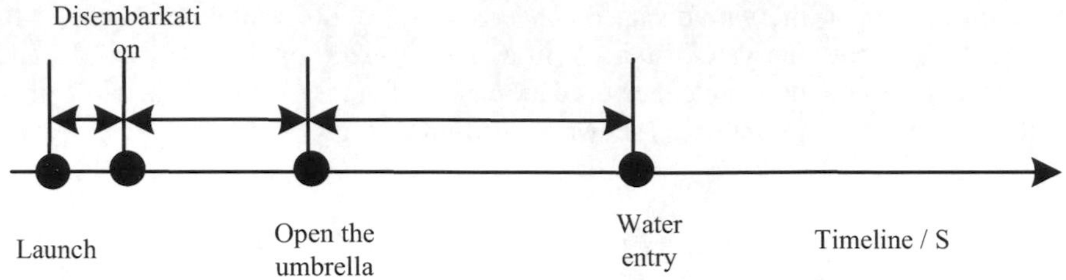

Fig. 6. Schematic of the buoy release time period

In the figure, t0 is the time from when the release button is pressed to when the buoy leaves the machine, that is, the response lag time; t1 is the time from when the buoy leaves the machine to when the buoy opens its parachute; t2 is the time from when the buoy is opened to the buoy enters the water [18]. After opening the parachute, the movement model of the buoy into the water is as follows: the buoy is opened about 1s after exiting the cabin. Regardless of the influence of wind speed, the movement of the buoy is affected by the combined action of gravity G and resistance Fs, as shown in the figure (Fig. 7).

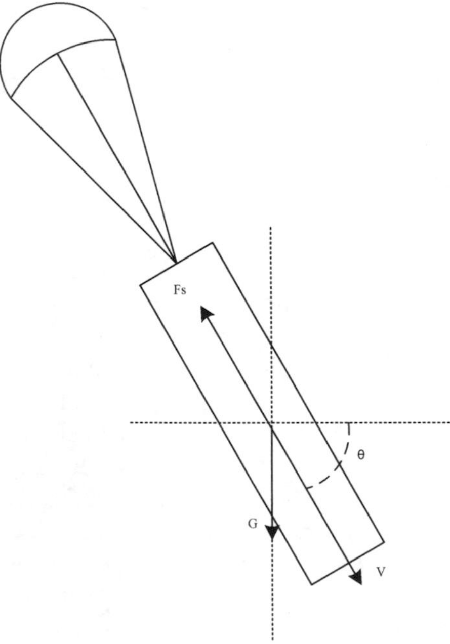

Fig. 7. Schematic of the air movement of the buoy

In the figure, Fs is resistance, G is gravity, and θ is the angle between the speed direction and the horizontal direction. After the parachute is opened, the air movement of the buoy is represented by the following equations

$$\begin{cases} \frac{dv}{dt} = g\sin\theta - \frac{c_d S \rho v^2}{2m} \\ \frac{d\theta}{dt} = \frac{g\cos\theta}{v} \end{cases} \tag{1}$$

In the formula, v is the air movement speed of the buoy, g is the acceleration of gravity, Cd is the drag coefficient, and S is the parachute drag area. ρ is the air density, generally 1.225 kg/m3; m is the mass of the buoy. The relationship between longitudinal displacement L and height H and time T is:

$$L = v_n t_1 + \int_0^{t2} v\cos\theta \frac{dv}{dt} \tag{2}$$

$$H = \frac{1}{2L} g t_1^2 + \int_0^{t2} v\sin\theta dt \tag{3}$$

$$T = H(t_1 + t_2) - \frac{d\theta}{dt} \tag{4}$$

In the formula, vn is the deployment speed of the carrier; t1 is the parachute-opening time; g is the acceleration of gravity; t2 is the time for the buoy to enter the water after the parachute is opened; v and θ are determined by the formula, and T is the time for the buoy to enter the water. According to the above model, the buoy shooting table under the typical launching conditions of 100 m launching height and 150 km/h launching speed is as follows (Table 2):

Table 2. Buoy projection radiation table

Launch height(m)	Water entry angle(°)	Water entry velocity(m/s)	Longitudinal displacement into water(m)	Water entry time(s)	Carrier speed(km/h)
100	87.2	17.5	97	7.4	150

The main sources of the longitudinal error σL are the position error σD of the aircraft at the time of launch, the displacement st0 produced by the response lag time error of the launch button, the displacement st1 produced by the time error of opening the parachute after the buoy is out of the cabin, the longitudinal displacement produced after the parachute is opened to the stable state of the buoy movement Error s, the influence of wind sw. It can be obtained from each error source:

$$\sigma_L = T\sqrt{\sigma_D^2 + \sigma_{i0}^2 + \sigma_{t1}^2 + \sigma_s^2 + \sigma_w^2} \tag{5}$$

Without considering the influence of wind, the longitudinal error is:

$$\Delta\sigma_L = T\sqrt{\sigma_D^2 + \sigma_{t0}^2 + \sigma_{t1}^2 + \sigma_s^2} \tag{6}$$

Assuming the values of each parameter are: the position error of the carrier is the navigation error of the carrier, set to σD = 30 m; the displacement error generated by the response lag time of the launch button is σt0 = v carrier Δt0, v carrier is the carrier speed; Δt0 is The response lag time error is set to Δt0 = 0.5 s; the displacement error generated by the parachute opening time after the buoy leaves the cabin is σt1 = v carrier Δt1, v carrier is the carrier speed; Δt1 is the parachute opening time error after exiting the cabin, set It is Δt1 = 1 s; the displacement error σ s generated after the parachute is opened to the steady state of the buoy movement is set to σs = 20 m. Under the above assumptions, for typical launch conditions, the longitudinal error of the buoy entering the water is 58.9 m. When the buoy is released, the longitudinal offset is controlled according to the buoy shooting table, which can realize the control of the release error. In actual deployment, in addition to the buoy shooting table, the influence of wind should also be considered. The influence of wind is the influence of wind speed and direction on the movement state t of the buoy after the parachute is opened, which leads to the deviation of the displacement of the buoy. At the time of delivery, the influence of wind can be eliminated by calculating the wind offset for real-time correction, but the error caused by the fluctuation of wind speed cannot be eliminated. Through the foregoing analysis, under different lateral error conditions, the buoy placement error is shown in the table (Table 3).

Table 3. Buoy release error

Longitudinal error(m)	Lateral error(1σ/m)	Total error(1σ/m)	Wind random error (m)	Accuracy requirements(1σ/m)	Error margin(m)
58.9	50	77.6	7	< 200	122.4
58.9	100	116.3	7		83.7
58.9	150	161.3	7		38.7
58.9	190	199.1	7		0.9

It can be seen from the table that reducing the lateral error can make the total error meet the maximum error requirement of the buoy formation and ensure that no alarm is missed. To launch the route for the square array of buoys, increase the turning guide points to reduce the yaw distance, control the lateral error, and make the total error meet the requirements.

2.3 Optimization of the Deployment Method of Small Buoys

The offshore marine buoy body is a kind of offshore floating body, and its basic design and theoretical research refer to the general theoretical research methods of the current offshore floating body structure. In order to better analyze the structure of the offshore buoy body, the article uses the basic analysis method of panel integration based on the three-dimensional potential flow theory to focus on the basic parameters of the target body structure that affect the buoy's motion response in the time domain and frequency domain.

When accelerating an object immersed in water, not only the object must be accelerated, but also the mass of a certain amount of water approaching the object or in front of it. The force F'required to accelerate an object in water is greater than the force F required to accelerate the same object in a vacuum. which is

$$F' = (m + m')a > F = ma \tag{7}$$

In the formula, m' is the attached mass added by attached water, and $m+m'$ is called virtual mass.

Internal analysis of the influence of these basic parameters on the force, motion performance and hydrodynamic parameters of the target body. Since the motion response of the buoy is mainly affected by the external environmental load, in order to better study the influence of the basic parameters of the target body on the motion performance of the target body, in this article, based on the analysis of the buoy body structure motion performance, we will focus on A more detailed analysis is carried out on the hydrodynamic parameters of the target body in a wave environment (Fig. 8).

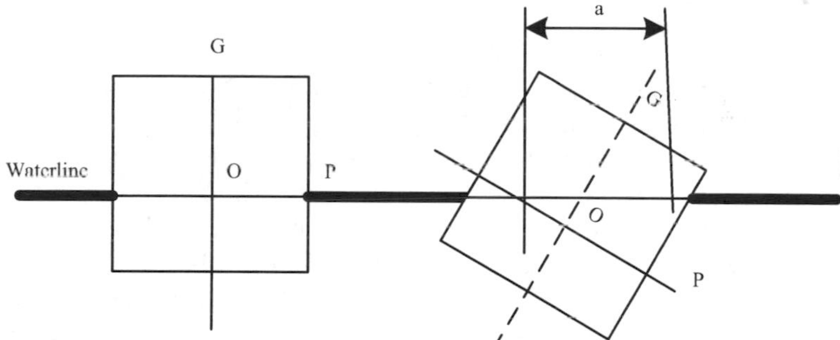

Fig. 8. Schematic diagram of static balance of floating body on water surface

When analyzing the motion performance and hydrodynamic parameters of offshore floating bodies, the offshore buoy body is usually considered as a rigid structure. The buoy body is a kind of rigid body. The marine buoy body can generally move freely in space. Therefore, when analyzing the motion performance and hydrodynamics of the marine buoy, it is necessary to establish a suitable structure for the marine buoy body under study. The coordinate system, otherwise there will be a lack of reference. For any object moving in space, its movement can be represented by six orthogonal direction vectors, as shown in the figure (Fig. 9):

The dynamic coordinate system (X', Y', Z') is established on the object, where the X'O'Y' plane is established on the vertical plane of the object, and the Z axis is on the vertical X'O'Y' plane and passes through the center of gravity of the object.; The fixed coordinate system (X, Y, Z) is based on a fixed reference system in space, where the XOY plane is established on the undisturbed stationary sea surface, and the Z axis is perpendicular to the sea level and upwards. For marine buoys, its movement

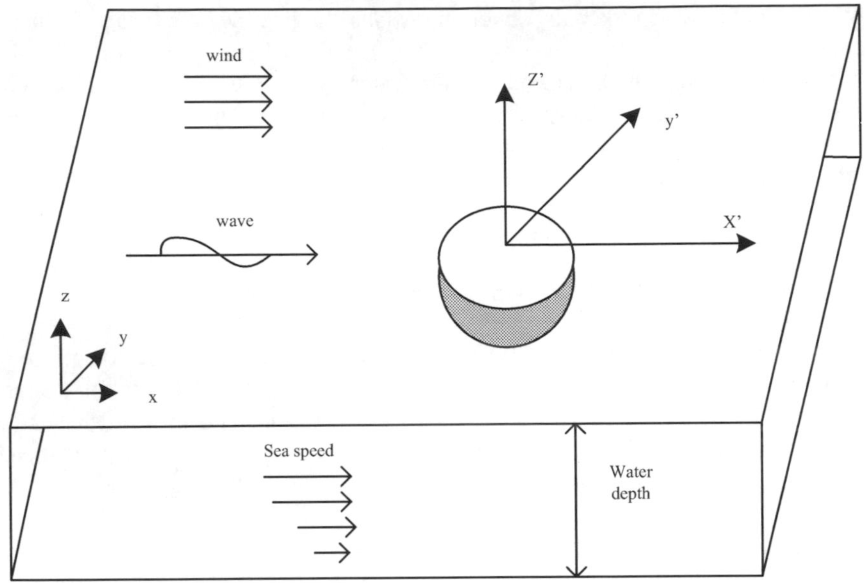

Fig. 9. The coordinate system of the rigid body movement of the buoy

in space can be described in terms of position and posture. Position is used to describe the geographic location of the marine buoy in the marine environment, including three degrees of freedom: heave, sway, and heave., Where x represents sway, Y' represents horizontal sway, z' represents heave; attitude is used to describe the appearance of marine buoys in the marine environment, such as tilt, rotation, etc., including pitch, roll, and sway For three degrees of freedom, the posture can be described by a posture matrix composed of the cosine values of the angles between the three coordinate axes of the coordinate system, where a1 represents pitch, a2 represents roll, and a3 represents first roll. Therefore, the movement state S of the buoy can be expressed as a formula.

$$S = x'i + y'j + z'k + w \times r \tag{8}$$

$$w = \Delta\sigma_L(a_1 i + a_2 j + a_3 k) \tag{9}$$

$$r = S(x'i + y'j + z'k) - \mathrm{w} \tag{10}$$

The natural period is an important indicator to measure the movement performance of a floating body. Since the main floating bodies of circular buoys are symmetrical to each other, when the mounted instruments are arranged, the mounted instruments are generally arranged symmetrically. Therefore, the natural periods of the buoys in the roll and pitch directions are roughly the same. In the calculation of the natural period of the buoy, It is only necessary to consider the natural period of the rolling direction of the buoy. According to the seakeeping theory, the approximate natural frequency of the free rolling of the floating body is:

$$f_\theta = \frac{1}{2\pi r}\sqrt{\frac{Dh}{J_{\prime\prime} + \Delta J_\omega}} \tag{11}$$

Among them, f is the natural frequency of free rolling of the floating structure, the unit is Hz; cënter, is the moment of inertia of the floating body rolling motion, the unit is kg.m3;4J}} is the additional moment of inertia of the floating body rolling motion, the unit is kg.m3; D is the displacement, kg; h is the horizontal stability of the center height, the unit is m; the calculation formula of the natural frequency of the buoy heave is as follows

$$f_z = \frac{1}{2\pi}\sqrt{\frac{\rho g S_w}{\frac{D}{g} + m_z}} - f_\theta \tag{12}$$

Among them, the natural frequency of the free heave of the floating body, Hz, mz are the additional mass of the free heave of the floating body, D is the displacement of the floating body; g is the acceleration of gravity, and the unit is m/s2. ρ is the density, the unit is kg/m3 Sw is the area of the load waterline. The movement performance of the buoy is related to whether the buoy equipped instrument can normally collect marine environment information and transmit data. In order to improve the movement performance of the buoy in the marine environment, this chapter proposes to improve the buoy vertical The design of the heave plate for the oscillating motion performance and the buoy anti-rolling design based on the TMD damping theory, and the feasibility calculation and analysis of the new design. In order to increase the additional mass of the buoy in the heave freedom, based on the design of the Spar platform, it is proposed to add a heave plate to the bottom of the hemispherical small marine buoy as shown in the figure (Fig. 10).

By analyzing the calculation formula of the natural period of the buoy, it can be known that the main parameters that affect the natural period of the buoy's roll and heave motion include moment of inertia, displacement, horizontal stability center height, and load waterline area, etc., and these parameters are usually determined by the buoy structure The weight of the equipment is determined by the weight and arrangement of the equipment. Through the analysis of the above formula, it can be seen that an appropriate increase in the weight of the buoy body can appropriately reduce the natural frequency of buoy heave and appropriately increase the natural frequency of buoy roll; under the premise of an appropriate increase in the area of the load waterline, it can be effective Improve the heave natural frequency of the buoy; appropriately reducing the buoy's moment of inertia and the additional moment of inertia can increase the natural frequency of the buoy's movement and reduce its natural period.

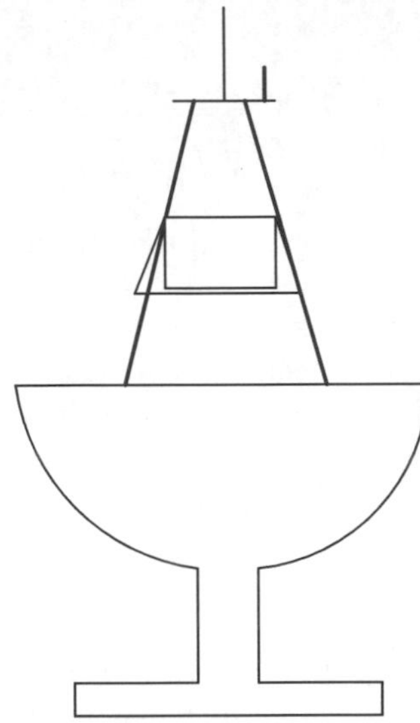

Fig. 10. Hemispherical marine buoy with heave plate

3 Analysis of Results

Two models of FVCOM and ROM were selected for this experiment. The FVCOM and ROMS models have many different sub-models. There are global and regional, climate and oceanic models, and there are many different models even in the northeastern waters of the United States.

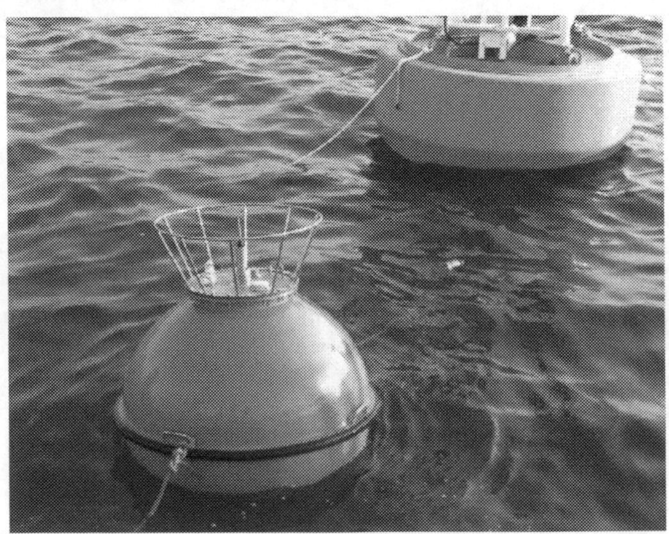

Fig. 11. Schematic diagram of sea trial

The buoy was transported to the Dalian Zhangzidao Wharf, where the temperature chain and the steel cable were connected and the anchoring was installed on the shore. Under the command and strong assistance of the marine technology research and development department of the Zhangzidao Fishery Group and the crew on the experimental ship, the overall installation went smoothly (Fig. 11).

The main purpose of this experiment is to compare and study various situations of drifting buoys at different depths, and to compare their influences. Since the result of each experiment is the latitude and longitude of each time point of the model or drifting buoy, it is necessary to convert the difference in latitude and longitude into a distance difference. This experiment uses the following formula:

$$a = \sin\left(^{(1at2-lat1)}/2\right)^2 + \cos(lat1) \times \cos(1lat2) \times \sin(lon2 - lon1)/2\Big)^2 \quad (13)$$

At the same time, when calculating the daily average error distance, the hourly distance error data is divided by the time weight, that is, the hourly distance error data is divided by 24th of the elapsed time. Every hour will get the error data of one kilometer per day, and divide this data by 24 to get the real error data for another day. Seawater can be divided into many layers in the horizontal direction, and the flow direction and speed of the seawater between different layers are different. It is not possible to simply replace the operation of the entire ocean current with one layer of ocean current. But in order to be able to better carry out the digital representation, but also for more convenient research, in many cases the sea water will be divided into multiple layers, the specific number of layers depends on the situation. This experiment uses drifting buoys to study different levels of sea and ocean currents. There are three types of buoys used, one is a four-rectangular page fan on the surface, and the other is a bucket-shaped submarine mark, as shown in the figure (Figs. 12, 13, 14, 15).

The heave and pitch motion RAOs of the hemispherical, truncated cone-shaped and cylindrical floating body models under ballast are obtained by numerical simulation.

The buoy is still in the testing state. The first step of the experiment is to filter the data. Choose a pair of buoys with a starting time interval of less than 1 h, a distance of less than 1 km and different depths, and compare the first three pairs of drifting buoys. Days of data. The data after three days is no longer of experimental significance and value because of the relatively large interval. Comparing the results of the experimental data, it is found that the results of the movement of the buoys at different depths under different conditions are different. Some directions are basically the same, while some are far apart, and some are more consistent at the beginning but inconsistent later. Comparing the data of this experiment, it is found that there are many errors. The first is the influence of the experimental data caused by the different styles of the drifting buoys on the upper and lower layers and the different responses after being stressed. Secondly, submersible targets often cross multiple water layers. Even if the buoy is small, the floating objects on the water and the line connecting the floating objects and the buoy will affect the movement of the submersible. Moreover, in order to achieve the production parameters of the submersible, the buoy itself It is not very small, and the direction of ocean currents between different water layers is not completely consistent, so this will inevitably affect the data results. In addition, it also includes the impossibility of placing buoys at exactly

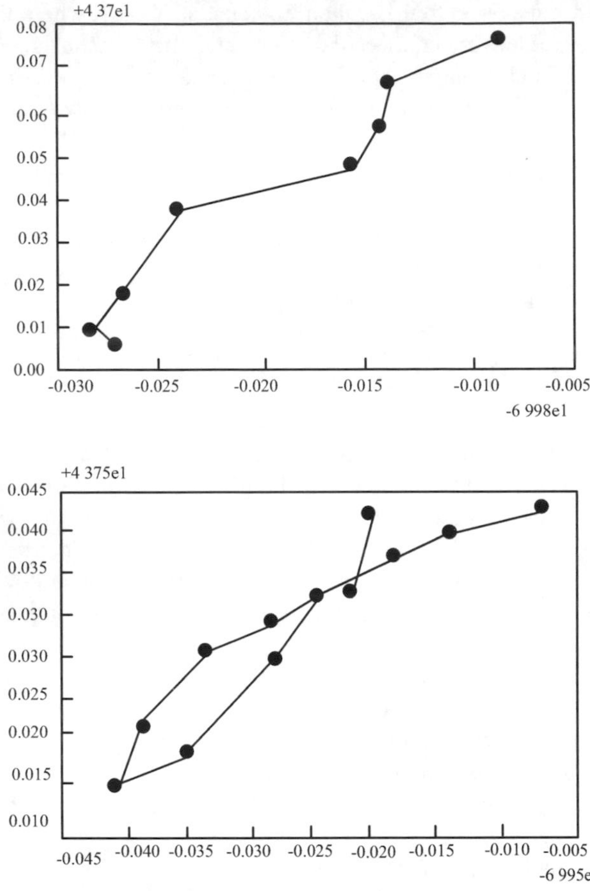

Fig. 12. Detection results of buoys at different depths

the same time and place. In addition to these errors, the data will also have various completely different problems. The red trajectory of the two trajectories on the left has only six data, which is probably caused by the damage of the buoy, which leads to the damage of the positioning system or the inability to send a signal for other reasons. And some data are extremely similar to the reason that the buoy is damaged, and the connection between the main body of the submersible target and the floating object storing the positioning system is disconnected, which causes the submersible target to become a kind of surface drifting buoy. The two types of drifting buoys are surface drifting buoys, so even if the styles are different, the data after running are very similar. Some similarities are caused by the similarity of ocean currents on the surface layer and the layer where the potential target is located. However, because the damage of the buoy cannot be determined in time, the real reason for the similar trajectory cannot be determined.

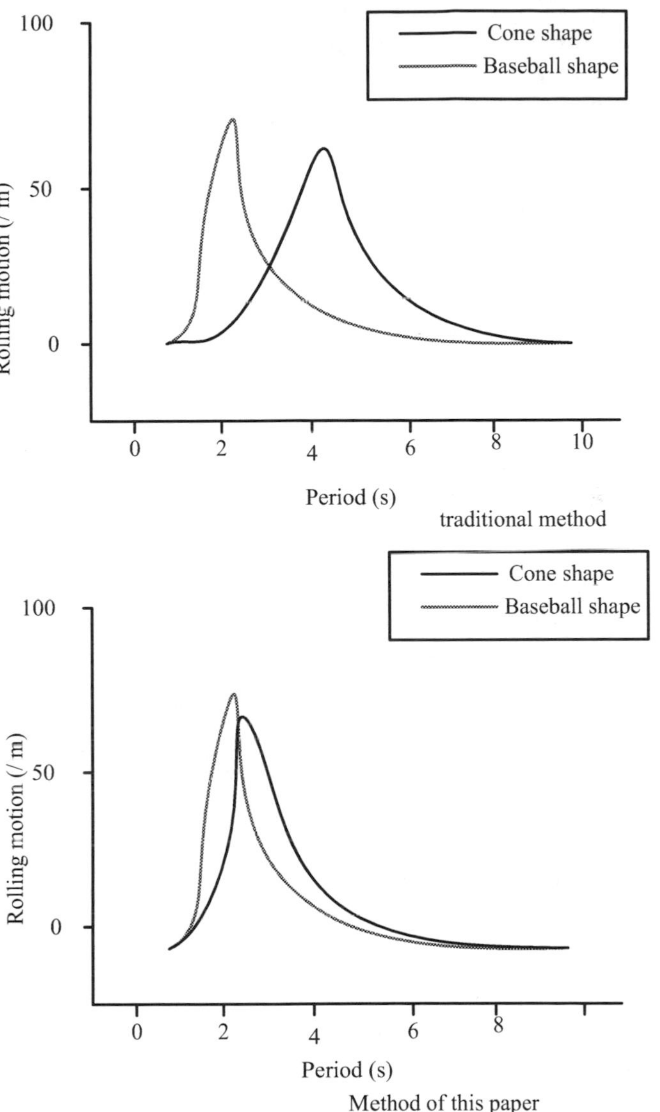

Fig. 13. Comparison results of roll motion response

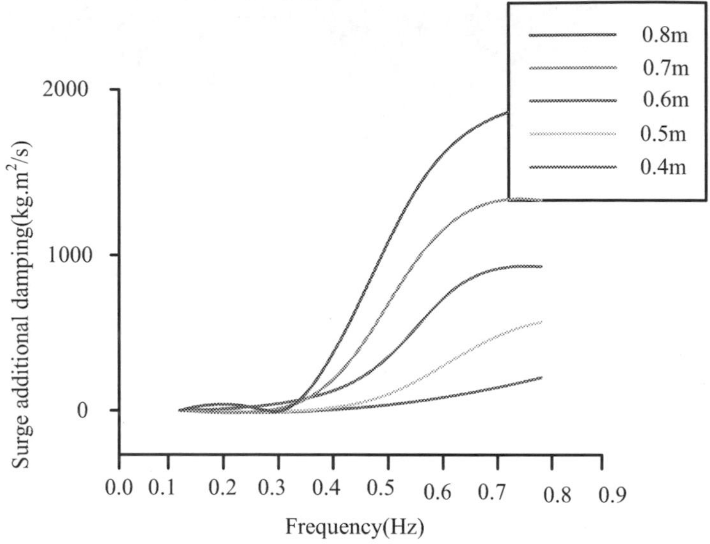

Fig. 14. The effect of draft on the additional damping of turbulence

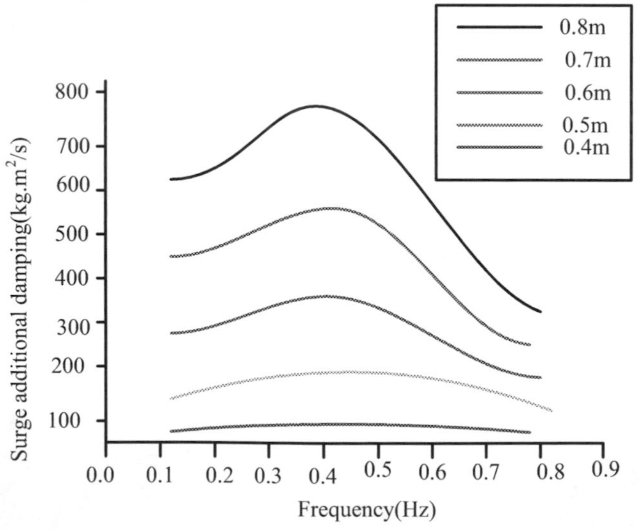

Fig. 15. The effect of draft on the additional mass of sloshing

4 Concluding Remarks

At this stage, the deployment conditions of small offshore buoys in China are relatively unfavourable, but if all parties can fully prepare and cooperate, the whole process can be orderly, safe and smooth. Buoy array anti-submarine is an important means of aviation anti-submarine. Controlling the accuracy of buoy placement is an effective measure to improve the anti-submarine efficiency of the buoy array. Planning the buoy placement route is the key to achieving high-precision buoy placement. The purpose of planning the buoy array route is to select a reasonable route that meets the performance index requirements of the buoy array and satisfies the aircraft deployment performance. At present, the research on the sonar buoy deployment formation simulation to improve the search probability of the buoy array is mainly carried out in this field in China. There is no report about the research on the influence of the buoy deployment error on the submarine search efficiency of the formation. In practical engineering applications, we found that the impact of buoy placement errors on the effectiveness of the buoy array cannot be ignored. Excessive placement error may lead to a decrease in the effectiveness of the formation's search for submarine, and even a large gap in the formation, resulting in missed alarms. From the perspective of engineering practice, this paper proposes a route experiment and error analysis method to reduce the error of the buoy array, which can achieve the purpose of effectively controlling the error of the buoy array.

References

1. Roostaei, J., Colley, S., Mulhern, R., et al.: Predicting the risk of GenX contamination in private well water using a machine-learned Bayesian network model. J. Hazardous Mater. **411**(10), 125075 (2021)
2. Ruggieri, A., Stranieri, F., Stella, F., et al.: Hard and soft EM in Bayesian network learning from incomplete data. Algorithms **13**(12), 329 (2020)
3. Zhao, C., Cao, F., Shi, H.: Optimisation of heaving buoy wave energy converter using a combined numerical model. Appl. Ocean Res. **102**(11), 102208 (2020)
4. Yang, J., Zhang, J.: Validation of Sentinel-3A/3B satellite altimetry wave heights with buoy and jason-3 data. Sensors **19**(13), 2914 (2019)
5. Sheridan, L.M., Krishnamurthy, R., Gorton, A.M., et al.: Validation of reanalysis-based offshore wind resource characterization using Lidar buoy observations. Mar. Technol. Soc. J. **54**(6), 44–61 (2020)
6. Venkatesan, R., Ramesh, K., Muthiah, M.A., et al.: Estimation of uncertainty in the atmospheric pressure measurement from the Indian ocean moored buoy systems. Mar. Technol. Soc. J. **55**(1), 137–146 (2021)
7. Meng, L., Wu, R., et al.: Experimental study on conversion efficiency of a floating OWC pentagonal backward bent duct buoy wave energy converter. China Ocean Eng. **33**(03), 50–61 (2019)
8. Mcallister, M.L., Van, D.: Lagrangian measurement of steep directionally spread ocean waves: second-order motion of a wave-following measurement buoy. J. Phys. Oceanogr. **49**(12), 3087–3108 (2019)
9. Venkatesan, R., Sannasiraj, S.A., Ramanamurthy, M.V., et al.: Development and performance validation of a cylindrical buoy for deep-ocean tsunami monitoring. IEEE J. Oceanic Eng. **44**(2), 415–423 (2019)
10. Yurovsky, Y., Dulov, V.A.: MEMS-based wave buoy: towards short wind-wave sensing. Ocean Eng. **217**(10), 108043 (2020)
11. Nielsen, U.D., Dietz, J.: Estimation of sea state parameters by the wave buoy analogy with comparisons to third generation spectral wave models. Ocean Eng. **216**(9), 107781 (2020)
12. Zhang, H., Zhou, B., Vogel, C., et al.: Hydrodynamic performance of a floating breakwater as an oscillating-buoy type wave energy converter. Appl. Energy **257**(8), 113996–113999 (2020)

13. Oikonomou, C., Rui, P., Gato, L.: Unveiling the potential of using a spar-buoy oscillating-water-column wave energy converter for low-power stand-alone applications. Appl. Energy **292**(11), 116835 (2021)
14. Zhang, H., Zhou, B., Vogel, C., Willden, R., Zang, J., Geng, J.: Hydrodynamic performance of a dual-floater hybrid system combining a floating breakwater and an oscillating-buoy type wave energy converter. Appl. Energy **259**, 114212 (2020)
15. Palm, J., Eskilsson, C.: Mooring systems with submerged buoys: influence of buoy geometry and modelling fidelity. Appl. Ocean Res. **102**, 102302 (2020)
16. Kumar, S., Nagababu, G., Kumar, R.: Comparative study of offshore winds and wind energy production derived from multiple scatterometers and met buoys. Energy **185**, 599–611 (2019)
17. Li, J., Huang, H., Chen, X., et al.: The small-signal stability of offshore wind power transmission inspired by particle swarm optimization. Complexity **2020**(13), 1–13 (2020)
18. Cao, Z., Liu, J., Lin, Z., et al.: Effects of inclined condition on LOCA for a small offshore reactor with OTSG. Nuclear Eng. Des. **375**(4), 111098 (2021)

Research on WiFi Indoor Positioning Error Correction Method Based on Adaptive Genetic Algorithm

Fa-yue Zheng(✉)

Mechanical and Electronic Engineering School, Beijing Polytechnic, Beijing 100176, China
fayue766@sohu.com

Abstract. In view of the shortcomings of traditional mean filtering and Kalman filtering in the process of receiving indoor WiFi positioning signal, a new method of correction of indoor positioning error based on adaptive genetic algorithm is proposed. In order to improve the accuracy of dynamic adaptive genetic signal processing, the algorithm uses adaptive genetic algorithm to modify the singular value of the average distance per hop and the error of measurement distance, and then the modified adaptive genetic value is processed again by weighted filter. The simulation results show that the algorithm has high accuracy, and compared with mean filter and Kalman filter, the algorithm has a great improvement in the influence of average ranging error and communication radius on the error.

Keywords: Adaptive genetic algorithm · Wifi; Indoor positioning · Error correction

1 Introduction

In this paper, the WIFI indoor positioning method is deeply studied, the whole positioning method is simulated and verified by MATLAB, the key parameters of the positioning system are determined, and the field test is carried out. Based on the field test results, an indoor positioning method based on adaptive genetic algorithm is proposed, By removing some reference points with small reference value in the database and the unsuitable nearest points selected in the precise positioning, the positioning accuracy is improved to a certain extent [1]. Literature [2] proposed a WiFi calibration method using BP neural network: using outlier detection algorithm to eliminate outliers in different mobile phone RSSI data pairs, and obtain relatively pure data to input to the BP neural network for training; and The weights and bias values of each layer of the network are repeatedly updated, so that the output value is close to the true value; when the square sum of the output layer error is less than the threshold, the training is completed, and the weights and bias values of each layer can be saved to get a more stable network The calibration model can be used to calibrate the observations of different models of mobile phones. However, this method has a poor correction effect of wifi indoor positioning errors. Based

S. Liu and X. Ma (Eds.): ADHIP 2021, LNICST 416, pp. 377–388, 2022.
https://doi.org/10.1007/978-3-030-94551-0_30

on this, this paper proposes a wifi indoor positioning error correction method based on adaptive genetic algorithm. The multi-user group cooperative localization algorithm based on WiFi and acoustic ranging is further studied, and the ranging method based on the sum of the amplitude of multi frequency audible acoustic wave is proposed. After the ranging is completed, the adaptive genetic algorithm is used to process the WIFI single point positioning results and acoustic ranging results, which effectively improves the positioning accuracy. In order to further improve the positioning accuracy, this paper uses Kalman filter to track the trajectory of mobile terminal equipment. In order to solve the problem that the tracking error of standard Kalman filter is large when the mobile terminal equipment turns, a Kalman filter algorithm combined with indoor map is proposed. The information of indoor map and the speed direction estimation of mobile terminal equipment are used to judge the turn, and the parameters of Kalman filter are adjusted when the turn occurs, The tracking accuracy of the corner is greatly improved. The client of indoor positioning system is implemented on Android platform, and the positioning management server of positioning system is implemented on Windows/Linux platform. The performance of the proposed positioning and correction algorithm is verified by field test.

2 WiFi Indoor Positioning Error Correction Method

2.1 Wireless WiFi Indoor Positioning Technology

Wireless indoor positioning technology refers to the use of Bluetooth, RFID, infrared and WIFI technology to estimate the location of people or objects in the indoor environment. Similar to the outdoor environment, the indoor environment also has the problems of wireless signal multipath effect, shadow fading and interference. However, due to the complex internal structure of buildings, different building materials and various indoor items, the indoor wireless environment is more complex than the outdoor environment. The indoor positioning technology based on WIFI can directly use the existing WIFI access points in the building, with low deployment cost, which is a major advantage of WIFI Positioning [3]. The location method needs at least three known IP points. Assuming their coordinates are $\left(x_1, y_1\right)$, $\left(x_2, y_2\right)$ and $\left(x_3, y_3\right)$ respectively, the time of wireless signal passing through the mobile terminal and each IP point can be measured, so as to calculate the distance a1, a2, and a3 from each IP point to the mobile terminal equipment. Taking the position of the three IP points as the center of the circle and the distance between the corresponding IP point and the terminal equipment as the radius, the three circles can be determined, The coordinates of their unique intersection point are the positioning results of mobile terminal devices. The formula is as follows:

$$\begin{cases} (x_1 - x)^2 + (y_1 - y)^2 = a_1^2 \\ (x_2 - x)^2 + (y_2 - y)^2 = a_2^2 \\ (x_3 - x)^2 + (y_3 - y)^2 = a_3^2 \end{cases} \tag{1}$$

Where x and y represent the x and Y coordinates of the mobile terminal device respectively, where they are unknowns. By solving the above equations, the positioning results of mobile terminal devices can be obtained. TOA based location methods need to ensure strict time synchronization between IP and mobile terminal devices [4]. Adaptive genetic algorithm (AGA) is also based on signal transmission time, but AGA does not measure the absolute time of signal arrival, but the time difference between signals from mobile terminal devices and different IP. Therefore, the location method based on AGA does not need time synchronization between mobile terminal devices and each IP, However, strict time synchronization is needed between IPS [5]. The positioning principle is shown in the figure (Fig. 1).

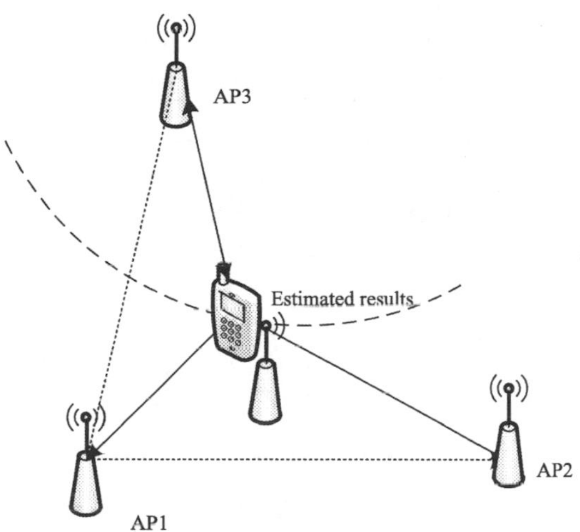

Fig. 1. Adaptive genetic algorithm indoor WiFi positioning principle

Adaptive genetic algorithm positioning requires at least three IP. During positioning, the time difference between the signal sent by the terminal equipment and each IP is measured:

$$\begin{cases} \sqrt{(a-x_2)^2-(b-y_2)^2}-\sqrt{(a-x_1)^2-(b-y_1)^2}=c(t_2-t_1) \\ \sqrt{(a-x_1)^2-(b-y_1)^2}-\sqrt{(a-x_3)^2-(b-y_3)^2}=c(t_1-t_3) \end{cases} \quad (2)$$

Among them, (a,b) represents the position coordinates of the mobile terminal to be located, and t_1, t_2 and t_3 respectively represent the time when the signal reaches the 3 IP. There may be two results calculated by the above formula, which will cause the positioning result to be blurred. At this time, the final positioning result can be determined by other information such as signal strength. When there are at least 4 IP, at least three hyperbolic equations can be obtained. At this time, the positioning result of the mobile terminal device can be uniquely determined [6]. If the position is known, the azimuth angle and length of the signal arriving at the IP are measured during positioning. Two straight lines can be obtained through the positions of the two IP and the azimuth angles of the two signals, and the intersection is the final positioning result. The position coordinate of the device to be located is (x, y), you can get:

$$\begin{cases} \tan\theta_1 = (y - y_1)/(x - x_1) \\ \tan\theta_2 = (y - y_2)/(x - x_2) \end{cases} \tag{3}$$

The positioning result can be obtained by solving the above-mentioned binary equations. The antenna array and other equipment are used to measure the angle of arrival of the signal, so the additional hardware equipment required is more complicated. The principle of indoor positioning using the wireless channel model is that the mobile terminal device estimates the distance between the terminal device and each IP by measuring the RSSI from multiple IP, and then uses the three-circle intersection positioning method to obtain the final positioning result. In the wireless channel model method, the most important thing is to have an accurate indoor wireless channel model. The accuracy of the channel model is directly related to the ranging accuracy using RSSI, thereby affecting the accuracy of the final positioning result [7]. When the transceiver device is in an open space and there are no obstacles between them, the received signal strength is inversely proportional to the square of the distance between the transceiver devices:

$$P_r \propto \frac{a_1^2+a_2^2+a_3^2-1}{d^2(\tan\theta_1+\tan\theta_2)} \tag{4}$$

Through a lot of theoretical and experimental research, a specific formula is obtained:

$$P_r(d) = \frac{P_t G_t G_r \lambda^2}{(4\pi)^2 d^2 L - P_r} \tag{5}$$

Where P_t represents the transmit power of the transmitting end, G_t and G_r represent the gain of the transmitting and receiving antennas respectively, λ represents the wavelength of the wireless signal, d represents the distance between the transmitting and receiving ends, and L represents the propagation loss [8]. Due to factors such as walls and other obstacles, people walking, door opening and closing, the indoor wireless environment is much more complicated than free space, so the free space attenuation model cannot be simply used for indoor positioning distance measurement. Through a large number of experiments, a logarithmic distance path attenuation model is summarized.

$$PL(d) = PL(d_0) + 10n\log_{10}\left(\frac{d}{d_0}\right) + X_\sigma \tag{6}$$

Among them, d_0 is the path attenuation index, the value of d_0 is different in different environments, and X represents the environmental noise, which obeys the normal distribution with the mean value of zero. d is the reference distance, usually 1 m. $PL(d_0)$ represents the loss at the reference distance; PL(d) represents the path loss value when the distance between the receiving and sending ends is d. The logarithmic path attenuation model does not include the penetration loss of the wireless signal from obstacles such as the floor and walls of the building, so the Keenan-Motley model is proposed:

$$PL(d) = L_0 + 10n\log_{10}(d) + \sum_{j=1}^{J} N_{wj}L_{wj} + \sum_{i=1}^{I} N_{fi}L_{fi} \tag{7}$$

Among them, L_{fi} represents the reference value of floor loss, and N_{fi} represents the number of penetrated floors. L_{wj} represents the reference value of through-wall loss. N_{wj} represents the number of through walls; other symbols have the same meaning as the logarithmic path attenuation model. As shown in the table, the penetration loss values of different building materials are different (Table 1).

Table 1. Penetration loss values of different building materials

Material type	Penetration loss (dB)
Ceiling management	1–8
Concrete floor	25–30
Ordinary brick concrete partition wall	10–15
Concrete wall	20–30
Elevator box top	30
Wooden furniture	3–6
Glass	5–8

During positioning, selecting an accurate indoor wireless attenuation model can obtain more accurate ranging results, thereby reducing the final positioning error.

2.2 Wifi Indoor Positioning Error

The quality of indoor positioning technology cannot be evaluated from a single aspect, but needs to be evaluated from various aspects such as positioning accuracy, deployment cost, and energy consumption. Indoor map is an important part and implementation basis of indoor positioning system. Without indoor map, the collection of location database and the display of final positioning results cannot be completed. The Arc GIS system is used to draw and publish the map. This geographic information system is developed by Esri [9]. The model structure of the Arc GIS system includes object classes, feature classes, and feature datasets. The Arc GIS indoor map of the realized indoor positioning system is deployed in the positioning and map server. First, the experimental area is measured geographically, and then the indoor map is drawn in the Arc GIS Desktop software based on the measurement information to generate the indoor map database, and finally in Arc GIS Publish the drawn indoor map in the Server and start the Arc GIS service to complete the drawing and deployment of the Arc GIS indoor map. The Android client can obtain the indoor map of the relevant area from the map server in real time, or convert the Arc GIS indoor map to an offline map and store it in the client. The client can display the indoor map on the screen, and zoom or zoom the map. Rotate, the final indoor positioning result can also be displayed on the map intuitively in real time. It should be noted here that if the client needs to obtain the Arc GIS indoor map from the map server in real time, the map server needs to close the firewall or set certain firewall rules, otherwise the client may not be able to obtain map data from the map server (Fig. 2).

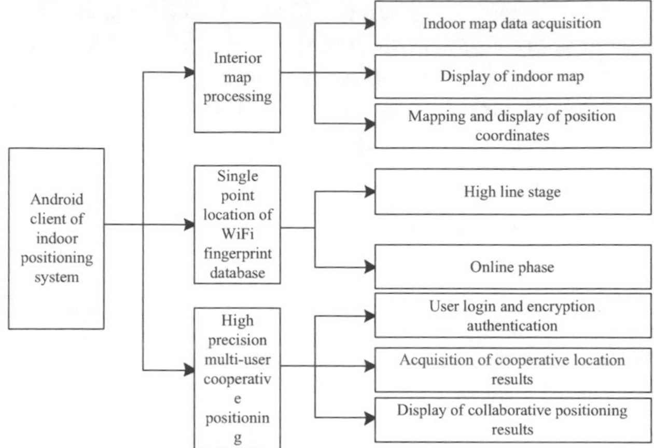

Fig. 2. Wifi indoor positioning data processing model

Positioning accuracy is the most important indicator, which depends on the positioning method and positioning algorithm. Generally, high precision means high deployment cost and high computational complexity. The positioning accuracy can be evaluated from the average positioning error, the cumulative distribution function of the positioning error, and the mean square positioning error. In a two-dimensional plane, if n single-point positioning experiments are performed, and the actual coordinates of the n points to be positioned are $(x_1, y_1), (x_2, y_2), \ldots, (x_n, y_n)$, then the average positioning error of the n times of positioning is:

$$ME = \frac{1}{n}\sum_{i=1}^{n}\sqrt{(x_i - \hat{x}_i)^2 + (y_i - \hat{y}_i)^2} \tag{8}$$

The mean square positioning error is

$$MSE = \frac{1}{n}\sum_{i=1}^{n}\left[(x_i - \hat{x}_i)^2 + (y_i - \hat{y}_i)^2\right] \tag{9}$$

Further, the root mean square positioning error is:

$$RMSE = \sqrt{MSE} = \sqrt{\frac{1}{n}\sum_{i=1}^{n}\left[(x_i - \hat{x}_i)^2 + (y_i - \hat{y}_i)^2\right]} \tag{10}$$

The CDF of the positioning error represents the probability of the positioning error within a certain range, and the CDF curve can more intuitively describe the error performance of the positioning system. The main work done in the offline phase is the construction of the WIFI location. When the positioning area is large, the number of reference points is large, and the amount of data may be very large. Therefore, when the positioning point is compared with the received signal strength of each reference point during online positioning, the calculation complexity is high, and the effect of real-time positioning cannot be achieved.. In order to solve this problem, a clustering algorithm is used to classify the reference points to form a certain number of clusters, which can greatly reduce the amount of calculation during positioning and improve the real-time performance of indoor positioning [10]. During the measurement process, the dynamic adaptive genetic algorithm often encounters some sudden singular values due to factors such as equipment or environment, and these singular values often bring great errors to the positioning. In the mean value and Kalman filtering methods, these singular values cannot be eliminated or corrected in real time, so errors caused by singular values cannot be eliminated. Accordingly, this paper proposes an improved dynamic adaptive genetic algorithm ranging model, as shown in the figure (Fig. 3):

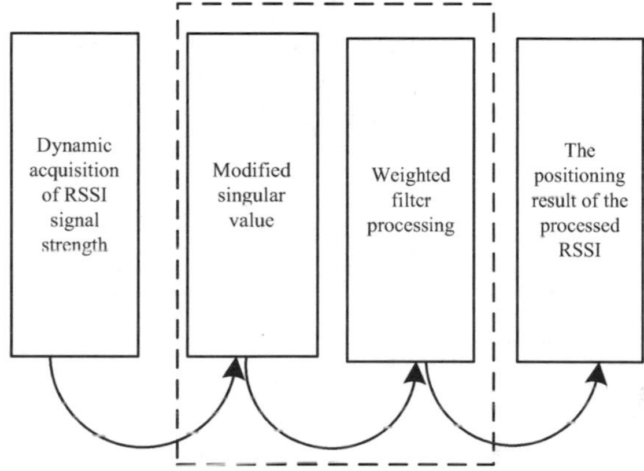

Fig. 3. Improved dynamic positioning algorithm

Specifically, it is divided into three parts: the first part is the dynamic acquisition of adaptive genetic signals; the second part is adaptive genetic signal processing, and this part is divided into two processes: singular value correction and weighted filtering; the third part is after processing Adaptive genetic and result output.

2.3 Wifi Indoor Positioning Error Correction

The online positioning phase completes the actual positioning process of the mobile terminal device, which is specifically divided into two processes, namely, rough positioning and precise positioning. The purpose of rough positioning is to narrow the to-be-positioned point from the entire positioning area to one or several clusters through cluster matching. Through rough positioning, the current position of the point to be located can be roughly determined, so there are fewer reference points to consider, which greatly reduces the number of reference points that need to be compared during the precise positioning stage and improves the positioning speed. When positioning, the mobile device first scans the RSSI of all access points it can currently detect:

$$\Phi_r = [\varphi_1, \varphi_2, \ldots, \varphi_L] \tag{11}$$

Among them, φ_i represents the RSSI from the i access point collected by the mobile terminal equipment in any direction, and L represents the total number of access points in the database. The co-location algorithm based on error diversity makes full use of the diversity of single-point positioning errors, and merges the received single-point position estimation results of surrounding users and itself to improve its own positioning accuracy. First, the user obtains his own single-point positioning result through the PDR positioning method, and then exchanges this result with the surrounding users, and processes the single-point positioning result of himself and the surrounding co-locators to obtain a coordinated positioning result with improved positioning accuracy.

This co-location method has designed a sound wave communication method, which can effectively exchange position estimation information of users by distinguishing the frequency of sound waves when they meet. In addition, experimental studies have proved that when users use PDR for positioning, the estimation error of the movement step size and the estimation error of the movement direction obey the normal distribution. Therefore, if the PDR positioning results of multiple users at the same or similar positions are averaged, different users can be averaged. The single-point positioning errors cancel each other out, resulting in higher positioning accuracy at this position. The block diagram of co-location based on error diversity is shown in the figure, and the specific implementation steps are as follows (Fig. 4):

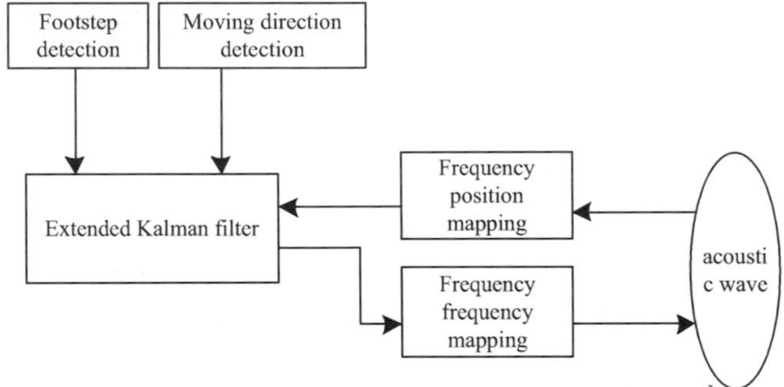

Fig. 4. Block diagram of co-location based on error diversity

As shown in the figure, the location area of interest is divided into different virtual grid areas. Each grid has a number as the grid label, which represents this grid area. As long as the grid size is determined, the entire grid can be covered by the grid. Indoor map. The disadvantage of the co-location algorithm based on error diversity is that due to the limited sound wave frequency range, each grid area corresponds to a sound wave frequency, so the total number of grid areas is limited, and the grid area is not fine enough, and users use sound waves. When the frequency exchanges position information, the center point of the regional grid is used as its own positioning coordinate, which brings about larger errors. In addition, this method only uses the diversity of single-point positioning errors for collaborative positioning, and does not make full use of the geometric relationship between the positioning results of multiple users, so that the improvement of positioning accuracy is relatively limited.

3 Analysis of Results

In order to study the ranging performance of the dual-transmit and dual-receive Chirp sonic ranging method used in this article, this article uses four mobile phones to conduct a ranging experiment indoors (Table 2).

Table 2. Basic configuration parameter list of four mobile phones

Basic configuration	HTC T328	Samsung S7568	Samsung I 9100	HTC G7
operating system	Android OS 4.0	Android OS 4.0	Android OS 2.3	Android OS2.2
CPU	Snapdragon MSM7227A	Snapdragon MSM7227A	Samsuing Exynos 4210	Snapdragon QSD8250
Wi Fi module	Broadcom BCM4329	Unknown	Samsung SWB-B23	Broadcom BCM4329

The statistical parameters of the RSSI sample data collected by different mobile terminals are compared, including the maximum and minimum values, the range of change, and so on. The maximum value is measured by measuring the mobile phone as close to a certain IP as possible. Similarly, the minimum value is measured by measuring the maximum distance from a certain IP but still being able to collect its RSSI data. As shown in the table (Table 3).

Table 3. Comparison of the maximum, minimum and variation range of RSSI data collected by different mobile terminals

Manufacturer	Mobile phone model	Minimum (dBm)	Maximum (dBm)	Range of change (dBm)
HTC	G7	−96	−38	61
HTC	T328	−97	−35	59
Samsung	GTS7568	−93	−13	80
Samsung	I9100	−95	−24	71

Samsung GTS7_568 has the largest variation range, which means that it converts the real signal energy in more detail and can recognize a larger range of signal changes. In contrast, the RSSI sample data collected by HTCG7 has the smallest change range. Therefore, this may cause the true signal energy within a certain range to be mapped to the same RSSI value, which makes the RSSI change range it can detect even smaller. For alignment, a larger RSSI variation means a higher degree of discrimination of signal energy, which is more helpful to distinguish two different positions and improve positioning accuracy. In the experiment, 200 different distance tests were selected, and the distance measurement error was counted. The figure shows the CDF curve of the ranging error (Fig. 5).

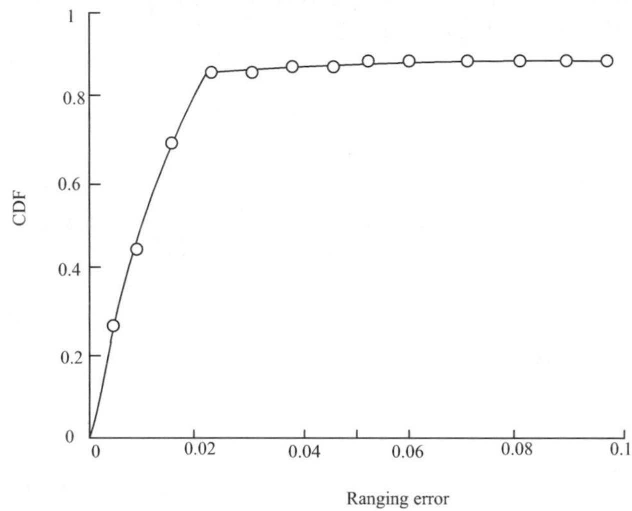

Fig. 5. CDF curve of signal ranging error

Obviously, the probability of ranging error within 3 cm is more than 90%, the probability of error within 8 cm is 100%, and the average error is 1.13 cm, which has reached an ideal ranging accuracy. Through experiments, the range of the dual-transmit and dual-receive Chirp sonic ranging can reach 8 m. In order to verify the feasibility of this algorithm, simulation comparisons are performed based on the mean filtering method, Kalman filtering method and improved dynamic adaptive genetic algorithm. The comparison results of average positioning error and node communication radius error are shown in the figure. The experimental parameters are set as follows (Figs. 6 and 7):

① The network area is set to a two-dimensional plane of 80 m*80 m;

② Randomly deploy 20 unknown nodes in this area;

③ In order to verify the influence of the distribution density of anchor nodes in the network on the positioning error of the algorithm, the number of anchor nodes in this area is 20–50 respectively. In order to simulate the actual situation, the anchor nodes are randomly distributed;

④ Communication radius: r = 50 m;

⑤ All experiments are simulated for 50 times, and the average value is taken as the experimental result;

⑥ Reference distance d0 = 1, mean square error of Gaussian distribution $X\sigma = 2$.

It can be seen from the figure that with the increase of the number of anchor nodes in the network, the average positioning error of the mean filtering method, Kalman filtering method and improved dynamic adaptive genetic algorithm proposed in this paper is generally decreasing. As the number of anchor nodes increases, the average positioning errors of the three algorithms tend to stabilize. When the number of anchor

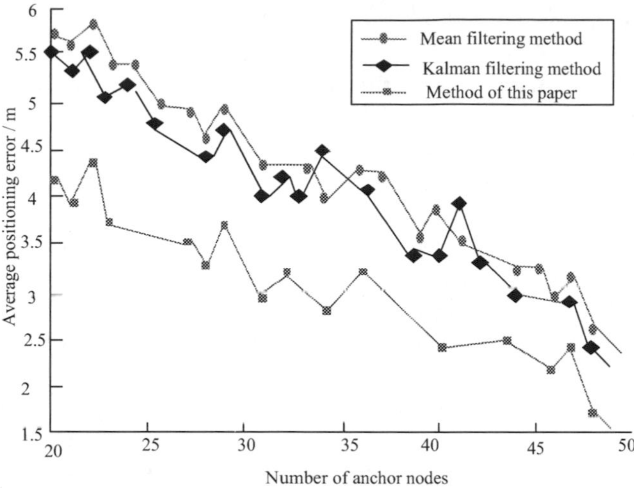

Fig. 6. Comparison of average errors of different algorithms

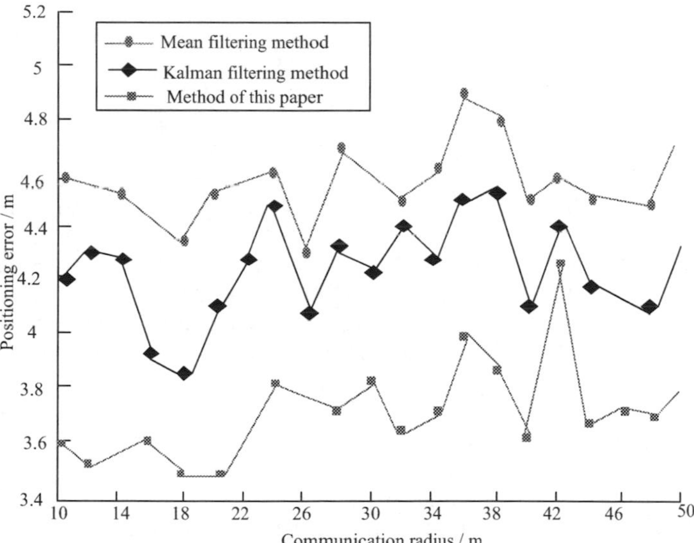

Fig. 7. The influence of node communication radius on error

nodes is 34, 41, and 46, the average positioning error of Kalman filtering is greater than that of average filtering. As the density of anchor nodes in the network increases, the average positioning error of the two methods tends to be equal; the average positioning error of the improved dynamic adaptive genetic algorithm is significantly smaller than that of the first two algorithms, indicating that the improved method can better suppress a single time The influence of the measured value on the average positioning error of the algorithm.

4 Conclusion

Aiming at the shortcomings of the existing mean filter method and Kalman filter method that the singular values cannot be eliminated and corrected in real time, a wifi indoor positioning error correction method based on adaptive genetic algorithm is introduced, and the weighted filter method is used to process the corrected self Adapt to genetic value. For multiple sets of adaptive genetic values measured by anchor nodes, a median strategy is adopted to select the most appropriate adaptive genetic values. Aiming at the problem that the target node and the anchor node are at different heights, a space compensation model is designed to reduce the impact of the target node and the anchor node on the positioning accuracy caused by the different planes. The simulation experiment results show that the algorithm improves the accuracy of dynamic adaptive genetic signal processing, has a higher calculation accuracy, and the ranging error after filtering has been greatly improved. Each parameter in the adopted environmental attenuation factor needs to be re-calibrated under the new environment, so the workload is relatively large. In future work, we can test several more model parameters in indoor environments, and then summarize a set of model parameters that can be used directly in indoor environments.

Fund Projects. Research on WiFi Based indoor positioning technology (Project No.: CJGX2016-KY-YZK040).

References

1. So, C., Ho, I.M., Chae, J.S., et al.: PWR core loading pattern optimization with adaptive genetic algorithm. Ann. Nucl. Energy **159**(9), 108331 (2021)
2. Song Binbin, Y., Min, H.X., et al.: A WiFi calibration method for indoor positioning based on BP neural network. J. Navig. Positioning **7**(01), 43–47 (2019)
3. Dbna, B., Jing, H.A., Vtt, A., et al.: An effective random statistical method for indoor positioning system using WiFi fingerprinting - ScienceDirect. Futur. Gener. Comput. Syst. **109**(8), 238–248 (2020)
4. Han K , Yu S M , Kim S L , et al. Exploiting User Mobility for WiFi RTT Positioning: A Geometric Approach[J]. IEEE Internet of Things Journal, 2021, PP(99):1–1
5. Al-Khaleefa, A., Ahmad, M.R., Isa, A., et al.: MFA-OSELM Algorithm for WiFi-based indoor positioning system. Inf. (Switzerland) **10**(4), 146 (2019)
6. Martin-Escalona, I., Zola, E.: Passive round-trip-time positioning in dense IEEE 802.11 Networks. Electronics **9**(8), 1193 (2020)
7. Madyatmadja, E.D., Hakim, L., Tannady, H., et al.: Use K-Nearest NEIGHBOR and FLOYD WARSHALL algorithms to determine store location and distance based on WIFI. Tech. Rep. Kansai Univ. **62**(5), 2379–2389 (2020)
8. Liu, S., Sun, G., Fu, W. (eds.): 2020. LNICSSITE, vol. 339. Springer, Cham (2020). https://doi.org/10.1007/978-3-030-63952-5
9. Liu, S., Liu, X., Wang, S., Muhammad, K.: Fuzzy-Aided Solution for Out-of-View Challenge in Visual Tracking under IoT Assisted Complex Environment. Neural Comput. Appl. **33**(4), 1055–1065 (2021)
10. Liu, S., Li, Z., Zhang, Y., Cheng, X.: Introduction of key problems in long-distance learning and training. Mob. Netw. Appl. **24**(1), 1–4 (2018). https://doi.org/10.1007/s11036-018-1136-6

AI System Research and Model Design

Minimum Redundancy Distributed Storage System of Enterprise Financial Information Based on Fuzzy Control

Yan Yu(✉) and Lin Chen

Wuhan Qingchuan University, Wuhan 430204, China

Abstract. In view of the problem that the load value of traditional storage system is too small, resulting in poor distributed burst effect, this paper designs a minimum redundancy distributed storage system of enterprise financial information based on fuzzy control. In the hardware part, the storage control board is designed, and the minimum redundant data acquisition circuit is constructed to control the operation load of each hardware branch. The software uses fuzzy control to process the minimum redundant information, and uses JavaScript to write distributed algorithm to realize the storage function of the storage system. After preparing the system test environment, two kinds of traditional storage system and design storage system are prepared for experiments. The results show that the design storage system has the largest load capacity and can carry a variety of storage tasks.

Keywords: Fuzzy control · Enterprise accounting information · Minimum redundancy · Distributed storage

1 Introduction

Financial information refers to the economic information used to reflect its characteristics in combination with other information. Compared with financial information, non-financial information refers to all kinds of information that appear in the form of non-financial information and have direct or indirect connection with the production and operation activities of enterprises. Nowadays, the business activities of enterprises are more and more complex and the expectation of shareholders for financial information is higher and higher. How to determine the scale of financial information disclosure has become one of the most difficult problems in accounting practice. The disclosure of financial information changes with the change of economic environment. The following factors affect the determination of disclosure content, and then affect the cost of financial information disclosure. The use of enterprise financial information is the top priority of enterprise financial information work. Only by making good use of enterprise financial information can we give full play to the role of "barometer" and "early warning device" of enterprise financial information.

At present, high-speed storage technology has become the key technology to be solved in the era of enterprise accounting big data. Fuzzy control can adjust the extremely

S. Liu and X. Ma (Eds.): ADHIP 2021, LNICST 416, pp. 391–402, 2022.
https://doi.org/10.1007/978-3-030-94551-0_31

high data transmission bandwidth, so its research has important practical value and application significance.

For this reason, relevant scholars have carried out research and made some progress. Literature [2] proposes a storage system of financial accounting information based on big data. Through the characteristics of big data, it plays an auxiliary role in the calculation of enterprise financial accounting work, the storage of accounting data, accounting prediction and other work, improves the quality of enterprise financial accounting work, optimizes the work process, and gives full play to the management function of accounting, But the system load of this method is too small.In reference [3], a financial information storage system based on UFIDA NC financial system is proposed. The system will store data information in the server of the enterprise headquarters. At the same time, each branch will set up an ERP system independently, and the headquarters will manage the ERP system of each branch by running UFIDA NC financial system, so as to improve the effect of financial information storage, But the effect of this system is not good.

Therefore, the research on the minimum redundancy distributed storage system of enterprise accounting information based on fuzzy control can provide reference for other high-speed storage systems. On the original hardware of the system, by adding other hardware, it can meet the needs of various functions [2]. High speed AD is added to the hardware of the system, and DSP is added to the back-end. High speed ad samples the data and transmits the data to DSP at high speed through the PCI bus. DSP is used to process the received data by compression, jigsaw puzzle and other algorithms. Based on this hardware platform, the design of high-speed data processing system based on PCI + DSP can be realized. Adding SATA hard disk on the basis of the original hardware of the system, because of the high data bandwidth and large capacity of SATA, the data can be stored into the SATA hard disk at high speed through the PCI, and the SATA hard disk can work in parallel to form a distributed storage structure.

2 Hardware Design of Distributed Storage System

2.1 Design Storage Control Board

The storage control board is composed of FPGA interface module, PCI hard core module, DDR3 memory module and six way tlk2711 interface module. The storage control board is designed as shown in the figure below (Fig. 1).

In the control board structure shown above, the FPGA is mainly used to realize the high-speed DMA read and write based on the PCIe hard core, the DDR3 controller, the design of the cache module and the interface design of the whole system. Xilinx virtex6 FPGA is adopted in the control board, which sets the working frequency to 600 MHz, supports DDR3 storage interface, provides DDR3 controller customization and rich digital clock management (DCM) resources [3]. Integrated with the PCIe hard core, it supports v2.0 protocol and X8 link width. GTX transceiver supports the transmission rate from 150mbit/s to 6.5 gbit/s. After the structure of the storage control board is built, the minimum redundant distributed data acquisition circuit structure is constructed.

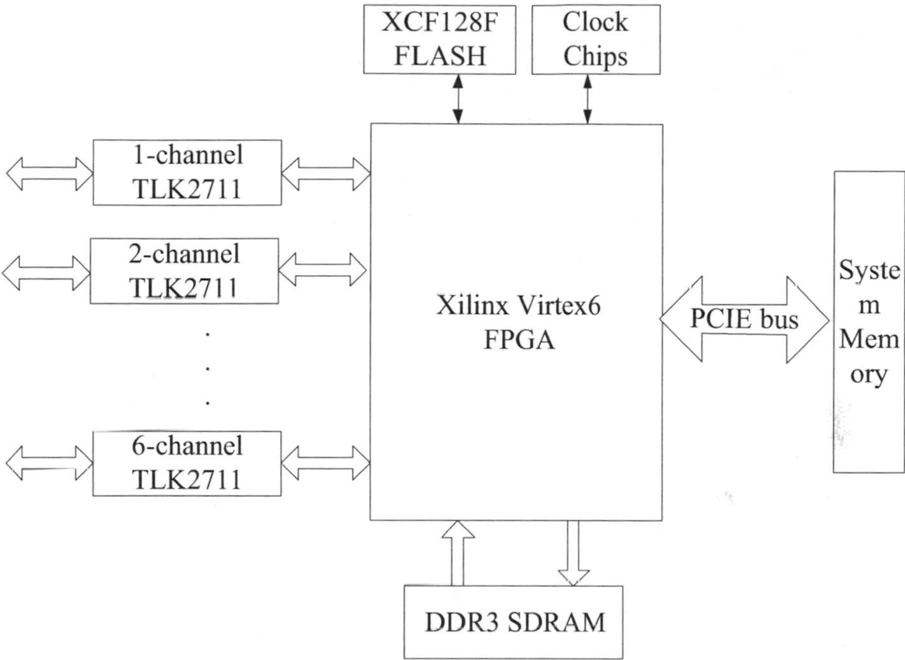

Fig. 1. Structure of storage control board

2.2 Construction of Information and Data Acquisition Circuit

Under the above hardware control, the serial port uses 422 interface chip to realize the conversion between TTL and serial port level. The control panel and serial communication select RX and TX cross connection mode [4] to realize the transmission of read-write data. The final connection circuit is shown in the figure below (Fig. 2):

According to the control circuit shown in the figure above, the plug-in serial port is used to connect the plug-in terminal, and the 9-pin serial port device terminal is controlled to run three interface units to control the voltage between the home appliance control board and the main controller to achieve the normal transmission of data. Four PNP triodes are used as the current driving unit of the relay, and four monochrome light-emitting diodes are set to monitor the working state of the furniture working unit [5]. Control MCU pin P1.0~p1.3 for low level, at this time the LED is on, diode interface Q1~Q4 is in the state of conduction, the internal coil of the relay is powered on, pull the internal switch of the relay, keep the relay in the closed state, synthesize the above circuit design, complete the hardware design of the storage system.

3 Software Design of Distributed Storage System

3.1 Using Fuzzy Control to Deal with Minimum Redundant Information

Using the distributed fuzzy controller structure, when processing the minimum redundant information, the clear physical quantity of the output and the output is converted into a

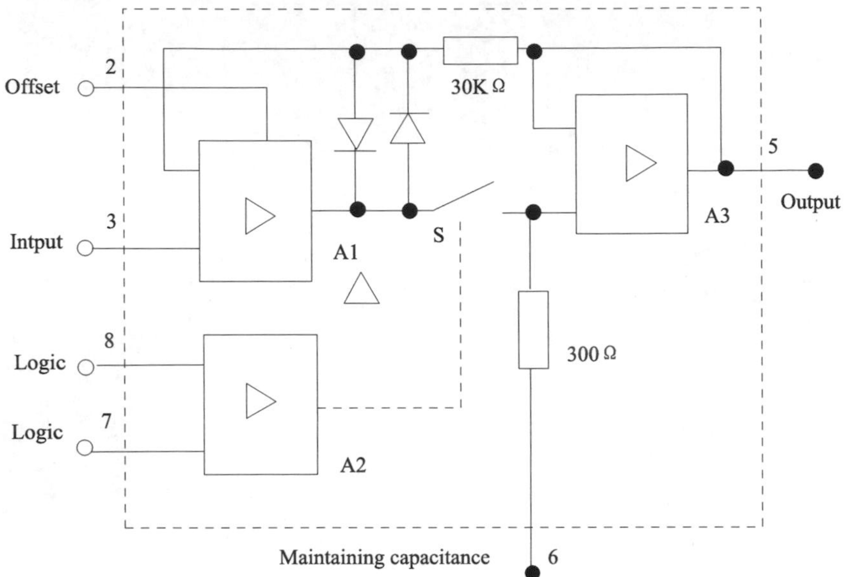

Fig. 2. Hardware control circuit

fuzzy quantity, and the physical theory domain of a component y_j of the known input variable Y_1 is:

$$Y_j = [-y, y](y > 0) \tag{1}$$

Obtain the fuzzy universe corresponding to Eq. (2) at this time as:

$$N_j = [-m_j, m_j](m_j > 0) \tag{2}$$

At this time, the fuzzy transformation coefficient k_j from the input variable Y to the variable m_j is the quantization factor, and the following conversion formula is obtained:

$$k_j = \frac{m_j}{y_j} \tag{3}$$

Among them, k_j is always positive and $k_j > 0$. According to (3), when the variable Y changes, in order to keep the control core of the fuzzy controller unchanged, only need to change k_j, so that the input quantity in the fuzzy processor can still fall into the original universe of discourse after being changed. In, the domain of discussion at this time is the use rule of fuzzy controller.

After formulating the use rules of the fuzzy controller, the approximate reasoning algorithm used by the fuzzy controller is selected [6]. Take the input element with the largest membership degree in the fuzzy set in the inference rule as the precise control quantity, adjust multiple elements to have the largest membership degree, and take the average of the membership degrees of several elements. The calculation formula is as follows:

$$\begin{cases} \Gamma = \frac{k_j}{2}L \\ \delta = \frac{L}{2}(\frac{d\beta_i}{d\beta_j} - \frac{\beta_j}{\beta_i}) \\ \beta = \frac{L(\beta_i - \beta_j)}{2} \end{cases} \tag{4}$$

Among them, Γ is the membership degree of the elements, δ is the precise control quantity, L is the number of elements, d is the applicable rules of the fuzzy set, and β_i and β_j are the maximum membership degrees of the two elements. The degree of membership is regarded as the minimum redundancy of corporate accounting information [7], which is used to realize the storage function of the system.

3.2 Storage Function Realization

When using JavaScript to write distributed algorithms and integrate them into different operation groups [8] according to the magnitude of the redundancy value, the operation grouping can be expressed as:

$$M(j) = \frac{P(j)}{Z(j)} \tag{5}$$

Among them, j represents the operation grouping, M represents the distributed function, P represents the signal power function, and Z represents the total number of enterprise accounting information operation groups. Choose a level of attribute as the storage index, and the storage process can be expressed as:

$$SE = \mu \frac{\langle E_t(t) - E_r(t)\rangle}{\langle E_t(t)\rangle} \tag{6}$$

Among them, μ represents the distribution parameter, $E_t(t)$ represents the information signal energy under the time scale, and r represents the signal data receiving parameter. When importing the storage algorithm [9], based on the system's receiving coefficient parameters, under the photon gain control, the algorithm import process can be expressed as:

$$G_r(t) = \frac{g(N_r(t) - N_0)}{(1 + sE_r)} \tag{7}$$

Among them, g represents the introduction rate, $N(t)$ represents the number of carriers, s represents the feedback coefficient, and N_0 represents the initial carrier number of the algorithm. After the distributed algorithm is imported, the storage algorithm is matched with the information attribute [10], and different distribution parameters are set according to the actual number of attributes to meet the needs of enterprise accounting information. Integrate the above hardware and software design, and finally complete the design of the distributed storage system.

4 Simulation

4.1 Experiment Preparation

Due to conditions, a pseudo-distributed cluster construction method was used when building a Hadoop cluster in the experiment, that is to say, a Hadoop cluster was constructed by creating multiple virtual machines on a single node. In the experiment, VMware is used to build a total of six virtual machine nodes, one of which is the Namenode name node, and the other five nodes are used as Datanode nodes, and the node virtual machines are built into a cluster. The hardware configuration of the experimental node is shown in the following table (Table 1 and 2):

Table 1. Hadoop cluster hardware configuration table

Serial number	Namenode node	Datanode node
Operating system	CentOS 7 64-bit system	CentOS 7 64-bit system
CPU	Intel(R) Core(TM) i5-2450 MCPU@2.S GHz	Intel(R) Core(TM) i5-2450 MCPU@2.S GHz
RAM	2 GB	1 GB
Hard disk	20 GB	20 GB

Under the control of the hardware node configured in the above table, prepare the experimental platform shown in the following table, and the parameters of the experimental platform are shown in the following table (Table 2):

Table 2. Experimental platform parameters

Serial number	Name of software	Version
1	operating system	CentOS 7 64-bit system
2	JDK	jdk-8u201-linux-x64.rpm
3	Hadoop	hadoop-2.9.0
4	Programming language	Java

Under the control of the platform parameters shown in the above table, the virtual machine is used to configure jps, and the storage system equipped with the host computer is continuously debugged. Under the control of the platform parameters shown in the above table, the virtual machine is used to configure JPS, and the upper computer is continuously debugged to carry the storage system.

In order to measure and verify the read-write performance of the onboard storage system, the DDR3 write request design interacts with the upstream FIFO, writes commands, addresses and data to the DDR3 controller, and configures the related interface timing, so as to complete the write request operation to the DDR3 memory. The sampling sequence of DDR3 memory interface write request is shown in Fig. 3.

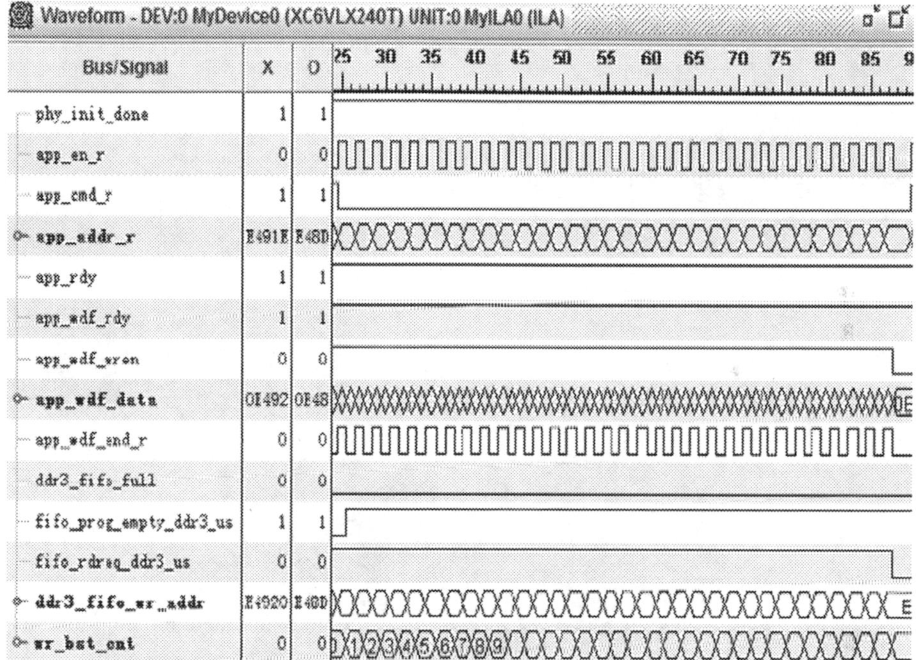

Fig. 3. DDR3 memory interface write request sampling sequence

In the process of request sampling sequence shown in the figure above, when the DDR3 controller completes the whole initialization process, it will send the signal PHY_ init_ Done set to valid. At this time, if the DDR3 memory is insufficient and the DDR3 uplink FIFO is not empty, the corresponding DDR3_ fifo_ Full signal and FIFO_ prog_ empty_ ddr3_ It's a signal. If command and address FIFO are ready to receive command and write data FIFO is ready to receive data, corresponding to app in the figure_ RDY and app_ wdf_ RDY signal, the user should initiate a write command operation, that is, the app_ Set CMD to low and give the write command address app_ addr。

When the write command is submitted successfully, the user should send the app_ Set WDF Wren to valid, and set the_ The data is submitted on WDF data, and the DDR3 memory write address is given. When the line is full, wr_ bst_ When CNT equals 32, the write operation stops. Through the analysis of the above figure, it can be seen that the sampling timing fully conforms to the DDR3 memory write design. In order to simplify the operation process, five switches are used to connect ZigBee devices, of which two switches are used as the core layer, and the remaining two switches are used as the access layer. After connecting the enterprise controlled area network, a firewall is set outside the enterprise LAN, and the final system test environment is as follows (Fig. 4):

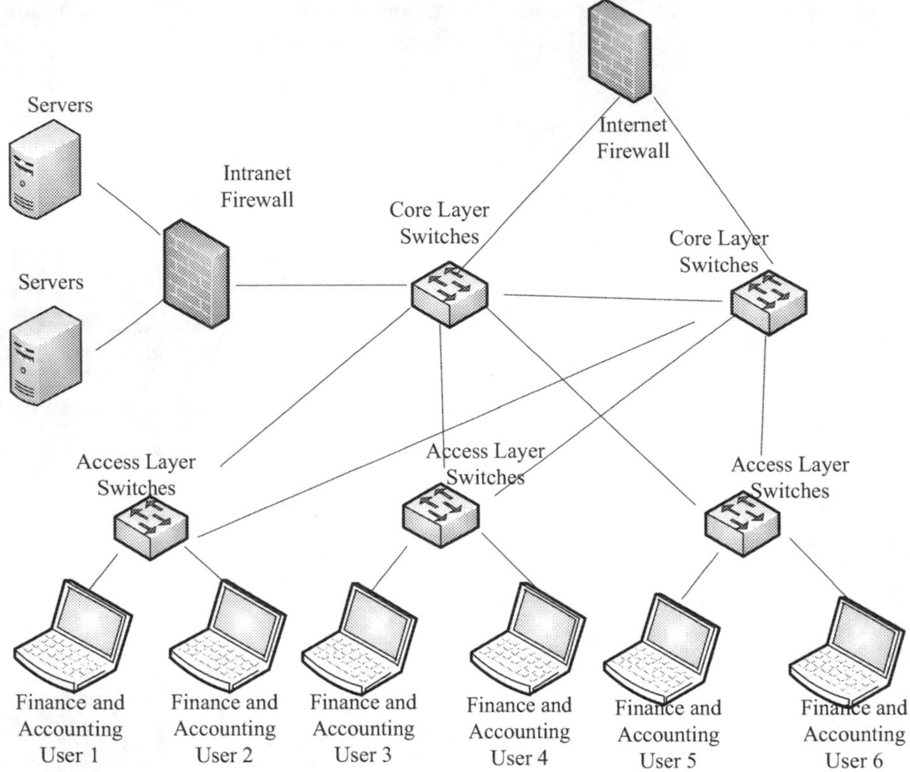

Fig. 4. Build the test environment of the system

In the test environment shown in the figure above, two ZigBee modules are used to continuously receive the information of the switch. The node of enterprise accounting information connects the minimum redundant information terminal node with the experimental PC through the hardware serial port. After starting the coordinator of the test environment, a distributed test network is built. In the above test environment, two traditional distributed storage systems and the designed reservation system are used to compare the performance of the three storage systems.

4.2 Structure and Analysis

Based on the above experimental preparation, the PCB antenna is used as the transmitting point of the minimum redundant information of the test node. The transmission power of the information is set to 5 dBm, the test data packet is 8 characters, the baud rate is set to 115000, and the minimum redundant accounting information is circularly sent for 1000 times (Table 3).

It can be seen from the packet loss rate results shown in the above table that with the increasing number of iterations, the packet loss rates generated by the three reservation systems show packet loss rates of different values. After the number of information cycles

Table 3. Packet loss rate of three storage systems

Number of times of sending information	Packet loss rate /%		
	Traditional storage system 1	Traditional storage system 2	Design storage system
50	8.6	4.5	0.4
100	8.7	4.8	0.4
150	8.7	4.8	0.6
200	8.7	4.8	0.6
250	8.7	4.8	0.8
300	8.7	4.9	0.9
350	8.8	4.9	1.1
400	8.9	5.1	1.1
450	8.9	5.2	1.2
500	8.6	5.9	0.4
550	8.6	5.1	0.7
600	8.7	5.7	0.8
650	8.8	5.5	1.0
700	8.5	5.2	0.7
750	8.2	4.2	0.8
800	8.8	5.7	0.8
850	8.9	5.8	1.0
900	8.6	5.6	1.0
950	8.6	5.8	0.4
1000	8.7	5.9	0.7

with the same accounting information number is fixed, the packet loss rate generated by the traditional storage system 1 is the largest, with a value of about 8.9%. The packet loss rate of traditional storage system 2 is about 5.2% under the same number of transmission, and the value of packet loss rate is small. However, the packet loss rate of the storage system designed in this paper is about 1.2% under the same number of transmission, and the value of packet loss rate is the smallest.

In the above experimental environment, simulate the data transmission phase of a static network without node join and exit behavior, and use three kinds of storage systems to test. If the number of information nodes participating in the regeneration process is d each time, for a single node storage process, it is necessary to start three storage processes, select d information nodes each time, and construct their transfer path to a new node Transmission path: the storage process of the next node will be started only after one node completes the storage. For a single storage information, the system storage process only needs to be started once. In this process, the transmission path from D information nodes and other two new nodes to this node is constructed for each new node. The actual available bandwidth of the three storage systems is simulated and counted. The available bandwidth values are shown in the figure below (Fig. 5):

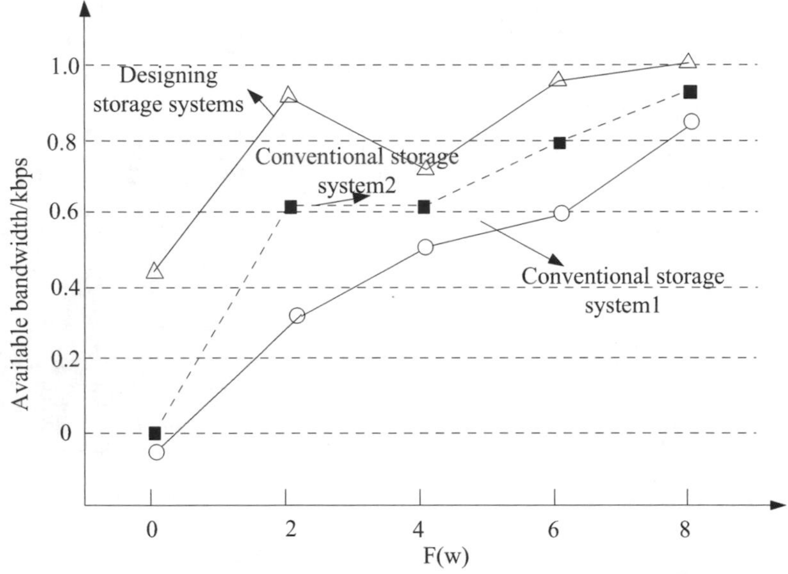

Fig. 5. Available bandwidth values of three storage systems

According to the experimental results shown in the figure above, under the control of the same information node, the available bandwidth value of traditional storage system 1 is about 0.8, and the actual available bandwidth value is the smallest. The available bandwidth value of traditional storage system 2 is about 0.9, and the actual available bandwidth of the system is larger. The available bandwidth value of the designed storage system is about 1.0, which is compared with the two traditional storage systems The designed storage system has the largest available bandwidth.

Keeping the above experimental environment unchanged, the performance runner software is used to test the three storage systems, and the test software is used to test the load, configuration, concurrency and other aspects of the performance of the three systems. The average response time of the evaluation software to the output of each module of the storage system is taken as the index of performance test, and the test results are shown in the following table (Table 4).

According to the experimental results in the table above, in the operation of each function of the system, the average reading and writing time of the storage system to the database is controlled within 1 s, and the response time of the hardware device is about 13 s. The average reading and writing time of the two traditional storage systems is more than 1 s, the response time of the traditional storage system hardware is about 17 s, and the response time of the traditional storage system hardware storage system is

Table 4. Response time test results

Storage function name User login	Average response time/ms		
	Traditional storage system 1	Traditional storage system 2	Design storage system
Video surveillance	1503.4	1268.2	731.5
Personnel positioning	864.3	659.7	452.8
Equipment information query	16083.9	14683.5	12834.5
Device delete operation	12523.4	1020.7	783.1
Issue	1010.2	869.7	616.3
Give an alarm	1120.5	986.4	817.8

about 15 s, which is larger than the storage system designed in this paper. Then, the first mock exam system is maintained, and three storage systems are tested. The number of users who store different online storage 50, 100 and 200 is simulated. The load capacity of the system is tested when the multi-user access the same module or data.

Table 5. Concurrent test results

Number of storage users	Load capacity /KBps		
	Traditional storage system 1	Traditional storage system 2	Design storage system
10	405.45	524.92	605.75
20	412.47	504.85	612.46
50	421.12	573.86	621.23
100	437.15	556.74	627.63
150	451.71	510.81	657.26
200	386.72	549.51	653.54

According to Table 5, when the number of storage users is 10, the load capacity of traditional storage system 1 can reach 405.45 kbps, the load capacity of traditional storage system 2 can reach 524.92 kbps, and the load capacity of designed storage system can reach 605.75 kbps; When the number of storage users is 150, the load capacity of traditional storage system 1 can reach 451.71 Kbps.The load capacity of traditional storage system 2 can reach 510.81 Kbps, and the load capacity of designed storage system can reach 657.26 Kbps; This method always has high load capacity.

5 Conclusion

With the rapid development of Internet and the acceleration of information digitization, the information age has come. The following is the rapid growth of data on the Internet. Traditional data storage methods are not ideal in terms of scalability, reliability and access delay, which can not adapt to the storage of massive data. Distributed storage system emerges as the times require. The designed storage system can improve the shortcomings of traditional storage system and provide theoretical support for the design and construction of storage system in the future.

References

1. Jing, W., Xuefei, Z., Shuxia, W., et al.: Construction of locally repairable codes based on fractional repetition cyclic codes in distributed storage systems. J. Electron. Meas. Instrum. **33**(04), 146–153 (2019)
2. Liang, P.: Research on the transformation from financial accounting to management accounting in the era of big data. Glob. Mark. **01**(03), 71–73 (2020)
3. Wen, H.: Financial information construction and fine management based on UFIDA NC financial system. Mod. Econ. Inf. **01**(21), 207–209 (2019)
4. Gou, Z., Zhang, X., Wu, D., et al.: Log analysis and workload characteristic extraction in distributed storage system. J. Comput. Appl. **40**(09), 2586–2593 (2020)
5. Guo, W., Xie, G., Zhang, F., et al.: Design and implementation of a mimic architecture for distributed storage system. Comput. Eng. **46**(06), 12–19 (2020)
6. Reyes-Ruiz, G., Hernández-Hernández, M.: Fuzzy clustering as a new grouping technique to define the business size of SMEs through their financial information. J. Intell. Fuzzy Syst. **40**(2), 1773–1782 (2021)
7. Popa, A., Safta, I.L., Miron, G.M., et al.: Manipulation of financial information through creative accounting: case study at companies listed on the romanian stock exchange. In: 18th RSEP International Economics, Finance & Business Conference – Virtual/Online 26–27 August 2020, Kadir Has University (2020)
8. ZHOU Yue, LI Gui-yang, JIANG Xiao-yu,et al. Piggyback Code Based on Distributed Storage Systems[J]. Journal of Chinese Computer Systems, 2020,41(05):1091–1097
9. Saghaeiannejad-Isfahani, S., Salimian-Rizi, N.: Assessment of success of financial information system in educational, health, and medical centers affiliated to Isfahan university of medical sciences. J. Educ. Health Promot. **9**(1), 128–135 (2020)
10. Frank, E.: An assessment of the conceptual linkages between the qualitative characteristics of useful financial information and ethical behavior within informal institutions. Ekonomski Horizonti **22**(2), 137–148 (2020)

Design of Abnormal State Monitoring System for Multi-channel Transmission of Social Network Information

Jun Li[1,2](✉), Hao-ding Murong[2], and Jun Xing[1]

[1] Shenzhen Academy of Inspection and Quarantine, Shenzhen 518045, China
zhengxiuli0831@aliyun.com

[2] Shenzhen Customs Information Center, Shenzhen 518045, China

Abstract. The security of social network communication is the most serious problem at present, because of the error of multiplex transmission, the risk of application security and the limited ability of most social network terminals. Based on this, the abnormal state monitoring system of social network information multiplex transmission process is optimized and designed. Through improving the system hardware and software functions, the accurate detection of multiplex transmission data is carried out. Finally, the experiment proves that the abnormal state monitoring system of social network information multiplex transmission process has higher monitoring accuracy and fully meets the research requirements.

Keywords: Social network · Information multiplexing · Abnormal state monitoring

1 Introduction

Social network is a kind of new technology energy, which is loved by people all over the world. With the improvement of information living standard, the challenge of information security is increasingly severe. Abnormal state monitoring technology of information multiplexing process has become the focus of attention. Communication security in social network environment is being studied by experts. The main problem is the security architecture of social network [1]. In the light of the information transmission process of social networks and the ability of the whole system to process information, the security framework is adjusted and constructed through three main ways: information collection, information processing and information transmission, the types of communication security threats under the social network environment are predicted, and an agreement on automatic reply by computers is signed [2]. In reference [3], aiming at the abnormal state monitoring of multi-sensors, the PCA model is established by using the principal component analysis adaptive reconstruction technology, and the multi-channel signals are detected by SPE statistics, thus realizing the fault detection and location. In reference [4], an abnormal state monitoring system is designed, which collects current and voltage information, and outputs it to the core chip operation module through the A/D

S. Liu and X. Ma (Eds.): ADHIP 2021, LNICST 416, pp. 403–415, 2022.
https://doi.org/10.1007/978-3-030-94551-0_32

conversion module for analysis, and constructs a system of multiplexing data collection and analysis.At present, there are few research methods on social network security monitoring, and the applied methods are all transmitted in one way, and the encryption intensity and authentication process are relatively small, so it has no security performance. Therefore, this paper puts forward the method of multiplex transmission, which has a great guarantee for communication security. By designing the overall hardware structure of the abnormal state monitoring system in the multiplex transmission process, the information acquisition module, data storage module and data transmission channel system are constructed. The hardware configuration and network port optimization were carried out to reduce the device drivers to achieve rapid response. Using the improved feature selection method to select abnormal traffic, the abnormal features of massive network traffic are extracted.

2 Multiplex Transmission Process Abnormal State Monitoring System

2.1 Hardware Structure of Multiplex Transmission Process Abnormal State Monitoring System

A monitoring and analyzing system for abnormal state of multiplex transmission process in social network is designed. This paper analyzes the main functions of the social network system and the possible abnormal conditions, designs the overall framework of the platform for monitoring the abnormal conditions of the multiplex transmission process, optimizes the hardware structure and interface of the system, and analyzes the overall technical scheme of the system, the system design specifications and the main functions of the system [5] The social network is a device module that integrates the collection of power consumption information, the monitoring of equipment operation status, intelligent control and communication, and is often installed at an important place where the collection of information and the monitoring data need to be collected from the distribution network, including places such as switches, ring network cabinets and

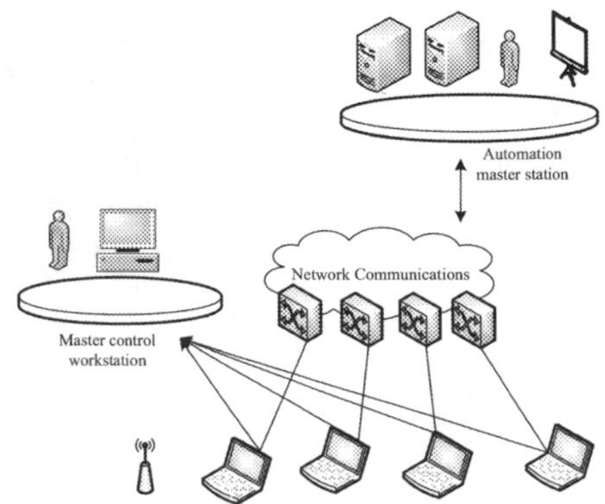

Fig. 1. Social network multiplex terminal architecture

feeder switches. According to the location of the distribution terminal, can be divided into terminal social network in the grid schematic diagram as shown in Fig. 1.

In order to meet the performance requirements of the network information security audit system, the device is developed based on CPU of X86 architecture, and the hardware platform NET-1711VD4N is adopted. The same hardware platform is used for the system device and the remote management device [6]. The NET-1711VD4N adopts the single board structure design of Intel 8456 V system chip, is a P4 gigabit network dedicated motherboard, and the motherboard integrates four Intel 825416 10M/100M/1000 Mbps gigabit network chip controllers, and directly leads out four RJ45 network interfaces and one serial interface on the board, which can be used to control the intranet, external network, ceasefire area, configuration network and control door, and there are also four network status indicator outputs on the trigger [7]. The Socket478 architecture supports 400/533 Delete Front-End Bus Pentium4/Celeron and Celeron D Series processors, storage aspects support DDR333/266, with VGA interfaces, PCI slots and CF card slots. The hardware parameters for this device are as follows (Table 1):

Table 1. List of hardware parameters

Processor	Intel pentium 4 2.8G
Memory	DDR 256M
CF card	256M
Hard disk	160G
Network interface	Four RJ45 network interfaces of Intel 82541
Serial port	1 COM interface
Supply voltage (V)	100–240 V AC, 6A, 50–60 Hz
Power supply (W)	520 W
Size	Standard 19 in. IU type, width × depth × height 430 mm × 355 mm × 44 mm
Weight	About 5 kg

Social network is the use of networking, information-based means to upgrade the traditional distribution terminal, its function is also more perfect. It is mainly capable of realizing the functions of monitoring the abnormal state of the multiplex transmission process, namely telemetry, telemetry, remote control and remote adjustment; realizing the short circuit fault detection function, realizing the fault recording and fault reporting, realizing the function of power quality monitoring, realizing the functions of intelligent charging, power management and remote communication, and realizing the distributed intelligent data exchange, etc. [8]. In the process of realizing the above functions, the key technologies include fault diagnosis and location technology, prejudgment and self-healing control technology, EMC adaptability under extreme conditions, compatibility and standardization of network communication protocol and feeder automation. In the hardware design, we can adopt two schemes: one is to use special embedded processor, using X86 structure of Intel processor. However, different options have different impact on the project, and need to consider the development costs, development cycle and other aspects.

The overall framework of the monitoring platform for abnormal state of multiplex transmission process of social networks mainly includes three parts: one is the data acquisition module composed of sensors, the other is the data storage module composed of an integrated system with data storage and management functions, and the third is the data transmission channel system [9]. The principle of the system is as follows: Firstly, the data in remote and intelligent terminal are collected by sensors, then the data are classified by data storage module, and the analog signals are converted to digital signals. Finally, the data are transmitted through multi-channels. The Control Center is the integrated system responsible for the processing and management of all data [10]. The integrated system can achieve centralized control of the whole platform, and has high robustness and controllability. The overall hardware configuration of the system is shown below (Fig. 2):

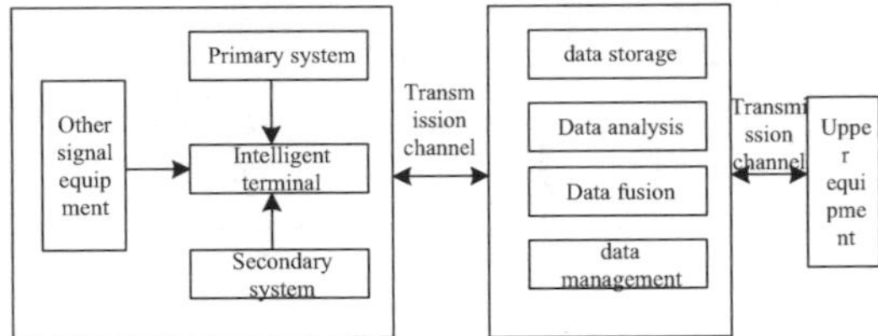

Fig. 2. System hardware configuration optimization

The hardware structure of automatic monitoring system for abnormal state of multiplex transmission process is composed of PC, switch, router and service. PC is the client of the system, also called the client. The main network monitors the data of the client through the access of the client. The monitoring system uses the C+ language, uses the ASP technology compilation dynamic monitoring code, monitors in the network each connection component the breakdown question.Switches exchange information between intranet and extranet, and exchange data between client and Internet database. To ensure the effectiveness of anomaly detection, the multiplex transmission network ports are optimized as follows (Fig. 3):

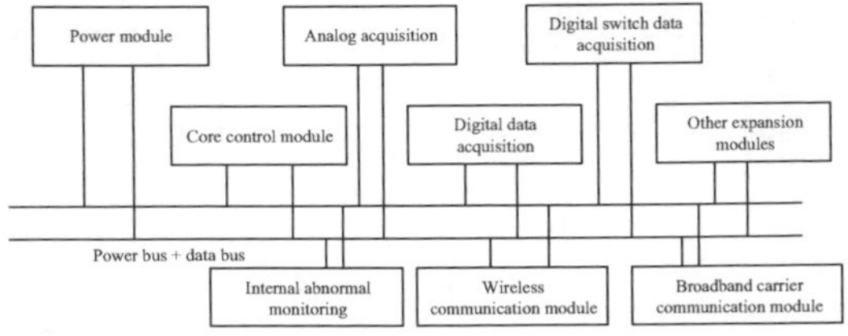

Fig. 3. Multiplex transport network port optimization

The system not only covers the normal functions, but also has the external interface to realize the docking with the external platform. The system needs login authentication, and can realize remote control. In the event of a state of emergency, an alarm signal will be issued and fault point information, images and data will be sent to the intelligent control terminal of the abnormal monitoring platform [11]. The platform can also provide historical data comparison and other functions to query and operate the corresponding control instructions. In addition, the system can also realize the analysis of the abnormal state inside the social network, that is, it is equipped with the monitoring module of internal abnormal.

System for network information monitoring applications, the key is to require the network function is perfect, compiling the kernel should be considered to meet this need. The key services provided by the standard kernel network subsystem are: data transmission, consistent device access interface, firewall and INET socket. These services are necessary for network applications, so they should be preserved when the kernel is clipped so that the system can provide complete network services. There are many network protocols and network hardware devices in the network subsystem. For the network security application, because it is used in the LAN environment to audit network information security, it needs the support of Ethernet communication protocol and Ethernet card device driver from the system kernel, and other network protocols and network hardware drivers can be simplified. Kernel configuration allows you to reduce kernel size by removing a large number of unused device drivers and modules from the kernel.

2.2 Optimization of Software Function of Multiplex Transmission Process Abnormal State Monitoring System

Aiming at the software function of abnormal state monitoring system in multiplex transmission process, the improved feature selection method is adopted. The aim of anomaly monitoring is to find the regular anomaly characteristics of network traffic based on the analysis of traffic data. Although the traditional analysis method is based on the general knowledge, it is only suitable for a small number of computing features. In the domain of network security, data mining will produce a large number of invalid data. Multiplex transmission process abnormal state monitoring system overall functions include packet capture, packet analysis and storage, query statistics and management 4 major modules [12]. The packet capture part includes initial filtering and capture. The data packet analysis section includes data packet analysis, keyword matching and storage; the query statistics section includes data query, data extraction analysis and data flow statistics; the management section includes the management of the keyword database and the management of account numbers of management personnel. The overall functional structure of the specific network information monitoring system is shown in the figure (Fig. 4).

Packet capture module can filter the original packet and catch the multiplex transmission model in the social network environment. The principle is that the information output path can be determined according to the stability of both sides of the communication network and the processing of the system terminal [13, 14]. Both the key of multiplexing and the process of multiplexing should follow this principle to select the

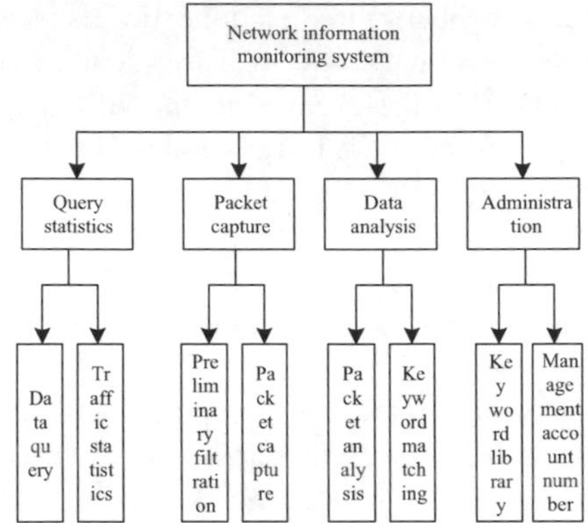

Fig. 4. System software module functional structure

path. The data packet capture module can filter and capture the original data packet. The module can get all the network adapters on the current monitoring system, set the preliminary filtering rules, and can capture the packets being transmitted in the current network. Data analysis module is the core module of the whole network information monitoring system [15]. Under this module, the administrator can analyze the content of the captured data, and use the keyword library to match the analyzed content, find out the information containing the keyword content, and store the captured and analyzed information for future reference. The management module can add, delete and modify the keywords in the keyword library, and encode the input keywords. Administrator account is divided into super administrator and ordinary administrator 2 permissions, super administrator can add, modify and delete the administrator account operation, account permissions can also be changed. Multiplex Traffic Monitoring Model in Social Network Environment is shown in Figure (Fig. 5).

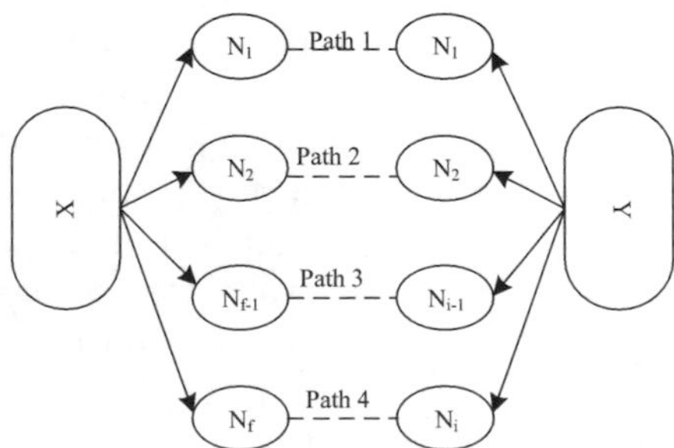

Fig. 5. Multiplex monitoring model in social network environment

As shown in the figure, x and y are the two terminals of communication. There are many paths of communication between these two terminals. N1,N2……Ni + 1 is the intermediate forwarding station on the path [16]. If x, y communicate by using one of the paths, the attacker can get all the information of x, y communication by attacking any intermediate forwarding station on the path. If x, y communicate by using more than one path, the attacker must attack at least one site on each path to get part of the information of x, y communication. The system can be divided into three layers: application layer, data processing layer and data acquisition layer. The system hierarchy diagram is shown in the figure (Fig. 6).

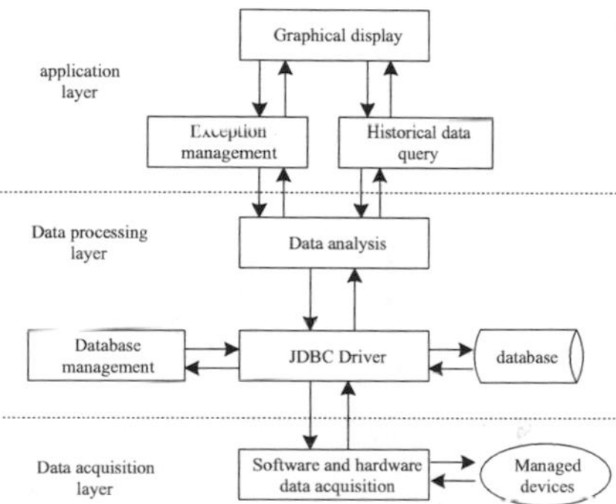

Fig. 6. Network state monitoring hierarchy

In the diagram, the data acquisition layer is responsible for the acquisition of the running state data of the equipment software and hardware (specific service running, CPU storage state, etc.). The client program deployed on the managed device acquires the running status of the host monitoring system and the status of the specific software through NETSNMP open source software and relevant Linux commands, and sends the collected device information to the monitoring server and stores the information in the database. In the data processing layer, the anomaly management module is responsible for analyzing the running state of the host computer monitoring system and the basic information of the equipment in the database. If any abnormality occurs, the abnormality information will be fed back to the application layer and alarm [17]. The application layer realizes the function of interacting with the user. This function will display the running state of the monitoring system and the running state of the managed devices to the user. If an exception occurs, the user is fed back the exception information. In addition, the user should set the operating state threshold of the managed device through the system management function of the application layer. Anomaly monitoring information collection is the necessary preparation for the distributed network, including information collection and processing. By analyzing the abnormal information, we can judge whether the client has accessed correctly and monitor the performance of the distributed network, so that the access information can be accurately monitored and the time consumed in network connection can be counted. During this period of time, the scope of network adaptation shall be predicted to obtain the information related to network security; the access mode of distributed network clients shall be selected to determine whether the information is abnormal. If the information is normal, it can be directly connected to the network client. Conversely, if the information is abnormal, it needs to be collected to provide data support for automatic monitoring database of subsequent access information.

2.3 Implementation of Abnormal State Monitoring in Multiplex Transmission Process

The general multiplex transmission process abnormal state monitoring system adopts the manager-agent model, and the manager-agent communicates with each other by standard protocol. The manager sends the order to the agent, the agent receives the order and collects the host status information from the controlled equipment and network equipment, and then feedback to the manager. The model diagram of network monitoring management is as follows (Fig. 7).

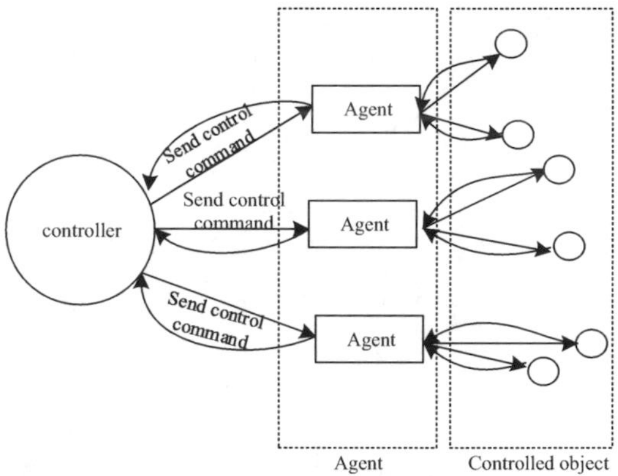

Fig. 7. Network monitoring management model

According to the definition of ISO, the management of abnormal state monitoring system for multiplex transmission process must include five major functions: anomaly management, configuration management, cost management, performance management, and security management. The collected information shall be pre-processed, and the relationship between information output and input components shall be analyzed according to the processing results. The specific calculation formula is as follows:

$$f(x_n, y_m) = \frac{\lambda(x_n, y_m)}{\sqrt{(x_n)(y_m)}} = \frac{\frac{1}{k}\sum_{m=1}^{k} x_{nm}y_m - \frac{1}{k^2}\left(\sum_{m=1}^{k} x_{nm}\right)\left(\sum_{m=1}^{k} y_m\right)}{\frac{1}{k}\sqrt{\sum_{m=1}^{k} (x_{nm} - x_n) \cdot \sum_{m=1}^{k} (y_m - \overline{y_m})^2}} \tag{1}$$

Where, $\lambda(x_n, y_m)$ represents the covariance between the input information x_n and the output information y_m; k represents the total amount of information. According to the formula, the input informationxn x_n and output information y_m can be normalized:

$$x' = \frac{x_n - x_{\min}}{x_{\max} - x_{\min}} \tag{2}$$

$$y' = \frac{y_n - y_{\min}}{y_{\max} - y_{\min}} \tag{3}$$

The formula is weighted to get the input data vector expression, as shown in the formula:

$$T = \left(f_1(x'_1, y')x', f_2(x'_2, y')x'_{i-1}, \cdots f_1\left(x'_j, y^+\right)x'\right)^{\mathrm{T}} \tag{4}$$

In the formula: T represents a period, from the formula can be obtained after weighting the input information vector. The monitoring system is divided into four modules: monitoring system state collection module, anomaly management module, database management module, data query module, the system module structure design as figure (Fig. 8).

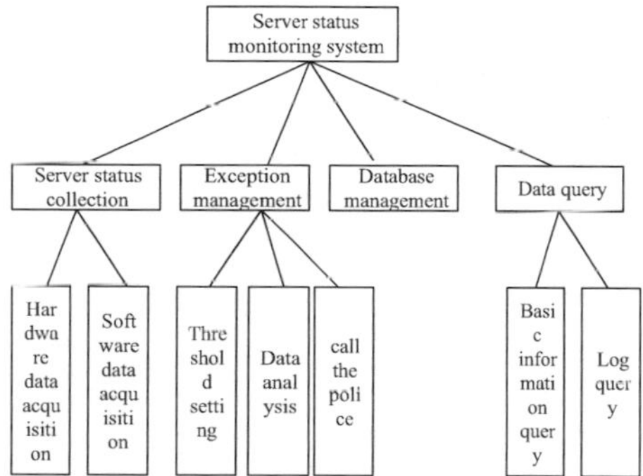

Fig. 8. System module structure optimization

The state acquisition module of abnormal state monitoring system in multiplex transmission process includes hardware data acquisition and software data acquisition. Anomaly management includes threshold setting, data analysis and alarm sub-functions. Data query includes basic information query and log query sub function. There are two kinds of data collected by monitoring system state acquisition module: fixed data and original data. Fixed data refers to data that does not need to be processed to determine whether it is abnormal, such as hardware data. Raw data refers to the data collected can not directly determine whether abnormal, need to be processed before judging, such as the number of secure access terminals online and site traffic data. Fixed data sent directly to the data analysis sub-function processing, if abnormal, then the alarm processing. The original data is stored in the temporary data table, after analysis and processing, the result data is obtained and sent to the data analysis sub-function. If there is an abnormality, alarm. Fixed data and original data are processed and the results are stored in the corresponding table in the database. The process of monitoring the abnormal state of the multiplexing process is shown in the figure (Fig. 9).

According to the schematic diagram of abnormal state monitoring system, the flow collection module, flow statistics module, anomaly monitoring module, alarm module and display module are designed. The flow collection module is mainly responsible for collecting all the flow data flowing through the collection point, and the flow statistics module is mainly responsible for splitting all captured network data packets and disassembling and storing them in the corresponding linked list structure. The abnormal

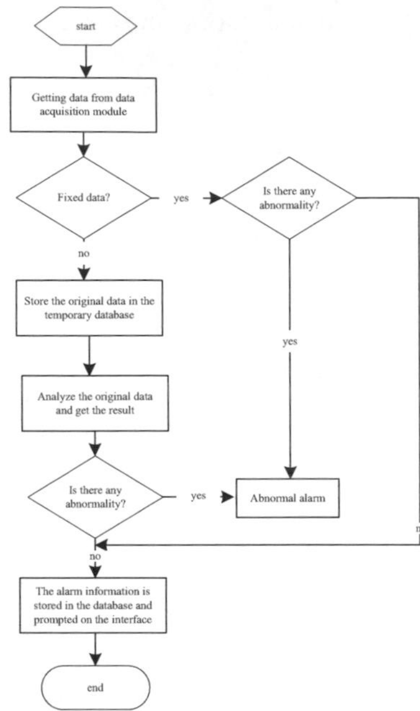

Fig. 9. Abnormal state monitoring process optimization

monitoring module is mainly responsible for monitoring the statistical results of abnormal data, the alarm module is mainly responsible for warning abnormal data, and the display module is mainly responsible for marking abnormal network information and displaying the monitoring results. By analyzing all the network traffic characteristics and selecting abnormal traffic by improved feature selection method, the abnormal traffic monitoring system of mobile communication network is designed.

3 Analysis of Experimental Results

In order to verify the operation effect of the abnormal state monitoring system in multiplex transmission process, the system equipment is connected to the internal LAN of the unit. The parameters of the tested equipment are shown in the table (Table 2).

The scale of the network on which the equipment under test is located is 20 hosts and the network segment is 192.168.18.0. The device is connected to the network via bypass mode, and the external Interact interface is connected to the hub, and the switch of the network shares the network data with the device under test through the hub. In this way, the device can monitor the Internet access information of all hosts in the network, and the Internet access information will be stored in the database of the system device. A host in the network, as the host of the management terminal, runs the management platform software under the Windows environment and connects to the database of the remote management terminal. In order to test the performance of monitoring technology, the traditional method is compared with this method, and 100 experiments are carried out to verify the information monitoring function of system equipment. Network packet

Table 2. Parameters of equipment under test

Processor	Intel Pentium 4 2.8G
Memory	DDR 256M
CF card	256M
Hard disk	160G
Network Interface	Gigabit network interface based on Intel 82541
Operating System	Embedded Linux based on 2.4.30 kernel

capture is the basis of the latter functions, and the results of protocol analysis and alarm are stored by database storage module to complete the data storage. The performance curve can be obtained as shown in the figure (Fig. 10).

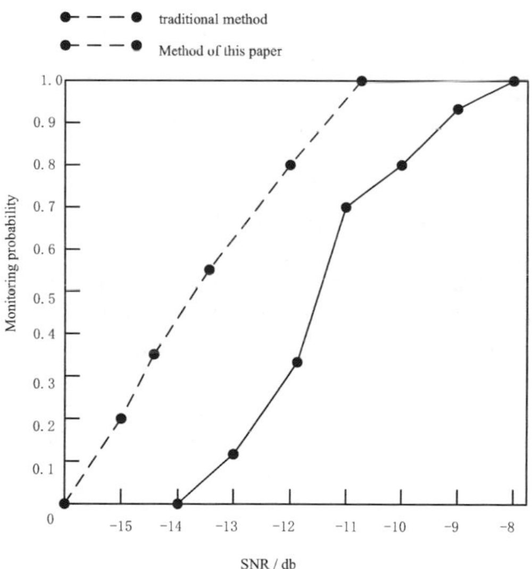

Fig. 10. Comparison of experimental results

It can be seen from the figure that, with the continuous improvement of the signal-to-noise ratio of the transmitted data, the monitoring probability of the traditional method is low and the response speed is slow, while the method in this paper can quickly reflect the abnormal data, and the monitoring accuracy of the abnormal data is high and the overall performance is good. Therefore, under the network delay, the monitoring system using the improved feature selection method is better than the traditional monitoring system to monitor the abnormal traffic of mobile communication network.

4 Conclusion and Outlook

At present, the level of social network technology is still in its infancy, the communication security design is not perfect, which brings great challenges to the communication security of social network. The security of communication system is the guarantee of every user's information, which has a direct impact on the effective operation of social network. Therefore, the design of communication security system should be strengthened. The design of secure communication system based on multiplex social network can make the system of computer terminal choose relatively simple decryption program and complex security performance, which greatly reduces the cost of network security prevention and control, and improves the processing capacity of terminal relatively. However, this paper has not been applied in the actual scene, and needs to analyze the needs of the actual environment in the future research, and further improve it to improve the perfection.

Fund Projects. National Key R&D Program of China (2018YFC0809105, 2018YFC0809100).

References

1. Olfat, M., Shokouhyar, S., Ahmadi, S., et al.: Organizational commitment and work-related implementation of enterprise social networks (ESNs): the mediating roles of employees organizational concern and prosocial values. Online Inf. Rev. **44**(6), 1223–1243 (2020)
2. Xu, S., Cao, J., Legg, P., et al.: Venue2Vec: an efficient embedding model for fine-grained user location prediction in geo-social networks. IEEE Syst. J. **14**(2), 1740–1751 (2020)
3. Chen, Y., Xu, P., Zhao, D., et al.: a fault detection and isolation algorithm for multi-sensor system. Comput. Sci. Appl. **9**(1), 8 (2019)
4. Jiang, S., Cheng, C., Wang, Q.: Design of voltage abnormal state detection system for power metering device. Comput. Meas. Control **28**(02), 44–47 (2020). No. 257
5. Shen, M., Zhang, J., Zhu, L., et al.: Secure SVM training over vertically-partitioned datasets using consortium blockchain for vehicular social networks. IEEE Trans. Veh. Technol. **69**(6), 5773–5783 (2019)
6. Salazar, J.J.R., Segovia-Vargas, M.J., Camacho-Miano, M.M.: Money laundering and terrorism financing detection using neural networks and an abnormality indicator. Exp. Syst. Appl. **169**(10), 114470 (2020)
7. Jang, H., Hou, J.U.: Exposing digital image forgeries by detecting contextual abnormality using convolutional neural networks. Sensors **20**(8), 2262 (2020)
8. Yan, F., Huang, X., Yao, Y., et al.: Combining LSTM and DenseNet for automatic annotation and classification of chest x-ray images. IEEE Access **7**, 74181–74189 (2019)
9. Lu, S., Wei, X., Rao, B., et al.: LADRA: log-based abnormal task detection and root-cause analysis in big data processing with spark. Future Gener. Comput. Syst. **95**, 392–403 (2019)
10. Vijayan, A., Tahoori, M.B., Chakrabarty, K.: Runtime identification of hardware Trojans by feature analysis on gate-level unstructured data and anomaly detection. ACM Trans. Des. Autom. Electron. Syst. (TODAES) **25**(4), 1–23 (2020)
11. Thiyagarajan, K., Kodagoda, S., Ranasinghe, R., et al.: Robust sensor suite combined with predictive analytics enabled anomaly detection model for smart monitoring of concrete sewer pipe surface moisture conditions. IEEE Sens. J. **20**(15), 8232–8243 (2020)

12. Chen, T., Liu, X., Xia, B., et al.: Corrections to unsupervised anomaly detection of industrial robots using sliding-window convolutional variational autoencoder. IEEE Access **8**(10), 117062 (2020)
13. Liu, S., Liu, D., Srivastava, G., Połap, D., Woźniak, M.: Overview and methods of correlation filter algorithms in object tracking. Complex Intell. Syst. **7**(4), 1895–1917 (2020). https://doi.org/10.1007/s40747-020-00161-4
14. Fu, W., Liu, S., Srivastava, G.: Optimization of big data scheduling in social networks. Entropy **21**(9), 902 (2019)
15. Liu, S., Bai, W., Zeng, N., et al.: A fast fractal based compression for MRI images. IEEE Access **7**, 62412–62420 (2019)

Construction of Digital Communication Platform of Yellow River Culture Based on Multimedia Communication

Yu-xuan Chen[1](✉), Yan-yan Chen[1], and Wu-lin Liu[2]

[1] Fuyang Normal University, Fuyang 236300, China
[2] Information and Communcition College National University of Defense Technology, Xi'an 710106, China

Abstract. The traditional digital communication platform construction method has a small number of functional nodes and high overall message complexity. In response to this problem, a construction method for the digital communication platform of the Yellow River culture based on multimedia communication is proposed. Transform data work into the development and application of the Yellow River culture, clarify the construction goals of the communication platform, use SBC to describe the platform architecture language, build the platform architecture hierarchy, use multimedia communication technology, allocate multi-point conference control servers, and realize cross-component data communication. And optimize the data storage algorithm. The experimental results show that the digital communication platform effectively reduces the network load of the Yellow River cultural information data transmission, reduces the complexity of message storage, and improves the query rate of information data.

Keywords: Multimedia communication · Digitization · Communication platform · Yellow River culture

1 Introduction

The Yellow River culture has a profound cultural heritage, but with the advent of the Internet era and the era of digital protection of intangible cultural heritage, the traditional mode of development and inheritance of the Yellow River culture has been unable to meet the cultural absorption needs of the public [1]. Therefore, in order to expand the power and efficiency of cultural communication, it is urgent to carry out the digital communication of culture.

In order to improve the status quo of the inheritance and protection of the Yellow River culture, it is imperative to use the digital communication platform to inherit the Yellow River culture. Through the collection and filing of data, it provides more and more possibilities for the protection, development and inheritance of culture. The popularity of the Internet has laid the foundation for the construction of the platform. The platform can realize the sharing, learning, communication and co construction of resources free of

S. Liu and X. Ma (Eds.): ADHIP 2021, LNICST 416, pp. 416–428, 2022.
https://doi.org/10.1007/978-3-030-94551-0_33

charge, break through the limitation of learning time and space, and accurately manage the storage, editing, visualization and other shared resources. It has a strong attraction for young groups and audiences who are interested in intangible cultural heritage. It's a digital broadcasting platform. The participants who actively participate in broadcasting also play an important role in the process of spreading and protecting culture.

Reference [2] applies digital management to cultural communication. This method considers that communication and sharing are the key to the completion of cultural inheritance. Therefore, starting from the form of digital information release, it connects the cultural information in the Internet to complete the construction of cultural information sharing platform. However, the communication efficiency of the platform still needs to be further improved. Reference [3] applies the cultural information transmission code to the design of cultural communication platform, and constructs the risk model of cultural communication from the perspective of communication risk theory. Based on the calculation results of risk model, the information decoding method is used to share and spread culture. But the transmission accuracy of the platform can not meet the demand.

There are three ways of digital protection of cultural works in China, namely, using digital technology to reconstruct the cultural environment, using cultural resources to build a remote sharing platform, and filing and preserving cultural resources in digital media, extending the preservation time of cultural resources through digitization and giving them practical value, and using automated 3D digital virtual machine to research cultural history. At the same time, it analyzes the significance of digital strategy on cultural protection and communication from the perspective of policy. On the basis of the above theory, this paper puts forward the construction of digital communication platform of Yellow River culture based on multimedia communication.

2 Construction of Digital Communication Platform of Yellow River Culture Based on Multimedia Communication

2.1 Platform Construction Goals

Establish a database of the Yellow River culture that connects the computer and the web, and use the database as the core of the entire digital communication work to transform the data-based work into the development and application of the Yellow River culture, bringing unlimited potential to it. The current development space of the traditional Yellow River cultural industry model is limited, and the main shortcomings are mainly reflected in: the diversified and innovative development mechanism of the Yellow River culture has not yet been formed; the lack of public service platforms that can take into account the protection of the original ecological culture, and the transformation of internationalization and industrialization; Independent innovation and technology development capabilities, and market expansion capabilities are limited, failing to form an influential brand and production base.

In view of the above shortcomings, the platform construction objectives and database platform construction objectives are clarified, and the important information support of digital informatization is provided for related industries based on the needs of the Yellow River cultural industry. Improve and enrich the diversification of the Yellow River culture industry chain, support the creation and development of the Yellow River culture brand, and build the Yellow River culture service platform for the whole industry [4]. The platform construction is conducive to the popularization, cultural creation, fashion and internationalization of the Yellow River culture. Using the Internet environment and digital technology, we have established an open, shared and interactive experience platform for serving the people and adhering to the people-oriented principle. The inheritance and protection of the Yellow River culture, rooted in all aspects of people's life, realizes the digital protection and international dissemination of the Yellow River culture. To a greater extent, promote the cultural transformation and industrial revitalization of the Yellow River, and realize the national cultural confidence. So far, the Yellow River Cultural digital communication platform construction goal is clear.

2.2 Constructing the Digital Communication Platform Architecture of the Yellow River Culture

Describe the Architecture Language of the Digital Communication Platform
Use the SBC architecture to describe the architecture language of the digital communication platform. The SBC architecture is an architectural theory that integrates "structural behavior". It has an integrated perspective. This theoretical system can completely describe and express multiple views, and master the backbone of the system. It can use the causal relationship between "structure" and "behavior". Logic formulas, graphical illustrations and other methods are clearly listed, and these relationships are all produced by the interconnection and interaction of "components". SBC architecture has guiding advantages. It is very practical to construct elements and define behavior functions of communication platform through architecture hierarchy diagram, framework diagram, component operation diagram, component connection diagram, structure behavior integration diagram and interaction flow diagram in architecture description language. With the help of architecture hierarchy diagram, structure behavior integration diagram and interaction flow chart, the media composition, function positioning and interactive behavior relationship of the Yellow River culture digital communication platform are constructed respectively. Compared with the non architecture oriented linearity and locality, SBC architecture can comprehensively sort out the "structure" and its corresponding "behavior" of the Yellow River culture digital communication platform, "Structure" can be understood as the internal form formed by all kinds of media, while "behavior" is the corresponding function, role and interactive relationship. Finally, a specific, targeted and interpretable analytic diagram is obtained. Based on this, in the practice of digital communication of the Yellow River culture, corresponding media strategies are formulated, and complementary advantages and multi-directional cooperation are paid attention to, so as to construct a pluralistic and multi-directional communication platform The communication platform system of the body [5]. So far, the description of digital communication platform architecture language is completed.

Constructing the Architectural Hierarchy of Digital Communication Platform
The digital communication of Yellow River culture should not only pay attention to the construction of "visual scene", but also pay attention to the construction of "public opinion scene", and build a platform architecture. The "visual scene" is mainly to build

a situation, so that people can experience the Yellow River culture personally, while the "public opinion scene" created by websites and social media can form multiple public opinion fields of the Yellow River culture in the network, so as to form a continuous and extensive information interaction and transmission [6]. The details are shown in Fig. 1.

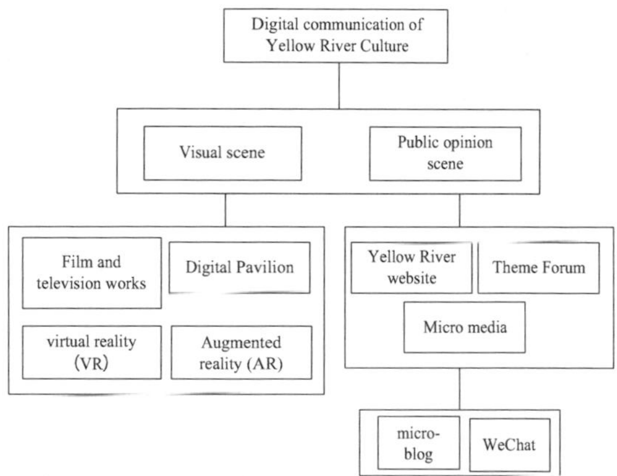

Fig. 1. Hierarchy of the communication platform architecture

As shown in the figure above, through the collaborative construction of multiple scenarios and multiple channels, the Yellow River culture has formed a diversified stimulus to the audience, prompting the people to continuously enrich and deepen the cognitive level, and shape a scientific and good cultural communication ecology. The digital communication platform of the Yellow River Culture must integrate its "structure" and "behavior". This is also an important feature that distinguishes the architecture orientation from other methods. After combing the components of the digital communication platform of the Yellow River culture, the "Behavior" refers to the clear positioning and description of functions, thereby enhancing the pertinence and effectiveness of cultural communication, allowing various media to play their strengths in their respective fields, diversifying and co-constructing, forming a three-dimensional red cultural digital communication network ecology. So far, the construction of the digital communication platform architecture has been completed, and the construction of the digital communication platform of the Yellow River Culture has been completed.

2.3 Establishing Platform Control Terminal Based on Multimedia Communication

SIP function is added to the platform architecture. By using multimedia communication technology, multipoint conference control server is assigned as the control end of the communication platform to complete the specific media processing. After users register for the digital communication platform, the platform carries out user management, including identity authentication and management. When users want to understand the

Yellow River culture, their SIP requests will be forwarded to the multi-point conference control server by the platform. The multi-point conference control server is used to complete the specific media negotiation. After the media negotiation is completed, the multi-point conference control server passes through xmlr PC interface, which interacts with multimedia communication server, including creating conference room, joining members and other operations. Finally, the media is negotiated through SIP protocol, and RTP transmission channel is established. Subsequent media processing is completed by media server [7]. The overall architecture of the media communication terminal is shown in Fig. 2

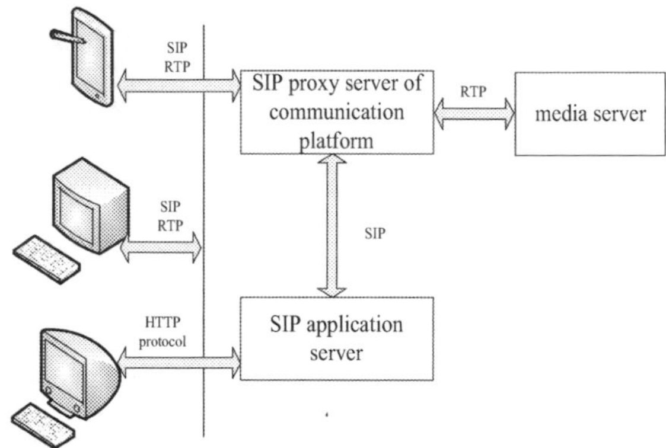

Fig. 2. Multimedia communication control terminal

As shown in the figure above, the multipoint conference control server is an important part of the multimedia communication platform. When more than two servers access the platform, the MCU must tandem, distribute and switch each communication signal, and complete the platform control. The MCU functions as a switch. The audio, video, data, signaling and other data of the platform terminal equipment must be processed and connected by the MCU and connected with other MCUs. The MCU separates the information flow of each terminal on the platform., Extract various information such as audio, video, data, signaling, and send different information to different modules for processing, complete audio mixing and switching, video information mixing or switching, data information broadcasting and routing, timing and signaling control, etc. Finally, the processed audio, video, data, and signaling are recombined and sent to each corresponding terminal of the platform [8]. The MCU of the Yellow River Cultural Digital Communication Platform is mainly composed of two parts. The first is terminal control, and the other is media processing. These two parts broaden the media processing capabilities of the digital communication platform, separate media processing from terminal control, and greatly improve The media processing capability of the dissemination platform also reduces the burden of terminal control, and the multipoint conference controller also has functions such as terminal control and management of media servers [9]. So far, the establishment of platform control terminal based on multimedia communication technology is completed.

2.4 Optimize the Communication Process of the Digital Communication Platform

SIP communication protocol is used as the transmission control protocol of multimedia communication to optimize the communication process of communication platform. The signaling and media transmission control protocol in the platform are the basis of platform communication development, as shown in Fig. 3

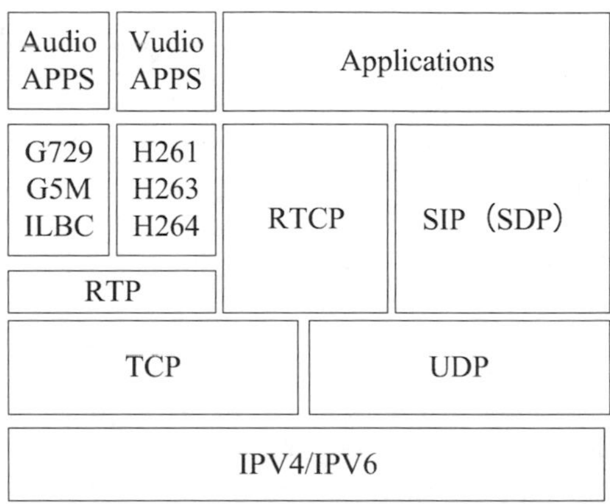

Fig. 3. Platform communication protocol and hierarchy

As shown in the figure above, the application of voice and video and SIP protocol is at the top level. SIP protocol can run independently. At the same time, for the convenience of SIP protocol, SDP protocol is used to modify the session attributes of SIP. SDP protocol is used to negotiate the media features in SIP signaling through offer answer mode. SIP signaling completes the establishment of session through UA. When the session is established, voice and video will play an important role The frequency media is encapsulated by RTP protocol, and the appropriate transmission mode is selected to interact on the network. RTCP is used to control RTP, UDP and TCP in transport layer to complete data transmission. At the same time, all the above protocols are located on the network layer. User agent is the logical network endpoint of SIP protocol, which is used to create, send and receive SIP messages, and manage SIP sessions. SIP user agent is divided into user agent client and user agent server. UAC is used to create and send SIP request messages. UAS is used to receive and process messages sent by UAC. The registration server also has a certain positioning ability. It forms a whole with proxy server and positioning server. The proxy server is located in sip on the network Between UAC and UAS, it is used for proxy routing messages between UAC and UAS. The proxy server is also used for routing control check, and the location server is used to store the location information of feedback users, which can be obtained from the registration server and related databases [10]. The LDAP protocol is used to locate the interaction protocol between the server and SIP server. The server accepts SIP requests, but does not send any SIP requests. It maps the original address in the request to an address list and returns it to the platform client. It can also return an empty table to indicate the rejection of the request. The whole message process is that when receiving the cancel request, the server replies to 200 OK, indicating successful response and response When the SIP transaction ends and a message other than cancel is received, all available routing addresses are replied. So far, the optimization of communication process of digital communication platform is completed.

2.5 Constructing Hierarchical Mechanism of SIP Communication Protocol

SIP is a layered protocol. Therefore, it is necessary to build a layered mechanism of SIP communication protocol so that when a function is described by the communication platform, it can easily span multiple components. The SIP protocol of the communication platform is divided into four layers, from top to bottom, they are transaction user layer, transaction layer, transport layer, syntax and coding layer. The SIP entity belongs to the transaction user layer, and the role of this layer is when the transaction user sends a request, It will establish a client transaction instance, transfer the request transmission mechanism, and its IP address and port number. The transaction layer is a basic element in the SIP protocol. From the client transaction to the server transaction request, the transaction layer processes the retransmission request of the application layer, matches the response request, and handles the timeout request of the application service layer. Any UAC request The completed tasks are composed of multiple transactions. The transport layer defines how to communicate requests and send responses between the client and the server on the network. The coding layer is the bottom layer of the SIP protocol to complete SIP protocol requests and response messages. The grammatical structure and analysis of SIP protocol, the message header field of the SIP protocol contains a lot of information in SIP signaling, and there are a total of 44 message header fields, each representing different information. The standard request types of SIP signaling are divided into six categories, which represent different functional descriptions, making it easy for users to distinguish the types of request messages. From the first line of the request message, you can see which type of request message belongs to. The message consists of message type, message network address and SIP version number. The first type of request is INVITE, which is used to invite users or services to join the session. The INVITE message contains a three-way handshake process. First, the client transaction initiates an INVITE request, and then serves. The client transaction returns a response, and finally the client transaction sends an ACK. The hop-by-hop mechanism is used to ensure that each signaling is successfully transmitted to the next hop. The INVITE message body contains SDP information to describe the media information in the SIP signaling.; The second type of request type is REGISTER, which is used by a licensed client to send registration information to the server, including requests for adding, querying, and deleting, so that the client's address can be recognized by the server; the third type of request type is ACK, using In the three-way handshake on the INVITE message, it confirms the INVITE message that has been received. If it is not received, it will cause the retransmission of the INVITE message; the fourth type of request is CANCEL, which is used to cancel a request, but CANCEL cannot be cancelled. End request message; the fifth type of request is BYE. BYE is used to send a request to the server to indicate that the communication platform wants to end the session. The BYE can be initiated by the calling party or the called party, and wait until the BYE request is received. After that, it will stop sending media streams to the other party; the sixth type of request type is OPTIONS, which contains the information used to request the performance of the SIP server, the methods supported by the SIP server, and the SDP protocol information. Finally, in response to the message sent by the communication platform, when the UAS receives the request information sent by the UAC, it will reply a response status code to the UAC, which is used to express the response to the request message. If you want to distinguish different reply status information, you can use three digit decimal numbers represent different status codes. From 100 to 699, the first digit of the status code indicates the type of the response message. The same first digit indicates that the type of the platform response message is the same. Combine the last two digits. Different response messages can be distinguished. So far, the construction of the layered mechanism of SIP communication protocol is completed.

2.6 Data Storage Platform Optimization Algorithm

Optimize the data storage algorithm of the platform, and transmit the stored data through the above communication protocol. Firstly, the routing protocols of the platform are divided into the following three categories: clustering based protocol, geographic location based protocol, and query based protocol. Clustering protocol divides the nodes in the network into multi-point clusters according to geographic location and signal strength, and selects a cluster head node to undertake the task of communication with other clusters, while other nodes in the cluster In order to save energy and wait for cluster head rotation, the non cluster head node has no task and can enter the sleep state. Location based routing protocol requires nodes to know their location information in the sensor network of the platform. The location information is the real geographic location, and it is also the virtual coordinates of the structure. The primary task of the protocol is to locate the node, build the location coordinate system, and then send the data in the network to the destination node. According to the set routing rules, select the next location of the data One hop node, a data centric storage method, uses geographic location information to map event data, and event attributes contain the location information of the collection node. Users can send query packets to network nodes, randomly find targets in network transmission, and find and determine the storage node of information data according to data storage rules. The probability of node k being selected as storage node is as follows:

$$P = \frac{e^{-\lambda|A|}(\lambda|A|)^k}{k!} \tag{1}$$

In the formula, P is the selection probability of nodes in the region, λ is the node density, and A is the area of the region. When routing protocol transfers information data for the purpose of data storage, it transfers the data from the source node to the storage node specified by the storage algorithm. Meanwhile, Lushan protocol also undertakes the task of discovering the storage node. According to the geographic mapping table, it maps the data to the geographic location:

$$x_{1i} \rightarrow \frac{k_1}{k_2} x_{1k}, \ldots, x_{mi} \rightarrow \frac{k_1}{k_2} x_{mk} \tag{2}$$

In the formula, x_{1k} and x_{mk} respectively represent the set of two mapping data tables. At the same time, considering the storage load balance and query path energy efficiency of the platform, the data packet is forwarded to the nearest target location in greedy forwarding mode:

$$P(V_i) = \lambda^{-1}\varsigma \frac{\sum_{i=1}^{m} x_i^2}{m\sum_{i=1}^{n} V_i^2} \frac{n!}{m!(n-1)!} f(V_i, V_k)^{-1} \tag{3}$$

In the formula, λ is the forwarding factor and ς is the relative error of data forwarding.

The expected number of hops between neighbor node and target node is calculated:

$$G = \frac{R}{d} \tag{3}$$

In the formula, G is the expected hop number of nodes, R is the transmission range of nodes, and d is the expected distance between any two points in the plane of network abstraction. Finally, the data files generated by the platform are stored in the HDFS file of Hadoop bottom layer, including hfile and Hlog file. Hfile is the data storage format of HBase, which is a binary format file. It is a lightweight packaging for the information data of the Yellow River culture, and Hlog file is used File is the wal storage format in HBase, and its data update is divided into the following processes: the platform client first initiates the operation of data update, encapsulates each operation, and then sends the data packet of information data through RPC remote procedure call. When the data arrives at the terminal, it will be sent to the specified key, and the data will be written into the wal and stored in the corresponding memstore to reach the specified destination After a certain period of time, the written data will be asynchronously persisted to the file, so the optimization of the platform data storage algorithm is completed. So far, we have completed the design of the construction method of the Yellow River culture digital communication platform based on multimedia communication technology.

3 Experiment and Analysis

This design method is recorded as experimental group A, and the two traditional digital communication platform construction methods are marked as experimental group B and experimental group C respectively. Comparative experiments are carried out to compare the number of functional nodes and storage methods of the communication platform sensor network Overall message complexity.

3.1 Experimental Preparation

According to the construction ideas of the above digital communication platform, the functional structure of the experimental group a communication platform mainly includes three aspects: Yellow River culture school, mass creation space and material library, highlighting the characteristic functions of personnel training and product innovation of the Yellow River culture. The platform also includes the function of "spreading the Yellow River culture", which helps the public to understand the latest information and information of the Yellow River culture in time Development trends. The digital communication platform of the Yellow River culture mainly includes the following contents: the Yellow River culture school carries out the communication education of the Yellow River culture simultaneously through the experience of online and offline schools, and the classroom includes master lecture, culture class, historical biography and other courses, so as to spread the Yellow River Culture, popularize the Yellow River culture knowledge of the platform readers, and implicitly cultivate the Yellow River culture talents The Yellow River culture school column undertakes the main content of the Yellow River culture autonomous learning section in the digital communication platform, which belongs to the functional characteristics of the platform. The public creative space includes such functional units as interactive communication, inspiration seeking, creative design and employment support. It combines innovation and entrepreneurship to provide a good innovation development space and entrepreneurial development platform for the majority of Yellow River culture lovers. By building an independent digital communication platform, it conforms to the development trend of "mass Entrepreneurship and innovation", and creates a good Yellow River culture environment. The column construction also belongs to the functional characteristics of platform construction. The material library mainly provides the electronic text library, online knowledge test database, video teaching and its materials, appreciation and download of pictures, etc. it is the main sharing column area of the platform, disseminating the latest information of the Yellow River culture, recording the story of the Yellow River culture, and developing innovative works of the Yellow River culture, including furniture, home decoration and clothing. At the same time, it also publicizes and introduces the excellent Yellow River cultural innovation team and R & D products.

The multimedia communication process of experiment group A is as follows: create multimedia communication data packets, obtain media streams through the channels, remove jitter, decapsulate data packets, and decode the media streams to complete the encoding of the media streams and encapsulate them into data packets. And finally send the data out through the channel, and the sent data is the transcoded data. The business starts with the creation of the video decoding end and the encoding end, completes the adjustment of the video resolution, creates a video mixer, mixes the media stream, and finally sends it to the access end of the platform. The initial processing capacity of the media control terminal is shown in the following table (Table 1):

Affirmative simultaneous interpreting of 1000 identical sensor nodes in the communication platform, and making them evenly distributed in the sensing area of the platform, so that all nodes can arrange a square area of 100 m * 100 m. The size of the packet is 70, the size of the query packet is 10, the index information size is 10, and the random location in the sensing monitoring area generates 10 events. Make the data packet larger than the index information size, select the number of hops between the storage node and the ordinary node as 3, and the number of hops between the storage node and the index node as 4.

Table 1. Processing capacity allocation of media controller

Serial number	Object type	Processing capacity
1	1080P	800
2	720P	330
3	SD	160
4	CIF	50
5	QVGA	30
6	QCIF	12
7	Audio frequency	10

3.2 Experimental Result

In the above experimental environment, the comparative experiments of three groups of digital communication platform were carried out. A total of 100 experiments were carried out, and the average experimental results of 100 experiments were taken for demonstration, so as to improve the reliability of the experimental results.

The comparison results of the number of function nodes are as follows (Table 2):

Table 2. Comparison of the number of functional nodes

Node transmission range/m	Number of functional nodes in group A	Number of functional nodes in group B	Number of functional nodes in group C
10	63	45	40
20	63	46	40
30	64	48	43
40	67	48	43
50	68	48	44
60	68	49	44
70	69	52	47
80	70	54	49
90	70	54	51
100	72	57	53

As shown in the above table, the average number of functional nodes of the experimental group A construction platform is 67.4, and the average number of functional nodes of the experimental group B and the experimental group C are 50.1 and 45.4 respectively. Compared with the experimental group B and the experimental group C, the group A The number of functional nodes increased by 17.3 and 22.0 respectively.

On the basis of the first set of experiments, compare the overall message complexity of the storage methods of the sensor networks of the three groups of communication platforms, and change the average query interval. When the message complexity is smaller, it indicates that the query rate of the communication platform is greater. The comparison results are shown in the following table (Table 3):

Table 3. Comparison results of message complexity

Average query interval/s	Group A message complexity	Group B message complexity	Group C message complexity
5	132	156	183
10	122	163	193
15	125	160	184
20	139	163	180
25	137	152	185
30	131	169	194
35	129	162	193
40	123	163	192
45	128	160	190
50	139	163	191

As shown in the above table, the average message complexity of the experimental group A construction platform is 130.5. The average message complexity of the experimental group B and the experimental group C are 161.1 and 188.5 respectively. Compared with the experimental group B and the experimental group C, A The message complexity of the group platform storage method has been reduced by 30.6 and 58.0 respectively. To sum up, this design method has given full play to the advantages of multimedia communication technology. Compared with the traditional digital communication platform construction method, it has increased the number of functional nodes of the sensor network, including index nodes, storage nodes, and spare nodes, effectively reducing The network load when the platform transmits the Yellow River cultural information data, at the same time, reduces the message complexity of the storage method, increases the query rate of the information data, and improves the competitiveness of the construction platform in the field of Yellow River cultural communication.

4 Concluding Remarks

The design method for the Yellow River culture, the construction of a digital communication platform, ease the load of the platform sensor network, improve the query speed of information and data. However, there are still some shortcomings in this research, which focus on the cross layer design of application layer data storage and network layer routing protocol. In the future research, we will provide different precision and multi-resolution query reply for data query.

References

1. Liu, S., Pan, Z., Cheng, X.: A novel fast fractal image compression method based on distance clustering in high dimensional sphere surface. Fractals **25**(4), 1740004 (2017)
2. Haotian, C.: Digital dissemination and information sharing framework of precise poverty alleviation policy list. J. Soc. Sci. Hunan Normal Univ. **47**(6), 148–155 (2018)
3. Yang, J.: The spread form of digital art and its risk avoidance. J. Hengyang Normal Univ. **39**(2), 152–155 (2018)
4. Qingqi, Z.: Research on strategies of digital construction and communication about Taohuawu woodcut new year pictures. Value Eng. **37**(26), 258–260 (2018)
5. Junkang, X., Peijun, C.: The reflections on the construction of digital transmission platform for marine intangible cultural heritage. J. Ningbo Univ. (Liberal Arts Edition) **31**(5), 127–132 (2018)
6. Ping, Z.: Architecture-oriented research on the construction of digital media platform of red culture. J. Sanming Univ. **36**(1), 83–88 (2019)
7. Liu, S., Fu, W., He, L., Zhou, J., Ma, M.: Distribution of primary additional errors in fractal encoding method. Multimedia Tools Appl. **76**(4), 5787–5802 (2014). https://doi.org/10.1007/s11042-014-2408-1
8. Yanxiang, H., Yunzhong, H., Fangxiao, Q.: The digital propagation paths of lu ban's craftsman culture. Urbanism Archit. **17**(27):80–81+116 (2020)
9. Mengye, L., Liu, S., Kumarsangaiah, A., et al.: Nucleosome positioning with fractal entropy increment of diversity in telemedicine. IEEE Access **6**, 33451–33459 (2018)
10. Wu, Q., Zhang, C., Zhang, M., et al.: A modified comprehensive learning particle swarm optimizer and its application in cylindricity error evaluation problem. Int. J. Performability Eng. **15**(3), 2553–2560 (2019)

Research on the Modeling of Regional Cultural Transmission Rate in Mobile Social Network Environment

Jing-chao Hua[1(✉)] and Heng Xiao[2]

[1] College OF Humanities And Communication, Sanya University, Sanya 572000, China
huajingchao@21cn.com

[2] Sanya University, Sanya 223003, China

Abstract. Regional culture belongs to the category of local characteristic culture, which is greatly influenced by local history, regional and human factors, so there are big differences. In the context of the era of converged media, the analysis of regional cultural dissemination and its discourse structure based on the perspective of media culture has certain practical significance. Analyze the regional cultural dissemination in the current media context, and put forward suggestions on how to realize the effective dissemination of regional culture. The rise and development of media culture in the new era is the epitome of cultural transformation. Based on regional cultural dissemination and the construction of discourse, it is hoped that the main cultural consciousness can be awakened, and the cultural autonomy and active position in the transformation process can be mastered. On the one hand, it is necessary to avoid the fate of assimilation of regional culture, on the other hand, under the condition of changing the current situation of homogenization of contemporary culture, analyze the harmony but difference of regional culture and strive for more space for its development.

Keywords: Mobile social network · Regional culture · Cultural communication

1 Introduction

With the development of information technology and the popularity of the network, the scale of data in the network is growing, and the audience's dependence on the network information is gradually increasing, but they have to face the problems of slow retrieval and difficult understanding of information [1]. Information visualization technology, supported by computer, realizes the visualization of abstract data through graphics and images, reveals the connotation and potential structure of information, makes the audience bid farewell to the irregular data heap, and reduces the cognitive burden. Guided by the idea of information visualization, this paper designs the information visualization model of regional cultural resources, and explores a new idea for the dissemination of regional culture [2]. Taking the regional cultural resources as the specific object, the

S. Liu and X. Ma (Eds.): ADHIP 2021, LNICST 416, pp. 429–439, 2022.
https://doi.org/10.1007/978-3-030-94551-0_34

visualization model of regional culture is designed from different angles, different levels and different dimensions, which provides practical experience for the application of information visualization technology in the field of regional culture communication.

Reference [3] puts forward a model of regional cultural transmission rate based on protocellular machine. The model considers that cultural competition and integration are inevitable under the background of globalization, and immigration significantly speeds up the process of cultural transmission and integration. Cultural change is an evolutionary process in a complex social system. According to the process of cultural transmission, the affected population can be divided into small rectangular units, and the time can be divided into multiple units to build a cultural transmission rate model. Reference [4] proposed a culture transmission rate model based on mobile social network. By forming a temporary group, the cultural information in mobile social network can be transmitted to others. And for practical considerations, it is expected to form a star network topology among mobile social network users to improve the speed of cultural transmission.

However, the traditional model of cultural communication rate still can not meet the needs, so this paper puts forward the modeling research of regional cultural communication rate in the mobile social network environment, extracts the characteristics of regional cultural information, clarifies the relationship of regional cultural information communication, constructs the regional cultural communication model, and realizes the high-speed transmission of regional culture.

2 Modeling of Regional Culture Transmission Rate

2.1 Characteristics of Regional Cultural Information Based on Multimedia

Once the research of regional cultural resources is separated from the initial environment of culture, it will cause the separation of the whole concept, which is a kind of damage to the characteristics of regional cultural resources. Therefore, it is necessary to study the regional cultural resources "region". Based on the reference information visualization reference model and map cube model, combined with the regional characteristics of regional cultural resources, a visualization model based on regional characteristics is proposed. The research of spatial geographic location is an important aspect of information visualization. The visualization of spatial location can clearly judge the spatial distribution of research information. At present, there are many researches on spatial geographic location. The typical research is that electronic map and virtual structure can be obtained. The combination of information visualization technology and geospatial information conforms to the thinking mode of the audience, so as to make the information visualization technology combined with geospatial information, so as to meet the thinking mode of the audience The visualization results have the advantages of large amount of information, intuitive image and various forms. The theory, method and technology of converting data into graphics or images on screen and interactive processing are made by using computer graphics and image processing technology. The main research fields are: scientific computing visualization, data visualization, information visualization and knowledge visualization. The objects, technologies, purposes, methods and application fields studied in these four fields are different, but they are not independent from each other. The four visualization technology relationships are shown in the figure. As a research field of visualization, information visualization includes many characteristics of data visualization, information graphics, knowledge visualization, scientific visualization and visual design, and develops and advances on this basis (Fig. 1).

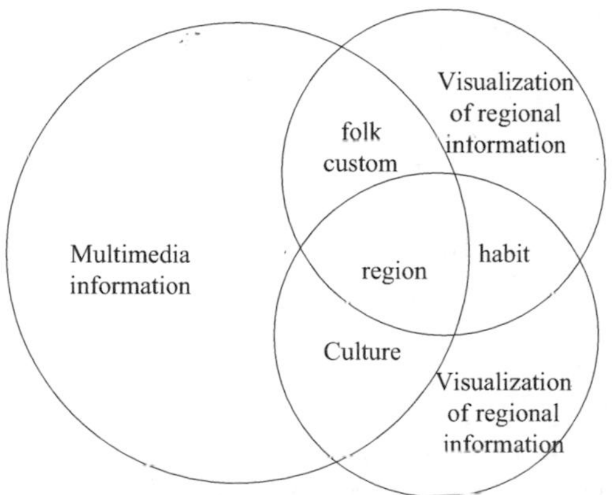

Fig. 1. Visualization of regional cultural information characteristics

Information visualization is an interdisciplinary field, which aims to study the visual presentation of large-scale non numerical information resources, and help people understand and analyze data by using graphics and image technology and methods [2]. The continuous upgrading of information software and hardware and the exponential growth of network resources and digital information have brought unprecedented opportunities to the development of information visualization. Information visualization as a discipline has gradually grown up. Information visualization concept map, information visualization is a kind of cross activity between data, computer and users, and its core is data exploration. Data exploration is mainly to achieve four purposes: to express information vividly, that is, to express abstract data in a visual way; to discover new knowledge; to identify various possible laws of information in terms of structure, pattern, trend, relationship, etc.; through exploration, human beings can obtain useful information from the vast ocean of data, so as to discover knowledge in information and learn from knowledge Seek a decision, as shown in the figure (Fig. 2).

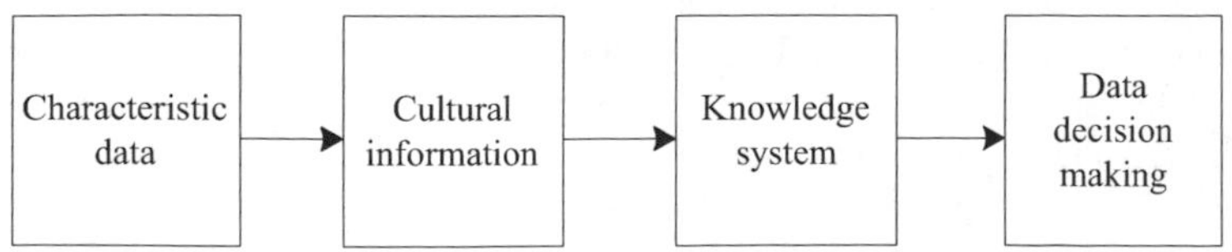

Fig. 2. Collection and processing process of regional cultural information characteristics

The research object of information visualization focuses on abstract data set, and often uses graphics to express the relationship of abstract data. Its purpose can be simply described as "interpretation with graphics". The relationship and common graphics expressed by information visualization are shown in the Table 1:

Table 1. Communication relationship of regional cultural information

Relationship	Features
Quantitative relationship	Bar chart, pie chart, curve chart and comparison chart
spatial relationship	Map, celestial map
Time relationship	Time line, flow chart
Organizational relations	Organization chart

According to different classifications, information visualization has different classification methods [5]. At present, the common classification standards include information type, information visualization technology and deep data arrangement. Information visualization classification information table, as shown in the table (Table 2).

Table 2. Classification of cultural information based on multimedia technology

Classification dimension	Classification results
Information type	One dimensional information, two-dimensional information, three-dimensional information, multidimensional information, time series information, hierarchical information, network information
Information visualization technology	Three dimensional technology, icon based technology, pixel positioning technology, hierarchical technology, chart based technology, hybrid technology
Deep data arrangement	Map type, information type, interactive and exploratory type

Starting from the filtering of the original data, effective information is analyzed and extracted from the original data; the mapping process is to map the screened effective information into standard data with dimensional structure; the visualization process is to complete the interactive view which is easy for the audience to operate; the feedback process reflects the impact of user behavior on the visualization process [6]. The information processing process of information visualization is the process of changing the information state, from the original scattered state to the final structured relational state. The processing structure of information visualization reduces the cognitive burden of the audience, so that users can quickly understand the information and find the rules, correctly interpret the information, master and apply the rules, and improve the cognitive level and insight [7].

2.2 Regional Culture Communication Model

Regional culture has rich connotation. According to the conditions of region, time, nationality, knowledge, art and human, we can analyze the following attributes of regional cultural resources. Cultural resources have their own geographical environment for generation, development and prosperity, and cultural objects are different due to their geographical location, climatic conditions, folk customs and living habits [8]. According to the subordination of location type, from high to low, it adopts "country", "province", "city", "county (District)", "township (town)" and "village". As shown in the figure. The design of regional cultural information visualization model needs a complete workflow as guidance [9]. This paper refers to the general steps of visualization data mining and design in the book information visualization and knowledge retrieval, and designs the workflow diagram of regional culture information visualization model construction, as shown in the figure (Fig. 3).

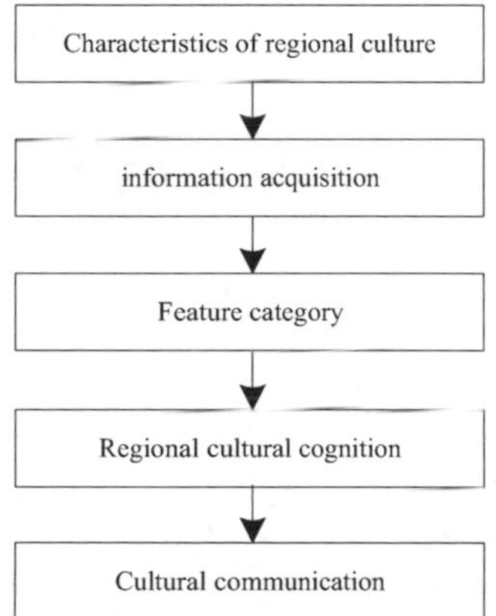

Fig. 3. Workflow of constructing the visualization model of regional cultural information

Every cultural resource is left over from history and has a distinct or hidden brand of the times. This paper analyzes the origin time of regional cultural resources, and understands the breeding, formation and development of culture according to the time sequence of feudal dynasties. Through cultural relics, economic conditions, production relations, social forms, major events and important figures to reflect the Era Connotation of cultural resources. China is a multi-ethnic country, because each nation has its own national background, national beliefs, national language, national habits and so on, there are great differences in regional culture. At the same time, due to the differences between nationalities, it also enriches the content of regional culture. To explore the regional culture from the perspective of geographic information and realize the interaction between human and image is in line with the audience's way of thinking. The audience can participate in controlling the flow of information and the way of visualization. This will help the audience to understand and infer the relationship between regional culture and information, and improve the cognitive level from initial knowledge to decision-making.

The content of regional research can be divided into three aspects: address data, including address number, address name and subordination, which is the attribute given by human beings to geographical space, and is helpful to the management and identification of geographical location; human geography, the special "regional" characteristics of regional cultural resources are mainly determined by its special initial environment, which includes not only macro resources, but also macro resources It also includes the micro food structure and crop varieties and other factors; the form of expression, information visualization technology for the expression of geospatial information, mainly using two parts of virtual space and electronic map (Fig. 4).

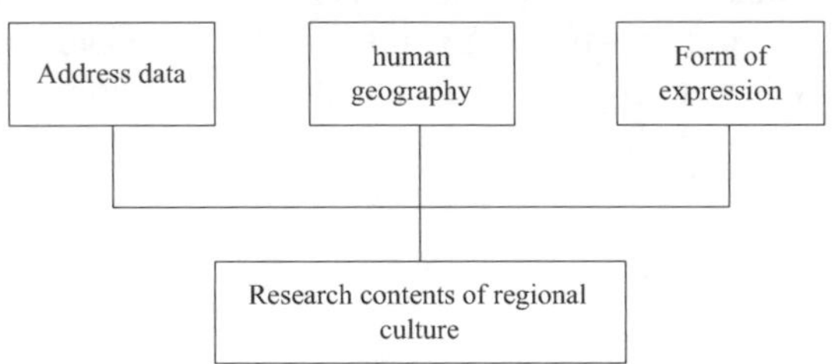

Fig. 4. Types of regional cultural communication information

The knowledge of regional culture can be divided into two kinds: one is the origin of regional culture, which is formed with the emergence of regional culture; the other is social science knowledge and natural science knowledge related to the formation, evolution, perfection and dissemination of regional culture. The knowledge of regional culture is extensive and profound. The other five attributes of regional culture can be used as a breakthrough in the study of knowledge. Taking "intangible cultural heritage" as an example, the forms of artistic expression of cultural objects are different. Culture can be divided into ten categories: folk story, folk literature, folk music, folk dance, opera, folk art, folk acrobatics, folk handicrafts, folk custom and traditional medicine. The artistry of regional cultural resources can be classified according to subordination, as shown in the figure (Fig 5).

The development of modern science and technology reflects a transformation process from "entity thinking" to "relationship thinking". From the perspective of cognitive law, logical information is easier for people to recognize. As a system, information visualization model is the reflection of information objects in various relationships. Information relationship is the association rules between research objects. Association analysis, as an important research in data mining, paves the way for the early research of information visualization. Analysis of regional cultural resources, information relations are mainly divided into the following two kinds. The internal information relationship of the object. This kind of relationship is mainly reflected in the hierarchical relationship within the object, which mainly includes: the subordination of geographical location, the inclusion of time attribute, the subordination of organizational structure, the relationship of characters, the generic relationship of cultural types, etc. These two locations belong to subordination in geographical location, and are the internal information relationship of objects.

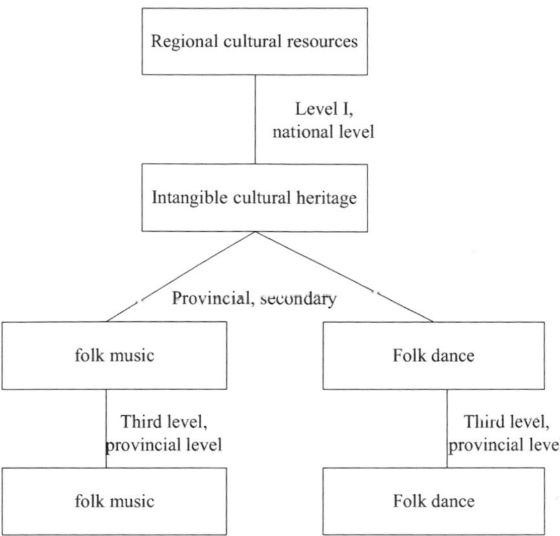

Fig. 5. Analysis of regional culture transmission rate

The cross information relationship between objects is expressed as the relationship under constraint binding, which mainly refers to the same genus relationship. It mainly includes: region, time and category, intangible cultural heritage also includes identification level and application batch.

2.3 The Realization of Regional Culture Communication

With the popularity of the network, network communication has become the mainstream of information communication. At present, there are many digital resources related to regional culture. However, in the design of traditional digital resources, the visual experience, interactive needs and cognitive burden of the audience on resources are not considered. The concept dynamic system of n participating enterprises can be given by the following equation:

$$D_1 x_i = F_{ii}(x_i, \theta_i) + \sum F_{ij}\left(x_i, x_j\right), i, j = 1, \ldots, n, \tag{1}$$

$$F_a(x_i, \theta_i) = -h_i(x_i - \theta_i) \tag{2}$$

$$F_{ij}\left(x_i, x_j\right) = \mu_{ij} B\left(x_j - x_i\right) \tag{3}$$

$$B(u) = u \cdot \exp\left(-(u/\lambda)^{\beta}/\beta\right) \tag{4}$$

The model is transformed into a social impact network model driven by linear resultant force

$$F_{ij}\left(x_i, x_j\right) \rightarrow \begin{cases} \mu_{ij}\left(x_j - x_i\right), \left|x_j - x_i\right| \leq \lambda \\ 0, \text{ otherwise} \end{cases} \tag{5}$$

The simplified analysis can be obtained:

$$D_r x_i = -(x_i - \theta_i) + \sum_{j=1, j \neq i}^{n} k_{ij} B(x_j - x_t), i = 1, \ldots, n \tag{6}$$

The linear dynamics of the evolution of the concept of cooperation are as follows:

$$\begin{cases} D_i x_1 = -h_1(x_1 - \theta_1) + \mu_{12}(x_2 - x_1) \\ D_1 x_2 = \mu_{21}(x_1 - x_2) - h_2(x_2 - \theta_2) \\ \theta_1, \theta_2, h_1, h_2, \mu_{12}, \mu_{21} \in (0, 1) \end{cases} \tag{7}$$

The steady-state values of the concept of cultural communication are as follows:

$$\begin{cases} \tilde{x}_1 = (\theta_1 + \theta_1 \mu_{21}/h_2 + \theta_2 \mu_{12}/h_1)/(1 + 2k) \\ \tilde{x}_2 = (\theta_2 + \theta_1 \mu_{21}/h_2 + \theta_2 \mu_{12}/h_1)/(1 + 2k) \end{cases} \tag{8}$$

The equilibrium value of concept difference is as follows:

$$\tilde{x}_1 - \tilde{x}_2 = \theta/(1 + 2k) \tag{9}$$

The difference of conceptive concession is as follows:

$$(\theta_1 - \tilde{x}_1) - (\tilde{x}_2 - \theta_2) = (\mu_{21}/h_2 - \mu_{12}/h_1)\theta/(1 + 2k) \tag{10}$$

Information visualization technology is applied to the dissemination of regional culture to improve the disadvantages of traditional culture dissemination, present an intuitive interactive interface for the audience, shorten the search time and reduce the cognitive burden. After fully analyzing the technical characteristics and visualization objectives of information visualization, a visualization model based on network dissemination of regional culture resources is designed. The basic architecture of the model is shown in Fig. 6:

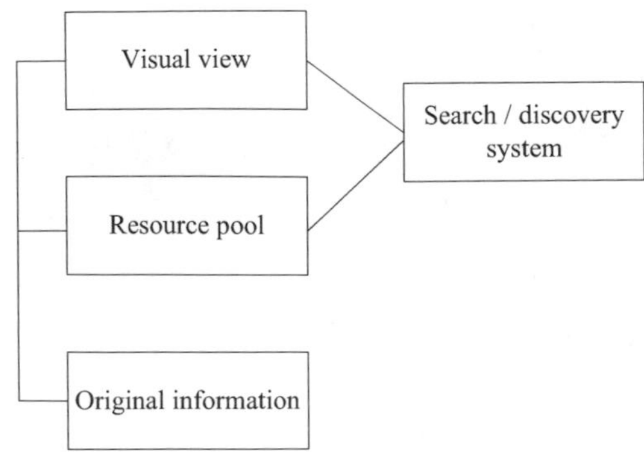

Fig. 6. Basic architecture of information network communication model

The main work of the model is simply summarized as the transformation process from original information to visual view. The model provides various services to the audience through visual view, and at the same time, the visual view can interfere with and affect the situation of resource base and original information; the resource base mainly stores and manages various digital objects, and the digital objects of the resource base are standard data obtained after the visual design of the original data, and can directly map the views; The main function of search / discovery system is to complete the retrieval request of the audience and realize interaction.

3 Analysis of Experimental Results

In order to verify the actual application of the regional culture propagation rate model in the mobile social network environment, the experiment is carried out in the following environment: Net Framework 3.5; operating system: Window XP; Web server: Internet information services 6; development language: C #, action script; database: SQL Server 2005; integrated development tools: Visual Studio 2008, flash. Based on the above experimental environment, comparative detection is carried out to analyze the cultural transmission rate of social networks in different geographical environments. The specific detection results are shown in the following figure (Fig. 7):

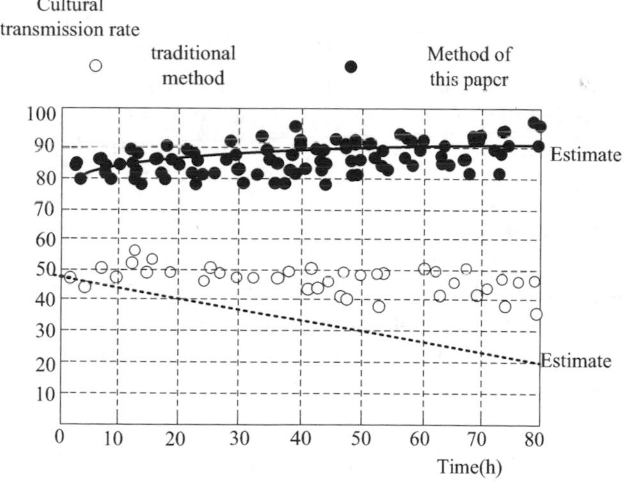

Fig. 7. Comparison of cultural communication efficiency under different methods

Based on the above comparative analysis results, the prediction model of regional culture propagation rate under this method can more accurately predict the data in the actual application process, and the prediction results are basically consistent with the actual distribution, which proves that the regional culture propagation rate model under the mobile design network environment proposed in this paper has a high accuracy in the actual application process, which is further improved The results are shown in the following figure (Fig. 8):

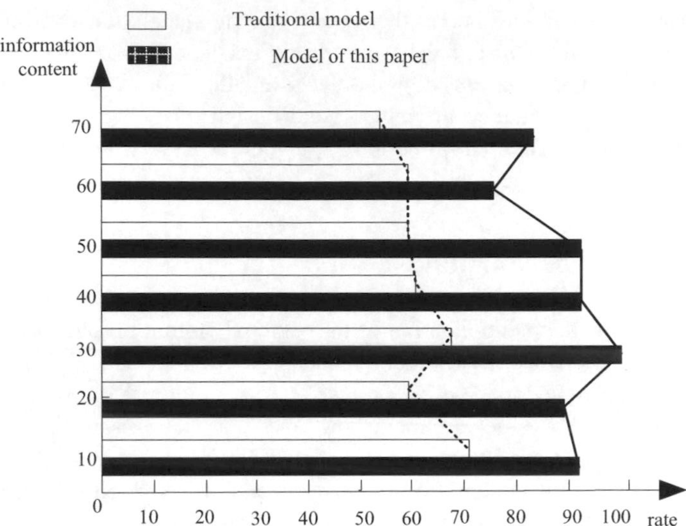

Fig. 8. Comparison results of regional culture rate estimation

Based on the above detection results, compared with the traditional methods, the regional culture propagation rate model in the mobile design network environment proposed in this paper has higher timeliness in the practical application process, which can more quickly carry out the culture propagation and control prediction, and fully meet the research requirements.

4 Concluding Remarks

Combined with the mobile design network environment, this paper designs and develops a cultural communication model that conforms to the cognitive law of the audience. The design process of the model is a combination of theory and practice, concept and technology. Under the comprehensive guidance of the visual model technology system, the research work determines the development and operation environment of the model and the technical system to be followed, processes the information structure of the intangible cultural heritage content, designs the attribute database according to the intangible cultural heritage information attribute model in the previous paper, and finally designs the attribute database The realization of client visual display system is described in detail.

Fund Projects. Hainan philosophy and social science planning project "Hainan fishermen history and South China Sea sovereignty research" (hnsk (ZC) 20–29).

References

1. Xu, Q., Su, Z., Lu, R.: Game theory and reinforcement learning based secure edge caching in mobile social networks. IEEE Trans. Inf. Forensics Secur. **15**(11), 3415–3429 (2020)

2. Liu, S., Fu, W., He, L., Zhou, J., Ma, M.: Distribution of primary additional errors in fractal encoding method. Multimedia Tools Appl. **76**(4), 5787–5802 (2014). https://doi.org/10.1007/s11042-014-2408-1
3. Shu, Z., Zhang, X., Jiang, H.: Computational model of cultural preservation based on cellular automata. J. Phys: Conf. Ser. **18**(4), 420–425 (2021)
4. Mao, Z., Jiang, Y., Di, X., et al.: Joint head selection and airtime allocation for data dissemination in mobile social networks. Comput. Netw. **166**(15), 106–109 (2020)
5. Zhao, Y., Zhang, J., Liu, Y., Sun, K., Zhang, C., Wenguang, W., Teng, F.: Numerical assessment of the environmental impacts of deep sea cage culture in the Yellow Sea, China. Sci. Total Environ. **706**, 135752 (2020). https://doi.org/10.1016/j.scitotenv.2019.135752
6. Zhao, Y., Yan, J., Hou, S.: "Scenes" updating the cultural landscape of villages along the grand canal under the all media communication. IOP Conf. Ser. Earth Environ. Sci. **75**(1), 121–125 (2021)
7. Liu, S., Liu, G., Zhou, H.: A robust parallel object tracking method for illumination variations. Mob. Netw. Appl. **24**(1), 5–17 (2018). https://doi.org/10.1007/s11036-018-1134-8
8. Sari, R.N., Dewi, D.S.: Development models of personality, social cognitive, and safety culture to work accidents in the chemical company. IOP Conf. Ser. Mater. Sci. Eng. **16**(1), 12201316 (2021)
9. Tvs, Jr A., Ldb, A.A., et al. Accurate prediction of blood culture outcome in the intensive care unit using long short-term memory neural networks. Artif. Intell. Med. **97**(7), 38–43 (2019)
10. Muslim, S.N., Ali, A.N.M., Fayyad, R.: A novel culture medium using lettuce plant (Lactuca Sativa) for detection of Pullulanase production by Paenibacillus Macerans isolated from agricultural wastes. J. Phys: Conf. Ser. **18**(1), 134–138 (2021)

Design of Cold Chain Logistics Information Real Time Tracking System Based on Wireless RFID Technology

Wang Li-feng[1(✉)], Huang Fei[2], and Guo-hua Zhu[3]

[1] BaiSe University, BaiSe 533000, China
wanglifeng1101@tom.com
[2] Macau University of Science and Technology, Macau 999078, China
[3] School of Artificial Intelligence, Jianghan University, Wuhan 430056, China

Abstract. In order to improve the proportion of cold chain distribution, effectively control transportation links and information, and ensure the quality of cold chain food, it is necessary to establish an effective monitoring and tracking mechanism in all aspects of cold chain transportation. Therefore, a real-time tracking system of cold chain logistics information based on wireless RFID technology is proposed. The tag data sensing terminal, reader data transfer terminal and background server data management terminal are used to construct the real-time tracking system framework of cold chain logistics information, and the wireless RFID card reader is used to obtain the real-time temperature of food, vehicle running position information, agricultural product information and other data, so as to realize the real-time tracking and monitoring of cold chain logistics information. The experimental test shows that the function of the system meets the needs of real-time tracking of cold chain logistics information, achieves the system design goal, and ensures the quality of cold chain food.

Keywords: Wireless RFID technology · Cold chain logistics · Information tracking system

1 Introduction

Under the wave of global economic integration, many industries in China, such as dairy products, cold drinks, raw and fresh meat products, have made great progress. The rapid development of these food industries has triggered and driven the development of cold chain logistics business [1]. Cold chain logistics is a kind of logistics system that keeps perishable and fresh food in the specified low temperature environment in all links of production, processing, storage, transportation, sales and final consumption, so as to ensure food quality and reduce food loss by the best logistics means. It is an essential safeguard measure to ensure that fresh food can be safely delivered to consumers from the production source. At present, China's annual demand for fresh agricultural products is about 100 million tons, but due to the inherent perishable characteristics of agricultural

S. Liu and X. Ma (Eds.): ADHIP 2021, LNICST 416, pp. 440–453, 2022.
https://doi.org/10.1007/978-3-030-94551-0_35

products, and the logistics distribution is usually carried out under normal temperature, resulting in excessive loss in the logistics process [2]. According to relevant data, the loss rate of agricultural products caused by improper production, transportation and storage is more than 100 million yuan, and the direct economic loss is about 100 million yuan [3]. Not only that, in recent years, malignant incidents of food quality and safety problems have also occurred constantly, such as the frequent occurrence of food decay, the discrepancy between the quantity of goods delivered and the quantity of goods delivered, and the inability to trace the source after the occurrence of food poisoning events, which lead to the safety of cold chain food has become the focus of public attention. Therefore, the demand for effective cold chain monitoring and tracking management of fresh products is becoming more and more obvious. On the background of this social demand, it is necessary to vigorously develop cold chain logistics to reduce unnecessary consumption of fresh products. At the same time, it is necessary to establish the whole process monitoring mechanism of cold chain supply chain and optimize and improve its tracking process. At present, scholars in related fields have carried out research on the cold chain logistics information real-time tracking system, and achieved certain research results. In reference [4], a real-time monitoring and traceability system of Tricholoma matsutake cold chain was designed. Using the Internet of things technology, a real-time monitoring traceability system is developed and tested to monitor the company's cold chain logistics environment parameters. The system can track and monitor the fluctuation of temperature, humidity, oxygen and carbon dioxide in the cold chain in real time.

Based on the above background, a cold chain logistics information real-time tracking system based on wireless RFID technology is designed. Using tag data sensing, reader data transmission and background server data management terminal, the real-time tracking system framework of cold chain logistics information is constructed. RFID tags with temperature sensors are placed to record and monitor the temperature in the cold storage, and labels are pasted on each container or paper bag. At the same time, record the freezing state change data, and immediately detect the temperature condition through visual inspection. If the temperature of the food in the car is abnormal, it should be detected immediately, and the problem should be stopped immediately. Realize the real-time tracking and monitoring of cold chain logistics information. So as to improve the distribution proportion of cold chain, effectively control the transportation links and information, and ensure the quality of cold chain food.

2 Key Technologies of Cold Chain Logistics Information Real Time Tracking System

The main research object of this paper is the design and implementation of cold chain logistics information real-time tracking system, which mainly uses satellite positioning technology, RFID technology and GPRS wireless transmission and other key technologies.

2.1 Satellite Positioning Technology

The logistics of fresh food products requires the whole process cold chain control, and the temperature environment is extremely strict, so the requirements of logistics tracking technology also need to be improved accordingly. The information system not only needs to obtain the geographical location information of the transported goods, but also needs to monitor the temperature information in real time as the guarantee of food safety and tracking evidence. Logistics tracking technology is usually applied to the following technologies: geographic information system (GIS) and global positioning system (GPS). GIS is an information system based on geographic spatial data, which can feed back spatial dynamic geographic information and provide decision-making service for relevant personnel through model algorithm. Combined with the corresponding software system, the distribution route model can be established to provide routes for transportation and positioning for monitoring, so as to better improve the logistics efficiency.

GPS uses satellites to realize the continuous and high-precision monitoring of ground distribution vehicles, which can effectively understand the location and trajectory of ground targets, and according to the needs, with the corresponding software, set a more reasonable geographical route, shorten the transportation time and reduce the transportation cost.

2.2 RFID Technology

RFID (radio frequency identification) radio frequency identification technology, also known as radio frequency identification, is a non-contact automatic identification technology. Through the radio frequency identification signal transmitted by the reader antenna, the target object can be automatically identified, so as to transmit the data from the electronic tag attached to the object, so as to achieve the purpose of automatic identification of the information in the electronic tag. The recognition process does not need manual intervention, and the reader can work in harsh environment, and can penetrate ice, snow, rain, fog and so on. At the same time, RFID technology can not only accurately identify high-speed moving objects, but also realize multiple tags reading at the same time, which is convenient to collect a large amount of data at one time, so as to improve work efficiency [5]. At present, long-distance RFID technology is widely used in airport security and intelligent transportation, such as its automatic toll system on expressway. Short range RFID products overcome the inconvenience of paper bar code in the dirty, dust and other harsh environment, can completely replace the paper bar code in this environment, and have a wide range of uses, such as the object tracking system on the factory assembly line.

RFID technology is widely used in logistics management. It mainly realizes the identification and management of goods information in the process of product transportation, so as to achieve the purpose of rapid, accurate and timely transportation of products. The electronic label is fixed on the surface of the product, and the whole process of goods tracking is realized in the process of goods delivery, loading and unloading, and transit.

2.3 GPRS Wireless Communication Technology

GPRS is the abbreviation of general packet radio service. It is a new packet data carrying service. It is the combination of GSM wireless access technology and internet packet switching technology. The most fundamental difference between GSM and the existing GSM Voice system is that GSM is a circuit switching system, while GPRS is a packet switching system [6]. GPRS service is suitable for most mobile Internet, and its data transmission characteristic is that it can be used for intermittent, sudden or frequent, small amount of data transmission.

The GPRS network covers a wide area. All the places where the GPS network can be used can use the GPRS network. It has high stability, powerful network functions, and relatively low cost in industrial applications. The specific advantages are as follows:

(1) Real-time online: Real time online means that GPRS service users can keep in touch with the network at any time. When users access the Internet, the mobile phone will send and receive data on the wireless channel. Even if there is no data transmission, the mobile phone will also keep the network unblocked.
(2) Billing by traffic: The charging method of GPRS service is calculated according to the number of packets received and sent by the user. Even if the user is online all the time, there is no need to pay any fee when there is no packet sent and received.
(3) High speed transmission: In theory, the data transmission of GPRS packet switching technology can reach 171.2 kb/s, but due to the limitation of coding and terminal function, the actual transmission rate will decrease. However, compared with the 9.6 kb/s of circuit switching data service, the transmission rate is significantly improved.
(4) Switch freely: Free switching means that users can simultaneously access the Internet and make voice calls on mobile devices, that is, data transmission function and voice call function can be switched at the same time or freely with each other.

3 Framework of Cold Chain Logistics Information Real Time Tracking System

The system consists of tag data sensing end, reader data transfer end and background server data management end as Fig. 1.

According to Fig. 1, the Tag data sensing terminal is to obtain the real-time temperature data of fresh food. The Tag is pasted on each packaging box to obtain energy from the RF signal sent by the nearby reader and work according to the instructions sent by the reader. The data transfer terminal of the reader should realize two functions: one is to control the Tag entering the reader communication range and obtain the sensing data. The second is to push the collected data to the server in real time. The data management end of the background server is to realize the centralized management of all monitoring data, provide user-friendly operation interface, realize the data tracking query function and simple data processing function. In addition, it also provides emergency alarm function. In case of an emergency, such as fresh food exposed to high temperatures for a long time. The background system will send a warning message to the designated staff according to the temperature overrun time.

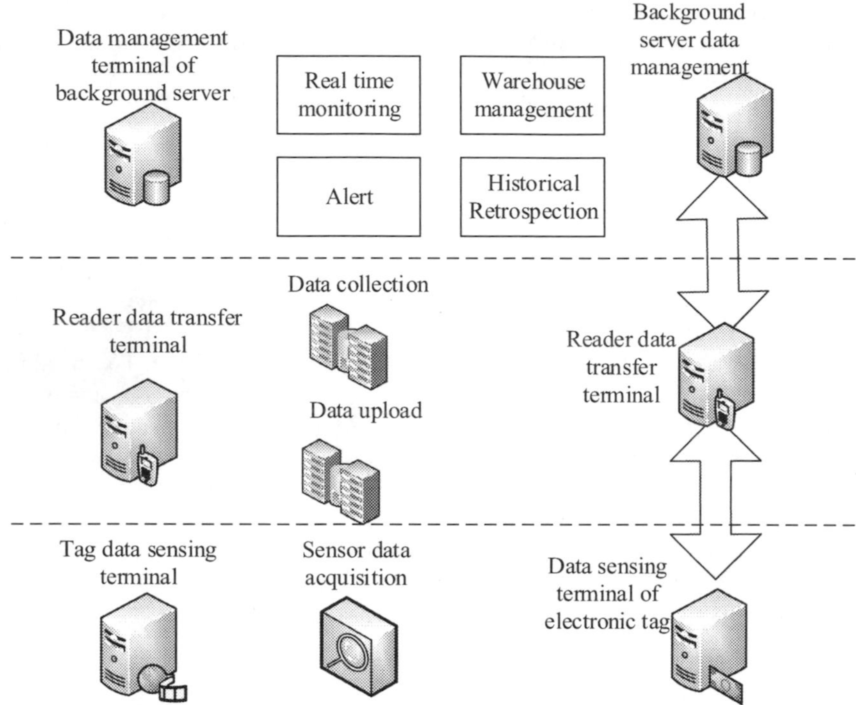

Fig. 1. Framework of cold chain logistics information real time tracking system

4 System Hardware Design

In the hardware part of the system, through the selection of microprocessor and the design of peripheral circuit, the main functions are the analysis and storage of GPS navigation message, the collection and storage of RFID logistics cargo information. Temperature and humidity data are collected and stored, and the collected data are sent to the server and mobile terminal of the monitoring center through GPRS network for users at all levels to query and process. At the same time, the collected information is displayed on the LCD of the vehicle terminal in real time [7]. The hardware structure is as Fig. 2.

4.1 MCU Processor

MCU (micro controller unit) processor, also known as single-chip microcomputer or single-chip microcomputer. It is to reduce the frequency and specification of the CPU, and integrate the peripheral interfaces such as memory, counter, USB, A/D conversion, UART and LCD driver circuit into a single chip to form a chip level computer, which can do different combination control for different applications.

The selection of processor chip is very important for any control system. For any type of processor, it is necessary to have corresponding peripherals and development methods to give full play to its maximum efficiency and bring the maximum efficiency to the whole control system [8]. The processor chip in this system is a 32-bit

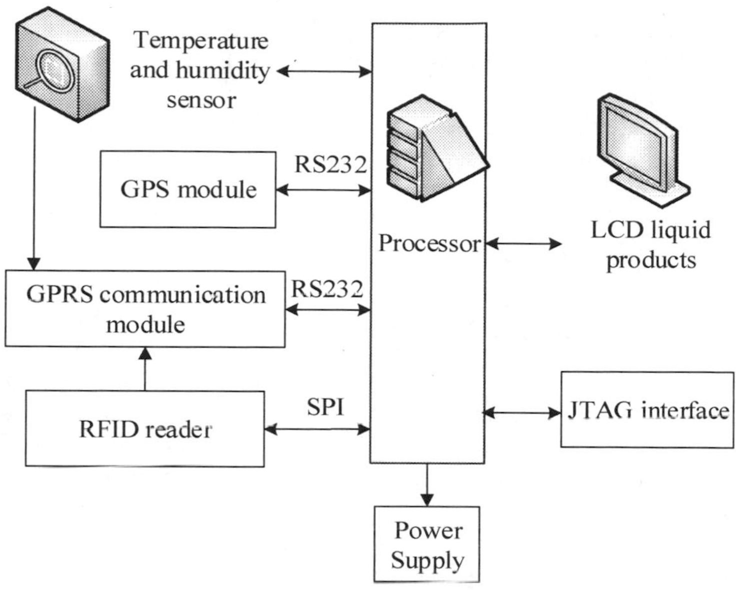

Fig. 2. Hardware structure

low-power, high-performance STM32F107VCT6 microprocessor produced by ST. Its standard peripherals include 10 timers, 2 AD analog-to-digital converters, 2 DA digital-to-analog converters, 5 USART interfaces, 3 SPI interfaces, 2 CAN2.OB interfaces, 2 full-speed USB (OTG) the interface, two I2C interfaces and Ethernet 10/100MAC module can fully meet the needs of industrial and medical products. Two USART interfaces of STM32F107VCT6 processor are used to complete the circuit design of GPS module interface and GPRS module interface. Use one of the pins of the microprocessor (here PA 11 pin) to complete the data transmission function with the temperature and humidity sensor. The SPI serial interface of the microprocessor is used to complete the data transmission function with the RFID reader module [9]. The hardware interface circuit design of these modules are introduced in detail in the follow-up content. According to the function design of the refrigerated cargo tracking system that the hardware design of the vehicle terminal needs to realize, the software flow of each module is developed. The experimental chapter verifies that the processor selection is correct, which can realize the functional design requirements of the vehicle terminal.

4.2 GPS Module Hardware

GPS receiving module is used to receive satellite signals, and send the information to the processor through the serial port for analysis, so as to obtain the required three-dimensional position coordinate information for subsequent storage, display and transmission. GPS module plays an important role in the vehicle terminal tracking system. According to the analysis of longitude, latitude, time and other information, we can know the location coordinates of the transport vehicles.

GPS positioning system consists of three parts: space part, ground monitoring part and user part. The space part is composed of 24 GPS satellites with an altitude of about 20200 km. The orbits are nearly circular, and the operation period is about 11 h and 58 min. They are evenly distributed on six orbital planes, with 4 satellites distributed on each orbital plane, and the orbital inclination is 55°. This distribution of satellites makes it possible for more than four satellites to be observed at any time and anywhere in the world at the same time, and can ensure good geometric figure of positioning calculation accuracy, so as to provide global navigation capability.

4.3 DHT11 Temperature and Humidity Sensor

DHT11 temperature and humidity sensor combines digital technology with sensing technology to ensure high reliability and excellent long-term stability. The sensor is composed of humidity sensing element and NTC temperature measuring element, and is connected with a high-performance 8-bit MCU [10]. Through the processor connection, the local humidity and temperature can be collected in real time. The communication mode between DHT 11 and STM32 adopts a single bus type, and only one I/O port is required. The 40-bit data of humidity and temperature data inside the sensor are transmitted to the processor at one time, and the data is verified by a checksum method, which effectively ensures the accuracy of data transmission. DHT11 has low power consumption, with an average maximum current of 0.5 mA under 5 V supply voltage. The main application areas include automobiles, home appliances, weather stations, humidity regulators, etc.

The humidity detection of DHT11 uses capacitive structure, and uses "micro structure" detection electrode system with different protection and polymer coating to form the capacitance of the sensor chip, which can not only ensure the original characteristics of the humidity sensitive parts of the capacitance, but also resist the interference from the outside world. Because it combines the temperature sensor with the humidity sensor to form a single individual, the measurement accuracy is high and the dew point can be accurately obtained, and the error caused by the temperature change can be avoided [11]. The sensor integrates signal amplifier, analog-to-digital converter, calibration data memory and standard IIC bus into one chip.

In order to achieve the purpose of accurate measurement, the calibration method of the sensor is different from the general sensor, which is calibrated in an extremely strict temperature and humidity closed room. It can fully meet the temperature monitoring requirements in the refrigerated cargo transportation, especially for fresh flowers whose fresh-keeping temperature requirements are between 0–10 °C, and the fresh-keeping temperature of general fresh flowers is 5 °C.

4.4 Wireless RFID Reader

The reason why the wireless RFID reader is called wireless is that it does not need to rely on the communication interface to exchange data between PC like the ordinary portable data collector, but can directly carry out real-time data communication with PC and server through the wireless network. Wireless RFID card reader has the advantage of online data acquisition. Its communication with the computer is realized by radio waves, which can transmit the data collected on site to the computer in real time [12]. Compared with the ordinary portable bar code scanner, it further improves the work efficiency of the operator, and makes the data from the original local verification and storage into remote control and real-time transmission.

The RFID reader module uses the MFRC522 contactless (13.56 MHz) reader card chip produced by NXP (NXP semiconductor) company. It uses the principle of modulation and demodulation, and integrates them into various contactless communication methods and protocols. The function of the RFID module is to record the quantity change and transit situation of goods in transit.

MFRC522 reader supports ISO 14443A/MIFARE working mode. The internal transmitter of MFRC522 can drive the reader's antenna to communicate with ISO 14443A/MIFARE card and transponder. The receiver part provides a powerful and efficient demodulation and decoding circuit to process the signals of ISO 14443A/MIFARE compatible cards and transponders, and the digital circuit part processes the complete ISO 14443A frame and error detection. MFRC_522 supports MIFARE Class devices and MIFARE's higher-speed non-contact communication, with a two-way data transfer rate of up to 424 kbit/s.

4.5 GPRS/ GSM Hardware

The SIM900A module is an Internet of Things communication product produced by SIMCom. The module is a compact GPRS/GSM module with built-in user applications. The module's working frequency bands are 900 MHz and 1800 MHz, which can be used normally all over the world. It adopts industrial-grade interface standards and can provide voice, short message and other service functions. The TCP/IP communication protocol is embedded, and the extended TCP/IP AT command makes it convenient for users to use the TCP/IP protocol. The main features of SIM900A module are as Table 1.

Table 1. SIM900A module parameters

Parameter	Description
Power supply	Single voltage: 3.2 V–4.8 V, low power consumption mode is only 3.5 mA
Working frequency band	Can automatically search for 900 MHz/1800 MHz working frequency band, or set working frequency band through AT command
GPRS data characteristics	GPRS data downlink transmission: up to 85.6 kbps GPRS data uplink transmission: up to 42.8 kbps
Short message SMS	MT, Mo, CB, Text and PDU modes SMS storage device: SIM card

5 System Function Design

The cold chain logistics tracking system is based on the Internet of things technology, based on electronic coding, through the electronic label technology to collect the attributes, status and physical environment parameters of cold chain food in real time, so as to realize the safety warning and product tracking of cold chain food logistics process, and support the monitoring of enterprises and government departments on cold chain food. From the above analysis, it is clear that the tracking system is located in the logistics among the supply chain enterprises, with the production and processing enterprises, logistics enterprises, sales enterprises and government management departments as the service objects, and mainly completes the data collection and monitoring management of the production, delivery, transportation and receipt of cold chain food [13]. Therefore, under the guidance of the system objectives of tracking and real-time monitoring, the functional objectives of the food cold chain logistics information tracking system mainly include:

(1) For the upstream and downstream enterprises to achieve seamless transmission of information, to solve the cold chain broken phenomenon;
(2) Improve the efficiency of cold chain operation, reduce the system cost and loss;
(3) In case of safety problems, it can quickly determine the source of products, provide convenience for product recall, and determine the responsibility of enterprises in the cold chain.

5.1 Logistics Information Collection Function

According to the flow of information in the cold chain supply chain, the completion of fresh product tracking must include three stages of key information, namely, the production and processing process of fresh products, the time information of transportation and sales stage, the location information, and the electronic label information of fresh products. Through the establishment of the relationship between the three, the data is organically combined to determine the tracking information needed by the final data query platform. Among them, the information of the electronic tag includes the unique identification number of the pig in the breeding stage, including the temperature and humidity information generated in each process. There are also such data information in fresh distribution center and sales terminal. The reader and the antenna are used to write the data in the village label. Here, writing the data to the electronic label means updating the information change of the electronic label in the database where the electronic label number is located in real time. The electronic label is the carrier of information, and the reader is the tool for updating the information of the electronic label. The following describes how to read and store the information generated in the supply chain by electronic tags and readers, and how to bind all kinds of data together to achieve tracking management.

Firstly, each tag and reader has three unique dimensions: number, time information and location information.

(1) Introduction to the attributes of electronic labels:

ID number: The unique identification of a fresh product.

Time information: in the process of production, processing and circulation, fresh products need to go through inspection, freezing, sorting, refrigeration, transportation, sales and other links. Each link is the update time point of RFID information. The update of RFID information is interval, which takes the completion of operation as the node. It does not mean that the data should be updated in real time in the slaughtering link.

Location information: Refers to the specific geographical location of fresh products in each link of the cold chain, and the update method is the same as the time information.

(2) Introduction to reader properties:

ID number: The unique identification of a reader.

Time information: the RFID reader is installed in some links of inspection, freezing, sorting, refrigeration, transportation and sales of fresh products. When the electronic tag passes through these links, the reader will actively update the information of the electronic tag within its reading and writing scope. The specific time point of information update is the time information of the reader.

Location information: the specific geographic location of some fresh products in the supply chain where readers are installed.

5.2 Information Tracking Management Function

This paper describes in detail how the system captures all kinds of information needed, and then mainly introduces how to use this information to realize the monitoring and tracking management of the supply chain. At present, there are mainly two kinds of information exchange between enterprises in the cold chain supply chain: transfer type and sharing type. Among them, the transitive interaction means that each enterprise independently collects and stores relevant product information, and according to the unified coding rules and storage form, it will be extended to the next enterprise along with the supply chain process. The shared interaction means that the supply chain information is transferred to the third-party monitoring center, and the enterprise has no right to change the specific information content, and the monitoring center uniformly manages and maintains it. The data handover method proposed in this paper is a combination of the two. Enterprises need to record and save relevant information, and also provide key logistics information and temperature and humidity information for the third-party monitoring center.

5.3 Real Time Monitoring and Management Function of Refrigerated Vehicle

In view of the current operating environment of the cold chain, the logistics distribution link is the most vulnerable part. Therefore, it is necessary to strengthen the monitoring and management of refrigerated vehicles, so as to solve the problem of fresh products in normal temperature transportation. In order to ensure the integrity of the cold chain, RFID scanning equipment and humidity sensors are required to be installed on the left, right and top of the refrigerator car door to identify the transportation environment and logistics status of goods. One or two cameras need to be installed inside the cargo door to track the door opening and closing, and to record the loading and unloading conditions; GIS, GPRS system and communication system need to be installed in the cockpit to realize real-time data transmission. RFID tags and readers need to be installed in front of the vehicle to indicate the vehicle identity information and read the terminal store information. Through the fixed reader installed in the car, the temperature data in the car is read in real time and transmitted to the third-party monitoring center, which is responsible for the information exchange, storage and forwarding with the vehicle terminal in the driving ship, and the data is transmitted to the supermarket fresh center; at the same time, the whole logistics network is monitored and managed.

The system adopts positioning technology and combines with the third-party visual monitoring platform, which can strengthen the real-time monitoring of the cargo transportation process and effectively solve the current closed logistics mode. When each refrigerated vehicle carries out loading business, the identity information of the refrigerated vehicle will be bound to the remote logistics real-time visual monitoring system, and automatically match the cargo Waybill Information, mainly including the cargo list, receiving address, contact information, scheduled arrival time, vehicle online landmark, on-way door opening process, duration and other data. If the order information is not transmitted to the monitoring center after the loading operation, the system will send a warning to the production of logistics enterprises from time to time. On the way of logistics, the monitoring center can query and understand the working status of refrigerated vehicles at any time according to the relevant information received. The actual location of refrigerated vehicles is displayed in real time on the electronic map of the monitoring center. In case of emergency, the monitoring center can issue relevant safety service instructions to the target vehicle.

6 System Test and Analysis

System testing is a very critical step after the completion of a system, testing can see whether the system has achieved the purpose of the initial design from the design to the final implementation. Through the test, we found some problems in the system and how to improve them. This paper will test whether the communication between the server and the vehicle terminal is normal, and whether the data transmission is normal. Finally, test whether the overall function of the system achieves the expected effect.

6.1 Test Environment

Due to the limited conditions, only the laboratory can test, in the early stage of the test, we need to build a good test environment. The environment is as follows:

Server: Alibaba Cloud Server, the system is Windows Server 2012 R2 Data Center Edition 64-bit Chinese version. Database: MySQL5.5, built on the server, and created various data tables as required. Hardware equipment: vehicle-mounted equipment (responsible for detecting data and sending it to the server). System operation carrier: This system runs on the computer side, and the computer system is the Win7 flagship system.

During the test, first simulate the order placement operation on the computer, that is, place an order to the carrier through the simulation entrustment, and the carrier then processes the order to simulate the transportation process of the refrigerated truck. Then put the vehicle-mounted equipment in the laboratory to detect the temperature and humidity in the air to simulate the temperature and humidity in the cabin. Finally, test the key functions of the system.

6.2 Vehicle Tracking Test

(1) Test conditions: the carrier successfully logs in to the system.
(2) Test steps: click the add vehicle information menu item in the main interface, fill in the relevant information in the pop-up interface, and click the confirm button to complete the addition of vehicle information; right click the selected vehicle, and click to modify and delete functions in the pop-up menu item. Fill in the modification information in the modification interface and click OK to complete the modification.
(3) Expected results: the functions of adding, modifying and deleting vehicles are successfully completed.
(4) Actual results: after the actual test, the system successfully achieved the expected results.

6.3 Real Time Monitoring Test

(1) Test conditions: the vehicle terminal equipment works normally, and the server data management function is turned on.
(2) Test steps: check whether the vehicle terminal of the server data management interface is connected successfully. If the connection is successful, check whether there are data records. Then, the client enters the main interface to observe the temperature and humidity data in the interface and view the vehicle position through the map interface. Change the environment of vehicle terminal and observe whether the alarm is given.
(3) Expected result: the vehicle terminal successfully connects with the server and starts to transmit data. The client can observe the temperature and humidity data change and position change through the main interface. After the environment of vehicle terminal changes, you can view the alarm information by clicking the alarm information button.
(4) Actual result: the connection between vehicle terminal and server is normal. Temperature and humidity data and position display are normal. The alarm information is checked successfully.

The function of the temperature and humidity sensor in the vehicle terminal is to monitor the temperature and humidity information in the refrigerator car in real time, display it through the LCD terminal, and finally transmit it to the remote terminal through the GPRS module.

6.4 Information Collection Test

The temperature and humidity sensor test mainly verifies whether DHT 11 can complete the indoor temperature and humidity data acquisition. Here, the processor is connected with the PC through the serial port, and the data acquisition results are displayed on the serial port terminal in real time. The temperature and humidity data acquisition results are as Table 2.

Table 2. Temperature and humidity data collection results

Acquisition times/time	Current temperature/°C	Current humidity/%RH
1	28	31
2	29	30
3	29	30
4	29	30
5	30	31
6	29	31

According to Table 2, when the acquisition time is 6 times, the current average temperature collected by DHT11 temperature and humidity sensor is 29 °C, and the current average humidity collected is 30.5% RH. The experimental results show that DHT11 temperature and humidity sensor can complete the collection of temperature and humidity data, and the accuracy of temperature and humidity acquisition is high.

7 Conclusion

At present, the construction of cold chain logistics can not meet the needs of development. The main problems include incomplete cold chain system, backward hardware and software equipment, complex types of supply chain enterprises, scattered information systems, etc. At the same time, due to the low proportion of cold chain distribution, excessive food loss caused by normal temperature transportation and uncontrollable operation links, the food quality and safety problems are also aggravating. In view of the above problems, this paper designs a cold chain logistics information real-time tracking system based on wireless RFID technology. After verification, the functions of the system meet the needs. By using tag data sensing, reader data transmission and background server data management terminal, the real-time tracking system framework of cold chain logistics information is established, and the real-time temperature of food, vehicle running position and agricultural product information is obtained by using wireless RFID card reader to realize the real-time tracking and monitoring of cold chain logistics information. It can effectively improve the proportion of cold chain distribution, control transportation links and information, and ensure the quality of cold chain food. However, due to the limitation of time, experience and other conditions, the cold chain logistics tracking system proposed in this paper is too idealistic. In theory, this system can largely solve the shortage of the existing tracking system in the field of cold chain. In the future, there are many areas that need to be improved. First, the implementation cost of the current system may be too high, leading to some difficulties in the later promotion. Second, when the designed tracking system is combined with the business system, it will change the existing business process of the enterprise. Here we need to do further study to minimize the loss.

Fund Projects. The research results of 2020 Guangxi Young and middle-aged University Teachers' scientific research basic ability improvement project (Humanities and Social Sciences) research on the docking and circulation mode of agricultural supermarket in Baise City from the perspective of agricultural products cold chain, Project No.: 2020ky19027.

References

1. Urbano, O., Perles, A., Pedraza, C., et al.: Cost-effective implementation of a temperature traceability system based on smart RFID tags and IoT services. Sensors **20**(4), 1163 (2020)
2. Anandarajah, P., Collins, D.: A review of chipless remote sensing solutions based on RFID technology. Sensors **19**(22), 1–50 (2019)
3. Liu, H., Yao, Z., Zeng, L., et al.: An RFID and sensor technology-based warehouse center: assessment of new model on a superstore in China. Assem. Autom. **39**(1), 86–100 (2019)
4. Li, X., Yang, L., Duan, Y., et al.: Developing a real-time monitoring traceability system for cold chain of tricholoma matsutake. Electronics **8**(4), 423–429 (2019)
5. Abdelkarim, M., Naoui, S., Latrach, L., et al.: Design of a compact size tag antenna based on split ring resonator for UHF RFID application. **29**(7), 1–12 (2019)
6. Li, X., Du, B., Li, Y., et al.: RFID-based tracking and monitoring approach of real-time data in production workshop. Assem. Autom. **39**(4), 648–663 (2019)
7. Li, J.: Optimal design of transportation distance in logistics supply chain model based on data mining algorithm. Clust. Comput. **22**(1), 1–10 (2019)
8. Liao, H., Chang, J., Zhang, Z., et al.: Third-party cold chain medicine logistics provider selection by a rough set-based gained and lost dominance score method. Int. J. Fuzzy Syst. **22**(6), 2055–2069 (2020)
9. Jani, D.B.: The cold chain and its logistics. Sci. Technol. Built Environ. **2**(9), 14–16 (2020)
10. Falcón, V.C., Porras, Y.V.V., Altamirano, C.M.G., et al.: A vaccine cold chain temperature monitoring study in the United Mexican states. Vaccine **38**(33), 5202–5211 (2020)
11. Liu, S., Liu, D., Srivastava, G., et al.: Overview and methods of correlation filter algorithms in object tracking. Complex Intell. Syst. **3**, 161–165 (2020)
12. Fu, W., Liu, S., Srivastava, G.: Optimization of big data scheduling in social networks. Entropy **21**(9), 902 (2019)
13. Liu, S., Bai, W., Zeng, N., et al.: A fast fractal based compression for MRI images. IEEE Access **7**, 62412–62420 (2019)

Design of Human Resource Big Data Parallel Classification System Based on Social Information Cognitive Model

Bo Sun and Cai-ming Zhang(✉)

School of Labor Relations and Human Resource, China University of Labor Relations, Beijing 100048, China

Abstract. In order to improve the classification function and operation performance of human resources, a new big data parallel classification system is designed. The parallel processor is installed to optimize the analog-to-digital converter, human resource data storage and wireless communication network. This paper constructs a social information cognitive model, under which human resource data can be obtained in real time, and preprocessed by data cleaning, Chinese word segmentation and stop word elimination. Human resource data features are extracted, and the similarity between the extracted data features and standard data features is calculated to realize the parallel classification function of human resource big data. Through the system test experiment, the conclusion is drawn: compared with the traditional classification system, the recall rate and accuracy rate of the design system are increased by 5.5% and 3.5% respectively, and it has more advantages in classification speed.

Keywords: Social information cognitive model · Human resources · Resource big data · Parallel classification

1 Introduction

Human resource management is the main component of enterprises and institutions, which has a great impact on the business contacts and development prospects of enterprises. In order to adapt to the change and development trend of the current economic structure of enterprises and improve the work efficiency and economic benefits of enterprises, the State applies big data technology to human resources management decision-making in all walks of life, and realizes a more comprehensive, systematic, efficient and realistic decision-making technology by using the software systems and frameworks in the big data platform, so as to enhance the core competitiveness of enterprises and public institutions and meet the long-term development goals of enterprises [1]. Human resource is an important part of enterprises and universities, which is mainly used to manage the information of employees in enterprises and facilitate personnel management. Because of the large number of employees in enterprises, universities and the society, in order to achieve effective utilization of human resources information and

S. Liu and X. Ma (Eds.): ADHIP 2021, LNICST 416, pp. 454–469, 2022.
https://doi.org/10.1007/978-3-030-94551-0_36

build human resources data, enterprises can acquire human resources information in big data, and employees can also upload real-time human resources data to big data to achieve resource sharing.

Due to the large number of social personnel, human resources data is relatively large. In order to achieve the efficient management of human resources big data, the human resources classification system came into being [2, 3]. Data classification is to combine the data with some common attributes or features, and distinguish the data by its category attributes or features. In other words, the same content, the same nature of information and information that requires unified management are gathered together, and the different information and information that needs to be managed separately are distinguished, and then the relationship between each set is determined to form an organized classification system. According to the research status at home and abroad, mature big data classification methods include the data classification system based on fractional-order derivative [4], and the big data classification system based on firefly and simulated annealing [5]. In reference [4], the fractional gradient descent method is proposed, which is an unconstrained optimization algorithm using support vector machine to train the classifier with convex problem. Compared with the classical integer order model, the fractional order model has a significant advantage in speeding up the calculation speed. In this study, in order to study the current situation that these new optimization methods can realize fractional derivative in the classifier algorithm, the problem is qualitatively studied. The main purpose of reference [5] is to introduce the ofs algorithm supported by meta heuristic algorithm of MapReduce paradigm. A new hybrid multi-objective firefly and simulated annealing (hmofsa) algorithm is proposed to select the optimal feature set. Therefore, as the first step, the original big data set is decomposed into sample blocks in the map phase. Hmofsaa algorithm is used to select the selected features from the examples. Part of the results are combined into the final feature vector in the reduce stage, and the kernel support vector machine classifier is used for evaluation. However, the above two traditional classification systems are lack of the construction of information cognitive model, and the acquisition of data to be classified is not comprehensive enough, resulting in the recall and precision of data classification can not reach the ideal effect.

However, the traditional classification system has the problem of low classification efficiency, so a parallel classification system of human resource big data based on social information cognitive model is proposed. Based on the traditional classification system, the system adjusts the classification mode, uses parallel processing mode to replace serial processing mode, and introduces social information cognitive model. Among them, parallelism means that two or more programs are executed in the same time period, with time overlap. The information cognitive model refers to the reaction law of employees' attitude and behavior to the social cues provided by others when they interact with others. Through parallel processing and the application of social information cognitive model, we hope to improve the classification efficiency of human resource big data classification system.

2 Hardware System Design of Human Resource Big Data Parallel Classification

The human resource big data classification system mainly uses text classification algorithm to classify various texts according to their characteristics. The system realizes the classification of large-scale unstructured text. The system has the following design goals: to classify large-scale unstructured human resource big data. The system can realize the function of feature extraction and classification of massive unstructured human resource big data, and the overall time consumption is controlled in an acceptable range. Test the system function and performance. Through the system test, the accuracy of system classification and the overall performance of the system are evaluated, and the improvement scheme is put forward for specific problems. Modular function. The main functional layers of the system are completely independent as different modules, and the loose coupling between each module is maintained as far as possible, so as to optimize, update and secondary development of the system in the future. Based on the overall goal of the system, the basic principles of advanced technology and usability, openness and extensibility, high stability and reliability, integration and fast response are followed in the system design, and the optimal design of the parallel classification system of human resources big data is realized from three aspects: hardware, database and software function.

2.1 Parallel Processor

Parallel processor is the core part of the computer, which plays a very important role in the overall operation of the computer. Parallel processor mainly includes arithmetic logic unit, instruction register, program counter, address calculator, data selector, controller and register group. The overall structure design block diagram of 32-bit parallel processor is shown in Fig. 1.

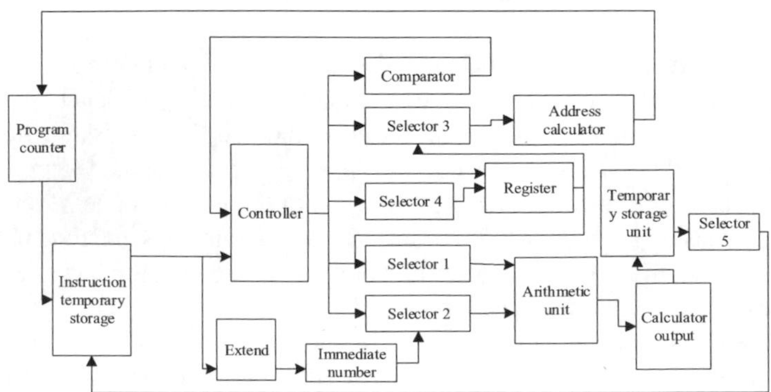

Fig. 1. Block diagram of parallel processor

The core idea of parallel processor pipeline is to divide each instruction into several smaller function segments, and each task is executed in parallel in time, so as to improve the efficiency of operation. Microprocessors are divided into five stages: instruction fetching stage, instruction decoding stage, execution stage, storage stage and write back stage. The instruction fetching stage includes instruction memory, instruction register, program counter and address calculator. In the value cycle, the computer takes out the data in the corresponding address from the memory according to the value of the program counter and sends it to the instruction register for temporary storage, so as to save the total time. Then, the program counter calculates the address of the next instruction to be executed. The address calculator is also a register whose function is to accept the address of the next instruction to be executed. The instruction decoding stage includes control unit, data selector and register group. During the decoding cycle, the instruction is sent from the instruction register to the decoder in the control unit, which is responsible for decoding the instruction and sending the generated control command to other corresponding parts of the processor. The data selector mux3 controls the output of the address calculator and sends the next address result to the program counter. Register group stores some operation data temporarily in register heap during CPU operation. Execution phase: including arithmetic logic unit and two data selectors. In the execution cycle, the decoding result of decoding phase is taken as the basis. The output data of the data selectors mux1 and MUX2 are sent to the arithmetic logic unit. Alu will select which two 32-bit operands to complete arithmetic, logic, shift and other arithmetic or logic operations. The results of other instructions are branch or jump instructions. After the operation, the output results are temporarily stored in aluout.

2.2 Analog to Digital Converter

The dual-channel analog-to-digital conversion chip ADS5402 of Texas Instruments is selected as the ADC chip in the hardware system. The ADC has a maximum sampling rate of 800MSPS, 12 bit resolution, data output interface mode supports LVDS level standard, and the maximum peak to peak voltage of input analog signal is 1 V. The dual channel design of ADS5402 chip fully ensures the synchronization of IQ two channel signals, and can ensure the signal performance of the whole system to meet the synchronization requirements. The analog-to-digital conversion chip ADS5402 supports LVDS level standards. DDR can be selected as the output signal format, which not only increases the anti-interference ability of D signal, but also greatly improves the reliability and transmission rate of D data. The input signal of ads5402 is differential form. The analog input signal received by the system is generally single ended. Therefore, before the input analog signal is connected to ads5402, the input single ended analog signal needs to be differential transformed and then connected to the analog-to-digital conversion chip. Two back-to-back 1:1 transformers are used to convert the single ended signal into differential signal, and the parasitic capacitance effect caused by the transformer is eliminated. For the two input channels of the ADC, the two circuits in the input conversion circuit are exactly symmetrical and the same.

2.3 Human Resource Data Storage

The total memory capacity of human resource big data parallel classification system is 64 kbit. Using 0.18 μm standard CMOS process, the working voltage of the chip is

3.3 V, the working temperature range is −55 °C–125 °C, and the key design indicators are read time and read threshold. Read time, defined as the time from the beginning of address change to the beginning of output read-out data, expressed as Access_ Time. Considering the requirements of the chip, different levels of reading time are designed in the circuit design process. The fastest reading time index of the chip is 45 ns at room temperature, and the rest are 55 ns/18.2 MHz and 65 ns/16.6 MHz respectively. The circuit parameters corresponding to the latter two different operating frequencies are all given corresponding options in the layout for verification during chip test. When the reading threshold is at least 20K, that is, when the equivalent resistance is below 20K Ω after anti fuse breakdown, it is considered to store "1", otherwise it is considered to store "0". There are 26 peripheral input/output signals, including enable control signal, address signal, data signal, power and ground signal. The internal function module is composed of address decoder, control logic, readout circuit, programming circuit, storage array and bidirectional data port module. The memory structure is shown in Fig. 2.

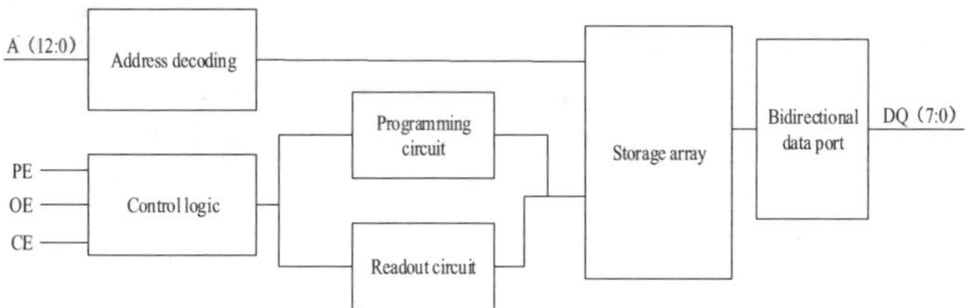

Fig. 2. Memory structure diagram

The 13 address signal lines A (12:0) of the memory are divided into three groups, which respectively control the block, row and column addressing of the memory array. The data signal line D (7:0) of the memory is a bidirectional port, that is, the programming data is input from this port, and the read-out data is output from this port. The number of ports indicates that the storage information width of the designed memory chip is 8 bits.

2.4 Wireless Communication Network

Wireless communication network adopts multi carrier modulation. Its basic principle is to divide high-speed data stream into low-speed data stream and transmit them on orthogonal subcarriers in parallel. In this way, the time dispersion of multipath fading is relatively reduced, and the frequency selective fading channel is transformed into several flat fading subchannels, which greatly reduces inter symbol interference. In wireless communication network, it is assumed that the system bandwidth is B, the number of subcarriers is N, the cyclic prefix is N_g sampling points, and the sampling period is T. If the transmission data of the first symbol period is X_i, then there are multiple carriers in the wireless communication network. Then the baseband equivalent signal can be expressed as:

$$s_l(t) = \sum_{k=0}^{N-1} X_{l,k}\phi_k(t - lT_s) \tag{1}$$

Where T_s is the length of the whole symbol including the cyclic prefix, and $\phi_k(t)$ is the carrier waveform function with frequency kB/N. According to the above communication principle, the human resource data of different regions are transmitted to realize the sharing of human resource data.

2.5 Component Chip Selection

FPGA Chip

FPGA is a kind of ASIC that allows users to configure and program independently. It supports the user to configure the system through the corresponding software. So through the control, we can complete the purpose of the system, and the field programmable gate array can also erase and write many times. Therefore, it greatly shortens the design cycle and R & D cycle of hardware products, and also greatly improves the flexibility of hardware implementation and reduces the cost of product design and R & D. The specific working principle of FPGA chip is based on the look-up table. Under the same hardware circuit conditions, the system related configuration files can be burned into the chip. Then, by querying the contents of the look-up table, we can achieve different logical purposes required by the system. Considering the data speed of ADC high-speed module and high-speed DAC module, XC5VLX330 chip is selected as FPGA chip.

Power Chip

The basic working principle of capacitive switching power supply is to make use of the energy storage characteristics of the capacitor, and store the input energy in the capacitor through the action of high-frequency switch through the controllable switch. When the switch is off, the electric energy is released to the load to provide energy. The output power or voltage is related to the duty cycle. Capacitive switching power supply can be used for step-up and step-down [6]. The internal FET switch array controls the charging and discharging of the fast capacitor in a certain way, so that the input voltage can be doubled or reduced by a certain factor, and the required output voltage can be obtained. Based on the traditional power supply chip, a linear regulator is embedded. The principle of the linear regulator is shown in Fig. 3.

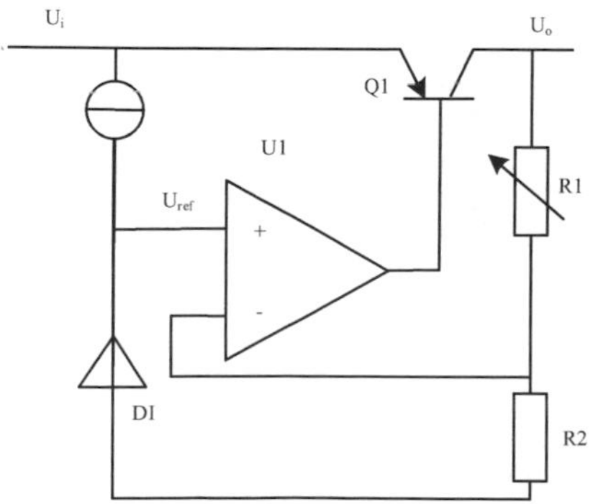

Fig. 3. Working principle of linear regulator

The sampling voltage is added to the non-inverting input of the comparator U1 and compared with the reference voltage U_{ref} added to the inverting input. After the difference between the two is amplified by the amplifier U1, the voltage drop of the series regulator is controlled to stabilize the output voltage. When the output voltage U_o decreases, the difference between the reference voltage and the sampling voltage increases, the driving current of the comparator amplifier increases, and the voltage drop of the series regulator decreases, so that the output voltage increases; If the output voltage U_o exceeds the required setting value, the front drive current of the comparison amplifier output is reduced, so that the output voltage is reduced.

3 Database Design of Human Resource Big Data Parallel Classification System

The system adopts ORACLE10G database, which is a large-scale database supporting JAVA. It is fully competent for the massive sharing of analysis platform data, and ORACLE10G's support for java is also satisfactory for the performance of system statistical queries. In the design of system database, the physical location of data storage should be determined first, then the data structure should be analyzed, and the relevant indexes and logs should be configured. After the above database analysis, the database concept and database table structure are analyzed [7]. The system is managed by SQL Server 2008, which can make the system more secure and reliable. In the human resource big data parallel classification system, the basic information management of employees is mainly to store the information of employees. If the information of employees needs tutors, the database will be updated in real time. The basic information of employees is shown in Table 1.

Table 1. Basic information of employees

Data name	Data description	Data type	Storage length
ID	Employee number (primary key)	Int	10
Name	Employee name	Varchar	20
Sex	Employee gender	Varchar	2
Date	date of birth	Date	8
ID number	ID number	Varchar	18
Place	Native place	Varchar	30
Tel	contact number	Int	20
Department	department	Varchar	30
Post	post	Varchar	20
Wages	base pay	Int	10

(continued)

Table 1. (*continued*)

Data name	Data description	Data type	Storage length
Form	Form of employment	Varchar	10
Education	Highest education	Varchar	20
University	University one is graduated from	Varchar	20
Date	Entry date	Date	8
State	On the job status	Varchar	10
Working years	working years	Int	5

In addition, the wage information table mainly stores the wage data information of employees, including the basic wage and performance wage of employees. The storage structure of salary information is shown in Table 2.

Table 2. Wage information

Data name	Data description	Data type
ID	Employee number (primary key)	Varchar(10)
Years	years	Int(10)
Bonus	bonus	Int(10)
Lunch	Lunch allowance	Int(10)
Transportation	Transportation subsidy	Int(10)

4 Software Design of Human Resource Big Data Parallel Classification System

4.1 Constructing the Cognitive Model of Social Information

Due to individual differences, that is, different work, life, learning environment and congenital differences, everyone's cognitive ability is in a relatively stable state. But when the brain is captured by scarcity, the focus will be relatively concentrated and the cognitive ability will be forced to decrease. Considering the changing process of users' cognitive ability. In the research of information retrieval cognitive model, it is necessary to analyze the whole cognitive process of users. In the design of information retrieval system, human intervention can make users' information needs more accurately expressed and interact with the system more tacit, so as to improve the utilization rate of information retrieval system. Figure 4 shows the dynamic change of users' cognitive ability affected by work and task situations.

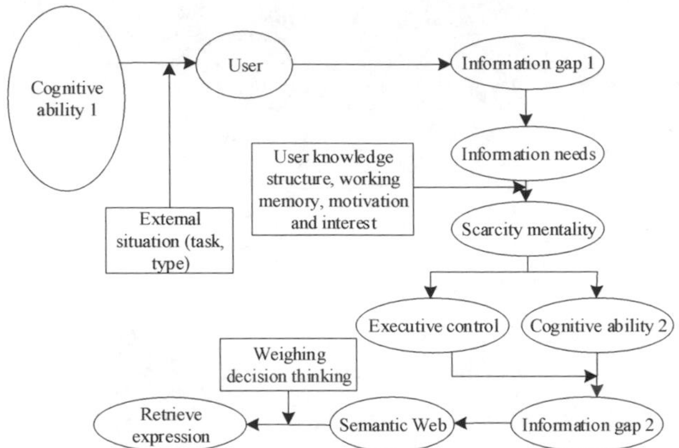

Fig. 4. Dynamic change of user cognitive ability

When users perform information retrieval, there will be an information gap between the user's knowledge structure and the task context. This difference is the sum of the knowledge that users objectively need to solve a certain problem [8]. Usually, when users are placed in the context of work tasks, they will have a scarcity mentality because they have less knowledge than they need. This mentality will capture the user's brain, making the bandwidth available to the user narrower, lacking insight and foresight. At this time, users use the external tool information retrieval system to acquire knowledge and make up for the lack of knowledge state when they solve the problem. When users express their needs to the information retrieval system, due to the capture of users' attention by scarcity mentality, the information needs expressed by users at this time will be less than the objective and actual needs needed to solve the problem. In this state, when the user interacts with the system, the information that the system feeds back to the user can not completely solve the current problems faced by the user. In order to solve this problem, it is necessary to analyze the user's cognitive ability and the ability to express their needs, find its root, and solve the impact of the dynamic change of user's cognitive ability on the search results. From the perspective of scarcity theory, this paper analyzes the dynamic change process of users' cognitive ability, and constructs a social information cognitive model. The model includes three layers: internal cognition layer, trade-off decision-making layer and human-computer interaction layer. The internal cognitive layer analyzes the change process from the user's objective knowledge needs to the user's cognitive needs. The trade-off decision-making layer is a process in which users form semantic network under the influence of weakened executive control and trade-off decision-making thinking. The internal cognitive layer and the trade-off decision-making layer constitute the user's cognitive process. The human-computer interaction layer is a process in which users express their needs in the form of retrieval after cognition and match with the information retrieval system, so as to obtain the required information.

4.2 Real Time Access to Human Resource Data

Multipoint data collection refers to the use of multiple servers at the same time from the same or multiple different data repository to obtain data in the form of web pages. For example, for human resource data, the storage structure and storage format of data in different resource databases are different, but the data will contain the following fields: company name, personnel name, position name, position description, position demand, job search keywords, address, publishing time, age, province, city, county, education background, etc. The collected data must undergo many structural and format transformations before further statistical research. The process of data collection is to directly save the web page in the form of text data locally, and then save it in the database in the form of binary. The purpose of multi-point data acquisition is to quickly collect the data from multiple remote data resource databases to the local database, saving valuable time. In order to dynamically obtain the data needed on the web page and download it in time, we should first set the corresponding conditions to process the data, then query through the trigger time of the database, and the database will immediately compare the relevant data to get the correct data. The data acquisition system processes the relevant pages through DOM technology, and controls the pages according to the corresponding requirements. For example: according to the theory of query data, the condition of the page is analyzed, and the data in the page is initialized according to the customer's needs. These values are processed dynamically in the background according to the needs. At the same time, various events on the page, such as button box link events, are set. According to the requirements of query events, start the query process to query the data effectively. Download the query data and save it to local in the form of web page. If the data is displayed in multiple pages, the corresponding new data can be found according to different page numbers.

4.3 Human Resource Big Data Processing

Human resource big data preprocessing includes three steps: data cleaning, Chinese word segmentation and stop word removal. In the process of data cleaning, we first get the data of the page to be cleaned to judge whether it needs to be cleaned. If there is data to be cleaned, we call the template framework of the page to be cleaned to get the structure description of the page header. Check the header structure and page control module, and output the initial processing results of HR big data after cleaning. Word segmentation is an important step in text preprocessing. In English, there are spaces between words for segmentation, but Chinese is composed of continuous strings without obvious separators, so it is necessary to perform word segmentation on strings [9]. After depunctuation, word segmentation and other operations are performed on all samples in the training text set, the word list of each sample is obtained. The set of the list and column of the obtained words constitutes the text feature dictionary. In order to reduce the data noise caused by the text representation dictionary in the feature vector, improve the ability of text discrimination, and reduce the search range of the representation dictionary in training. To improve the classification accuracy and training efficiency of the classifier, before representing the sample as feature vector, the text representation dictionary will generally implement operations such as "taking root" and "removing stop words". Stop words usually refer to high-frequency words that appear frequently in various categories of documents at the same time, and are considered to be unfavorable for classification. Generally, they are some prepositions, pronouns, conjunctions, etc. Filtering out the words in the stop-word list from the text representation dictionary is the ultimate goal of de-stopping.

4.4 Extracting Human Resource Data Features

The human resource data feature extraction prepared by the data simulation layer, and the selection of human resource feature words to construct the human resource data feature vector in the language vector model is the core part of the classification process. There are three steps for feature extraction: word frequency statistics, word weight calculation and feature word sorting. Count the frequency of a given word in a given file, that is, calculate the word frequency of the given word. At the same time, count the number of documents in which the word appears in all documents, that is, calculate the inverse word frequency of the word. The calculation formula of word frequency statistics is as follows:

$$\begin{cases} tf_{i,j} = \frac{n_{i,j}}{\sum_k n_{k,j}} \\ idf_i = \lg\left(\frac{D}{\{d:d \in t_i\}}\right) \end{cases} \tag{2}$$

In the above formula, $n_{i,j}$ is the number of times that the word appears in the human resource big data file d_j, and the denominator is the sum of the number of times that all the words appear in the file jd. $|D|$ is the total number of files in the corpus, and $\{d : d \in t_i\}$ is the number of files containing the word t_i. Word weight is used to evaluate the importance of a word in an article. In TF-IDF feature extraction algorithm, word weight is calculated by using the product of tf and idf. As shown in Formula 3.

$$Weight_i = tf_i \times idf_i \tag{3}$$

Taking into account the differences of the domains in the document, the vector normalization process is performed to convert the feature vector into a unit vector, which is convenient for the subsequent similarity calculation.

4.5 Realize the Parallel Classification Function of Human Resource Big Data

The collected human resource big data features are expressed as a central vector, which is obtained by the arithmetic average method or the weighted arithmetic average method. In the process of classification, no training is needed. Given an unknown class of text, the distance between its center vector and the center vector of each known class is calculated. Calculate the center vector of each category of text, the center vector calculation method is generally used to calculate the arithmetic average [10–12]. After the arrival of new text, the center vector is calculated by the same method. Before calculating the center vector, the text preprocessing such as word segmentation and removing stop words is needed. The similarity between the new text and each class is calculated by the following formula:

$$Sim(d_i, d_j) = \frac{\sum_{k=1}^{m} w_{ik} \times w_{jk}}{\sqrt{\left(\sum_{k=1}^{m} w_{ik}^2\right)\left(\sum_{k=1}^{m} w_{jk}^2\right)}} \tag{4}$$

Where d_j is the center vector of class j, d_i is the new text feature vector, and the dimension of the feature vector is. The similarity between the new text and each known category is calculated, and the category with the highest similarity is regarded as the classification result of the new text.

5 System Test

The design of human resource big data parallel classification system based on social information cognitive model is to provide auxiliary tools for human resource management of enterprises. In the process of system testing, we need to follow the following aspects. The system is an enterprise human resource system. Testers should include developers and users of the system, so that errors can be easily detected and corrected in time. When testing the human resource management system, the probability of finding errors is proportional. In the process of system development, it is inevitable that there will be errors, so we need to test the system. In addition to developers, system testers also need users to test. When designing test cases, we should not only design effective test cases, but also set unreasonable data for testing.

5.1 System Test Environment

According to the basic requirements of developing software information system, the minimum configuration requirements of test environment are determined. Two IBM high-performance servers, application server and database server, need to be installed on the server side. The operating system and database version are Windows Server 2003 and SQL Server 2005 respectively. The hardware configuration of the client part is p43.0, 4G or more memory, 160 g hard disk, and the operating system and communication network are Windows XP and 100M Ethernet LAN respectively.

5.2 Describe the Test Data Set

The data set used for training in the system test is to use the web crawler to crawl the news information in the human resource management website which has been manually edited and labeled. And through the system unstructured simulation preprocessed data set. It includes six categories: basic information of employees, salary, social insurance, enterprise position, labor cost and training work. The size of human resource data of each category is 5 GB. The documents in each category are manually marked to indicate their category. The data set used for big data classification test is the classification document downloaded from the laboratory. The data comes from the news corpus edited and manually sorted by all human resource management websites in a certain region, corresponding to the six classifications in the training document.

5.3 System Test Process

First of all, the functional test of the system should make the test user, and then test the function of the system, and use the actual user to log in and operate the test. The performance test of the system is to quickly complete the system function when the system is used. System testing is a link before the system test run. System testing is also a relatively important part of the system development process. System testing can ensure the quality of the system. The main methods of system testing include black box testing and white box testing. The system is carried out under the enterprise human resource management system, using black box testing methods. Through the black box test, a black box can be regarded as a running program after it is opened, and the test can be carried out without considering the internal environment. Through the operation of relevant hardware equipment, the main operation interface of the human resource big data parallel classification system is obtained.

In order to form the experimental comparison, the big data system based on fractional derivative, the big data system based on firefly and simulated annealing were used as the experimental control group. The experimental results are compared with those of the system designed in this study. Among them, the traditional human resource data classification system mainly uses the serial classification method to classify and store the input human resource data one by one. The classification system based on multi-classifiers uses multi-classifier equipment on the basis of traditional systems to achieve parallel processing of human resource data to a certain extent.

5.4 System Test Results

This system test experiment mainly conducts specific tests from two aspects: classification function and system performance. The evaluation of the classification results can reflect the quality of the classification algorithm from one side, and at the same time reflect the overall performance level of the classification system. Among them, the system function test evaluates the quality of the system text classification by calculating the recall rate and precision rate. The recall rate refers to the ratio of the number of samples correctly judged by the classifier as the class to the total number of samples belonging to the class. Accuracy refers to the proportion of samples that the classifier judges to be of this class, which actually belong to the class. Calculated as follows:

$$\begin{cases} recall = \frac{A}{A+C} \\ precision = \frac{A}{A+B} \end{cases} \tag{5}$$

In the formula, A represents the number of documents that belong to this category in the original manual classification standard and are also classified into this category by the classifier. B represents the number of data that did not belong to this category in the original manual classification standard but were classified into this category by the classifier. C represents the number of data that should have been classified into this category but were classified into other categories. After the operation of the three classification systems, the quantitative test results are obtained and substituted into formula 5 to get the test results that can reflect the classification function of the system, as shown in Table 3.

Table 3. System classification function test results

Experiment number		1	2	3	4	5
Big data system based on fractional derivatives	A	4.47	4.54	4.62	4.48	4.55
	B	0.38	0.25	0.27	0.28	0.24
	C	0.15	0.21	0.11	0.24	0.21
Big data system based on firefly and simulated annealing	A	4.82	4.77	4.83	4.85	4.79
	B	0.09	0.11	0.07	0.11	0.12
	C	0.09	0.12	0.1	0.04	0.09
Design of human resource big data parallel classification system	A	4.97	4.96	4.94	4.98	4.95
	B	0.02	0.02	0.03	0.01	0.02
	C	0.01	0.02	0.03	0.01	0.03

Through the statistics of multiple experimental data, it is found that the classification recall rate and accuracy rate of big data system based on fractional derivatives are 94.1% and 96.1%, respectively. The recall and accuracy of the big data system based on firefly and simulated annealing are 98.0% and 98.2% respectively, while the average recall and accuracy of the designed classification system are 99.6%. In addition, the statistical results of system classification time are shown in Table 4.

Table 4. System performance test results

Experiment number	Classification time of big data system based on fractional derivatives/s	Classification time of big data system based on firefly and simulated annealing/s	Design of human resource big data parallel classification system classification time/s
1	2.33	1.57	0.18
2	2.35	1.63	0.53
3	2.17	1.49	0.49
4	2.28	1.52	0.32
5	2.36	1.55	0.26

According to Table 4, the average classification time of the three data classification systems is 2.3 s, 1.6 s and 0.4 s respectively. To sum up, the classification function and operation performance of the designed human resource big data parallel classification system based on social information cognitive model are higher than the two comparison systems, that is, the designed classification system has more application advantages.

6 Conclusion

The human resource big data parallel classification system based on social information cognitive model is easy to operate. Data processing is simple and effective, and its association rules provide an important scientific method for students to understand the development of different echelons of human resources, establish a scientific talent evaluation mechanism, and clarify the age structure of the teaching staff. Of course, the application of data mining technology in the enterprise human resource management system needs to be improved, such as optimizing the algorithm of data mining, data mining from different angles such as multi-layer and multi-dimensional, so as to develop and optimize the allocation of social and enterprise human resources more reasonably and scientifically.

References

1. Ramadhany, S.E., Azwir, H.H.: Developing database system for managing medical insurance claims in the human resource department. PERFORMA Media Ilmiah Teknik Industri **19**(2), 213–230 (2020)
2. Gurmu, A.T., Ongkowijoyo, C.S.: Predicting construction labor productivity based on implementation levels of human resource management practices. J. Constr. Eng. Manage. **146**(3), 04019115.1–04019115.13 (2020)
3. Fu, W., Liu, S., Srivastava, G.: Optimization of big data scheduling in social networks. Entropy **21**(9), 902 (2019)
4. Hapsari, D.P., Utoyo, I., Purnami, S.W.: Unconstrained optimization based fractional order derivative for data classification. J. Phys. Conf. Ser. **1613**(1), 012066 (2020)
5. Gayathri Devi, S., Sabrigiriraj, M.: A hybrid multi-objective firefly and simulated annealing based algorithm for big data classification. Concurrency Comput. Pract. Exp. **31**(14), e4985.1–e4985.12 (2019)
6. Yuan, Y.: Research on the reconstruction of human resource management based on computer big data and artificial intelligence technology. J. Phys. Conf. Ser. **1574**(1), 012075 (2020)
7. Bragina, D., Molodchik, N.: Big data in human resource management. Manage. Pers. Intellect. Res. Russ. **9**(3), 76–80 (2020)
8. Novianti, W., Qadri, F.: Design of information system human resource management. IOP Conf. Ser. Mater. Sci. Eng. **879**, 012126 (2020)
9. Liu, S., Li, Z., Zhang, Y., Cheng, X.: Introduction of key problems in long-distance learning and training. Mob. Netw. Appl. **24**(1), 1–4 (2019). https://doi.org/10.1007/s11036-018-1136-6
10. Devarriya, D., Gulati, C., Mansharamani, V., Sakalle, A., Bhardwaj, A.: Unbalanced breast cancer data classification using novel fitness functions in genetic programming. Exp. Syst. Appl. **140**, 112866 (2020). https://doi.org/10.1016/j.eswa.2019.112866

11. Mousa, S.K., Othman, M.: The impact of green human resource management practices on sustainable performance in healthcare organisations: a conceptual framework. J. Cleaner Prod. **243**, 118595 (2020)
12. Liu, S., Bai, W., Zeng, N., et al.: A fast fractal based compression for MRI images. IEEE Access **7**, 62412–62420 (2019)

Speech Signal Feature Extraction Method of Tibetan Speech Synthesis System Based on Machine Learning

Ze-guo Liu(✉)

Key Lab of China's National Linguistic Information Technology, Northwest Minzu University, Lanzhou 730000, China

Abstract. In order to improve the accuracy of Tibetan speech synthesis, a feature extraction method of Tibetan speech synthesis system based on machine learning is proposed. Based on the analysis of Tibetan speech text content, the construction of speech synthesis system is realized. By judging the level of Tibetan prosody, a synthetic encoder is designed to realize the feature extraction of Tibetan speech signal. According to the experimental results, under the condition of normal speaking speed and identical Tibetan speech content, the Tibetan speech synthesized by the speech signal feature extraction method of Tibetan speech synthesis system based on machine learning is more accurate.

Keywords: Machine learning · Tibetan · Synthesis system · Signal extraction

1 Introduction

Tibetan language is a national language with a long history. Tibetan language is mainly divided into: Wei Tibetan dialect, Anduo and Kangxi dialect. Tibetan language is not only used in Tibet, Qinghai, Gansu, Sichuan, Yunnan and other parts of China, but also used in Nepal, India and other countries. Tibetan Lhasa dialect is a kind of Wei Tibetan dialect, which is mainly used by Tibetan people in Lhasa city and its surrounding areas [1]. Lhasa dialect is the most used and influential Tibetan dialect in Tibetan areas. Therefore, Tibetan Lhasa dialect is also known as Tibetan "Putonghua", and its pronunciation has some basic characteristics, such as: there are no voiced initials and blocking initials in Lhasa dialect, and complex consonant initials are relatively rare [2, 3]. In reference [4], a speech signal feature extraction method based on EMD for Tibetan speech synthesis system is proposed. Firstly, the speech signal is decomposed into several intrinsic mode components by using empirical mode decomposition (EMD), and then these components are processed by FFT to obtain more detailed signal division, The obtained speech sequence of Tibetan speech synthesis system is mixed with the first-order difference and short-term energy features of Tibetan speech synthesis system, and then used in the following experiments. The experimental data show that the speech recognition rate of this algorithm is significantly higher than that of traditional Tibetan speech synthesis system

S. Liu and X. Ma (Eds.): ADHIP 2021, LNICST 416, pp. 470–479, 2022.
https://doi.org/10.1007/978-3-030-94551-0_37

under different test environments. However, the effect of Tibetan speech synthesized by this method is not good.

In view of the above problems, this paper proposes a speech signal feature extraction method based on machine learning for Tibetan speech synthesis system. Tibetan speech synthesis system based on machine learning is a technology that uses computer and some devices to realize text to speech. The aim of the feature extraction method of Tibetan speech synthesis system based on machine learning is to endow human machine with the ability of speech, so as to realize the speech communication between human and machine. Speech signal feature extraction of Tibetan speech synthesis system is an interdisciplinary subject, which involves linguistics, psychology, machine learning and many other fields. The typical Tibetan speech synthesis system mainly includes two parts: the front-end and the back-end.

2 Speech Signal Feature Extraction Method for Tibetan Speech Synthesis System

2.1 Tibetan Text Analysis

Based on machine learning Tibetan speech synthesis system, a prototype system with Tibetan speech synthesis is designed in the embedded platform, and Tibetan text content is input to the system, and the output will be continuous and fluent Chinese Tibetan speech with a certain degree of naturalness [5]. The feature extraction of Tibetan speech synthesis system based on machine learning adopts Xscale px255 of inter company. The specific structure of Tibetan speech synthesis system based on machine learning is shown in Fig. 1.

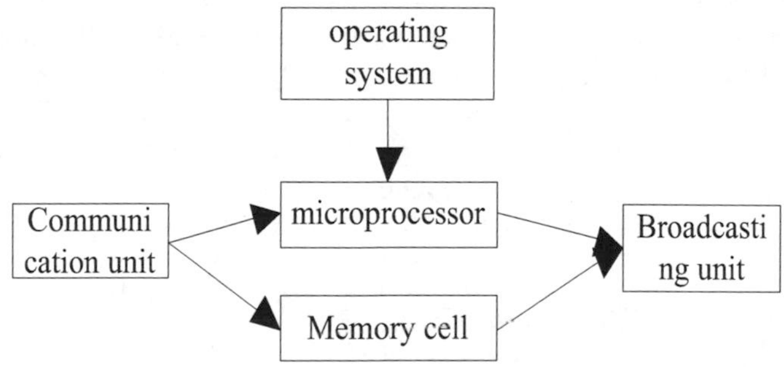

Fig. 1. Tibetan speech synthesis system

Based on machine learning Tibetan speech synthesis system, the feature extraction of speech signal is realized by arm 5te kernel technology. It provides 16 DMA channels to provide data for peripheral devices, and each channel has a special FIFO. When more than half of the FIFO data is received, DMA is triggered for data transmission, which can not only transfer the data to the memory quickly, but also greatly improve the efficiency [6].

The feature extraction of typical Tibetan speech synthesis system mainly includes two modules: front-end text analysis and back-end speech synthesis [7]. The front-end of Tibetan speech synthesis system is mainly responsible for analyzing the input Tibetan text, and then extracting the information needed by the back-end modeling. Therefore, the back-end of Tibetan speech synthesis system builds an acoustic model according to the front-end text processing results. The speech signal feature extraction method of Tibetan speech synthesis system based on machine learning. The specific speech signal feature extraction structure of Tibetan speech synthesis system is shown in Fig. 2.

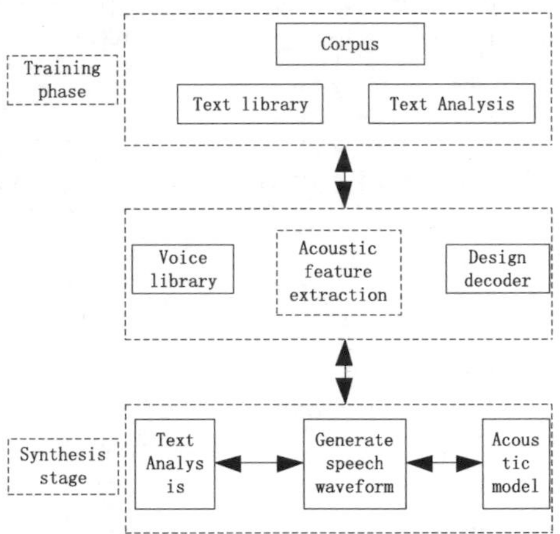

Fig. 2. Structure diagram of speech signal feature extraction in synthesis system

Based on machine learning Tibetan speech synthesis system, speech signal feature extraction is designed into two stages: training and synthesis [8]. The training stage is to analyze the Tibetan text in the corpus, send it to the encoder module, extract the acoustic characteristic parameters from the speech database, and then send it to the decoder; get the acoustic model through the decoder; in the Tibetan speech synthesis stage, send the text to be synthesized into the acoustic model through text analysis, and then generate the speech waveform corresponding to the text sequence through the acoustic model.

2.2 The Judgment of Tibetan Prosodic Level

Prosody is mainly an auditory feature and a psychological quantity. Prosody contains the speaker's intention information and the hearer's perception information, which is very useful in helping the hearer understand the language and intention [9]. The speech signal feature extraction of Tibetan speech synthesis system based on machine learning is described by its corresponding acoustic features, such as fundamental frequency, duration, amplitude and frequency spectrum, and four speech auditory features, such as pitch, duration, intensity and timbre. In addition, the appropriate pause in Tibetan speech synthesis is also a very important component of prosody [10]. The basic structure of Tibetan texts is shown in Table 1.

Table 1. Classification of Tibetan characters

Positive addition			Add after		
Character character	Timbre	Pronunciation	Character character	Timbre	Pronunciation
Positive	Strong	Airless sound	Positive	Strong	Airless sound
Neutral	Neutralization	Voiceless aspirated	Neutral	Neutralization	Voiceless aspirated
Negative	Weak	Voiceless consonant	Negative	Weak	/
	Very weak	Secondary voiceless		/	/

The prosodic feature of Tibetan speech synthesis system is not only to complete the pronunciation of consonants and vowels, but also to pay attention to the factors of Tibetan tone, strength and duration. However, these factors can not exist alone, but are attached to the consonants and vowels in Tibetan speech synthesis system [11]. The prediction and judgment of Tibetan prosodic level is to better realize the feature extraction of Tibetan speech synthesis system based on machine learning.

The characteristics of Tibetan prosody are to convey information about fundamental frequency, duration change and amplitude [11]. It is difficult to measure, model and simulate the intonation and stress in Tibetan speech synthesis system. In the speech signal feature extraction of Tibetan speech synthesis system based on machine learning, the prosodic levels from small to large are: last position, syllable, step, phonological copula, attached morpheme phrase, phonological phrase, intonation phrase and prosodic sentence.

Generally, they are simplified into prosodic words, prosodic phrases, intonation phrases, and prosodic sentences [12]. A smaller prosodic component is contained in a larger prosodic component, which forms the prosodic hierarchical structure of Tibetan

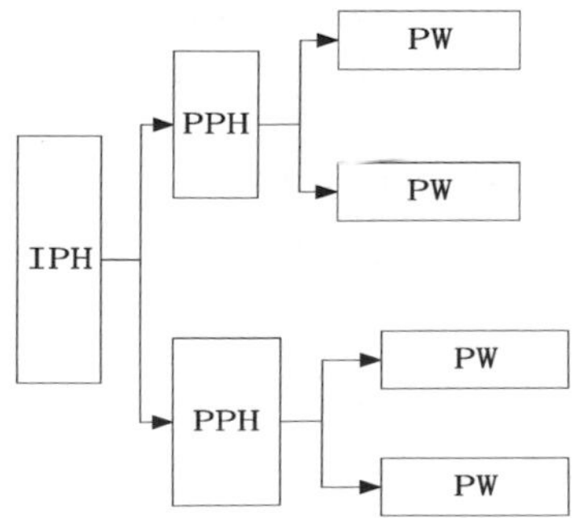

Fig. 3. The prosodic level of Tibetan speech

speech. PW is a prosodic word; PPH is a prosodic phrase; IPH is a intonation phrase. The specific prosodic hierarchy is shown in Fig. 3.

The speech signal feature extraction of Tibetan speech synthesis system based on machine learning is from the point of view of the pronunciation position and pronunciation method of Tibetan natural speech, but the prosodic structure and grammatical structure of Tibetan are not completely consistent. In the Tibetan speech synthesis system, there is a certain close relationship between the components of Tibetan prosodic words and the part of speech features, between the pronunciation position of stress and the syntactic structure, and between the pause of prosodic boundary and the characteristics of syntactic structure and the part of speech [13]. Therefore, grammar in prosodic structure plays an important role in feature extraction of Tibetan speech synthesis system based on machine learning.

In the speech signal feature extraction of Tibetan speech synthesis system based on machine learning, with the continuous development of computer, the in-depth exploration of Tibetan text information is also developing. Similar to Chinese speech synthesis information processing, Tibetan word segmentation also plays an important role in Tibetan information processing [14]. The speech signal feature extraction of Tibetan speech synthesis system based on machine learning can analyze and process the words, words, phrases, sentences, semantics, even the whole text or language in Tibetan prosody. It can also realize the automatic segmentation of Tibetan prosody. It can not only mark some grammatical features of grammatical words such as part of speech and word length automatically, but also do not need too much manual repair The efficiency is greatly improved.

2.3 Feature Extraction of Speech Signal Based on Machine Learning

The speech signal feature extraction of Tibetan speech synthesis system based on machine learning, the feature of Tibetan speech signal obtained by its encoder is linear prediction residual signal approximated layer by layer, and the synthesized speech quality increases with the rate. The Tibetan speech synthesis system based on machine

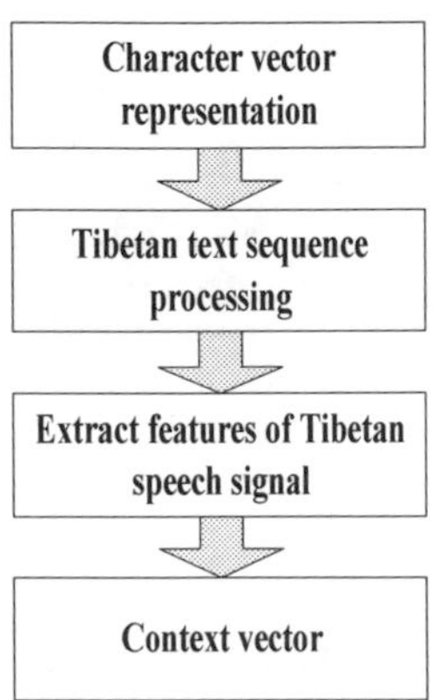

Fig. 4. Framework of Tibetan speech synthesis encoder

learning transforms a variable length Tibetan text sequence into a fixed length feature vector [15]. At the same time, the encoder in the system obtains the output value context vector of the encoder by inputting the character vector, preprocessing the Tibetan text sequence in the hidden layer, and extracting the text feature network (Fig. 4).

In the Tibetan speech synthesis system based on machine learning, the design of adaptive codebook structure plays an important role in improving the Tibetan speech synthesis encoder. The previous Tibetan speech synthesis algorithm only uses the excitation signal of the core layer to update the adaptive codebook buffer, thus abandoning the excitation signal of the enhancement layer 1 or the enhancement layer 2 with better quality [16]. Therefore, the new algorithm for feature extraction of Tibetan speech synthesis system based on machine learning is the core layer, enhancement layer 1 and enhancement layer 2, and a buffer is set respectively to update the corresponding adaptive codebook with the excitation signal of each layer, so as to achieve the best matching of coding parameters of each layer.

In order to reduce the complexity of the algorithm, the first level target vector of the core layer is analyzed by closed-loop pitch analysis, and then the best integer and fractional pitch delay are obtained. According to the analysis results of common core layer, enhancement layer 1 and enhancement layer 2, it is clear that the optimal pitch delay used in the three layers of interpolation is the same in the Tibetan speech synthesis system based on machine learning.

According to the different adaptive codebook buffers of each layer, the excitation vectors of each layer are obtained by interpolating the past excitation at the optimal pitch delay, which are recorded as vn, vn_{12} and vn_{16} respectively. At the same time, the feature extraction gain of Tibetan speech synthesis system based on machine learning is calculated. In order to save the number of coding bits, the adaptive codebook gain of the core layer is taken as the common gain of the adaptive codebook of each layer of the current subframe, which is recorded as g_p. Vector xn, xn_{12}, xn_{16}, get the specific expression, see Formula 1:

$$\left\{ \begin{array}{l} Xn_2(n) = xn(n) - g_p \cdot y_1(n) \\ Xn_{2-12}(n) = xn_{12}(n) - g_p \cdot y_{1-12}(n) \\ Xn_{2-16}(n) = xn_{16}(n) - g_p \cdot y_{1-16}(n) \end{array} \right\} \tag{1}$$

Where, $y_1(n) = vn * h(n)$ is the convolution of the core layer's adaptive codebook excitation vector and the perceptual weighted synthetic filter's unit impulse response; $y_{1-12}(n) = vn_{12} * h(n)$ is the convolution of the enhancement layer's adaptive codebook excitation vector and the perceptual weighted synthetic filter's unit impulse response; $y_{1-16}(n) = vn_{16} * h(n)$ is the convolution of the enhancement layer's adaptive codebook excitation vector and the perceptual weighted synthetic filter's unit impulse response.

Therefore, in the new algorithm of speech signal feature extraction of Tibetan speech synthesis system based on machine learning, the adaptive codebook structure makes full use of the feature that the Tibetan speech synthesis coding can generate multiple excitation signals; the excitation signals of different quality adaptive codebooks are obtained, and are respectively used for searching algebraic codebooks at different levels [17]. In the Tibetan speech synthesis coder based on machine learning, the algorithm complexity of the candidate coder is measured by the execution of millions of operations per second, and the operations such as addition, subtraction, multiplication and division are weighted according to different weights. The specific statistical results are shown in Table 2.

Table 2. Storage complexity of candidate encoders

/	/	Storage complexity (kwords)
ROM	Coding end	14.92
	Analytic end	1.247
	Total	16.177
RAM	Coding end	13.185
	Analytic end	6.273
	Total	19.458

In the feature extraction of Tibetan speech synthesis system based on machine learning, in order to prevent over fitting of acoustic parameters in Tibetan speech synthesis system, a series of nonlinear transformations are carried out for each input frame through calculation. In the Tibetan speech synthesis coder based on machine learning, there are two hidden layers, and the layers are completely connected. Therefore, when the Tibetan speech output in the decoder is not directly converted to audio, it is necessary to introduce the machine learning output into the form of waveform expression.

3 Simulation Experiment Analysis

3.1 Experiment Preparation Activities

Speech signal feature extraction of Tibetan speech synthesis system based on machine learning, in which Tibetan speech synthesis refers to converting text information into speech output. The construction of Tibetan speech synthesis corpus plays an important role in the design of the system and the realization of feature extraction of Tibetan speech signal based on machine learning. The construction process of the corpus in the Tibetan speech synthesis system is shown in Fig. 5.

The corpus of the Tibetan speech synthesis system aims to study, develop and evaluate the synthesis and recognition of Tibetan. In this experiment, the feature extraction experiment of Tibetan speech synthesis system based on machine learning is carried out. Based on the initial manual segmentation of the half syllable boundary in the phonetic database, the HTK toolkit is used as the basic unit. After the special training, the personnel have manually checked and adjusted. At the same time, the basic frequency parameters of each syllable are modified, and the whole sound bank is marked with Tibetan prosody structure and stress level.

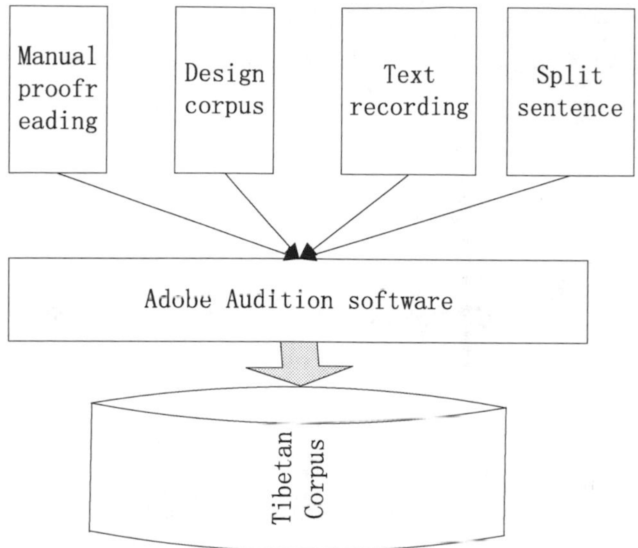

Fig. 5. Construction process of Tibetan speech synthesis corpus

3.2 Experimental Process

The front-end part of Tibetan speech synthesis system based on machine learning mainly includes text analysis module. It mainly extracts the information needed for the back-end modeling of the system from the text through the process of analyzing the language features of Tibetan text, checking the grammar rules, and searching the semantic data. Therefore, the back-end part of Tibetan speech synthesis system uses machine learning or neural network technology to build acoustic model of speech according to the front-end text analysis results, and generates target speech by using text information and trained acoustic model.

In the Tibetan speech synthesis system based on machine learning, the corpus used is the human-computer speech interaction group. In the process of traditional Tibetan corpus construction, every step of text design, text recording and sentence segmentation needs careful manual supervision and proofreading, which requires a lot of energy and time. In the Tibetan speech synthesis system based on machine learning, the application of modern technology shortens the synthesis time and improves the work efficiency.

3.3 Experimental Result

Under the condition of normal speaking rate and the same Tibetan speech content, through praat's speech analysis software, the traditional method 1 and traditional method 2 and the method of this paper are used to extract the voice signal characteristics of the Tibetan speech synthesis system, and the two are verified. The accuracy of Tibetan speech synthesis of the method, the specific comparison result is shown in Fig. 6:

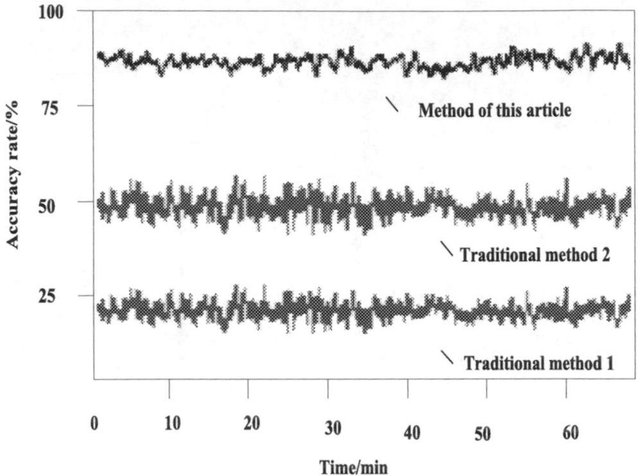

Fig. 6. Tibetan speech synthesis comparison results under different methods

According to the comparison results in the figure, the traditional method 1 extracts the features of the voice signal of the Tibetan speech synthesis system, and the accuracy of Tibetan speech synthesis is within 30%, and the traditional method 2 extracts the features of the voice signal of the Tibetan speech synthesis system. The accuracy is less than 60%. The Tibetan speech synthesis accuracy after extracting the speech signal features of the Tibetan speech synthesis system with the method in this paper is less than 90%, indicating that the Tibetan speech synthesized by the method in this paper is more accurate. Therefore, the method for extracting features of the speech signal of the Tibetan speech synthesis system based on machine learning, on the basis of ensuring the quality of Tibetan speech synthesis, better completes the content of Tibetan speech synthesis, which has more important practical significance.

4 Conclusion and Outlook

In traditional Tibetan speech synthesis, the prosodic control of Tibetan speech synthesis is not considered, so the control effect of the range of Tibetan speech synthesized according to timbre is not obvious. With the extension of Tibetan sentences and the narrowing of the range, the duration of vocal cord vibration is limited. Especially when the speech is close to the end of the sentence, the acoustic characteristics of natural speech are obviously affected by its physiological factors. However, the feature extraction of Tibetan speech synthesis system based on machine learning can improve this point, make Tibetan speech synthesis more flexible, and greatly meet the changing requirements of Tibetan speech rhythm. However, the method in this paper does not consider the calculation time, which leads to a longer time for the speech signal feature extraction of the Tibetan speech synthesis system. Therefore, the following research will focus on the calculation, aiming to improve the speech signal feature extraction efficiency.

Fund Projects. Funded by Graduate Research Innovation Project of Northwest Minzu University (Granted No. YXM2019010).

References

1. Müezzinoğlu, T., Karaköse, M.: An intelligent human–unmanned aerial vehicle interaction approach in real time based on machine learning using wearable gloves. Sensors **21**(5), 1766 (2021)
2. Yao, Y., Ding, J., Wang, S.: Soil salinization monitoring in the Werigan-Kuqa Oasis, China, based on a three-dimensional feature space model with machine learning algorithm. Remote Sens. Lett. **12**(3), 269–277 (2021)
3. Subramani, P., Srinivas, K., Kavitha Rani, B., Sujatha, R., Parameshachari, B.D.: Prediction of muscular paralysis disease based on hybrid feature extraction with machine learning technique for COVID-19 and post-COVID-19 patients. Pers. Ubiquitous Comput. (2021, prepublish)
4. Yilin, F., Yanming, H., Feng, J.: Research on MFCC speech signal feature extraction algorithm based on EMD. Electro. World **008**, 23–25 (2019)
5. Su, H.: Design of the online platform of intelligent library based on machine learning and image recognition. Microprocess. Microsyst. **82**, 103851 (2021)
6. Mei, Y., Ye, D.-P., Jiang, S.-Z., Liu, J.-R.: A particular character speech synthesis system based on deep learning. IETE Tech. Rev. **38**(1), 184–194 (2021)
7. Lei, Y., Zhang, B., Li, R.: Detection of unusual targets in traffic images based on one-class extreme machine learning. Traitement du Signal **37**(6), 1003–1008 (2020)
8. Gajurel, A., Chittoori, B., Mukherjee, P.S., Sadegh, M.: Machine learning methods to map stabilizer effectiveness based on common soil properties. Transp. Geotech. **27**, 100506 (2021)
9. Silvello, G.C., Bortoletto, A.M., Costa, M., de Castro, A., Alcarde, R.: New approach for barrel-aged distillates classification based on maturation level and machine learning: a study of cachaça. LWT **140**, 110836 (2021)
10. Yao, Y., Yu, L., Chen, Y.: Feature Extraction Method of Radiation Source in Deep Learning Based on Square Integral Bispectrum. J. Phys. Conf. Ser. **1678**(1), 012074 (2020)
11. Kumar, S., Singh, S., Agarwal, P., Acharya, U.K., Sethy, P.K., Pandey, C.: Speech quality evaluation for different pitch detection algorithms in LPC speech analysis–synthesis system. Int. J. Speech Technol. **24**(3), 545–551 (2020)
12. Xinyi, Y., Boyu, S., Qingyun, M., Kailin, H.: Design of the speech tone disorders intervention system based on speech synthesis. J. Phys. Conf. Ser. **1617**(1), 012078 (2020)
13. Zhu, X., Xue, L.: Building a controllable expressive speech synthesis system with multiple emotion strengths. Cogn. Syst. Res. **59**, 151–159 (2020)
14. Yue, W.: Research on feature point extraction and matching machine learning method based on light field imaging. Neural Comput. Appl. **31**(12), 8157–8169 (2019)
15. Li, G., Li, J., Zhaojie, J., Sun, Y., Kong, J.: A novel feature extraction method for machine learning based on surface electromyography from healthy brain. Neural Comput. Appl. **31**(12), 9013–9022 (2019)
16. Liu, S., Bai, W., Liu, G., et al.: Parallel fractal compression method for big video data. Complexity **2018**, 2016976 (2018)
17. Liu, S., Fu, W., He, L., Zhou, J., Ma, M.: Distribution of primary additional errors in fractal encoding method. Multimedia Tools Appl. **76**(4), 5787–5802 (2014)

Research on Tibetan Speech Endpoint Detection Method Based on Extreme Learning Machine

Ze-guo Liu(✉)

Key Lab of China's National Linguistic Information Technology, Northwest Minzu University, Lanzhou 730000, China

Abstract. The traditional method of Tibetan speech endpoint detection will reduce the detection accuracy in low SNR environment, so a new method based on limit learning machine is proposed. Firstly, speech signal is preprocessed. By analyzing the human discourse generation model, the speech signal is filtered and processed by frame, the high frequency interference signal in the signal flow is filtered, and the signal flow is decomposed into multiple frames by using the short-term stationary characteristics of the speech signal, so that the characteristics of the speech signals in each frame are kept constant; the sentence analysis is optimized and the word association in the sentence is analyzed by using line graph syntax The system forms an independent semantic block, introduces the pattern template based on pseudo matrix, generates pseudo points through the original feature vector structure and inserts it into the original speech features for protection. Finally, the classification algorithm of limit learning machine is introduced, the optimal configuration of i-h-o is selected, and the end detection method of Tibetan speech based on limit learning machine is completed. The simulation results show that the accuracy of the speech endpoint detection results is significantly higher than that of the traditional method under different SNR and ambient noise.

Keywords: Limit learning machine · Tibetan speech · Speech endpoint detection

1 Introduction

Tibetan has always played an important role in the past and present. It is an indispensable tool for Tibetan compatriots to communicate with the outside world, and also the carrier of Tibetan culture, recording and inheriting the splendid historical culture. With the development of society, ethnic culture is more and more inclusive, and minority characters are gradually loved by people. Tibetan language, as an integral part of minority languages, has also played an important role in the international world, not only by the recognition of international standard words, but also by the recognition of the global highway [1, 2]. Therefore, whether from the actual development of the country or from the perspective of cultural inheritance, Tibetan studies have very important significance. The structure of Tibetan mainly depends on syllables to realize the expression of words and sentences. The phonetic segment is closely related to syllable. When the

S. Liu and X. Ma (Eds.): ADHIP 2021, LNICST 416, pp. 480–490, 2022.
https://doi.org/10.1007/978-3-030-94551-0_38

phonetic segment is found, the end point of syllable is easy to find. The study on the characteristics of Tibetan speech is the protection of Tibetan culture and helps to better use Tibetan characters. In the process of speech detection, the fluctuation of vocal cord appears gradually, which makes the whole syllable promote the implementation of the detection. When the syllable appears the main vowel, the audio is the maximum. At this time, the speech components can be analyzed according to the language characteristics of Tibetan language, and the speech can be detected according to the time frequency or the frequency of the voice vibration. Analogy is the process of hearing detection in the natural noise environment. Therefore, it can achieve the purpose of separating Tibetan speech signals. As far as speech signal is concerned, the clear part and the voiced part can be distinguished from the spectrum characteristics of speech. The spectrum of the former voice is rising at more than 4 kHz, while the latter is declining and above 4 kHz.

With the continuous progress of society, great changes have taken place in the way of daily communication. Voice communication is an important form of people's emotional interaction. At the same time, speech signal processing technology has developed rapidly. The purpose of speech signal processing is to determine the starting and ending point of speech signal in the presence of noise. When the effect of endpoint detection is good, the first thing we can see is that the processing time decreases significantly, and then there will be noise interference, which needs to be eliminated by certain technology, so as to increase the performance of the detection system. There are many endpoint detection methods, among which the short-term energy method was invented in the earliest period. With the passage of time and the maturity of the theory, the form of zero crossing rate method appeared again. The premise of these methods to achieve good results is to carry out experiments in the environment of high signal-to-noise ratio. When the environment becomes the environment of natural noise and low signal-to-noise ratio, its effect will be significantly weakened. Reference [3] combines spectral subtraction with energy zero ratio method. With the help of spectral subtraction, the signal-to-noise ratio of speech signal is improved, and the endpoint detection of processed speech is carried out by energy zero ratio method. Finally, the experimental analysis of Tibetan speech is needed. However, the accuracy of speech endpoint detection result of this method is low.

In view of the problems of the above methods, this paper proposes a Tibetan speech endpoint detection method based on extreme learning machine, and verifies the effectiveness of this method through simulation experiments, which solves the problems of traditional methods.

2 Research on Tibetan Speech Endpoint Detection Method Based on Extreme Learning Machine

2.1 Speech Signal Preprocessing

Speech signal is a kind of nonlinear time-varying and discrete complex signal. From the whole, the speech signal is a nonstationary signal. But the local analysis shows that the signal has the short-term stability characteristic when the signal is consistent in about 30 ms. To realize the analysis of the signal, first of all, we should understand the sound generating principle of the voice signal. The voice signal is an acoustic signal sent out by the cooperation of multiple organs of the human body. It has high frequency part and low frequency part, and also has the clear sound and the voiced component [4]. According to these characteristics, we will preprocess the language signal. Human voice is produced by the coordination of multiple functional organs. The speech signal is produced by these processes. When the direct air flow from the lungs flows through the complex and changeable channels composed of trachea, throat, upper jaw and lip, different sounds are produced due to the changing channels. In short, what kind of speech to send out is determined by the combination of the tongue, upper jaw and lip. Third, the radiation system, through the above two systems, has been produced, and it needs to be radiated

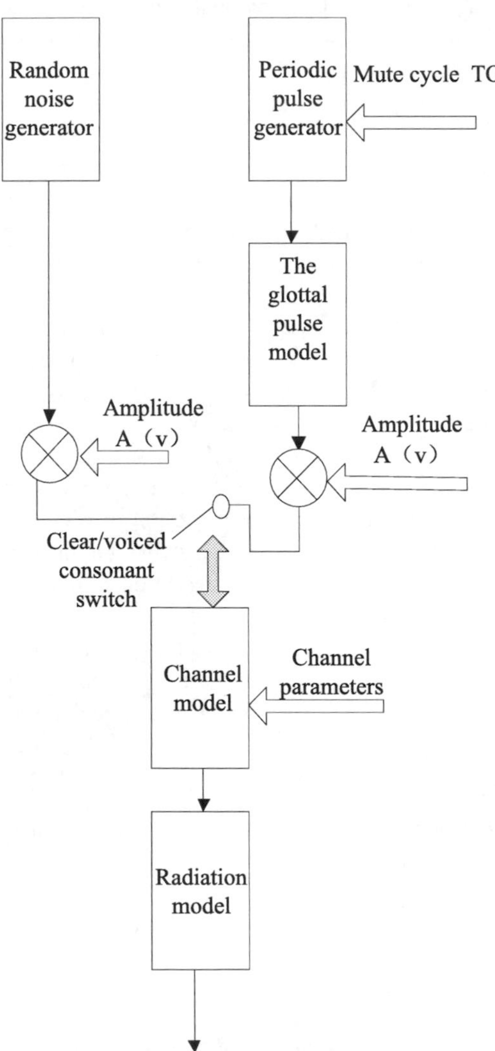

Fig. 1. Speech signal generation model

outside the lip and nasal cavity. So the human discourse we hear is just like this. The specific model diagram is as follows (Fig. 1):

The purpose of speech signal filtering is to filter out the high-frequency interference signal in the signal stream, which is usually bounded by half of the sampling frequency f_S, and the band larger than that frequency will be isolated [5, 6]. The reason why the filter is set as a band-pass filter is that the 50 Hz power frequency interference must be filtered at the same time. f_H , f_L will be set as the upper and lower limit of the filter frequency. Generally speaking, the upper and lower limit frequencies are $f_H = 3.44$ Hz, $f_L = 60$–100 Hz, and f_S is usually 8 kHz. For the design of speech recognition, the upper and lower limits of intermediate frequency rate are $f_H = 4.5$ Hz or 8 kHz, $f_L = 60$ Hz, and the selection of f_S is usually 10 kHz. Then, according to the sampling theorem, f_S is selected, and the value of sampling frequency f_S must meet the standard of more than 2 times of the frequency of the sampled signal. A/D converter generally selects 8-bit or 12 bit, and then performs A/D conversion on speech signal.

After the completion of the filtering process, it also needs to carry out the frame processing. From the overall time-domain waveform of the speech signal, the signal characteristics are in a non-stationary state, but from the micro point of view, in a very short period of time, its characteristics are in a stationary state, almost unchanged, which we call the short-term stationary characteristics of the speech signal. This feature brings a breakthrough for the complex and changeable speech signal processing, which can decompose the signal flow into several small segments, that is, a frame, in which the characteristics of the speech signal remain constant [7, 8]. Therefore, the frame processing of speech signal is the beginning of speech analysis. The signal can be decomposed into multiple parts of a frame of 10 ms–30 ms, and then each frame of speech signal can be analyzed as the research object. However, if the whole speech information is simply cut into multiple speech segments, the feature parameters between the adjacent frames will change too much, which is not conducive to signal analysis and processing.

In order to make the speech feature of adjacent frames have the characteristics of smooth transition, the signal of the next frame is often moved forward for a distance, called frame shifting, which makes the adjacent frames cross overlap, so that the feature parameters of adjacent frames have relative consistency [9, 10]. In the design of speech recognition, the length N of each frame is generally within 10 ms–30 ms, while the size of frame shift M is generally more than 1/3 of frame length N. It is important to select the frame length. If it is too long, it violates the short-term stability of speech signal, which makes the feature parameters of a frame different and can not find the correct feature value. If the selection is too short and the number of frames is too many, the calculation amount will be greatly increased and the operation time will be prolonged. At this point, the speech signal preprocessing is completed.

2.2 Analysis of Optimized Sentences

In the original endpoint detection method, the detection of simple phrases is relatively easy, but multiple backtracking will affect the efficiency of the algorithm, so this paper chooses the line graph syntax. The algorithm involves three data structures: line graph, process table and active edge set. These three data structures include more information.

In the analysis process of the algorithm, when the process table is blank, in order to obtain the part of speech tags of the next language input, the start and end positions will be stored in the process table together, and an element will be extracted from the process table and recorded as $X(i, i')$, where i represents the start position, i' represents the end position, and according to certain rules, the rule set will not be changed The matching points are marked and added to the active edge set together with the extracted element $X(i, i')$, and then the extension subroutine is called. The analysis results are as follows (Fig. 2):

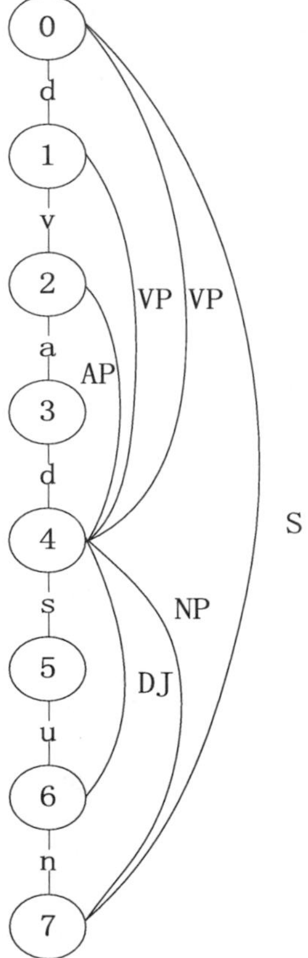

Fig. 2. Analysis result curve

The symbols in the figure above are marked with reference to the classification of symbols and the definitions in the marking Grammar Library. After using syntactic analysis and applying it to the original input string of Tibetan speech, we can get the algorithm result graph as shown in the figure above. From the result graph, we can see the relationship and connection between words in sentences, and further form independent semantic blocks on the basis of these relationships. After morphology and analysis, new rewriting rules can be obtained. The original input sentences of Tibetan speech can be processed to get a limited number of semantic blocks. After processing, these semantic blocks show the characteristics of speech endpoint, and they are identified and classified. In order to improve the recognition accuracy of this method, this paper introduces the voiceprint template based on the chaff matrix, generates the pseudo points (CPS) through the original feature vector elements, and inserts the CPS into the original speech features for protection. Compared with the existing system based on Fuzzy vault (FV), this method has stronger security. The feature extracted in the registration process is X_0, different authentication feature $X_I, X_J \in R^{m\times p}$, where m is the feature length of audio and p is the order of Mel cepstrum coefficient. The pseudo matrix of chaff matrix is expressed as C, $C \in R^{m\times r}$ and r, which indicate the number of CPS added. The feature of positioning matrix after adding C can be expressed as follows:

$$X_{C0}, X_{CI}, X_{CJ} \in R^{m\times (p+r)} \tag{1}$$

In the above formula, X_{C0} is the voiceprint template designed in this paper, which is composed of codebook inserting matrix C, and all audio features after adding C matrix are collectively referred to as X_C. In order to simulate the perception of human ear to sound in real situation, a sound signal scale converted into pseudo matrix is selected to reflect the frequency range of spectrum coefficient. Firstly, the linear spectrum of the sound frequency is selected by Fourier transform, then the transverse axis of natural spectrum is scaled nonlinearly, and a new energy spectrum in the cepstrum coefficient of Mel frequency is obtained. The Mel frequency cepstrum is obtained by the inverse spectrum analysis.

The linear transformation of horizontal spectrum axis is to convert the speech spectrum into a spectrum range. Mel spectrum coefficient module is used to convert the audio spectrum module into an audio spectrum range. Then the transfer function of each triangle filter is as follows:

$$H_m(k) = \begin{cases} 0, & k < f(m-1) \\ \frac{k-f(m-1)}{f(m)-f(m-1)}, & f(m-1) \le k \le f(m) \\ \frac{f(m+1)-k}{f(m+1)-f(m)}, & f(m) \le k \le f(m+1) \\ 0, & f > f(m+1) \end{cases} \tag{2}$$

In the above formula, $1 \le m \le M$ and M represent the number of triangular filters, and $f(m)$ represents the center frequency of the m-th filter. Through the filter, the characteristics of speech can be obtained from the linear natural spectrum to the nonlinear Mel spectrum, so that a frame of speech sample can be represented by a 22 dimensional vector. It can be used as input to the back-end identification part of the system.

2.3 Introducing the Classification Algorithm of Limit Learning Machine

Extreme learning machine (ELM) has been effectively applied in different pattern recognition applications. Many applications show that extreme learning machine has better accuracy and lower time consumption than SVM in small data, sparse data to medium large data. Therefore, elm is used as Tibetan speech endpoint detection classifier in this paper. Elm is a variant of artificial neural network. It can solve the problem of too long training time by randomly selecting hidden nodes and analyzing the weight of output layer. Therefore, it is of great significance to quickly solve the regression and classification model. Elm has two important parameters, the number of hidden nodes and the regularization coefficient. The default number of neurons in the output layer is equal to the number of classes, and the regularization coefficient can also be changed. The standard range is 0.1–10. The function of elm mainly includes two steps: feature mapping and elm learning. For training sample set $\{x_i, t_i\}, i = 1 \ldots, N$, there are a total of N samples. Each sample x_i is a d dimensional column vector, and t_i is the output label. In the feature mapping stage, the input is transformed into a hidden layer, and the output function of elm is as follows:

$$f_L(x) = \sum_{j=1}^{L} \beta_j G(a_j, b_j, x_i) = h(x_i)\beta \tag{3}$$

In the above formula, $\beta = [\beta_1 \ldots \beta_L]^T$ represents the output weight vector between the hidden layer (L nodes) and the output layer:

$$h(x_i) = [h_1(x) \ldots h_L(x)]^T \tag{4}$$

In the above formula, the function of $h(x_i)$ is that x_i maps the d dimensional space to the L dimensional feature space, $G(a, b, x)$ is a continuous piecewise nonlinear function, and (a_i, b_i) is the parameter of the ith hidden node:

$$G(a, b, x) = \frac{1}{1 + \exp(-(a \cdot x) + b)} \tag{5}$$

In the above formula, parameter a, b is generated randomly. In the learning stage, the main goal is to obtain the minimum training error and norm output weight β. the general approximate values are as follows:

$$\lim_{L \to \infty} \left\| \sum_{i=1}^{L} \beta_j h_j(x) - f_L(x) = 0 \right\| \tag{6}$$

This paper uses i-h-o configuration for elm, which corresponds to input layer, hidden layer and output layer respectively. The number of neurons in I is equal to the number of input features, the number of neurons in O is the number of output classes (in this paper, there are two, that is, speech frames and noise frames), and the number of neurons in H is between 50 and 600. Based on the experiment, the most appropriate value is 360. So far, the research of Tibetan speech endpoint detection method based on extreme learning machine is completed.

3 Simulation Experiment

3.1 Setting Simulation Experiment Parameters

In order to extract the speech signal characteristics required by limit learning machine, the software mainly uses MATLAB r2014a. All experiments are carried out in Windows 10 professional edition. Core (TM) i5-7300hq processor is used, and the physical memory is 16 GB. By comparing the results of traditional Tibetan speech detection, the validity and performance of the method based on limit learning machine are determined. In the parameter setting of simulation experiment, the performance of the endpoint detection system is tested by using true positive rate (TPR), false positive rate (FPR) and the time and precision parameters of feature extraction, modeling, training and testing of classifier.

Endpoint detection is a binary classification problem, because any frame either contains speech signal or does not contain it. Assuming a given speech signal, the parameters of the classifier after classifying the frame are defined as follows:

$$\begin{aligned} TPR &= \frac{L_1}{N_1} \\ FPR &= \frac{L_2}{N_2} \\ ACC &= \frac{L_1+N_2-L_2}{N} \end{aligned} \tag{7}$$

In the above formula, L_1 represents the true positive parameter, L_2 is the false positive parameter, and ACC represents the detection accuracy. N represents the total number of voice frames, N_1 represents the total number of frames containing voice signals, and N_2 represents the total number of frames without speech signals, so there are:

$$N = N_1 + N_2 \tag{8}$$

In the data set of simulation experiment, TIMIT standard voice database and noise-92 standard noise database are used to train and test the VAD method proposed in this paper. In order to create a real simulation environment, the short sentences in the database are merged (the total time is 5–10 s) for simulation, in which the ratio of talking frame and non talking frame is kept between 1:3 and 3:1. The sentences in the dataset are divided into six different types of real-world noise: babble noise, street noise, room noise, restaurant noise, car noise and train noise. The signal-to-noise ratio is set to 0 dB, 5 dB, 10 dB and 15 dB. The feature vector of the frame is composed of low-frequency denoising energy, MFCC and formant frequency. In order to extract these features, the input speech signal is divided into frames with frame length of 25 ms and frame shift of 10 ms. In the experiment, *TPR*, *FPR* and *ACC* obtained by the trained extreme learning machine classifier are compared and analyzed with the existing detection methods, and the total time of the two detection methods in classification training and testing is calculated, and the results are compared and analyzed.

3.2 Result Analysis

Under the above experimental conditions, the true positive rate and false positive rate of this method under different SNR and noise types are obtained (Table 1).

The above table shows the test results of this method. It can be seen that *TPR* and *FPR* are not the same for different types of noise conditions and SNR levels, which indicates that the sensitivity of the improved features to different noises is not the same. Through calculation, the average *TPR* and *FPR* values of the two methods under different noise types are obtained as follows (Table 2):

Table 1. Statistical results of true positive rate and false positive rate of this method at 15 dB

Noise type	*TPR*	*FPR*
Babble	98.64	2.13
Street	96.25	0.58
Room	88.14	1.35
Restaurant	94.25	0.64
Car	98.64	0.00
Train	99.22	0.00

Table 2. Average *TPR* and *FPR* values of the two methods

Signal to noise ratio		Method	
		Method of this paper	Traditional method
15 dB	*TPR*	97.04	73.52
	FPR	0.84	10.36
10 dB	*TPR*	94.56	70.74
	FPR	1.32	13.59
5 dB	*TPR*	83.51	61.37
	FPR	1.58	20.35
0 dB	*TPR*	78.02	50.67
	FPR	3.11	28.31

The table above shows the comparison between the average *TPR* and *FPR* of the proposed method and the traditional method under six different noise types. The proposed method has obvious improvement compared with the traditional method.

The comparison of accuracy results is shown in the figure below (Fig. 3):

The figure above shows the average accuracy of the proposed method compared with the traditional method for endpoint detection. Under low SNR, the proposed method can greatly improve the accuracy of endpoint detection.

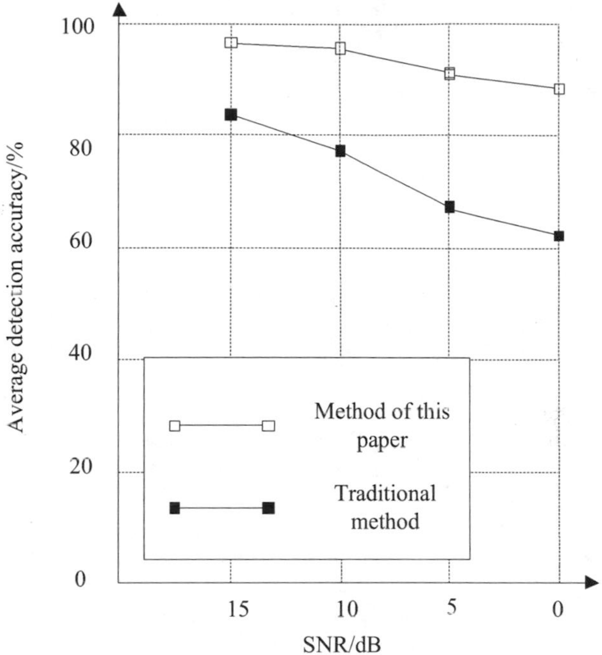

Fig. 3. Comparison of the accuracy of the two methods

4 Conclusion and Outlook

An excellent speech endpoint detection method can maintain high detection accuracy in the complex low SNR noise environment, and escort other steps in the speech recognition system. In this paper, the speech signal is taken as the research object, and the speech endpoint detection in low SNR environment is realized by integrating extreme learning machine and other methods. The simulation results show that the method designed in this paper is effective. Due to the limited research time and ability, the selection of noise signal in this paper is single noise, because of the lack of mixed noise library, such as white noise and pink noise mixed noise, car noise and human noise mixed noise. Therefore, in order to verify the endpoint detection effect of the proposed method in mixed noise, we need to enrich the noise database and supplement the corpus of various mixed noise.

Fund Projects. Funded by Graduate Research Innovation Project of Northwest Minzu University (Granted No. YXM2019010).

References

1. Zhang, T., Liu, Y., Ren, X.: Voice activity detection based on long-term power spectrum variability. J. Front. Comput. Sci. Technol. **13**(09), 1534–1542 (2019)
2. Xiangfeng, W., Yi, Y., Quan, Z., et al.: A dataset of Mongolian, Tibetan and Uyghur speech fragments based on voice activity detection. Sci. Data China (Chin. Engl. Online) **4**(04), 112–122 (2019)
3. Xie, X.: Research on endpoint detection method of Tibetan speech. Inf. Comput. (Theor. Ed.) **32**(450(08)), 106–108 (2020)
4. Zhou, B., Wei, S., Tang, Y., et al.: ELM network intrusion detection algorithm based on rough set attribute reduction. Transducer Microsyst. Technol. **38**(01), 122–125 (2019)
5. Yang, X., Ma, Z., Shen, H., et al.: Fault diagnosis of airflow jamming fault in double circulating fluidized bed based on multi-scale feature energy and KELM. CIESC J. **70**(07), 2616–2625 (2019)
6. Wei, J., Zhou, B., Tang, H., et al.: Transformer fault diagnosis with the combination of RapidMiner-modified particle swarm optimization-Extreme Learning Machine algorithm. Proc. CSU-EPSA **31**(03), 133–138 (2019)
7. Han, T., Zhang, H., Zheng, Z., et al.: Auditory perception speech signal endpoint feature detection based on temporal structure. J. Jilin Univ. (Eng. Technol. Ed.) **49**(01), 313–318 (2019)
8. Liu, S., He, T., Dai, J.: A survey of CRF algorithm based knowledge extraction of elementary mathematics in Chinese. Mob. Netw. Appl. https://doi.org/10.1007/s11036-020-01725-x
9. Liu, S., Pan, Z., Cheng, X.: A novel fast fractal image compression method based on distance clustering in high dimensional sphere surface. Fractals **25**(4), 1740004 (2017)
10. Liu, S., Fu, W., He, L., Zhou, J., Ma, M.: Distribution of primary additional errors in fractal encoding method. Multimedia Tools Appl. **76**(4), 5787–5802 (2014). https://doi.org/10.1007/s11042-014-2408-1

Design of Data Visualization System for Mobile Social Web Front End Development Based on HTML5

Jian-hong Wei[1], Yun-li Cheng[1], Xiao-ru Chen[1], Yang-hong Mao[1], Yu Gao[1], and Jia Yu[2](✉)

[1] Department of Software Engineering, Software Engineering Institute of Guangzhou, Guangzhou 510990, China

[2] School of Information Engineering East, China Jiaotong University, Nanchang 330013, China

Abstract. In view of the low efficiency of data visualization and the slow speed of database reading and writing when the current data visualization system is dealing with the front-end development data of mobile social Web, a data visualization system for front-end development of mobile social Web based on HTML5 is designed. The data collector is used to collect and store the front-end development data of mobile social Web, and the visual controller is used to communicate with the visual external devices to complete the hardware design of the system. Preprocess the Web front-end development data to ensure the data dimension differences within a reasonable range, divide the data into HTML5, JavaScript, CSS three different forms of files, classify low dimensional data and high-dimensional data, design two types of data visualization interface, complete the system software design. The experimental results show that the reading and writing speed of the database of the design system is accelerated, the data insertion time, query time and visualization time are effectively shortened, and the efficiency of information visualization is improved.

Keywords: HTML5 · Mobile social · Web front end · Development data · Visualization system

1 Introduction

As a new subject, data visualization plays an important role in people's intuitive observation of data, insight into the connotation of data, understanding the law of data, and exploring the knowledge behind data. At present, Internet technology, represented by Web front-end technology, provides technical support for data visualization on display platforms such as PC and mobile terminals. With the help of Web front-end technology, people can visually display and analyze data on computers, mobile phones, tablets and

S. Liu and X. Ma (Eds.): ADHIP 2021, LNICST 416, pp. 491–503, 2022.
https://doi.org/10.1007/978-3-030-94551-0_39

other terminal devices, improving the accessibility and comprehensibility of data. Therefore, it is of great practical value and practical significance to study the data visualization system of mobile social Web front-end development.

Since the establishment of the Department of visual chemistry, great progress has been made in the research of data visualization in theory, technology and application. Reference [1] analyzes the semantics of high-dimensional time-varying data, allowing users to change the color, brightness, contrast and other attributes of visual graphics according to their own requirements. It meets the needs of developers and commercial companies for multiple data types visualization, and solves the visualization problems caused by high dimension and large amount of data in parallel coordinate visualization, but the efficiency of data visualization is low. Reference [2] according to the parallel coordinate visualization technology, a parallel coordinate visualization method based on principal component analysis is proposed. This paper defines the parallel coordinates of principal component analysis and cluster analysis, aiming at the problem that the dimension of multi-dimensional data is too high, resulting in the limited width plane, the space between coordinate axes is too narrow, and the visualization lines are dense and crowded. Principal component analysis is used to reduce the dimension of the data, and the parallel coordinates of the processed data are visualized, but the reading and writing speed of the database is slow.

In order to solve the problems of low efficiency of data visualization and slow speed of database reading and writing, a data visualization system based on HTML5 is designed. Through the data collector of visualization system to collect and store the front-end development data of mobile social network, a web front-end data visualization controller is designed with PLC controller as the core hardware. RFID technology is used to identify multi-dimensional web front-end development data. Based on HTML5, the visualization development process of Web front-end is optimized. Taking the web front-end development data of HTML5 file as low dimensional data, the multi-dimensional visualization prototype interface and low dimensional visualization prototype interface are designed. It solves the problems of low efficiency of data visualization and slow speed of database reading and writing.

2 Mobile Social Web Front-End Development Data Visualization System Hardware Design

2.1 Data Collector

The data collector of the visualization system is designed to collect and store the front-end development data of mobile social Web. The data acquisition channel, communication interface, system operation conditions and human-computer interaction are the main components of the data collector. The overall framework of the data collector is as Fig. 1.

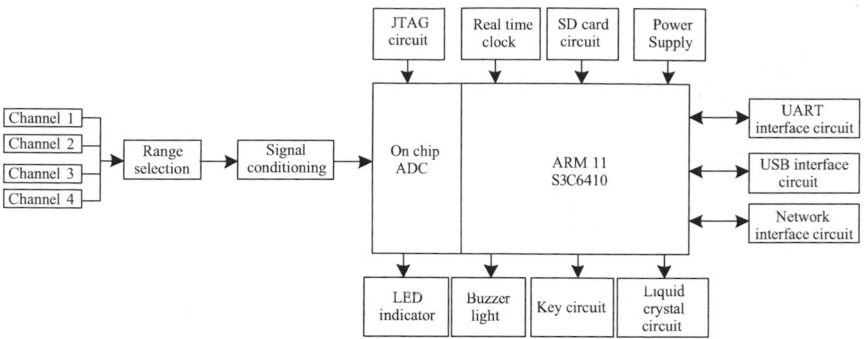

Fig. 1. Overall framework of data collector

As shown in Fig. 1, before the voltage signal is input into the ADC interface, the internal ADC of S3C6410 is used to carry out analog-to-digital conversion on the data acquisition channel, so as to complete signal conditioning, range selection and isolation protection, so that each channel can be compatible with current and voltage. SD card circuit, JTAG circuit, power supply and real-time clock are selected to provide clock source, current and voltage, Linux kernel image and bootloader for the visualization system [3]. Human-computer interaction selects key circuit, liquid crystal circuit, buzzer and LED indicator, and communication interface selects USB interface circuit, UART interface circuit and network interface circuit to provide support for LXI bus LAN specification, system function interface, and peripheral equipment. The circuit of the data acquisition channel is designed by using the backplane and core board. The core board uses two Samsung K4X1G163PE chips, omitting external components, using DDR SDRAM memory, and reading the startup code through the reset address [4]. The power circuit structure of the data collector is as Fig. 2.

After the voltage is introduced through the backplane, after voltage stabilization, denoising, filtering, and DC/DC isolation, the Web front-end development data and user stack are stored in DDR SDRAM, the input voltage is reduced to a fixed limit, and a low-level reset signal is output [5]. Furthermore, it provides a stable input voltage and realizes the design of a data collector for Web front-end development data.

2.2 Visual Controller

Taking PLC controller as the core hardware, the Web front-end data visualization controller is designed. The model of the SCM is selected as F8LE52-0S, which uses Ethernet interface, communication mobile phone, computer and other visual display external devices to provide output signals for external devices, and control the conversion of high and low signal levels, so as to realize the DC input and output of the system power circuit [6]. The visual controller develops multiple peripheral interfaces. Through the FA-M3 PLC controller, it programs each visual scheduled task, outputs the visual control instructions of the Web front-end development data, gradually issues the visual operation commands, communicates the signal transmission circuit, responds to the transmitted command signals, and then drives the visual equipment [7]. The basic peripheral circuit structure of the visual controller is as Fig. 3.

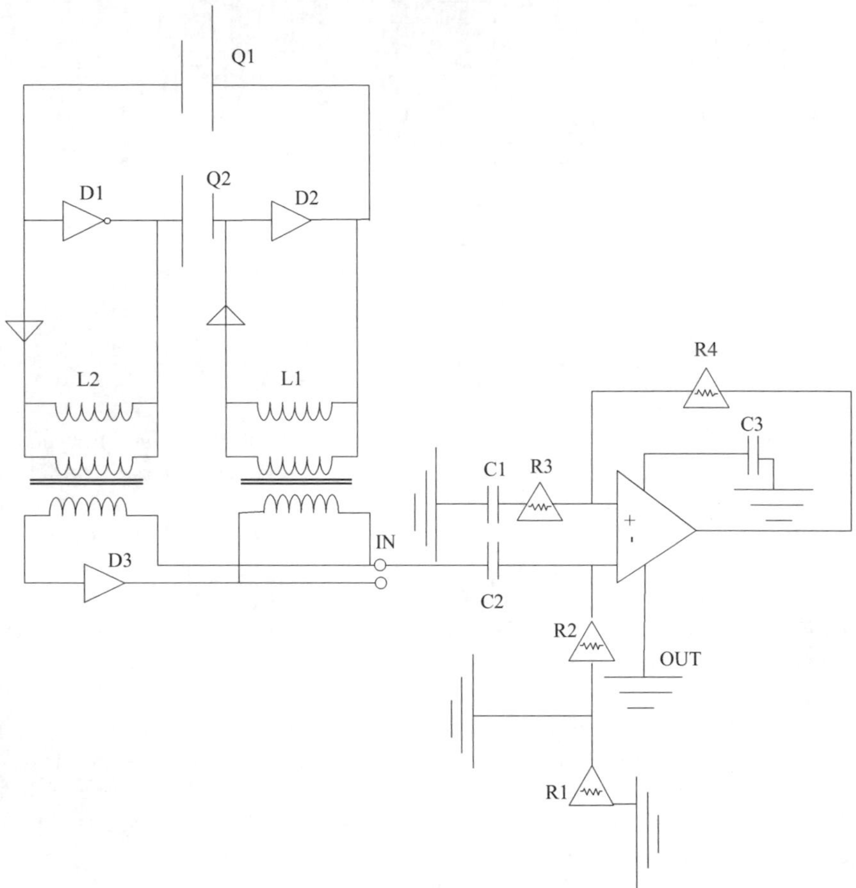

Fig. 2. Power circuit structure of data collector

The circuit is equipped with multiple analog comparators, supports multiple communication modes, and centrally processes the communication data of visualization devices. The driver chip of the circuit is TH83872S, which turns the reset signal to the low level to control the power switch, amplify and shape the collected data of the sensor, and respond to the visualization information after A/D conversion to ensure the normal operation of the visualization controller. So far, the design of data visualization controller for Web front-end development is completed, and the hardware design of the system is completed.

3 Software Design of Data Visualization System for Mobile Social Web Front End Development

3.1 Preprocessing Mobile Social Web Front End Development Data

On the basis of hardware design, the software is designed. RFID identification technology is used to identify the multi-dimensional Web front-end development data, and

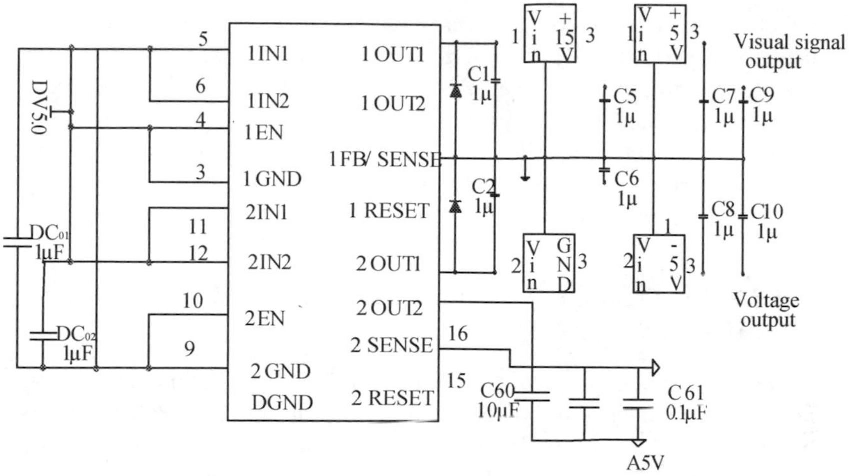

Fig. 3. Basic peripheral circuit structure of visual controller

then mobile information technology, computer, network and mobile devices are used to establish a computer supported collaborative environment, form a sensor network, and screen the Visual Web front-end development data, including the state quantity of development data and the quantity measurement of development data [8]. The error square sum criterion function is used as the clustering criterion function. The Web front-end development data set is divided into clusters composed of multiple types of objects, and then the connection rules between the development data are mined to achieve the accurate division of data dimensions. The clustering formula is as follows:

$$E = \sum_{j=1}^{K} \sum_{x \in C} \left| x_j - \overline{x_j} \right|^2 \tag{1}$$

In formula (1), E is the clustering function of Web front-end development data, x_j is the data set of the j cluster, $\overline{x_j}$ is the mean of the data set, C is the cluster center of the data set, and K is the cluster set. Through formula (1), the same set of Web front-end development data is divided into different sets. This paper uses statistics to clean the Web front-end development data, uses k-nearest neighbor algorithm to remove the default value in the original Web front-end development data, and counts the information data with the highest frequency as the replacement value of the default value [9]. The formula for calculating the encoding width B of substitution value is as follows:

$$B = \frac{V}{2(\xi \times F)} \tag{2}$$

In formula (2), V is the range available for Web front-end development data, F is the selectable gain for data acquisition, and ξ is the resolution of the collector. Through smooth filtering, the high-frequency part corresponds to the noise component, and the Web front-end development data is denoised. Through linear interpolation technology, the lost Web front-end development data is recovered. Finally, the data comparison work is carried out to detect the retention, error and expiration rate of Web front-end development, so as to ensure that the overall state of Web front-end development data set is good, so as to ensure that the dimension difference of Web front-end development data is within a reasonable range, so that a large number of Web front-end development data can be transformed into an easy to understand mode. So far, the preprocessing of mobile social Web front-end development data is completed.

3.2 Optimize Web Front End Visualization Development Process Based on HTML5

Optimize the Web front-end visual development process, and divide the preprocessed Web front-end development data into three different forms of files: HTML5, JavaScript and CSS. When developing Web front-end and data visualization Web page, we first write HTML structure document, add header, navigation, content and footers, etc. In W3C style specification, the use of abbreviated attributes, including font, background and border, etc., in which the background location of HTML file can obtain images, which can reduce the number of HTTP requests, save bandwidth, but also speed up the loading process, reasonably plan the image background, and facilitate the current visual file design and maintenance. Then add more detailed HTML tags, before completing all the document structure, write the performance code to ensure that the HTML page structure is good, and can be displayed through any style.

CSS belongs to both design and development. Like other languages, its syntax also needs to be standardized, and the parsing performance of CSS needs to be improved while maintaining clear structure [10]. First of all, the Web front-end development data is abbreviated, CSS attribute abbreviated can effectively reduce the data style, through the repeated use of style class, can control the number of ID and class, flexible use of HTML tags for image merging, deep understanding of CSS priority rules, mining Web front-end development data deep style attribute, ensure the data style attribute mode. In order to ensure the writing of standard CSS, we need to verify it. W3C provides an online verification tool to ensure that the writing of verification style meets the standard and solve the problem of Web standard.

JavaScript file format is compatible with the Web development environment, which needs to be comprehensively considered in combination with the technical level of developers, the applicability of visualization requirements, and the applicability of Web front-end development data [11]. JavaScript Mobile is used to assist the development of visualization, such as sliding, click events, AJAX requirements and other visual effects. For the applicability of Web front-end development data, A-level browser is used to provide support to detect whether the document quality is perfect, the framework scalability, framework size, visual performance, execution speed, visual code modularity, code reusability, etc. In view of the good code structure of the visualization module, the ready method is used to initialize it, and the data document of the Web front-end development and the use method of the visualization module are described in detail, so as to improve the maintainability of the front-end development. So far, the Web front-end visualization development process optimization based on HTML5 has been completed.

3.3 Design Web Front End and Develop Data Visualization Interface

Taking the Web front-end development data in the form of HTML5 file as the low dimensional data, and the development data in the form of JavaScript and CSS file as the high dimensional data, the multi-dimensional visual prototype interface and the low dimensional visual prototype interface are designed respectively. According to different data dimensions, different data visualization processing methods are adopted [12].

When the Web front-end data is low dimensional data, enter the low dimensional data visualization module to analyze and study the data. The low dimensional data visualization prototype interface is as Fig. 4.

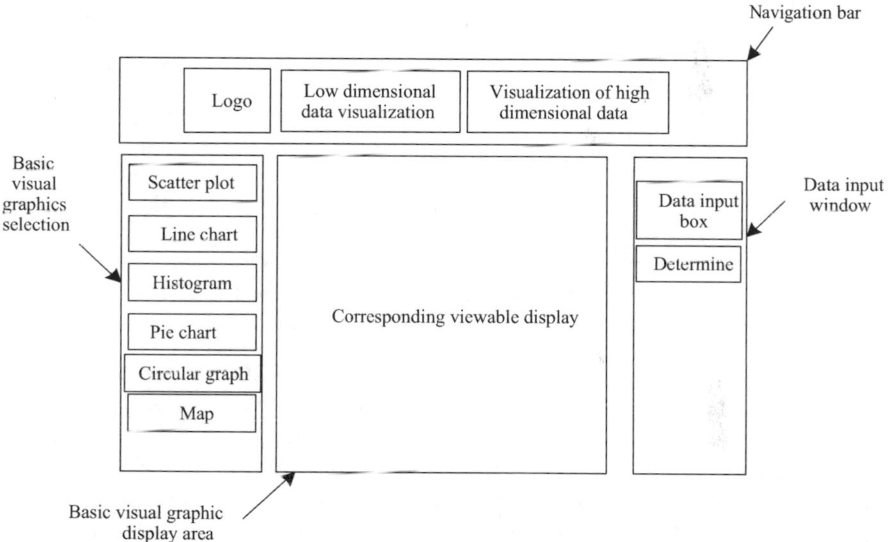

Fig. 4. Low dimensional data visualization prototype interface

The Web interface mainly includes four parts: navigation bar, basic visual graphics selection bar, graphics display area bar and data input window bar. In the visualization graph selection area, there are basic types of data visualization charts, and different types of basic charts can be selected. The graph display area is used to display the visualization results of low dimensional data, and the data input window area is used to input low dimensional data [13].

For multi-dimensional web front-end development data, K-means clustering algorithm is used to process the data. A large amount of data is divided into groups, that is, the data is divided into multiple categories, and the parallel coordinates of the clustering results are visualized. The multi-dimensional data visualization prototype interface is as Fig. 5.

The interface is composed of four parts: navigation bar, interaction bar, graphic display area and data input window. The upper side of the interface is the navigation bar, which has a logo, low dimensional data visualization and multi-dimensional data visualization module selection button [14]. Click the button to switch to the corresponding

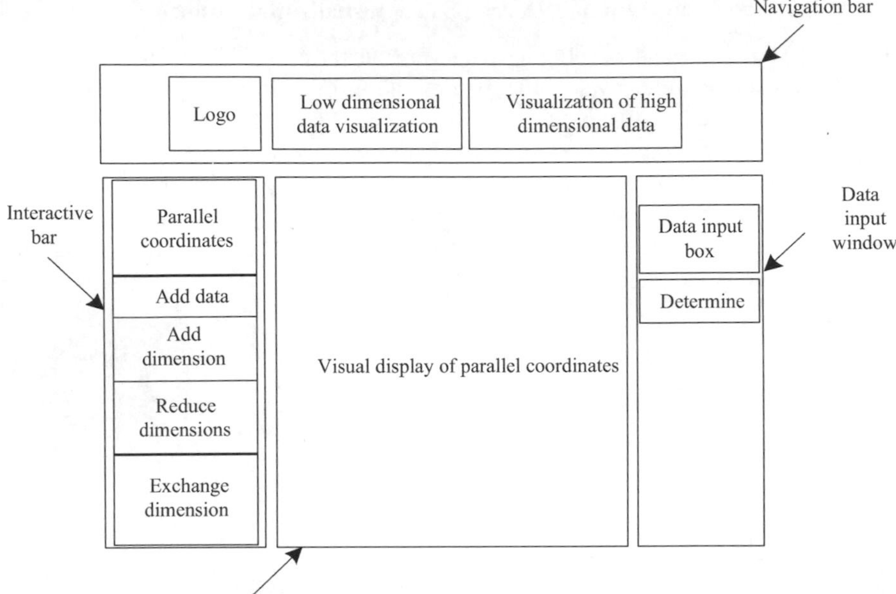

Fig. 5. Multi dimensional data visualization prototype interface

function module. The interactive column area mainly includes parallel coordinate graph, add data, add dimension, reduce dimension and other interactive operation types. The graphic display area is mainly used to display the parallel coordinate map. The parallel coordinate map is obtained through interactive operation and displayed in this area. The data input window is responsible for data input [15]. In the web front-end interface of visualization system, the navigation bar of low dimensional data and high-dimensional data has the same function and can be regarded as a public module, while the other three parts are slightly different according to the specific function division. So far, the design of data visualization interface for web front-end development is realized, and the system software design is completed. Combined with hardware design and software design [16], the design of data visualization system for mobile social web front-end development based on HTML5 is completed.

4 Experiment and Analysis

The design system is compared with reference [1] and reference [2] mobile social web front-end development data visualization system, and the visualization efficiency and database reading and writing speed of the three groups of system Web front-end development data are compared.

4.1 Experimental Preparation

Set up the experimental environment as the client computer is the s530i3-41304G1T host, the T410 G645 host, the hard disk is 1000G, the memory is 60 GB, the server host is a 2G/300AN rack-mount host, the host device CPU is PowerEdge 2900, and two GHSK 7283 processing are configured. The CPU cache is 8M, the frequency is 1700 MHz, the hard disk is 400 GB, and the network environment is 150 Mb/s of the power internal private network. The deployment architecture built is as Fig. 6.

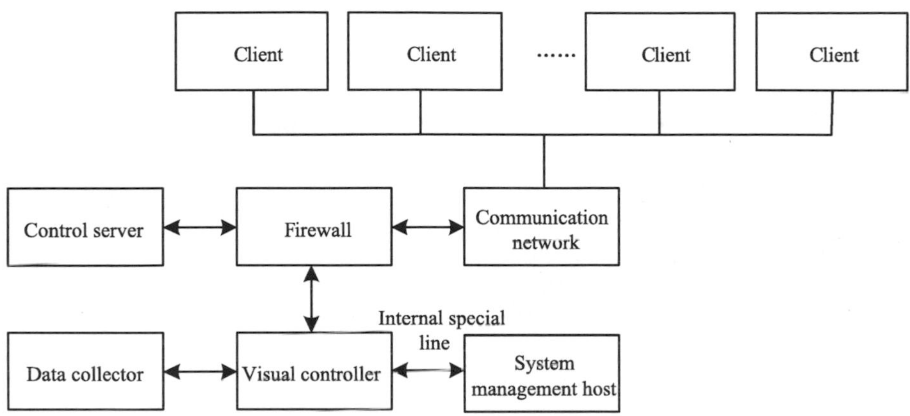

Fig. 6. Visualization system deployment architecture

Select the appropriate load balancer, forward the front-end development data of mobile social web to the front-end cluster, buffer and send it to the visualization system for data analysis and processing, and the manager completes the management operation of the system through the host.

4.2 Experimental Results

Data Visualization Efficiency Test

Set the amount of information data to 5000 MB, compare the data visualization efficiency of the three groups of systems, set the data insertion to single thread execution, change the number of data import, and compare the data insertion time of the three groups of systems. The experimental results are as Table 1.

It can be seen from Table 1 that with the increase of information data, the data insertion time of three groups of systems increases. When the amount of information data is 5000 MB, the data insertion time of the design system is 8.16 s, the data insertion time of reference [1] system and reference [2] system is 29.35 s and 34.92 s respectively. Therefore, compared with the reference [1] system and the reference [2] system, the data insertion time of the design system is shorter. Because the data collector of the visualization system of the system can speed up the speed of data collection and process data faster. This method is obviously superior to the other two data acquisition methods.

On this basis, the visualization time of the three groups of systems is further counted, and the experimental comparison results are as Table 2.

Table 1. Comparison results of data insertion time of three groups of systems

Information data volume/MB	Design system data insertion time/s	Reference [1] System data insertion time/s	Reference [2] System data insertion time/s
500	2.83	10.37	12.74
1000	3.38	11.73	13.74
1500	3.94	12.06	14.02
2000	4.15	12.74	14.72
2500	4.65	13.17	15.38
3000	5.05	13.84	15.73
3500	6.73	16.63	17.18
4000	7.25	20.39	23.73
4500	7.74	24.04	28.03
5000	8.16	29.35	34.92

Table 2. Comparison results of visual display time of three groups of systems

Information data volume/MB	Design the visual display time of the system/s	Reference [1] system visualization time/s	Reference [2] system visualization time/s
500	257	521	622
1000	261	532	643
1500	289	557	672
2000	302	593	702
2500	328	624	731
3000	362	667	769
3500	381	681	791
4000	399	709	827
4500	437	746	872
5000	458	781	906

It can be seen from Table 2 that with the increase of information data, the visualization time of three groups of systems increases. When the amount of information data is 5000 MB, the visual display time of the design system is 458 s, the visual display time of the reference [1] system and the reference [2] system is 781 s and 906 s respectively. Therefore, compared with the reference [1] system and the reference [2] system, the visualization time of the design system is shorter. The system uses RFID technology to identify multi-dimensional web front-end development data, which can quickly transfer information and shorten the visual display time of the system.

Set the number of thread queries to 100, read the front-end development data of mobile social web in the system at the same time, use the server to query the corresponding data in the system according to the visualization request of the client, and compare the data query time of three groups of systems. The experimental comparison results are as Table 3.

Table 3. Comparison results of data query time of three groups of systems

Number of thread queries/pcs	Design system data query time/s	Reference [1] system data query time/s	Reference [2] system data query time/s
10	0.82	3.78	3.98
20	0.89	3.82	4.27
30	0.93	3.91	4.42
40	0.96	4.02	4.68
50	0.99	4.17	4.92
60	1.05	4.33	5.27
70	1.14	4.49	5.74
80	1.22	4.62	5.92
90	1.26	4.89	6.21
100	1.27	5.71	6.39

It can be seen from Table 3 that with the increase of the number of thread queries, the data query time of the three groups of systems increases. When the number of thread queries is 100, the data query time of the design system is 1.27 s, the data query time of the reference [1] system and the reference [2] system is 5.71 s and 6.39 s respectively. Therefore, compared with the reference [1] system and the reference [2] system, the data query time of the design system is shorter. Based on the above analysis, the time of data insertion, query and visualization of the design system is short, which can effectively improve the efficiency of data visualization. Because the system uses the circuit driver chip TH83872S to turn the reset signal to the low level to control the power switch, amplify and shape the data collected by the sensor, so that the data transmission is faster and the data query time is shorter.

Database Reading and Writing Speed Test

On the basis of data visualization efficiency test experiment, unlimited web front-end development data sampling points are set. The databases of the three groups of systems can support the sampling of data points, and the maximum number of records is more than 15 million. 10 experiments are conducted respectively to compare the read-write performance of the database and the read-write speed of direct positioning. The experimental comparison results are as Table 4.

Table 4. Comparison results of database reading and writing speed of three groups of systems

Number of experiments/pcs	Read and write speed of design system database (10000 times/s)	Read and write speed of reference [1] system database (10000 times/s)	Read and write speed of reference [2] system database (10000 times/s)
1	201.3	176.2	162.1
2	200.1	175.9	163.2
3	200.9	175.2	163.0
4	201.5	175.4	163.7
5	200.6	176.2	162.9
6	200.5	176.1	162.5
7	201.8	175.9	162.1
8	200.9	175.9	162.6
9	201.1	176.3	162.7
10	201.3	176.2	162.5

It can be seen from Table 4 that the average reading and writing speed of the database of the design system is 2.01 million times/s, and the database can save 200000 data sampling points within 1–2 s, while the average reading and writing speed of the reference [1] system and the reference [2] system are 1.759 million times/s and 1.627 million times/s respectively, and the database can only save 175000 and 160000 data points within 1–2 s. Therefore, compared with the reference [1] system and the reference [2] system, the reading and writing speed of the design system database is faster. To sum up, the design system shortens the data insertion time, query time and visualization time, improves the efficiency of information visualization, speeds up the reading and writing speed of database, and ensures the throughput performance of Web front-end development data. Compared with the other two systems, the system uses mobile information technology, computers, networks and mobile devices to establish a computer supported collaborative environment, forming a sensor network. The information transmission speed of the system is faster, the transmission content is more accurate, and the reading and writing speed is faster.

5 Conclusion and Outlook

This research designs a visualization system for mobile social web front-end development data, which effectively improves the visualization efficiency and reading and writing speed of Web front-end development data. It can greatly improve the speed of network information circulation, further promote the development of games and e-commerce, speed up the refinement of Internet industry, and strengthen network marketing. However, there are still some shortcomings in this study. The system is not necessarily suitable for non numerical data. Moreover, it has poor applicability in the fields of text, image, sound and other computer applications. In the future research, the initial clustering center will be randomly selected from the data set to realize the visualization of hierarchical data and network log data.

Fund Projects. 2020 Scientific research, education and teaching research projects of South China Institue Of Software Engineering.GU: Application of GIS-based intelligent environmental protection IoT system construction technology (ky202001).

References

1. Quanyong, S.H.A.O.: Development of data visualization technology platform. Microcomput. Appl. **36**(1), 144–148 (2020)
2. Bai, B.: Analysis of data security protection in heterogeneous network based on visualization and data fusion technology. Electr. Des. Eng. **28**(13), 137–140+146 (2020)
3. Zhang, X., Pang, X.: The design and research of big data visual analysis system in library. Digit. Technol. Appl. **38**(8), 138–139 (2020)
4. Cong, H., Li, H., Song, X.: Research on visualization mode of sports data news in the age of media convergence. J. Xi'an Inst. Phys. Educ. **37**(4), 449–456 (2020)
5. He, Y., Chen, L., Li, F., et al.: Research and implementation of visualization of text data fed back by e-commerce users. Packag. Eng. **41**(10), 228–234 (2020)
6. Wu, Y., Cai, Y., Li, J., et al.: Exploration on the prototype development method of data visualization display system. Mod. Inf. Technol. **4**(8), 107–108 (2020)
7. Ma, N., Yuan, X.: Tabular data visualization interactive construction for analysis tasks. J. Comput. Aided Des. Comput. Graph. **32**(10), 1628–1636 (2020)
8. Han, L.: Visualization method of complex multidimensional data in multiple heterogeneous networks. Comput. Simul. **37**(11), 299–303 (2020)
9. Chen, S., Sun, Z.: Method for motion data set visualization based on two-layer auto-encoders. J. Nanjing Univ. Posts Telecommun. (Nat. Sci.) **40**(3), 22–30 (2020)
10. Cao, Q., Shi, S.: Technical research and system design of data mining analysis and visualization in spatiotemporal big data. Jiangsu Sci. Technol. Inf. **37**(3), 45–47 (2020)
11. Liu, S., Liu, D., Srivastava, G., et al.: Overview and methods of correlation filter algorithms in object tracking. Complex Intell. Syst. **3**, 161–165 (2020)
12. Fu, W., Liu, S., Srivastava, G.: Optimization of big data scheduling in social networks. Entropy **21**(9), 902 (2019)
13. Demaree, D., Kruse, A., Pennestri, S., Russell, J., Schlafly, T., Vovides, Y.: From planning to launching MOOCs: guidelines and tips from GeorgetownX. In: Vincenti, G., et al. (eds.) E-Learning, E-Education, and Online Training. LNICSSITE, vol. 138, pp. 68–75. Springer, Cham (2014). https://doi.org/10.1007/978-3-319-13293-8_9
14. Tanaka, H.: Development of automatic analysis and data visualization system for volcano muography. J. Disaster Res. **15**(02), 203–211 (2020)
15. Enevoldsen, P.: Estimating wind conditions in forests using roughness lengths: a matter of data input. Wind Eng. **44**(2), 142–155 (2019)
16. Hermida, R.C., Mojón, A., Fernández, J.R.: Comparing the design of the primary-care based Hygia chronotherapy trial and the internet-based TIME study. Eur. Heart J. **41**(16), 1608–1608 (2020)

Design of Bearing Alloy Intelligent Batching System Based on Electrical Control

Ming-fei Qu(✉), Dan Zhao, and Yi Xin

College of Mechatronic Engineering, Beijing Polytechnic, Beijing 100176, China

Abstract. In order to realize the optimization and intelligent control of the burden ratio of bearing alloy, the intelligent burden system of bearing alloy based on electrical control was constructed with the minimum error of the formula as the index, in order to complete the work of alloy ratio and intelligent burden independently. The feature of the system is to realize the automatic and accurate batching of bearing alloy by using fuzzy algorithm, and to configure electrical control to supply alloy raw materials quantitatively. Experimental results show that the system can basically control the error of bearing alloy batching within ±1%, which proves that the system can meet the requirements of batching precision.

Keywords: PLC electrical control · Bearing · Alloy batching system · Fuzzy algorithm

1 Introduction

The forming quality of materials is closely related to the accurate proportion of charge composition. It is of great help for lean management of raw material consumption, reducing the difficulty of composition adjustment in front of furnace and improving the casting characteristics of alloy to accurately match the alloy composition and control the charge composition. At present, there are two common matching methods: artificial experience calculation and algorithm optimization derivation. The former has low efficiency and heavy workload, and the proportioning accuracy is affected by subjective factors. With the help of computer algorithm optimization and programming derivation, it is easier to realize the accurate proportion of alloy composition.

High quality sliding bearing alloy should have good antifriction, compressive strength and fatigue resistance. Accurate control of alloy composition is the key to produce high quality bearing materials.

The electrical control system is generally called the secondary control loop of electrical equipment. Different equipment has different control loops, and the control mode of high-voltage electrical equipment and low-voltage electrical equipment is not the same. Specifically, the electrical control system refers to the combination of several electrical components, used to achieve the control of a certain or some objects, so as to ensure the safe and reliable operation of the controlled equipment. Therefore, an intelligent

S. Liu and X. Ma (Eds.): ADHIP 2021, LNICST 416, pp. 504–516, 2022.
https://doi.org/10.1007/978-3-030-94551-0_40

batching system for bearing alloys based on electrical control is designed in this study to reduce batching errors through electrical control.

The innovative design process of the system is as follows:

(1) Based on the minimum formula error, the fuzzy algorithm was used to realize the automatic and accurate batching of bearing alloy.
(2) Configuration of electrical control to ration the supply of alloy raw materials.

2 Design Requirements of Intelligent Batching Control System

The design of bearing alloy intelligent batching control system must first meet the actual process requirements, and then design a reasonable technical scheme on this basis, so that the design of the system and customer requirements to achieve the optimal unity. Combined with the project contract and the actual situation of the enterprise, this paper puts forward the following process requirements and technical requirements.

2.1 Process Requirements of Batching Control System

In order to achieve the best effect and requirements of proportioning, the design of proportioning control system has the following process requirements:

(1) The system has comprehensive functions, stable and reliable work, accurate weighing, and can meet the requirements of batching production.
(2) It should be able to adjust the formula data of various raw materials online to achieve higher batching accuracy.
(3) It has monitoring screen to know the running state of the system at any time.
(4) It has good adaptability and can adapt to different requirements of environment and process form.
(5) On the premise of meeting the batching accuracy, the system has clear structure, reliable use, convenient maintenance, high performance and price [1].

2.2 Technical Requirements of Batching Control System

According to the actual situation of bearing production equipment and the experience requirements of bearing proportioning, the following technical requirements are put forward for the bearing proportioning control system.

(1) The system has automatic, automatic and manual functions. No matter in which way, it has a kind of compulsive function. This function can start or stop an operation immediately [2].
(2) The system accurately weighs all kinds of raw materials, and the error value is not more than 0.1%.
(3) The system shall be able to display, store and print the current formula and production report.
(4) The system can display and print the batch, shift, daily and monthly ingredients.

(5) Automatic diagnosis function of the whole system.
(6) The system has good man-machine interface function.
(7) Out of tolerance alarm and fault printing function.
(8) The system has good communication function.

3 Framework of Bearing Alloy Intelligent Batching System

A complete batching control system is mainly composed of feeding system, weighing system, mixing system and control system [3].

(1) Feeding System

For weighing bulk materials, often use solenoid valve feeding, screw feeding, material door feeding, electromagnetic vibration feeding, belt feeding and motor vibration feeding and other feeding methods.

In order to accelerate the speed of ingredients, generally choose a slightly larger feed caliber silo, but it will produce a large and unstable drop, so it is difficult to improve the accuracy of ingredients. If in order to solve the precision problem and choose a slightly smaller feed caliber silo, although the ingredients are stable, but the speed of the ingredients can not meet the requirements, so that the production efficiency is low. Therefore, this paper uses double - gate batching method to solve the problem of batching speed and precision.

Double-door batching method is a silo with a raw material, but each silo has two feeding doors, the system determines the switch state of the two feeding doors according to the set value. After starting the batching, the two feeding doors of each bin are opened. In a short time, once the weight value reaches the set value, one of the feeding doors will be closed, and the remaining one will be closed to complete the fine feeding. When the weight value is the rated value, the remaining feeding doors will be closed. Double-gate batching method not only reduces the error caused by unstable drop, solves the problem of batching speed and precision, but also makes the control system more flexible.

(2) Weighing System

The key of the batching control system is the weighing system, and the weighing mode quickly transits from the mechanical scale stage to the electromechanical electronic scale stage and the sensor electronic scale stage.

The mechanical scale composed of scale bucket and lever system has slow weighing speed and low efficiency, so it is not suitable for automatic production. The electronic scale is mainly composed of load-bearing, force transmission reset system and load cell. Among them, the load-bearing and force transfer reset system is the mechanical connection between the measured object and the conversion element. Because of the good sealing of the load cell, the sensor electronic scale can replace the mechanical scale and the electromechanical scale to complete the work under bad conditions. The development of electronic scale is a process from simple to complex, from rough to precise, from mechanical to mechanical and electrical integration, and finally to full electronic, from single function to multi-function.

(3) Hybrid System

At present, the domestic mixer design technology is very mature, such as the horizontal screw belt mixer used by most manufacturers. Horizontal screw belt mixer is a new type of mixing equipment with high efficiency, high uniformity, high loading coefficient, low energy consumption, low pollution and low crushing. In this paper, the motor and inverter are selected for simulation experiment.

(4) Control System

The main function of the control system is to make the field equipment complete all the batching work according to the preset formula proportion.

Nowadays, most batching control systems adopt computer control method, because this control method not only makes the whole process automatic and improves the productivity, but also provides users with a large number of ratios that can be easily called and modified. At the same time, the computer can carry out real-time online monitoring on the field control system, and realize the function of storing and viewing historical data, so as to achieve the purpose of scientific management.

Based on the composition of the above batching control system, the framework of bearing alloy intelligent batching system is designed, which consists of user, controller and equipment. The control computer and touch screen are oriented to the technologists, and it is convenient to set the bearing alloy grade, mass, element mass fraction, etc. The controller includes industrial computer and programmable logic controller (PLC). The industrial computer participates in the alloy proportioning optimization and sends the proportioning parameters to PLC module through industrial Ethernet.

PLC is responsible for the process control of the batching machine. The field equipment communicates with PLC through CC link to complete a series of actions such as weighing, lifting, transferring and charging according to the alloy proportion. The alloy proportioning is a semi-automatic process, which requires the participation of the technologists: PLC displays the proportioning processed by the industrial computer on the touch screen, and the technologists place the burden in the weighing area of two stations in order. When the quality and weight meet the proportioning requirements, the carrier rises, and the frequency conversion speed regulating conveyor belt loads the raw material into the furnace.

4 System Hardware Design

4.1 Electrical Control Hardware—PLC

PLC is a programmable memory which can realize the functions of logic operation, sequence control, timing and counting. It is also a kind of digital or analog input and output to control various production processes.

In the running process, the CPU of the programmable controller repeatedly executes three working stages of input sampling, user program execution and output refresh in a single scanning cycle according to a certain scanning speed [4, 5]. PLC has the following advantages:

The hardware is complete and adaptable;
Programming is simple and easy to use;
The control program can be changed with good flexibility;
It has high reliability and strong anti-interference ability;
Abundant input/output interface modules;
The workload of system design, installation and debugging is less.

In modern control engineering, Siemens 57-200 series PLC is used as the ideal controller. Because of its compact structure, good expansibility, powerful command function and low price, it can fully meet the system requirements.

4.2 Load Cell

The Weighing Sensor
Load cell is a kind of force sensor, which measures the mass by converting the measured mass into another one. The load cell is an important part of the electronic scale, which converts the weight of the weighed object into electrical signal and indicates it after processing [6]. Therefore, its advantages and disadvantages largely determine the accuracy of the electronic scale.

Combined with the structural characteristics and working principle of the hopper, the resistance strain type tension weighing sensor is selected. The resistance strain type tensile load cell mainly has the following characteristics:

(1) High precision and wide measuring range.
(2) Good frequency response.
(3) The utility model has the advantages of simple structure, small size, light weight and simple installation.
(4) It can work under high temperature, low temperature, high pressure, strong vibration, strong magnetic field and nuclear radiation.

The resistance strain gauge is made of resistance wires with a diameter of 0.025 mm and a high resistance string. In order to obtain high resistance, the resistance wires are arranged into a grid network, called sensitive grid, and pasted on the insulating substrate. The two ends of the resistance wire are welded with leads. A protective covering layer is pasted on the sensitive grating. When measuring with strain gauge, paste it on the surface of the object to be measured. When the measured object is deformed by force, the sensitive gate of the strain gauge is also deformed, and its resistance value changes accordingly. It is converted into the change of voltage or current through the conversion circuit, which is used to directly measure the strain.

Weighing Indicator
Panther weighing terminal is a high quality weighing terminal product designed by METTLER TOLEDO company for chemical industry, pharmaceutical industry, food industry and other process industry applications. Panther uses sigma delta analog-to-digital converter (ADC) and digital processing technology to provide a variety of applications for

industrial weighing demand: weighing display, weight checking, sorting, setting control, etc.

The other weighing terminal adopts traxdsp digital filtering patent technology, which can obtain stable weight value in real time. Even if there is motion equipment on the connected weighing body, the real weight data can be obtained by adjusting the parameters of the weighing terminal.

(1) Easy to install Panther can choose different voltage power supply. Optional AC voltage is 100 V AC, 120 V AC, 230 V AC. The panel instrument can be fixed on the control cabinet only by opening a rectangular port on the control cabinet. Sealed (desktop/wall type) instrument wiring harness connection adopts sealed joint, which has dust-proof and waterproof effect (in line with IP65 standard). With bracket structure, Panther can be placed on the platform or installed on the wall with adjustable angle.
(2) Easy to use, fluorescent display, high brightness, touch film keyboard, durable resin panel, anti physical and chemical damage.
(3) The flexibility adopts modular tree structure parameter setting, which is convenient for users to find setting parameters.
(4) Panther is designed, manufactured and tested in Mettler Toledo's IS09001 certified factory. All parts (including options) of panther are installed and debugged in the factory. Built in diagnostic program, easy to find and solve problems. The design of standard Panther conforms to relevant international standards of metrology, electrical safety and electromagnetic compatibility.

Panther weighing terminal function the functions of panther instrument are as follows:

(1) The Panther meter with weighing function can be connected with up to eight 350 Q analog sensors, its maximum display graduation is 10 000 D, and its a/D conversion speed is 30 times/s. The functions of automatic peeling, automatic clearing, unit conversion, automatic tracking and dynamic detection are embedded in the system.
(2) Storage function: the storage function of the four target weight values in the excessive, normal and insufficient sorting functions: the storage function of the two preset points and the advance value.
(3) Communication interface function Panther has a RS232 serial communication interface 1:21 and parallel I/O interface, which can be used to connect serial printer or computer.
(4) PLC interface function Panther meter has Profibus-DP Field bus connected to El; Rio of Allen Bradley PLC connected to 121: Modbus Plus interface: 16 bit D/a precision analog output interface.

4.3 Batching Controller

The batching controller is the foundation and key of the system. It is directly connected with the batching actuator to realize the detection and control of the batching actuator, such as the belt weighing, the start and stop of the conveyor, the switch of the mixer and so on. The batching controller can work independently, that is, when other structures of the system stop working, the batching controller can still independently control the batching actuator to realize the basic automatic batching function, so as to improve the reliability of the system work.

Each batching controller is connected with each other through RS.485 communication interface, and connected to industrial computer through data concentrator. A data concentrator can be connected with 32 control modules. Each control module connected to RS.485 bus can be distinguished from each other by setting different address identification by micro switch. The upper computer can access each control module individually by addressing, and detect and control its status in real time. The industrial control computer in the system mainly completes man-machine dialogue and system management functions, such as formula generation, management and download, remote control and real-time status display of batching unit, statistics of raw material consumption and finished product production, etc. The lower batching controller mainly completes the real-time control function of batching actuator, such as the real-time control of feeding speed.

Each set of batching belt weigher in the system is in the same logistics line, and there is an interlocking relationship between them. The whole batching process is programmed by PLC to realize automatic control. The communication between PLC and IPC can realize remote operation and control of IPC, and realize equipment alarm (such as belt deviation alarm). In the power control design, not only through the PLC internal software to achieve the interlocking, but also in the electrical circuit also increased the electrical signal interlocking, increased double the reliability.

The hardware structure of batching controller includes MCU, keyboard and interface, LCD and interface, analog/switch input processing circuit, analog/switch output

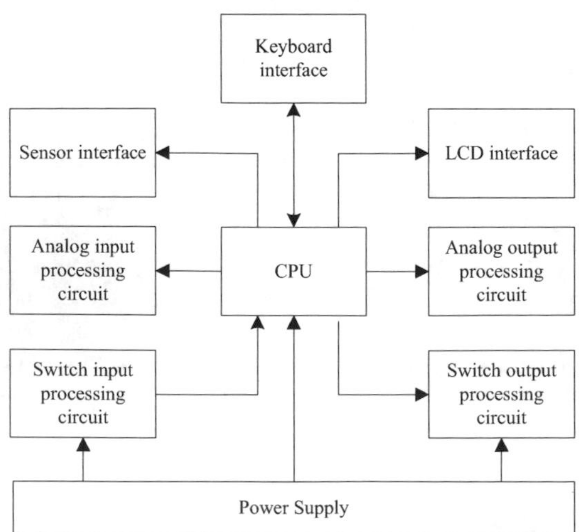

Fig. 1. Hardware structure block diagram of batching controller

processing circuit, LED status indication and interface, RS-232/rs.485 serial communication interface, etc. The hardware structure block diagram of batching controller is shown in Fig. 1.

4.4 Programming Cable

If we want to realize the effective control of the system, we need a programming cable to complete the upload and download of PLC program. The serial communication interface of the computer is different from that of the PLC. The computer generally uses USB interface, while the 57-200 series PLC uses RS485 interface. Therefore, the communication between the computer and the PLC can only be realized by converting the serial interface. Siemens programming cable can realize the conversion of serial interface, such as MPI (multiple point interface) card or PC/PPI (point to point interface) programming cable, and special usb-rs485 converter can also realize the conversion of serial interface. This paper uses the special communication cable designed by Siemens for 57-200 series PLC, namely PC/PPI programming cable.

5 System Software Design

5.1 General Procedure of System Batching

The system is divided into two parts: feeding and batching, and at the same time. Among them, the automatic feeding process can be realized only by using the proximity switch. In comparison, the batching part is slightly complicated, and its program flow is shown in Fig. 2.

In this process, the upper computer first transmits the relevant parameters to the field control system, so that the PLC can quickly calculate the required closing advance and closing weight. In the batching process, the batching weighing control system continuously transmits the weighing data to the PLC; when the real-time weight value is far from the rated weight value, the PLC opens two feeding doors at the same time to improve the batching speed; in order to reduce the impact caused by the falling of raw materials on the scale bucket, once the real-time weight value is equal to the fast casting weight value, the PLC closes one of the feeding doors to improve the batching accuracy; when the real-time weight value is far from the rated weight value, the PLC opens two feeding doors to improve the batching speed When the weight value is equal to the closing weight value, PLC will close the remaining feeding door and delay for a period of time, so that all the remaining materials in the air fall into the weighing bucket. The program of batching control algorithm updates the relevant parameters after a single batching. The stop condition of unloading process is that the real-time weight value read by PLC is the expected residual weight value W, but the weighing system fault will make the unloading process unable to stop, so the timer control program is specially added, that is, once the set unloading time is exceeded, the control system will force to stop the unloading process.

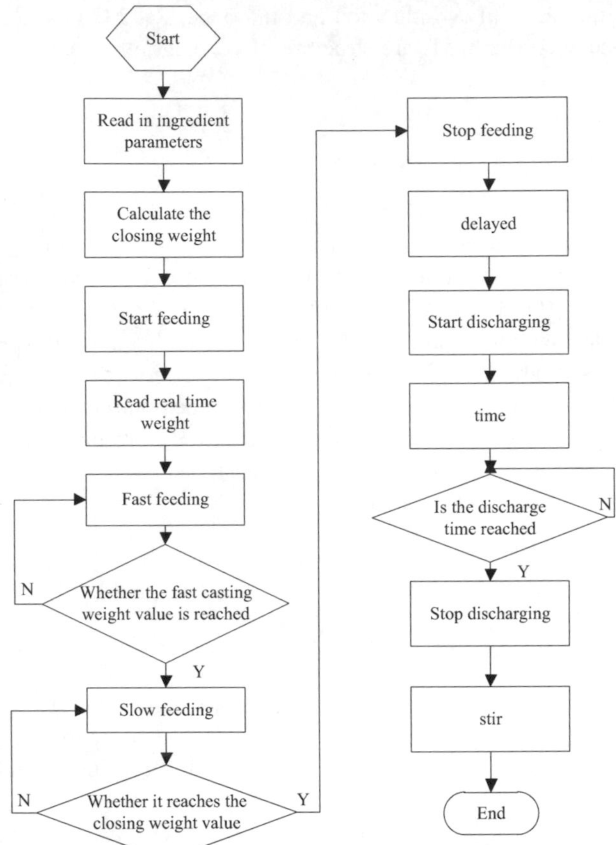

Fig. 2. Workflow of bearing alloy intelligent batching system

5.2 Algorithm Program of Electrical Control for Batching

Fuzzy control is a kind of computer control which takes human control experience as knowledge model and is realized by computer. Its basic idea is to summarize the control strategy of human experts for a specific controlled object or process into a series of control rules expressed in the form of "if (condition) then (action)", and obtain the control action set through fuzzy reasoning, which acts on the controlled object or process [7]. Therefore, fuzzy control is suitable for the control of complex systems without mathematical model or difficult to establish mathematical model. The control action set is a group of conditional sentences, and the state condition and control action are a group of quantified fuzzy language sets. Therefore, what fuzzy control should do is: first, summarize human control ideas, that is, the knowledge of experts or the experience of operators into fuzzy control rules, and then turn these control rules into algorithms that can be executed by computers, and then control the controlled object [8].

The fuzzy control system usually consists of five parts: fuzzy controller, I/O interface, actuator, controlled object and measuring device.

(1) Fuzzy controller is the core of fuzzy control system. It is a language type "fuzzy controller" based on fuzzy control knowledge representation and rule reasoning. The fuzzy controller stores the fuzzy control algorithm derived from rules, which is generally implemented by computer program or hardware [9].
(2) The D/A fuzzy controller takes the difference between the given value and the feedback value of the controlled quantity as the input, which is synthesized by the fuzzy control algorithm to obtain the corresponding control quantity. Because the control quantity is digital quantity and the actuator accepts analog quantity, D/A conversion is needed between fuzzy controller and actuator. Sometimes a level conversion circuit is needed after D/A conversion.
(3) The actuator includes AC/DC motor, stepping motor, hydraulic motor, pneumatic control valve, etc.
(4) The controlled object can be a kind of equipment or device and its group. They work under certain constraints to achieve people's goals. These controlled objects can be deterministic or fuzzy, univariate or multivariable, with or without delay, or linear or nonlinear, steady or time-varying, as well as strong coupling and interference.
(5) Sensor is a kind of device that converts various non electric quantities of the controlled object, such as flow, temperature, pressure, speed and concentration, into electrical signals [10]. When selecting sensors, we should pay attention to the accuracy of sensors.
(6) The A/D sensor converts the information of the controlled quantity into an electrical signal, which is then fed back to the computer.

According to the above analysis, the fuzzy control model of the weighing batching process can be established, as shown in Fig. 3.

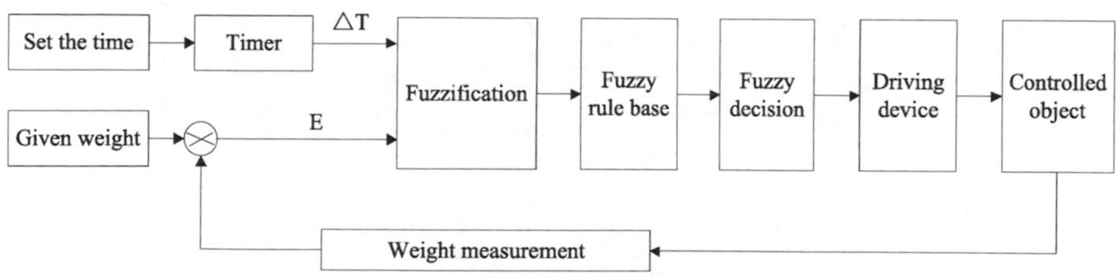

Fig. 3. Electrical paste control principle of bearing alloy intelligent batching

Two inputs of the fuzzy control system are weight deviation E and time deviation ΔR, and one output is control voltage U. The principle of the fuzzy control system is: the system compares the actual weight of the material with the set value to produce the weight deviation E, at the same time, the timer compares with the set time to produce the time deviation at, and sends these two variables into the fuzzy controller. According to the value of the weight deviation E and the time deviation ΔT, the fuzzy controller conducts fuzzy reasoning according to the fuzzy rule base of the system to produce a fuzzy control system a control output voltage U controls the feeding speed of the electromagnetic vibration feeder, so as to realize the control of the controlled system. In this system, when the deviation e is large, the constant speed feeding of PID control is adopted to ensure the weighing time; when the deviation is small to a certain value, the fuzzy control is adopted to make the control quantity U decrease with the decrease of weight deviation E and time deviation Δt until the weighing is completed to ensure the weighing accuracy.

6 System Test Experiment

Using MATLAB software and its fuzzy logic toolbox, the design and performance test of bearing alloy intelligent batching system based on electrical control can be realized conveniently and quickly.

6.1 Proportioning Scheme of Bearing Alloy

Cast a batch of copper base bearings for pumps (zcupb2sn5, Brinell hardness no less than 560 HBW, tensile strength no less than 150 MPa), prepare 500 kg furnace charge, require Pb no less than 20%, Sn no more than 6% (mass fraction), raw materials are Cu-CATH-1 (Cu no less than 99.99%), sn99.95 and pb99, corresponding burning loss mass fraction are 1%, 1%, 1.4% respectively.

6.2 Experimental Platform

In order to study the control effect of batching control algorithm, this paper designs and builds a batching weighing experimental platform to simulate the fine blanking process, as shown in Fig. 4.

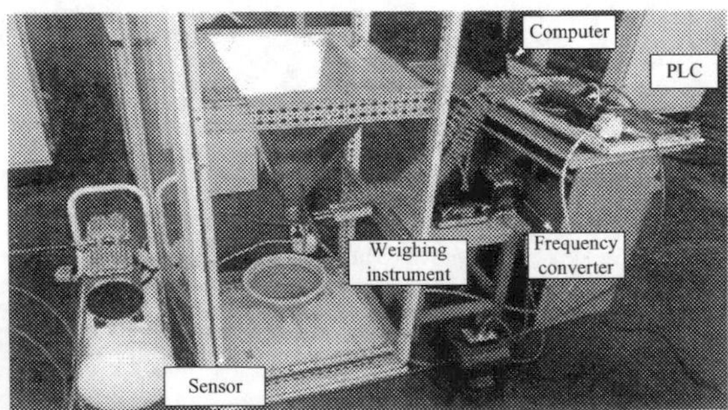

Fig. 4. Diagram of batching and weighing experimental platform

6.3 System Test Index

The batching error is calculated by formula (1): the batching error is calculated by formula (2).

$$Z = X - Y \tag{1}$$

In formula (1), Z is the error value of batching, X is the instrument display value, and Y is the formula setting value.

$$P = \left|\frac{Z}{Y}\right| \times 100\% \tag{2}$$

In formula (2), P is the batching error.

6.4 System Operation

The batching system has been in good condition since it was put into operation in June, and there is no major batching error accident. It shows the advantages of friendly interface, stable operation, high control precision and strong function. Some production commissioning records are shown in Table 1.

Table 1. Production record of copper base bearing for pump

Time	Recipe set point (Kg)	Instrument display value (Kg)	Batching error (Kg)	Batching error (%)
9:55	500	510.0	10.0	2.02
10:00	500	507.2	7.2	1.44
10:02	500	503.6	3.6	0.72
10:05	500	500.4	0.4	0.08
10:10	500	501.3	1.3	0.26
10:15	500	498.8	−1.2	0.24
10:20	500	499.5	−0.5	0.2

From the analysis of the results shown in Table 1, it can be seen that the bearing batching error was relatively large at the beginning of the debugging. After constant modification of the system in this paper, the batching error is gradually reduced and basically controlled within $\pm 1\%$, which proves that the system in this paper has a high batching accuracy.

7 Conclusion

In this paper, an intelligent batching system based on electrical control process is designed and developed independently. The system is suitable for the proportion optimization of bearing alloy and intelligent feeding control, and has the advantages of low efficiency, convenient application and easy control of alloy cost. In the process of operation of the system, the minimum error is taken as the control index, and the batching control is completed based on the fuzzy control algorithm, which meets the requirements of bearing technology for batching accuracy, and proves that the system in this paper has a good application prospect.

Fund Projects. Key topics of Beijing Polytechnic, Development and design of bearing alloy intelligent batching system based on electrical control (2017Z004-006-KXZ).

References

1. Shad, Z.M., Ghavami, M., Atungulu, G.G.: Occurrence of aflatoxin in dairy cow feed ingredients and total mixed ration. Appl. Eng. Agric. **35**(5), 679–686 (2019)
2. Capellades, G., Wiemeyer, H., Myerson, A.S.: Mixed-suspension, mixed-product removal studies of ciprofloxacin from pure and crude active pharmaceutical ingredients: the role of impurities on solubility and kinetics. Cryst. Growth Des. **19**(7), 4008–4018 (2019)
3. Hassani, S.A.M.: Simultaneous determination of active ingredients in multicomponent common over the counter tablets in the present of parabens and 4-aminophenol by HPLC. Chromatographia **83**(7), 791–805 (2020).
4. Dash, S.P., et al.: Performance analysis of coherent PLC With MPSK signaling in Nakagami-*m* noise environment. IEEE Trans. Veh. Technol. **69**(3), 3057–3067 (2020)
5. Deng, Y., Liu, J., Li, D.: Development of a thermal compensator based on PLC for Fanuc CNC system. Int. J. Adv. Manuf. Technol. **112**(7), 1885–1902 (2021)
6. Carek, A.M., Jung, H., Inan, O.T.: A reflective photoplethysmogram array and channel selection algorithm for weighing scale based blood pressure measurement. IEEE Sens. J. **20**(7), 3849–3858 (2020)
7. Yuvaraja, T., Kumar, K.A.R.: Fuzzy control in H-bridge MLI for solar photovoltaic system to enhance load sharing. Int. J. Electr. Eng. Educ. **57**(1), 64–72 (2020)
8. Liu, S., Liu, D., Srivastava, G., et al.: Overview and methods of correlation filter algorithms in object tracking. Complex Intell. Syst. **11**(3), 66–78 (2020)
9. Liu, S., Lu, M., Li, H., et al.: Prediction of gene expression patterns with generalized linear regression model. Front. Genet. **10**(6), 120–131 (2019)
10. Liu, S., Bai, W., Zeng, N., et al.: A fast fractal based compression for MRI images. IEEE Access **33**(7), 62412–62420 (2019)

Research on Cost-Benefit Evaluation Model of Social Security Project Based on Fuzzy Entropy

yi-huo Jiang[1](✉), Wen-bing Jiang[2], and Hong-liang Ni[1]

[1] City College of Dongguan University of Technology, Dongguan 523419, China
[2] Lanzhou City University, Lanzhou 730070, China

Abstract. At present, there is no uniform standard for the cost estimation of social security projects for reference, and the cost composition and estimation contents of projects under the self-management mode of all regions are different. In order to do a good job in the basic data collection and information submission for cost estimation, it is necessary to deeply study the cost efficiency of social security projects and realize the high precision assessment thereof. Aiming at the shortcomings of current cost-benefit evaluation model of social security project, such as large error, long time-consuming, low sensitivity to change of inspection value, etc., this paper studies the relevant reference, and puts forward a cost-benefit evaluation model of social security project based on fuzzy entropy. This paper uses analytic hierarchy process to construct the cost-benefit evaluation index system of social security project. By comparing the two indexes, the importance of each index is judged. The final evaluation index is determined according to the consistency test results of the judgment matrix. Combining with the calculation of fuzzy entropy, the indexes are empowered, and the social security project is established by synthesizing the weight of each index, the relevant data of investment budget and project demand. The cost-benefit evaluation model of the project is integrated to realize the benefit evaluation. The experimental results show that the evaluation efficiency of the proposed model is higher, the sensitivity of the model to the changes of the observed values is significantly higher than that of the experimental comparison model, and the evaluation results are closer to the reference values.

Keywords: Fuzzy entropy · Social security project · Cost- benefit · Evaluation

1 Introduction

With the treaty-based development of information technology and the wide application of computer systems in production, life and business activities, social security projects as an independent whole have gradually become independent and developed rapidly. Social security project is an interdisciplinary subject based on computer technology, system science, management science and communication technology. Social security project is a cross-disciplinary subject, which is oriented to technology and management, and

S. Liu and X. Ma (Eds.): ADHIP 2021, LNICST 416, pp. 517–532, 2022.
https://doi.org/10.1007/978-3-030-94551-0_41

pays attention to the combination of engineering methods and human subjective analysis methods [1]. In today's era of knowledge economy, social security projects will play a more important role. While seeing the rapid development of social security projects, we must clearly see the problems in the construction of social security project. Whether in developed countries or in developing countries, the success rate of social security project construction is far less than other construction projects. One of the important reasons for this situation is lack of timely and necessary evaluation. Compared with the rapid development of information industry, social security project evaluation is lagging behind seriously. Strengthening the research of social security project evaluation theory and method is one of the important contents in this field [2, 3]. The cost-benefit of social security project has a strong lag and recessiveness, which can be reflected only after a considerable period of time, and there are hidden costs, such as maintenance costs and a large proportion of investment, etc. There are many factors affecting the cost-benefit of social security project, which increase the difficulty of evaluation [4].

Now some scholars have put forward better cost-benefit evaluation models. For example, in reference [5], a project cost-benefit evaluation model based on multi-level extension evaluation method was proposed. According to the characteristics of social security project, an index system for evaluating the economic benefit of social security project was constructed from three aspects: operation economic benefit, enterprise financial benefit and social economic benefit. The cost theory in whole life cycle was introduced into the analysis and calculation of enterprise's financial benefit evaluation index. From the long-term economic benefit, the cost of different stages of the system project was considered comprehensively. The comprehensive weight of each index was determined by interval analytic hierarchy process. By introducing matter-element extension theory, an social security project benefit evaluation model based on multi-level extension evaluation method was established. The evaluation results of the model were accurate and sensitive to the changes of the observed values, but it took a long time. Reference [6] presented a cost-benefit evaluation model of social security project based on multi-objective optimization. According to the time dimension and influence, the multi-objective optimization mathematical model of project's cost-benefit evaluation was established. By queuing and filtering, a portfolio solution satisfying the objective and constraint conditions was generated, and the characteristic parameters of the solution were used as the constraint conditions for linear optimization. The local optimal solution of the portfolio was obtained, and the feasible solution was obtained by combining genetic algorithm. The evaluation efficiency of the model was high, but the sensitivity of the model was low. Reference [7] proposed a cost-benefit evaluation model of social security project based on regression analysis. The main influencing factors of project benefit were analyzed, and the benefit evaluation index system was constructed from two aspects of operation efficiency and investment benefit. Through the regression analysis model, the grading interval of indicators for different project characteristics was to determine, an improved matter-element extension model based on A.J. Klee method was proposed to realize the combination of subjective and objective evaluation, and the benefit evaluation model was constructed to calculate the comprehensive evaluation value of each project benefit. The computational complexity of the model was small, but the sensitivity of the model to the variation of the observed values was low. Reference [8] presented a cost-benefit evaluation model of social security project based on marginal benefit. The optimization framework of a project was constructed to describe the overall idea of optimal decision-making, and the marginal benefit of each index to project decision-making optimization was calculated based on the reliability calculation results of the impact index system. The weight distribution of the index was determined according to the marginal benefit of the index, and the comprehensive decision-making value of the project was calculated by combining the corresponding relationship analysis between the project and the index. Combining with the total investment quota, the optimal project was determined and the cost-benefit of social security project was evaluated. Reference [9] taking the power grid cost as an example, based on the importance evaluation, a hierarchical backbone grid construction method is proposed. Based on the life-cycle cost theory and considering the equipment grade, service life and depreciation value, a cost calculation model for differentiated construction investment is established. The typical users of various loads are taken as representatives to establish a load-side outage loss model, which constitutes the disaster-resistant benefit of the hierarchical backbone network with the benefits of the generation side and the power grid side. When the proportion of investment was high, the evaluation result of the model was more accurate, but the sensitivity of the model to the change of the observed value was lower.

To solve these problems, a cost-benefit evaluation model of social security project based on fuzzy entropy is proposed. The main research work is as follows:

(1) Using analytic hierarchy process (AHP), the evaluation index system is constructed and compared. After obtaining the judgment matrix, according to the consistency test results, the matrix is revised to determine the final evaluation index system.
(2) Combining the theory of entropy and the calculation of fuzzy entropy, the comprehensive weight of each index is determined.
(3) The cost-benefit evaluation model of social security project is constructed based on the integration of investment budget and demand of social security project. The model is used to evaluate the cost-benefit.

2 Cost-Benefit Evaluation Model of Social Security Project Based on Fuzzy Entropy

2.1 Cost-Benefit Evaluation Index System of Social Security Project

Before the cost-benefit evaluation of social security project, it is necessary to construct the cost-benefit evaluation index system of social security project, and to complete the construction of evaluation index system by using analytic hierarchy process.

Analytic Hierarchy Process (AHP) was proposed by Saaty, an American operational researcher at the University of Pittsburgh, in the 1970s. The main idea is as follows: divide the evaluation object index into several layers according to the membership relationship, invite experts in relevant fields to compare the importance of each level and factor, establish two or two judgment matrices, and finally calculate the relative weight of the index based on matrix knowledge. The above ideas can be summarized into the following steps:

(1) Establishing an index system;
(2) Obtaining weight vectors of indexes at all levels;
(3) Determining the evaluation value of the relative evaluation index of each scheme;
(4) By synthesizing the evaluation values at all levels, the index weight and the comprehensive evaluation value relative to the total objective are obtained.

This method is a multi-criteria decision-making method combining qualitative method and quantitative method, which has the characteristics of conciseness, systematicness and practicability. Now it has been widely used in complex decision-making systems.

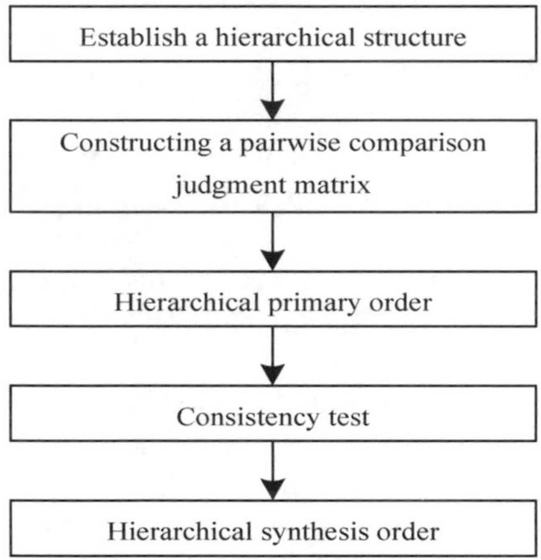

Fig. 1. Basic process of analytic hierarchy process

The analytic hierarchy process is shown in Fig. 1. As can be seen from Fig. 1, the following four aspects of work need to be solved in the analytic hierarchy process:

(1) Establishment of the index system: that is, dividing the constituent elements into different levels and constructing a multi-level hierarchical structure model;
(2) Construction of pairwise comparison judgment matrices: using the upper indexes as the basic criteria, invite experts in the field compare the importance of different indexes at the same level, and construct pairwise comparison judgment matrices according to the corresponding judgment scales;
(3) Hierarchical single ranking and consistency test: on the basis of the judgment matrix, the single ranking of the elements at each level, i.e. the weight vector, is determined. The consistency of the pairwise comparison judgment matrices is tested according to the relevant formulas.
(4) Hierarchical whole-ranking: the ultimate goal of evaluation is to obtain the ranking of the underlying indicators relative to the total objective. This step mainly uses the single-ranking results to obtain the whole-ranking of the evaluation objectives.

The establishment of index system is the first step of analytic hierarchy process (AHP). Its main function is to construct a hierarchical model of evaluation index according to a certain degree of orderliness. The quality of its construction directly affects the effectiveness of the evaluation results. Under this hierarchical model, the complex evaluation object can be decomposed into a hierarchical model composed of many index elements. According to the different attributes of elements, elements of different attribute categories can be divided into different levels. Usually, before implementing this step, we should have a comprehensive, systematic and scientific understanding of the evaluation objectives. It satisfies the principles of scientificity and advancement, comprehensiveness and systematicness, combination of qualitative and quantitative analysis, as well as feasibility and operability [10].

According to the typical hierarchy shown in Fig. 2, each layer element is not completely dominated by the upper element, it is dominated by at least one element of the upper layer. Among them, the hierarchical structure can be macroscopically divided into target layer, intermediate layer and scheme layer.

(1) Target layer: This layer is the highest level, which contains only one element and is the target or ideal state of the evaluation object.
(2) Intermediate layer: This layer, also known as criteria layer, contains relevant intermediate steps and criteria to achieve the intended targets of the object to be evaluated. It often includes a main criteria layer and a sub-criteria layer.
(3) Scheme layer: This layer is the lowest level, representing the decision-making scheme provided to achieve the predetermined evaluation objectives in the target layer.

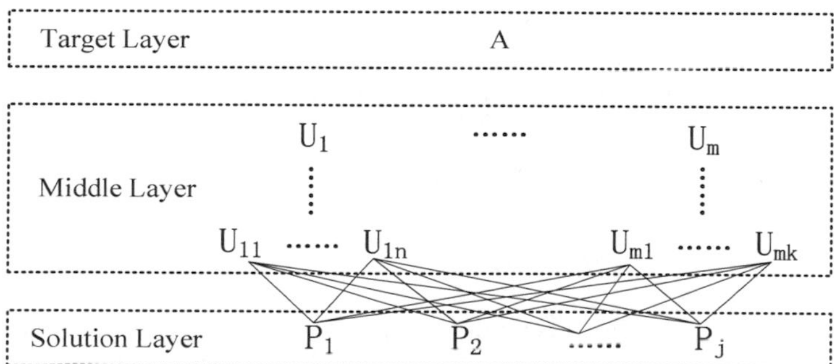

Fig. 2. Hierarchical diagram

In these three macro levels, the target level and the scheme level are fixed for different evaluation objects. In the target layer, no matter what the evaluation object is, the element is always fixed. In the scheme layer, the number of schemes can be taken from 1 to infinite in theory, and the specific number is related to the actual evaluation object. When there is only one scheme, the evaluation will evolve into a single scheme evaluation. For multi-scheme evaluation, it can be seen as a multiple realization of the single scheme evaluation. Among them, the order of the middle layer is at least one layer. The number of specific layers is related to the complexity of the evaluation object and the detailed degree of demand analysis. There is no limit to the number of layers in theory. However, the single-layer construction is better to satisfy the requirement of having less than nine elements, so as to reduce the difficulty of constructing a pairwise judgment matrix. According to the above analysis, a complete hierarchical structure contains at least three levels of order. With the increase of the complexity of the evaluation object, the order may also increase.

An important feature of analytic hierarchy process (AHP) is to express the importance degree of two indexes relative to the previous level in the form of the ratio of two importance degrees for each index [11, 12]. Next, we will explain how to build a pairwise comparison judgment matrix.

If in the cost-benefit evaluation index system of social security project, the index elements U_{k1}, U_{k2},$\cdots$, U_{kn} are the lower indicators with U_k as the basic criteria, now we can get the relative weights of U_{k1}, U_{k2},$\cdots$, U_{kn} based on U_k. When comparing the relative importance of U_{ki} and U_{kj}, a numerical value to represent them is selected. This value can be given directly by the decision maker or obtained through some kind of technical consultation. According to this method, the pairwise comparison judgment matrix $A = \left(A_{ij}\right)_{n\times n}$ can be obtained finally. Among them, A_{ij} represents the important value of factor i relative to factor j.

The matrix descriptions corresponding to the comparison results of the relative importance of pairwise elements are given by using formula (1):

$$A = \begin{pmatrix} A_{11} & A_{12} & \cdots & A_{1n} \\ A_{21} & A_{22} & \cdots & A_{21} \\ \vdots & \vdots & \cdots & \vdots \\ A_{n1} & A_{n2} & \cdots & A_{nn} \end{pmatrix} \tag{1}$$

The matrix shown in the above formula has the following properties:

(1) $A_{ij} \neq 0, (i, j = 1, 2, \cdots, n)$;
(2) $A_{ij} = \frac{1}{A_{ji}}, (i \neq j)$;
(3) $A_{ii} = 1, (i = 1, 2, \cdots, n)$.

Matrix satisfying the above conditions is called positive and negative matrix. For both positive and negative matrix A, matrix A is called a consistent matrix if both of them satisfy $\forall i, j, k(i, j, k = 1, 2, \cdots, n)$ and $A_{ik} \cdot A_{kj} = A_{ij}$ is tenable. According to the above properties, when applying analytic hierarchy process (AHP) to evaluate an object containing n elements, the decision maker needs to carry out pairwise comparisons for at least $n(n-1)/2$ times, and quantifies the relative important values of the elements according to a certain ratio scale to form a numerical judgment matrix. The commonly used comparison scales are shown in Table 1.

Table 1. Scaling table of judgment matrix

Element comparison results	Scale
Element i is as important as j	1
Element i is slightly more important than j	3
Element i is obviously more important than j	5
Element i is intensely important than j	7
Element i is extremely important than j	9
Element i is slightly less important than j	1/3
Element i is obviously less important than j	1/5
Element i is intensely less important than j	1/7
Element i is far less important than j	1/9
The comparison results of the importance of the two elements are between the above conditions	2, 4, 6, 8, 1/2, 1/4, 1/6, 1/8

Ideal judgment matrix is consistent, but due to the lack of awareness of the evaluation object and the complexity of the evaluation object itself, the established judgment matrix often fails to meet the consistency. In order to ensure the use of analytic hierarchy process to obtain reasonable and effective evaluation results, it is necessary to test the consistency of two judgment matrices. The method of consistency test is introduced below.

From the knowledge of matrix theory, we can draw the following conclusion: if $\lambda_1, \lambda_2, \cdots, \lambda_n$ is the characteristic root of matrix A, it satisfies:

$$Ax = \lambda x \tag{2}$$

And if A satisfies $a_{ii} = 1 (i = 1, 2, \cdots, n)$, then:

$$\sum_{i=1}^{n} \lambda_i = n \tag{3}$$

When matrix A satisfies complete consistency, there is $\lambda_1 = \lambda_{\max} = n$, then $\lambda_i = 0 (i = 2, 3, \cdots, n)$; when matrix A does not satisfy complete consistency, $\lambda_1 = \lambda_{\max} > n$, and the sum of the remaining eigenvalues satisfies:

$$\sum_{i=2}^{n} \lambda_i = n - \lambda_{\max} \tag{4}$$

The formula for calculating the consistency test index CI is as follows:

$$CI = \frac{\lambda_{\max} - n}{n - 1} \tag{5}$$

The smaller the calculated value of CI is, the better the consistency is; when $CI = 0$, the judgment matrix has complete consistency.

After completing the consistency test of hierarchical single-ranking, the consistency test of hierarchical whole-ranking is carried out from top to bottom. Assuming that the consistency test index value $CI_j^{(k)}$ is based on the element j at the $k-1$th level, the comprehensive index $CI^{(k)}$ at the kth level is:

$$CI^{(k)} = \left(CI_1^{(k)}, CI_2^{(k)}, \cdots, CI_{n_{k-1}}^{(k)}\right)w^{(k-1)} \tag{6}$$

When $CI^{(k)}$ is less than the set threshold, it is considered that all judgement matrices above the kth level of the hierarchical structure have overall satisfactory consistency.

According to the consistency test results, the judgment matrix that does not meet the needs is revised to determine the final evaluation index system.

2.2 Weighting Computation

After determining the evaluation index system, it is necessary to empower the indicators, describe the proportion coefficient of the corresponding indicators in the cost-benefit evaluation of social security project, and improve the accuracy of the evaluation results. The specific process is as follows:

Let $(i' = 1, 2, \cdots, n', \ j' = 1, 2, \cdots, m')$ be the observation data of the j'th index of the i'th evaluated object, and n' be the number of evaluated objects. m' is the index number of the the i'th evaluated object. For any j', the greater the difference between the observed data $x_{i'j'}$ is, the greater the comparative effect of the index on the system is, and the more the information it contains and transmits are. Entropy is used to measure the uncertainty of information. If entropy increases, information decreases, and vice versa. This method of measuring information with entropy is the Entropy Value Method [13]. The steps to determine the weight of the index by the method of entropy are as follows:

(1) Under calculating the i'th object, the characteristic proportion $f_{i'j'}$ of the j'th index is calculated, and the formula is as follows:

$$f_{i'j'} = \frac{x_{i'j'}}{\sum_{i'=1}^{n'} x_{i'j'}} \tag{7}$$

The constraint condition is $x_{i'j'} \geq 0$, and $\sum_{i'=1}^{n'} x_{i'j'} > 0$.

(2) According to the calculation formula of the entropy value, the entropy value $e_{j'}$ of the j'th evaluation index is:

$$e_{j'} = -\frac{1}{\ln n'} \sum_{i'=1}^{n'} f_{i'j'} \ln(f_{i'j'}) \tag{8}$$

In the formula above, $n' > 0$, $e_{j'} > 0$.

(3) Determining weights.

When given j', the difference of $x_{i'j'}$ becomes smaller, $e_{j'}$ will increase. For a given j', when $x_{i'j'}$ is all equal, $f_{i'j'} = 1/n'$, $e_{j'} = e_{\max} = 1$, then $x_{j'}$ will not play any role in comparing the indicators between systems; conversely, when the given difference of $x_{i'j'}$ increases, $e_{j'}$ will decrease, and then the comparative effect of index $x_{j'}$ on the system will increase. Based on the above analysis, $1 - e_{j'}$ is as the difference coefficient of index $x_{i'j'}$, the greater the value of it is, the more the attention should be paid to the comparative role of index is. Let the index entropy weight set $w' = [w'_1, w'_2, \cdots, w'_{m'}]^T$, and the entropy weight of the j'th evaluation index is as follows:

$$w'_{j'} = \frac{1 - e_{j'}}{m - \sum_{j'=1}^{m} e_{j'}} \tag{9}$$

Assuming the fuzzy number $r_{i'} = [a_{i'j'}, b_{i'j'}, c_{i'j'}]$, the corresponding membership function is:

$$f_{r_{i'}}(x) = \begin{cases} 0, & x \le a_{i'j'} \\ \frac{x - a_{i'j'}}{b_{i'j'} - a_{i'j'}}, & a_{i'j'} < x \le b_{i'j'} \\ \frac{x - c_{i'j'}}{b_{i'j'} - c_{i'j'}}, & b_{i'j'} < x \le c_{i'j'} \\ 0, & x > c_{i'j'} \end{cases} \tag{10}$$

Where, $x \in R$, $a_{i'j'} < b_{i'j'}$ and $c_{i'j'}$ are upper and lower bounds respectively, indicating the degree of ambiguity. The larger the values of $a_{i'j'}$ and $c_{i'j'}$ are, the higher the degree of ambiguity is. Among them, $a_{i'j'}$, $b_{i'j'}$ and $c_{i'j'}$ represent the most conservative, most likely and optimistic evaluation values given by the i'th expert on the importance of index j', respectively.

The expert evaluation weight set $E = [e_1, e_2, \cdots, e_k]$ is obtained, in which e_k is the proportion of the evaluation value of the kth expert to the evaluation object in the comprehensive evaluation.

The triangular fuzzy number complementary judgment matrix [14] is established by several experts according to the purpose of evaluation and the relevant information of evaluation index. Then the weight set of expert evaluation is used to calculate the triangular fuzzy comprehensive judgment matrix T of evaluation index.

$$T = [\alpha_{i'j'}e_{i'}, b_{i'j'}e_{i'}, r_{i'j'}e_{n'}] \tag{11}$$

Where, $\alpha_{i'j'}e_{i'}$, $b_{i'j'}e_{i'}$ and $r_{i'j'}e_{n'}$ represent the most conservative, most likely and optimistic estimates of index j', respectively.

If the weight vector of the fuzzy number to be determined is v, the formula for calculating the fuzzy score of the j'th index is as follows:

$$v_{j'} = (a_{j'} + 5b_{j'} + c_{j'})/7 \tag{12}$$

Let the index fuzzy set be $w'' = [w''_1, w''_2, \cdots, w''_{m'}]^T$, the j'-th index fuzzy weight is obtained by normalization.

$$w''_{j'} = \frac{v_{j'}}{\sum_{j'=1}^{m'} w''_{j'}} \tag{13}$$

In the index combination weight set $W = [w_1, w_2, \cdots, w_{m'}]^T$, $w_{j'}$ is the j'th index weight after the above weighted combination. The $w'_{j'}$ and $w''_{j'}$ are replaced by $w_{j'}$ by linear combination.

$$w_{j'} = \theta w'_{j'} + (1 - \theta) w''_{j'} \tag{14}$$

Where, θ is the proportion of objective preference coefficient weight to combination weight; $1-\theta$ is the proportion of subjective preference coefficient weight to combination weight.

According to the linear weighted comprehensive evaluation method, a comprehensive score of $D_{j'}$ is obtained:

$$D_{j'} = \sum_{j'=1}^{m'} d_{i'j'} w_{j'} \tag{15}$$

Where, $d_{i'j'}$ is the evaluation index score, $w_{j'}$ is the combination weight of the j'th index.

2.3 Establishment of Cost-Benefit Evaluation Model

In order to meet the market demand, social security project expansion needs to adapt to the market development through certain adjustments. In the process of project expansion decision-making, decision makers must consider the cost of project expansion and subsequent operational benefits. With this as the objective function, an evaluation model is constructed, and the project expansion is carried out according to the results of calculation and analysis to maximize the benefits [15]. The concrete process of constructing cost-benefit evaluation model of social security project is as follows:

Assuming that in the stage t, for the economic benefits produced by the part x_{ij} of the social security project, α'_{ij} represents the cost consumed, β'_{ij} represents the depreciation cost of the project investment in that stage, y_{ij} represents the Boolean variable, b_t represents the investment budget in the stage t, B represents the total investment amount, d_j represents the project demand amount in this stage, M_{ij} represents the expansion restriction of the project in this stage, and then the demand is determined and the social security project is constructed. The mathematical model of project's cost-benefit evaluation is as follows:

$$BM = \max \sum_{i=t_1}^{t_s} \sum_{j=1}^{n} (\gamma'_{ij} x_{ij} - \alpha'_{ij} x_{ij} - \beta'_{ij} x_{ij}) \tag{16}$$

The constraints are:

$$\begin{cases} s.t.\ 0 \le x_{ij} \le M_{ij} y_{ij} \\ \sum_{i=t_1}^{t_s} \sum_{j=1}^{n} (\alpha'_{ij} x_{ij} + \beta'_{ij} x_{ij}) \le B \\ \sum_{i=1}^{n} x_{ij} \ge d_j \\ y_{ij} \in \{0, 1\} \end{cases} \tag{17}$$

For the above model, the project demand is known, and the objective function of benefit evaluation requires the maximum of the net benefit obtained by the project. The constraints are the restriction of the increment of the project in each stage, the restriction of the investment cost budget, and the restriction of the demand of the project quantity.

In order to simplify the calculation and reduce the complexity, the model expression is transformed into the following form:

$$BM' = \min \sum_{i=t_1}^{t_s} \sum_{j}^{n} -(\gamma'_{ij} x_{ij} - \alpha'_{ij} x_{ij} - \beta'_{ij} x_{ij}) \tag{18}$$

For uncertain demand, in order to describe uncertain demand, the probability p'_i of different changes in demand is introduced. In order to avoid increasing the complexity of model constraints, the model is decomposed by scenario tree method, and the cost-benefit equilibrium analysis of uncertain demand is established on the description of new model. After splitting and reconstructing, the mathematical expression of the model is as follows:

$$BM^* = \min \sum_{j=1}^{n} P'_j \left(\sum_{i=t_1}^{t_s} \left(-\gamma'_{ij} x_{ij} + \alpha'_{ij} x_{ij} + \beta'_{ij} x_{ij} \right) \right) \tag{19}$$

The constraints of the model are the same as above. For the project's cost-benefit evaluation model with certain demand and uncertain demand, after model transformation, it is solved by lot-sizing heuristic algorithm.

3 Experimental Simulation

In order to prove the comprehensive effectiveness of the proposed cost-benefit evaluation model for social security project based on fuzzy entropy, a simulation experiment is needed. The experimental environment is as follows:

Windows 7, 32-bit operating system platform, Intel Core i7 8700K processor, 3.7 GHz main frequency, 4.7 GHz turbo boost, six cores.

Experiments are carried out with the proposed model, the cost-benefit evaluation model based on multi-level extension evaluation method, the cost-benefit evaluation model based on multi-objective optimization, the cost-benefit evaluation model based on regression analysis, and the cost-benefit evaluation model based on marginal benefit for social security project. The models are compared when the proportion of investment is different. The results of the overall benefit evaluation are shown in Table 2. In Table 2, MOA, MOB, MOC, MOD and MOE represent the proposed model, the cost-benefit evaluation model based on multi-level extension evaluation method, the cost-benefit evaluation model based on multi-objective optimization, the cost-benefit evaluation model based on regression analysis, and the cost-benefit evaluation model based on marginal benefit for social security project.

Table 2. Comparison of the total benefit evaluation values of the funds of each model

Evaluation model	Proportion of expenditure %			Relevance
	60%	30%	3%	
MOA	0.9995	0.5213	0.0433	Strong
MOB	0.9986	0.7547	0.5436	Stronger
MOC	0.9987	0.9277	0.8251	Weak
MOD	0.9979	0.9168	0.8709	Weak
MOE	0.9962	0.9205	0.8324	Weak

Analyzing the data in Table 2 and comparing the changes of the evaluation value of each model with the proportion of investment, when the proportion of investment is 60%, the evaluation value of each model is close, but when the proportion of investment is reduced to 3%, the change of the evaluation value by using the cost-benefit evaluation model of social security project based on multi-objective optimization, the cost-benefit evaluation model of social security project based on regression analysis and the cost-benefit evaluation model of social security project based on marginal benefit are small, which shows that the sensitivity of the three evaluation models to the change of the evaluation value is small. The evaluation value of the cost-benefit evaluation model based on multi-level extension evaluation method decreases to 0.5436. It shows that the sensitivity of the model to the change of the evaluation value is high, and the evaluation value of the proposed model decreases to 0.0433, indicating that the sensitivity of response to change of value is very high.

The proposed model and the project's cost-benefit evaluation model based on multi-level extension evaluation method are used to carry out the experiment. The short-term net income evaluation results of the two models are compared and the experimental results are shown in Fig. 3. In Fig. 3, RA, RB and RC represent the evaluation results of the proposed model, the evaluation results of the project's cost-benefit evaluation model based on multi-level extension evaluation method and the reference evaluation values, respectively.

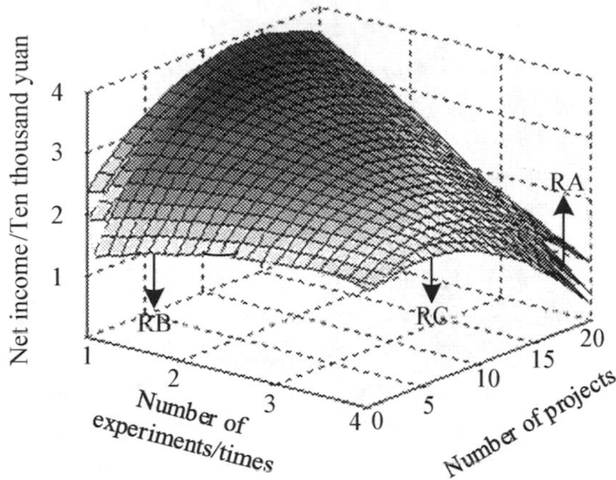

Fig. 3. Comparison of short-term net income evaluation results

According to Fig. 3, when the number of projects increases from 0 to 15, the net income of social security project increases gradually. When the number of projects increases from 15 to 20, the net income of social security project decreases gradually. Compared with the cost-benefit evaluation model based on multi-level extension evaluation method, the short-term net income evaluation results of the proposed model are closer to the reference value, indicating that the result of benefit evaluation of the proposed model is more accurate.

The cost-benefit evaluation model based on the proposed model, the cost-benefit evaluation model based on multi-level extension evaluation method, the cost-benefit evaluation model based on multi-objective optimization, the cost-benefit evaluation model based on regression analysis and the cost-benefit evaluation model based on marginal benefit for social security project are used to carry out experiments, and the models are compared with each other in the case of the number of projects changing. The time required for cost-benefit evaluation is shown in Table 3. In Table 3, MOA, MOB, MOC, MOD and MOE have the same meanings as Table 2.

Table 3. Time-consuming comparison of various models

Evaluation model	Number of projects		
	5	10	20
MOA	5.45 s	10.21 s	20.05 s
MOB	6.27 s	13.56 s	27.33 s
MOC	5.82 s	12.63 s	25.71 s
MOD	6.71 s	14.25 s	28.34 s
MOE	6.03 s	13.67 s	27.16 s

By analyzing the data in Table 3, the time needed for the benefit evaluation of each evaluation model increases with the increase of the number of projects. By calculating, it can be seen that when the amount of project increases, the evaluation time of the proposed model increases the least. By comparing the evaluation time of the same number of projects, it can be seen that the cost-benefit evaluation time of the proposed model is the shortest.

Based on the above experimental results, in order to further verify the effectiveness of the proposed model, an ablation experiment is designed below. The cost-benefit evaluation model based on multi-level extension evaluation method, the cost-benefit evaluation model based on multi-objective optimization, the cost-benefit evaluation model based on regression analysis, and the cost-benefit evaluation model based on marginal benefit for social security project are selected as the control group. The experimental results are compared with those of the designed model. The short-term net income and time consumption of the above different models are set to be the same. The short-term net income is 30,000 yuan, the time consumption is 20 s, and the number of experiments of different models is 70 times. The evaluation accuracy of different models is compared. The comparison results are as follows:

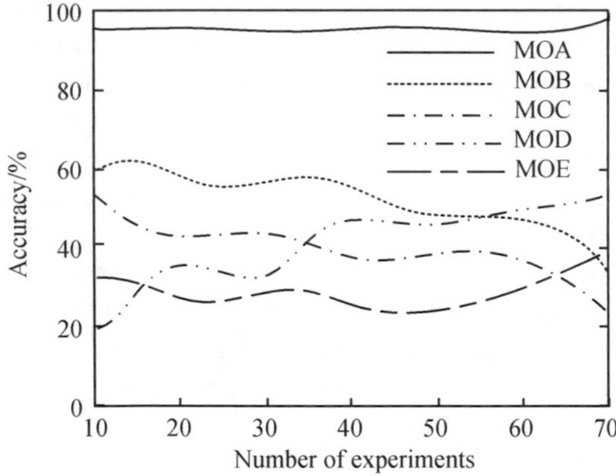

Fig. 4. Comparison results of accuracy of different models

According to Fig. 4. The experimental results, compared with the cost-benefit evaluation model based on multi-level extension evaluation method, the cost-benefit evaluation model based on multi-objective optimization, the cost-benefit evaluation model based on regression analysis, and the cost-benefit evaluation model based on marginal benefit for social security project, the proposed model has higher accuracy.

4 Conclusions

In order to improve the accuracy and decision-making efficiency of cost-benefit evaluation of social security project, a cost-benefit evaluation model of social security project based on fuzzy entropy is proposed in view of the problems existing in the current cost-benefit evaluation model of social security project. The cost-benefit evaluation index system of social security project is established by analytic hierarchy process to provide basis for cost-benefit evaluation. The evaluation matrix is obtained by comparing the importance of the index. According to the consistency test results of the judgement matrix, the judgement matrix that does not meet the consistency needs is revised to determine the final evaluation index system, reduce the number of indicators and improve the operation efficiency. Through the corresponding calculation results of fuzzy entropy, the comprehensive weight of each index is determined to improve the accuracy of cost-benefit evaluation results of social security project. Based on the above evaluation index system and weight calculation, the cost-benefit evaluation model of social security project is constructed by synthesizing the related data of social security project demand and investment budget. The simulation results show that the proposed model has good robustness, less error and higher evaluation efficiency.

The simulation results validate the cost-benefit evaluation performance of the proposed model, but some problems have been found in the whole research process, which still need to be further studied. The prospects for future research are as follows:

Cost-benefit model in social project expansion has always been a research hotspot. Especially in recent years, the benefit of social security project has attracted more attention and more extensive research. In the future research, there are still several problems to be further studied:

(1) Cost-benefit optimal equilibrium model of project expansion constrained by stochastic factors.

In recent years, stochastic programming and algorithms have been applied to the study of network flow planning theory, these models mostly involve stochastic demand, and there are many ways to represent stochastic demand. The optimal cost equilibrium model of stochastic social project construction is the maximum capacity model of social security project under stochastic constraints, the maximum net profit model of social security project, the minimum cost model of project construction and the profit generated by the maximum unit cost. This kind of model has a wide application background in real life, and the description of this kind of model is very close to many uncertain situations in real life.

(2) Cost-benefit optimal equilibrium model for multi-stage project expansion.

Multi-stage project expansion model has always been a very important model in the theory of social security project construction. The study of cost-benefit optimal equilibrium model for multi-stage project expansion is also complicated. Because it is built in stages, the decision-making of project expansion in the latter stage must be based on the former stage. The expression of operational benefit function of social security project will be more complex, and the expression of cost function will vary greatly in each stage, not only piecewise linear function. When we study the optimal adjustment capacity and maximum net profit model of multi-stage social security project expansion, there will be more changes, and the algorithm will be completely different from the single-stage project construction model. The cost-benefit equilibrium model of multi-stage project expansion can also be considered together with the stochastic project expansion model.

Fund Projects. Major scientific research and cultivation project (2021), City College of Dongguan University of technology.

References

1. Kühnast, S., et al.: Evaluation of adjusted and unadjusted indirect comparison methods in benefit assessment. A simulation study for time-to-event endpoints. Meth. Inf. Med. **56**, 261–267 (2017)
2. Wang, G.P., Chan, C.Y.: Efficient processing of enumerative set-based queries. Inf. Syst. **55**, 54–72 (2016)
3. Harris, A., Lim, S.L.: Temporal dynamics of sensorimotor networks in effort-based cost-benefit valuation: early emergence and late net value integration. J. Neurosci. **36**, 7167–7183 (2016)
4. Heydt, G.T.: A probabilistic cost/benefit analysis of transmission and distribution asset expansion projects. IEEE Trans. Power Syst. **32**(5), 4151–4152 (2017)
5. Chen, Y., Lu, L., Huang, Y.Z., et al.: Evaluation model of economic efficiency of distribution network planning based on multi-level extension evaluation method. China Electr. Power **49**, 159–164 (2016)
6. Hua, B., Chen, Y.Q., Mu, L.X., et al.: A hybrid optimization method for overseas project investment optimization combination. Fault Block Oil Gas Field **24**, 238–242 (2017)
7. Fang, Y.J., Wang, X.L., Shi, J., et al.: Evaluation of efficiency and efficiency of provincial power grids with new energy access. Power Grid Technol. **41**, 2138–2145 (2017)
8. Zhou, X.M., Zhang, Q., Yang, W.H., et al.: Optimization of medium voltage distribution social projects based on reliability marginal benefits. Power Grid Clean Energy **33**, 24–30 (2017)
9. Lin, X., Liu, Y., Xu, L.X., Ma, C.X., et al.: Construction of graded backbone grid and its economic assessment. J. Chongqing Univ. (Nat. Sci. Edn.) **2**, 31–41 (2019)
10. Lee, D.H.: Cost-benefit analysis, LCOE and evaluation of financial feasibility of full commercialization of biohydrogen. Int. J. Hydrogen Energy **41**, 4347–4357 (2016)
11. Li, J., Deng, C., Li, Y., Li, Y., Song, J.: Comprehensive benefit evaluation system for low-impact development of urban stormwater management measures. Water Resour. Manage **31**(15), 4745–4758 (2017)
12. Khodashenas, P.S., Thouénon, G., Rivas-Moscoso, J.M., et al.: Benefit evaluation of all-optical subchannel add-drop in flexible optical networks. IEEE Photonics J. **8**, 1–9 (2016)
13. Cao, Z., Lin, C.T.: Inherent fuzzy entropy for the improvement of EEG complexity evaluation. IEEE Trans. Fuzzy Syst. **26**, 1032–1035 (2018)
14. Christensen, V., Steenbeek, J., Failler, P.: A combined ecosystem and value chain modeling approach for evaluating societal cost and benefit of fishing. Ecol. Modell. **222**, 857–864 (2017)
15. Bai, C., Sarkis, J.: Supplier development investment strategies: a game theoretic evaluation. Ann. Oper. Res. **240**(2), 583–615 (2014)

Research on Evaluation Model of Social Network Information Management Based on Asymmetric Fuzzy Algorithm

yi-huo Jiang[1(✉)], Wen-bing Jiang[2], and Hong-liang Ni[1]

[1] City College of Dongguan University of Technology, Dongguan 523419, China
[2] Lanzhou City University, Lanzhou 730070, China

Abstract. The existing evaluation model of the index system and the weight of each index is not perfect, leading to the final evaluation result is not ideal. Therefore, the evaluation model of social network information management based on asymmetric fuzzy algorithm is established. Based on the theory of analytic hierarchy process (AHP), by comparing two indexes of the same level, the judgment matrix is constructed, and the weights of indexes of different levels are obtained. On this basis, the process of determining the security level of MIS by using fuzzy asymmetric closeness degree is discussed, and the specific steps of system security early warning evaluation are given. The practical application of the evaluation model shows that the social network information management system developed by the enterprise is generally satisfactory, especially in the aspect of performance level index. For the independent business system and special business system put into use in the early stage of the company, the development level and performance level of the integrated social network information management system are satisfactory. On the other hand, the technical performance, system characteristics, operation efficiency and operation use of the system need to be improved.

Keywords: Asymmetric fuzzy algorithm · Social network information management · Evaluation model · Analytic hierarchy process · MIS safety level

1 Introduction

In the Internet age, the social communication mode is broken, and the people begin to accept the communication in the form of words, pictures, videos and so on. This online form of social interaction is becoming more common and is gradually replacing some of the traditional face-to-face social interaction. Social networks allow users to freely generate content, voluntarily share content, and spontaneously distribute content on their platforms. This makes the user's offline social relations in the online can be divergent, transfer and extension. Users' behavior of posting, replying, commenting, forwarding and so on not only enrich the content of social network, but also make the potential influence of social network huge. SNS (Social Networking Services) is a Social network-oriented Internet service established by people according to the six degrees of

S. Liu and X. Ma (Eds.): ADHIP 2021, LNICST 416, pp. 533–550, 2022.
https://doi.org/10.1007/978-3-030-94551-0_42

segmentation theory. Generalized says, everything will "relationships" function as the interaction between the core network products can be attributed to under the large concept of SNS, typically including Facebook, renren and kaixin, domestic well-known micro bobby such as sina weibo, in addition to information release, also fully integrated all sorts of social functions, such as chatting, micro group, classification of community, etc., so also belong to the category of social networking applications.

In recent years, with the rapid development of modern social network information system technology, more and more attention has been paid to the use of social network information systems in all walks of life. Compared with the traditional mode of work, social network information system can reduce the working link, speed up the transmission of information, and greatly enhance the storage of data security and confidentiality. At the same time, data and information processing, query, statistics, and analysis will be more convenient. With the improvement of people's daily work efficiency, social network information system is vigorously promoting the development of social productive forces. However, due to the rapid development of social network information systems, the realization of their functions is often the focus of scholars and users' research and application, but the application quality of the system is easy to be ignored. Therefore, although many social network information systems are in operation and use, their actual efficiency is difficult to be satisfactory [1]. In order to adapt to the development situation, enterprises will continue to develop new social network information systems or constantly improve the existing systems, which requires us to conduct a comprehensive evaluation of the social network information system, so as to find out the existing problems of the system, and also to avoid the same problem again in the newly developed system [2].

The evaluation of social network information system is a very large and complex project. First of all, the construction of social network information system is a high-tech field. On the one hand, it has practical significance and is worth studying. On the other hand, the risks associated with it are relatively large. Secondly, the construction of social network information system is a long-term investment, not only visible hardware facilities, soft investment (development, maintenance, training, etc.) in the system construction and operation is bound to account for a large proportion of the total investment, but this part of the cost cannot be clearly reflected. Thirdly, the benefit which the social network information system brings is difficult to manifest in the system movement in short time, which has the lagging characteristic. Fourthly, the enterprise management foundation, the rule Institutions, user's awareness level, knowledge level and other factors have a direct bearing on the size of the role played by social network information systems [3, 4]. The above intertwined factors will affect the quality of social network information systems and the evaluation of project implementation. These factors include political, economic, social, technical and so on. Therefore, enterprise social network information system project evaluation is actually a very complex and comprehensive subject. At present, there is a groping stage in China and abroad in the three major research fields of evaluation basic theory, evaluation index system design and specific evaluation implementation methods.

Reference [5] puts forward and establishes a social network information management evaluation model based on integration perspective. From the perspective of technology-service-business support integration, Delphi method is used to conduct two rounds of expert consultation on 17 experts to establish the evaluation index system of social network information management system construction effect, and AHP is used to determine the index weight. The recovery rates of the two rounds of expert consultation were 94% and 88%, respectively. More than 50% of the 17 experts were enterprise social network information system managers who had worked in related fields for more than 10 years. They were direct users of the social network information system, and their authority was 83%. The final index system consists of 6 first level indicators and 24 two level indicators. The index system reflects the measurement of different levels of the social network information management system, and lays a foundation for the comprehensive analysis of the problems and improvement direction of the social network information management system. Reference [6] puts forward and establishes a social network information management evaluation model based on expectation-satisfaction evaluation. From the perspective of "user view" and based on the method of user expectation-satisfaction evaluation model, it reveals the informatization effect of enterprises at the micro-level by analyzing the correlation between users' expectation effect and actual performance of enterprise social network information system. In reference [7], aiming at the problem of resolution limitation in the new method of network key node identification h-index, based on the consideration of network information propagation probability, a method of social network key node identification based on adjacent h-index is proposed to realize the evaluation of social network information management, This method can accurately measure the real influence level of the nodes to be evaluated. Sir (stochastic infectious recovered) model is used to evaluate the actual communication ability of nodes in real social networks, and the effectiveness of the proposed method is verified. The characteristics of knowledge sharing management in social networks are analyzed theoretically in reference [8], The evaluation of social network information management is completed by visual tools. Firstly, the questionnaire is designed, and the matrix is drawn through the survey results, and then the organizational knowledge sharing and communication chart is drawn by using social network analysis method.

The above-mentioned evaluation model is not comprehensive enough to establish the evaluation index system of MIS and assign the weights of the indicators at all levels, so that the final evaluation results cannot find the existing problems in MIS in time. A social network information management evaluation model based on asymmetric fuzzy algorithm is established in this study. The overall framework of the study is described as follows:

(1) According to the demand of enterprise for MIS and the characteristics of MIS, the evaluation index system of MIS is established on the basis of consulting experts' opinions.
(2) Based on the theory of analytic hierarchy process (AHP), a judgment matrix is constructed by comparing two indexes at the same level, and the weights of indexes at different levels are calculated by combining the mathematical operation of the judgment matrix, then the consistency test is carried out.

(3) On this basis, the fuzzy asymmetric closeness degree is used to determine the security level of MIS, and the concrete calculation steps of system security early warning evaluation are given.
(4) The practical application results of the evaluation model show that the social network information management system developed by an enterprise is generally satisfactory, especially in the aspect of performance level indicators. For the independent and special business systems that have been put into use in the early stage of the company, the development level and performance level of the integrated social network information management system are both satisfactory. On the other hand, the technical performance, system characteristics, operational efficiency and operational use of the system still need to be improved.

The spread speed of social network information is very fast, which makes social network become a bigger influence. The current situation of huge user base and long time investment determines the influence of social network in our country. Therefore, the analysis of social network information management methods and related evaluation is to promote the healthy development of social networks necessary premise. Therefore, this research has established the social network information management appraisal model based on the asymmetric fuzzy algorithm.

2 Research on Social Network Information Management Evaluation Model Based on Asymmetric Fuzzy Algorithm

2.1 Construction of Evaluation Index System for Social Network Information Management

On the basis of consulting and summarizing the existing domestic and foreign literature, combining with the enterprise's demand for social network information management system and the characteristics of enterprise social network information management system itself, and fully consulting experts and scholars' opinions, the evaluation index system of social network information management system is finally obtained. The specific process is as follows:

(1) According to the domestic and foreign references, it summarized the existing social network information management system evaluation index system situation, and listed different evaluation index system one by one;
(2) The existing evaluation index system is comprehensively analyzed, the indicators with the same meaning are summarized, the different indicators are classified, and a comprehensive evaluation index system of social network information management system based on literature is obtained.
(3) According to the demand of the public transportation industry and the enterprise for the social network information management system and the characteristics of the enterprise social network information management system, the index is screened, optimized, recombined and designed, and the preliminary evaluation index system of the enterprise social network information management system is obtained.

(4) The social network information system technical experts, the enterprise social network information management system developers, the managers of the enterprise departments and the university teachers (altogether 8 persons) were invited to put forward their opinions on the preliminary evaluation index system of the enterprise social network information management system.
(5) After collecting the opinions of experts and scholars, the evaluation index system of enterprise social network information management system is optimized.
(6) It consulted experts and scholars again to optimize the evaluation index system of the enterprise social network information management system.

After two consultation and optimization, the enterprise social network information management system evaluation index system was finally obtained. The index system consists of five primary indicators, including technical performance, system characteristics, operational efficiency, performance level and operational use. Each primary index contains several secondary indcxes, and there are 20 secondary indexes in the evaluation index system. The specific indicators are as follows:

Technical Performance Index

(1) The degree of realization of goal function, which indicates the degrce to which the system achieves the planning objectives proposed in the planning, design and analysis phases.
(2) Progressiveness, which indicates whether the social network information management system integrates the leading scientific management knowledge, and whether the design of the system is scientific and has strong applicability.
(3) Standardity, which indicates whether to set up social network information system platform according to relevant international standards, national standards, industry standards, database standard format and metadata standard format, and whether documents and materials are standardized for archiving.
(4) The quality of software, which includes the operability, scalability and practicability of the system software.

Characteristics of Social Network Information Management System

(1) Functional integrity, which indicates whether the system covers information collection, processing, storage, archiving, organization, utilization and other functions.
(2) Work efficiency, which mainly refers to the efficiency characteristics related to time and speed, including reaction time, operation speed and so on. It can be embodied in the time of opening the system, searching for information, importing and exporting data, implementing various functions, opening various links and so on.
(3) Security and confidentiality, which indicates whether the system software and hardware are reliable, whether it meets the national security requirements for credit user identification, authentication, anti-illegal intrusion, firewall, information management software, system data backup and so on.
(4) Maintainability, maintenance personnel maintenance system difficulty level, it is determined by the system modularization, simplification and standardization and other factors.

(5) Reliability, which indicates the basic guarantee of the system's continuous operation according to the functional requirements, the system can be quickly repaired after failure, while the data can be protected in this case.

Operation Benefit

(1) Direct economic benefits and direct benefits gained from the application of social network information systems. It includes the increase of passenger traffic, the increase of passenger revenue and the increase of government subsidies.
(2) Indirect economic benefits and indirect benefits gained from the application of social network information system. It includes the expansion and optimization of line resources, the improvement of brand value, the promotion of corporate culture, the promotion of corporate image and so on.
(3) Social benefits, which providing better services for the public and scientific utilization of public resources.

Performance Level

(1) The degree of realization of the planning objectives, which indicates the phased slogans of the strategic planning, the overall objectives and the degree of realization of the specific indicators.
(2) Management efficiency, which indicates whether the application of the system brings about the reform and innovation of organizational structure, management mode and business process; it includes the ability of information collection, transmission, processing speed, operation efficiency, responding to user needs, improving the level of decision-making and so on.
(3) Cost savings. It can be measured by saving office expenses, reducing the number of workers, reducing the intensity of work, and optimizing the use of resources.
(4) User satisfaction, which indicates whether the functions, characteristics, services and interfaces of the system meet the diversified and personalized needs of users, whether users are satisfied or not.

Based on the above analysis, we set up an evaluation index system for social network information management, as shown in Table 1.

Table 1. Evaluation index system of social network information management

Evaluation index system	Primary index U_i	Secondary index U_{ij}
Social network information management evaluation index system (U)	Technical performance (U_1)	Target function implementation (U_{11})
		Progressiveness (U_{121})
		Standardization (U_{13})
		Software quality (U_{14})
	System characteristics (U_2)	Functional integrity (U_{21})
		Work efficiency (U_{22})
		Security and confidentiality (U_{23})
		Maintainability (U_{24})
		Reliability (U_{25})
	Operation benefit (U_3)	Direct economic benefits (U_{31})
		Indirect economic benefits (U_{32})
		Social results (U_{33})
	Performance level (U_4)	Planning target realization degree (U_{41})
		Management efficiency (U_{42})
		Cost savings (U_{43})
		User satisfaction (U_{44})
	Operation and use (U_5)	Input standardization (U_{51})
		Output standardization (U_{52})
		Operation safety (U_{53})
		Friendly and convenient interface (U_{54})

2.2 Determining the Weight of Risk Indicators

In the index system, the importance of each index to the target is different, when measuring the contribution of each index to the target, different weights should be given, the important person should be given greater weight. Index weight is a quantitative way to reflect the proportion of various indicators in the comprehensive evaluation. Whether the weight is determined scientifically and rationally will directly affect the accuracy of evaluation, which is an important factor in the evaluation process. The weight can be determined individually, but in order to make the result more authoritative and reasonable, the expert investigation method is adopted, and the AHP method is used to determine the weight on the basis of the experts' scoring according to China's national conditions.

(1) The General Procedure of AHP (AHP) is as Follows:

First, determine the evaluation problem. The analytic hierarchy process (AHP) is used to define the objectives and problems to be evaluated in order to clarify the purpose of decision-making.

Second, list the elements of evaluation. In view of the problem to be evaluated, it collects the opinions of experts and scholars, and the elements to be evaluated will be listed one by one.

Third, establish a hierarchical structure model. After in-depth analysis of the problems faced, the factors included in the problem are divided into different levels (such as target level, criterion level, index level, scheme level, measure level, etc.). Then the hierarchical structure of the hierarchy and the subordinate relationship of the factors are illustrated in the form of block diagram. When there are many factors in a certain level, the level can be further divided into several sub-levels, the index system has three levels: the highest level (target W), the intermediate level (primary evaluation index U_i, $i = 1$, 2, 3, 4, 5) and the lowest level (secondary evaluation index U_{ij}, $i = 1, 2, 3, 4$, $j = 1, \cdots, 6$).

Fourth, use the 22 comparison method to construct judgment matrix. The judgment matrix represents the comparison of the relative importance of the factors at the upper level. The values of the elements of the judgment matrix reflect people's understanding of the relative importance of the factors. Generally, the scaling method of the number 1–9 and its reciprocal is adopted (as shown in Table 2). When the importance of the comparative factors can be explained by a meaningful ratio, the value of the judgment matrix can take this ratio.

Table 2. Scale values of relative importance of indicators

Scale	Extremely important	Very important	Important	A little more important	Equally important	A little less important	Unimportance	Very unimportant	Extremely unimportant
b_{ij}	9	7	5	3	1	1/3	1/5	1/7	1/9

Fifth, calculate the weight of each judgment matrix and make consistency test. Calculate the product of every row element in the judgment matrix M_i

$$M_i = \prod_{j=1}^{n} b_{ij},\ i = 1, 2, \cdots, n \tag{1}$$

The n square root $\overline{W}_i n$ of M_i can be obtained as follows:

$$\overline{W}_i = {}^n\sqrt{M_i} \tag{2}$$

Normalized vector $\overline{W} = (\overline{W}_1, \overline{W}_2, L, \overline{W}_n)^T$ can be obtained:

$$W_i = \frac{\overline{W}_i}{\sum_{j=1}^{n} \overline{W}_i} \tag{3}$$

Then $W = (W_1, W_2, \cdots, W_n)^T$ is the eigenvector.

The largest eigenvalue $\lambda_{\max}$ of the judgment matrix is calculated.

$$\lambda_{\max} = \sum_{i=1}^{n} \frac{(BW)_i}{nW_i} i \tag{4}$$

Among them, $(BW)_i$ represents the i element of vector BW, and its formula is as follows:

$$BW = \begin{bmatrix} b_{11}\ b_{12} \cdots b_{1n} \\ b_{21}\ b_{22} \cdots b_{2n} \\ \vdots \quad \vdots \quad \ldots \quad \vdots \\ b_{n1}\ b_{n2} \cdots b_{nn} \end{bmatrix} \begin{bmatrix} W_1 \\ W_2 \\ \vdots \\ W_n \end{bmatrix} = \begin{bmatrix} (BW)_1 \\ (BW)_2 \\ \vdots \\ (BW)_n \end{bmatrix} \tag{5}$$

Consistency Check:

When the consistency check coefficient CR of the judgment matrix satisfies the following formula (6), the judgment matrix can be considered to have satisfactory consistency; otherwise the element values of the matrix must be adjusted [9, 10].

$$CR = \frac{CI}{RI} < 0.01 \tag{6}$$

$$CI = \frac{\lambda_{\max} - n}{n - 1} \tag{7}$$

Among them, RI represents the average random consistency index; CI represents the consistency index.

For the 1–9 judgment matrix, the value of RI is shown in Table 3.

Table 3. *RI* value

n	1	2	3	4	5	6	7	8	9
RI	0	0	0.58	0.90	1.12	1.24	1.32	1.41	1.45

Finally, the weight set of the first level evaluation index of social network information management is $W = (W_1, W_2, W_3, W_4)$, $W \geq 0$, $\sum_{i=1}^{5} W_i = 1$. The weight set of the secondary evaluation index is $W_i = (W_{i1}, W_{i2}, \cdots, W_{ij})\ U_{ij}$, $W_i \geq 0$, $\sum_{j=1}^{n} W_{ij} = 1$.

Analytic Hierarchy Process (AHP) synthesizes the advantages of a certain aspect of venture capitalists' personal experience and quantitative analysis tools. The data of AHP comes from the judgment of the venture capitalists and related personnel based on their personal experience. The calculation results also require the venture capitalists to handle flexibly according to specific circumstances. It can be said that the process of determining the weight of indicators by AHP is the unification of science and art.

(2) Determining Membership Degree Matrix

The percentage statistics method is used in the study. The method is to count the percentage of the evaluation results of the evaluation objects directly, and take the results as the membership degree of the index. The method of determining membership is as follows [11]:

There are m elements in the evaluation factor domain and 5 grades in the evaluation level domain. The evaluation result is r_{ij} $(i = 1, 2, \cdots, m,\ j = 1, 2, \cdots, nk)$ and H is the expert to participate in the social network information management evaluation. For the evaluation result $u_{i1}^k, u_{i2}^k, \cdots, u_{in}^k k$ $(k = 1, 2, \cdots, H)$ of the expert k to the evaluation object i, one component of the result is 1 and the other component is 0, as shown in Table 4. Only one of each element in the Chinese comment element is 1, and the rest is 0. The determination of membership matrix is obtained from the following calculation.

$$r_{ij} = \sum_{k=1}^{H} u_{ij}^k \tag{8}$$

Table 4. Evaluation results of qualitative indicators given by an expert

Grade evaluation object	1	2	3	...	m
1	0	0	1	0	0
2	1	0	0	0	0
3	0	0	1	0	0
...	0	0	0	0	1
m	0	1	0	0	0

The judgment matrix constituted by r_{ij} is:

$$R = \begin{vmatrix} r_{11} & r_{12} & r_{13} & \cdots & r_{1n} \\ r_{21} & r_{22} & r_{23} & \cdots & r_{2n} \\ r_{31} & r_{32} & r_{33} & \cdots & r_{3n} \\ \vdots & \vdots & \vdots & \cdots & \vdots \\ r_{m1} & r_{m2} & r_{m3} & \cdots & r_{mn} \end{vmatrix} \tag{9}$$

(3) Using Fuzzy Symmetric Closeness Degree to Carry Out Comprehensive Evaluation to Determine the Evaluation Set

In order to effectively carry out the fuzzy comprehensive measurement and prevent the loss of effective information, the weighted average operator can be used to replace the traditional large and small operators [12, 13]. Set the fuzzy comprehensive evaluation set as B, that is:

$$B = W \circ R = (b_{j'}),\ j' = 1, 2, 3 \cdots, c \tag{10}$$

$$b_{j'} = \sum_{i=1}^{m} W_i u_{ij} \tag{11}$$

Among them, u_{ij} indicates the membership of evaluation index. Each element in B can be obtained according to fuzzy operation rules.

Definition of Asymmetric Closeness Degree
Closeness is a measure of the degree of approximation of two fuzzy subsets, which is divided into symmetric and asymmetric degree of approximation. Previous studies have proved that the degree of asymmetry evaluation is effective. The definition of asymmetric degree of closeness is:

$$N(A, B) = 1 - \frac{1}{n}\sum_{k=1}^{n}\left|u_A^{\frac{1}{p}}(u_k) - u_B^{\frac{1}{p}}(u_k)\right|^k \tag{12}$$

If $b_{i'} = \max\limits_{1 \le j' \le c}(b_{j'})$ is satisfied, $D_i = (0, \cdots, 0, 1, 0, \cdots, 0) = (d_1, \cdots, d_{i-1}, 1, d_{i+1}, \cdots, d_c)$ is called the characteristic fuzzy subset of component $V = \{v_1, v_2, \cdots, v_c\}$ in the fuzzy universe $v_{i'}$ of social network information management evaluation level.

Fuzzy Comprehensive Evaluation Algorithm Based on Asymmetric Closeness Degree
Step 1: standardize B

Vector $b_i(i \in i_c = \{1, 2, \cdots c\})$ is ranked last, for any $i_1, i_2 \in I_c = \{1, 2, \cdots, c\}$, for any $i_1, i_2 \in I_c$, if $|i_1 - i| > |i_2 - i|$, b_{i1} is placed before b_{i2}; if $|i_1 - i| = |i_2 - i|$, and $i_1 > i_2$, b_{i1} is placed behind b_{i2}, the standardized B is recorded as:

$$B^{(1)} = (b_1^i, b_2^i, \cdots, b_c^i) = (b_{i+2}, b_{i-2}, b_{i+1}, b_{i-1}, b_i) \tag{13}$$

Accordingly, standardization of D_i can be obtained:

$$D_c = (d_1^c, d_2^c, \cdots, d_{n-1}^c, d_n^c) \tag{14}$$

Step 2: calculate the degree of asymmetric closeness:

$$N(B, D_i) = N(B^{(1)}, D_c),\ i \in I_c \tag{15}$$

$$N(B^{(1)}, D_c) = 1 - \frac{1}{c}\sum_{r=1}^{c}\left|(b_r^i)^p - (d_r^c)^p\right|^r \tag{16}$$

Among them, r is a constant; p means probability.

Step 3: if the following conditions are met:

$$N(B, D_k) = \max_{1 \le j \le c} N(B, D_i) \tag{17}$$

It indicates that the evaluation results of social network information management system belong to grade $v_{i'}$.

3 Analysis on Implementation Effect of Social Network Information Management Evaluation of an Enterprise

3.1 Evaluation Process

(1) Quantitative Evaluation

In the evaluation system constructed in the study, quantitative evaluation refers to the information collected through various ways, the design of the index system in the evaluation object through the above series of operations, get a specific quantitative value according to the evaluation model, the value is the result of quantitative evaluation. Therefore, quantitative evaluation reflects the direct evaluation results of the evaluation objects through the index system,. and it is also the most important process in the evaluation of the whole social network information system.

In the process of quantitative evaluation, considering the authoritativeness, authenticity, objectivity and scientificity of the evaluation, and considering the characteristics of the enterprise to be evaluated and the characteristics of the social network information management system itself, the evaluation data of the design evaluation model comes from three types of personnel: users of the social network information system, company leaders and information technology experts. At the same time, the evaluation value of these three types of personnel is given a certain proportion of weight, and the final evaluation result of the system is decided by them. The actual users of the system are the users of the social network information system. Their experience is the most authentic and direct. Their evaluation results are also the most credible and objective. The real situation in the process of using the system can be reflected by their evaluation. Therefore, the user rating of the social network information system in this evaluation model is set up. The weight of the value in the overall system score is 60%; compared with the social network information system users, the company leaders use the system less frequently, and they generally only care about some important functions and indicators, but at the same time, the system development experience can also give professional evaluation of the system, and easier to find the system exists.

Quantitative evaluation is divided into four steps: social network information system user rating, corporate leadership rating, information technology expert rating and data aggregation, in which social network information system user rating, corporate leadership rating and information technology expert rating can be synchronized. The final score of users, leaders and experts in each evaluation index is the average score of all users, leaders and experts under the index. The evaluation table of enterprise social network information management system is designed, which is scored by social network information system users, company leaders and information technology experts.

In the analysis of evaluation questionnaires, the corresponding scoring criteria and criteria should be worked out according to the different grades of each evaluation. These criteria can control the objective evaluation, reduce the subjective factors of the evaluators, and also provide convenience for the data processors, so that the quantitative analysis can be based on. Five qualitative ratings (excellent, good, general, poor, very poor) are assigned to each of them, as shown in Table 5.

Table 5. The corresponding table of qualitative evaluation grades and scores

Evaluation grade	Excellent	Good	General	Poor	Very poor
Corresponding score	100	80	70	60	50

Note: 1) Score of each secondary indicator = User score * User score weight + Leader score * Leader score weight + Information Technology Expert score * Expert score weight.
2) The score of each primary index = $\sum$(the score of each secondary index under the primary index * the corresponding weight of each secondary index under the primary index).
3) The total score of the system = $\sum$ (index of each level * index weight of each level).

(2) Qualitative Evaluation

Quantitative evaluation results obtained from the quantitative evaluation step should be between 50 and 100 points, and a quantitative fractional section should be designed to correspond to the qualitative evaluation criteria, so as to qualitatively judge the construction and operation of the system.

Since the corresponding scores of each quantitative evaluation grade are excellent (100 points), good (80 points), general (70 points), poor (60 points), very poor (50 points) and so on, the corresponding criteria of quantitative evaluation results and qualitative evaluation are obtained, as shown in Table 6.

Table 6. Relationship between quantitative evaluation results and qualitative evaluation

Qualitative evaluation grade	Quantitative evaluation score
Excellent	$85 \leq$ System quantitative evaluation total score ≤ 100
Good	$75 \leq$ System quantitative evaluation total score < 85
General	$65 \leq$ System quantitative evaluation total score < 75
Poor	$55 \leq$ System quantitative evaluation total score < 65
Very poor	System quantitative evaluation total score < 55

(3) Feedback from Evaluation Results

The last link of the evaluation implementation process is the feedback of evaluation results, which is the direct expression of the evaluation results of social network information systems. It mainly reflects the evaluation methods, basis, indicators, standards, means and the final evaluation level, etc. According to the final evaluation situation, the improvement suggestions for social network information systems are given.

3.2 Evaluation and Implementation

(1) Data Collection

Questionnaire design and questionnaire release. The third party is responsible for the issuance of the questionnaire. The social network information system users and enterprise leaders are sent to them by the third party within the enterprise. The information technology experts are invited by the third party to conduct the survey outside the enterprise. At the same time, the third party will guide the evaluators to fill in the relevant questionnaires, including the specific meaning of the relevant questions in the questionnaire, scoring criteria, filling criteria and so on. The problems encountered by the evaluators in the scoring process will be explained one by one by the third party, so as to ensure the quality of the questionnaire.

(2) Data Processing

A total of 120 questionnaires were sent out in this survey. Among them, 48 users of the social network information system, 7 leaders of the company and 5 experts of information technology were given questionnaires. According to statistics, 60 questionnaires were collected, 120 of which were valid and the effective recovery rate was 100%.

Processing the data of the collected questionnaires, the comprehensive total score of each secondary index of the evaluation system of enterprise social network information management system to be evaluated is obtained, as shown in Table 7.

(3) Model Calculation

Table 7. Comprehensive evaluation of the secondary indicators of the enterprise social network information management evaluation system

Evaluation index system	Primary index	Secondary index	User score (60%)	Leader score (25%)	Expert score (15%)	Total score (100%)
Social network information management evaluation index system (U)	Technical performance	Target function implementation (U_{11})	78	88	86	85
		Progressiveness (U_{121})	74	84	81	81
		Standardization (U_{13})	77	94	81	86

(*continued*)

Table 7. *(continued)*

Evaluation index system	Primary index	Secondary index	User score (60%)	Leader score (25%)	Expert score (15%)	Total score (100%)
		Software quality (U_{14})	73	84	76	80
	System characteristics (U_2)	Functional integrity (U_{21})	75	84	71	80
		Work efficiency (U_{22})	80	88	86	87
		Security and confidentiality (U_{23})	76	88	76	83
		Maintainability (U_{24})	80	88	86	87
		Reliability (U_{25})	79	94	81	87

According to the weight value of each index in the evaluation index system of enterprise social network information management system, the overall score of each index and the evaluation index system is obtained, as shown in Table 8.

Table 8. Total score of the enterprise social network information management evaluation index system

Evaluating indicator	Score	Primary index	Weight value	Score	Secondary index	Weight value	Score
Social network information management evaluation index system (U)	84	Technical performance (U_1)	0.11	83	Target function implementation (U_{11})	0.28	85
					Progressiveness (U_{121})	0.51	81
					Standardization (U_{13})	0.10	86
					Software quality (U_{14})	0.15	80
		System characteristics (U_2)	0.29	84	Functional integrity (U_{21})	0.49	80
					Work efficiency (U_{22})	0.29	87
					Security and confidentiality (U_{23})	0.07	83
					Maintainability (U_{24})	0.11	87

(continued)

Table 8. *(continued)*

Evaluating indicator	Score	Primary index	Weight value	Score	Secondary index	Weight value	Score
					Reliability (U_{25})	0.09	87

(4) Evaluation Results

According to the corresponding relationship between quantitative evaluation results and qualitative evaluation, the technical performance index score is 83 and the evaluation grade is "good"; the system characteristic index score is 84 and the evaluation grade is "good"; the operation benefit index score is 80 and the evaluation grade is "good"; the effective level index score is 87, which evaluation grade is "excellent"; operational use index score is 82, which evaluation grade is "good"; system total score is 84, which evaluation rating grade is "good".

(5) Analysis and Suggestion of Evaluation Results

Through this evaluation, the social network information management system developed by the enterprise is evaluated quantitatively and qualitatively.

Through this evaluation, we can see that the social network information management system developed by the enterprise has reached a good level on the whole, especially in the performance level indicators, reached the "excellent" level, for the company has been put into use earlier, independent, special business systems, the performance level have been greatly improved.

However, through the evaluation results, it can be seen that the system can only achieve "good" level in technical performance, system characteristics, operational efficiency and operational use, indicating that there are still some shortcomings in the above aspects. The details are as follows:

(1) In terms of technical performance, the enterprise social network information management system is developed entirely by the technical team of the company. The advantage of this is that the developers are familiar with the company and the developed system is closer to the actual situation of the company. On the other hand, the technical level of the technical team inside the company is still higher than that of the professional developers outside the company. There is a certain gap, resulting in some shortcomings in the technical performance of the system, especially in the advanced and software quality. Therefore, it is suggested that the enterprise should introduce external technical team in the process of system follow-up improvement and maintenance, and let the external professional team give guidance from the technical level, so as to continuously improve the technical performance of the system.

(2) In the aspect of system characteristics, the reliability of data is not guaranteed to a certain extent due to the lack of necessary data entry and verification mechanism in the system design; the functional integrity score is low; and the different requirements of different business personnel on system functions also lead to the absence of special security management in the system social network information system. Institutions,

security management system, input verification mechanism, to a certain extent, also affect the safety score. To this end, it is proposed to develop enterprise social network information management system safety management system or set up system safety management posts, standardize the use and management of the system.

(3) In terms of operational efficiency, MIS improves work efficiency and reduces costs, which will inevitably bring direct and indirect economic benefits to enterprises. However, the economic benefits of MIS are lagging behind. In the short term, the benefits may not be obvious, but in the long run, it will bring great economic and social benefits to enterprises.

(4) Operational use, as the enterprise belongs to the traditional public transport enterprises, most of the managers of the traditional public transport enterprises are promoted from front-line drivers, stationmasters and technical workers, the level of computer operation is not high, therefore, more attention should be paid to the social network information management system development such as easy to operate, simple and other aspects. Therefore, it is suggested to further simplify and unify the system interface so as to make the interface display clearer and the information can be obtained more easily and quickly. At the same time, some auxiliary and prompting functions, such as help prompt, step prompt and so on, can be added to the system interface to reduce the difficulty of system operation and improve the work efficiency.

4 Conclusions

Under the background of information age, MIS has become an indispensable part of the development of enterprises. The advantages and disadvantages of MIS, to a certain extent, determine the competitiveness of enterprises. Therefore, it is very important to evaluate the social network information management system for the level of enterprise information and the development of enterprises. According to the requirement of enterprise to MIS and the characteristics of MIS, the evaluation index system of MIS is established on the basis of consulting experts' opinions. Based on the theory of analytic hierarchy process (AHP), a judgment matrix is constructed by comparing two indexes of the same level. Combined with the mathematical operation of the judgment matrix, the weights of indexes at different levels are calculated, and the consistency test is carried out. On this basis, the fuzzy asymmetric closeness degree is used to determine the tube. The process of safety classification of social network information system is discussed, and the specific steps of system safety early warning evaluation are given. Through the above theoretical analysis and practical research, the enterprise has a clearer understanding of the success and shortcomings of the development and application of the social network information management system. For the enterprise's information construction, this study is not only a summary of experience, but also a perfect and improved process for the enterprise's future information development.

The evaluation of enterprise social network information management system is a difficult but highly practical research. However, due to the limitations of research time, the quantity and scope of literature reading and the level of research ability, there are still some shortcomings.

(1) With the development and application of the system, the emphasis of evaluation in different periods will be different, and the evaluation index system needs to be improved and verified to meet the evaluation objectives in different periods.
(2) It is subjective to assign the weights of evaluation indicators and evaluate the system by the evaluators, which may lead to some deviations in the evaluation results. The next step is to have professional quality and level of the evaluators, and to adhere to an objective, conscientious and responsible attitude in order to make more effective evaluation of enterprise social network information management system.

Fund Projects. Major scientific research and cultivation project (2021), City College of Dongguan University of technology.

References

1. Daniel, G., Enrique, G.A., Manuel, C., et al.: Centrality and power in social networks: a game theoretic approach. Math. Soc. Sci. **1**, 27–54 (2020)
2. Abrahams, B., Sitas, N., Esler, K.J.: Exploring the dynamics of research collaborations by mapping social networks in invasion science. J. Environ. Manage. **1**, 27–37 (2019)
3. Brown, M.E., Cooper, M.W., Griffith, P.C.: NASA's carbon monitoring system (CMS) and arctic-boreal vulnerability experiment (ABoVE) social network and community of practice. Environ. Res. Lett. **11**, 115–126 (2020)
4. Zhang, Y., Xiang, Y., Wang, L.: Power system reliability assessment incorporating cyber attacks against wind farm energy management systems. IEEE Trans. Smart Grid **99**, 1–15 (2016)
5. Guo, M.J., Liu, H.Y., Wang, B.Q., et al.: Research on evaluation model of provincial new rural cooperative information system construction based on integration perspective. Chin. J. Health Stat. **6**, 1071–1074 (2016)
6. Yue, H.J., Xu, H.: Research on the performance evaluation of university information system based on the expectation and satisfaction. Mod. Inf. **7**, 51–57 (2016)
7. Liu, C.H., Liu, Q., Qi, R.H.: Vital node identification technology of social networks using neighbors' H-index values. China Sciencepaper **3**, 296–302 (2019)
8. Liu, Y.Y., Jiang, S.: Research on social network analysis and knowledge sharing management based on Big Data. Inf. Sci. **4**, 109–115 (2019)
9. Liu, X., Shahidehpour, M., Li, Z., et al.: Power system risk assessment in cyber attacks considering the role of protection systems. IEEE Trans. Smart Grid **99**, 1 (2017)
10. Zhang, X.S., Su, X., Wang, Y.M.: Study on evaluation method for military information systems interoperability. J. China Acad. Electron. Inf. Technol. **6**, 649–654 (2016)
11. Abdo, H., Flaus, J.M.: Uncertainty quantification in dynamic system risk assessment: a new approach with randomness and fuzzy theory. Int. J. Prod. Res. **19**, 1–24 (2016)
12. Jeon, H., Seo, K.-K.: A new performance assessment modeling and development of a performance assessment system for a cloud service. Wirel. Pers. Commun. **89**(3), 795–818 (2015)
13. Mehdizadeh, M., Ghazi, R., Ghayeni, M.: Power system security assessment with high wind penetration using the farms models based on their correlation. IET Renew. Power Gener. **8**, 893–900 (2018)

Human Resource Social Insurance Data Remote Reporting System Based on Big Data Technology

Cai-ming Zhang(✉) and Bo Sun

School of Labor Relations and Human Resource, China University of Labor Relations, Beijing 100048, China

Abstract. In order to cooperate with the China Insurance Regulatory Commission's national economic census work, meet the various needs of the Insurance Regulatory Commission for insurance statistical information work, and increase the transmission speed of social insurance data. Therefore, this paper designs a remote reporting system for human resources and social insurance data based on big data technology. Upload the data file structure through the web method and manually enter the data report to optimize the C/S structure in the system hardware. Improved the system software module functions and operating steps. The statistical data collected from the business system, system, and personnel system is reported from low-level (prefecture-level) to high-level (head office level). The data is summarized and sorted at all levels. A file containing all off-site statistical data is generated. Finally, the actual application of the human resources and social insurance data reporting system in different places is investigated and analyzed. The experimental results confirm that the system has a high reporting rate and fully meets the research requirements.

Keywords: Big data technology · Human resources · Insurance data

1 Introduction

It has the function of economic security and social security, and can promote the development of insurance and financing. In recent years, with the rapid development of China's insurance industry as a whole, the situation of increasing spatial differentiation is more and more obvious, which actually affects the sustainable development of the insurance industry and the effective play of its function [1]. At the same time, the theoretical research on the insurance industry is constantly enriched, while the academic research on the insurance industry space is relatively backward. The theoretical framework of insurance spatial differentiation is still in the exploratory stage, which leads to a lack of in-depth understanding of the trend and degree of insurance industry differentiation, differentiation mechanism, influencing factors and change trend. The practice of policy design and spatial layout of insurance supervision lacks theoretical guidance. In order to cooperate with the China Insurance Regulatory Commission's national economic census work, meet the various needs of the Insurance Regulatory Commission for insurance

S. Liu and X. Ma (Eds.): ADHIP 2021, LNICST 416, pp. 551–561, 2022.
https://doi.org/10.1007/978-3-030-94551-0_43

statistical information work, and increase the transmission speed of social insurance data. Therefore, this paper designs a remote reporting system for human resources and social insurance data based on big data technology. Upload the data file structure through the web method and manually enter the data report to optimize the C/S structure in the system hardware. Improved the system software module functions and operating steps. The statistical data collected from the business system, system, and personnel system is reported from low-level (prefecture-level) to high-level (head office level). The data is summarized and sorted at all levels. A file containing all off-site statistical data is generated. Finally, the actual application of the human resources and social insurance data reporting system in different places is investigated and analyzed. The experimental results confirm that the system has a high reporting rate and fully meets the research requirements.

2 Human Resource Social Insurance Data Remote Reporting System

2.1 Hardware Structure of Social Insurance Data Remote Reporting System

The development and design of social insurance data remote reporting system is realized by combining Java servlet, JSP, HTML and JavaScript [2]. This model is page based, with each JSP file or servlet file linked to another. Clicking on the link will send a request to the server. The server will process the request and replace the browser's page. Each user's behavior will lead to an HTTP request. Data is provided by each business system of each central branch company, uploaded to the statistical information reporting system by means of file or input and ETL automatic data. After the summary, review and

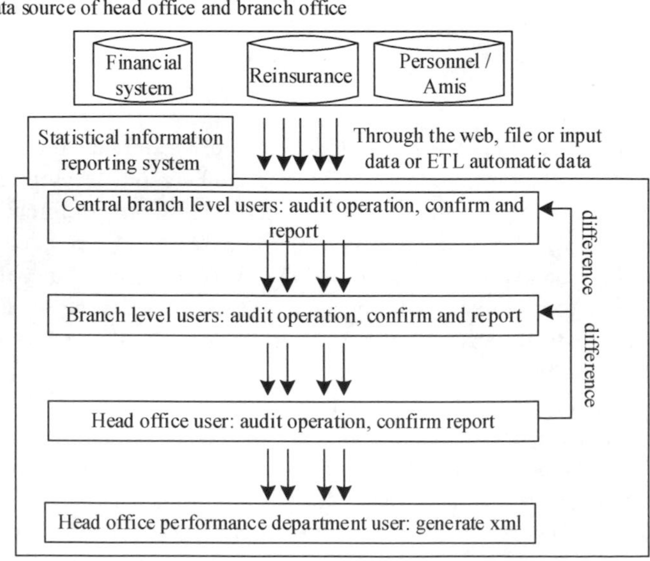

Fig. 1. Hardware configuration of social insurance data remote reporting system

confirmation of provincial and head office companies, it is submitted to the performance Department of the head office. The performance department uses the system to generate the overall data file of the whole company in XML format submitted to CIRC [3]. In order to test the effect of data processing, the hardware configuration of social insurance data remote reporting system is optimized, as shown in Fig. 1.

As shown in Fig. 1, server-side control technology simulates stateful connection based on stateless HTTP, and realizes state maintenance and event processing [4]. Write B/s program in the way of desktop program. Some desktop programs that used to be very complex can now be easily implemented in browsers, and even code migration is very easy [5]. The client does not need any additional plug-ins. Upload the data file through the web, input the data manually and report the data. Generate the data file in XML format submitted to CIRC. System management subsystem: used to maintain the data needed by the system operation. Further optimize the function of c/s system structure of social insurance data remote reporting system, as shown in Fig. 2.

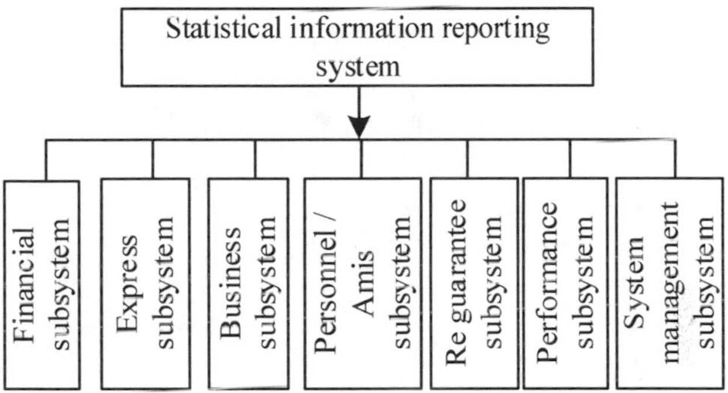

Fig. 2. C/S architecture of social insurance data remote reporting system

For the C/S architecture can not meet the needs of the development of distributed enterprise application system, developers constantly put forward new software architecture to improve the difficulties in the development and maintenance process [6]. In the current enterprise and application system development field, there are two camps of software architecture, they can provide a powerful and complete development platform. These two platforms are known as. Net architecture and J2EE architecture.

2.2 Software Flow of Insurance Data Remote Reporting System

From the business point of view, the information reporting system can be divided into the following subsystems.

Business subsystem: including data upload, data modification, account verification, report generation, summary, data audit and other functions.

Personnel and Amis subsystem: including data input, data modification, account verification, report generation, summary, data audit and other functions.

Reinsurance subsystem: including data upload, data modification, account verification, data reporting, single account query, data prompt, report generation, summary and report audit, excel upload, data modification, account verification and report generation.

Performance subsystem: including data upload, data modification, account verification, export XML file and download functions.

System management: including role management, organization management, system user management, function module management and data source configuration. The subsystem reporting process is shown in Fig. 3.

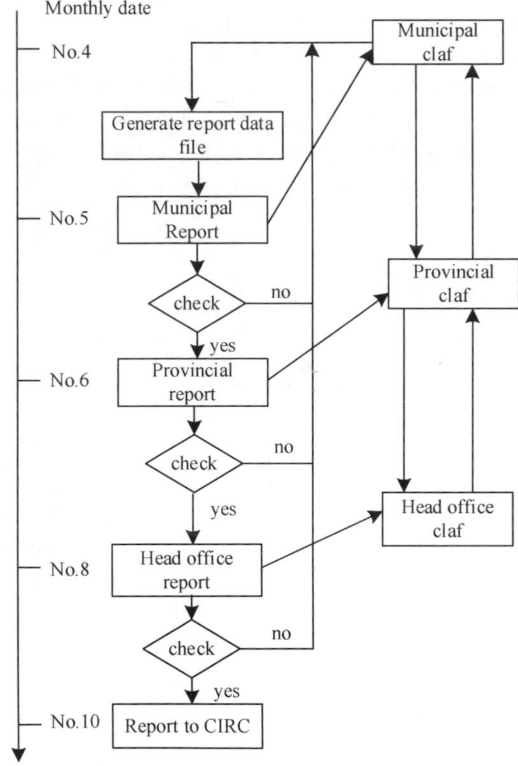

Fig. 3. System software operation process optimization

Municipal users generate data submission documents through core business systems and systems, and then upload data submission files to the information submission system [7]. The system will upload data for analysis and generate city level reports, and then the user will verify the report items. Submit data to provincial branch after verification is successful. The report of prefecture and city level is in a paragraph.

After the provincial users log in, the data of their subordinate prefectures and cities are collected and sorted, and the provincial reports are generated. After the report item is verified successfully, it shall be submitted to the head office.

After the head office user logs in, the data of all its branches will be summarized and sorted, and the head office report will be generated. After the report items are verified successfully, the XML file containing all the submitted subjects will be generated. This module is the main module for each city branch and business data upload [8]. Support to upload the TXT text of account information separated by "," to the server. At the same time, check the uploaded account information. The system administrator and the business role can configure the header code of the uploaded file to determine the name of the uploaded file.

Table 1. Function point description of upload module

Function point	Describe
File name check	Check whether the upload file name matches the configuration name of
Subject information check	Check the contents of the uploaded file, including whether the account ID belongs to the current role. Whether the report period is correct. Whether the account data meets the specification, such as whether the account data should not appear in English, etc.
File transfer	Upload files from client to server
File deletion	Delete unnecessary files
File data storage	Put the data of the file uploaded to the server into the database, and delete the file on the server after the database is successfully put into the database

Based on the data in Table 1, the check file name method is further called to check the file name. The naming rule of upload file name is: GF +6 digits of institution +4 or 5 digits of system name +1 digit of report period (M/Q/h/y, etc.) +6 digits of report period + generation date (8 digits). TXT file [9]. After passing, the file data should be checked, and the file data should be put into storage after checking the data items. So as to effectively ensure the effect of data processing, system efficiency.

2.3 Realization of Human Resource Social Insurance Data Reporting in Different Places

Through the detailed investigation and analysis of the social security fund data acquisition system, the functional requirements of the product are described in detail. This section describes the functional requirements of each module and its subordinate function points from the perspective of business function according to the order of function module structure [10]. Data conversion is one of the main modules of social security fund data acquisition system. Its main subordinate functions include conversion scheme management, setting account correspondence, data conversion and data deletion. Two disequilibrium measure indexes are introduced. Based on the research results of regional financial differences, the Gini coefficient formula (1) reflecting the differences in the history and current situation of China's regional insurance development is

$$\mathrm{GiNi} = \frac{2}{n^2 u_y} \sum_i i y_i - \frac{n+1}{n} \tag{1}$$

In formula (1), n represents the total number of insurance regions in China, which can be divided into 3 or 31 according to different research levels. y_i is the number of the i-th regional indicators after ranking according to the indicators from high to low, and u_y is the mean value of the selected indicators. In the process of specific calculation, the difference degree of relevant indicators is based on different regions or different income classes. The calculation formulas of Theil index and logarithmic mean deviation are listed in (2) and (3).

$$\mathrm{GE}_0 = \frac{1}{N}\sum_i LN\left(\frac{u}{y_i}\right) \tag{2}$$

$$\mathrm{GE}_1 = \frac{1}{N}\sum_i \frac{y_i}{u} LN\left(\frac{y_i}{u_i}\right) \tag{3}$$

In the formulas (2) and (3), N is the total number of regions, u is the mean value of regional extraction index, and y_i is the measurement index of regional insurance development level of the i-th region. Logarithmic mean deviation and Theil index. On the basis of the above formula, the operation process is shown in formula (4).

$$\begin{aligned}\mathrm{GE}_0 &= \frac{1}{N}\sum_i LN\left(\frac{u}{y_i}\right)\\ &= \frac{1}{n}\sum_{t=1}^{k}\sum_{i\in N_t} LN\frac{u}{y_i}\\ &= \sum_{t=1}^{k}\frac{n_t}{n}\frac{1}{n_t}\sum_{i\in N_t} LN\frac{u_t}{y_i} + \frac{1}{n}\sum_{t=1}^{k}\sum_{i\in N_t} LN\frac{u}{u_t}\\ &= \sum_{t=1}^{k} Z_t GE_0(y^t) + \sum_{t=1}^{k} Z_t LN\frac{u}{u_t}\end{aligned} \tag{4}$$

On the basis of formula (4), the operation of formula (5) is carried out.

$$\begin{aligned}\mathrm{GE}_1 &= \frac{1}{N}\sum_i \frac{y_i}{u} LN\left(\frac{y_i}{u_i}\right)\\ &= \frac{1}{n}\sum_{t=1}^{k}\sum_{i\in N_l} \frac{y_i}{u} LN\frac{y_i}{u}\\ &= \sum_{t=1}^{k}\frac{n_t}{n}\frac{u_t}{u}\frac{1}{n_t}\sum_{i\in N_i} \frac{y_i}{u_t} LN\frac{y_i}{u_t} + \frac{1}{n}\sum_{t=1}^{k}\sum_{i\in N_t} \frac{u_t}{u} LN\frac{u_t}{u}\\ &= \sum_{i=1}^{k} Z_t\frac{u_t}{u} GE_1(y^t) + \sum_{t=1}^{k} Z_t\frac{u_t}{u} LN\frac{u}{u_t}\\ &= \mathrm{I} + \mathrm{O}\end{aligned} \tag{5}$$

Due to the different sensitivities of the indicators, the mean logarithmic deviation and Theil index are very sensitive to the changes of the indicators of the ethnic group with the lowest proportion and the highest proportion, respectively. Thus, it can make up for the deficiency the Gini coefficient is sensitive to the changes of indicators in the medium level, so that we can more comprehensively analyze the degree of regional differentiation of China's insurance. Each city formulates its own accounting subjects according to its specific situation. The accounting subjects of each city are different, which needs to be transformed into a unified accounting subject. In the data deletion function, the data management organization users delete the problematic and incomplete data submitted by the lower level data center users or municipal agency users through this function.

The amount of data transmission between reporting task r_i and information resource r_j is recorded as $F\left(r_\text{i}, r_j\right)$. after obtaining all resources, task r_i can perform the communication function. After the completion of the task, all resources are released to become available resources. The execution time of reporting task r_i on all resources E_k is $T_\text{c}(r_\text{i}, E_\text{k})$, and the traffic process mainly depends on the calculation time and the transmission time between resources. For the part where calculation and reporting cannot overlap, $T_\text{c}(r_\text{i}, E_\text{k})$ can be defined as:

$$T_\text{c}(r_\text{i}, E_\text{k}) = R_c(r_\text{i}, E_\text{k}) + \sum_{r_j}\left(F\left(r_\text{i}, r_j\right) \times Q_c\left(r_{j\text{i}}, E_\text{k}\right)\right) \tag{6}$$

In formula (6): $\sum\limits_{r_j}\left(F\left(r_\text{i}, r_j\right) \times Q_c\left(r_{j\text{i}}, E_\text{k}\right)\right)$ is the total time of resource transmission; $R_c(r_\text{i}, E_\text{k})$ is the time of calculation. The average execution time of reporting task r_i is shown in formula (7).

$$\overline{T}_\text{c}(r_\text{i}) = \sum_{m=1}^{n} T_\text{c}(r_\text{i}, E_\text{k})\big/n \tag{7}$$

If the escalation task r_i is mapped to all resources E_k, then make $T_\text{a}(r_\text{i}, E_\text{k})$ and $T_\text{b}(r_\text{i}, E_\text{k})$ the start and end time of the task execution, and $T_\text{a}(r_\text{i}, E_\text{k})$ can be expressed as formula (8).

$$T_\text{a}(r_\text{i}, E_\text{k}) = \max\left\{E_\text{a}(E_\text{k}), D_{pred}(r_\text{i}, E_\text{k})\right\} \tag{8}$$

In formula (8), $D_{pred}(r_\text{i}, E_\text{k})$ is the time of all data obtained by the reporting task from the start task r_i set. And $T_\text{b}(r_\text{i}, E_\text{k})$ can be expressed as formula (9).

$$T_\text{b}(r_\text{i}, E_\text{k}) = T_\text{a}(r_\text{i}, E_\text{k}) + T_\text{c}(r_\text{i}, E_\text{k}) \tag{9}$$

According to the start time and end time of the reporting task, the task is mapped to the resource target, and the matching time between the task and the computing resource is determined. According to the current situation reported, parallel distribution scheme implementation. According to the requirement analysis mentioned above, entities are summarized in each subsystem, and the relationship between entities is determined. Based on this, the system structure function points are described, as shown in Table 2. Submit the verified data to the superior company. If the superior company returns the submitted data, the system will display "re submit data". The returned data must go

Table 2. System structure function point description

Function point	Describe
Entrance	Is the access to the system
Legitimacy check	Determine whether the user exists
Safety inspection	Judge whether the user password is correct
Authority check	Check the permission that the user has

through the verification step again before it can be submitted again. Note: once the data is submitted to the superior company, it cannot be submitted again, and the system will automatically control the data after it is submitted, and it is not allowed to upload and modify the data. If errors are found after submitting to the superior company, the data status of the company shall be released to the "not submitted" state through the operation of the users of the superior company at the provincial level, so as to re-operate. See Table 3 for the function points of data reporting module.

Table 3. System module function point description

Function point	Describe
Data verification status check	Check whether the data of the organization passes the verification
Data reporting status check	Check whether the data of the organization is reported
Data summary status check	Check whether the data is "full summary" or "temporary summary"
Data number check	Check for data

Select the data item to be summarized in the interface of displaying summary status and summary records, click summary and select the subordinate, and STS sum action will process the data summary request of provincial and head office. After judging the summary type, accumulate the submitted data item by item and return to the summary result display page. So as to effectively guarantee the operation effect of the system and improve the operation efficiency of the system.

3 Analysis of Experimental Results

The configuration requirements for the experimental platform to be built are: a database server with a simulated minicomputer, 8 $\times$ 4.0 GB, 64-bit dual-core or above. Memory $\geq$ 128 GB, support memory mirroring and protection. 4 pieces of 8 TB capacity hard drives, hard drive speed $\geq$ 16 000 r/min, SCSI8068, support Raid-0, 1, 5, 6, 10. Application-side server 1 emulation minicomputer, processor, 4 GB, 36 MB cache. $\geq$32 GB fully buffered DDR4 memory, which can be expanded to 128 GB, supports memory mirroring and memory protection. Two hard drives with 4 TB capacity, rotating speed $\geq$ 20000 r/min. Support Raid-0, 1, 5, 6, 10. In the above-mentioned experimental environment, the method of literature [2] is used as a traditional comparison method, and the experiment is compared with the text method. All the test data are collected into the virtual branch; the business system is being tested, and the data of new order, renewal, preservation, surrender and insurance payment are generated; the system imports the relevant data generated by the business system. System input capital investment, in the reporting system, new institutions: branch, number of documents: 440300. Operators of new branch company. To generate business system report data, execute business system "data report" program; click "generate business report data" to output file. In the system, the corresponding responsibilities are used to execute the "operation report", select the accounting period and rewrite section "40300...", and the report output is "report submitted by CIRC (balance sheet)", "report submitted by CIRC (monthly profit and loss statement)" and "report submitted by CIRC (cash flow statement)". The report output is processed in Excel, and the specific steps are shown in Table 4.

Table. 4 Test process

Serial number	Test Case Specification
1	$1 \rightarrow 9 \rightarrow 13 \rightarrow 14 \rightarrow 15 \rightarrow 16\ldots$ Business system data is reported separately
2	$10 \rightarrow 11 \rightarrow 12 \rightarrow 13 \rightarrow 14 \rightarrow 15 \rightarrow 16\ldots$ Financial system data is reported separately
3	$1 \rightarrow 8 \rightarrow 11 \rightarrow 12 \rightarrow 13 \rightarrow 14 \rightarrow 15 \rightarrow 16\ldots$ Report through the interface

In order to analyze the data upload processing performance of this traditional system and this system more intuitively, the paper compares and analyzes the efficiency of file processing, and the results are shown in Fig. 4.

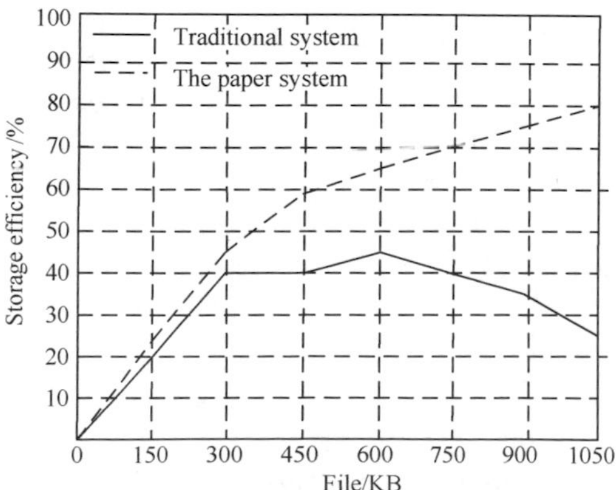

Fig. 4. Comparison results of data processing performance of two systems

Compare the time of file reporting between the traditional system and the system in this paper, and verify the stability of the system in this paper. The results are shown in Fig. 5.

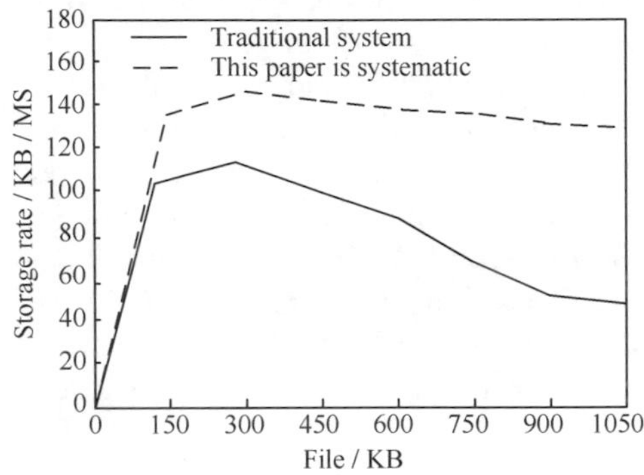

Fig. 5. Comparison results of reporting rate between two systems

Based on the above test results, it can be seen that the human resources social insurance data remote reporting system based on big data technology has high timeliness and stability in the practical application process, which fully meets the research requirements.

4 Conclusion

Based on big data technology, the paper designs the remote reporting system of human resources social insurance data. The paper analyzes the mechanism of the spatial differentiation of insurance industry from the measurement level, expands the research on the influencing factors of the differentiation of insurance industry from analytical description to measurement level, and improves the theory of insurance space to some extent. The results of the application of the remote reporting system of human resources social insurance data based on big data technology are proved by experiments. The research shows that the system of this design has a theoretical research and empirical test on the evolution trend of the spatial differentiation of insurance industry in China, and discusses and forecasts the trend of the differentiation of insurance industry in a long period.

References

1. Dhieb, N., Ghazzai, H., Besbes, H., et al.: A secure AI-driven architecture for automated insurance systems: fraud detection and risk measurement. IEEE Access **8**, 58546–58558 (2020)
2. Wu, J., Wang, J., Liu, Y.: Design and research of insurance survey claims system based on Big Data analysis. In: 2019 International Conference on Virtual Reality and Intelligent Systems (ICVRIS), pp. 211–214. IEEE (2019)
3. Li, Z., Xiao, Z., Xu, Q., et al.: Blockchain and IoT data analytics for fine-grained transportation insurance. In: 2018 IEEE 24th International Conference on Parallel and Distributed Systems (ICPADS), pp. 1022–1027. IEEE (2018)

4. Bo, Z.A., Cdb, A., Tz, C., et al.: An IGAP-RBFNN-based secondary control strategy for islanded microgrid-cyber physical system considering data uploading interruption problem. Neurocomputing **397**(6), 422–437 (2020)
5. Liu, D., Xu, Y., Xu, Y., et al.: Opportunistic data collection in cognitive wireless sensor networks: air-ground collaborative online planning. IEEE Internet Things J. **7**, 8837–8851 (2020)
6. Usman, M., Jan, M.A., Jolfaei, A., et al.: A distributed and anonymous data collection framework based on multi-level edge computing architecture. IEEE Trans. Ind. Inf. **16**, 6114–6123 (2019)
7. She, W., Gu, Z.H., Lyu, X.K., et al.: Homomorphic consortium blockchain for smart home system sensitive data privacy preserving. IEEE Access **7**(6), 62058–62070 (2019)
8. Liu, S., Fu, W., He, L., et al.: Distribution of primary additional errors in fractal encoding method. Multimedia Tools Appl. **76**(4), 5787–5802 (2017)
9. Honeycutt, W.T., Kim, T., Ley, M.T., et al.: Sensor array for wireless remote monitoring of carbon dioxide and methane near carbon sequestration and oil recovery sites. RSC Adv. **11**(12), 6972–6984 (2021)
10. Liu, S., Bai, W., Liu, G., et al.: Parallel fractal compression method for big video data. Complexity **2018**, 1–16 (2018)

Multi Cultural Music Remote Assistant Teaching System Based on Mobile Internet of Things

Yang-lu Ren[1(✉)], Jun Zhou[1], and Yang-bo Wu[2]

[1] College of Art, Xinyu University, Xinyu 338000, China
[2] College of Mathematics and Computer, Xinyu University, Xinyu 338000, China

Abstract. In view of the imperfection of the current music teaching mode system in Colleges and universities, this paper puts forward the design of the multi-cultural music remote auxiliary teaching system based on the mobile Internet of things. Through the combination of emerging mobile Internet of things technology and teaching, optimize the system hardware structure and configuration parameters, further optimize the software function of multi-cultural music remote assistant teaching system based on mobile Internet of things, improve the collection and selection of teaching materials, so as to better improve students' interest in learning and ensure the quality of teaching.

Keywords: Mobile Internet of things · Multi culture · Music teaching

1 Introduction

In the process of music teaching, through the accumulation and summary of teachers' experience, we can find that the problems students encounter in learning have some similarities to a certain extent [1].With the help of the distance assistant teaching system, the assistant teaching is carried out. Generally speaking, all the music equipment used in music classroom teaching activities and extracurricular music activities in the form of the Internet of mobile animals can be regarded as the music teaching system of the remote assistant teaching system. Music system is a series of mobile Internet of things system, which can stimulate students' interest and improve their memory.

Relevant scholars have studied this. Literature [2] proposes a music privacy information protection incentive mechanism based on perceptual task allocation, which allocates tasks by perceiving mioT music information, constructs a crowd perception framework according to cloud and edge cooperation, and realizes music privacy information protection through greedy algorithm. This method can improve the quality of teaching, but the correlation between music teaching content and students' needs is poor. Literature [3] puts forward the music remote auxiliary teaching method based on flipped classroom teaching, which improves the learning experience of music students through flipped classroom teaching method, and measures the learning effect by students' academic performance. This method can enhance students' learning interest, but the teaching quality is poor.

S. Liu and X. Ma (Eds.): ADHIP 2021, LNICST 416, pp. 562–574, 2022.
https://doi.org/10.1007/978-3-030-94551-0_44

The traditional music teaching mode system is not perfect, and the teaching quality is not good. This paper proposes the design of multi-cultural music remote assistant teaching system based on mobile Internet of things. Optimize the system hardware structure and configuration parameters, improve the collection and selection of teaching materials, improve students' interest in learning, and ensure the quality of teaching.

2 Multi Cultural Music Distance Assistant Teaching System

2.1 Hardware Structure of Music Remote Assistant Teaching System

The overall design of the system includes function structure design and system structure design, which can be simplified as modular design. In the modular design, it is not only necessary to divide the components of the whole system, but also to design the communication between modules, module continuity, module protection, module solvability and module combination [4]. The original framework of music teaching hardware system starts from the identity of system administrator, and every login needs to log in as an administrator. After login, users can upload music learning information, music teaching plan and music teaching resources. This paper introduces the technology of remote assistant teaching system, improves the system framework, retains all the functions of the original system, and on this basis, adds two modules: user message and user online music answering. Through the two modules of user's message and user's online music Q & A, learners can not only have a better understanding of music knowledge, but also effectively shunt the teaching video watching [5]. Some users will have questions in the process of watching the teaching video. Through the function of leaving a message and answering questions, users' doubts can be solved. In the stage of original users' leaving a message, new users can enter the system to watch other music teaching videos. The frame structure of the improved system is shown in the figure (Fig. 1).

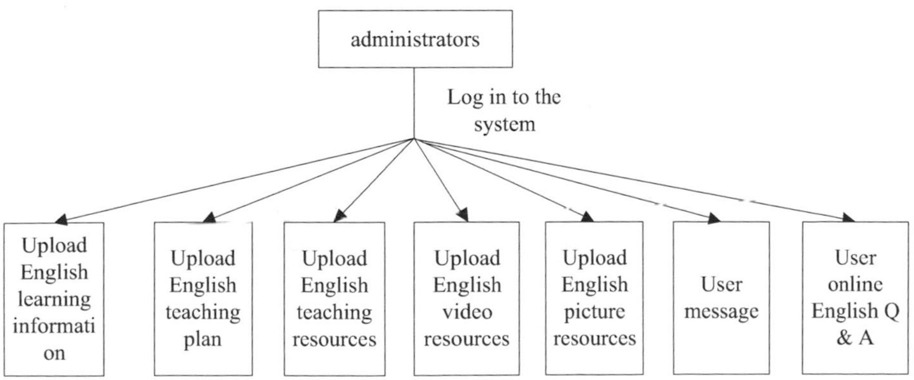

Fig. 1. The frame structure of the improved system

The core of the hardware equipment of the system is PC. The hardware equipment of other parts includes MIDI keyboard, monitor speaker, professional audio card, sound source, microphone, etc. the software includes all kinds of music software. The function of the whole system is to carry out music teaching through MIDI production and digital audio recording and editing [6]. MIDI keyboard is used to input information in this system, which is a bridge for human-computer interaction; microphone is used as audio input equipment to input the voice of songs or other original musical instruments into the Internet of mobile animals. The Internet of things and all kinds of music software are responsible for recording and storing, no matter the information input by MIDI keyboard or the audio input by microphone, they are all stored in it. When the modular design of the system is carried out, it needs to meet the high cohesion characteristics within the module [7]. Therefore, the system equipment module must be able to run independently to complete the design function, but need to control the scale of the module. The properties of modules can be summarized as interchangeability, pluggable parallelism and boundedness.

(1) When the internal requirements of the module change, the changes will not affect the normal operation of other modules; (2) when the module needs to be deleted, only the functions handled by the module will be affected; (3) if a new module realizes the same functions and has the same operation interface, the operation of the whole system will not be affected after replacement [8]. Based on this, the system equipment structure framework is optimized as follows (Fig. 2):

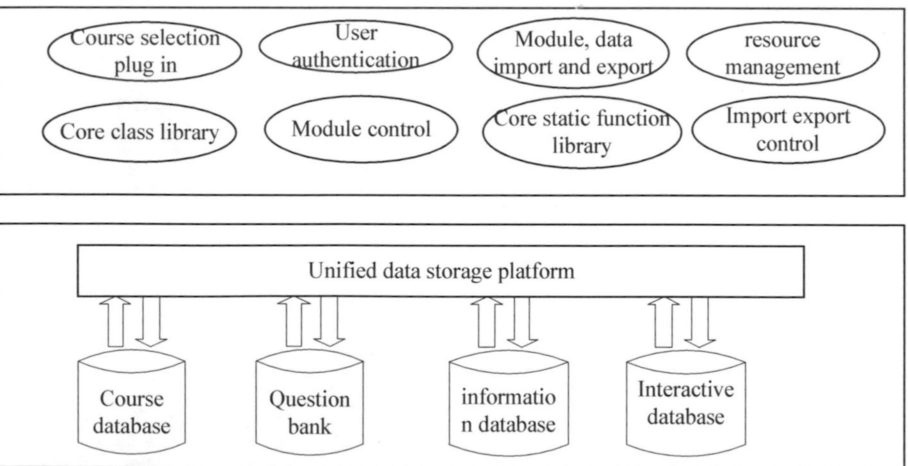

Fig. 2. Optimization of system equipment structure framework

The design of music teaching environment includes: the design of "situation" in music teaching, the design of "cooperative learning" environment in music teaching, the design of learning environment (non teaching environment) and the design of external information resources [9]. Based on the theory of music teaching design under the guidance of multicultural theory, this paper designs and refines the specific curriculum design of music teaching system based on mobile Internet of things, and applies the teaching strategies under multicultural theory to the specific curriculum design. As shown in the figure, this paper presents the specific application of teaching strategies in curriculum creation (Fig. 3).

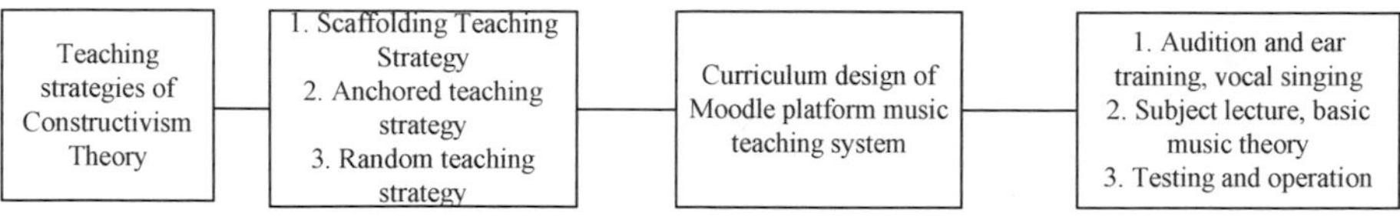

Fig. 3. Optimization of system equipment operation standard

The use of mobile Internet of things and remote auxiliary teaching system in music theory teaching can reduce the difficulty of teaching and enrich teaching means. Students should make comprehensive use of hand, brain, vision and hearing senses in learning, which can not only stimulate students' interest in learning, but also play a great role in promoting the efficiency of classroom learning, so as to realize the goal of classroom teaching "Double main teaching mode", so as to improve the teaching quality and efficiency [10]. Before the music theory class, the teacher first makes the music examples needed for the class, and saves them in the corresponding music software, so that in the class, the teacher only needs to move the mouse, and the students can see the music examples and hear the sound at the same time, and can listen to some part of the music examples repeatedly to strengthen the understanding [11]. At the same time, teachers can also use mobile Internet of things software to lead students to play music theory games, with the help of some staff learning software to teach basic music theory. On the one hand, it reduces the time for teachers to prepare lessons, on the other hand, it reduces the difficulty and dullness of music example explanation, and stimulates students' interest in learning.

2.2 Function Optimization of System Software

Music assisted teaching management system is mainly composed of information release, document management, teaching resource management, assisted teaching management and system management modules. The information release module is divided into school news, notice, school news and BBS. Document management is divided into document management, receiving document management and filing management; teaching resource management includes music video courseware management, homework management and music knowledge management; auxiliary teaching management is divided into course management, online examination, question answering discussion and score analysis; system management is divided into teacher and student management, system setting management, log management and auxiliary decision-making management. The function structure of music assistant teaching management system is shown in the figure (Fig. 4).

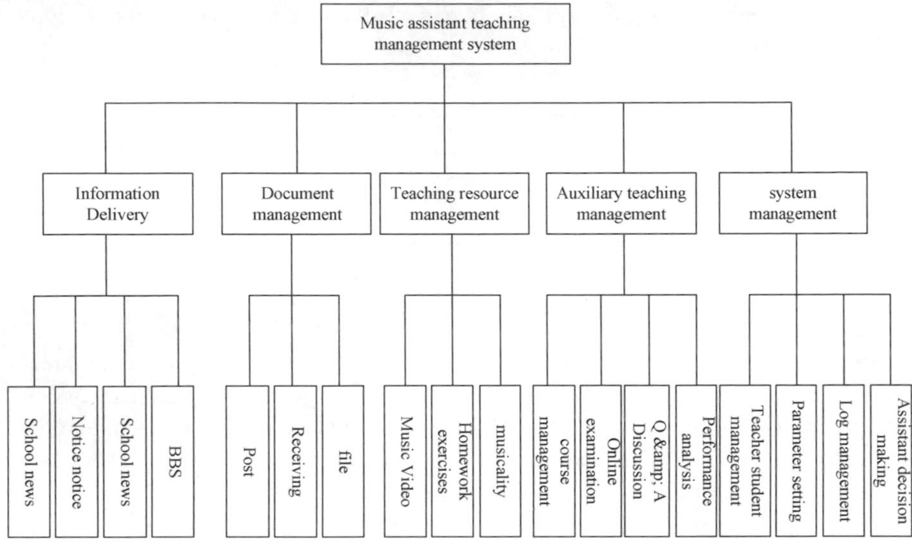

Fig. 4. Optimization of software function structure of music teaching system

In the process of information service, information management and resource development, the system users and designers summarize an important experience that database technology is the most effective way to manage data [12]. With the progress of network technology and mobile Internet of things technology, data management through database has become an important consensus. In order to complete the sharing, integrity and consistency of system data, both large management systems and small transaction processing are using database technology to complete data management [13]. Based on this, the top-level data flow of the database is shown in the figure (Fig. 5).

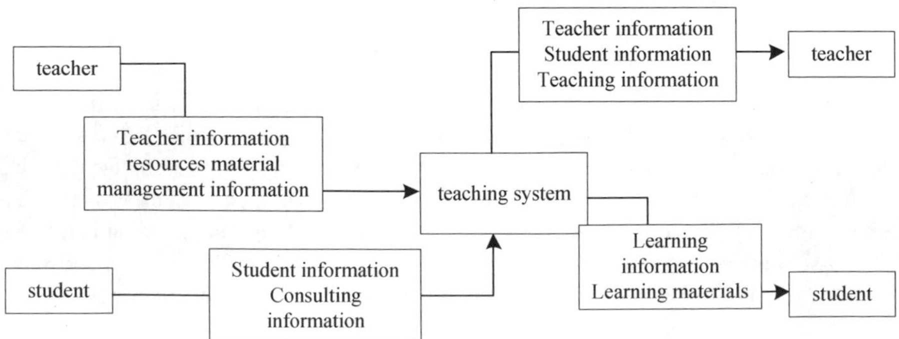

Fig. 5. Top level data flow processing steps of database

The analysis of data model is mainly carried out by means of data flow chart, which can systematically and comprehensively depict the data logic of music teaching assistant management system. In the process of information storage, processing and flow, data flow chart is mainly realized with the help of centralized common symbols. The data flow chart consists of the following four basic elements: external entity, data flow, processing (function) and data storage. Each module uses the data flow chart to represent the source of data and the relationship between data.

2.3 The Realization of Music Remote Assistant Teaching

The database design of music assisted instruction management platform mainly serves for the business knowledge base of music assisted instruction management. Through the management of the database, the processing of knowledge data and business data is optimized [14]. In the process of database design, the following aspects are taken as important standards: verification is based on the database design specification, and the data structure is designed in the form of standardization to ensure the consistency and normal operation of data operation. The naming standard is helpful to the later unified management and upgrade maintenance. Therefore, in the design of database and table, the naming standard must be strictly implemented, and all column information must be annotated. Data redundancy and normalization of data paradigm will affect the retrieval speed of later data. Therefore, in the design of data table, we need to master various degrees in order to achieve the highest retrieval value and reduce the system response time. The implementation of strict identity authentication management, users with different permissions access to different data and operation degree, improve the security of data. Through the use of triggers and stored procedures, we can strictly control the simultaneous operation of tables, ensure the control of simultaneous modification and access, and reduce the inconsistency of data. The detailed data flow of music teaching material management module is shown in the figure (Fig. 6):

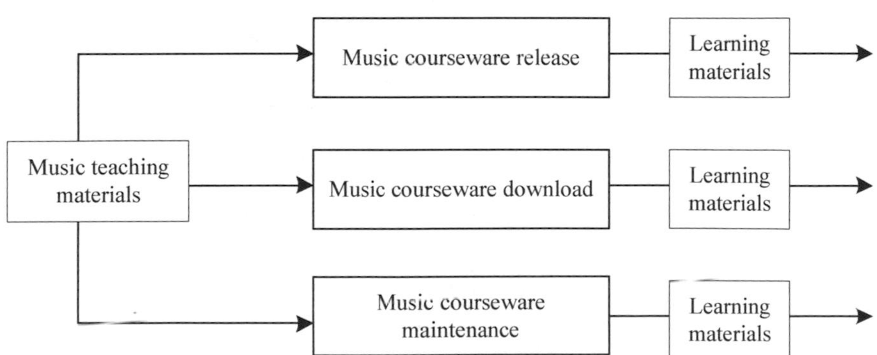

Fig. 6. Optimization of collaborative operation steps of teaching system

In the process of system operation, in order to better realize the management of music video, homework exercises, music knowledge and other teaching resources. In order to better ensure the realization of the functions of adding, modifying, deleting and querying music teaching video materials, and optimize the function of uploading and previewing music videos, students can download different music teaching videos according to their permissions, and realize the editing management and online evaluation of homework exercises. Homework exercises editing includes adding, modifying and deleting homework exercises In addition to, query and release; exercise online evaluation function is similar to online examination evaluation function. Effective management of music text knowledge, including the addition, modification, deletion and release of music knowledge. In order to ensure the effective processing of the above requirements, further optimize the teaching resource management business process, as shown in the figure (Fig. 7).

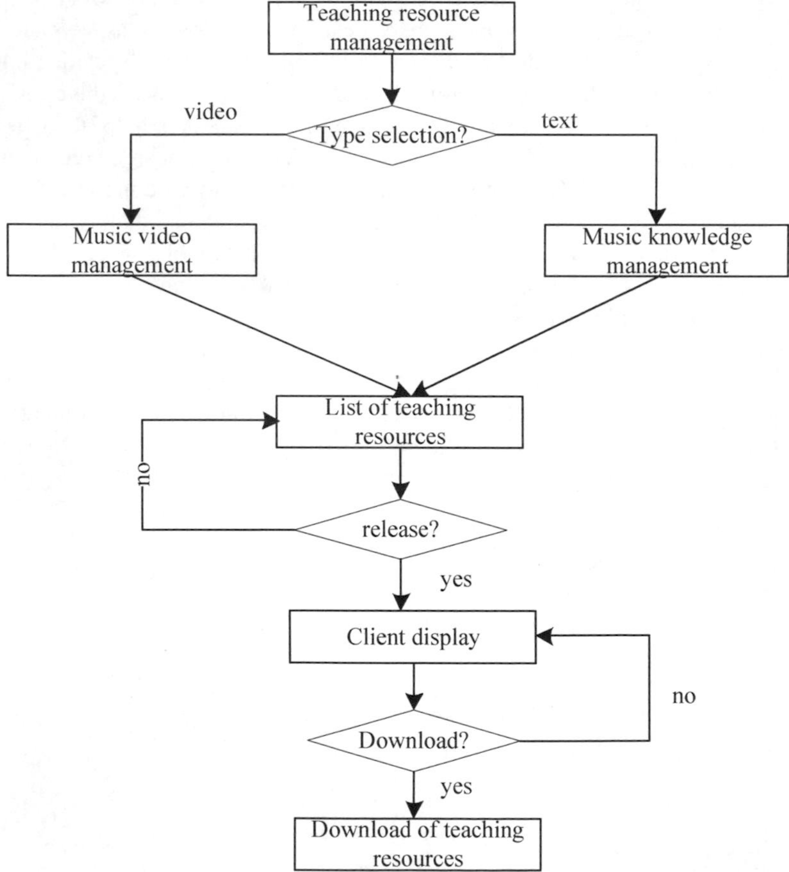

Fig. 7. Operation process optimization of auxiliary teaching system

Based on the optimization steps, the main functions of each module of the system are further optimized. The auxiliary teaching management module includes teacher-student communication module management, music video management, examination management and courseware management.

The management module of music question bank mainly realizes the classification management, question management and answer management of music related examination knowledge.

Classification management includes the functions of adding, modifying and deleting test questions or knowledge types. The types of classification include selection, filling in blank and objective questions. The management of test questions includes the functions of adding, modifying and deleting test questions. When adding test questions, you must select the corresponding type of test questions, score, answer and other detailed information, so that the parameter setting and automatic statistics of score analysis can be realized when generating test papers.

Answer management includes the functions of adding, modifying and deleting answers. Each answer must correspond to a specific question, otherwise it will not be operated.

The videos of music courseware and related materials are placed in the Resource Center for unified management, which provides convenience for students to browse related materials, expand their learning knowledge and deepen their understanding. At the same time, it helps to improve learning interest and enthusiasm. The materials and knowledge related to music teaching are the results of music teachers' accumulation and collection. There are many forms, including pictures and text, and remote auxiliary teaching system, such as video, audio or animation.

The authority to add related data is mainly given by the system administrator to teachers, and the related operations include data upload, modification and maintenance. At the same time, according to the students' views, this module automatically records the students' views, and teachers can analyze the popularity of relevant knowledge.

The music video courseware is released by the corresponding teachers, and the corresponding permissions are set. The students with the corresponding permissions can download the courseware materials.

Courseware video and information release: teachers release electronic handouts and teaching videos online.

Music video and material download: students with permission can download the material and video.

Maintenance of courseware and related information: the administrator can regularly maintain and manage the expired materials.

3 Analysis of Experimental Results

System testing is an integral part of system development, which is not the content of system implementation. It is mainly to verify the functions of the system to ensure the quality of software. The process of software testing is to complete the test of the program or the whole system and evaluate the test according to certain limited conditions. The limited conditions can be abnormal or normal, and there is no complete limit. In software testing, we need to simulate errors by various means to ensure that any unexpected situation can be handled normally. The test of music teaching assistant management platform is an integrity test process, which includes not only various functions and modules, but also performance parameters, robustness, security and verification tests.

In the test process, the overall test is mainly carried out through the establishment and verification of test cases. The foundation and purpose of establishing test cases are from the requirements analysis document to verify the credibility of the satisfaction.

The system test must be implemented as quickly as possible, find the defects or deficiencies of the system, and propose solutions for the deficiencies, and finally make the system to meet the actual requirements of users. The purpose of music teaching assistant management system test is the same.

Hardware environment:

Server: the CPU of the application server requires Intel Pentium 2 GHz or above, and the data service and application service are configured separately. The memory configuration of each server is no less than 4G. The data server adopts the form of disk array for automatic increment, and the application server space is no less than 50 g.

Client:

The requirement of client is low, the CPU is Intel Pentium 3 or more, and the memory is 1 g or more.

Software environment:

Server: the operating system is Windows Server 2003 or higher. The database is SQL Server 2000 relational database.

Client: unlimited client operating system, recommended for windows, browser recommended IE6.0 and above, report and file access, install prompt software.

The effect evaluation of courseware recommendation can be analyzed by the change chart of difficulty coefficient and ability. The evaluation of learners' ability is dynamic. Every time learners submit their own learning results, the system will automatically evaluate their learning ability and save relevant records, and recommend new learning courseware according to the difficulty coefficient. The comparative relationship between

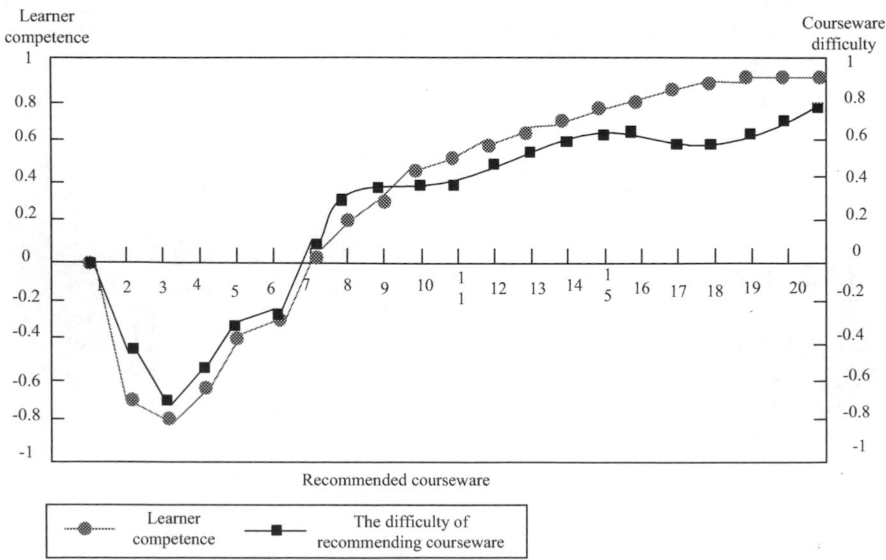

Fig. 8. Operation effect of teaching system in this paper

the difficulty coefficient of recommended courseware and learners' own ability is shown in the figure.

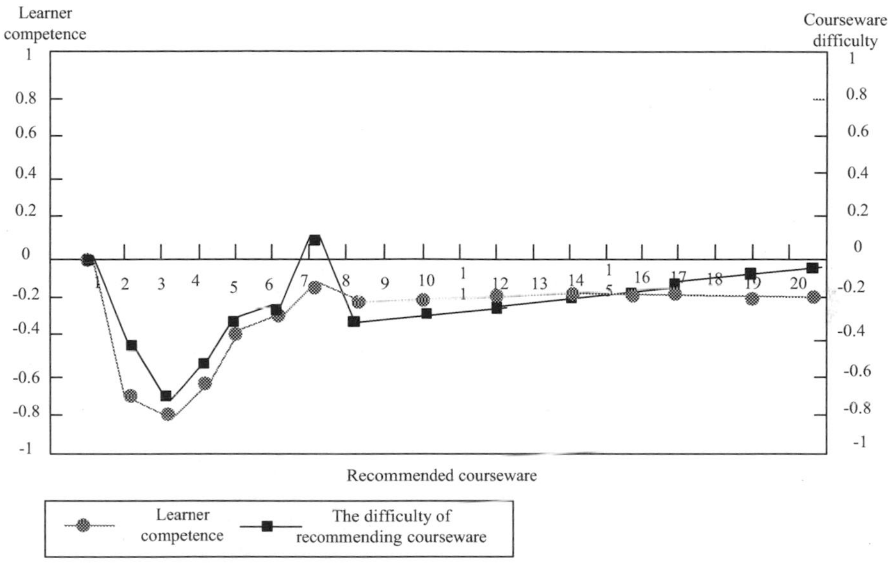

Fig. 9. Operation effect of traditional teaching system

It can be seen from the analysis of Fig. 8 and 9 that the operation effect of the teaching system under different methods is different. When the number of recommended courses is 12, the recommended difficulty of the designed system is 0.48 and the learner ability is 0.57. The recommended difficulty for the traditional teaching system is 0.30 and the learner's ability is 0.20.Comparing the two figures, it can be seen that the design system can effectively improve learners' ability. Moreover, compared with the traditional system, the multicultural distance assisted teaching system based on the mobile Internet of Things can provide more reasonable teaching information for learners in the practical application process, effectively improve the learning quality of students, and have a high specific effectiveness.

Further test the friendliness of the system, when there is a problem or input error, to give an error prompt. The function test is the core and focus of the system. Here we mainly introduce the function test cases of the system. The results of the function friendly survey are shown in the table (Table 1).

Through the test of the system, the current music teaching assistant management system can meet the design requirements, the system has good operation efficiency and stability; the human-computer interaction interface is friendly, all kinds of errors and friendly prompt information is perfect, which can guide the user to operate correctly and start quickly; Through the establishment of business use case and data model, online learning and teaching assistant business can be well realized, and the system has high security and reliability. To sum up, the music teaching assistant management system

Table 1. Function friendly use case table

User actions	Test case	Result	Result analysis	Is it up to expectations
Any input, and reset operation	Enter "123" and click "Refresh"	Reset input field	Through the browser client operation, the original state of the system is restored, and the interface is refreshed	Yes
Input 18 or 15 teacher ID numbers and fill in other information correctly	Enter "310005198504040611" and press "save"	Display "saved successfully"	1. The relevant information is saved to the database through the browser client; 2. After saving successfully, the client prompts "save successfully"	Yes
Fill in the correct form again and submit it	Enter "310005198504040611" and press "save"	Display "saved successfully"	1. Prompt operation success; 2, complete the database update; 3, do not need to repeat the input	Yes
Enter illegal characters in the client of browser, such as special characters of system database operation	Enter "@@@"	1. Display "input nonstandard" 2. Display "do not enter illegal characters"	1. In the client or smart client input is not standardized prompt; 2, the client prompt error, and display the correct way to fill in; 3, information return does not save	Yes

(*continued*)

Table 1. (*continued*)

User actions	Test case	Result	Result analysis	Is it up to expectations
When filling in the question information, fill in the non numerical form and submit it	Enter "one two three"	1. Please fill in the question information 2. Display "e.g.: 300158"	1. It is suggested to fill in the grade form of the type of the question; 2. Examples of correct filling; 3. Information is not saved	Yes

based on mobile Internet of things can well adapt to the actual requirements of teachers, students and managers for real-time management and learning of music teaching information.

4 Conclusion and Prospect

With the continuous development of economy and the acceleration of social progress, people pay more and more attention to quality education and art education. At the same time, higher requirements are put forward for music teaching. At present, although China's middle school music teaching has made some progress in teaching conditions, teaching staff, teaching level and so on, there is still a certain gap between middle school music teaching and students' aesthetic needs. In traditional middle school music teaching, simple blackboard writing is often adopted. In the past, this teaching mode can not really meet the current middle school music teaching needs. Therefore, it is necessary to improve the quality of middle school music teaching. This paper puts forward the design of multi-cultural music remote assistant teaching system based on mobile Internet of things, so that students can feel the charm of music more intuitively, stimulate students' enthusiasm for learning music, and let students better experience the beauty of music.

References

1. Huang, C., Yu, K.: Research on the innovation of college music teaching mode based on Artificial Intelligence. J. Phys. Conf. Ser. **1915**(2), 022–028 (2021)
2. Xu, Q., Su, Z., Dai, M., et al.: APIS: privacy-preserving incentive for sensing task allocation in cloud and edge-cooperation mobile internet of things with SDN. IEEE Internet Things J. **7**(7), 5892–5905 (2020)
3. Gren, L.: A flipped classroom approach to teaching empirical software engineering. IEEE Trans. Educ. **63**, 155–163 (2020)
4. Yuan, L., Xiaofei, Z., Yiyu, Q.: Evaluation model of art internal auxiliary teaching quality based on artificial intelligence under the influence of COVID-19. J. Intell. Fuzzy Syst. **39**(6), 8713–8721 (2020)

5. Yang, L.: Comprehensive evaluation of music course teaching level based on improved multi-attribute fuzzy evaluation model. Int. J. Emerg. Technol. Learn. (iJET) **15**(19), 107 (2020)
6. Strong, J.T., Giegold, C.: A multidisciplinary music and visual arts center at Earlham College. J. Acoust. Soc. Am. **145**(3), 1739 (2019)
7. Al-Smadi, M.H.: The effect of using songs on young English learners' motivation in Jordan. Int. J. Emerg. Technol. Learn. (iJET) **15**(24), 52 (2020)
8. Taheri, A., Meghdari, A., Alemi, M., Pouretemad, H.: Teaching music to children with autism: a social robotics challenge. Sci. Iranica **26**(1), 40–58 (2019)
9. Lin, P.-H., Chen, S.-Y.: Design and evaluation of a deep learning recommendation based augmented reality system for teaching programming and computational thinking. IEEE Access **8**, 45689–45699 (2020)
10. Cano Ortega, A., Sánchez Sutil, F.J., De la Casa Hernández, J.: Power factor compensation using teaching learning based optimization and monitoring system by cloud data logger. Sensors **19**(9), 2172 (2019)
11. Fu, W., Liu, S., Srivastava, G.: Optimization of big data scheduling in social networks. Entropy **21**(9), 902 (2019)
12. Liu, S., Bai, W., Zeng, N., et al.: A fast fractal based compression for MRI images. IEEE Access **7**, 62412–62420 (2019)
13. Liu, S., Li, Z., Zhang, Y., et al.: Introduction of key problems in long-distance learning and training. Mob. Netw. Appl. **24**(1), 1–4 (2019)
14. Martinovi-Bogojevi, J.: Aspekti kolaborativne kreativnosti u nastavi glazbe u osnovnoj koliCollaborative creativity in elementary school music teaching. Metodički Ogledi **27**(1), 127–148 (2020)

Design of Visual Listening Music Teaching System in Local Colleges and Universities Based on Mobile Augmented Reality

Jun Zhou[1(✉)], Hui Lin[1], Yang-lu Ren[1], and Yang-bo Wu[2]

[1] College of Art, Xinyu University, Xinyu 338000, China
[2] College of Mathematics and Computer, Xinyu University, Xinyu 338000, China

Abstract. The design of audio-visual music classroom teaching system should be based on the characteristics of music classroom teaching, and the design and implementation of music classroom teaching system are completed. According to the current situation of music classroom teaching, based on the commonly used music teaching software, combined with the characteristics of the actual music classroom teaching, the overall structure design of the visual listening music classroom teaching system is completed. Through experiments, it is proved that the visual listening music teaching system based on mobile augmented reality in local colleges and universities has high practical value and fully meets the research requirements.

Keywords: Mobile augmented reality · Audio visual · Music teaching

1 Introduction

With the rapid development of the information age, multimedia technology teaching has been widely recognized. Through the continuous popularization and application of computer technology, colleges and universities put forward higher requirements for music teaching. Visual listening to music classroom teaching is one of the key points of music teaching at this stage. Various music teaching software have appeared one after another, but in the public's view, there are few system software that can really meet the actual needs of music classroom teaching.

Some scholars put forward the design and application of music teaching system based on Moodle platform, using B/S structure to design Moodle platform module and database module, analyzing the actual needs of music teaching. Although it improves the interaction between teachers and students in the classroom, the storage capacity of music knowledge resources is low, which affects the actual experience. Some scholars have proposed a music teaching system based on virtual reality, which realizes the interactivity and immersion of virtual technology through audio processing and analysis. Although the music information resources meet the needs of daily teaching and have a new mode of interactive experience, it takes a long time to connect in the information interaction and has a poor sense of experience.

S. Liu and X. Ma (Eds.): ADHIP 2021, LNICST 416, pp. 575–586, 2022.
https://doi.org/10.1007/978-3-030-94551-0_45

In order to solve the above problems, this paper proposes a visual listening music teaching system based on mobile augmented reality technology, which makes the music teaching software and hardware configuration updated and improved. The proposed design can meet the actual needs of music classroom teaching, and has important practical significance for music teaching system.

2 Visual Listening Music Teaching System in Local Colleges and Universities

2.1 Hardware Configuration Optimization of Visual Listening Music Teaching System in Local Universities

The overall architecture design of music assistant teaching system based on mobile enhancement can be realized with modular design idea. In the process of modular design, not only the division of the whole system architecture should be considered, but also the comprehensive design of communication, continuity, retention and protection between modules should be carried out [3]. In the modular design, the main aspects of attention are as follows: when the internal requirements of the module change, the changes of the module should be carried out without affecting the normal work of other modules; when deleting the module, the deleted part should only act on the deleted module itself, without affecting the functions of other modules; when deleting the module, the deleted part should not affect the functions of other modules; When the new module has the function of the original module, it is necessary to ensure that the interface of the module is consistent, and the function of the whole system will not be affected after replacement [4]. From the functional structure and architecture, the architecture of music assistant teaching system based on mobile enhancement is designed. The system architecture will directly determine the future operation of the system. Based on the comparative analysis of the current C/s and B/C modes, the B/S mode is adopted. The main reason is that the maintenance and development cost of the three-tier mechanism system is lower than that of the C/S mode [5]. The specific system architecture design is shown in the figure (Fig. 1):

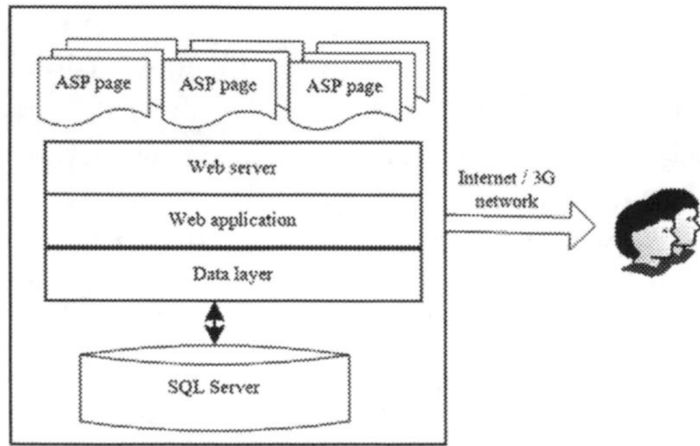

Fig. 1. Overall architecture design of the system

In the hardware system of the server, the CPU is required to be intel5, and the memory is more than 2G. The data service and application service are configured separately, and the minimum memory should not be less than 4G. The data hard disk adopts the way of disk array, so as to improve the speed of data storage, and the hard disk size should not be less than 500 g. In the development tool, visual studio is selected as the development tool [6]. The application server adopts IIS 6.0, and the server operating system adopts Windows Server 2003. SQL Server 2010 is used as the database management system. In terms of client requirements, the CPU of the computer is Intel Pentium III or above, and the memory is 1g. The minimum version of IE is 6.0. The system is mainly composed of user management, music knowledge learning, music appreciation, homework management, resource sharing, online examination and online question answering modules. The user management module includes the following functions: add users, edit the basic information of users, delete the basic information of users [7]. The function of music knowledge learning module includes: display music knowledge learning materials, display music teacher's materials, browse and consult music materials, reply to submit materials. The main function of music appreciation module includes: video information classification display, video detail information display, video information retrieval, video information upload and download. The function of homework management module includes: Homework submission, score check Inquiry, job release management, job online review. The main functions of the resource sharing module include: resource upload, music resource download, music resource evaluation; the main functions of the online examination module are: question editing and uploading, students' online examination, submitting papers, marking papers, score entry and score query. The main functions of the online answering module include: asking questions, answering online, and submitting feedback.

The system adopts the structure of B-S mode, that is, browser server structure mode. Users only need to use any browser on their own computer to access the server, all the data processing work is completed by the server, and the browser is only responsible for presenting the processing results of the server, which greatly reduces the workload of the user's computer. For users, a browser can access, convenient and simple [8]. For the system design, a lot of energy can be spent on the server program design, without considering the specific situation of the client, which is conducive to better improve the

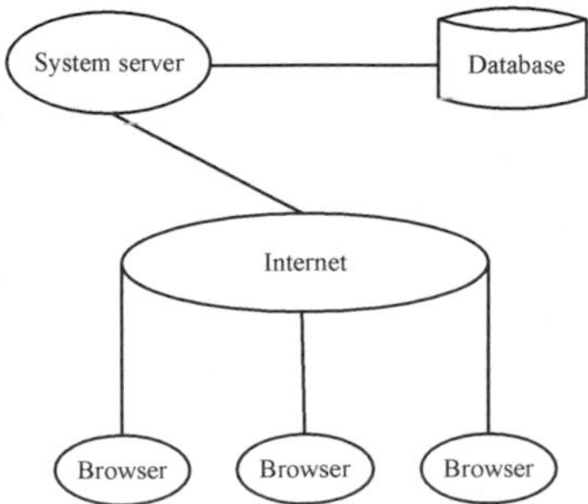

Fig. 2. Structure of audio visual music teaching management system

data processing and security of the server. The figure below shows the structure of the network-based excellent audio-visual education resource management system (Fig. 2):

Because B/S mode concentrates many core parts of system functions on its server, it can unify the clients together. Because of this structure, the system is easier to maintain and upgrade in use, and its operation is more concise [9]. B-S system architecture generally adopts three-tier structure, that is, adding an intermediate structure between the user data server layer and the user layer, and the intermediate structure includes different hierarchical structures such as presentation layer, logic layer and data processing layer. The middle layer builds a bridge between the server and the client, which enables the effective data communication between the user's browser and the server. However, each layer of the three-tier structure is independent of each other, and the modification of one layer does not affect the functions of the other layers and the whole. The design of the excellent audio-visual teaching resource management system is inseparable from the consideration of the technical factors that support its realization [10]. Based on the investigation of the current teaching management system and the development of network technology, Moodle is selected as the basic system architecture. Augmented reality itself is a global development project, its function is more comprehensive, at the same time, its support characteristics of constructivism teaching concept is very clear, in line with the teaching needs of the system. Moodle provides users with flexible modification freedom, which can meet the personalized needs of different users. Moodle itself is an application based on B-S mode, which is consistent with the original idea of the system.

2.2 Software Function Optimization of Audio Visual Music Teaching Management System

The excellent audio-visual teaching resource management system based on network is mainly used to assist teachers and learners in the management and use of teaching resources, in order to provide a reliable way to make better use of audio-visual resources. From the perspective of the performance of the whole system, the educational resource management system needs to be flexible and convenient in use, and can meet the needs of human-computer interaction. The flexibility and popularization of the system is the key. It can provide timely feedback when processing command requests, so as to avoid the situation of users' anxiety and waiting [11]. Finally, the educational resource management system must be safe and reliable, because the number of resource visitors may be large, the system should be able to ensure the concurrent processing of multiple commands. At the same time, because the formation of educational resources requires a lot of manpower and financial resources, it is more likely that some resources are accumulated by teachers' painstaking efforts for many years. Therefore, it is necessary to ensure the safety and reliability of the database itself, strengthen the protection function and timely backup work, and avoid data loss caused by loopholes. The overall goal of the system is to coordinate the smooth human-computer interaction between different users, and then make rational use of the excellent audio-visual teaching resources on the network to serve their own learning and Teaching [12]. In order to make different users use the system according to their own needs, four user roles with different permissions are set up, which are system administrator, teacher, student and visitor. The system administrator has the highest authority and is responsible for the initialization and data maintenance of the whole system, realizing the functions of user management, data management, information management and resource management. Teachers are responsible for the update and release of teaching resources and the interaction with students. After successful registration, students log in to the system and use the system resources for online learning and questioning. Visitors can only browse audio-visual resource information.

Combined with the reality of the current music teaching course, the role of the system is divided into teachers, administrators and college students. At the same time, through the analysis of the specific work in the music assisted instruction system, the overall use case of the system is as shown in the figure. Through the above use case analysis, the function of the system is designed as shown in the figure (Fig. 3).

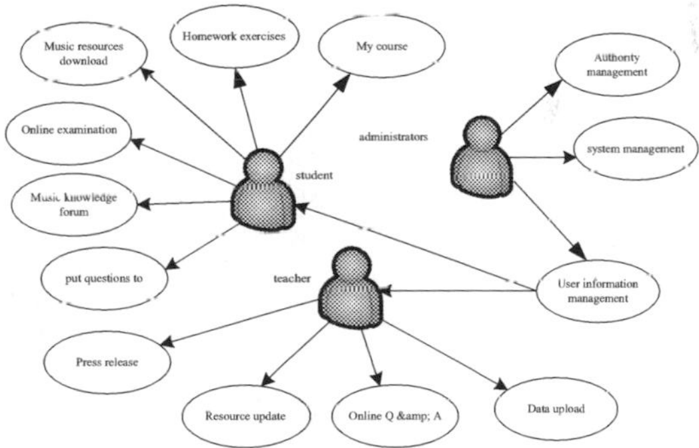

Fig. 3. Software function case of audio visual music teaching management system

Information release in the system mainly includes news release and announcement release. The news release is mainly for all users who log in to the system, including news editing, news maintenance and news release. Announcements are mainly for registered users of the system, including the management, release and maintenance of announcements. File management is mainly used for the management of system files, including file submission, file audit, file deletion and other functions, so as to provide the paperless office of the system and improve the overall work efficiency. Teaching resource management mainly realizes the management of music video resources, homework system and classroom courseware. Auxiliary teaching includes online examination, my courses, homework exercises, questions and answers, performance analysis, learning resources, etc. Through this way, we can make clear the main significance of computer teaching system and traditional teaching methods, make clear the teaching objectives, improve the teaching quality, and make clear what is the key link to improve the efficiency. In order to further improve the advantages of system implementation. In the design of music assistant teaching system, the main functions are summarized from the actual investigation. First of all, for a system, the most important and basic function is the identification function. The main task of students using the auxiliary teaching system is to learn better, so it should fully reflect the content of music learning, the completion of music homework and other links, and need the learning and homework submission function of network classroom. And in the process of students' learning, they may encounter problems that can not be solved, which requires timely communication with teachers. There should be a teacher's question answering module on the system to realize the real-time communication between teachers and students. Thirdly, from the teacher's point of view, to be able to verify the students' learning results, we need to submit homework online. In order to better complete the teaching task, the system needs to have the function of lesson preparation management. Finally, from the perspective of managers, any system should have the function of user management and resource management. Among them, the online examination is used to test the basic knowledge of music through teachers' uploading of test questions; my courses are mainly used to record personal learning, including time, course type, etc.; score query and analysis are mainly used to query the test results; question answering is mainly used to provide space for teachers and students to communicate (Fig. 4).

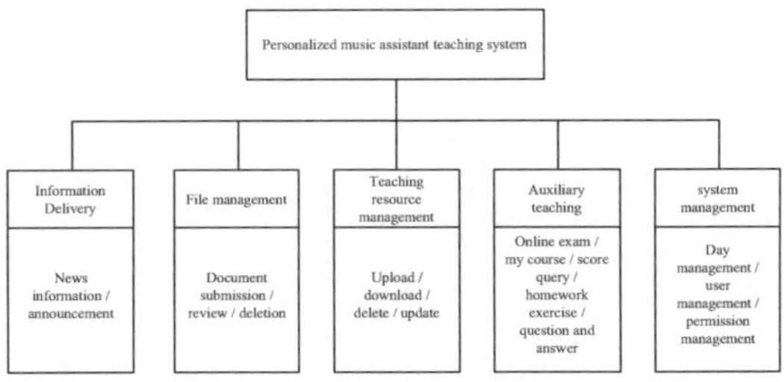

Fig. 4. System function structure optimization

System management mainly includes log management, user basic information management, authority management and so on. Log management is mainly used to record the system operation records and processes; user basic information management is mainly used to modify and delete the basic information of registered users; authority management is mainly used to assign roles to users of the system. The main function of audio-visual education resource management system is to achieve teacher assisted teaching and students' autonomous learning. Each registered student user has a "personal space" module. Remind learners of the gap between the current learning progress and their own learning plan, or provide comparison with other learners' progress, so that learners can adjust their pace. The recommendation part mainly recommends relevant learning resources according to learners' learning preferences, so that learners can learn more efficiently according to their own interests and habits. The main learning resources of this system are video resources, and the personal learning space also records the progress of the last learning for learners. The system will automatically remind students to choose to learn from the beginning or continue to learn from the last time when they continue to learn the next time, so as to humanize the learning process.

2.3 The Realization of Audio Visual Music Teaching Management

In view of the problem that the visual listening music classroom teaching system involves multiple functional modules, on the premise of ensuring the overall function of the system, the overall design of the system is carried out, including functional structure design and functional module design, in order to ensure the stable operation and scalability of the system. According to the actual needs of the visual and auditory music classroom teaching system, the overall function of the system is divided into three parts: simulation teaching template, music teaching resource editing module and music demonstration module. The structure of the system operation module is optimized as shown in the figure below (Fig. 5):

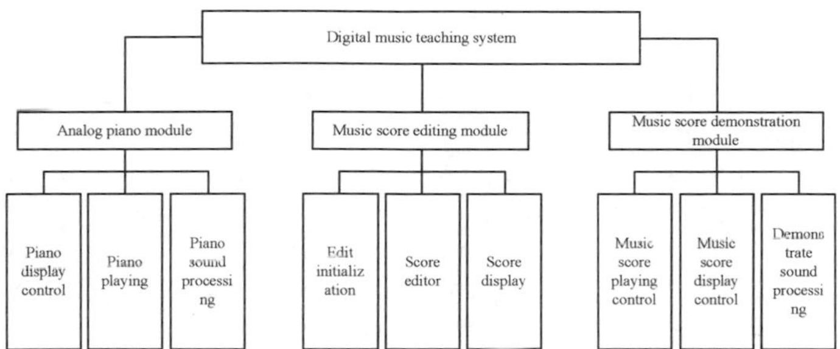

Fig. 5. Structure design of system operation module

The three functional points of the overall design of the functional structure are: the music editing module needs to realize the basic open or new music, edit and preview music, store music and other functions; the music demonstration module needs to realize. In music classroom teaching, playing demonstration is the core component. In the process of music learning, the corresponding music can not only promote the interaction between teachers and students, but also make students understand music more deeply. In order to improve the combination of the system and the classroom, and ensure the playback effect, the system combines the actual situation of music classroom teaching, further enriches the classroom playback demonstration methods, including single track/multi track playback, multi instrument selection playback, up and down tone playback, specific section playback and other methods, and achieves high-quality playback effect through special processing. Further, the demonstration module of visual listening music course is designed as follows (Fig. 6):

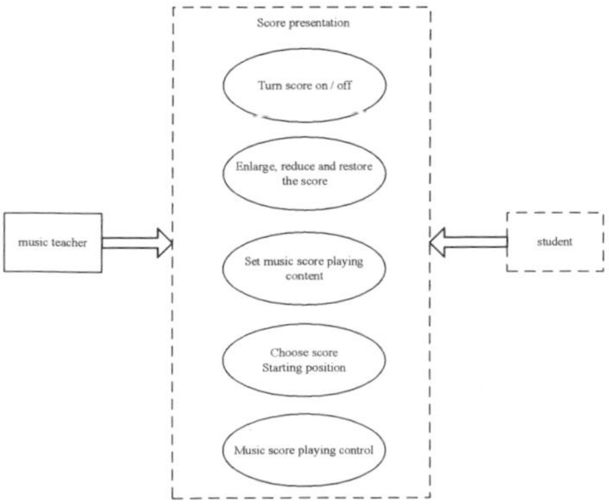

Fig. 6. Design of audio visual music course demonstration module

This module is mainly responsible for the display of teaching courses, such as determining the specific position of piano keys in the process of piano teaching, pressing and releasing multi post piano keys at the same time, realizing the processing of sliding piano keys, etc. As the core module of analog piano, the implementation of piano playing sub module is the most difficult, and the performance effect of analog piano depends on the quality of its implementation. In order to meet the needs of better interactive friendliness, it is realized by using mobile augmented reality technology combined with multi touch interactive technology. Excellent audio-visual education resource management itself needs regular maintenance and management, including account management, database management and version information management. In addition to the login user name, each registered user also has a serial number configured according to the corresponding order. The allocation of the serial number is specified according to different permissions as the unique identification of different users. The effective serial number is sorted regularly and the expired serial number is deleted. Only scientific and reasonable serial number management can ensure the normal operation of the system. The database connected with the server stores user information, audio-visual resource information, learning process records and other important information, so the database itself also

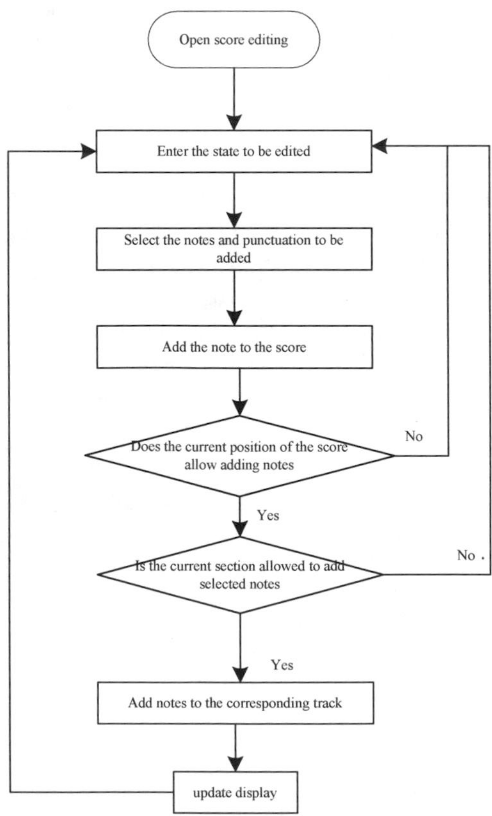

Fig. 7. System operation process

needs to be constantly updated and maintained to ensure its normal work. With the progress of the times and the development of technology, it is bound to put forward more new requirements for the education management system, so the system also needs to be constantly updated, each update will be a new version of the system, the smooth transition between the systems is needed, to ensure the compatibility of each version in the use process is also an important part of the system management.

The music editing sub module is mainly responsible for the editing process of music, including dragging to modify the staff score, key and time sign; adding or deleting music punctuation, bars and notes, etc. As the core module of music editing, music editing sub module involves many complex music rules. The accuracy and stability of music editing depend on the quality of its implementation. In order to further improve the effect of music editing and presentation, WPF technology is adopted to complete the design of the module. The specific flow of music editing sub module is shown in the figure (Fig. 7):

As shown in the figure, editing music can be realized by editing menu and symbol panel. For example, adding or deleting notes and dots can also be realized by editing menu for newly built music with two tracks. Music clef, key and time can be modified by symbol panel on the left. The resource management module of the network-based excellent course teaching resource management system is mainly in the charge of teachers. Teachers can add and update audio-visual resources at any time, delete obsolete and useless resources, and modify the classification and storage mode of resources. At the same time, with the collection and increase of resources, it will become more and more difficult to find information, so the system provides search function, users can query and retrieve resources through keywords. It should be noted that querying resources is applicable to users with all permissions. Excellent audio-visual teaching resources are the core of learning, and its management is also the most important part of all management services. At the same time, in order to avoid unnecessary loss caused by the loss of resources, teaching materials should be backed up safely in time. Especially in the network environment, audio-visual resources can be stored in the cloud server, which is safe and convenient, and can reduce the workload of the local server, so as to improve the operation effect of the system.

3 Analysis of Experimental Results

In order to verify the practical application effect of the teaching system, the experimental setup is carried out. The test of the system includes function test, overall test and performance test. In order to ensure the test effect, the test environment is preferred to be standardized. In terms of software environment suggestions, the system is based on Web service architecture, so it can adapt to most of the operating systems, mainly involving the following aspects:

Operating system: windowserver2008;
Java virtual machine: sunj2sdk5.0/beajrockit5.0;
Application server: tomcat7. X;
Database: sql2008.
Hardware environment: server.

Configuration: Intel Xeon es-2600, 2 gigabit network cards. Memory: 8g.
Hard disk: above 2506, use disk array to improve the speed and security of the system.

Unit test cases of system login module are shown in the table:

Table 1. Unit test of login module

Test entry	User name	Anticipate result	Test result
Click menu: user login	Admin	Enter the user login interface	Normal
Fill in user name	Admin	The system will query according to the user name	Normal
Fill in the password	Admin	The system will query with the password according to the input	Normal
Choose roles	Admin	The system will check according to the input role	Normal
Click login	Admin	The system will check according to the user name and password	Normal
Login successful	Admin	The page shows successful login	Normal
Login failed	Admin	The page shows that the user does not exist. Please register	Normal
Click the register button	Admin	Enter the registration interface	Normal
Click finish	Admin	Prompt that the registration is successful and jump to the login interface	Normal

As can be seen from Table 1, the average response time of the login module of the design system is less than 10 s, and the response time of most things is less than 7 s, which is short. When the login module is running, the utilization rate of CPU, memory and process is low, and the function of login module can meet the requirements of system running. The unit test cases of music information retrieval module are as follows:

Table. 2 Unit test of music teaching information retrieval management module

Test entry	User name	Anticipate result	Test result
Click the menu: music knowledge learning system input keywords	Admin	Enter the music knowledge learning interface	Normal
Input keywords	Admin	Enter the music knowledge retrieval interface	Normal
Key words of input music information	Admin	The system will query according to the key words	Normal

(*continued*)

Table. 2 (*continued*)

Test entry	User name	Anticipate result	Test result
Display music knowledge information	Admin	Display the music knowledge information directory of query and keyword matching	Normal

As can be seen from Table 2, the average response time of music knowledge retrieval is less than 10 s, and the response time of most things is less than 7 s. When the music knowledge information reaches the peak, the utilization of CPU, memory and process is low, and the storage capacity of music knowledge resources is large, which can meet the needs of the system. It can be proved that the function of music knowledge learning module can meet the requirements of system operation. The unit test cases of the system music appreciation module are shown in the table (Table 3):

Table 3. Unit test of music information browsing and playing management module

Test entry	User name	Anticipate result	Test result
Click menu: music appreciation system	Admin	Enter the music appreciation interface	Normal
Input basic information of music appreciation	Admin	Display response music appreciation resources	Normal
Click the play button	Admin	Music appreciation resources begin to play	Normal
Click the stop button	Admin	End of play	Normal

The ability of the system music appreciation system can meet the needs. From the above, we can see that the function of the music appreciation module designed at present can meet the needs of the system operation from the performance.

4 Conclusion and Prospect

Under the background that multimedia system is widely used in classroom teaching, this paper puts forward the design of visual listening music teaching system in local colleges and Universities Based on mobile augmented reality. According to the characteristics of music classroom teaching, this paper designs the overall architecture of music assistant teaching system, increases the system memory, establishes user management module, knowledge learning module, music appreciation module, homework management module, resource sharing module, online examination module, online answer module, and constructs effective communication between browser and server through B/S mode, At the same time, it ensures the security of education system and provides technical support for music education classroom. Through the experimental verification of the actual teaching effect of the system designed in this paper, it can be confirmed that the storage capacity of the visual listening music teaching system based on mobile augmented reality is high, and the average response time is less than 10 s, which can meet the needs of music classroom education, improve the level of music classroom teaching, and reflect the progress of music education system.

In the next research, we plan to start with online interaction, focus on the online learning function of the Internet, strengthen the visual design of online learning module, understand the user preferences through real-time feedback, develop personalized pages, enrich the use experience of music teaching system, and improve the teaching atmosphere of music classroom.

Fund Projects. The research project is: the research results of the 2021 project of Jiangxi education science "the 14th five year plan" – the application research on the infiltration of Ganpo music culture into "aesthetic education" teaching in local colleges and universities.

References

1. Yuan, F.: Design and application of music teaching system based on Moodle platform. Microcomput. Appl. **314**(06), 136–138 (2019)
2. Rui, H.: Implementation of music teaching system based on virtual reality. J. Suzhou Univ. Sci. Technol. (Nat. Sci.) **036**(001), 80–84 (2019)
3. Jin, Y.H., Hwang, I.T., Lee, W.H.: A mobile augmented reality system for the real-time visualization of pipes in point cloud data with a depth sensor. Electronics **9**(5), 836 (2020)
4. Zhang, D.: Application of audio visual tuning detection software in piano tuning teaching. Int. J. Speech Technol. **22**(1), 251–257 (2019)
5. Bogach, N., Boitsova, E., Chernonog, S., et al.: Speech processing for language learning: a practical approach to computer-assisted pronunciation teaching. Electronics **10**(3), 235 (2021)
6. Neto, L.P., Godoy, I.R.B., Yamada, A.F., Carrete, H., Jasinowodolinski, D., Skaf, A.: Evaluation of audiovisual reports to enhance traditional emergency musculoskeletal radiology reports. J. Digit. Imaging **32**(6), 1081–1088 (2019)
7. Lin, P.-H., Chen, S.-Y.: Design and evaluation of a deep learning recommendation based augmented reality system for teaching programming and computational thinking. IEEE Access **8**, 45689–45699 (2020)
8. Cano Ortega, A., Sánchez Sutil, F.J., De la Casa Hernández, J.: Power factor compensation using teaching learning based optimization and monitoring system by cloud data logger. Sensors **19**(9), 2172 (2019)
9. Yang, L.: Comprehensive evaluation of music course teaching level based on improved multi-attribute fuzzy evaluation model. Int. J. Emerg. Technol. Learn. (iJET) **15**(19), 107 (2020)
10. Liu, S., Bai, W., Zeng, N., et al.: A fast fractal based compression for MRI images. IEEE Access **7**, 62412–62420 (2019)
11. Liu, S., Li, Z., Zhang, Y., et al.: Introduction of key problems in long-distance learning and training. Mob. Netw. Appl. **24**(1), 1–4 (2019)
12. Vincenti, G., Bucciero, A., Helfert, M., Glowatz, M. (eds.): e-Learning, e-Education, and Online Training. LNICSSITE, vol. 180. Springer, Cham (2017). https://doi.org/10.1007/978-3-319-49625-2

Integrated Management System of Instrument and Equipment in University Financial Media Laboratory Based on Hybrid Cloud Architecture

Zhen-bin Huang(✉)

Art Design and Media College, Sanda University, Shanghai 201209, China

Abstract. Drawing lessons from previous research results, design an integrated management system for instruments and equipment in the university's media converged laboratory based on a hybrid cloud architecture. The device basic information management module mainly defines the basic information of the device, and registers all the information of the device or changes in all the information. The equipment scrap management module realizes early warning of scrap information and processing of scrap applications for scrap equipment. The function of the equipment maintenance management module is to realize the recording of information about the periodic quality inspection and maintenance of test equipment and other equipment in the laboratory. Equipment accessories management module is mainly to add, delete, maintain and query equipment accessories information. Register the information of Equipment Distribution Department through the equipment distribution management module. The system management module is responsible for the basic user information and other basic settings. The system uses firewall technology to isolate and protect the user network from the server. Test results show that the system function and performance are good, can be put into use.

Keywords: Hybrid cloud architecture · Financial media · Data migration service · Laboratory equipment · Information dispersion algorithm · Integrated management

1 Introduction

For the university financial media laboratory, the traditional equipment management mode can not meet the current needs because of its backward equipment management mode and slow processing speed. The demand of modern equipment management information system is particularly urgent [1]. Modern equipment management information system adopts advanced computer technology and network technology, and integrates modern management concepts and methods for equipment management. Through the equipment management information system, ensure that the equipment is always in a good state of operation, give full play to the potential of the equipment, and improve the

S. Liu and X. Ma (Eds.): ADHIP 2021, LNICST 416, pp. 587–600, 2022.
https://doi.org/10.1007/978-3-030-94551-0_46

operation efficiency and economic efficiency of the equipment [2]. With the improvement of production management informatization, the informatization modernization of equipment management needs to realize the informatization of basic information of equipment first. In daily management, the equipment information is unified planned into the management system. With the help of the real-time characteristics of the computer system, the equipment information data can be quickly and accurately counted, and the summary and analysis reports can be automatically generated. It provides an effective means for the integrated management of laboratory equipment.

After more than 20 years of development, the laboratory information management system has made great progress in both design and application technology. Especially with the development of today's network society and the acceleration of information exchange, more and more new international standards have been determined, more and more laboratory information management systems have been put on the market, some products have also appeared the development trend of "homogenization", and more and more manufacturers have turned to the competition of price and service. Users are more willing to accept suppliers with good service and strong company strength. More and more manufacturers begin to pay attention to the development of products from the perspective of customer needs, based on cost saving, and from the details to improve the user experience. The development of laboratory information management system is a special management system developed with the development of computer information technology. In the aspect of system architecture, it has experienced different stages from C/S architecture to B/S architecture. C/S architecture is mainly used in the developed regions of local area network. With the development of Internet technology, the B/S architecture model based on Web server has become popular. Through web application, the cost of client application is reduced, and the number of user groups is expanded, and the user area is no longer restricted. With the contact of customers to computer information technology, more and more customers no longer focus on the function, and the requirements for user experience are also higher and higher. Many LIMS system suppliers also value the R & D investment of related technologies. The laboratory information management system based on C/S architecture has experienced the development and evolution mode from two-tier architecture, three-tier architecture to multi-layer architecture. Traditional C/S architecture system, fat client software is not easy to deploy, thin client has complex development, need to use the network, users lack of good experience. But with the development of C/S architecture technology, the problems such as data synchronization and deployment are gradually eliminated or become non main contradictions. Based on these two architectures, an integrated management system of instrument and equipment in university financial media laboratory is designed based on hybrid Cloud Architecture.

2 Design of Instrument and Equipment Integrated Management System of University Financial Media Laboratory Based on Hybrid Cloud Architecture

2.1 System Architecture Design

Based on the hybrid Cloud Architecture, this paper designs the instrument and equipment integrated management system architecture of university financial media laboratory. The core of the architecture mainly includes five parts: main service, data storage service, data restoration service, data migration service and access statistics service, and cloud storage access component [3].

The main service provides a friendly, easy-to-use and flexible user interface for applications and clients to complete the functions of data file access and resource management. The main service forwards the specific data processing request to other services for completion. The data control module responds to the user's request for file storage and acquisition, and forwards the request to the data storage service or data restore service to complete the operation. In addition, the data migration service can be called by the data control module to complete the migration of data files. The cloud storage management module interacts with private cloud or public cloud storage service providers based on cloud storage access components, manages the joining and exiting of cloud storage resources, configures the parameters of cloud storage service providers, monitors the load status and connection status of storage resources, and feeds back to other modules in time. The data management module is used to manage the information of system data files. The functions between the main service and other services are fully integrated. According to the scale of the hybrid cloud storage system, different ways such as program interface, WEB service or message queue can be used for communication coordination.

Data storage service: the data splitting module uses IDA algorithm to block the data, and records the metadata of the data block for data query and recovery. The data security module can be used to encrypt and sign the data symmetrically to ensure the confidentiality and integrity of the data [4]. The storage scheduling module determines the data storage strategy and calls the cloud storage access component to complete the storage function.

Data acquisition service: it is used to obtain data from private cloud or public cloud. Data security module is responsible for data decryption and integrity verification, restore scheduling module queries data block meta information, determines data file acquisition scheme, and data recovery module restores original data according to IDA algorithm.

Data migration service and access statistics service: the implementation of data migration service is to solve the problem that the cloud service provider replacement and data storage solution are not optimal, and run automatically according to a certain cycle or directly called by the main service. Access statistics service makes statistics on the access of data files in cloud service providers, providing assistance for data storage and migration.

Cloud storage access component: it is called by other services in the form of component. The private cloud access module encapsulates the interface of the private cloud, and the public cloud access module provides a unified access interface for all cloud service providers in the system to complete identity authentication, access, storage management and other operations [5].

Data processing in hybrid cloud architecture is completed by data storage, data acquisition, data migration and access statistics services. Data storage: the user uploads the data to the main service through the service interface. After the latter is delivered to the data storage service, the data storage service first determines whether the storage space occupied by the private cloud has reached the threshold set by the system. If not, it indicates that the private cloud has enough space to store the user's data. At this time, the system directly stores the data file and generates metadata to write to the server database, otherwise it indicates that the data needs to be uploaded to the public cloud space. If the user sets the encryption mechanism for the data stored in the public cloud, the data encryption is completed first. Then the system uses the cloud service provider selection algorithm to determine the storage scheme, calls the information dispersion algorithm to divide the data into blocks, and generates data files and metadata of data blocks, including file size, storage location and other information. After the metadata is written to the database, the system puts the data block into the upload queue to wait for uploading to the public cloud storage space.

When the data acquisition service responds to the user's data download request, it first obtains the metadata of the requested data file from the database, so as to know where the data file is stored: if it is stored in the private cloud, the system directly accesses the private cloud to obtain the data and returns it to the user; When the data needs to be obtained from the public cloud, in order to improve the access performance, the system sets a certain size of cache space in the server hard disk. If the data file is in the cache space, it can be directly obtained and returned to the user. Otherwise, the system calculates the current best data block download scheme and accesses the public cloud to obtain the corresponding data block. Finally, the merged complete data file will be cached to the local cache and returned to the user. The existence of public cloud cache space enables some frequently accessed data to be obtained directly without acquiring data blocks from the cloud and then merging them, which improves the performance of data acquisition to a certain extent [6].

Data migration and data access statistics: in a reasonable storage architecture, the access history of data files should be established, which can be queried by users and system administrators. Based on this, the data access statistics service classifies and statistics the data file access, and provides support for cloud service provider selection and data migration implementation in public cloud data storage. The binary relation of file type and file size only identifies a kind of access statistics data. When the file type is the same and the integer value after rounding is the same, the two files are considered to be the same classification. The access statistics service maintains the statistics in the following ways: when the user stores the data files to the system, the total cumulative quantity of the matched classification statistics increases accordingly; Access statistics service regularly scans metadata information of data files in the system, and updates the total storage time and total number of visits in the statistics. Based on the access statistics, the system can make a simple prediction of the data file storage when the data is stored. The predicted storage time and the number of downloads in the storage time can be set as the average storage time and the average number of visits. The more data in the system, the closer the access statistics are to the actual storage situation, which makes the selection strategy of cloud service providers more reasonable when the system is storing data. After the system uploads data to the public cloud, the storage scheme composed of cloud service providers may not be optimal in actual situation, or there is a certain difference from the expectation, and the data migration service will be adjusted automatically. Migration service scans the data files stored in public cloud periodically according to the cycle, recalculate the storage scheme for the data files that have reached the predicted storage time, and when the migration cost is relatively small, the cloud service provider can be replaced for data migration.

In the system architecture, the Vandermonde matrix in the information dispersion algorithm is as follows:

$$P = \begin{bmatrix} 1 & 1 & \cdots & 1 \\ 1 & 2 & \cdots & n \\ 1 & 4 & \cdots & n^2 \\ \vdots & \vdots & \vdots & \vdots \\ 1 & 2^{m-1} & \cdots & n^{m-1} \end{bmatrix} \tag{1}$$

In formula (1), n and m represent natural numbers.

The identity matrix in the information dispersion algorithm is as follows:

$$R = \begin{bmatrix} 1 & 0 & 0\cdots & 0 \\ 0 & 1 & 0\cdots & 0 \\ \vdots & \vdots & \vdots & \vdots \\ 0 & 0 & 0\cdots & 1 \\ 1 & 1 & 1\cdots & 1 \\ 1 & 2 & 3\cdots & n \\ \vdots & \vdots & \vdots & \vdots \\ 1 & 2^{m-1} & 3^{m-1} & n^{m-1} \end{bmatrix} \tag{2}$$

2.2 Design Equipment Basic Information Management Module

Equipment basic information management module mainly defines the basic information of the equipment, and registers all the information of the equipment or the changes of all the information. It mainly manages the classification and abstract properties of the device definition, and the main design operations include adding, deleting and modifying. This module is part of the daily management of the most commonly used devices. The requirement of equipment information management is to record the information of the whole process from adding equipment to abandoning equipment. Specifically, management involves the steps of registering storage devices, establishing and maintaining device card information. Warehousing registration is the storage of registration information of warehousing equipment or materials, the storage of equipment information and equipment log, and the establishment of equipment account information. Device information modification is to modify the error information, which is generally the error in the process of device input registration and distribution, in order to ensure the correctness of data information. The query module can be used to query system data and generate business reports [8]. It mainly includes equipment inventory statistics, equipment usage statistics, equipment storage statistics, equipment scrap statistics and equipment maintenance statistics.

Design and implement the equipment basic information management module to realize the management of fixed assets of the laboratory. Through the construction of equipment information files and use logs, it is convenient for the equipment management personnel to allocate the equipment. At the same time, the management personnel can understand the risk of equipment management in time, and achieve risk early warning and control in time. In the process of equipment information query statistics, we can query according to different conditions, and set the priority of the query, so as to filter the equipment information.

The activity diagram design is shown in Fig. 1.

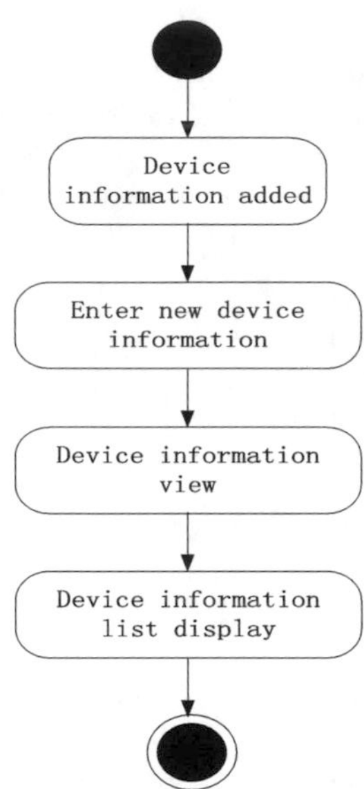

Fig. 1. Activity diagram of equipment basic information management

2.3 Design Equipment Scrap Management Module

The equipment scrapping management module realizes the early warning of scrapping information and the processing of scrapping application of scrapped equipment. Scrap management realizes the creation, maintenance and approval of scrap application information. After the approval of scrap application, scrap information is recorded, including time, reason and operator. Database to save the corresponding data [9]. In the system design, the scrapped database is designed and implemented to mark the scrapped equipment. It is convenient to query scrap information.

The activity diagram of equipment scrap management is shown in Fig. 2.

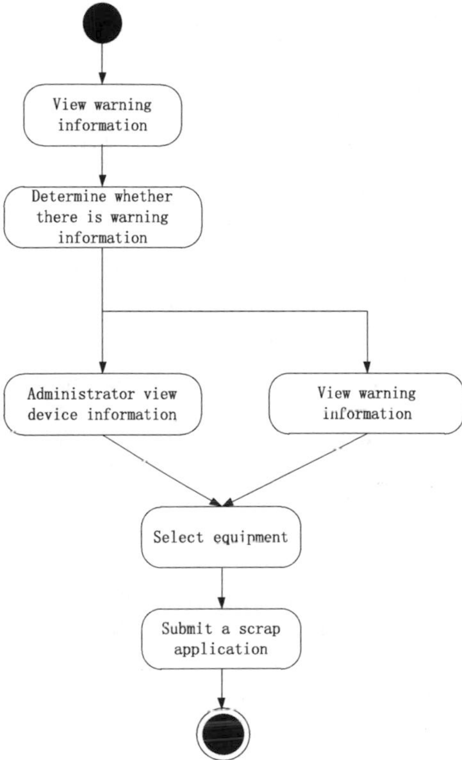

Fig. 2. Equipment scrap management activity chart

The function of this module is to manage the equipment scrap information. The process flow is as follows: the system will make an early warning for equipment scrapping, the equipment administrator will put forward an application for equipment scrapping, the authorized personnel will approve it, and the financial personnel can view the scrapping information for corresponding financial processing.

2.4 Design Equipment Maintenance Management Module

The function of equipment maintenance management module is to record the regular quality inspection and maintenance information of test equipment and other equipment in the laboratory. The main record information includes the setting of maintenance period and the basic management of maintenance date. During equipment maintenance, change the equipment status, create and modify equipment maintenance records, and delete equipment maintenance records. In this module, it also supports the view of equipment maintenance record information, and allows the view of equipment maintenance record information in the way of log view.

2.5 Design Equipment Accessories Management Module

Equipment accessories are the parts information needed in the maintenance or use of the equipment. The equipment accessories management module is mainly used to add, delete, maintain and query the information of equipment accessories. For the purchased accessories, the information can be added and modified according to the specifications and models, and the quantity of used accessories can be maintained. Provide the corresponding query conditions to query the information of equipment accessories. Equipment accessories management is similar to the design of equipment basic information management module, which will not be repeated here.

2.6 Design Equipment Allocation Management Module

The equipment purchased from the laboratory is used by special departments or personnel or departments, which needs to be realized through the equipment allocation management module in the system. When the equipment is allocated, the equipment allocation department, person in charge, equipment number, operator, equipment use function and other information are registered through the equipment allocation management module. For equipment such as consumables, record the number of users and the name of consumables. The function of equipment allocation management module includes the registration of equipment allocation information and the management of equipment allocation application [10]. The system approval user can process the equipment allocation application, and the applicant can view the application status.

2.7 Design System Management Module

As the basic management function of the whole laboratory equipment management system, system management is operated by the system administrator to complete the basic settings of basic user information, role authority matching and organization allocation. It is mainly used for system user data, system security, organization management and user authority control. User management registers the users of the system, and combines with the authority control subsystem to determine the user's operation authority and department ownership. It includes: user basic information management; User account: the account and password of the person logging in to the system; User role: determine the role permissions of personnel. In the process of business authorization, users can be grouped. A user can belong to multiple groups at the same time. By authorizing to groups, business function permissions can be authorized to users at the same time. The main task of department management is to set up and maintain the organization of the company. The goal of the system's authority management is to complete the authority assignment of the user role, and the purpose is to enable the personnel with the corresponding role to have the corresponding business operation authority according to the established user needs. In the design of the system, the business operation is realized through the business operation module, that is to say, the assignment of permissions is the operation control of the business module, that is, users with corresponding user roles are required to be able to modify, add, delete and query the module. Users without corresponding permissions can only operate some modules in the system. In the permission design of this system, in order to facilitate the management, the permission group management is set. The modification and addition and deletion of design basis are the basic permissions, and then the approval and query are set as permission groups. Each group is defined as a permission when the administrator sets it. Role management is mainly designed for the user role in the second chapter of the system. The role of the user in this system is horizontal management. Through the definition and management of the role, the effective allocation of the role corresponding to the authority is realized, which greatly reduces the workload of the system administrator. It is also convenient for the control of authority.

2.8 Design System Network Topology

In order to improve the robustness and security of the system, the system uses firewall technology to isolate and protect the user network and the server, so as to ensure the information security. And the use of dual database to ensure the reliability and stability of the system, the backup database will use hot backup method for synchronous backup, considering that the data transmission volume of the system will not be large, so the hot backup method will not have a significant impact on the efficiency. In addition, in order to improve the system performance, the application server and database server are combined to separate the data management and processing from the client. The use of database server can reduce the amount of programming, it provides a standard interface API for data manipulation, and can better ensure the security of the database. It provides monitoring performance, concurrency control and other tools for the database server. The system uses J2EE platform for development, uses web logic as the application server, considers the dynamic function of Java in network development and deployment, and introduces the standard of Java enterprise. Through the combination of these methods, the network security is greatly improved.

2.9 System Hardware Design

The basic configuration of the system is as follows:

CPU: Intel(R) Pentium(R) M 786 MHz 2.13 Gz;
Hard disk space: 200 GB;
Memory: 2 GB.

3 System Test

After the completion of the system design, through professional testing methods and tools, the completion of the test system design is consistent with the expected requirements. This chapter describes the purpose, scope, tools, process and results of the system.

3.1 Test Overview

The system test is to test the function and performance of the equipment management system. Through the system test, we can obtain the throughput, load capacity, response capacity, stability, reliability, security and scalability of the system, verify whether the system function meets the design requirements, and optimize the system design and improve according to the problems in the test process. Test the accuracy, tolerance and recovery ability of the database operation of the system. First of all, the test of the system is not simply because of looking for errors, but to analyze the error of the system, find the defects of the system, improve the software, and error analysis can improve the efficiency of the test. In addition, when there is no error in the test, it is also very important to test the quality of the system. The object of system testing needs to cover the hardware and software of the whole system, including the software and its dependent hardware, peripherals and so on.

The flow chart of the system test is shown in Fig. 3. The final purpose of the test is to verify whether the system meets the requirements and specifications and achieves the system indicators.

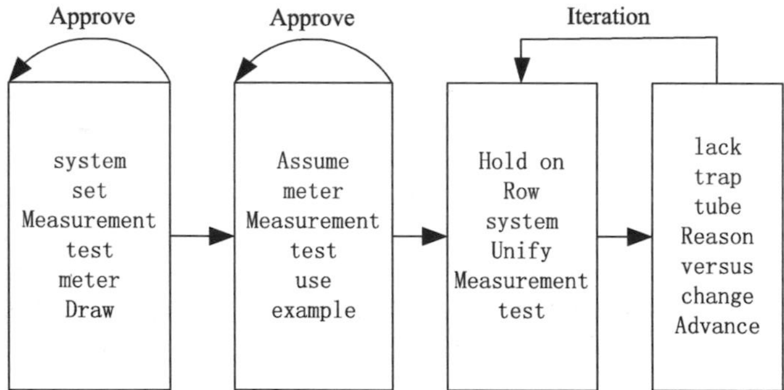

Fig. 3. Flow chart of system test

3.2 Test Purpose

The purpose of this test is to verify the functions and system performance of the laboratory equipment management system, mainly to obtain the following test indicators. Correctness: the system should be able to ensure 100% accuracy of data input, output and data transmission. Robustness: robustness requires the system to have a certain degree of fault tolerance. It requires the system to be able to remind rather than crash when illegal data is input. It requires the software to prompt in time when there is a problem. When the hardware is abnormal, it is required to be able to save the data at the first time, and the fault recovery can continue to work. Reliability: the system is required to have good reliability, and the operation failure rate caused by the system is required to be less than 0.5%; It is required that the average repair time of the system should be less than 5 h, and the number of unexpected downtime should be less than 2 times/year.

Performance, efficiency: the average page response time is less than 5 s. Security: the system is required to have good design in data backup, security prevention and security control. Scalability: the system can allow technology updates and new business modules to be added in the next 2–3 years.

3.3 Test Tools and Test Environment

Test Tools

The system is divided into two parts: function test and performance test. Among them, the function test is completed by making test cases and then executing them manually, while the performance test needs to be implemented with the help of tools. The performance testing tool is LoadRunner developed by mercury company, which is used to test the performance of the system.

Testing Environment

The test environment is shown in Table 1.

Table 1. Test environment

Environment	To configure	Operating system/software
Application server	CPU: 4*3.0 GHz Memory: 8 GB Hard disk: 500G network card: 100 Mbps/1000 Mbps	Windows Server 2003 SP3 Tomcat6.0
Database server	CPU: 2*3.0 GHz Memory: 4G Hard disk: 250G network card: 100 Mbps/1000 Mbps	Windows Server 2003 SP3 Oracle11g
Client	CPU: 2*2.8 GHz Memory: 2G Hard disk: 750G network card: 100 Mbps	Windows 7/ Windows XP SP3 Windows IE Windows Office 2007

3.4 Functional Testing

The function test of the system is used to verify whether the functions of the laboratory equipment management information system meet the needs of users. Test cases will be designed for all functions of the system, and then tested one by one.

After testing, it is found that the system functions meet the needs of users and design.

3.5 Performance Testing

This system will use LoadRunner's virtual user generator engine to build system load, through which virtual users can be generated, so as to simulate the business operation behavior of real users. The engine first records the business process, and then converts it into a test script. Considering that there will not be too many visitors to the laboratory equipment management information system at the same time, we will simulate the situation of 10 users online at the same time to test the concurrent operation pressure of the system.

The results of the performance test are shown in Table 2. Through the additional test of the system performance under the conditions of 5, 10, 50 virtual users respectively, and comparing with 100 virtual users, we can find the change rate of the system performance with the increase of the number of users.

Table 2. Test results of system concurrent operation pressure

Serial number	Number of virtual users (people)	Average response time (S)
1	5	0.86
2	10	0.98
3	15	1.19
4	20	1.46
5	30	1.83
6	50	2.17
7	60	2.69
8	70	3.17
9	80	3.82
10	90	4.35
11	100	4.71

It can be seen from the test result table that when there are 100 concurrent users operating on the system at the same time, the average response time is 4.71 s, which is in line with the expected results and meets the performance indicators of the system requirements.

3.6 Software Test

All functions of the system software are tested, and it is found that all functions of the software have been realized. Now the quality of the software is evaluated in several aspects: after three rounds of testing, the execution rate of test cases is 100%, the test execution is full, the pass rate of test cases is 99.54%, and the defect remaining rate is 1.38%. The defect rate and the number of residual defects have met the quality control requirements of the company, so we can end the testing stage and enter the next stage. The main purpose of this test is to test the software functions. All the software functions in the requirements have been realized. In the last test, the new functions have been fully tested, and the defects have been corrected and the final verification test has been carried out.

4 Concluding Remarks

The design system has been formally put into operation. During the operation of the system, it is found that the system runs stably and the business process is easy to operate. Therefore, the further research of the next step of the system will mainly focus on solving the problem of printing functional defects, and develop functional modules more suitable for the new business of laboratory equipment management.

With the continuous increase of test instruments, the amount of in-vehicle test data will increase by at least an order of magnitude. Therefore, it is necessary to consider the big data and data visualization of test data to further improve the efficiency and readability of data.

References

1. Shibu, J.A.: Asset inspection management system as a reliable inspection/monitoring tool. Mater. Eval. **78**(12), 1276–1285 (2020)
2. Zenginis, I., Vardakas, J.S., Abadal, J., et al.: Optimal power equipment sizing and management for cooperative buildings in microgrids. IEEE Trans. Industr. Inf. **15**(1), 158–172 (2019)
3. Cai, C., Wang, et al.: Resonant wireless charging system design for 110-KV high-voltage transmission line monitoring equipment. IEEE Trans. Ind. Electron. **66**(5), 4118–4129 (2019)
4. Toyama, T., Nagamine, N., Omori, T., et al.: Structure gauge measuring equipment using laser range scanners and structure gauge management system. Quarterly Rep. RTRI **60**(1), 40–45 (2019)
5. Mitoma, T., Yu, W.S., Chen, D.: New RBI Tool- ERAS (Equipment Risk Analysis System). Int. J. Comadem **22**(4), 21–27 (2019)
6. Eltyshev, D.K., et al.: Intelligent diagnostic control and management of the condition of electrotechnical equipment. Russian Electr. Eng. **90**(11), 741–746 (2019)
7. Yao, L., Shang, D., Zhao, H., et al.: Medical equipment comprehensive management system based on cloud computing and internet of things. J. Healthc. Eng. **2021**(4), 1–12 (2021)
8. Zhang, H., Li, Q., Meng, S., et al.: An adaptive industrial control equipment safety fault diagnosis method in industrial internet of things. Secur. Commun. Networks **2021**(5), 1–8 (2021)
9. Ssg, A., Nb, B., Ey, B., et al.: The impact of an integrated electronic immunization registry and logistics management information system (EIR-eLMIS) on vaccine availability in three regions in Tanzania: a pre-post and time-series analysis. Vaccine **38**(3), 562–569 (2020)
10. Goble, B.J., Mackay, C.F., Hill, T.R.: Design, development and implementation of a decision support info-portal for integrated coastal management, KwaZulu-Natal, South Africa. Environ. Manage. **64**(1), 27–39 (2019)
11. Toyama, T., Nagamine, N., Omori, T., et al.: Structure gauge measuring equipment using laser range scanners and structure gauge management system. Quarterly Rep. RTRI **60**(1), 40-45 (2019)

12. Negri, E., Berardi, S., Fumagalli, L., et al.: MES-integrated digital twin frameworks - ScienceDirect. J. Manuf. Syst. **56**, 58–71 (2020)
13. Jaouhari, S., Palacios-Garcia, E., Anvari-Moghaddam, A., et al.: Integrated management of energy, wellbeing and health in the next generation of smart homes. Sensors **19**(3), 1–24 (2019)

Research on Experimental Teaching Information Collection Method of Visual Communication Specialty in Hybrid Universities Under Intelligent Terminal

Zhen-bin Huang(✉)

Art Design and Media College, Sanda University, Shanghai 201209, China

Abstract. In order to enhance the efficiency of information collection in experimental teaching of visual communication specialty and improve students' art design ability, this paper proposes a method of information collection in experimental teaching of visual communication specialty in hybrid Universities under intelligent terminal. The delay model of experimental teaching information collection is established, and the user delay under different modes is obtained. According to the residual value of all the information in the experimental teaching information processing node, the dynamic weight of each experimental teaching information is calculated, and the residual load capacity of the experimental teaching information processing node is estimated to achieve the balanced distribution of the experimental teaching information of visual communication specialty, This paper constructs the experimental teaching information collection mode of visual communication specialty in compound university to realize the experimental teaching information collection. The experimental results show that the method can meet different reliability requirements, has high acquisition accuracy and the effectiveness of redundant information filtering, and can promote the further improvement of students' art design ability.

Keywords: Intelligent terminal · Mixed colleges and universities · Visual communication major · Experimental teaching information · Collection method · Art design

1 Introduction

In recent years, with the continuous implementation of the enrollment expansion plan of colleges and universities in China, the improvement of teaching quality in Colleges and universities has gradually become the focus of extensive social discussion. The reason why we want to further improve the teaching quality of colleges and universities is that from the perspective of social employment demand, many college students still have some prominent problems of difficult employment, and the fundamental way to solve this problem is to improve the teaching quality of college education and make it more scientific and reasonable for effective education and teaching, And further close

S. Liu and X. Ma (Eds.): ADHIP 2021, LNICST 416, pp. 601–614, 2022.
https://doi.org/10.1007/978-3-030-94551-0_47

to the rigid needs of social talent demand [1]. In improving the quality of teaching in Colleges and universities, many colleges and universities have done a lot of work, and measuring the quality of college work also needs to be processed and analyzed through the standardized process, which involves the collection and processing of teaching related data, because only the real data can effectively reflect the real teaching situation of the school.

In the past, in the process of collecting experimental teaching information of visual communication specialty in hybrid universities, although it also includes measuring the quality of teaching and so on, due to the low degree of informatization, many hybrid universities use office software such as Excel or word to collect experimental teaching information of visual communication specialty, which leads to the slow submission of experimental teaching information data of visual communication specialty, Moreover, the possibility of human intervention in the form is large, which is easy to lead to the inconsistency of the data format of the experimental teaching information of visual communication specialty [2]. In addition, the transmission process of the form needs to be issued level by level, and the recovery progress is often relatively long.

The ultimate purpose of information collection of visual communication experiment teaching in hybrid universities is query and statistics. Because of manual collection and statistics of visual communication experiment teaching information data in hybrid universities, it is difficult to achieve effective statistical work only through the existing office software, Sometimes it takes a lot of computing time to get the experimental teaching information data of visual communication specialty in hybrid universities.

At present, the teaching quality of visual communication specialty in hybrid colleges is more diversified. It is necessary for the more comprehensive collection of experimental teaching information of visual communication specialty to reflect the teaching level of visual communication specialty, such as the collection of comprehensive data such as curriculum arrangement of visual communication, teacher-student ratio, cooperation between school and enterprise. Therefore, every region of our country is also exploring and practicing the corresponding evaluation standards of colleges and universities. With the rapid development of information technology, the office efficiency of enterprises and institutions can be effectively improved through software system platform. For the current mixed university to solve the visual communication professional experimental teaching information collection related problems, can also be solved through information. As long as the application of advanced technology, design in line with the design of business needs, so that the system has a high performance and key problem solving ability, it can be widely recognized.

The early 21st century is the beginning of the research and development and application of information acquisition software system for experimental teaching of visual communication major in mixed universities. With the synchronous development of computer operating system, computer network technology and computer hardware technology, all walks of life and units began to use relevant software systems one after another, Its purpose is to help enterprises solve problems such as office efficiency [4]. For China's colleges and universities, this stage is just at the time of expanding the enrollment plan, which increases the management pressure of colleges and universities at the same time. Many colleges and universities have new tentative choices in score processing, course selection management, book use, etc., that is to introduce the corresponding score management system, course selection management system, book management system and other software systems, At this stage, the system mostly appears in C/S mode, that is, the operation mode of direct interaction between client and server is realized through LAN deployment. With the continuous development of computer technology, the growing maturity of Internet technology, and the increasing complexity of enterprise business, the management system of colleges and universities is constantly enriched and improved. On the one hand, the needs of business are considered in the type, It further enriches the types of the system; On the other hand, advanced technology architecture mode is adopted in technology application. At present, China's colleges and universities have a relatively high degree of education informatization in running schools. From the level of education departments, students' diploma information can be identified and viewed through xuexin.com. From the inside of the school, although most of the original teaching systems are retained, they are mostly through technology upgrading and system business integration [5]. However, at present, with the requirements of our country's education authorities for colleges and universities in the new situation of personnel training, colleges and universities pay attention to the management and teaching quality on the one hand, and actively prepare historical data and data on the other hand, so that the school itself or the higher authorities can have a comparison of the current school running state, and the ultimate goal is to promote the improvement of its school running quality. At present, some provinces in China have begun to brew and study such systems, and strive to collect and summarize the teaching information of colleges and universities in time, so as to ensure the first-hand regional teaching quality.

According to the national conditions and current development status of each country, there are great differences in the level of information education in foreign countries. Developing countries generally lag behind developed countries, some underdeveloped countries even have some problems in universal education, and the construction of education informatization is even less [6]. The development of educational informatization in developed countries such as the United States and Germany benefits from their advanced information technology level and educational teaching mode. Their computer software and hardware levels are at least 10 years ahead of those in developing countries. The development of educational informatization in developed countries also benefits from their advantages in educational resources.

With the terminal innovation of communication and network technology, it is easy to produce too much redundant information when collecting experimental teaching information of visual communication specialty in hybrid universities, resulting in the consequence that the information collection is not accurate enough. Therefore, this paper proposes a method of information collection for experimental teaching of visual communication specialty in hybrid universities under the intelligent terminal.

2 Design of Experimental Teaching Information Collection Method

In order to solve the problem of low collection rate of traditional experimental teaching information collection methods, this paper proposes a hybrid experimental teaching information collection method of visual communication specialty in colleges and universities under intelligent terminal. The overall framework of this method is as follows:

Firstly, from the perspective of a pair of primary users and a pair of secondary users in the experimental teaching system of visual communication specialty in hybrid colleges and universities, an experimental teaching information acquisition delay model is established to obtain the user delay under different modes.

Secondly, according to all the information residual values of the experimental teaching information processing node, the dynamic weight of each experimental teaching information is calculated, the residual load capacity of the experimental teaching information processing node is estimated, and the balanced distribution of the experimental teaching information of visual communication specialty is completed.

Finally, the experimental teaching information collection model of visual communication specialty in hybrid colleges and universities is constructed to realize the experimental teaching information collection.

2.1 Calculate the User Delay in Different Modes

In the process of collecting experimental teaching information of visual communication specialty in hybrid universities, with the support of intelligent terminal, the delay model of experimental teaching information collection of visual communication specialty in hybrid universities is established to obtain the user delay in different modes. The specific steps are as follows:

Considering that there are a pair of primary users and a pair of secondary users in the experimental teaching system of visual communication specialty in hybrid universities, it is assumed that λ_p and λ_s represent primary users and secondary users, the Poisson distribution of the experimental teaching information of visual communication specialty obeys the parameters, L_p and L_s represent the corresponding packet length, pU_s represents the sending node of the primary user, and pU_D represents the receiving node of the primary user, R_p and R_{PS} denote the channel rate at which the secondary user sends the experimental teaching information node SU_S of visual communication specialty, R_{SP} and R_S denote the channel rate at which the secondary user sends the experimental teaching information node SU_S of visual communication specialty to the primary user receiving node pU_D and the secondary user receives the experimental teaching information node of visual communication specialty, respectively. If the secondary user participates in the process of collecting the experimental teaching information of visual communication specialty for the primary user, the equivalent packet transmission rate of the primary user represented by R_{CP} can be calculated

$$R_{CP} = \frac{pU_S \oplus pU_D}{SU_S[R_S + SU_D]} * R_{SP} \tag{1}$$

According to the above, the model of information collection delay of experimental teaching for visual communication specialty is established by using formula (2) (Fig. 1)

$$q(R_{CP}) = \frac{SU_S * R_{PS}}{R_{SP}} SU_D \tag{2}$$

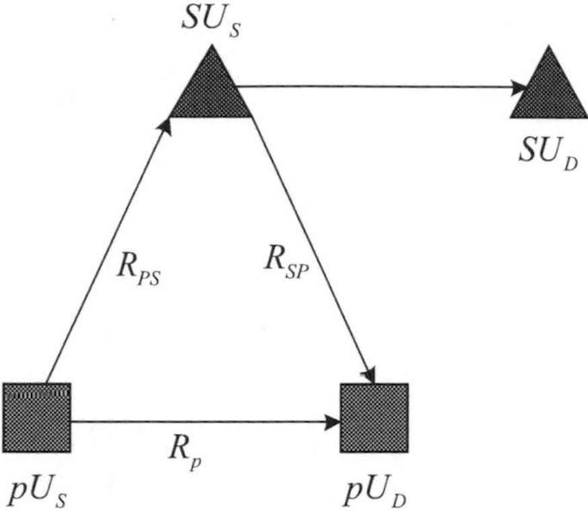

Fig. 1. Network virtual model

In the non acquisition mode, the primary user has the priority to use the experimental teaching information of visual communication specialty in hybrid universities. When the primary user no longer uses the experimental teaching information of visual communication specialty, the secondary user can use the visual communication information, And when the primary user needs to use the visual communication professional experimental teaching information, the user should give up the right to use the visual communication professional experimental teaching information to the primary user in time

$$E\|T_{NC-P}\| = E\left[X_{NC-P}\right] + \frac{\lambda_P E\left[X_{NC-P}^2\right]}{2(1-\rho_{NC-P})} \tag{3}$$

In the above formula, $E\left[X_{NC-P}\right]$ represents the average service time of the primary user in the non acquisition mode, λ_P represents the utilization rate of the experimental teaching information of the visual communication specialty of the primary user, and ρ_{NC-P} represents the ratio of the utilization rate of the experimental teaching information of the visual communication specialty of the primary user to the departure rate of the primary user in the non acquisition mode.

Then the average delay of stimulating users can be obtained by using formula (4)

$$E\|T_{NC-S}\| = \frac{E\left[X_{NC-S}\right]}{(1-\rho_{NC-P})} + \frac{\lambda_S E\left[X_{NC-S}^2\right]}{2(1-\rho_{NC-P})\left(1-\rho_{NC-p}-\rho_{NC-S}\right)} \tag{4}$$

In the above formula, $E\left[X_{NC-S}\right]$ represents the average service time of secondary users, ρ_{NC-S} represents the ratio between the utilization rate of experimental teaching information of visual communication specialty of secondary users and the departure rate of primary users in non acquisition mode, and λ_S represents the utilization rate of experimental teaching information of visual communication specialty of users.

In the process of the utilization of professional experimental teaching information in the acquisition mode, the secondary user provides information service for the main user until the main user is listed as empty. Since the visual communication professional experimental teaching information allocation protocol [7] between the main and secondary user, the main user will not interrupt the professional experimental teaching information and formula (5) can get the average delay of the main user in the acquisition mode, namely:

$$E\|T_{BE-P}\| = \frac{E\|T_{NC-P}\| \times E\|T_{NC-S}\|}{E\|X_{BE-P}\| \otimes E\|W_{BE-P}\|} \tag{5}$$

In the above formula, $E\|X_{BE-P}\|$ represents the average service time of the primary user and $E\|W_{BE-P}\|$ represents the average waiting time of the main user.

The average delay of the secondary user is obtained by formula (6), namely:

$$E\|T_{BE-S}\| = \frac{E\left[X_{BE-S}\right] \oplus E\left[W_{BE-S}\right]}{E\left\|T_{BE-p}\right\|} \tag{6}$$

In the above formula, $E\left[X_{BE-S}\right]$ represents the average service time of secondary users, and $E\left[W_{BE-S}\right]$ represents the average waiting time of secondary users.

To sum up, it can be explained that in the process of collecting experimental teaching information of visual communication specialty in hybrid universities, from the perspective of the existence of a pair of primary users and a pair of secondary users in the experimental teaching system of visual communication specialty in hybrid universities, a delay model of experimental teaching information collection of visual communication specialty in hybrid universities is established to obtain the user delay under different modes, And take it as the basis of experimental teaching information collection of visual communication specialty in hybrid universities.

2.2 Balanced Distribution of Experimental Teaching Information of Visual Communication Specialty in Hybrid Universities

In the process of balanced distribution of experimental teaching information of visual communication specialty in hybrid universities, the proportion of each dimension of experimental teaching information of each experimental teaching information processing node in its own information is calculated [8], On this basis, the ratio of the allocated experimental teaching information in each dimension to the total allocated experimental teaching information in each node is obtained, and the residual value of experimental teaching information in each node is obtained, According to the residual value of all information in the information processing nodes of the experimental teaching of the visual communication specialty, the dynamic weight of each experimental teaching information of the visual communication specialty is calculated, and the residual load capacity of the experimental teaching information processing nodes of the visual communication specialty is obtained, so as to complete the balanced distribution of the experimental teaching information of the visual communication specialty [9]. The specific steps are as follows:

Suppose c represents the number of experimental teaching information processing nodes of visual communication specialty, and $R(r_1, r_2, \ldots, r_m)$ represents the m dimensional information of each experimental teaching information processing node of visual communication specialty

$$\gamma(g) = \frac{R(r_1, r_2, r_m \ldots) \otimes c}{m \otimes \theta(\iota)} \odot \mu(K) \times o(f) \tag{7}$$

In the above formula, $\theta(\iota)$ represents ι disjoint subsets of selected task form in task set, $\mu(K)$ represents K dimensional experimental teaching information of information processing node of visual communication specialty, and $o(f)$ represents the experimental teaching information of visual communication specialty required by each task.

Suppose r_{ij} represents the experimental teaching information allocated by the information processing node of visual communication specialty, $\mu(g)$ represents the balance of experimental teaching information of visual communication specialty, $j(g)$ represents the load of experimental teaching information of visual communication specialty, and $\hbar(y)$ represents the amount of experimental teaching information allocated to the y dimension by the i information processing node.

Assuming that $\phi_{(f,j)}$ represents that the f experimental teaching information processing node of visual communication specialty has been allocated to the j experimental teaching information, the residual value of each experimental teaching information processing node of visual communication specialty is obtained, and the dynamic weight of each experimental teaching information of visual communication specialty is calculated

$$RS = \frac{\phi_{(f,j)} \otimes W'}{E''} \otimes \frac{\eta(S) \otimes V(\sigma)}{\varsigma(s)} \tag{8}$$

In the above formula, W' represents the dynamic weight of the residual experimental teaching information, E'' represents the residual value of the experimental teaching information of visual communication specialty, η represents a constant, ϕ represents the load weight of the node, $V(\sigma)$ represents the impact of the same experimental teaching information of visual communication specialty on the residual load capacity of the node, and $\varsigma(s)$ represents the balance of the experimental teaching information of visual communication specialty after the node is connected to task s.

Assuming that ϕ represents the total number of request tasks, the remaining load capacity of the experimental teaching information processing node of visual communication specialty is obtained

$$\psi(\varpi) = \frac{\phi(\overline{\gamma})\mu''(x)}{\vartheta(E)} \oplus \nu(d) \times N''' \times \wp(\iota) \tag{9}$$

In the above formula, $\mu''(x)$ represents the number of task sets requested by the outside world, $\vartheta(E)$ represents the number of experimental teaching information required by each task in each dimension, $\nu(d)$ represents the capacity of experimental teaching information in each dimension, N''' represents the number of tasks in each set in the task set, and $\wp(\iota)$ represents the information in the experimental teaching system of visual communication specialty.

Assuming that $\theta(f)$ is a random non negative positive integer, the balanced distribution of experimental teaching information for visual communication specialty can be completed

$$\mu(S) = \frac{\theta(f)(\psi)\psi(w)}{\gamma(g)} \tag{10}$$

To sum up, it can be explained that the principle of balanced distribution of experimental teaching information of visual communication specialty in mixed universities is used to complete the balanced distribution of experimental teaching information of visual communication specialty in mixed universities.

2.3 Construction of Experimental Teaching Information Collection Model of Visual Communication Specialty in Hybrid Universities

The experimental teaching information collection model of visual communication specialty in hybrid universities is mainly based on the intelligent terminal technology background, through the selective energy perception of the experimental teaching information of visual communication specialty. The intelligent terminal is used to make the model parameters balance the total network energy consumption. At the same time, based on the expected delay requirements, the optimal route with great probability to effectively collect the experimental teaching information of visual communication specialty is selected, and the performance of the experimental teaching information collection model of visual communication specialty is maximized according to the service level agreement [10]. By mapping the requirements of service level agreement, acquisition accuracy and energy consumption control to the network flow of WSN, it can be solved. Figure 2 shows the mapping diagram of WSN.

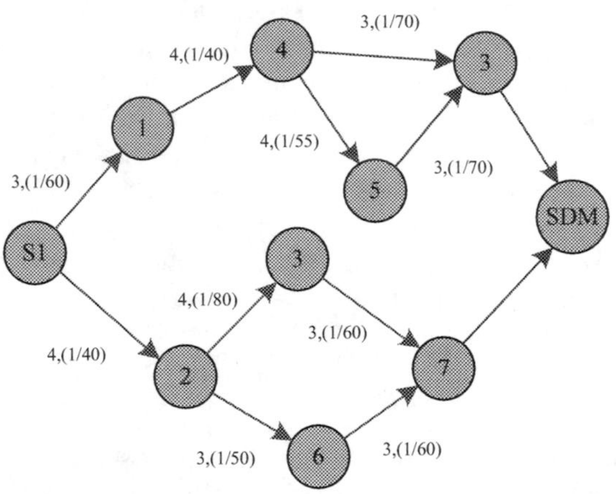

Fig. 2. Diagram of network flow

Each node in Fig. 2 above represents a sensor. The initial information source is node s and the target node is SDM. In the original Ping stage of the placement process, all nodes can create the area view of the nearest node, and the view will not be updated until the next maintenance or update password appears [11]. The update frequency of all nodes is an important factor to be considered when creating the acquisition model. Because the frequency is a controlled factor, it can be set through the most suitable level of the network. If a pre-set fixed frequency is used to maintain and update the window in the model setting stage, the optimal processing strategy of the model will not face the whole level, but all sub graph levels.

The edge in Fig. 2 represents the connection of two nodes, the volume of each connection shows the connectivity of the receiving node, and the expenditure of each edge is equivalent to the reciprocal of the battery volume of the receiving node. For example, the edge volume of node 4 to node 8 shown in the figure is 3, so the connectivity of node 8 is 3. The expenditure on the edge represents the percentage value of the remaining battery capacity of node 8 (the remaining battery capacity of node 8 is 70%). Therefore, after combining the heuristic algorithm of connectivity and cost function, the routing selection can balance the energy consumption and improve the routing reliability of experimental teaching information collection of visual communication specialty [12].

Figure 3 shows a class of simple energy improvement goals of routing clustering.

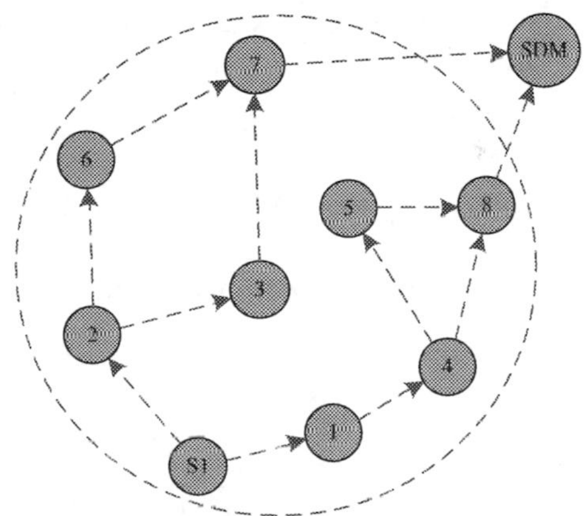

Fig. 3. Route selection

The experimental teaching information of visual communication specialty, which comes from the expected node and finally arrives at the target SDM through node 1, node 4 and node 8, will be clustered through the selected route. The above nodes present the maximum value of the current energy feasibility, because the remaining power of the route will gradually reduce, which is less than other feasible routes, and its feasibility will also decline with the extension of time, eventually making the route infeasible. In a certain stage, the routing process with the least energy consumption will break the balance of different routing energy, so as to maximize the life expectancy of WSN.

The establishment of information collection model of experimental teaching for visual communication specialty balances the energy consumption, while the information collection of experimental teaching for visual communication specialty can be combined with SLA to maximize the number of experimental teaching information selection for visual communication specialty and balance the energy consumption based on a reasonable degree. If the increase of the boundary points does not exceed the reasonable threshold, the experimental teaching information collection model of visual communication specialty will select the route with the maximum connectivity. For example, if the threshold is set to 35%, there will be two routes to access the next node. If the power difference between the two routes is less than 35%, then the model will enforce the selection of the route with the highest connectivity.

The information collection model of visual communication experiment teaching is not only realized by a simple maximization strategy, but also by an ideal route selection method. As long as it is in a reasonable energy balance target area, it will maximize the amount of information collection of visual communication experiment teaching until it exceeds the preset threshold, Then it is converted to the next more matching route, where the termination condition of the conversion is that the energy threshold difference between the maximum route and the secondary route can be converted into the maximum route. If the delay of experimental teaching information collection of visual communication specialty is regarded as the incentive object and integrated into the intelligent terminal, the closer to the SDM, the larger its volume, which reflects the approximation with the target node from the side. Therefore, as long as the reasonable performance consumption threshold between routes is not exceeded, the experimental teaching information of visual communication specialty can be effectively collected through the intelligent terminal [13].

3 Experimental Comparative Analysis

3.1 Experimental Environment and Data Information

In order to test the feasibility of the design of experimental teaching information collection method of visual communication specialty in hybrid Universities under the intelligent terminal, simulation is set to verify. The experimental environment is i5-3337U1.8 GHz, memory 8G, Windows 7–64 bit system, Matlab 2013 version.

In the above experimental environment, four Chinese search engines are used to collect the experimental teaching information of the same visual communication major, and the amount of feedback displayed is different. The experimental information is shown in Table 1.

Table 1. Data information table

Category	Search engines	Training information	Test information
Text message	Baidu	20	100
	Sogou	20	100
Audio-visual materials	Baidu	20	100
	Sogou	20	100
Machine readable information	Baidu	20	100
	Sogou	20	100

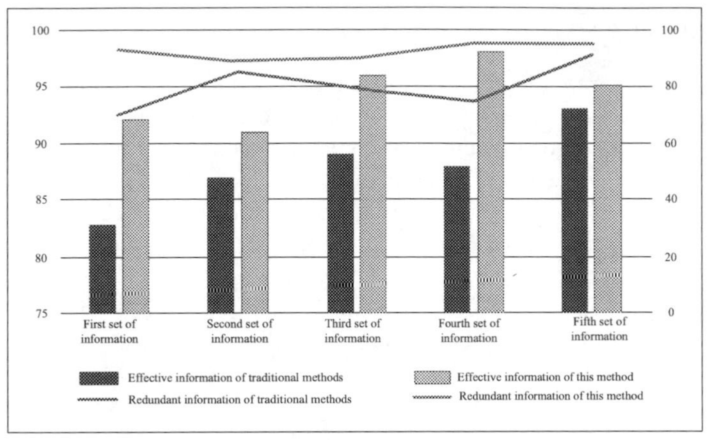

(a) collection results based on text information category

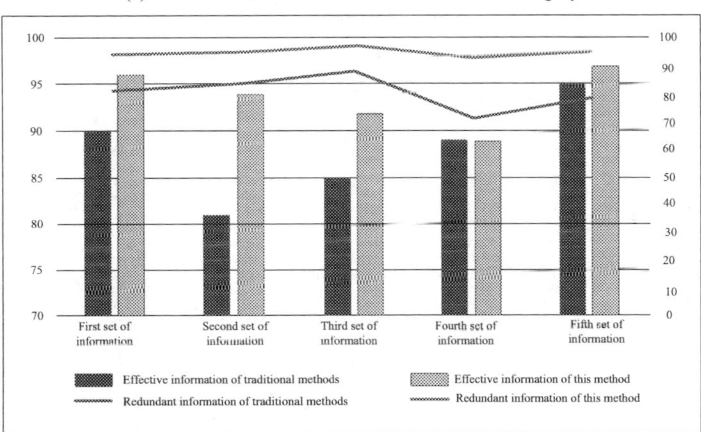

(b) collection results based on the category of audio-visual data

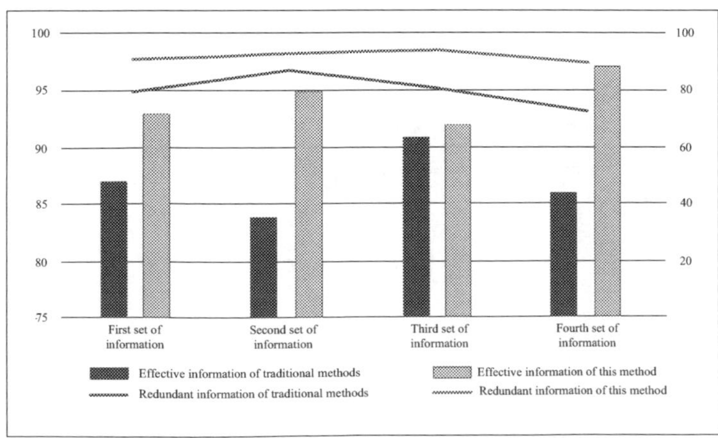

(c) collection results based on machine-readable information category

Fig. 4. Schematic diagram of information collection results comparison

3.2 Experimental Architecture and Analysis

20 keywords used for training in each category in the information table of visual communication experiment teaching were uploaded to the designed model, and all the information feedback from search engine was combined according to the location arrangement method to create meta set. If the search results have the same URL, the information feedback from the search engine is repeated about the keyword. If multiple search engines' feedback information coincides, the search results ranking is top selected, and the quadratic of the repetition times is taken as the weight value.

10 keywords in each category are randomly selected from the results, and 1000 available information and 1000 conditional redundant information are selected from the results of the classification of each keyword. The simulation is carried out for the 200 network information. The following Fig. 4 shows the comparison of the traditional information acquisition model and the designed model.

As can be seen from the Fig. 4 above, the average collection rate of experimental teaching information of visual communication specialty in hybrid Universities under the intelligent terminal is 94.08%, while the traditional collection method is only 87.67%. The reason is that the method designed in this paper constructs the experimental teaching information collection model of visual communication specialty in hybrid universities. Therefore, in the process of filtering redundant information, the recognition rate of traditional collection methods will fluctuate greatly, while under the intelligent terminal, the experimental teaching information collection method of visual communication specialty in hybrid universities fluctuates more gently and has a more stable identification performance.

Integrate all the information together. Under the same experimental environment, collect the experimental teaching information of visual communication specialty in hybrid universities by using traditional methods and this method respectively, and compare the information collection time of different methods. The specific comparison results are shown in Table 2.

Table 2. Data collection time/s

Number of experiments	Traditional method	Method of this paper
5	1.96	1.48
10	2.33	0.42
15	1.89	0.38
20	2.58	0.57
25	2.47	0.54
30	2.67	0.61
35	2.58	0.51
40	1.96	0.46

By analyzing the data in the above table, it can be seen that compared with the traditional methods, the experimental teaching information collection time of visual communication specialty in hybrid colleges and universities in this method is shorter and more efficient, which can realize the rapid and accurate collection of experimental teaching information of visual communication specialty in hybrid colleges and universities.

4 Concluding Remarks

In the current experimental teaching information collection process of visual communication specialty in hybrid colleges and universities, the user delay calculation and balanced distribution of teaching information under different modes are not taken as the research focus, which leads to the problem of low information collection rate. Therefore, this paper focuses on solving the problems existing in traditional methods, This paper puts forward the research on the experimental teaching information collection method of visual communication specialty in hybrid colleges and Universities under the support of intelligent terminal, calculates the user delay under different modes, and constructs the experimental teaching information collection model of visual communication specialty in hybrid colleges and universities by evenly distributing the experimental teaching information of visual communication specialty in hybrid colleges and universities. The experimental teaching information collection of visual communication specialty is realized. The results show that the information collection method designed in this paper has higher collection efficiency and can promote the further improvement of students' art design ability. The key research direction in the future is to comprehensively improve the comprehensive performance of the designed information collection method and provide an effective basis for more ideal construction.

References

1. Shestopalova, O.L.: Forecasting obsolescence of information collection and processing systems of space complexes according to the criterion of information support efficiency. Izvestiâ vysših učebnyh zavedenij Priborostroenie **64**(2), 81–89 (2021)
2. Huang, Z., Chen, C., Pan, M.: Multiobjective uav path planning for emergency information collection and transmission. IEEE Internet Things J. **7**(8), 6993–7009 (2020)
3. Tenqchen, S., Su, Y.J., Chen, K.P.: Using the cvp traffic detection model at road-section applies to traffic information collection and monitor -the case study. Artif. Intell. Eng. **1**(2), 38–43 (2019)
4. Liu, X., Yin, J., Zhang, S., Xiao, B., Ou, B.: Time-efficient target tags information collection in large-scale rfid systems. IEEE Trans. Mob. Comput. **PP**(99), 1 (2020)
5. Tang, X.: Research on logistics information collection based on image recognition. IEEE Access **PP**(99), 1 (2020)
6. Yan, B., Wu, C., Yu, R., Yu, B., Yu, Y.: Big data-based e-commerce transaction information collection method. Complexity **21**(2), 1–11 (2021)
7. Duan, R., Du, J., Jiang, C., Ren, Y.: Value based hierarchical information collection for auv enabled internet of underwater things. IEEE Internet Things J. **PP**(99), 1 (2020)
8. Luis, F.L., Agostinho, A., Rosa, F.: Mixed integer formulations for a routing problem with information collection in wireless networks - sciencedirect. Eur. J. Oper. Res. **280**(2), 621–638 (2020)

9. Eytan, A., Nichita, A.: Maintenance service platform (MSP) for maintenance information collection and sharing. IFAC-PapersOnLine **53**(3), 330–335 (2020)
10. Dong, G.L.: A case study on elementary and middle school teacher's method of information collection in the curriculum. J. Educ. Res. Math. **29**(3), 363–387 (2019)
11. Qin, Y., Wei, X.G., Wang, X.R., Zhang, Y.: Research on information collection and evaluation of learning intelligence media classroom teaching of red culture group based on mobile teaching. Wirel. Internet Technol. **16**(6), 154–155 (2019)
12. Zhong, Z.: Design of web based remote monitoring system for ship resistance teaching network information. Ship Sci. Technol. **41**(4), 224–226 (2019)
13. Xin, J.S., Liu, Z.H., Li, H.: Study on big data collection and analysis platform for education. J. Guangdong Ind. Tech. Coll. **19**(1), 11–17 (2020)

Fault Diagnosis of Stepping Motor PLC Control Loop Based on Fuzzy ADRC

Cai-yun Di[1,2(✉)]

[1] State Grid Jibei Electric Power Company Limited Skills Training Center, Baoding 071051, China
[2] Baoding Technical College of Electric Power, Baoding 071051, China
zz20221145@yeah.net

Abstract. The method of multivariable statistical monitoring and parameter estimation based on analytic model relies too much on the model, which leads to the low accuracy of fault diagnosis of three-phase current. According to the loop fault diagnosis structure based on fuzzy ADRC, any phase of the three-phase current in the current measurement loop is taken as the analysis object to obtain the fault feature vector. Through off-line and on-line modeling and identification, the control limits of statistics at a certain confidence level are determined. Define the branch current measurement loop as the fault criterion, and design the loop fault diagnosis process. The experimental results show that the method has little difference from the actual waveform, the maximum error is 10A, and has accurate fault diagnosis results.

Keywords: Fuzzy ADRC · Step motor · PLC control · Loop fault · Diagnosis · Contribution rate

1 Introduction

Stepper motor is widely used in various motion control systems because of its high precision, small inertia and reliable operation, and it can realize high precision and quick open-loop control [1]. There is a close relationship between the PLC control loops of stepper motor. The fault of a control loop often causes chain reaction, which leads to the abnormal operation of many loops, and even causes the paralysis of the whole production process. Therefore, the reliability and safety of the control loop seem particularly important [2]. Real-time monitoring of the operation status of the control loop can quickly and accurately detect the faults in the control loop, provide accurate, necessary and timely reference information for staff, quickly deal with the faults in a short time, reduce the waste of funds and resources, and effectively reduce the occurrence of catastrophic accidents [3]. Because of the particularity of the structure of the control loop, the output signals of all kinds of equipments show a large range of uncertainty, high correlation and non-gaussian characteristics, and a large number of data show that the SNR of data collected in the production process is sometimes low, because of these

S. Liu and X. Ma (Eds.): ADHIP 2021, LNICST 416, pp. 615–626, 2022.
https://doi.org/10.1007/978-3-030-94551-0_48

characteristics, the work of establishing accurate mathematical model of the control loop becomes very complicated. In general working environment of stepper motor, there are strong nonlinearity, uncertainty and correlation between data, which makes it difficult for engineers to obtain a complete and accurate mathematical model, and increases the probability of misinformation. Therefore, a fault diagnosis method of stepping motor PLC control loop based on fuzzy ADRC is proposed. Firstly, the basis of loop fault diagnosis based on Fuzzy ADRC is determined. Then the fault diagnosis method of stepping motor PLC control circuit is studied from two aspects: control circuit fault identification and circuit fault diagnosis process design, and the effectiveness of the proposed method is verified by experiments.

2 Loop Fault Diagnosis Based on Fuzzy ADRC

ADRC is a kind of nonlinear robust control technology. It uses the extended state observer to transform all the unknown nonlinear uncertain objects into the series type of integrator by nonlinear state feedback, and uses the state error feedback to design the ideal controller [4, 5]. It does not depend on the precise mathematical model of the controlled plant, the algorithm is simple, and it has good control effect and strong robustness.

Its structure and principle are shown in Fig. 1:

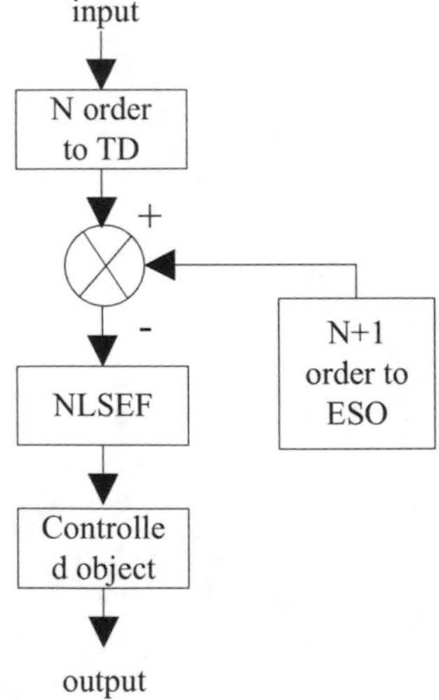

Fig. 1. Circuit fault diagnosis structure and schematic diagram based on fuzzy ADRC

In Fig. 1, the Tracking Differentiator (TD) is responsible for solving the problem of measuring signals from discontinuous or random noises [6, 7], reasonable extraction of continuous signals (given tracking) and differential signals. According to the differential output and the steepest synthesis function, the transition process of the closed-loop system can be arranged; and the extended state observer (ESO) expands the disturbances affecting the output of the controlled plant into a new state variable, and observes the expanded total disturbance signal through a special feedback mechanism [8, 9]. Through the inputs and outputs to construct the total disturbance as a state variable, the second-order system, whose extended observer reaches the third order, adds the state of the total disturbance. The total disturbances include internal disturbances and external high-frequency noise disturbances. The output of the extended state observer is the observed total disturbance of the system. The nonlinear state error feedback control law (NLSEF), based on the given signal derived from the tracking differentiator (TD) and the derivative errors of the system output and output observed by the differential and state observers of the given signal, further compensates the control and disturbance. The nonlinear control method is constructed by the fall or the fastest control synthesis function Fhan [10, 11].

The fault of relay protection current measurement loop is characterized by the abnormality of generalized ratio of variation caused by the increase of comprehensive error. Although this error may vary with ambient temperature and operation over time, it may be considered constant for a short monitoring period of time, so the monitoring of abnormal or faulty current measurement loops may translate into a mathematical approach to the solution of the generalized ratio of variances for each relay protection loop [12].

The current conversion ratio of each link of the measurement loop is defined as a variable to identify, and the dynamic change of the ratio is identified to diagnose whether the measurement loop is abnormal. The primary current, the secondary current and the GVR are constrained by each other, and the secondary current is known. Therefore, it is only necessary to solve the GVR to know the primary current, i. e. only 1 variable needs to be solved. The primary current must satisfy the Kirchhoff's law at the bus node of substation, so the variables can be solved by the constraint equation of Kirchhoff's law. So it is not necessary to get the primary current directly, and the generalized ratio of variation can be solved by using the constraint equation of primary current.

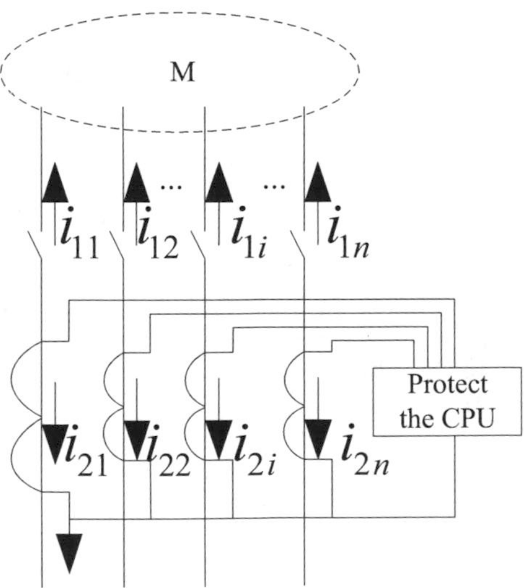

Fig. 2. Protection system for the current comparison principle

With any phase of the three-phase current in the current measurement loop as the analysis object, the protection system of the current comparison principle is shown in Fig. 2.

As shown in Fig. 2, $i_{11}, i_{12}, \cdots, i_{1i}, \cdots i_{1n}$ is a primary current flowing through a certain phase of node M, which is converted into a corresponding protective secondary sampling current which is converted through a current measurement loop to a corresponding protective secondary sampling current $i_{21}, i_{22}, \cdots, i_{2i}, \cdots i_{2n}$, according to the definition of generalized ratio of variation:

$$i_{1i} = N_{gi} \cdot i_{2i} \tag{1}$$

In formula (1), N_{gi} is the generalized ratio of variation of the current measurement loop of the i branch; i_{1i} is the primary current at the time of i; i_{2i} is the secondary current sampling value protected at the time of i [13, 14].

Set the monitoring time period of the measurement loop as $t_1 \sim t_m$, in which the protection CPU calculation value of section j branch of i. time period is $i_{2j}^{(i)}$, and the calculation value of each group shall satisfy the formula (1), then the calculation value of $m(m > n)$ hours may be listed as follows:

$$\mathbf{A} = \begin{bmatrix} i_{21}^{(1)} & i_{22}^{(1)} & \cdots & i_{2n}^{(1)} \\ i_{21}^{(2)} & i_{22}^{(2)} & \cdots & i_{2n}^{(2)} \\ \vdots & \vdots & \vdots & \vdots \\ i_{21}^{(m)} & i_{22}^{(m)} & \cdots & i_{2n}^{(m)} \end{bmatrix} \tag{2}$$

If the number of equations in formula (2) is greater than the number of unknown quantities, A may be decomposed into:

$$\mathrm{A} = \mathrm{SVD} \tag{3}$$

In formula (3), S and D are orthogonal matrices of order $m \times m$ and $n \times n$, respectively; V is diagonal matrix, whose diagonal elements are singular values of A and arranged in descending order [15, 16].

In practical calculation, the sampling group m is much larger than the current measurement loop n, and the column vector of A is linearly correlated. When m is large enough, $\mathrm{rank}(A) = n - 1$ can be obtained. At this time, there is only one basic solution system for the over determined homogeneous equations. In order to eliminate the zero solution, the constraint condition $\|n_v\| = 1$ is added. Substitution of type (3) into type (2) may be converted as follows:

$$A^{\mathrm{T}} A n_{\mathrm{g}} = (SVD)^{\mathrm{T}} (SVD) n_{\mathrm{g}} = D^{\mathrm{T}} \left(V^{\mathrm{T}} V \right) D \mathbf{n}_{\mathrm{g}} = 0 \tag{4}$$

In formula (4), D^{T} represents the orthogonal invertible matrix of unit; D is the orthogonal matrix; and $D\mathbf{n}_{\mathbf{g}} = (0, 0, \cdots, 0, 1)^{\mathrm{T}}$; the eigenvector corresponding to the minimum eigenvalue of $A^{\mathrm{T}} A$.

3 Research on Fault Diagnosis of Stepping Motor PLC Control Loop

3.1 Control Loop Fault Identification

In the PLC control loop of stepper motor, the input and output of controller, output of control valve, output of controlled object and output of transmitter should be selected as signal observation points of fuzzy ADRC to form mixed signal matrix [17–19].

By analyzing and calculating the collected signals, the statistics can be calculated from the decomposed independent components and separation matrix, and the running state of the control loop can be monitored in real time. During the operation of the control loop, if some devices fail, their output signals will be abnormal, which will affect the fuzzy ADRC supervisory statistics calculated from the data signals. Then, the statistical value obtained from online monitoring is compared with the statistical limit. If the control limit is exceeded, the control loop is normal. If the control limit is exceeded, the control loop is abnormal or fails.

Offline Modeling Identification

In the off-line modeling, the observation data of the above five observation points in the normal operation state of the control loop are selected to form observation matrix as historical data. The main aim of the data preprocessing of the observation matrix is to remove the outliers [20]. Because some signals are disturbed instantaneously, some inaccurate measurement values may be produced. When modeling off-line, removing these erroneous data can obviously improve the parameter estimation and performance parameters of statistics, and improve the accuracy of model training. According to the fuzzy ADRC method, the independent component s(k) and separation matrix W are obtained, and two statistics are calculated by s(k) and W.

There are two ways to deal with the historical data in off-line modeling of control loop monitoring. One is that the modeling data does not contain the change of the set value. For off-line modeling without setting value change, the effect of noise of different variance in control loop on fault diagnosis is mainly studied. The data selected in the modeling are the historical data of the normal operation of the control loop under different noise forms and different variances. Because the form and variance of noise in historical data are unknown, the variance of control loop output can be calculated to replace the variance of noise. The historical data under normal conditions are divided into two groups. The mean and variance of the first group are calculated [21]. The mean and variance of the other group are subtracted by the mean and divided by the variance of the first group. The statistical control limit of the normalized fuzzy ADRC method is used.

The fault diagnosis ability of fuzzy ADRC to control loop before, after and during the change of set point is studied for off-line modeling with change of set point in modeling data [22, 23]. When modeling offline, the historical data selected shall be the same as the change of the set value of the monitored data. The selected historical data are divided into two parts, and the data pretreatment and statistic control limits are calculated according to the modeling method with the set values unchanged.

In fuzzy ADRC method, because the distribution of independent components does not conform to a common distribution law, the control limit of statistics can not be determined from the confidence interval of a specific distribution. In most statistical process monitoring, it is usually assumed that the statistics follow the Gaussian distribution, but if the two statistics are treated as Gaussian distribution, the final result is obviously inaccurate and the final monitoring effect is not credible. In view of this situation, the method of fuzzy ADRC estimation is generally adopted to estimate the fuzzy ADRC function of the statistics, and then the confidence interval of the statistics is determined by the estimated fuzzy ADRC function, which is defined as:

$$f_n(x) = \frac{1}{nh}\sum_{i=1}^{n} K\left(\frac{x - X_i}{h_n}\right) \tag{5}$$

In formula (5), n represents the number of samples; X_i represents the independent sample-distribution samples drawn from population X ;$f(x)$ represents the density function of X ; $K(\cdot)$ represents a given probability density function over the real number field, called the kernel function; and h_n represents a positive number related to the number of samples, called the smoothing parameter or window width.

Through the definition of fuzzy ADRC estimator, the performance of kernel density estimator can be greatly improved by choosing proper kernel function and window width. The key factor in the estimation is the choice of window width h, which increases the effect of randomness if the window width is too large. The choice of window width is so large that the small properties of density function are easily neglected. Usually, the principle of evaluating the minimum window width is to get the minimum mean square error, and then the control limit of the statistics is determined at a certain confidence level.

Online Modeling Identification

Firstly, the monitored data are preprocessed, then the normalized data are decomposed into fuzzy ADRC, and the statistics are calculated and compared with the control limits obtained from off-line modeling. If the statistics are not overrun, indicating that the normal operation of the system, if the statistics overrun, indicating that the system has failed, the need for fault handling. By calculating the statistical contribution diagram of the fault time, we can determine which observation variable has the problem. By looking up the fault device of the problem variable, we can work out the solution to the fault of the control loop.

The variables that contribute the most to the failure graph are the main variables that cause the failure, as shown in Table 1.

3.2 Circuit Fault Diagnosis Process Design

The fault of the current measurement loop can be judged by comparing the calculated GVR L_{gi} of the current measurement loop with the ideal VR L_i of the CT. Thus, the expression of the fault criterion T_i for the current measurement loop of the branch circuit i is defined as:

$$T_i = \frac{L_{gi} - L_i}{L_i} \tag{6}$$

Table 1. Failure elements corresponding to the variables in the contribution diagram

Variable name in contribution diagram	Faulty device in control loop
Controller input	Transmitter failure or change of setting
Controller output	Controller failure
Control valve output	Control valve failure
Output of controlled object	Control object exception
Transmitter output	Transmitter failure

Under normal operation, the comprehensive error ε of the current measurement loop shall be less than 10%, and the expression of the fault criterion eigenvalue Q_i of the i branch current measurement loop shall be:

$$Q_i = \frac{-\varepsilon}{1 + \varepsilon} \quad (7)$$

When the measurement loop is normal, the error is within the specified range, i.e. p; E $Q_i \in (-0.0909, 0.1111)$; when the measurement loop is abnormal, the generalized

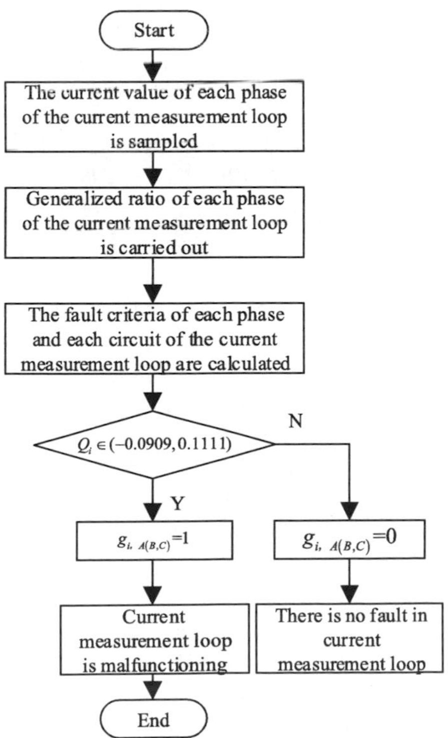

Fig. 3. Loop fault diagnosis process

variation ratio exceeds the above range. Therefore, according to the value range of Q_i, it can be judged whether there is a fault in the i-branch current measurement loop. The method is applied to the three-phase current measurement circuit of A, B and C respectively, and then the abnormity of the whole circuit can be identified by logic decision.

In order to clarify the fault criteria, the following definitions are given:

If $Q_{i,A(B,C)} \in (-0.0909, 0.1111)$, then $g_{i,A(B,C)}=1$ means that there is no fault in the A(B,C) phase measurement loop of the current protection device i; if $Q_{i,A(B,C)} \notin (-0.0909, 0.1111)$, then $g_{i,A(B,C)}= 0$ means that there is a fault in the A(B,C) phase measurement loop of the current protection device i, then the flow of fault diagnosis of the step motor PLC control loop based on fuzzy ADRC as shown in Fig. 3 can be obtained.

As can be seen from Fig. 3, the above criteria can be integrated when a bus current differential protection is installed in a substation. There are many ways to use this method in the station area and wide area protection. When a branch exits, the branch object in this method must be changed accordingly. In practice, the station and wide area protection have the function of network topology identification, so the network topology identification result can be used in the research method at the same time. However, when the number of branches changes, the identification equation variable shall be changed and the data shall be switched temporarily, but the number of branches included shall be unlimited.

4 Experiment

In order to verify the fault diagnosis method of stepping motor PLC control loop based on fuzzy ADRC, the experimental verification analysis is carried out.

4.1 Experimental Environment

A parallel fault diagnosis experimental platform of power equipment state information is established; the experimental environment includes 10 ordinary PCs with the same configuration, one as NameNode and the other as datanode. Each computer has a dual-core Interi5-2400 CPU with 4.00 GB of memory and 80 GB of hard disk. Build a parallel diagnostic test platform for power equipment status information, which includes 10 ordinary PCs with the same configuration, one for NameNode, and one for datanode. 4.00 GB of memory, 80 GB of hard disk, each computer has CPU dual-core Interi5-2400.

4.2 Experimental Results and Analysis

The life signal is divided into 4000 data segments, and the non-zero feature values and their corresponding feature vectors are determined as the comprehensive feature indices for the first time. Figure 4 shows the schematic diagram of fault diagnosis of stepper motor PLC control loop.

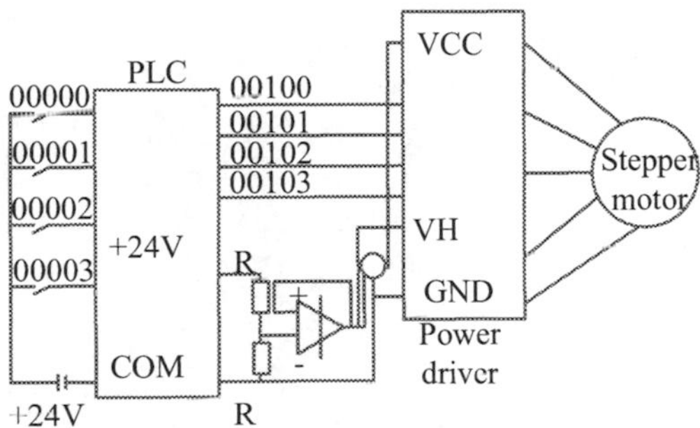

Fig. 4. Schematic diagram of fault diagnosis of stepping motor PLC control loop

According to the bearing faults shown in Fig. 4, the three-phase short circuit of the PLC control loop of the stepping motor is diagnosed by using multivariable statistical monitoring, parameter estimation based on analytical model and fault diagnosis based on fuzzy ADRC, as shown in Fig. 5.

Through the comparison of the results shown in Fig. 5, it can be seen that there is little difference between the waveform of the fault diagnosis method based on fuzzy ADRC and the actual waveform after the three-phase short circuit fault occurs, and the waveform change at other time points is basically the same as the actual curve, in which the maximum error is 10A, while the waveform difference of the method based on multivariable statistical monitoring and parameter estimation based on analytical model is large.

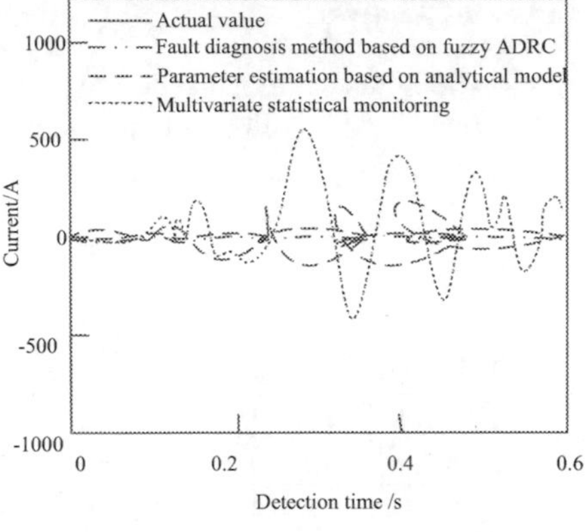

(a) Phase A

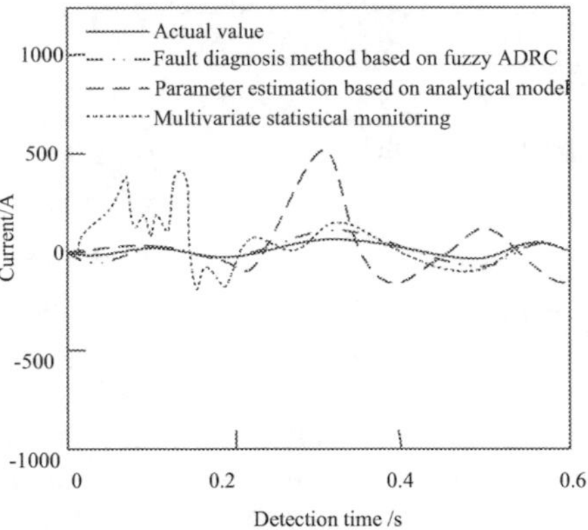

(b) Phase B

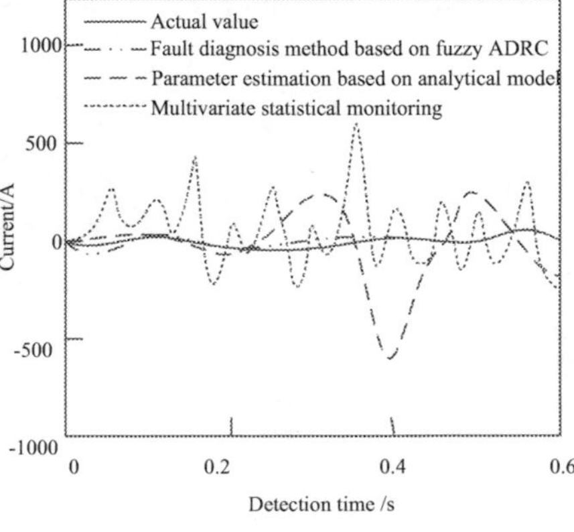

(c) Phase C

Fig. 5. Comparative analysis of three fault diagnosis methods

5 Conclusion and Prospect

Based on fuzzy ADRC, the fault diagnosis method of PLC control loop of stepping motor is studied. The input and output of PLC controller in the loop are selected as signal observation points. When a component in the control loop fails, it can be monitored through changes in statistics. This method can fully consider the problem of signal correlation in the control loop, and can monitor all parts of the control loop at the same time, not just a single device in the control loop. Compared with multivariate statistical monitoring and fault diagnosis method based on analytic model parameter estimation, this method has the advantages of fast operation speed and strong self-learning ability. The simulation results show that the method can detect the fault accurately and locate the fault accurately.

At present, the research on fuzzy ADRC control loop fault diagnosis method is still preliminary, and many related problems need to be further studied. Fuzzy ADRC has strict restrictions on the historical data needed for monitoring the transition process when the set point is changed, but it is difficult to obtain a large number of historical data with the same set point change method in the actual continuous process. At present, the method only involves a single loop control system, and there is no research on more complicated control loops such as cascade control system and proportional control system.

References

1. Zhang, D., Wang, J., Qian, L., et al.: Stepper motor open-loop control system modeling and control strategy optimization. Arch. Electr. Eng. **68**(1), 63–75 (2019)
2. Chen, T., Pan, Y., Xiong, Z.: Fault diagnosis scheme for single and simultaneous open-circuit faults of voltage-source inverters on the basis of fault online simulation. J. Power Electron. **21**(2), 384–395 (2021)
3. Khil, S.E., Jlassi, I., Cardoso, A.M., et al.: Diagnosis of open-switch and current sensor faults in PMSM drives through stator current analysis. IEEE Trans. Ind. Appl. **55**(6), 5925–5937 (2019)
4. Li, X., Xu, J., Chen, Z., et al.: Real-time fault diagnosis of pulse rectifier in traction system based on structural model. IEEE Trans. Intell. Transp. Syst. **PP**(99), 1–14 (2020)
5. Zhao, S., Wang, E.: Fault diagnosis of circuit breaker energy storage mechanism based on current-vibration entropy weight characteristic and grey wolf optimization-support vector machine. IEEE Access **7**(99), 86798–86809 (2019)
6. Wassinger, N., Penovi, E., Retegui, R.G., et al.: Open-circuit fault identification method for interleaved converters based on time-domain analysis of the state observer residual. IEEE Trans. Power Electron. **34**(4), 3740–3749 (2019)
7. Song, G., Wang, T., Hussain, K.S.T.: DC line fault identification based on pulse injection from hybrid HVDC breaker. IEEE Trans. Power Deliv. **34**(1), 271–280 (2019)
8. Liu, S., Liu, X., Wang, S., Muhammad, K.: Fuzzy-aided solution for out-of-view challenge in visual tracking under IoT assisted complex environment. Neural Comput. Appl. **33**(4), 1055–1065 (2021)
9. Palmer, K,A., Bollas, G.M.: Analysis of transient data in test designs for active fault detection and identification. Comput. Chem. Eng. **122**, 93–104 (2019)
10. Liu, S., Liu, D., Srivastava, G., et al.: Overview and methods of correlation filter algorithms in object tracking. Complex Intell. Syst. **2020**(3), 1-23 (2020)

11. Chen, T., Zeng, C., Wang, C.: Fault identification for a class of nonlinear systems of canonical form via deterministic learning. IEEE Trans. Cybern. **PP**(99), :1–12 (2021)
12. Shen, Q., Yue, C., Yu, X., et al.: Fault modeling, estimation, and fault-tolerant steering logic design for single-gimbal control moment gyro. IEEE Trans. Control Syst. Technol. **PP**(99), 1–8 (2020)
13. Wu, F., Sun, J., Zhou, D., et al.: Simplified fourier series based transistor open-circuit fault location method in voltage-source inverter fed induction motor. IEEE Access **8**, 83953–83964 (2020)
14. Huang, W., Du, J., Hua, W., et al.: Current-based open-circuit fault diagnosis for PMSM drives with model predictive control. IEEE Trans. Power Electron. **PP**(99), 1 (2021)
15. Chen, T., Pan, Y., Xiong, Z.: A hybrid system model-based open-circuit fault diagnosis method of three-phase voltage-source inverters for PMSM drive systems. Electronics **9**(8), 1251 (2020)
16. Zhang, Z.F., Wu, Y., Qi, S.Y.: Diagnosis method for open-circuit faults of six-phase permanent magnet synchronous motor drive system. IET Power Electron. **13**(15), 3305–3313 (2020)
17. Yang, H., Zhou, Y., Zhao, J.: Current covariance analysis-based open-circuit fault diagnosis for voltage-source-inverter-fed vector-controlled induction motor drives. J. Power Electron. **20**(2), 492–500 (2020)
18. Zhang, J., Zhan, W., Ehsani, M.: Fault tolerant control of PMSM with inter-turn short circuit fault. IEEE Trans. Energy Convers. **PP**(99), 1 (2019)
19. Chen, C.L., Hung, S.K.: Visual Servo Control System of a Piezoelectric2-Degree-of-Freedom Nano-Stepping Motor. Micromachines **10**(12) (2019)
20. Zhou, H., Xu, J., Chen, C., et al.: Disturbance-observer-based direct torque control of five-phase permanent magnet motor under open-circuit and short-circuit faults. IEEE Trans. Ind. Electron. **PP**(99), 1 (2020)
21. Zhao, M., Liu, G., Chen, Q., et al.: Fault-tolerant control of a triple redundant PMA-SynRM driven under single-phase open-circuit by mono-inverter. IEEE Trans. Power Electron. **PP**(99), 1 (2021)
22. Nedelec, S., Martinez-Arias, A.: In vitro models of spinal motor circuit's development in mammals: achievements and challenges. Curr. Opin. Neurobiol. **66**, 240–249 (2021)
23. Guo, M., Pan, S., Chen, L., et al.: Improvement in step resolution and response time of ultrasonic motor by using a piezoelectric resonant shunting circuit as damping control. Rev. Sci. Instrum. **91**(12), 125008 (2020)

Design of Encryption System for Urban Cultural Heritage Protection Database Based on Blockchain Technology

Cai-ting Peng, He-qing Zhang(✉), and Yi-lin Feng

Management (Tourism), School of Guangzhou University, Guangzhou 510006, China

Abstract. Aiming at the problems of long data encryption time and low encryption efficiency in the traditional urban cultural heritage protection database encryption system, an urban cultural heritage protection database encryption system based on blockchain technology is designed. In terms of hardware design, database encryption services were added, and database encryption system hardware was deployed. In terms of software design, based on hardware deployment, design the system application architecture; design system interfaces from the perspectives of encryption and decryption; generate and manage system encryption keys; use blockchain technology to reduce encryption principles. Language composition, adding symbols and terminology, data encryption structure, designing a mixed encryption and decryption flowchart, and realizing the encryption of the urban cultural heritage protection database. The experimental results show that under different tuple pairs, the average time of Setup data encryption is 38 s and 41 s less, and the average dynamic update time consumption is less than 5.77 s, 10.48 s. Under the condition of different number of keywords, Setup data encryption The average time is shorter by 34 s and 96 s, and the average dynamic update time is shorter by 45.19 s and 98.72 s. Has a high encryption database efficiency.

Keywords: Blockchain technology · Urban culture · Heritage protection · Database encryption

1 Introduction

Since the reform and opening up, China's economy has shown an unprecedented high-speed development trend, and the urban population has become larger and larger, followed by "old city reconstruction", real estate development, infrastructure reconstruction, etc., which has led to the serious destruction of many historical and cultural heritage. At the same time, with the improvement of people's material living standards, people's eager desire for cultural needs has transformed into a growing nostalgia complex. It is this complex that directly stimulates the economy that leads to the phenomenon of "antique" replacing the real ancient. City is not only the carrier of history and culture, but also the cultural landscape of social economy. To maintain the continuity of urban landscape, to protect the local characteristics of local architecture and to retain the memory of street space are the needs of the development of modern human civilization [1].

S. Liu and X. Ma (Eds.): ADHIP 2021, LNICST 416, pp. 627–644, 2022.
https://doi.org/10.1007/978-3-030-94551-0_49

Nowadays, social network technology has developed rapidly and has been widely used in all walks of life. Everyone obtains the information he needs on the Internet, and information has gradually become an important resource in the online world. Based on this, the protection method of urban cultural heritage data has also changed from traditional paper, photo and other protection methods to network technology protection, which stores all urban cultural heritage data in the heritage database [2]. However, with the continuous development of the network, various network attacks have also occurred. Therefore, the security of the urban cultural heritage protection database is highly valued by people. The core of network security is database security, so the discussion on database security is of great significance.

In the database management system, the administrator has great authority, which can not only manage what the system wants to do, but also query any information he wants to know. As long as the database administrator wants to know the user's information, he can do it. In addition to hacker attacks, part of cybercrime comes from the inside of the system. Since internal attacks on the database are simpler than hackers, in order to prevent insiders from unauthorised querying of data to view the urban cultural heritage protection data, and record them, Under this kind of situation, the information leaked out makes the protection of urban cultural heritage more difficult [3]. In order to avoid this problem, the urban cultural heritage protection data can be encrypted. After the information is encrypted, the urban cultural heritage protection data can be decrypted by the private key of the relevant personnel. In this way, even if the management personnel or hackers obtain the information, they can not decrypt the data and make the act of leaking the urban cultural heritage protection data, In this way, the confidentiality of urban cultural heritage protection data is guaranteed.

At home and abroad, database encryption technologies such as time control encryption technology, timed release proxy re-encryption scheme, specific time encryption, TSE scheme, TRE scheme, information security timed release secret sharing scheme, and access control model have been developed to solve malicious intrusion by hackers, Database security issues such as stealing and data leakage [4]. On the basis of domestic and foreign research, literature [5] chooses the way of encrypting the data in the database, so that important data information is stored in the form of ciphertext in the database. Even if the database is attacked, it can effectively ensure that the data will not be leaked, so as to ensure the security of the data in the urban cultural heritage protection database. Literature [6] uses a symmetric encryption algorithm to improve the efficiency of multi-keyword encryption, combines blockchain technology to solve the problem of dishonest search in cloud servers, and uses a linear index structure to achieve multi-keyword search while improving search efficiency. Blockchain is proposed. A searchable encryption scheme that supports multiple keywords on the Internet to ensure the security of data in the urban cultural heritage protection database.

Based on the above research, this research uses blockchain technology to encrypt the data in the urban cultural heritage protection database, through the client PC, based on the browser, remote configuration management system hardware, and from the encryption and decryption two In terms of designing the system interface, ensuring the security of the database encryption system through the system encryption key, using blockchain technology to realize the encryption of the urban cultural heritage protection database, enhancing the encryption performance of the urban cultural heritage protection database encryption system, and proposing a blockchain-based technology The design of the encryption system of the urban cultural heritage protection database.

2 Deploy the Hardware of the Database Encryption System for the Protection of Urban Cultural Heritage

In this study, considering the data of urban cultural heritage protection and the demand for database encryption system, the database encryption service is added on the basis of the hardware of database encryption system determined by predecessors. Through the PC of the client, based on the browser, the database encryption system hardware is remotely configured and managed, as shown in Fig. 1.

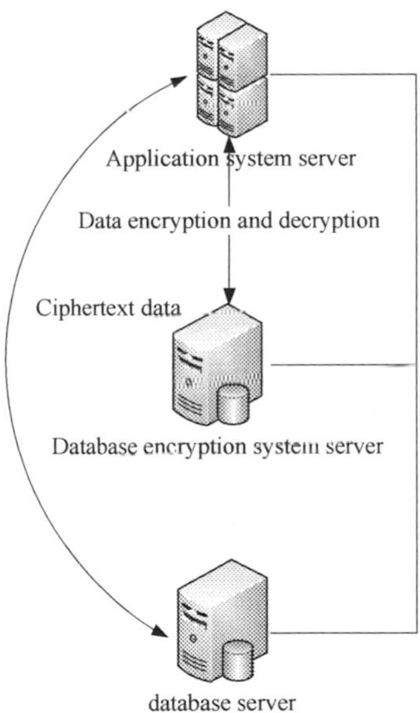

Fig. 1. Database encryption system deployment

The database encryption system deployment method shown in Fig. 1, the selected database encryption server is a high-performance dedicated all-in-one server, and the key management system of the data encryption system, the main control service system and the management control platform are deployed inside.

In Fig. 1, the database encryption system is located between the application system and the application system database. The specific implementation of data encryption and decryption is carried out between the application system server and the database encryption system server, completely outside the application system database, so it is transparent to the application system database, and will not affect the original architecture and performance of the application system.

3 Software Design of the Encryption System for Urban Cultural Heritage Protection Database Based on Blockchain Technology

3.1 Design System Architecture

Based on the deployment mode of the database encryption system shown in Fig. 1, the application architecture of the designed database encryption system is shown in Fig. 2.

As shown in Fig. 2, the system application architecture diagram includes three parts: user's database application system, database encryption server and user's application system database.

Among them, the user's application system is the same as an ordinary database application system. It can be a client in a C/S structure, a browser in a B/S structure, or an independent database application. It is just a client who deploys a database encryption server. interface. The database encryption server is the core part of the whole database encryption system. It provides security services to the user application system, receives encryption and decryption requests from the user application system, and performs encryption and decryption of database information, key management, etc. The application system database can be all kinds of relational data, such as Oracle, DB2, mysql, MS SQL server, etc. the database encryption server is transparent to the application system database.

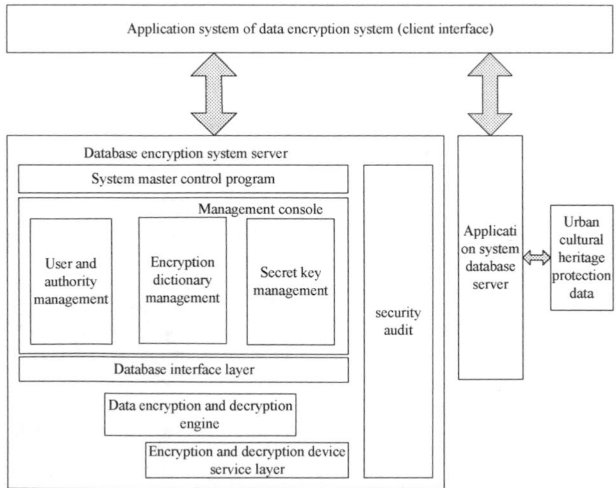

Fig. 2. System application architecture diagram

The data encryption and decryption processing is completed between the application system and the database encryption server, and the database encryption server is not directly related to the application system database. In this way, it is compatible with the application system database to the greatest extent, and does not need to change the user's logic of operating the original database. At the same time, for application developers, they can use various ways of operating the application system database (such as JDBC, ODBC, hibernate, special database operation language).

From a security perspective, the database security architecture of this study uses a four-level security layer to achieve the security of the entire system, namely the outer layer, the middle layer, the inner layer and the analysis layer. The function of the outer layer is to verify the legitimacy of the user's identity. Only having a legal login account (together with password) can access the database encryption server and use the database encryption server for encryption and decryption; The middle layer is a restrictive management security measure based on RBAC policy. Its function is to verify the user's rights and ensure that legitimate users can only encrypt and decrypt the data within their rights. The inner layer is a data encryption layer that includes a data encryption and decryption engine and key management. This security measure converts the original data into an encrypted form that is not directly readable, and the confidentiality, tamper-proof and reliability of the data are guaranteed. The analysis layer is a security analysis layer including security audit control. Since it is impossible to build data that is not attacked, the security system must provide a mechanism to detect illegal access to the database. In the four security layers, the outer layer, middle layer and inner layer provide active protection for database, while the analysis layer provides passive protection for database.

3.2 System Interface Design

When providing the interface to the application system, we can consider using static global class to save some global setting information and authentication status of users, and consider caching mechanism and timing mechanism to ensure the efficiency of authentication and verification. In addition, the connection pool can be used to improve the efficiency of the system, and the reserved interface can be designed for intelligent automatic processing of user SQL requests.

Encrypted Interface

Parameters to be provided by the application system ("database name, table name, field name", "data to be encrypted"); The interface first determines whether it has been authenticated. If the system automatically obtains the hidden parameters (user ID, password), it connects to the server for identity authentication. After the authentication is passed, the system needs to set the authentication ID [7]. If the authentication is passed, the encryption process with the server is automatically completed, and the encryption result is obtained and returned to the caller. The interface description is shown in Table 1.

Table 1. Data encryption interface description

Serial number	Encrypted representation	Paraphrase
1	Prototype	String db-Encrypt (string strKeyHash, byte[] byPlainText)
2	Description	Encrypt the data and return the encrypted data after Base64 encoding

(*continued*)

Table 1. (*continued*)

Serial number	Encrypted representation	Paraphrase
3	Parameter	StrKeyHash [in]: Data belongs to library name, table name, field name byPlainText [in]: Plaintext data (hexadecimal string)
4	Return value	base64Encrypted data after encoding: success Null: fail

As shown in the data encryption interface description in Table 1, the data encryption interface access system process is as follows: 1. Import the user ID and encrypted password, and configure the file by the client; 2. The background automatically establishes a connection with the server and completes the identity Authentication; 3. Determine whether the plaintext data is greater than the maximum number of bytes configured in the system;4. When the plaintext data is less than the maximum number of bytes configured by the system, the function is used to encrypt the data. When the plaintext data is greater than the maximum number of bytes allowed, the user is prompted to convert the packet and encrypt it in sections; 5. Automatically release the connection and clear the authentication mark after completion.

Decryption Interface

The parameters (database name, table name, field name, data to be solved) that the application program needs to provide; The interface first determines whether it has been authenticated. If not, the system automatically submits the implicit parameters (user ID, password) to the server for authentication. After the authentication is passed, the system needs to set the authentication ID [8]. If the authentication has been passed, the decryption process with the server is automatically completed, and the decryption result is obtained and returned to the caller. The interface description is shown in Table 2.

Table 2. Description of data decryption interface

Serial number	Encrypted representation	Paraphrase
1	Prototype	Byte[] db_Decrypt (string strKeyHash, string strEncryptData)
2	Description	Decrypt the data and return the decrypted data (plaintext)
3	Parameter	strTablename [in]: Data belongs to library name, table name, field name strEncryptData [in]: Ciphertext data (base64 encoding)
4	Return value	Data decryption: successful Null: fail

As shown in Table 2, the process of data decryption interface accessing the system is as follows: 1. Import the user ID and encrypted password, and configure the file by the client; 2. The background automatically establishes the connection with the server and completes the identity authentication; 3. Determine whether the encrypted data is greater than the maximum number of bytes configured by the system; 4. When the data is less than the maximum number of bytes configured by the system, use this function to decrypt the data (automatically break the sub-packages of suitable size and return to the caller automatically Group package); 5. After completion, the connection will be automatically released and the certification mark will be cleared.

3.3 Encryption Key Management

The management of data encryption keys directly affects the security of the database encryption system, so key management occupies a very important position in the database encryption system. For this reason, the encryption key management module of the system that conforms to this design is studied.

Generate Encryption Key
To generate the encryption key of the urban cultural heritage protection database encryption system this time, a user-specific data will be selected to participate in the hash operation to generate the data encryption key, so this design uses the following method to generate the data encryption key:

$$D = H(u, p, s, r) \tag{1}$$

D represents the last data encryption key; H represents hash algorithm; u is the user name of the user; p is the user's private key; s is the name of the field to be encrypted; r represents the random number generated by the system for different table structures [9].

Using this data encryption key generation method can make the same user generate different keys for the same-named fields in different table structures, which increases the difficulty of key cracking.In addition, the security of this key generation method completely depends on the security of the user's password. Even if the attacker can get the user's database password, only the user's corresponding urban cultural heritage protection data will be affected. The attacker cannot get the encryption key of other users' corresponding urban cultural heritage protection data through the attacked user's encryption key.

Design Encryption Key Management
When the database administrator creates a table, it will store the field information that needs to be encrypted in the table in kmdb, including the encryption algorithm used by the field and a random number generated when the table is generated.When the user stores data in the table, DECO's encryption judgment module will determine that this operation needs to be encrypted based on the information in the KMDB. At this time, DECO calls the key management module to generate the final data encryption key based on the user's user name, password, field name, and a random number generated by the key management module.Then, when the user reads the data in the encrypted field,

DECC obtains the corresponding encryption algorithm and random number from kmdb through the same steps, generates the data encryption key according to the user name and password, decrypts the field, and finally returns the information to the user. The process is shown in Fig. 3.

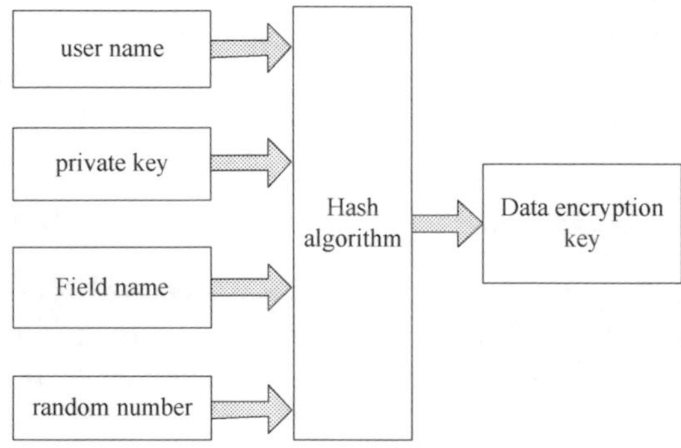

Fig. 3. Data encryption key generation process

3.4 Basic Representation of Data Encryption Based on Blockchain Technology

Determine the Basic Encryption Primitives of the System

Based on the encryption key generated by formula (1), the user's private key encryption scheme is composed of three polynomial time algorithms

$$S = (G, E, D) \tag{2}$$

(2) In the formula, S represents a time algorithm; G represents a conceptual algorithm, which requires a security parameter k and returns a key K; E also represents a probability algorithm, which requires a key K and a message m as parameters, and returns the ciphertext c; D represents a deterministic algorithm, enter the key K (if K is the key to generate c) and the cipher text c and return m [10].

Informally, if the output of the ciphertext does not reveal any part of the plaintext information to the adversary, the private key encryption scheme is secure against chosen plaintext attack. Let's suppose that a scheme outputs ciphertexts that are computationally indistinguishable, And from random to adversary, it can query Encryption o adaptively. It is said that this scheme is safe for random ciphertext under the selected plaintext attack. In addition, the pseudo-random function P is a polynomial-time computable function, and it cannot be distinguished from a random function in polynomial time with any probability.

Basic Symbols and Terms of the System

The set of all binary strings of length n is denoted as $\{0, 1\}^n$, and the set of all binary strings of finite length is denoted as $\{0, 1\}^*$. And $[n]$ represents the integer set $\{1, \cdots, n\}$, and $2^{[n]}$ represents the corresponding power set. And $x \leftarrow X$ indicates that the element x

is sampled from the distribution X, and $x \leftarrow A$ indicates that the output of the algorithm A is x. Given a sequence o of n elements, the i element is denoted as o_i or $o[i]$. If T is a set, then $\#T$ is its base. In addition, $\langle x, y \rangle$ or xy represents the concatenation of the character strings x and y.

d is the plaintext structure, while e is the ciphertext structure constructed by the client using K as the key. If the scheme is stateful, the setup program will also output state st. Then, the client sends the encrypted structure e to the untrusted server and keeps the state st and the key K private. Among them, the client can use supported query operations on e.

Data Structure

The design of the urban cultural heritage protection database encryption system is adjusted to the original data decryption structure. The blockchain will generate a block Merkle tree based on the on-chain data, and generate the block header during the consensus process, redesigned as Fig. 4 shows the block data structure.

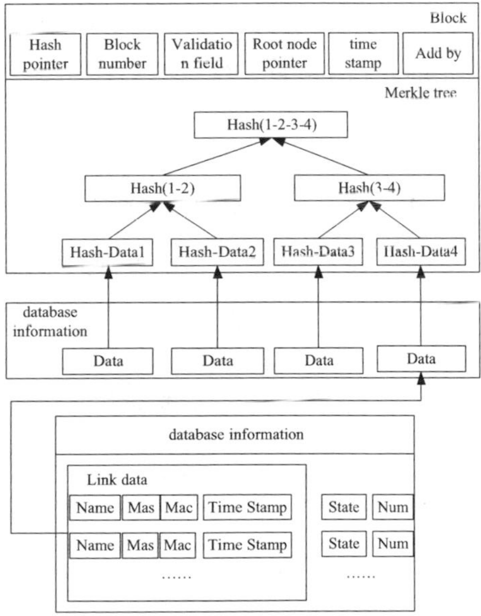

Fig. 4. Block data structure

In Fig. 4, the hash pointer is a pointer recorded with a hash value. Pointing to the predecessor block of the block, the entire block chain can be traversed through the hash pointer. The verification field of the predecessor block is stored. During the traversal process, the verification field of the block can be checked to determine whether the block is attacked.

The block number records the number of the block in the whole network.

The root node pointer is a pointer to the root node of the Merkle tree in this block, which is used to find the root node in the search process. Save the hash value of all root nodes.

The verification field is generated by hash pointer and root node pointer through specific hash function. In this way, all the blocks are concatenated with hash values from the beginning to the end.

The timestamp is used to record the adding time of the block, which is generated locally by each node, and small differences are allowed in the network. However, the timestamp of a block is not allowed to be earlier than the predecessor node.

The adder tag is used to record the adder of a block. When a node successfully applies to add a block, its public key is recorded on the block as a token.

The Merkle tree node is similar to the most basic tree structure. When the Merkle tree node points to the pointer of the child node, the pointer of the leaf node is empty. Used to store the hash value calculated by the child node or the content of the Mas field of the message, which is used to solidify the data.

3.5 System Encryption Process

The database encryption system designed this time will meet the following functions in actual use:

1. The storage period of the data in the database is relatively long. When using the encryption algorithm, the attacker should not be able to crack the algorithm in a short time and then attack the database.
2. After the data in the database is encrypted, the storage space it occupies cannot be clearly larger than that when it is not encrypted. The speed of data encryption and decryption must be fast enough to be transparent to the user, and the user should not be clearly aware of the encryption and decryption. The resulting delay operation.
3. The database encryption system has flexible authorization mechanism and reliable key management function.
4. After the database is encrypted, try not to affect the original functions, and at the same time, allow users to access the database at the level of their own authorization.

Based on this, a hybrid encryption algorithm is used to encrypt the urban cultural heritage protection database. The flow chart of hybrid encryption and decryption is shown in Fig. 5.

As shown in Fig. 5, the user encrypts the urban cultural heritage protection data to be stored with an encryption algorithm to generate the key K, and then obtains the public key K_1 generated by the database from the key server, Encrypt K with K_1 to get the encryption key K_2, and then the user will need to store the urban cultural heritage protection data, send the ciphertext and the encryption key K_2 to the database, The database uses its own private key to unlock the encryption key K_2 to get the encryption key K_1, and then uses K_1 to unlock the ciphertext of the stored urban cultural heritage protection data to get the plaintext.

4 System Test

Two groups of conventional database encryption systems were selected. By means of comparative test, the urban cultural heritage protection database was used as the experimental object of comparative test. The data was drawn into a chart by MATLAB software

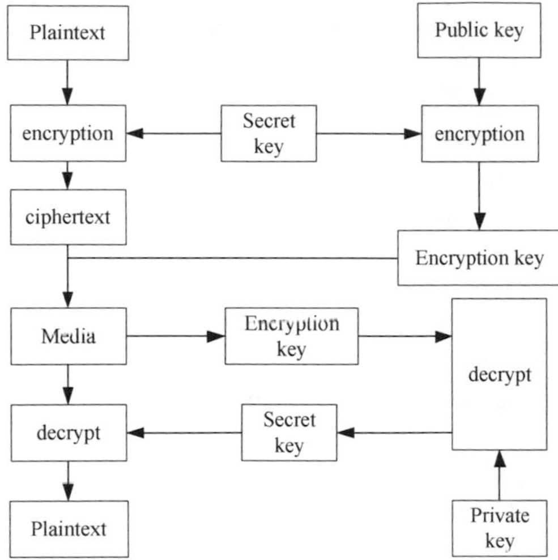

Fig. 5. Flow chart of hybrid encryption and decryption

to verify the encryption system of the research. Compare the data encryption time of three groups of database encryption systems for the same data.

4.1 Experimental Preparation

In order to avoid the failure of the three groups of system software and hardware selected in this experiment, which will affect the system test results, the various hardware circuits in the three groups of systems will be tested on the PCB board; for the system software, the build as shown in Fig. 6 System software testing environment, testing system software.

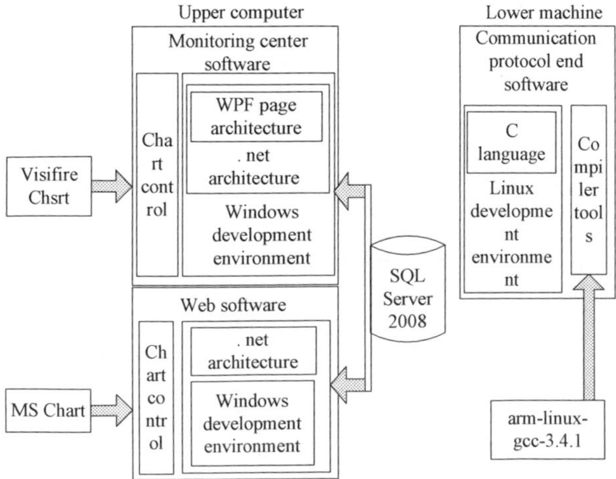

Fig. 6. System software testing environment

The system software and hardware testing process is as follows:

1. Using PCB circuit board, check whether there are circuit faults, short circuit, loose welding, open circuit, load, poor contact and other problems in the system circuit. 2. Check whether the three groups of system hardware equipment are complete and whether the detection equipment is in normal working condition.3. When there is no problem with the system hardware, the operating system and system communication protocol are embedded in the system software operating environment shown in Fig. 6. 4. Separately test all the drivers in the system, focusing on the system encryption program; 5. Use the database encryption module to control the system to encrypt the urban cultural heritage protection database, and check whether the encryption module is operating normally;6. When it is confirmed that the encryption module of the system is in normal operation, Download all the programs of the system to the corresponding chips of the system.

In this experiment, the system test hardware environment set up is shown in Fig. 7.

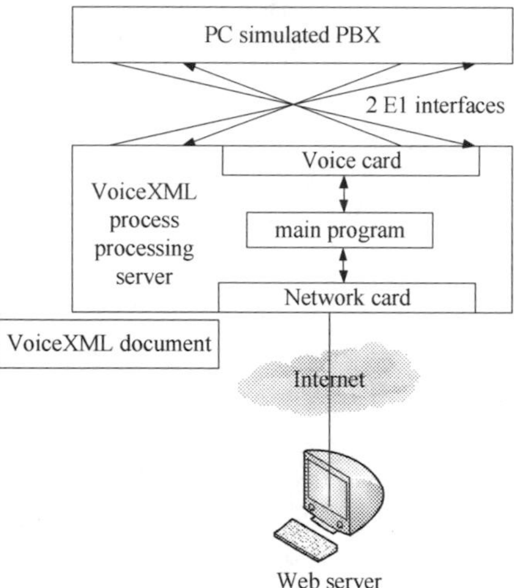

Fig. 7. System test environment

In this experiment, the selected urban cultural heritage protection database, in txt format, stores a total of about 500,000 urban cultural heritage data. Using each urban cultural heritage data as an edge, 36,692 points and 183,831 edges can be obtained. The data set size is about 1.4G.

According to the database selected for this experiment, the experimental data is divided as shown in Table 3, so that the number of (W, ID) pairs in the database is from $1 * 10^6 - 20 * 10^6$, and the setup time, search voucher generation time, query time and dynamic update time are tested in different numbers, Test each quantity 10 times, take the average value, and test a total of 10 different quantities.

Table 3. Pairs logarithm and number of keywords in the data set

Pairs (10^6)	Keyword	Pairs (10^6)	Keyword
1	1108	2	5680
3	7650	4	9320
5	18653	6	30492
7	51116	8	61255
9	72594	10	76594
11	91254	12	179830
13	210346	14	257364
15	307812	16	353155
17	395713	18	434379
19	496193	20	500000

In Table 3, pairs (106) represents tuple number pairs after database data partition; Keyword is the number of keywords. It is divided into 1–20 blocks according to the key value logarithm, and the key value logarithm of each block is $i * 10^6$, where i represents the block.This paper uses this data set as the algorithm verification data set, and has carried out the setup data encryption time, token formation time, query time, update time and other perspectives to verify the efficiency of the system encryption database.

4.2 Experimental Results

The First Set of Experimental Results

The setup data encryption time includes the structure of security index and data encryption operation. Experiments are carried out under the pairs logarithm and the number of keywords of the dataset shown in Table 3, in which the number of different inverted index key value pairs and the number of unique keywords are taken as variables respectively, Comparing the Setup data encryption time of the three schemes, the experimental results are shown in Table 3 and Table 4.

Table 4. Comparison of setup data encryption time under different tuple numbers (s)

Pairs (10^6)	Design system	Conventional system 1	Conventional system 2
1	5	43	46
2	10	48	51
3	13	51	54

(continued)

Table 4. (*continued*)

Pairs (10^6)	Design system	Conventional system 1	Conventional system 2
4	20	58	61
5	25	63	66
6	30	68	71
7	35	73	76
8	40	78	81
9	45	83	86
10	50	88	91
11	55	93	96
12	60	98	101
13	65	103	106
14	70	108	111
15	75	113	116
16	80	118	121
17	85	123	126
18	90	128	131
19	95	133	136
20	100	138	141

It can be seen from Table 4 that in different tuple number pairs of three groups of database encryption systems, the encryption time of setup data increases linearly with the increase of data volume, and basically maintains a positive relationship with non duplicate tuple data pairs. Among them, the conventional system 2 encrypts the tuple pair data in the urban cultural heritage protection database, and the average time of Setup data encryption is 93.4 s, which takes the longest time; the conventional system 1 encrypts the tuple pair data in the urban cultural heritage protection database, The average time of Setup data encryption is 90.4 s; The design system encrypts the tuple number pair data in the urban cultural heritage protection database. The average time of setup data encryption is 52.4 s, which is 38 s and 41 s less than that of the two conventional systems.

As can be seen from Table 5, the experimental comparison under different number of keywords shows that the data encryption time of setup also shows a linear growth with the increase of the amount of data, and basically maintains a positive relationship with the number of non duplicate keywords.Among them, the conventional system 2 encrypts the tuple pair data in the urban cultural heritage protection database, and the average time of Setup data encryption is 117 s, which takes the longest time; the conventional system 1 encrypts the tuple pair data in the urban cultural heritage protection database, Setup data encryption average time is 55 s; The design system encrypts the tuple number

Table 5. Comparison table of Setup data encryption time under different number of keywords (s)

Keyword	Design system	Conventional system 1	Conventional system 2
1108	2	36	98
5680	4	38	100
7650	6	40	102
9320	8	42	104
18653	10	44	106
30492	12	46	108
51116	14	48	110
61255	16	50	112
72594	18	52	114
76594	20	54	116
91254	22	56	118
179830	24	58	120
210346	26	60	122
257364	28	62	124
307812	30	64	126
353155	32	66	128
395713	34	68	130
434379	36	70	132
496193	38	72	134
500000	40	74	136

of data in the urban cultural heritage protection database. The average time of the setup data encryption is 21 s, which is 34 S and 96 s less than the average time required by the two groups of conventional systems.

It can be seen that in this design system, the time efficiency of index construction and data encryption module has been significantly improved compared with other solutions. The system designed this time has a high encryption database efficiency.

Results of the Second Group

The urban cultural heritage protection database encryption system, the stored data will also be adjusted with the discovery of cultural heritage, and there will be deletion changes, which involve the dynamic update operation of database encryption, so the elements shown in Table 3 Group data pairs and keywords, as dynamic update time verification data, Verify three groups of systems, and dynamically update the time consumption when encrypting the urban cultural heritage protection database. The experimental results are shown in Table 6 and Table 7.

Table 6. Comparison table of dynamic update time consumption under different tuple number pairs (s)

Pairs (10^6)	Design system	Conventional system 1	Conventional system 2
10	0.05	5.82	10.53
20	0.10	5.87	10.58
30	0.03	5.8	10.51
40	0.02	5.79	10.5
50	0.25	6.02	10.73
60	0.30	6.07	10.78
70	0.35	6.12	10.83
80	0.04	5.81	10.52
90	0.45	6.22	10.93
100	0.05	5.82	10.53
110	0.55	6.32	11.03
120	0.06	5.83	10.54
130	0.65	6.42	11.13
140	0.07	5.84	10.55
150	0.75	6.52	11.23
160	0.08	5.85	10.56
170	0.85	6.62	11.33
180	0.09	5.86	10.57
190	0.95	6.72	11.43
200	0.97	6.74	11.45

It can be seen from Table 6 that in different pairs of tuples, the dynamic update time of conventional system 2 fluctuates between 10–11 s, and its average dynamic update time consumption is 10.813 s, and the resulting dynamic update time consumption is relatively large.; Conventional system 1 dynamic update time consumption fluctuates between 4–7 s, its average dynamic update time consumption is 6.103 s, the dynamic update time consumption generated is too large; The dynamic update time consumption of the design system fluctuates between 0 and 1 s, and the average dynamic update time consumption is 0.333, which is 5.77 s and 10.48 s less than the two conventional systems respectively.

Table 7. Comparison of dynamic update time consumption under different number of keywords (s)

Keyword	Design system	Conventional system 1	Conventional system 2
1108	80.14	125.33	178.86
5680	98.79	143.98	197.51
7650	98.96	144.15	197.68
9320	99.99	145.18	198.71
18653	97.18	142.37	195.90
30492	96.47	141.66	195.19
51116	95.31	140.50	194.03
61255	97.81	143.00	196.53
72594	90.16	135.35	188.88
76594	95.41	140.60	194.13
91254	97.16	142.35	195.88
179830	85.41	130.60	184.13
210346	83.14	128.33	181.86
257364	89.16	134.35	187.88
307812	94.31	139.50	193.03
353155	96.71	141.90	195.43
395713	87.46	132.65	186.18
434379	85.31	130.50	184.03
496193	89.77	134.96	188.49
500000	94.56	139.75	193.28

It can be seen from Table 7 that the dynamic update time consumption of conventional system 2 fluctuates between 170 and 200 s, with an average dynamic update time consumption of 191.38 s, resulting in a large dynamic update time consumption; The dynamic update time consumption of conventional system 1 fluctuates between 120–150 s, and its average dynamic update time consumption is 137.85 s, and the generated dynamic update time consumption is too large;The time consumption of dynamic update of the design system is between 80 and 100 s, and the average dynamic update time consumption is 92.66, which is 45.19 s and 98.72 s less than the two groups of conventional systems.

It can be seen that the system designed this time encrypts the urban cultural heritage protection database when updating data, and still has a high encryption database efficiency.

5 Conclusion

In this research, an encryption system is designed on the server side, which is the periphery of the database, to ensure the security of the database and encrypt the important information in the database. Therefore, this design makes full use of blockchain technology to design a data encryption structure, and realizes the design of various functional modules of the encryption system through both system hardware and software, and completes the design of an urban cultural heritage protection database encryption system based on blockchain technology. Through experiments, the encryption efficiency of the system designed this time was verified, and the problems existing in the traditional system were solved.

Fund Projects. National Planning Office of Philosophy and Social Science Foundation of China: Research on Cultural Heritage Conservation and the Activation of South China Historical Trail. (NO:19FSHB007).

References

1. Shufen, N., Yaya, X., Pingping, Y., et al.: Cloud-assisted attribute-based searchable encryption scheme on blockchain. J. Comput. Res. Dev. **58**(4), 811–821 (2021)
2. Bingsheng, D.I.N.G.: Research on database encryption technology. J. BeiBu Gulf Univ. **35**(2), 46–51 (2020)
3. Liu, S., Liu, X., Wang, S., Muhammad, K.: Fuzzy-aided solution for out-of-view challenge in visual tracking under IoT assisted complex environment. Neural Comput. Appl. **33**(4), 1055–1065 (2021)
4. Liu, S., Li, Z., Zhang, Y., Cheng, X.: Introduction of key problems in long-distance learning and training. Mob. Networks Appl. **24**(1), 1–4 (2018). https://doi.org/10.1007/s11036-018-1136-6
5. Shuai Liu, Guanglu Sun, Weina Fu. e-Learning, e-Education, and Online Training, pp. 1–386. Springer, Cham. Doi:https://doi.org/10.1007/978-3-030-63952-5
6. Ma, Y.: Design of inter satellite communication network security encryption control system based on blockchain. Comput. Measur. Control **29**(3), 171–175 (2021)
7. Nie, M., Pang, X., Chen, W., et al.: Fair searchable encryption scheme based on ethereum blockchain. Comput. Eng. Appl. **56**(4) (2020)
8. Wei, Y.: Research on encryption optimization of data base access information transmission. Modern Ind. Econ. Informationization **9**(4), 65–66 (2019)
9. Liu, T., Chen, Q., Zhang, Y.: Research on data encryption technology for computer network security protection. Modern Inf. Technol. **3**(15), 153–154, 157 (2019)
10. Xianlong, T.: Design and implementation of encryption module for geographic information field acquisition system. Land Resour. Herald **16**(3), 31–33 (2019)

Network Resource Personalized Recommendation System Based on Collaborative Filtering Algorithm

Gang Qiu[1,2](✉) and Jie Cheng[2]

[1] School of Software, Shandong University, Jinan 250101, China
[2] Changji University, Changji 831100, China

Abstract. Several existing personalized recommendation systems for network resources do not specifically classify user history information personalized classification criteria, resulting in a low degree of matching between system personalized recommendations and user interests. In order to improve the accuracy of personalized recommendations for network resources, we designed a collaborative filtering based Algorithm-based personalized recommendation system for network resources. In the hardware design, design the overall circuit module, configure the bus timing, and improve the operating speed of the system hardware. In software design, calculate the time weight function to construct the user's implicit scoring matrix; calculate the similarity between network resources and user browsing history items, design personalized classification criteria for user history information based on collaborative filtering algorithms; calculate predicted item scores and actual scores Establish a personalized recommendation model for network resources.In the experiment, the system is compared with several existing systems, and the average absolute error in different adjacent sets is tested. According to the data results, in the five data test sets, the average absolute error of the system is less than that of other systems, so the personalized recommendation system based on collaborative filtering algorithm has better recommendation accuracy.

Keywords: Collaborative filtering algorithm · Overall circuit module · Bus timing · Time weight function · Recommendation model

1 Introduction

Modern e-commerce has the main characteristics of personalization, which is not possessed by the traditional business model. E-commerce has an urgent demand in the dynamic provision of precise marketing "comparative shopping" services and real-time provision of optimized distribution services. With the emergence of e-commerce, it provides an unprecedented opportunity to achieve this demand, making it a new direction of e-commerce development, "personalized" business model came into being [1].

Literature [2] analyzes the user's historical behavior data based on the system dynamics algorithm, mines the user's interest preferences, classifies users according to different

S. Liu and X. Ma (Eds.): ADHIP 2021, LNICST 416, pp. 645–655, 2022.
https://doi.org/10.1007/978-3-030-94551-0_50

interest preferences, and recommends items with similar preferences. This kind of system requires a combination of qualitative and quantitative methods, and the calculation process is more complicated and the calculation speed is slow. Literature [3] designed a personalized recommendation system for college books and bibliographies in view of the poor effect and poor accuracy of the current college bibliographic personalized recommendation system combined with association rules. However, the hardware of this system is not up to the standard and cannot realize fast calculation and processing.In reference [4], a personalized recommendation system is designed based on tensorflow. The recurrent neural network module of the system can build a model for time series, fully tap the changing interests of users, and save training time to a great extent through multi super parameter adjustment, and can significantly reduce the error rate. But this kind of system needs a lot of parameters to build the model, otherwise the accuracy is difficult to reach the standard. In this paper, we design a new personalized recommendation system based on collaborative filtering algorithm.

2 Hardware Design of Network Resource Personalized Recommendation System

2.1 Circuit Module Design

The structure before the expansion of the network port of the ultrasonic flowmeter only includes three parts: TMS320F28335 as the control core; an analog front end composed of a transmitter circuit that transmits ultrasonic probe drive signals, a receiver circuit that receives ultrasonic probe signals, and a switch circuit that switches the transmit and receive states [5]; The ADC sampling module that samples the received signal.

DM9000 Ethernet control chip provides three kinds of bus connection modes: 8-bit, 16-bit and 32-bit. The 16-bit bus connection mode is chosen in this article. In the actual connection process, what needs to be connected includes 16-bit bus SD0~SD15, chip select signal *AEN*, processor read command signal *IOR* and processor write command signal *IOW*. It should be noted that the data bus and address bus are multiplexed in DM9000, and the actual access type of the bus is determined by the input signal of the *CMD* pin. When the input signal of *CMD* pin is a high level signal, the data bus is accessed; When the input signal of *CMD* pin is low level, the address bus is accessed. Therefore, in TMS320F28335, a signal is needed to control the *CMD* pin. The *CMD* pin in this paper is controlled by the SA2 pin in the address bus of TMS320F28335. In this way, TMS320F28335 can access dm9000 data bus and address bus by accessing different addresses.

2.2 Bus Timing Configuration

In the process of using xinf bus of TMS320F28335, the key point is to configure the timing of xinf [6]. In the internal clock signal of TMS320F28335, xtimclk and xclkout are used by xinf. The relationship between these two clock signals and CPU clock sysclkout is shown in Fig. 1.

Access to XINTF is based on XTIMCLK. XTIMCLK can be configured as SYSCLKOUT or SYSCLKOUT/2 through XINTFCNF2 register, and its default value

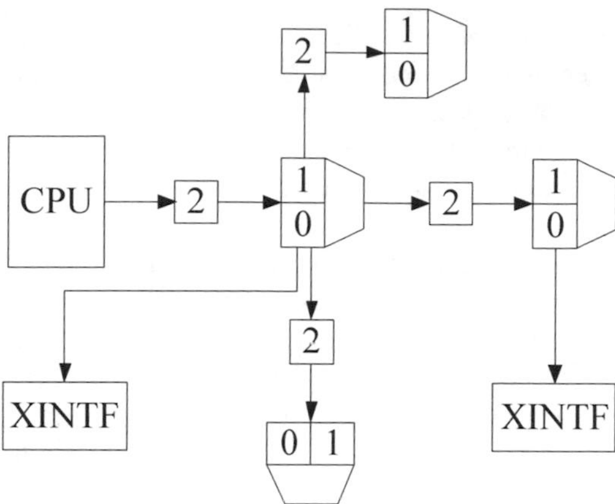

Fig. 1. Bus configuration relationship

is SYSCLKOUT/2. XCLKOUT is the output clock provided by XINTF module. In the process of DM9000 accessing Zone0 memory area of TMS320F28335 through XINTF bus, the process of accessing Zone0 includes three stages: establishment, activation and tracking.The timing of the three stages can be configured in the register XTIMING. In the setup phase, XZCSO is set to low level, and then the address is placed on the address bus. In the activation phase, if data is read from DM9000, the read signal is set to low level again; if data is written to DM9000, the write signal XWE is set to low level. In the tracking period, when the read and write signal becomes high, it will continue to be maintained for a period of time to ensure that the read and write process can be completed.The establishment, activation and tracking cycle of each XINTF can be configured through the XTIMING register. Through the design of the circuit module and the configuration of the bus timing, the operating speed of the network resource personalized recommendation system can be improved, so that it can search for network resources and sort them within the allowable range of the software [7].

3 Software Design of Network Resource Personalized Recommendation System Based on Collaborative Filtering Algorithm

3.1 Constructing the User Implicit Rating Matrix

The system's mining of user's implicit information is mainly through the log files of the website, where the main analysis and use of user behavior records, which mainly include user name, target IP, number of visits, visit duration, application type, etc., can be used for user interests [8]. It is expressed by the type of item visited, that is, the type of website visited. The interests of users are reflected by the websites they browse, and are classified according to the application types of the websites, and each application type includes at least one item. Since the log data has been classified according to application types, the realization of user interest classification is relatively simple. According to the record information of users visiting learning website, the user interest is mapped, and the user interest degree is transformed into implicit scoring of project resources by users, and the implicit scoring matrix of users is established. Considering the timeliness of the system recommendation, the user's latest browsing record is the most favorable for the recommendation effect, which makes the recommendation result very accurate, so the system selects the magazine file data of the last month as the analysis object [9]. Because this article analyzes and mines the user's implicit rating information, the rating information is indirectly reflected by the user's behavior when browsing the page. In order to intuitively describe the user's interest in the project, the interest is converted into an interest value, the data weight based on the user's visit time is introduced, and the user's interest value is generated in combination with the number of visits. Based on the time weight function, the calculation formula of $W_i(m, n)$ can be obtained as:

$$W_i(m, n) = (1 - \beta_x) + \beta \times \frac{D_{mn}}{T_m} \tag{1}$$

Where, $W_i(m, n)$ is the time weight function of a user accessing a network resource; β_x is the index to adjust the weight growth, that is, to change the speed of weight change; D_{mn} represents the access time of user m to resource n; T_m is the time interval for user m to access such resources. Considering that in a period of time, some users' interests may be more scattered or concentrated, which will have a certain impact on the recommendation effect to a certain extent, the appropriate correction should be made according to the number of types of users' interests in the quantitative scoring. The user scoring value formula is as follows:

$$H_{mn} = \frac{\beta_m}{C_i} \times K_{mn} \tag{2}$$

In the formula, H_{mn} represents the value of user m's interest in network resource n; β_m represents the adjustment parameter of network personalized information classification, generally taking a value of 1; C_i represents the standard threshold of user i for the item; K_{mn} represents the user's interest in a certain network resource The value of the score.The behavior records of a user in the network are obtained from the log file, and then these behaviors are classified as $C_i = \{C_1, C_2, \cdots, C_n\}$ according to the user's interest, and then the implicit scoring matrix of the user is output. The length and times of user I's access to item J are extracted from the log file and recorded as D_{mn} and T_m respectively. The time weight $W_i(m, n)$ is obtained by formula (1), and then multiplied by the number of visits to obtain the user interest value. According to formula (2), the user's rating value is obtained, denoted as H_{mn}, and then the user's rating value of the item is divided into 6 standards from 0 to 5, and a threshold value is set for each standard, denoted as $CL_i = \{CL_1, CL_2, \cdots, CL_n\}$, and then the relevant threshold value is passed Evaluate the corresponding interest value, repeat the above steps, calculate the user's score for all access items and the score for all application types, and finally form the user's implicit score matrix.

3.2 Design User History Information Personalized Classification Standard Based on Collaborative Filtering Algorithm

Firstly, preprocess the historical query information on the web, and use a vector composed of multiple search keywords to represent the vector model of the similarity of historical query information on the web to improve the accuracy of retrieval. On this basis, the temporal semantic classification of the information in the network historical query information database is carried out, and the temporal semantic feature vector is extracted. At this point, a vector model should be established first, so that the vector can represent network history query information in this feature space, and all the temporal semantic sentences related to the network history query information in the database are extracted to establish this vector model. Every feature in the vector model has its eigenvalue, that is the weight of the feature. Each temporal semantic keyword is one of the feature dimensions. Assuming that there are I temporal semantic keywords in the database, the feature dimension in the database is also I. For a web history query information H_i, we can use the frequency of temporal semantic keyword H_j in the database to calculate the weight of temporal semantic keywords, and take the similarity as its continuous measurement index. The formula for calculating the frequency of temporal semantic keywords is as follows:

$$H_{ij} = \frac{\mu_{ij}}{\gamma_j} \tag{3}$$

Among them, H_{ij} represents the frequency of temporal semantic keywords, μ_{ij} represents the frequency of temporal semantic keywords appearing in historical online query information, and γ_j is the most frequent occurrence of temporal semantic keywords in historical online query information. The number of times. This formula is mainly used to calculate the relative frequency of a temporal semantic keyword in the online historical query information. The temporal semantic keywords appearing in the network history query information are transformed into a vector according to the frequency order. The content of the network history query information is extracted, the attributes and attribute values are extracted, and the temporal semantic vector describing the content of the network history query information is characterized. Each web history query information has different temporal semantic features, so when calculating the similarity of web history query information, we need to first consider the concept attribute.Assuming that the concept G has an instance g, the instance g can be represented as $g_y = W_\eta[C_L]$, where $D = (D_1, D_2, \cdots, D_m)$. At this point, the similarity of instance G and Q can be calculated. First, the attribute vector of instance G and Q is a common attribute vector through the above method, and then the similarity of the historical query information of the network is calculated according to the attribute value, and the comparison of the two instances [10]. The formula for attribute value and similarity is as follows.

$$\alpha_G(m, n) = \sum_{t=1}^{n} \frac{\mu_{ij} + \gamma_j}{2} \cdot \alpha_G(m_1, n_1) \tag{4}$$

Among them, μ_{ij} and γ_j are the weight coefficients of attributes m_1 and n_1 in each vector, which are preset parameters. They are usually the statistical values obtained after preprocessing the network history query information. The final weight value of the network history query information is determined by this statistical value, and its value range is [0.1]. By substituting the above similarity into the semantics of the network history query information, the similarity between the network history query information and the standard semantics can be calculated, and then the personalized classification standard of the user's network history query information can be obtained [11].

3.3 Establish a Personalized Recommendation Model for Network Resources

When building the recommendation model, the set of neighboring users of the target item is obtained through the above formula, and then the target user's score of unknown items is predicted according to the score of neighboring users, and the items with higher prediction score are recommended to users [12]. The algorithm for predicting, scoring and recommending unknown items is shown in Fig. 2:

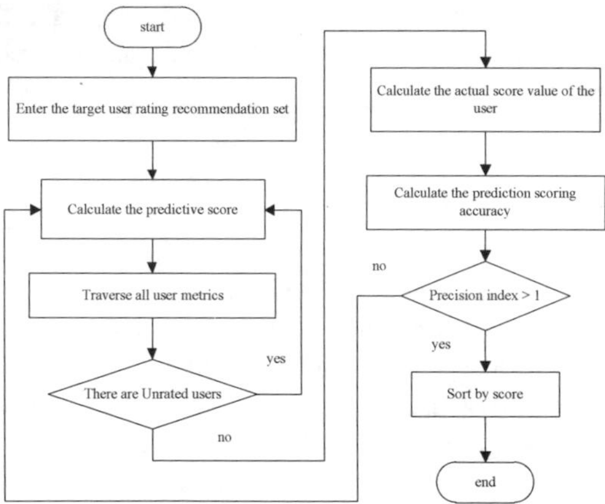

Fig. 2. Algorithm flow

As shown in Fig. 2, it is first necessary to input the nearest neighbor sets of various types of target users in the algorithm, such as $C_i = \{C_1, C_2, \cdots, C_n\}$ above, and then use formulas to calculate the target user's predicted scoring value for unknown items and the ranking of recommended items. Assuming that the project user is C_i, the unrated projects of the network resource X_n can be roughly summarized in a set of projects. Then use the evaluation formula to calculate the predicted accuracy value:

$$G_{m,n} = \frac{\sum_{i=1}^{m} sim(m, G_i) \times H_M}{\sum_{j=1}^{n} \left| sim(n, G_j) \right|} \tag{5}$$

Where, $G_{m,n}$ is the precision index between the predicted value and the actual value of user m in network resource item n; $sim(m, G_i)$ represents the predicted value of a user relative to a network resource; $sim(n, G_j)$ the actual Recommendation Index of a user relative to a network resource; H_M represents the user's own rating of the item.If the accuracy does not meet the standard, repeat the calculation steps of the user's recommended predicted value for the network resource until the accuracy of the scoring items of all users reaches the standard value. Finally, the score values of the unknown items in all types of the target user are calculated, and the predicted score values are sorted from high to low, and the top X items are selected as recommended items.The algorithm can recommend the users who first visit the network resources, classify the interests according to the types of network resources accessed by users, mine the implicit information of users in the system described in this paper, and calculate the similarity to produce the nearest neighbor, and finally recommend the similar items. With the increase of access to network resources and the extension of time, the number of recommended projects will be more and more, and the accuracy is also higher and higher.

4 Experimental Research

4.1 Experiment Preparation

The evaluation of personalized recommendation system is also an important content. If the results of recommendation can not meet the personalized needs of users, it will reduce the user's trust in personalized recommendation service, and recommendation is meaningless. On the contrary, if the personalized recommendation results are consistent with the user's personalized needs, it will enhance the user's sense of trust in the system. More importantly, it can help users solve the problem of "information overload" and lack of personalized services, so that users can reduce the time cost of finding the resources they need.Therefore, it is very necessary to evaluate the recommendation effect of the algorithm. Use the SPM method to calculate the similarity between the predicted score and the user's real score in the recommendation results of the personalized recommendation system to measure the recommendation accuracy of the recommendation system, calculate the average absolute error (MAE), and detect the items recommended by the personalized recommendation system Whether it is appropriate, statistics recommend the frequency of correct or incorrect. The MAE value is used to calculate the error value between the predicted score of the recommendation algorithm and the actual score, so MAE only calculates the resource items that the user has scored in the test data set.The smaller the MAE value, the higher the recommended accuracy of the proposed algorithm. If the score set of I users in n network resources can be expressed as n_i, the user score predicted by the system can be expressed as $ky_{i,j}$ and the real score of the user can be expressed as $kz_{i,j}$, then the calculation of the average absolute error can be expressed as:

$$\eta_i = \frac{\sum_{j-1}^{n_i} \left| ky_{i,j} - kz_{i,j} \right|}{n_i} \tag{6}$$

In the formula, η_i represents the absolute error of the resource recommendation accuracy of the network resource personalization system. Take the average of the MAE values of the selected i users to get the overall average absolute deviation of the algorithm:

$$\overline{\eta}_i = \frac{\sum_{i=1}^{n} \eta_i}{i} \tag{7}$$

Where $\overline{\eta}_i$ is the average absolute error of the recommendation accuracy of the network resource personalization system; i is the number of users who scored. Through the previous article, we know that the selection of similar user groups has a great impact on the final recommendation results. In the case of selecting different neighbor sets, the stability of recommendation accuracy is a big index to measure the efficiency of collaborative filtering recommendation algorithm. Therefore, this experiment will design experiments to test the algorithm.Taking movie scoring as an example, a data set on network resources is established, and the Movie Lens data set provided by the Group Lens project research team of the University of Minnesota is used as the data set of this experiment. This includes nearly 200,000 rating data on 2,249 movies from 10,484 users, and each Web user has rated at least 19 of them. In the experiment of this article, the training set and the test set are divided into 5 cross-validation, and the data set is randomly divided into 5 groups. Each test set occupies about 20% of the total data. Therefore, it can be known that the total data of the data set The amount is about 80% to form 5 sets of test sets and training sets.

4.2 System Personalized Recommendation Test

Under the same experimental conditions and experimental environment, five rounds of cyclic experiments were carried out on the experimental group and the control group with different training sets and test sets, and their recommended results were compared with the test set data, and five sets of experiments were calculated. In, set up the experimental group and the control group respectively. The experimental group is the recommendation system based on the collaborative filtering algorithm designed in the article, and the control group is the personalized recommendation system based on system dynamics proposed in the literature [2], literature [3] The personalized recommendation system based on association rules mentioned in [4] and the personalized recommendation system based on TensorFlow proposed in [4]. In the same test set, different sizes of neighbor sets are selected to test and compare the traditional collaborative filtering recommendation algorithm with the optimized recommendation algorithm. In this experiment, we increase the number of neighbors from 1 to 200, and the test results are shown in Fig. 3.

According to the experimental data of the five test sets in Fig. 3, the size of the neighbor set has a significant impact on the average relative error value, that is, the accuracy of the system recommendation. With the increase of the neighbor set, the average relative error value will gradually decrease, and the accuracy of the recommendation will be improved. In the comparative test of the four systems, the network resource personalized recommendation system based on collaborative filtering algorithm is smaller than other systems, and has higher recommendation accuracy in the five test sets. From the experimental data in Fig. 4, we can know that the personalized recommendation system based on collaborative filtering algorithm designed in this paper has better recommendation accuracy than several existing systems.

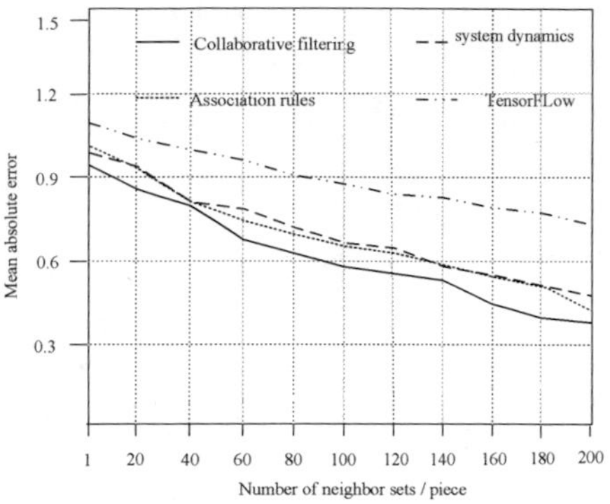

(A) Test set 1

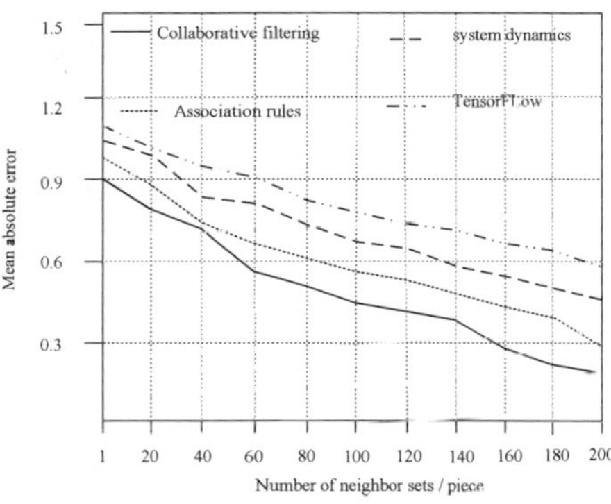

(B) Test set 2

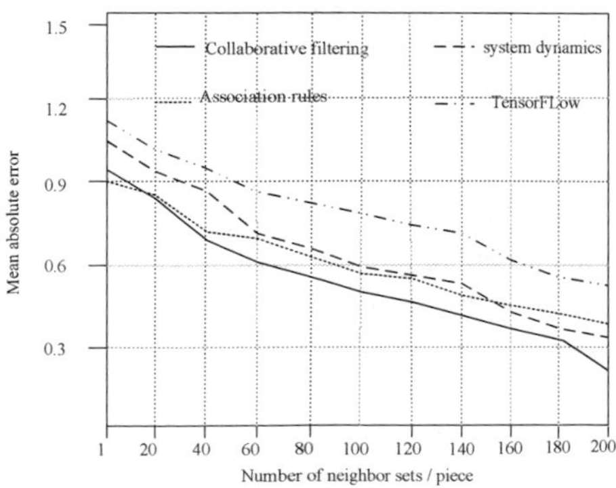

(C) Test set 3

Fig. 3. System performance comparison test

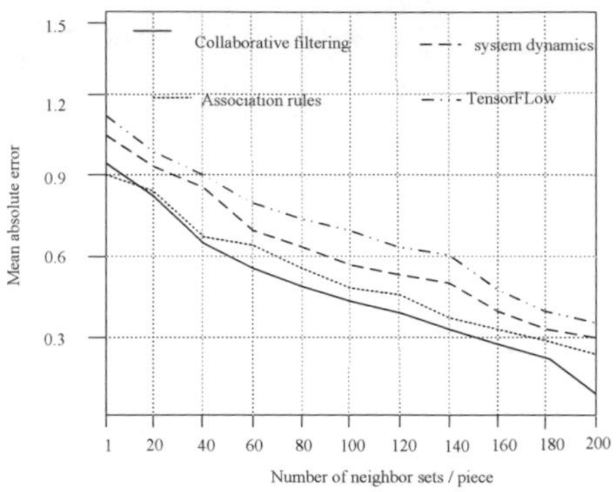

(D) Test set 4

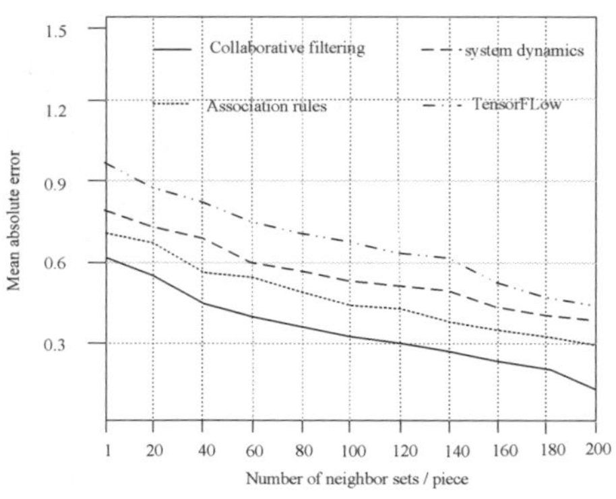

(E) Test set 5

Fig. 3. continued

5 Conclusion and Prospect

The collaborative filtering recommendation system was born in the field of e-commerce and has been widely and successfully applied. Traditional websites can only recommend unified network resources to customers, but are unable to meet the individual needs of customers. In order to provide them with personalized network resource recommendation, we learn from the successful application of collaborative filtering recommendation technology in the field of e-commerce, and try to introduce this technology into the network environment design in this research.Based on the three systems of literature [2], literature [3] and literature [4], this paper optimizes them. Then, based on the optimized collaborative filtering recommendation algorithm, a personalized network resource recommendation model is proposed. Finally, a personalized network resource recommendation system is designed and implemented to meet the personalized needs of customers and improve the efficiency of the network.

Although this method seems to have achieved some results, there are still many deficiencies in the research. In the future, research will be conducted on the category and amount of network data to improve the effectiveness of recommended resources.

Fund Projects. The work was supported by the 2021 Autonomous Region Innovation Environment (Talents, Bases) Construction Special-Natural Science Program (Natural Science Foundation) Joint Fund Project (2021D01C004) and the 2019 Xinjiang Uygur Autonomous Region Higher Education Scientific Research Project (XJEDU2019Y057, XJEDU2019Y049).

References

1. Zhang, Y-J., Dong, Z., Meng, X,-W.: Pesearch on personalized advertising recommendation systems and their applications. Chin. J. Comput. **44**(3), 531–563 (2021)
2. Wang, Z., Si, L., Liu, B., et al.: Study on information personalized recommendation based on system dynamics. J. Hebei Univ. Sci. Technol. **42**(2), 170–179 (2021)
3. Xu, B.: Design of university library bibliography personalization recommendation system based on association rules. Modern Electron. Tech. **44**(8), 68–72 (2021)
4. Yang, H.: Design of personalized recommendation system based on TensorFLow. Adhesion **41**(2), 166–169 (2020)
5. Liu, S., Liu, D., Muhammad, K., Ding, W.: Effective template update mechanism in visual tracking with background clutter. Neurocomputing (2020). https://doi.org/10.1016/j.neucom.2019.12.143
6. Li, X., Liang, H., Feng, J., et al.: Design of personalized learning resource recommendation system for online education platform. Comput. Technol. Dev. **31**(2), 143–149 (2021)
7. Liu, S., Li, Z., Zhang, Y., Cheng, X.: Introduction of key problems in long-distance learning and training. Mob. Networks Appl. **24**(1), 1–4 (2018). https://doi.org/10.1007/s11036-018-1136-6
8. Chen, Y., Geng, X.: Recommendation of personalized product-service system scheme based on improved collaborative filtering. Comput. Integr. Manuf. Syst. **27**(1), 240–248 (2021)
9. Ou, D.: Research and implementation of personalized travel recommendation system based on data mining[J]. Hubei Agric. Sci. **60**(9), 123–126.(2021)
10. Gao, P., Li, J., Liu, S.: An introduction to key technology in artificial intelligence and big data driven e-learning and e-education. Mob. Networks Appl. (2021). doi: https://doi.org/10.1007/s11036-021-01777-7
11. Niu, Q., Yang, R., Tian, H.: Optimization scheduling simulation of network high coverage resource based on AGV communication. Comput. Simul. **38**(6), 410–414 (2021)
12. Chin, H.H., Varbanov, P.S., Kleme, J.J., et al.: Enhanced cascade table analysis to target and design multi-constraint resource conservation networks. Comput. Chem. Eng. **148**(3), 107262 (2021)

Design of Public Opinion Research and Judgment System for Network Hot Events Based on Data Mining Technology

Qing-mei Cao[1(✉)] and Hui-fang Guo[2]

[1] Department of Computer Technology and Information Management, Vocational and Technical College of Inner Mongolia Agricultural University, Baotou 014100, China

[2] School of Transportation and Municipal Engineering, Inner Mongolia Vocational and Technical College of Architecture, Hohhot 010010, China

Abstract. The traditional network hot event public opinion research and judgment system in the face of excessive operating pressure, there are a large number of blank database connections, leading to the quality of the system running worse and worse. In order to solve this problem, this paper proposes the design of network hot event public opinion research and judgment system based on data mining technology. In hardware design, the serial communication module is designed. With the support of embedded processor, it is combined with the designed Ethernet transceiver to realize real-time communication; In the software design, it collects public opinion data, uses data mining technology to find network hot events, calculates the public opinion level of hot events, and designs corresponding early warning function to realize the judgment of public opinion. So far, the overall design of the system is completed. The experimental results show that: in the case of increasing operating pressure, the designed public opinion research and judgment system based on data mining has high operating efficiency, good public opinion recognition ability, and high operating quality.

Keywords: Data mining · Network hot events · Public opinion collection · Public opinion research and judgment

1 Introduction

Born in the 1960s, the Internet has been applied far beyond its original intention in today's society. With the rapid development of the Internet, it has become an important part of people's life, work and study and an important carrier of the spread and development of human civilization [1, 2]. As the main carrier of network communication, the Internet has penetrated into economy, politics, culture, social life and other aspects, changing people's way of communication and thinking [3].

S. Liu and X. Ma (Eds.): ADHIP 2021, LNICST 416, pp. 656–671, 2022.
https://doi.org/10.1007/978-3-030-94551-0_51

Due to the characteristics of Instant Internet communication, uncontrollability of communication content, relative equality of discourse power and long-term retention of information, Internet communication is completely different from traditional communication mode. First of all, in terms of communication speed, the traditional communication mode has to go through the audit of each editorial level before releasing an information, but it is different on the Internet, It saves the cumbersome audit of traditional communication mode and greatly improves the speed of communication; Secondly, in terms of communicators, the communicators of traditional communication mode are only a few traditional media, and there are a large number of blogs, various types of websites, forums, various chat rooms and communication tools on the Internet, which can be used as communicators; Secondly, in the aspect of equal right of discourse, in the traditional media environment, only a few media and individual groups have the right to speak, and the information spread is spread after screening. In the Internet environment, everyone can say that all kinds of information can appear in front of the Internet people at the same time; Finally, in terms of information retention, traditional media, such as radio and television, disappear after a while, and few people go to read the previous materials for newspapers and magazines. On the Internet, information will be retained for a long time. Even if the problem is solved, negative information will be left on the Internet [4–7]. Therefore, in the Internet era, Internet communication can have a huge impact on economy, politics, culture, social life and other aspects [8].

With the continuous development of the Internet, public opinion plays a more and more important role in government administrative departments or industries. At present, there are certain developments in the theory and software products of network public opinion at home and abroad.

There are two main parts involved in network public opinion, one is web information mining, the other is data collection, that is, web crawler [9, 10]. Web information mining, network public opinion is the mapping of social public opinion in the Internet space, is a direct reflection of social public opinion, the main data of network public opinion comes from the web. The other part is web crawler, which is more based on text content and can realize the transparency of public opinion data [11]. But in the face of the above two forms, the traditional public opinion research and judgment system is mainly for the collection and analysis of various unstructured data, such as text data, audio and video data, graphics, image data and other data fusion multimedia data. In the practical application, there are some disadvantages, such as a large number of blank requests and blank database connections in the application process, which lead to limited utilization of resources and poor quality of system operation.

Therefore, this paper proposes the design of network hot event public opinion research and judgment system based on data mining technology to solve the problems existing in the traditional public opinion research and judgment system. From the aspects of serial communication module, Ethernet transceiver, public opinion acquisition module, hot event discovery function and public opinion information early warning, the public opinion research and judgment system of network hot events is optimized and designed by using data mining technology, and the effectiveness of the design system is verified by experiments.

2 Hardware Design of Network Hot Event Public Opinion Research System Based on Data Mining Technology

2.1 Design of the Serial Communication Module

In the system design, serial communication refers to RS-232 communication, and serial communication can be divided into synchronous communication and asynchronous communication. In the process of data transmission of asynchronous communication system, receiver clock and transmitter clock are not synchronized, and are transmitted in independent byte mode [12]. Now most microprocessors provide UART interface. The program can receive and receive data only by operating the relevant registers. Full duplex communication is generally supported. After the level conversion, RS-232 serial communication with PC can be carried out, or chip level communication can be directly conducted with other processors. Hardware driver module is always closely related to hardware, and different hardware will have different influence on program processing [13, 14]. In the serial communication module of the design system, the USART hardware interface of ATmega128 is used.

ATmega128 has 2 USART serial interfaces. It is an enhanced UART interface. Besides the traditional UART interface, it also has some more reliable and flexible features, such as supporting high-speed asynchronous communication, independent high-precision baud rate generator, etc.; In the stop bit detection, AVR is optimized, which makes the detection error smaller.

For data receiving and receiving, interrupt driven mode is adopted to receive or send data in interrupt service program, and the upper program will be further processed. For data receiving, in interrupt mode, a byte received will produce an interrupt. Reads bytes from the data register in the interrupt service program and clears the receive interrupt flag. At this time, the received bytes can be processed directly in ISR or put them in the buffer for background tasks. In the project development, if the data package is relatively simple, this kind of processing is very simple [15]; If the packet is complex, the ISR is quite complex. When the background task is processed, the software level is particularly clear, and the interrupt is only responsible for receiving, while the background task is responsible for parsing the data content. However, the size of its buffer should be

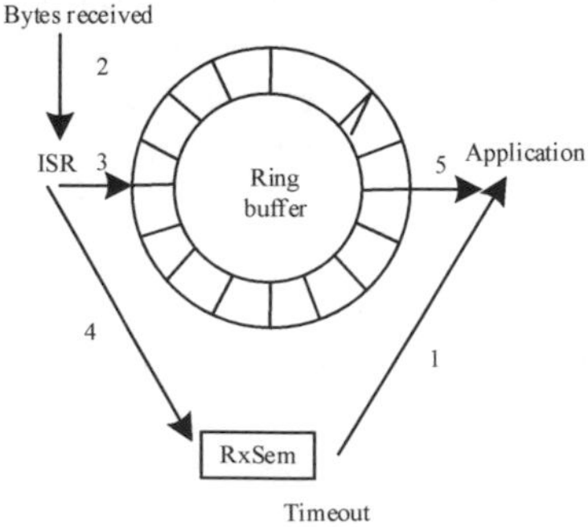

Fig. 1. Serial buffer receiving process with Signal Volume

determined reasonably to meet the worst-case scheduling situation, otherwise data will be lost. The response to input data depends on the execution speed of background tasks. If the frequency of execution is fast, the response time is fast. For real-time kernel like IBC/OS-II, the speed of processing input data is almost as fast as that of ISR only receiving data without processing [16]. Therefore, in the management of the ring buffer, a semaphore rxsem is added, as shown in Fig. 1.

The application task waits for the arrival of the semaphore (1). After receiving a byte, an interrupt is generated. The ISR reads the byte (2) from the data register and stores it in the ring buffer (3). Then the ISR sends a semaphore to tell the task that it has received a byte (4). After the application gets the semaphore, it fetches the byte from the ring buffer and performs the required task (5).

The whole process of data receiving is as follows: the application program calls commgetchar() or other functions to read data from the buffer. Generally, the timeout value should be specified to prevent the application task from being suspended permanently. When a semaphore is received, the required processing is performed. After receiving a byte, an interrupt is generated to enter the interrupt service program. Call the middle layer function commputrxchar() to put the data into the ring buffer. If the receive buffer is full, the data will be discarded [17, 18]. Therefore, when setting the buffer size, the processing speed of the task and the maximum amount of data actually received should be considered to prevent data loss. When the byte is put into the buffer, commputrxchar () decides when to send a signal to the received semaphore according to the actual needs. Is it sent every time, or when a specific character is encountered.

Compared with data receiving, data sending is simpler. Because the receiving interrupt will happen at any time, it is completely asynchronous, sometimes it will involve complex packets, and the receiver will have a special task to process the data. When sending data, generally the data to be sent can be packaged and sent directly, and the background still adopts buffer and interrupt drive mode.

2.2 Ethernet Transceiver Design

In order to ensure the reliable and timely communication of the public opinion research and judgment system, the embedded processor is used as the core processor of the system to support the normal communication of the system. STM32F407 is a 32-bit RISC chip based on ARM Cortex M4F. Its main frequency is 168 mhzo STM32F407. It has two APB enhanced I/O and peripheral buses, three AHB buses and a 32-bit AHB bus matrix, as shown in Fig. 2.

STM32F407 is a high-performance ARM chip. Its main features are as high as 1 Megabyte flash memory, flexible static memory controller, high-performance AD converter, general DMA controller and 17 timers. With the support of the processor, the Ethernet transceiver is designed.

Each node in the system communicates through Ethernet. Because the embedded processor STM32F407 has Ethernet controller, it only needs one Ethernet PHY chip. The nodes in the Internet of things use LAN8720 as the Ethernet PHY layer chip [19].

LAN8720 is an IEEE 802.3–2005 standard transceiver. Embedded processor STM32F407 communicates with it through standard RMII interface. LAN8720 supports auto negotiation mode, which can be used to determine the optimal network speed

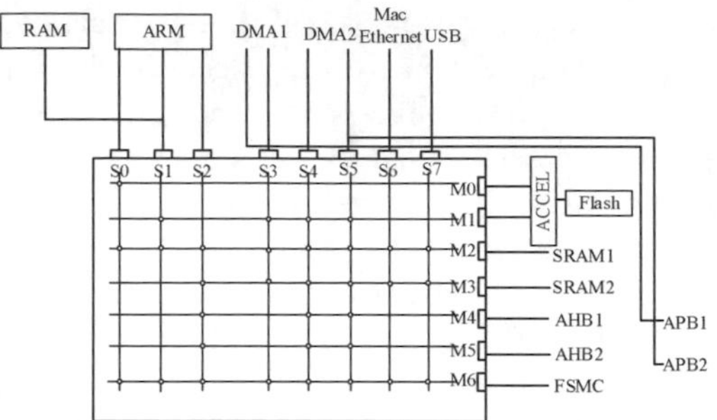

Fig. 2. STM32F407 Bus architecture diagram

and single and duplex communication modes. LAN8720 supports the operation of IEEE 802.3–2005 specific registers. This kind of operation does not need to access registers. The initial configuration can be selected through the configuration pin, and the function of transceiver can be further defined by using the register configuration option [20, 21]. All digital interface pins of LAN8720 can withstand 3.6 V. The LAN8720 uses an integrated 3.3 V to 1.2 V linear regulator, which can be selectively disabled.

LAN8720 is connected with the main control chip through RMII interface. RMII interface (reduced MII interface) is a simplified MII interface. It has two modes of MAC and PHY, and the nodes of Internet of things use PHY mode. The definition of RMII interface of LAN8720 chip in PHY mode is shown in Table 1.

Table 1. RMII interface definition in PHY mode

Signal name	Number	Direction	Describe
CLK_REF	1	Input	Reference clock
TXD[0:1]	2	Input	Transmit data
TX_EN	1	Input	Send enable
RXD[0:1]	2	Output	Receive data
RX_ER	1	Output	Acceptance error
CSR_DV	1	Output	Conflict detection
MDC	1	Input	Management clock
MDIO	1	I/O	management data

By combining the designed Ethernet transceiver with the serial communication module [22, 23], the internal and external communication of the public opinion research and judgment system can be realized, and the real-time public opinion data can be obtained. With the support of this technology, the software part of the system is designed.

3 Software Design of Network Hot Event Public Opinion Research System Based on Data Mining Technology

3.1 Design of the Public Opinion Collection Module

Public opinion collection is the crawling and storage of all media data and information on the Internet, including text content and other content. Media forms include forum, news, microblog, etc. Public opinion information must be timely, accurate and complete, and the information and attributes of the sender should be classified, counted and analyzed, which requires that the design of public opinion collection module should consider the characteristics of comprehensiveness, accuracy and stability. Therefore, the design of public opinion system should consider the storage space and database efficiency.

Data comprehensiveness includes that the content of data collection must include all URLs and keywords of all preset website pages, and cover large websites with large traffic and rich content, and the website content and keywords can be expanded. The timeliness of public opinion collection is to ensure the final update status of public opinion information. Considering the storage space and data validity of public opinion, the system only collects and saves information within one month. Data stability is to consider the collection efficiency and stable operation of the collection system, without process crash, downtime and other special circumstances. The architecture of data acquisition module is shown in Fig. 3.

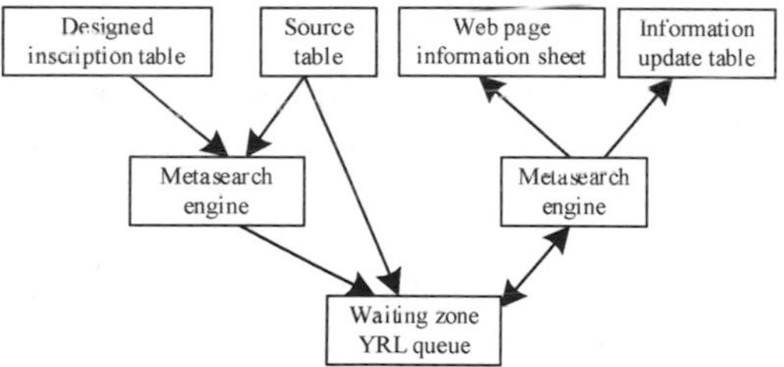

Fig. 3. Architecture of the public opinion data collection module

Public opinion data collection adopts three-level thread structure: the rule making and operation of search engine and web crawler are classified as one level thread, including the definition of website, crawler rules and keywords. The search engine collects data by keyword judgment, filters and discards the data, which is a secondary thread. Its main work is to complete the data collection through the current mainstream search engine. The data collected by the search engine is analyzed and stored locally as a three-level thread. Its main work is to store and write the processing results to the database. After collecting public opinion data, the hot event discovery function is designed to further analyze public opinion.

3.2 Design of Hotspot Event Function

Hot spot discovery is to discover the influential public opinion events in a certain period of time by automatically grabbing the information and identifying the topic of the target website set by users. At present, the commonly used target websites are mainstream news websites, large-scale commercial websites and popular BBS.

The traditional network public opinion analysis system of public opinion discovery basically adopts two methods: first, manually input keywords to search, the second method is based on the word frequency statistics of web page. The first method has low statistical efficiency, manual input of subjective interference factors, it is not easy to grasp the public opinion information in time: the statistical accuracy based on the word frequency of the web page is low, and the theme of the web page is not clear, which will lead to the distortion of the relevant statistical data. The design of the software part of the system is based on the original public opinion discovery method, and designs a new hot public opinion discovery function.

Through the web crawler module to capture the required information of news web pages, multiple network data sources are continuously monitored, and then the web page parser starts to clean up the web page. The main function is to remove the "noise" in the web page, and retain the web page links, title, time, first-class title, and related document fields, so as to complete the data cleaning. Through the word segmentation module, the feature words of the standardized document are segmented. Randomly select a part of web pages, use them as test samples for feature extraction to obtain a group of feature words, then the feature extraction module evaluates the feature words of the sample web pages, and extracts the common feature vectors of the test samples. The TFIDF value of the feature vector is used to express the text topic, and the VSM construction module establishes the vector space model. The hot spot discovery module clusters the text relation matrix to find new hot spots.

The similarity between the report and the topic is calculated, and the maximum similarity and the topic with the maximum similarity are recorded to determine the topic closest to the current report; The theme is expressed by several feature words with the highest comprehensive weight in all the news within the theme itself; The similarity between news report and topic is calculated by Formula 1. If the calculated value is greater than the set closed value, it is considered that the new topic is included in the calculated page. The calculation formula is as follows:

$$score(x) = 1 - max\left\{1 - \frac{k}{m}\right\} \times sim(\vec{x}, \vec{c}_i) \tag{1}$$

$$sim(x, c) = \frac{\sum_{j=1}^{M} a_{jx} \times a_{jc}}{\sqrt{\left(\sum_{j=1}^{M} a_{jx}^2\right) \times \left(\sum_{j=1}^{M} a_{jc}^2\right)}} \tag{2}$$

$$a(r, \vec{w}) = \frac{rf(r, \vec{w}) \times \log\left({}^{N}/_{n_i} + 0.01\right)}{\sqrt{\sum_{r \in \vec{w}} \left[rf(r, \vec{w}) \times \log\left({}^{N}/_{n_i} + 0.01\right)\right]^2}} \tag{3}$$

Where $a(r, \vec{w})$ represents the weight of word r in text w, $rf(r, \vec{w})$ refers to the word frequency of word r in text w, N represents the total number of training text, n_i represents the number of r text in the training text set, $sim(x, c)$ represents the similarity of the new file x for a certain time cluster set, a_{jx} represents the weight of word j in cluster c, M

table shows the total number of words in the file set, and x represents the information of new file, c_i is the i cluster in the time interval, i is the number of files contained in the time interval, and k represents the number of files increased between the latest file collection time in cluster c_i and the arrival time of the new file x. In the setting threshold, as long as score is greater than the set value, the new file is considered as a new topic.

Cycle the above process. After a class processes a fixed number of new topics, it compares the topics in the class. If the similarity of the two topics is greater than the merge closed value, it merges them. Two clustering similarity methods can be used to calculate the similarity between topics

$$Sim(P_1, P_2) = \frac{\sum_{b_i \in P_1} \sum_{b_j \in P_2} sim(b_i, b_j)}{|P_1| \cdot |P_2|} \tag{5}$$

Where P_1 and P_2 are the monitored news topics, b_i and b_j are the news reports in P_1 and P_2 respectively, and $|P_1|$ and $|P_2|$ are the number of news reports in the two topics respectively. When a class processes a fixed number of new topics determined by users, the text information in each topic is eliminated, and the similarity between news reports and the topic is recalculated; When the similarity is lower than the closed value of clustering, or does not meet the restrictions (such as whether the report is within 30 days or not), the news report will be eliminated; Then, the internal representation and its weight are recalculated; Finally, the monitoring results are output according to the needs of users. For all the current topics in the category, combined with the time characteristics of the topic and the quantity characteristics of the text information in the topic, several topics with the highest score are selected from all the categories as the hottest information of the category.

3.3 Warning Design of Public Opinion Information

The degree of netizens' attention to a specific network public opinion event is the popularity of network public opinion, which can be used to measure the development stage of a network public opinion event, and also can be used to investigate the attitude and behavior of the participants in the network public opinion event. The popularity of Internet public opinion depends on the appeal of public opinion to post and follow a news or event. In the system design, the average number of follow-up posts in a certain period of time is used to express the appeal and the degree of attention of the theme post.

$$NH = \left[\frac{g_i}{n_i} + \sum_{j \in u_i} \frac{\sum\limits_{k \in v_j} \frac{g_k}{n_k}}{m_j} \right] \tag{6}$$

Where g_i is the number of post nodes, n_i is the number of posts of post node i, u_i is the reply set of post node i, v_i is the reply set of post node j, m_j is the number of user posts of post node j, T is the delay, and N is the number of replies of current post node. The larger the value calculated by the formula, the higher the degree of attention of the post corresponding to the news or event. It can not directly determine the severity of public opinion, but it can be used as a reference for the classification and early warning of public opinion events.

According to the severity and level of network public opinion (very serious, serious, general and slight), systematic early warning can be carried out, corresponding to red warning, orange warning, yellow warning and green warning respectively. In addition to the popularity of Internet public opinion, the severity of the classification is also related to the spread speed, scope and political tendency of public opinion. Generally speaking, public opinion against the party, the state and ethnic conflicts can be classified as very serious public opinion. For the early warning parameter setting of public opinion, the system needs to carry out the classification and weight score calculation of comment posts, and the calculation method is shown in formula 7.

$$NW = \frac{2n_1 + n_2 + 0.5n_3}{N} \tag{7}$$

Where, N represents the number of Posts related to public opinion, n_1, n_2 and n_3 represent the number of Posts related to level 1, level 2 and level 3 public opinion. Finally, the value calculated by formula 7 is used for early warning of public opinion. The score between 1–1.5 indicates serious public opinion, the score between 0.5–1 indicates general public opinion, and the score below 0.5 indicates slight public opinion. So far, the design of network hot event public opinion research and judgment system based on data mining technology is completed.

4 Experimental Research on Public Opinion Research System of Network Hot Events Based on Data Mining Technology

4.1 Experimental Data Preparation

The main carriers of online public opinion are news websites, blogs, forums, post bars, etc. (Baidu Post Bar, zhonghua.com, Sohu Community, Sina forum, Tianya community, people.com, xinhua.com, cnews.com, Chinanet, sina.com, sohu.com, Netease and Tencent), Several representative news websites and forums are selected as shown in Table 2. The selected websites are very representative, with top click through rate and rich in network public opinion.

Table 2. Network public opinion data collection sources

Serial number	Website name	Home page URL address
1	Baidu Post Bar	http://tieba.baidu.com/
2	China.com	http://club.china.com/
3	Sohu Community	http://club.sohu.com/
4	Sina Forum	http://bbs.sina.com.cn/
5	Tianya community	http://www.tianya.cn/

(continued)

Table 2. (*continued*)

Serial number	Website name	Home page URL address
6	People's daily	http://www.people.com.cn/
7	Xinhuanet	http://www.xinhuanet.cn/
8	China News Network	http://www.chinanews.com.cn/
9	China.com	http://www.china.com.cn/
10	Sina	http://www.sina.com.cn/
11	Sohu	http://www.sohu.com/
12	Netease	http://www.163.com/
13	Tencent	http://www.qq.com
……	……	……

Supported by the data sources in Table 2, several groups of comparative experiments are designed to verify the actual performance of the proposed public opinion research and judgment system compared with the public opinion research and judgment system based on information awareness in reference [2] and the public opinion research and judgment system based on SNA in reference [7]. In order to ensure the scientific and fair experiment.

The comparative experiment is designed according to the application requirements of the system, and the experimental results are analyzed.

4.2 Application Performance Performance Analysis

In the experiment, two traditional public opinion research and judgment systems are introduced. Under the same experimental environment, the application performance of the system under pressurized conditions is tested. The concurrent users for each system are set to 300 and the number of experiments is 50. The experimental results are shown in Table 3.

Table 3. Experimental results of performance performance of different public opinion system

Public opinion research and judgment system	Project	Unit	Result value
Public opinion research and judgment system of reference [2]	Total duration	Hours: Minutes: seconds	1:19:37
	Maximum number of running Vuser		180
	Total throughput	Byte	156872253
	Average throughput	Byte	37547
	Total hits	Second	16643
	Average hits per second	Second	3983

(*continued*)

Table 3. *(continued)*

Public opinion research and judgment system	Project	Unit	Result value
Public opinion research and judgment system of reference [7]	Total duration	Hours: Minutes: seconds	00:56:49
	Maximum number of running Vuser		180
	Total throughput	Byte	128379926
	Average throughput	Byte/s	28136
	Total hits	Second	15348
	Average hits per second	Second	3048
Public opinion research and judgment system of this paper	Total duration	Hours: Minutes: seconds	00:36:15
	Maximum number of running Vuser		240
	Total throughput	Byte	276584359
	Average throughput	Byte	92491
	Total hits	Second	18231
	Average hits per second	Second	4134

According to the data in Table 3, the experimental data of the public opinion research and judgment system in reference [2] shows that with the increasing pressure, the number of users who send out requests has reached 180, the overall response time of the system has decreased, the average throughput is low, the system can not respond to customer requests in time, nor can it meet the needs of users, and the overall running time is long. The experimental data of the public opinion research and judgment system in reference [7] shows that the running time is reduced by a part compared with the previous experimental result, and the total number of clicks is similar to the previous result, but the overall response time of the system is also reduced, and the average throughput is lower, which can not meet the user's request; Compared with the previous two groups of experimental results, the designed public opinion research and judgment system has a short duration, an average throughput of 92491 bytes, which is much higher than the previous two groups of results, and the number of users processed reaches 240, indicating that the system can maintain a better response level and application performance in the case of increased pressure, and can meet various user requests.

4.3 Experimental Analysis of System Middleware Resources Use

When the concurrent users of the system reach a high level, the database connection pool and thread pool will run out of resources, which indicates that the resource allocation is unreasonable and idle connections exist when the system uses middleware resources. To solve this problem, the middleware resource usage experiment is designed. The number of concurrent users is set to 100, and the idle connection of the system database is observed. The specific experimental results are shown in Fig. 4.

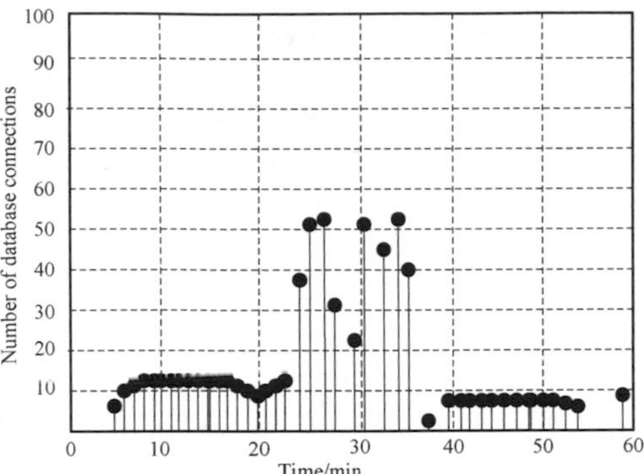

(a) Experimental results of public opinion research and judgment system in reference [2]

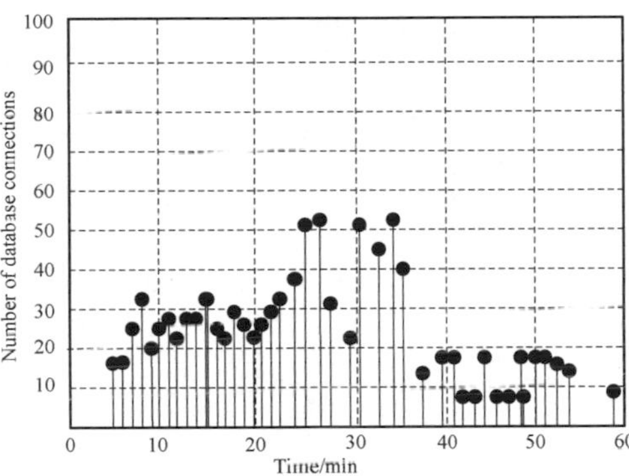

(b) Experimental results of public opinion research and judgment system in reference [7]

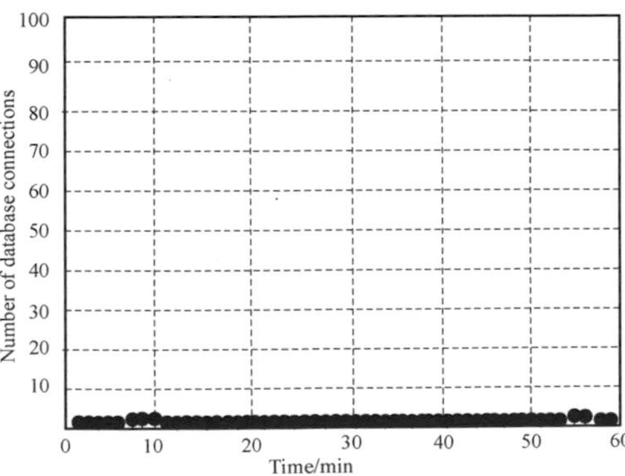

(c) Experimental results of public opinion research and judgment system in this paper

Fig. 4. Experimental results of middleware resources of different public opinion research system

According to the experimental results of each public opinion research and judgment system in Fig. 4, when the system is faced with the concurrent access of 100 users, the number of idle connections of the two traditional public opinion research and judgment systems is high and low, up to 50, which indicates that there are a lot of idle connections in the process of system operation, resulting in excessive consumption of resources and reducing the operation efficiency of the system; In contrast, in the effective experimental time, the number of idle connections in the database of the designed public opinion research and judgment system is always relatively low, approaching 0, which indicates that the system has high efficiency, and there is no excessive consumption of resources. The main reason is that the design system uses data mining technology to compare the two topics in a class after a fixed number of new topics are processed. If the similarity of the two topics is greater than the combined closed value, they will be combined to improve the clustering similarity and further optimize the resource utilization. To sum up, the design of public opinion research and judgment system based on data mining middleware resource use effect is good, high efficiency.

4.4 Experimental Analysis of Public Opinion Identification

The public opinion recognition experiment mainly evaluates the public opinion recognition module in the system, and evaluates the performance of the system by using the missing rate, error rate and normalized recognition cost. The calculation formula of missing rate and error rate of public opinion is defined as follows:

$$miss_i = \frac{p_i}{S_i} \tag{8}$$

$$F_i = \frac{e_i}{E_i} \tag{9}$$

Based on the above formula, the average miss rate, average error rate and normalized identification cost of the system are calculated.

$$R_{miss_i} = \sum\nolimits_i \frac{miss_i}{N} \tag{10}$$

$$R_{F_i} = \sum\nolimits_i \frac{F_i}{N} \tag{11}$$

$$C_{norm} = \frac{C_{miss} \cdot R_{miss} \cdot R_t + C_{miss} \cdot R_F \cdot R_{-t}}{N} \tag{12}$$

Where $miss_i$ is the missing rate, p_i is the number of unrecognized texts related to public opinion i, S_i is the total number of texts related to public opinion i, F_i is the error rate, e_i is the number of recognized texts not related to public opinion i, E_i is the total number of texts not related to public opinion i, R_{miss} is the average missing rate, R_F is the average error rate, C_{norm} is the normalized recognition cost, C_{miss} is the missing cost, C_{miss} is the cost of the error, R_t is the prior probability of the current event public opinion, $R_{-t} = 1 - R_t$. After the above calculation, the evaluation results of different

Table 4. Public opinion identification experimental results of different research and evaluation systems

	Number of network hot events	Missing rate	Error rate	Normalized cost
Public opinion research and judgment system of reference [2]	20	0.351	0.05632	0.0711
	50	0.409	0.07936	0.0793
	80	0.482	0.08413	0.0886
	100	0.594	0.11237	0.0922
Public opinion research and judgment system of reference [7]	20	0.692	0.09547	0.0118
	50	0.749	0.09846	0.0512
	80	0.811	0.10114	0.0831
	100	0.829	0.10236	0.1057
Public opinion research and judgment system of this paper	20	0.062	0.00006	0.0007
	50	0.084	0.00011	0.0015
	80	0.092	0.00015	0.0019
	100	0.109	0.00018	0.0024

public opinion research and judgment systems are obtained, and the specific contents are shown in Table 4.

According to the data in Table 4, when the two traditional public opinion research and judgment systems face different number of hot events, the missed check rate and error rate are increasing, and the normalization cost is increased. Compared with the previous two experimental results, the designed public opinion research and judgment system error rate and missed check rate based on data mining technology are also increasing, but the increase is not obvious, The values are always low and the normalization overhead is low. According to the analysis of the above two experimental results, the design of the network hot event public opinion research and judgment system based on data mining technology has high efficiency, accurate public opinion recognition and good overall application performance, which is superior to the traditional two public opinion research and judgment systems. The reason for this result is that the public opinion research and judgment system of network hot events based on data mining technology optimizes the public opinion collection module, crawls and stores all media data and information on the Internet, including text content and other content, and optimizes the performance of public opinion recognition.

5 Conclusion and Prospect

Due to the increasing impact of the Internet on people's lives, the increasing number of Internet users, and the more complex and changeable network environment, the establishment of an effective network public opinion monitoring system is of great significance for coping with network emergencies, maintaining social stability, and helping the government make democratic and scientific decisions.

Based on the characteristics of rapid formation, rapid change, large amount of information and great social impact of network public opinion, as well as the requirements of current government work for network public opinion monitoring, combined with the research results of data mining, this paper proposes to establish a public opinion research and judgment system based on Data Mining, and studies the key technologies in it, It provides a method for further semantic analysis of network public opinion information for public opinion control. After the experimental demonstration, the practical application ability of the system proposed in this paper has been proved, which shows that the design of the system effectively solves the problems existing in the traditional system, and realizes the more effective control of network public opinion.

However, in the research process, there are still some shortcomings, such as the supervision of network hot events, including text form, picture form, video form, etc. in this paper, there is no one by one verification of multiple forms. Therefore, in the follow-up development, research and analysis will be carried out from this aspect to further improve the public opinion research and judgment system.

Fund Projects. Science and Technology Plan Project of Inner Mongolia Autonomous Region Project Description: Research and Development and Application of Key Technologies of Smart Agriculture in Inner Mongolia Yellow River Region (2020GG0033).

References

1. Wei, D., Wei, X.: Design of network experimental teaching evaluation system based on data mining. Modern Electron. Tech. **43**(03), 142–145+149 (2020)
2. Zhang, S., Wang, L., Lou, G.: Research on network public opinion study and early-warning system based on information awareness. Inf. Stud. Theory Appl. **43**(12), 149–155 (2020)
3. Xu, D.: Research on the judgment system of network public opinion on the emergency initiated by the major epidemic based on spatio-temporal big data. Modern Inf. **40**(04), 23–30+81 (2020)
4. Liu, J., Shi, M., Liu, C.: Semantic graph of net citizens' information perception on network group emergencies. Inf. Stud. Theory Appl. **42**(02), 158–163 (2019)
5. Shi, J., Guo, J.: Public Opinions on "The US-China Trade War": features and strategies based on the study of Youtube platform. J. Intell. **38**(08), 105–112+135 (2019)
6. Liu, J., Li, L.: Analysis of intelligent early warning mechanism of network public opinion in the background of big data. J. Intell. **38**(12), 92–97+183 (2019)
7. Cheng, J., Zhang, S., Ji, Q.: Analysis of micro-blog public opinion periodic propagation network in social hotspot events based on SNA—an empirical study on "Yu Huan". Manage. Rev. **31**(03), 295–304 (2019)
8. Gao, P., Li, J., Liu, S.: An introduction to key technology in artificial intelligence and big data driven e-Learning and e-Education. Mob. Networks Appl., 1–4 (2021)
9. Chen, Z., Wang, J.: Analysis of network public opinion events based on bayesian network. Inf. Sci. **38**(04), 51–56+69 (2020)
10. Zhang, C., Ma, X., Zhou, Y., et al.: Analysis of public opinion evolution in COVID-19 pandemic from a perspective of sentiment variation. J. Geo-Inf. Sci. **23**(02), 341–350 (2021)

11. Liu, S., Liu, D., Srivastava, G., et al.: Overview and methods of correlation filter algorithms in object tracking. Complex Intell. Syst. **3**, 1–23 (2020)
12. Li, Z., Huang, S.: Analysing on network public opinion based on LDA topic model. J. Syst. Sci. Math. Sci. **40**(03), 434–447 (2020)
13. Liu, S., Liu, X., Wang, S., Muhammad, K.: Fuzzy-aided solution for out-of-view challenge in visual tracking under IoT assisted complex environment. Neural Comput. Appl. **33**(4), 1055–1065 (2021)
14. Lei, L., Wu, Y., Zhang, Y., et al.: Time+user dual attention based sentiment prediction for multiple social network texts with time series. IEEE Access **PP**(99), 1 (2019)
15. Peng, L.J., Shao, X.G., Huang, W.M.: Research on the early-warning model of network public opinion of major emergencies. IEEE Access **PP**(99), 1 (2021)
16. Gu, Q., Li, L.: Research on process optimization and unified system content in big data network information sharing environment. J. Phys. Conf. Ser. **1952**(3), 032039 (9pp) (2021)
17. Li, K., Jing, M., Tao, X., et al.: Research on online management system of network ideological and political education of college students. Int. J. Electr. Eng. Educ. (2021), 002072092098370
18. Gao, J., Niu, S., Li, H., et al.: The reliability research of V2V based on sample judgment and balanced modulation. Int. J. Internet Protoc. Technol. **12**(4), 229–235 (2019)
19. Xie, X., Ma, Z., Ye, J., et al.: Research and development of sound quality in portable testing and evaluation system based on self-adaptive neural network. Appl. Acoustics **154**, 138–147 (2019)
20. Yue, H.A., Lw, A., Tz, A.: Area definition and public opinion research of natural disaster based on micro blog data - ScienceDirect. Procedia Comput. Sci. **162**, 614–622 (2019)
21. Durso, G., Haws, K.L., Way, B.M.: Drug influences on consumer judgments: emerging insights and research opportunities from the intersection of pharmacology and psychology. Mark. Lett. **31**, 19–23 (2020)
22. Graf, H., Menter, M.: Public research and the quality of inventions: the role and impact of entrepreneurial universities and regional network embeddedness. Small Bus. Econ., 1–18 (2021)
23. Chen, M.Y., Liao, C.H., Lughofer, E.D., et al.: Informetrics on social network mining: research, policy and practice challenges. Libr. Hi Tech. **38**(2), 273–275 (2020)

Design of Marine Buoy Profile Information Monitoring System Based on Machine Vision

Huan-Yu Zhao[1,2,3](✉) and Xing-kui Yan[1,2,3]

[1] Institute of Oceanographic Instrumentation, Qilu University of Technology (Shandong Academy of Sciences), Qingdao, China
[2] Shandong Provincial Key Laboratory of Ocean Environmental Monitoring Technology, Qingdao, China
[3] National Engineering and Technological Research Center of Marine Monitoring Equipment, Qingdao, China

Abstract. Based on the comprehensive investigation of the existing marine information platform, this paper puts forward an architecture model of the electronic information system of the marine platform, which optimizes the system hardware structure, system configuration model and system software function by combining with machine vision, so as to realize the effective monitoring of the marine buoy profile information, This paper describes the functional composition of the marine buoy profile information monitoring system, and comprehensively analyzes the reliability, safety and other characteristics of the marine buoy profile information monitoring system, which provides the basis for the design modeling, index decomposition and development integration of the electronic information system of the offshore platform.

Keyword: Machine vision · Ocean buoy · Information monitoring

1 Introduction

In the field of marine environmental monitoring, marine production and aquaculture, long-term, fixed-point and real-time vertical profile monitoring of seawater parameters in specific areas is often needed. The existing marine environment monitoring platform mainly obtains the monitoring data of the vertical profile of the underwater environment through the chain sensor buoy method in the specific sea area. Although this method can accurately monitor the water quality data of the same vertical profile at different depths, it needs to work multiple sensors at different depths to get more comprehensive underwater environment data, This is bound to increase the cost of the whole underwater buoy profile information monitoring system [1].

Aiming at the problems of high cost and poor reliability of traditional system monitoring, this paper introduces machine vision technology to design the marine buoy profile information monitoring system. The specific research ideas of this paper are as follows:

Using the lifting control method, only a single sensor can realize the accurate measurement of different depths of the same vertical profile. According to the function of

S. Liu and X. Ma (Eds.): ADHIP 2021, LNICST 416, pp. 672–688, 2022.
https://doi.org/10.1007/978-3-030-94551-0_52

the ocean buoy profile information monitoring system, analyze the system operation principle, realize the function optimization of the system software, and greatly reduce the cost of the long-term continuous monitoring system of different depths of underwater vertical profile.

2 Design of Ocean Buoy Profile Information Monitoring System

2.1 Hardware Structure of Marine Buoy Profile Information Monitoring System

The whole marine buoy profile information monitoring system is mainly composed of environmental monitoring sensor group, winding coil, underwater electronic cabin, anchoring device, underwater information acquisition device and surface floating ball [2]. The environmental monitoring sensor group is placed inside the underwater floating ball. It is used to complete the acquisition of temperature and deep salt physical quantity, and the anchoring device and winding coil are used to fix the system and realize the lifting function of underwater floating ball [3]. The main function of the underwater electronic cabin is to control the rotation of the winding coil, realize the actual number of turns and speed control operation, and display various configuration functions of the system hardware equipment, as follows (Fig. 1):

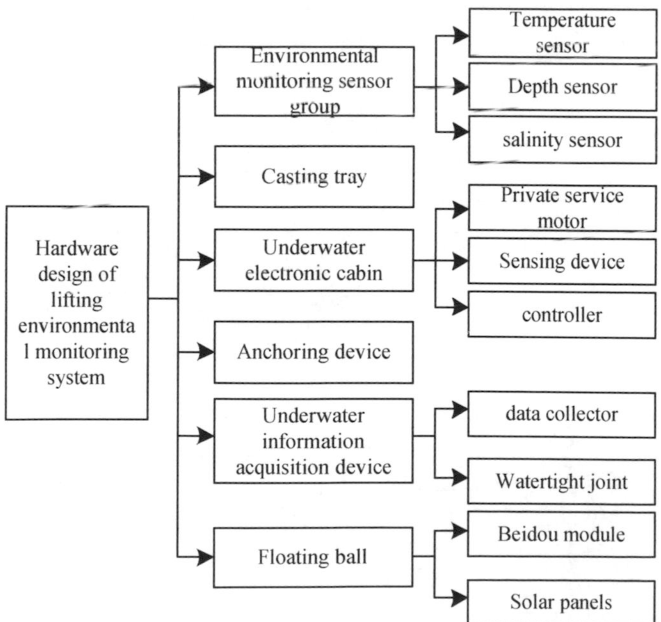

Fig. 1. System hardware configuration function

The profile information monitoring system of marine buoy in shallow water vertical section is mainly composed of underwater floating ball, underwater electronic cabin, winding coil and anchoring device. The underwater electronic cabin is mainly composed of servo motor, motor controller, voltage converter, watertight interface and other modules [4]. The underwater electronic cabin is placed in the inside of the winding coil. The underwater floating ball and the winding coil are connected by armored cable. One end of the armored cable is connected with the wave meter, temperature sensor and salinity sensor inside the underwater floating ball through a watertight connector to provide the working voltage and monitoring data transmission function. The other end of the armored cable is wound on the winding coil, and the armored cable joint is connected with the underwater cable from the underwater information acquisition device through the underwater connector, and the collected underwater environment data is sent to the underwater information acquisition device through the cable, and returned to the shore station system through the Beidou module for data processing, In addition to providing power for the underwater sensors, the underwater cable from the underwater information acquisition device also connects its branch connector into the underwater electronic cabin through the underwater connector to provide working voltage for the motor and motor controller in the underwater electronic cabin [5]. The marine buoy profile information monitoring system in this paper is different from the ordinary information system. The design, construction and implementation of the whole system are based on the special background of marine environment and resources. Therefore, in order to make the system run normally, the system function, process and composition design must follow certain requirements [6]. According to the system design objectives and planning from the system integration level, the marine buoy profile information monitoring system is composed of data acquisition subsystem, remote control subsystem, data management

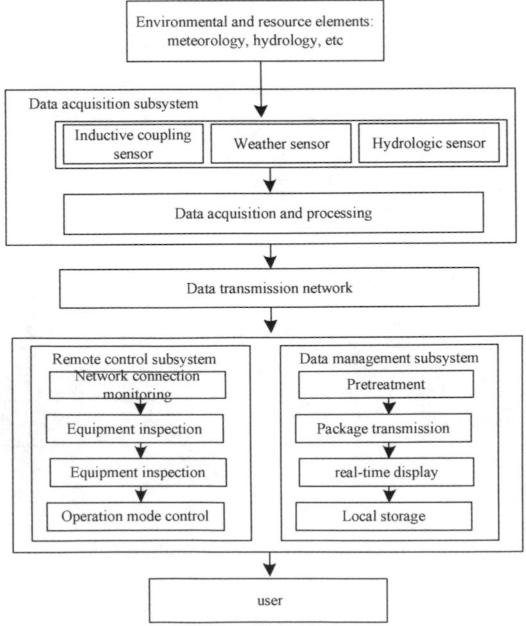

Fig. 2. Hardware configuration of ocean buoy monitoring system

subsystem, etc. the overall structure of the hardware configuration of the marine buoy monitoring system is shown in the Fig 2.

The architecture of the system can be divided into four layers: support layer, access layer, processing layer and information transmission layer, and the standard system, security system and guarantee system run through all layers of the system architecture. The architecture is as follows (Fig. 3):

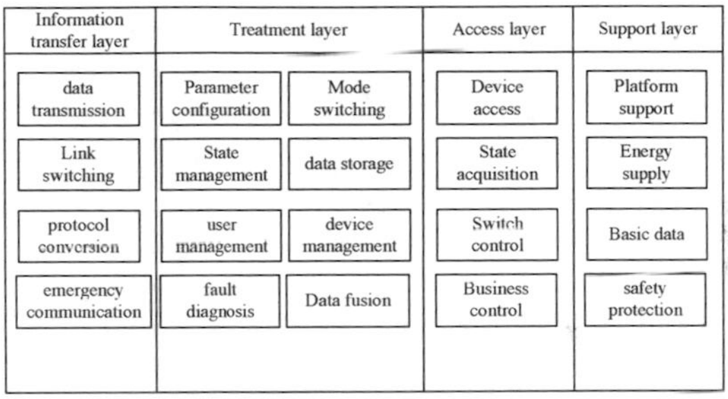

Information transfer layer	Treatment layer		Access layer	Support layer
data transmission	Parameter configuration	Mode switching	Device access	Platform support
Link switching	State management	data storage	State acquisition	Energy supply
protocol conversion	user management	device management	Switch control	Basic data
emergency communication	fault diagnosis	Data fusion	Business control	safety protection

Fig. 3. System structure framework

The support layer is responsible for providing basic guarantee conditions for the system operation, which mainly includes platform support, energy supply, basic data, security protection and other functions [7]. Among them, support provides installation space and interface for carrying equipment, energy supply supplies power for carrying equipment through various power generation means, basic data provides various benchmark data for carrying equipment, and security protection provides environmental monitoring and intrusion protection for carrying equipment. The support layer is responsible for providing basic guarantee conditions for the system operation, which mainly includes platform support, energy supply, basic data, security protection and other functions [8]. Among them, platform support provides installation space and additional interface for carrying equipment, energy supply supplies power for carrying equipment through various power generation means, basic data provides various benchmark data for carrying equipment, and security protection provides environmental monitoring and intrusion protection for carrying equipment [9]. The processing layer is responsible for the processing of all kinds of state data and perception business data, as well as the management and control of the system. Among them, the state management is responsible for analyzing the state information of the access system equipment and forming the system work log; The fault diagnosis is based on the system work log and real-time equipment status information, and the fault solution suggestions are given. Data storage is responsible for saving all kinds of business data, status data and work logs of the system [10]. Data fusion is responsible for data fusion of access services to form a unified situation information; According to the configuration instructions issued by the information transmission layer, the software and hardware parameters of the equipment are configured; User management is to assign the permissions of different login users; The mode switching completes the system working mode switching according to the control instruction; The equipment management completes the business mode management and on-off control of the equipment according to the control instructions. The electronic information system of offshore platform is composed of floating platform system, integrated electronic information system and multi energy complementary intelligent power supply system. The system consists of intelligent control system, basic support system, multi-source target processing system and information transmission system. The system establishes communication links with users through different wireless transmission equipment to realize information exchange on the other side. In the system, the communication network is established in the form of switch to realize the internal information transmission of the system, and separate the internal and external network to ensure the information security of the system. At the same time, the internal real-time data bus is established to provide real-time data such as attitude and position to the user equipment. The working principle of the system is shown in the Fig. 4.

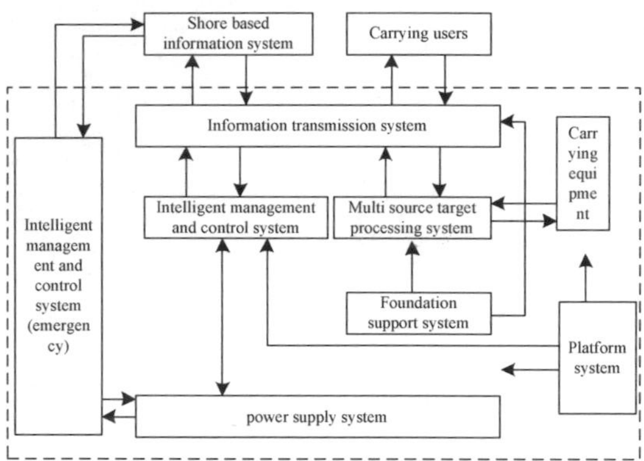

Fig. 4. System operation principle

The system provides stable marine physical carrier for the whole system equipment through platform and anchoring device, and is equipped with lighting, ventilation, sewage and fuel storage equipment, so as to provide physical environment guarantee for maintenance personnel to go on the stage for maintenance and diesel engine operation of energy subsystem, and is equipped with ballast water regulation equipment to realize platform draft regulation capacity under different water depth conditions. Through the platform monitoring, the status monitoring of the above equipment and local control of some equipment are completed, and the status monitoring data is reported to the intelligent management and control system through the internal network. The power supply system realizes the independent power generation on the sea through the complementary form of diesel engine, wind energy, photovoltaic and ocean energy. The power storage is completed by the battery to ensure the continuous operation of the system when the power generation means are interrupted, and provide time buffer for the system maintenance and repair in the case of lack of power generation capacity. The power supply system completes the external power transmission through the two bus forms of AC 220 V and DC 48 V, and uses the diesel engine to generate power to realize the power guarantee for the high-power equipment of the platform subsystem. At the same time, the power supply system reports all kinds of state information of its own power generation, storage and distribution to the intelligent management and control system through the internal network. The barocap sensor is proprietary to Vaisala. It is a kind of silicon capacitive absolute pressure sensor with high precision, high reliability and many other excellent characteristics. With its outstanding long-term stability, it minimizes the field calibration requirements in many application fields (Table 1).

Table 1. Performance parameters of hardware equipment

Parameter	Specifications
Accuracy (20 °C)	±0.3 hPa
Resolving power	0.1 hPa
Measuring range	500–1100 hPa
Working temperature	−40 + 60 °C
Voltage output range	0–5 V
Supply voltage	10–30 VDC
Current consumption	4 mA

In the working process of the system, because the stator of the servo motor in the underwater electronic cabin is fixedly connected with the transmission device, and the winding coil can be driven by the rotation of the stator of the servo motor relying on the engagement of the gear, the whole lifting control system can realize the constant speed winding in the positive and negative directions of the winding coil through the rotation of the motor. Because the winder is connected with the underwater anchoring device through the steel cable and is in a positive buoyancy state, the winder is always in a stable equilibrium state. When the winding coil winds the cable at a constant speed, the length of the cable between the winding coil and the underwater floating ball changes, so that a single group of sensors can monitor the seawater environment at different depths in the same vertical section. Because the vertical section ocean buoy profile information monitoring system designed in this paper uses a single sensor group to measure the underwater environment through lifting control mode, which replaces the traditional fixed chain measurement mode of multiple sensors, so that the whole system only needs to supply power to a single set of sensors of each underwater measurement device, It avoids the traditional way of multi-sensor group power supply. Therefore, the power consumption of the lifting underwater buoy profile information monitoring system is much smaller than that of the chain underwater monitoring system. In practical application, servo motor control winding coil rotation is not frequent operation, so it will not cause the overall power consumption burden of the system. Based on this, the system uses solar panels and battery power supply, and through the Beidou satellite, the underwater environment information data is transmitted to the shore station data center for post-processing. This kind of system power supply and communication mode replaces the traditional underwater cable power supply and data transmission mode, which not only saves the cost of power supply and transmission cable, but also avoids the maintenance cost required by long-term wear of underwater cable, and ensures the safety and stability of data transmission.

2.2 Function Optimization of System Software

The marine information system is deployed on the sea and unattended. After the failure, it needs to be repaired at sea, and the preparation time for a single maintenance is long and the maintenance cost is high. Therefore, the maintenance work should be carried out immediately only when the system has a serious failure that affects the task. When there is a general failure of the system that does not affect the task, it is generally not necessary to carry out the maintenance immediately, but try to carry out the maintenance together with the maintenance of important failures or regular maintenance. Therefore, the system reliability work should focus on the task reliability of the system, try to avoid serious failures, reduce the frequency of sea maintenance, and improve the availability of the system. According to gjb813–1990, the task reliability model of the system is established. Fault criteria of mission reliability: the system can not provide necessary survival guarantee, energy supply, intelligent control and data communication services for the carrying equipment during the mission, resulting in the carrying equipment unable to work, so the fault maintenance must be carried out immediately. Among them, the reliability of supporting equipment such as ventilation and exhaust, pressure transformation and oil supply for oil turbine power generation in the platform subsystem will be considered in the power supply subsystem. The block diagram model of mission reliability is shown in the figure below (Fig. 5):

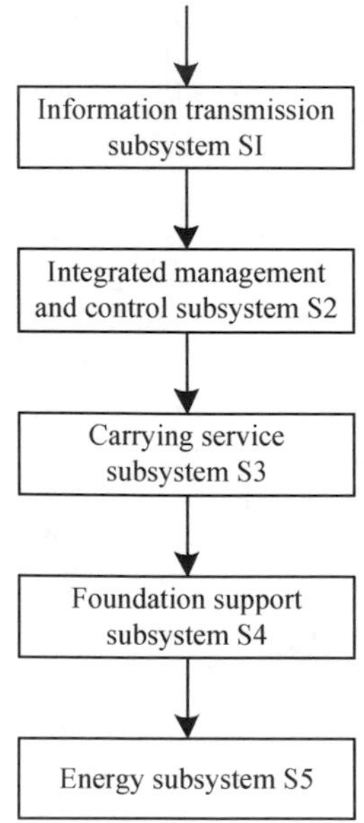

Fig. 5. Information monitoring task processing flow

Because the whole shallow water environment vertical profile monitoring system is composed of multiple underwater test units, when multiple monitoring systems send data to the underwater information acquisition device at the same time, the data acquisition device will not receive the underwater environment data transmitted by each set of marine buoy profile information monitoring system orderly, resulting in data loss or error. There are many kinds of PC software development languages, such as VB, C, C++, C #, Java and so on. We know C and C #, but the main software development tools for these two languages are VC++ and vs. the main differences between them are shown in the Table 2.

Table 2. System performance comparison results

	VC++ 6.0	VS
Development mode	The application is dialog based MFC	Windows form application
Development language support	C, C++	C, C++, C#
Development efficiency	Faster	Fast
Help system	Preferably	Fast
Code execution efficiency	Very high	Higher
Database support	Weak	Strong
Easy to learn	It's hard	Commonly

The multi parameter acquisition system of marine profile environment needs to realize the measurement of various marine environment parameters, store and analyze the data, and at the same time, it can control the sensor remotely and monitor the operation status in real time. The system software mainly includes the software of the lower computer, the communication protocol software and the PC user software. The system software structure is shown in the Fig. 6.

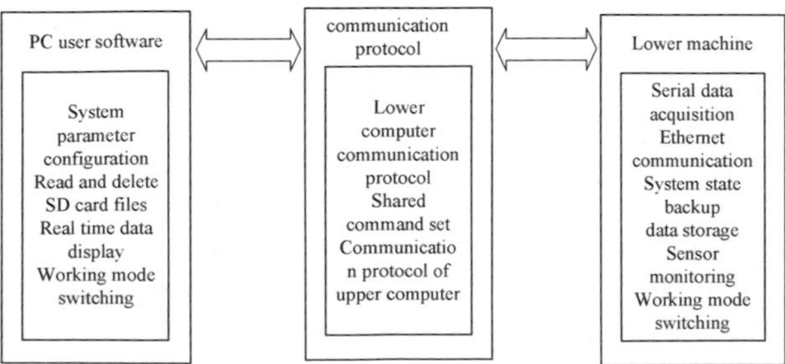

Fig. 6. System software structure

The lower computer software mainly includes the subroutine of initialization module, serial data acquisition module, Ethernet data sending and receiving module and real-time data storage module; The communication protocol software mainly realizes the normal communication between the upper computer and the lower computer by making software rules. PC user software mainly realizes the extraction and deletion of SD card files, underwater sensor parameter configuration and operation control, underwater node network parameter configuration, real-time data display, system status monitoring and other functions.

2.3 The Realization of Ocean Buoy Profile Information Monitoring

In recent years, with the rapid development of marine science and technology, more and more monitoring data have been accumulated. There are many important information hidden behind the surge of data. How to analyze it at a higher level in order to make better use of the data becomes more and more important. However, the original data often has many problems that affect the quality of data use. There are many reasons for these problems, as shown in the figure. Therefore, effective data preprocessing technology is an effective method to solve the problem of data quality and improve the accuracy of data decision-making (Fig. 7).

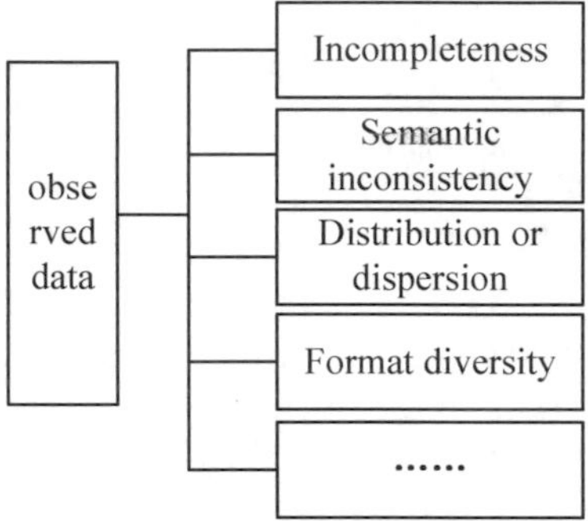

Fig. 7. Information management model of observation data monitoring

Through data preprocessing, the incomplete data can be complete, the wrong data can be corrected, the redundant data can be removed, the required data can be selected and integrated, the unsuitable data format can be converted to the required format, and the redundant data attributes can be eliminated, so as to achieve the same data type, data format consistency, data integration Data information refining and data storage centralization can improve data quality, data service accuracy and decision accuracy.

Data preprocessing is to clean, transform and select the original data before data application, so as to make it reach the minimum specification and standard required by the application algorithm for knowledge acquisition. In a word, after preprocessing, we can not only get the data set required by the mining system, but also reduce the cost of the application system and improve the effectiveness and comprehensibility of knowledge. The main tasks of data preprocessing are as follows.

Data cleaning: such as filling missing data, eliminating noise data, etc. The principle of data cleaning is to use the existing technical means and methods to clean the "dirty data" by analyzing the causes and existing forms of "dirty data", and transform the "dirty data" into data that meets the data quality or application requirements, so as to improve the data quality of the data set.

Data conversion: the data will be stored in the database, data warehouse or file to form a complete data set, and the redundant data should be eliminated in this process. It is mainly to standardize the data, such as limiting the data value to a specific range. For some application patterns, the data needs to meet a certain format. Data conversion can convert the original data to the format required by the application pattern to meet the needs.

Data selection: eliminate the attributes that can not describe the key characteristics of the system, so as to get a refined set of attributes that can fully describe the applied object. For the mining system that needs to deal with discrete data, we should first quantify the continuous data so that it can be processed.

After the system is powered on, the control chip must first carry out the necessary configuration work, and then start the data acquisition work. The sampling period is set according to the density requirements of various meteorological factors in scientific research. Different sensors have different sampling frequency. The system samples the underwater sensor CTD every half an hour, collects the weather sensor every one hour, and obtains the GPS information every three hours, as shown in the figure. These processes are initiated by the buoy control system actively, and each sensor is powered on regularly or waked up by the controller for data communication, which is not the party initiating the communication actively. As the host of the system, the controller needs an accurate clock to ensure accurate data sampling and communication time. In the process

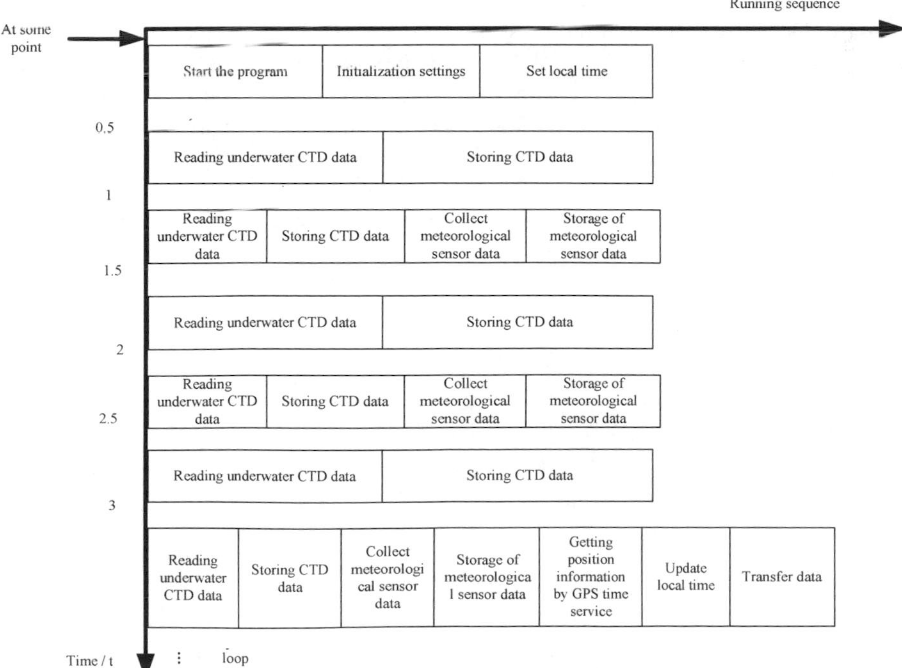

Fig. 8. Buoy information management system operation process

of communication between the controller and the GPS module, the satellite time at the moment will be obtained to calibrate the buoy time (Fig. 8).

According to the influence of wind when sound wave propagates in the atmosphere, windsonic wind speed measurement process is as follows. At the same time, the time of the ultrasonic wave from transducer s to N and the time of the sound wave from n to s can be expressed by the relationship between speed, time and distance:

$$\begin{cases} T_1 = \frac{L}{c+V} \\ T_2 = \frac{L}{c-V} \end{cases} \tag{1}$$

The results of solving the quadratic equation are as follows

$$V = \frac{L}{2}\left(\frac{1}{T_1} - \frac{1}{T_2}\right) \tag{2}$$

$$C = \frac{L}{2V}\left(\frac{1}{VT_1} + \frac{1}{VT_2}\right) \tag{3}$$

L is the distance from transducer s to N, Ti and TA are the time of ultrasonic from transducer s to N and from n to s respectively, *C* is the speed of sound, and Y is the component of wind speed to be measured in the N direction of transducer. The other two ultrasonic transducers on the sensor can measure the component of wind speed in E or W direction with the same principle, which is the process of anemometer measuring wind speed and direction.

When the system transmits data, it adopts the control mode of CAN bus, and conducts data arbitration through the protocol of data acquisition unit, so as to ensure that only one set of system can use the channel to transmit environmental data in each time period. Based on the use of can technology, the following gives the process of fuzzy data assimilation of ocean observation information data, as shown in the Fig. 9.

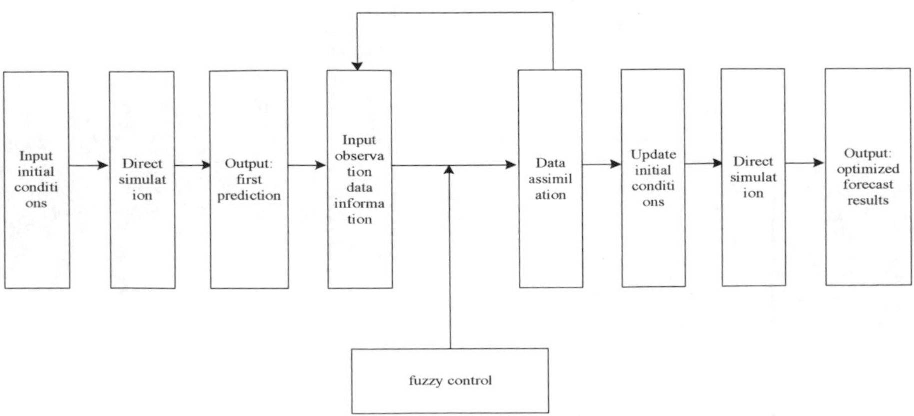

Fig. 9. The process of data assimilation of buoy information fuzziness

In fact, the key to solve the problem of data is the purpose of application. Here, many definite cases can be described by fuzzy conditional statements. For example, if two sets of data X and Y are interactive, we can consider them to be related, which can be described by the following statement:

$$R_i : IfX^i isA^i(n)ThenY^i isB^i(n) \tag{4}$$

The result obtained from the previous step is the formalization of fuzzy relation, and the de fuzzy formula is as follows:

$$y = \frac{R_i \sum_{i}^{L} y_i}{L - C} \tag{5}$$

According to the fuzzy basic representation, the central average method is calculated.

$$P(x) = \frac{\Pi_{i-1}^{M} \mu A_{ij}(x_i)}{\sum_{j=1}^{L} \prod_{i=1}^{M} \mu A_{ij}(x_i)} \tag{6}$$

$$\Delta y' = f(x) = P(x) \sum_{j=1}^{L} p_j(x) y_j \tag{7}$$

In order to obtain effective fuzzy prediction, there must be an overall measure of individual errors in the form of average error. Ape's formula is as follows:

$$APE = \frac{\sum_{i}^{n} |x_i - \hat{x}_i|}{\sum_{i} |x_i| - \Delta y'} \times 100 \tag{8}$$

Therefore, in a more general form, it is defined as the minimization of the following cost function:

$$J(x) = (x - x_b)^T B^{-1}(x - x_b) + \sum_{i=0}^{n} (y_i - H_i[x_i])^T R_i^{-1}(y_i - H_i[x_i]) \tag{9}$$

Under the premise of ensuring the stability and reliability of data transmission, the focus of the whole system software design is to complete the accurate winding operation of the winding coil by controlling the rotation of the motor. The specific system software flow chart is shown in Fig. 10:

After the control system is powered on and initialized, the motor controller writes the working mode to the servo motor and waits for the instruction from the upper computer. After receiving the upper computer instruction, judge the specific content of the instruction. If it is a speed instruction, write the corresponding command word after judging the motor speed and direction, and then write 0f, then the motor starts to work. If the received command is 0x1111 or 0x0000, it means that the motor holds the brake and continues to work after the motor stops holding the brake. If the motor controller receives the instructions from the host computer, it will send the working state information of the motor to the host computer, so as to achieve the research goal of real-time and accurate monitoring of the ocean buoy.

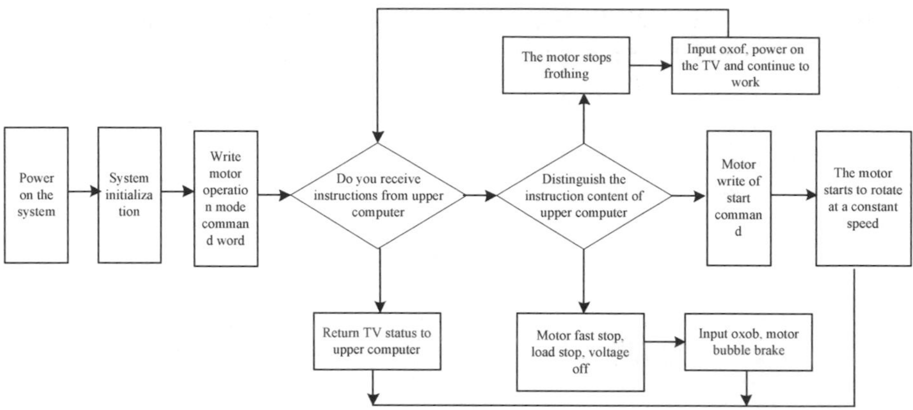

Fig. 10. System operation process optimization

3 Analysis of Experimental Results

The system performance was tested in depth in laboratory environment, and the conversion accuracy of AD was verified by using NAV meter and high precision voltage source. A series of joint commissioning experiments have been carried out in the National Laboratory of marine science and technology pilot. The calibration process of the system is as follows: the output of high-precision constant current source is taken as the voltage to be measured, and the measurement result of the system is taken as the measurement value VMS; At the same time, the nanovoltmeter measured the voltage and took the result as the standard value vstd. Because the NAV meter selected in the calibration process is of high precision and resolution of 1 nv. Even if the range is increased to 10 V, the resolution can be kept within 1 uv, and the accuracy is better than 3 ppm. Therefore, the results of Nav table can be used as the accurate value for system reference. Then, the fitting curve and fitting equation of vstd VM and each channel are obtained by using the least square method. After the measurement process, the measurement results need to be calibrated by this fitting equation to obtain more accurate measurement values. Several voltage values are selected for measurement, five values are tested for each voltage value, and the average value is obtained as the measurement result of this point. The least square fitting is carried out with origin. The least square fitting function and its coefficients of each analog conversion channel are listed in the Table 3.

Table 3. Parameters of fitting curve

Passageway	Slope	Loading distance
Pressure	1.0012	−5.72341E-6
Precipitation	1.00184	−9.81E-3

(*continued*)

Table 3. (*continued*)

Passageway	Slope	Loading distance
Wind speed	1.00013	−9.16752E-6
Temperature	1.00034	−8.08157E-5
Wind direction	1.00012	−5.42655E-5
Compass	1.00033	−1.82208E-4
Long wave	1.00391	−6.46393E-4
Shortwave	1.00287	−1.46E-3

Through the VI editor, the underwater data acquisition program and the communication program of the buoy platform are written respectively. Through GDB debugging tool, you can easily debug programs in the Linux system of the host, and set breakpoints for debugging. Compile to generate executable file, write makefile file.

The underwater data acquisition system mainly tests the data receiving of multiple serial ports and the storage of SD card. At the same time, open and receive 6 channels of serial port data, and each channel sends channel * string to test the bit error rate. Where * is the channel number, debug the serial port to connect the Linux system of PC. Save the collected data to the file/MNT/sdcard/collect.txt. For several consecutive data transmission, each test time is 1 h and 3 h, and there is no bit error, so the reliability of the system is high.

In order to analyze the monitoring effect of the designed system and compare the measurement error of the marine buoy profile information monitoring system before and after use, the mean value of the measurement error under 30 iterative experiments is obtained, as shown in Fig. 11.

According to the analysis of Fig. 11, after the calibration of the designed system, the measurement error of analog quantity is greatly reduced and the measurement accuracy is significantly improved.

Furthermore, the accuracy and stability of single point voltage measurement are studied. In the same environment, the actual application performance of the traditional monitoring system and the detection system in this paper are compared and analyzed. The horizontal part in the figure is the expected value, and the specific detection results are shown in the following Fig. 12.

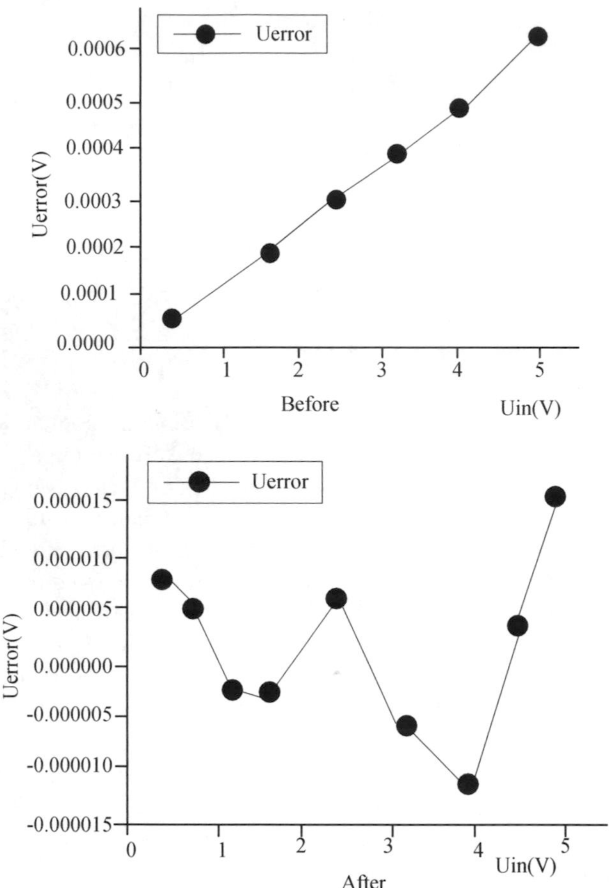

Fig. 11. Deviation between measured value and standard value

Based on the monitoring results above, the proposed information monitoring system of buoy profile has higher accuracy in the practical application process and fully meets the research requirements.

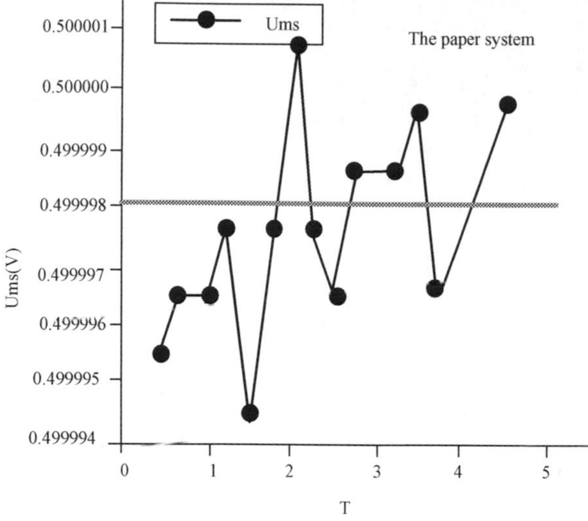

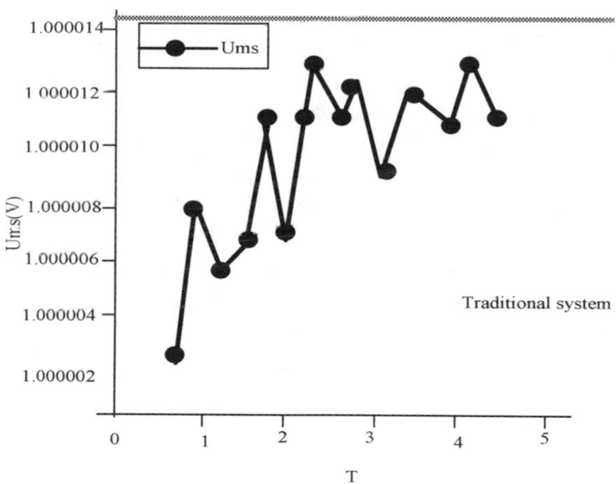

Fig. 12. Monitoring results of system performance comparison

4 Concluding Remarks

This paper designs an ocean buoy profile information monitoring system based on machine vision. The hardware structure of the marine buoy profile information monitoring system is designed, the fuzzy data assimilation process of buoy information is optimized, and the fuzzy prediction method is used to realize the stable transmission of the data of the information monitoring system and the effective monitoring of the marine buoy profile information. It provides scientific design ideas and methods for design modeling, index decomposition, development integration and function joint test of offshore platform electronic information system.

References

1. Kumar, R., Patil, O., Karthik, N.S., et al.: A machine vision-based cyber-physical production system for energy efficiency and enhanced teaching-learning using a learning factory. Procedia CIRP **98**(1), 424–429 (2021)
2. Liu, S., Liu, X., Wang, S., Muhammad, K.: Fuzzy-aided solution for out-of-view challenge in visual tracking under IoT assisted complex environment. Neural Comput. Appl. **33**(4), 1055–1065 (2021)
3. Sergiyenko, O., Tyrsa, V.: 3D optical machine vision sensors with intelligent data management for robotic swarm navigation improvement. IEEE Sens. J. **PP**(99), 1 (2020)
4. Liu, S., Liu, D., Muhammad, K., Ding, W.: Effective template update mechanism in visual tracking with background clutter. Neurocomputing (2020) https://doi.org/10.1016/j.neucom.2019.12.143
5. Sergiyenko, O., Flores-Fuentes, W., Rodríguez-Quiónez, J.C., et al.: Control theory and signal processing in machine vision for navigation. Int. J. Adv. Robot. Syst. **17**(4), 172988142092647 (2020)
6. Sheridan, L.M., Krishnamurthy, R., Gorton, A.M., et al.: Validation of reanalysis-based offshore wind resource characterization using lidar buoy observations. Marine Technol. Soc. J. **54**(6), 44–61 (2020)
7. Oikonomou, C., Rui, P., Gato, L.: Unveiling the potential of using a spar-buoy oscillating-water-column wave energy converter for low-power stand-alone applications. Appl. Energy **292**(11), 116835 (2021)
8. Salcedo-Bosch, A., Rocadenbosch, F., Gutiérrez-Antuano, M.A., et al.: Estimation of wave period from pitch and roll of a lidar buoy. Sensors **21**(4), 1310 (2021)
9. Imzilen, T., Chassot, E., Barde, J., et al.: Fish aggregating devices drift like oceanographic drifters in the near-surface currents of the Atlantic and Indian Oceans. Progress Oceanography **171,** 108–127 (2019)
10. Gao, P., Li, J., Liu, S.: An introduction to key technology in artificial intelligence and big data driven e-learning and e-education. Mob. Networks Appl. (2021). https://doi.org/10.1007/s11036-021-01777-7

Design of Electromechanical Measurement and Control System for Marine Mobile Buoy Based on Neuron PID

Xuan-qun Li[1,2,3](✉) and Ji-ming Zhang[1,2,3]

[1] Institute of Oceanographic Instrumentation, Qilu University of Technology (Shandong Academy of Sciences), Qingdao 266000, China

[2] Shandong Provincial Key Laboratory of Ocean Environmental Monitoring Technology, Qingdao 266000, China

[3] National Engineering and Technological Research Center of Marine Monitoring Equipment, Qingdao 266000, China

Abstract. In order to overcome the shortcomings of buoys that can only observe the marine environment at fixed stations, a neuron PID-based electromechanical measurement and control system design for marine mobile buoys was developed. The overall structure of the hardware equipment of marine mobile buoys was introduced, and a mobile buoy was constructed. Designed a mobile buoy measurement and control system, described the data communication, motion control, and the man-machine interface of the mobile buoy measurement and control system, and gave an example of testing the mobile buoy in the lake. Marine mobile buoys have the ability to autonomously conduct marine environmental observations in a larger area, and are of great value for obtaining large-scale marine environmental information.

Keywords: Neuron PID · Marine mobile buoy · Electromechanical measurement and control

1 Introduction

Among various three-dimensional ocean observation systems, the buoy system is an important marine environment observation equipment. It has the advantages of continuous, long-term and automatic monitoring of marine meteorology and hydrological environment in the harsh marine environment. It is the expansion of marine observation shore base stations, survey ships and survey aircraft in space and time, and has any other surveys. The irreplaceable advantages of the method play a huge role in ocean observation [1]. However, after the buoys are deployed, observations can only be carried out at fixed stations to obtain ocean observation elements near the stations. Relocation and deployment requires a lot of consumption of personnel and ships. The autonomously controlled mobile marine monitoring platform is a powerful tool for in-depth research on the marine environment. Recently, domestic and foreign scholars have proposed the

S. Liu and X. Ma (Eds.): ADHIP 2021, LNICST 416, pp. 689–707, 2022.
https://doi.org/10.1007/978-3-030-94551-0_53

concept of a mobile marine buoy [2]. Designed a marine mobile buoy electromechanical measurement and control system based on neuron PID, using the control of the propeller and rudder and neuron PID technology to realize the monitoring task of the mobile buoy at different stations, and using the digital radio to realize the marine mobile buoy and the shore The control system performs wireless data transmission and communication, and collects various data required [3].

2 Electromechanical Measurement and Control System of Marine Mobile Buoy

2.1 Hardware Structure of Electromechanical Measurement and Control System for Marine Mobile Buoy

The marine mobile buoy can realize the maneuverability of the buoy through the propeller and rudder, and realize the environmental monitoring task of a given sea area through navigation control. It uses solar energy to provide energy and charge the battery, so that the mobile buoy can work stably and reliably for a long time [4]. The marine mobile buoy system consists of the marine part of the mobile buoy and the remote control center. The structure of the marine mobile buoy system is shown in the Fig. 1.

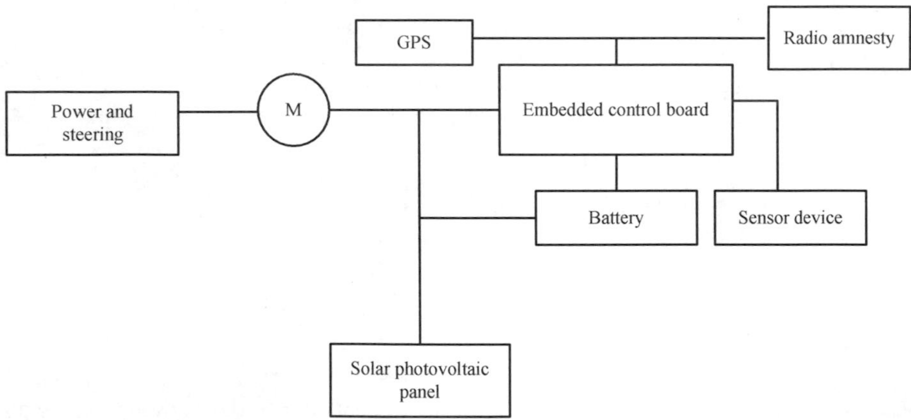

Fig. 1. The structure of the marine mobile buoy maneuvering system

The remote control center is called the upper computer, which can directly issue the control command [5]. The control system of the marine part of the mobile buoy is called the lower computer, which is controlled by the upper computer, and directly controls the function of the mobile buoy. The marine part of mobile buoy is mainly composed of buoy body and embedded control board. By controlling the movement of a hydraulic single degree of freedom experimental platform, the frequency of pressure change of the system is changed. By controlling whether the energy storage is connected to the hydraulic system, the influence of accumulator on different motion frequencies of hydraulic system is studied, and the effect of accumulator on the heave compensation system is verified. The hardware system of the measurement and control system is mainly composed of the upper computer, adlinkpci-8132 motion control card and magnetic grating ruler. The details are shown in the Fig. 2:

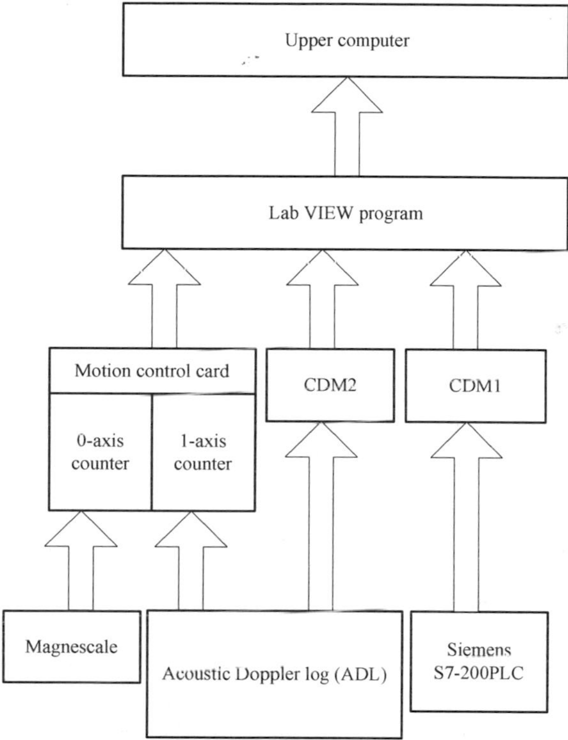

Fig. 2. Overall design block diagram of the system

The single-degree-of-freedom platform is mainly composed of three parts, from left to right: hydraulic station, single-degree-of-freedom actuator, and electronically controlled experimental platform [7]. By operating the electronic control test bench, the movement of the servo cylinder is controlled. The hydraulic station is the power supply unit of the single-degree-of-freedom test bench, which is mainly composed of a motor, a plunger variable pump, an electromagnetic overflow valve, an accumulator, and oil pipes; the single-degree-of-freedom actuator is mainly composed of a servo cylinder and an electro-hydraulic servo valve Composition [8]. The cylinder barrel is fixed, and the movement of the piston rod of the servo cylinder can be controlled by adjusting the voltage of the servo valve [9]. The accumulator is installed at the pump outlet. The composition and principle diagram of the entire hydraulic system are shown in the Fig. 3

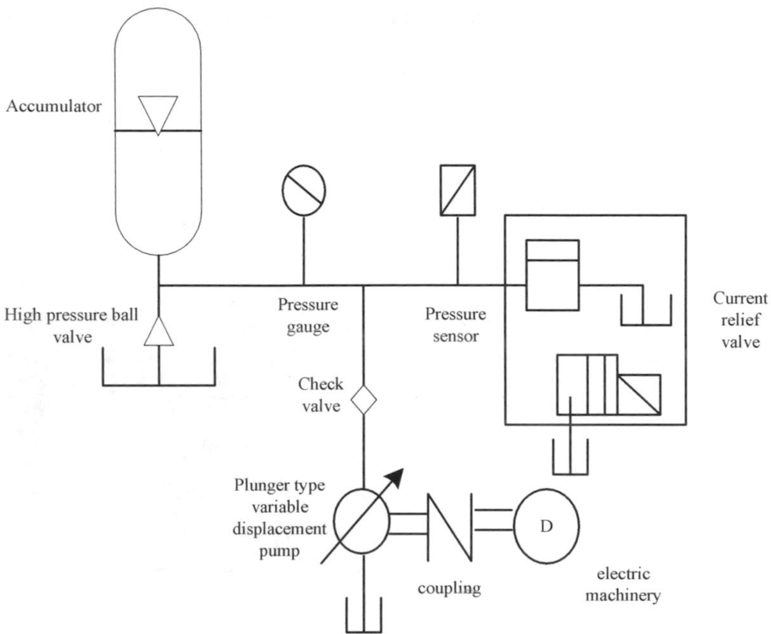

Fig. 3. Hydraulic system power equipment structure

Among them, the power module, motion and direction control module, sensor module, communication module, positioning module and data processing module are integrated on the embedded control main board. Propellers, rudders, solar panels and various sensors are directly installed on the buoy body. The floating part of the marine mobile buoy plays the role of data collection, processing and transmission, and various information collected will be transmitted to the remote control center through wireless communication [10]. The remote control center includes man-machine interface module, data storage module and data communication module. The communication module plays the role of receiving and processing the data sent by the ocean buoy, and transmits the collected data of the ocean buoy to the upper computer in real time to store and display the data. The data displayed include the precise position, heading, speed, sensor data, measurement time, etc. of the marine mobile buoy. In the host computer, by establishing a friendly man-machine interface to process and display the data, the user can conveniently operate the host computer [11]. The electric control experiment platform is mainly composed of a strong current control cabinet, a weak current control cabinet and a computer control system. The strong current control cabinet is used to control the start and stop of the motor and the switch of the electromagnetic overflow valve; the weak current control cabinet and the computer control system are located in the same control cabinet, which are respectively used for the manual and computer control operations of the servo cylinder, and the computer control system realizes the Control of cylinder movement [12]. The computer control system mainly includes computer, PCI-6221 multi-function acquisition card, Moog controller, displacement sensor, pressure sensor, etc. PCI-6221 is used to complete the acquisition and buffering of input signals.

The data is sent to the computer through PCI bus, and the data is processed under the control of application program. At the same time, the command of application software is received, and the electrical signal is sent out to control the action of servo cylinder through MOOG controller. Moog controller as a hardware closed-loop makes the output voltage of PCI-6221 change linearly with the displacement range of the cylinder. The principle of the measurement and control system is shown in the Fig. 4

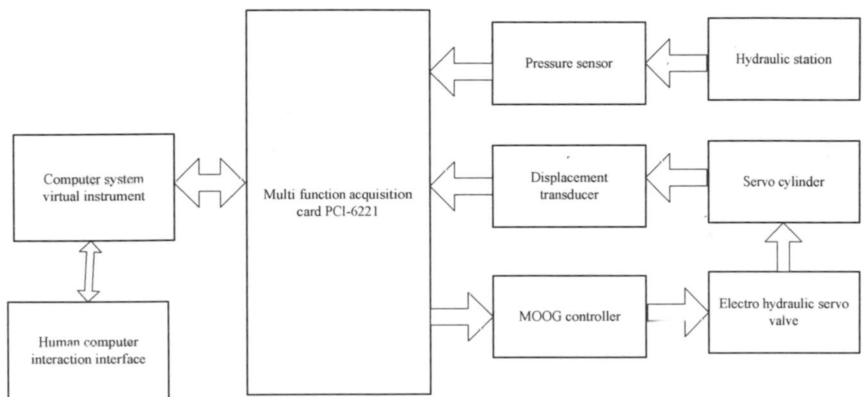

Fig. 4. Perfect function structure of measurement and control system

The specific use process of the measurement and control system is as follows: the displacement signal of the hydraulic cylinder collected by the displacement sensor is input to the controller through the A/D conversion module, and the control voltage signal of the hydraulic cylinder is obtained under a certain control strategy after processing in the controller. The control voltage is converted through the D/a module, and then output to the electronic amplifier, and then output to the electro-hydraulic proportional directional valve, To control the telescopic movement of the hydraulic cylinder. The modular design of the system is essentially the same as the mechanical modular design in the conventional sense, which realizes the overall function of the system through the combination of relatively independent functional modules [13]. However, the combination of mechanical modules is generally guaranteed by mechanical rigid connection, such as bolts, etc., and the combination and Realization of functions are guaranteed by the geometric related conditions between each other. The connection between different modules of the measurement and control system mainly depends on the electrical interface (software) connection to realize the information exchange between modules and complete the relevant control tasks [14]. Therefore, in the modular division and design of the measurement and control system, in order to ensure the reliable operation between modules and improve the reliability of the system, we must follow several criteria: the connection mode and information exchange between modules are simple and reliable. Module functions should be relatively centralized and independent; Module combination should have clear purpose, greater flexibility and good economy [15]. The hardware composition of the six hydraulic cylinders is exactly the same, and the control strategy is responsible for coordinating the telescopic motion of the six hydraulic cylinders to realize the motion of the upper platform. The system is composed of local master station and upper computer, which can realize remote control and data sharing in LAN.

2.2 Function Optimization of System Software

With the continuous development of artificial intelligence control, the artificial intelligence method is used to store the adjustment experience of the operator in the actual operation as knowledge into the computer, and according to the actual situation on the site, the computer can automatically adjust the PID parameters and control, intelligent control Fuzzy control theory in is one of the effective methods to solve this problem [16]. The software system is developed using the LabVIEW graphical programming language of National Instruments. LabVIEW is the most representative graphical programming software in the field of virtual instruments. It is widely used in the fields of testing, process processing and control. The motion control card is used on the LabVIEW development platform. Realizing motion control, data collection is simple and easy (Fig. 5).

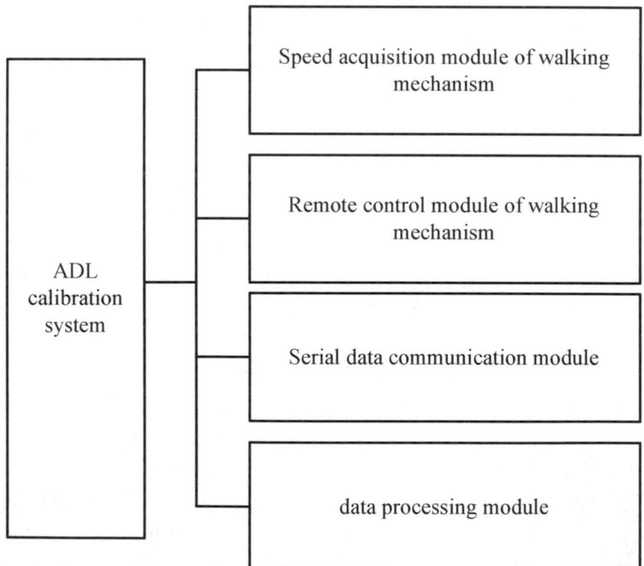

Fig. 5. Software structure block diagram of remote measurement and control system

According to the requirements of the mobile buoy system for the upper computer, it has four main function modules: data communication module, human-computer interface module, data processing module and motion control module. The human-computer interface module is to provide friendly human-computer interface, which is convenient for users to operate. The motion control module mainly controls the motion of Marine Mobile buoy in real time by sending the command to control the marine mobile buoy. The data communication module mainly completes the communication function with the lower computer. The main function of data processing module is to process and store the received data. This paper mainly introduces the human-computer interface module. The man-machine interface module of the upper computer mainly includes the display of the position of the buoy, the track of motion and the parameters in the course of the motion. The position of the motorized buoy is displayed by receiving the GPS information transmitted by the serial port and calling the map. Use the map data as the base map of the map, and use the WebBrowser control in C# to call the JavaScript based program to obtain the map data. The user performs human-computer interaction on the location of geographic information in the electronic map presented by the Web program to obtain the latitude and longitude coordinates of the data object, find the XML information through Xpath, and then update it to the user's special application data [18]. In this software, the latitude and longitude information returned by GPS will be displayed on the map on the left as shown in the figure after being analyzed. GPS keeps sending back data, and will display the track and label on the map in real time. Its specific application will be introduced in the test (Fig. 6).

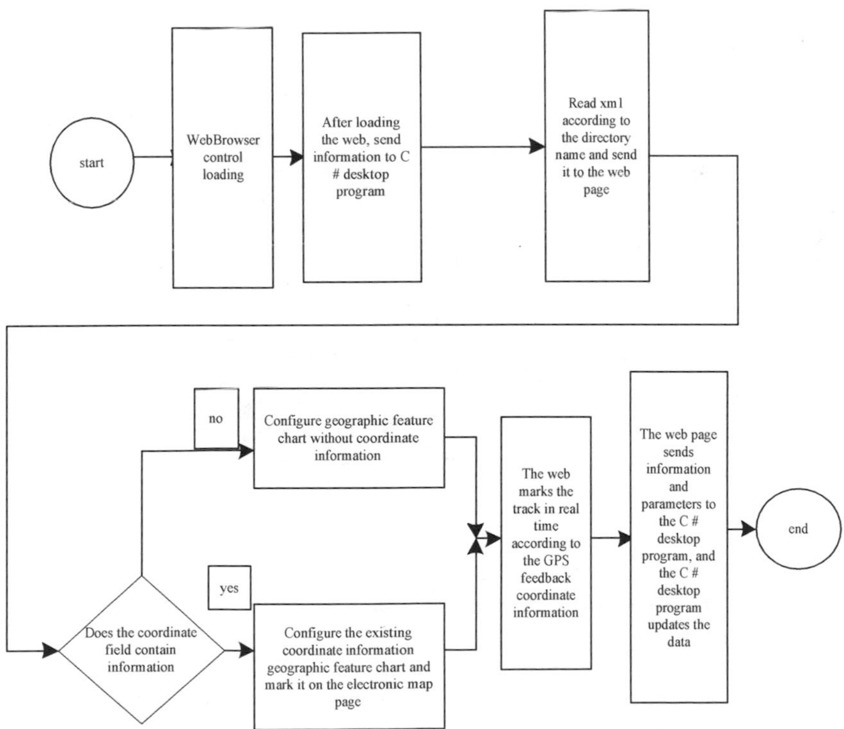

Fig. 6. Optimization of electromechanical measurement and control information call steps

Due to the particularity of marine monitoring and measurement, the mobile buoy must be able to cruise along the given track, reach the designated coordinate position, provide enough energy supply, carry the necessary monitoring equipment, and reach the designated position as fast as possible for various test and monitoring work. The overall size of the motorized buoy is: 2000 mm × 800 mm × 600 mm, and the floating body on the water surface is made of foam floating body material covered with thin steel plate. The power source of the whole device is propeller propulsion, which mainly relies on the rudder to control the direction, and the battery is charged by the solar panel to provide electric power to drive the propeller and rudder board. The motion control module controls the speed of the propeller, the start and stop and the swing angle of the rudder, so that the motion of the maneuverable buoy can be accurately controlled, and the motion of the established trajectory can be realized. In order to clearly display the navigation direction and rudder position of the mobile buoy, a friendly man-machine interface is designed to display the above information. The rudder angle data and the absolute course of the mobile buoy sent back by the lower computer through the data transmission radio will be displayed on the upper computer. At the same time, the upper computer can send commands to the lower computer to control the movement of the mobile buoy. The basic principle of conventional PID control system is shown in the Fig. 7.

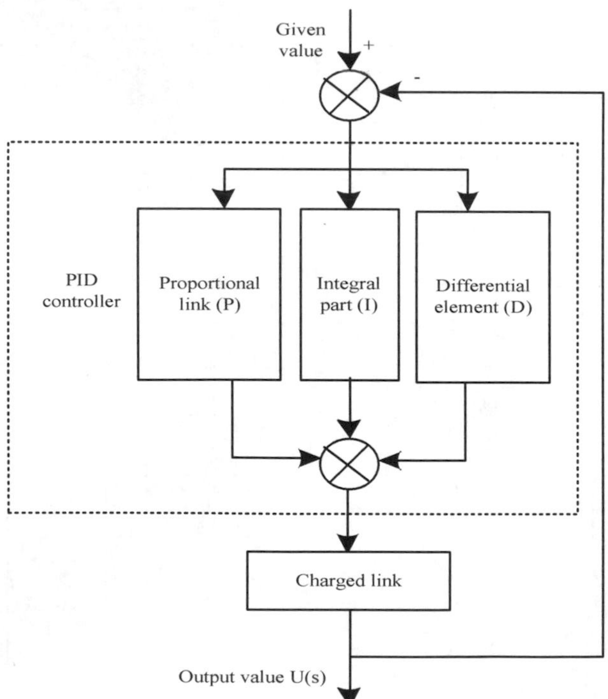

Fig. 7. Basic principle of PID control system

The function of each correction link of PID controller is briefly described as follows: the proportional link reflects the deviation signal of the control system in proportion immediately. Once the deviation is generated, the controller immediately produces a control effect to reduce the deviation. The integral part improves the error free degree of the control system by eliminating the static error; The differential part reflects the variation trend of the deviation signal, and introduces an effective correction signal before the deviation signal becomes too large, so as to speed up the reaction speed of the system and reduce the adjustment time. The selection of PID control parameters should be based on the specific characteristics of the controlled object and the performance requirements of the control system. The basic concept of fuzzy control is mainly used for macro control of the system. It is a computer intelligent control based on fuzzy set theory, fuzzy linguistic variables and fuzzy reasoning. Its basic principle is shown in the figure, and the core part is the fuzzy controller in the dotted diagram (Fig. 8).

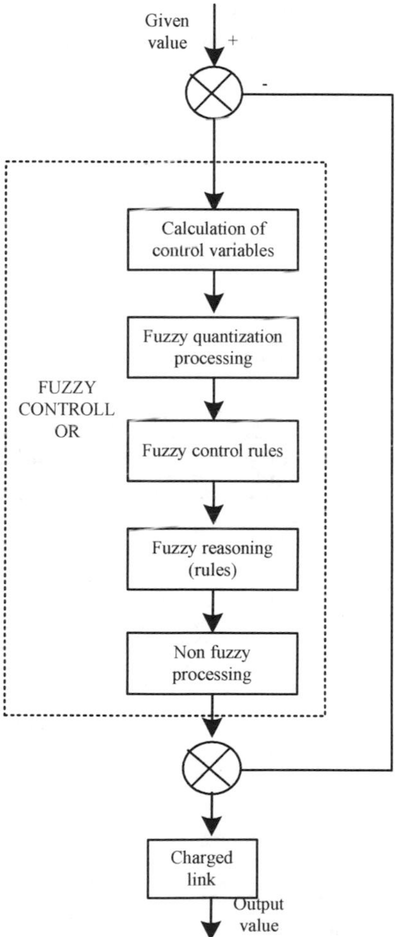

Fig. 8. Measurement and control principle of fuzzy control system

The control law of the fuzzy controller is realized by the computer. The realization process of the fuzzy control algorithm is as follows: First, the computer obtains the precise value of the controlled quantity through sampling, and then compares the precise value with the given value to obtain the deviation signal e, and takes the deviation signal e as the FC An input quantity is fuzzified to become a fuzzy quantity, which is expressed in the corresponding fuzzy language to obtain a subset E of the fuzzy language set of the deviation signal e, which is then obtained by reasoning from E and the fuzzy control rule R (fuzzy relationship) The synthetic rule makes fuzzy decision-making and obtains the fuzzy control U = ER. Finally, the fuzzy quantity U is unfuzzified and transferred to the controlled object for further control. In summary, the fuzzy control algorithm can be briefly summarized as the following four steps:

According to the output value of the system obtained this time, the input variable selected by the system is calculated.

Fuzzy the precise value of the input variable into a fuzzy value.

According to the fuzzy value of the input variable and the fuzzy control law, the fuzzy inference (rule) is performed to calculate the fuzzy value of the control variable.

Unfuzzy the fuzzy control variables obtained, calculate the precise control quantity, control the system, and loop through it.

The overall function diagram of the platform measurement and control system software is as shown (Fig. 9):

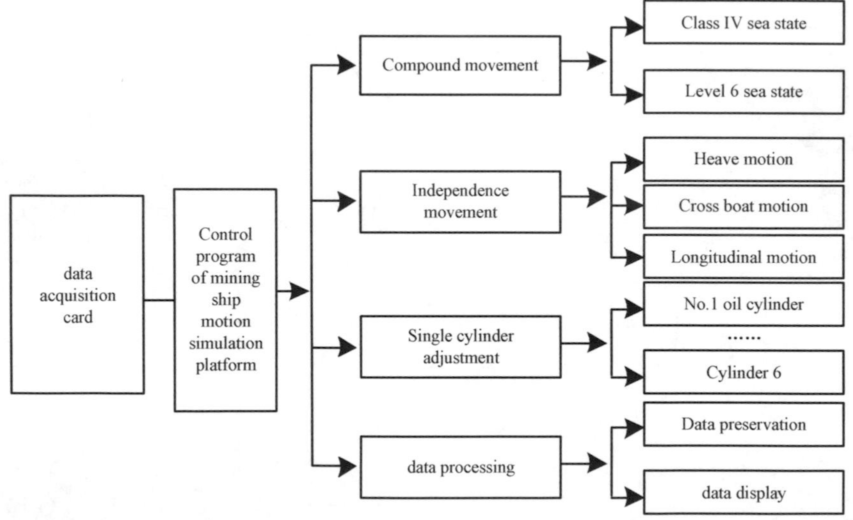

Fig. 9. The overall function of the platform measurement and control system software

In order to complete the various functional requirements of the mobile buoy, the lower-level computer mainly includes the following modules: the position of the mobile buoy is determined by GPS, and the buoy is controlled by the position of the mobile buoy, and the information is collected by the high-precision GPS module. The GPS latitude and longitude information is returned to the embedded control board, and then the information is returned to the upper computer control terminal through the digital radio, which is displayed on the map and the geographic location information is stored. The motorized buoy will be powered by solar panels, and the voltage must be sufficient and usable. The power management module will detect whether the voltage meets the

14 V working voltage and perform various actions when it meets the requirements. At the same time, different voltage values need to be configured to meet the working voltage of each module. The steering gear part and the data transmission station supply 12 V voltage, and the GPS and system voltage supply 5 V voltage, and then the system voltage is distributed to 3 V, 3.3 V, etc. Including propeller motor, steering gear and encoder. The propeller motor can control the start and stop of the marine mobile buoy, and the steering gear and encoder are used to precisely control the steering gear's angle, and then control the steering of the mobile buoy so that it can reach the specified position. The communication module must realize the long-distance.

2.3 Realization of Electromechanical Measurement and Control Management of Marine Mobile Buoy

The waves in my country's deep-sea mining areas are mainly divided into swells and waves. Waves are waves that directly act on the surface of the ocean by wind, and surges are waves that propagate to static waters when the wind and waves leave the wind area. The average annual wind speed in my country's mining areas is equivalent to Grade 4. Tropical storms and typhoons mainly occur in summer. According to foreign meteorological statistics, the average annual wind speed in the three months of July, August and September is 3.6–4.4 and 2.9, and the annual average is 15 times, each lasting 1–2 days. Therefore, tropical storms and typhoons will affect mining operations for 30–40 days a year. According to a 1994 survey by the Ministry of Geology "Ocean No. 4", the average wind wave height during exploration in China's mining areas was 0.7 m. Small waves account for 52%, light waves 40%, medium waves 8%, and big waves only 0.3%. The frequency of mid surge is 86%, the maximum surge wave height is 2.5 m, and the probability of occurrence is very small. The higher the wave height of a sea wave, the longer its period and wavelength, and vice versa. For example, the period of medium waves is 3.647 s; strong waves are 5.8–7.2 s. The sea surface water velocity in the mining area is about 0.33 m/s, the water temperature is about 250 C, the seabed water velocity is 0.18 m/s, and the water temperature is 1–20 C. Since the sea wave condition is the most concerned issue during the mining operation, the statistics of the surge condition in the C-C area are made as shown in the (Tables 1 and 2).

Table 1. Sea state characteristics survey statistics table

Sea state	Wave height (m)	Frequency of occurrence (%)
Maximum surge	2.5	Minimum
Chung Chung	0.819	86
Big wave	>2	0.3
Mid wave	1.319	8
Light waves	0.812	40
Xiaolang	0.307	52

Table 2. Wave characteristics of sea conditions of class 4 and 6 **H1/3**

Sea state	Mean wave height (m)	Average wind speed (m/s)	Meaningful wave height $H_{1/3}$ (m)	Period (s)
Level 6	4.9	18	6	8.8–10
Level 4	2	9	2.4	4.7–5.8

From the analysis of the table, we can see that the overall surge conditions in my country's mining areas are good, with small light waves and moderate waves as the main ones, and severe sea conditions are rare. Therefore, the prerequisite for the design of my country's deep-sea mining system is that the system can perform normal operations under four-level sea conditions and stop operations under six-level sea conditions. That is, the mining ship simulation system needs to simulate the movement characteristics of the mining ship under four-level and six-level sea conditions. The wave height in the table refers to the vertical height between the peak of a separate wave and the adjacent trough; the average wave height refers to the average value of all wave heights and the average value of all wave heights observed through the stationary observation point during the observation point observation period; Significant wave height refers to the average value of the maximum wave height observed through a stationary observation point. Assuming that the propagation direction of ocean waves is constant, and the crest and trough lines are parallel to each other and perpendicular to the propagation direction, such ocean waves are called binary irregular waves, or long-peak waves, and are denoted as ε_i. The modeling and simulation of binary irregular long peak waves have been widely used in ship motion control systems. The Longuet-Higgins model is used to analyze the two-dimensional irregular long-peak wave ocean wave spectrum. It is believed that it can be described by the superposition of countless cosine waves with different periods and initial phases. The expression of the wave equation at a certain fixed position is:

$$\xi(t) = \sum_{i=1}^{N} \alpha_i \cos(\omega_i t + \varepsilon_i) \tag{1}$$

The relationship between the amplitude a of each harmonic wave and the wave spectrum ω_i is as follows:

$$a_i = \sqrt{2S(\omega_i)\Delta\omega_i - \xi(t)} \tag{2}$$

Therefore, the wavefront equation of the long peak wave can be obtained as:

$$\xi(t) = \sum_{i=1}^{N} \sqrt{2S(\omega_i)\Delta\omega} \cos(\omega_i t + a_i \varepsilon_i) \tag{3}$$

It can be obtained from the derivation of the three basic parameters x of the proportional valve flow-pressure equation P, the hydraulic cylinder flow continuity index V and the hydraulic cylinder dynamic equation. The forward and reverse motion parameters β

of the asymmetric cylinder need to be selected according to the direction of motion A. Corresponding equations, in mathematical modeling, cannot reflect the characteristics of the sudden change in pressure C and asymmetry w of the two chambers of the hydraulic cylinder when the valve-controlled cylinder system commutation movement, which is not conducive to simulation calculations.

$$\begin{aligned}&\frac{C_d w_1 x_v}{2}\left[(1+\text{sign}(x_v))\cdot\text{sign}(P_s-P_1)\sqrt{\frac{2}{\rho}|P_5-P_1|}-(1-\text{sign}(x_v))\cdot\text{sign}(P_1-P_0)\sqrt{\frac{2}{\rho}|P_1-P_0|}\right]\\&-C_i(P_1-P_2)-C_eP_1=A_1\dot{x}_5+\frac{V_1}{\beta_e}\dot{P}_1\end{aligned} \tag{4}$$

Further establish the nonlinear equation of the system as:

$$\begin{cases}\dot{X}_1=X_2\\ \dot{X}_2=\frac{A_1}{M}X_3-\frac{A_2}{M}X_4-\frac{B}{M}X_2-\frac{K}{M}X_1+\frac{1}{M}F_L\\ \dot{X}_3=g_1U-\frac{\beta_\epsilon A_1}{V_1}X_2-\frac{\beta_eC_i}{V_1}(X_3-X_4)-\frac{\beta_eC_e}{V_1}X_3\\ \dot{X}_4=g_2U+\frac{\beta_eA_2}{V_2}X_2-\frac{\beta_eC_i}{V_2}(X_3-X_4)-\frac{\beta_eC_e}{V_2}X_4\\ x_s=X_1\end{cases} \tag{5}$$

In order to maintain the integrity of the physical meaning of the entire system, in the co-simulation system constructed in this paper, the second working mode of the hydraulic cylinder is adopted, that is, the speed and displacement of the piston rod relative to the hydraulic cylinder are calculated by the ADAMS dynamic model. Output to the EASYS model; in this way, the EASYS hydraulic system calculates the pressure in the upper and lower oil chambers of the hydraulic cylinder and the pressure in the accumulator according to the given speed and displacement at each step of the simulation iteration; at the same time, EASYS will calculate the pressure in the upper and lower oil chambers of the hydraulic cylinder according to the given speed and displacement. The difference and the damping force caused by the speed, calculate the hydraulic driving force acting on the piston by the hydraulic system. The ADAMS model uses the hydraulic driving force as an input signal for dynamic calculations. This completes the data exchange of the co-simulation process. The simulation results of the hydraulic model of the co-simulation can be viewed in EASYS, and the dynamic model, which is the simulation result of the platform movement, can be viewed in ADAMS. The co-simulation schematic diagram and the co-simulation process are shown in the Fig. 10.

The three software used by the system, namely EASYS, ADAMS and MatLAB, need to be installed in the root directory of the hard disk, such as C:}IVIATLAB701, and the installation directory must not contain spaces and Chinese, otherwise EASYS cannot find the necessary during the compilation process The file is compiled. When installing EASYS and ADAMS, choose CompaqVisualFortran6 as the software compiler. During the software installation process, the software compiler settings for EASYS and ADAMS have been carried out. After the installation is complete, you need to set up the MatLAB compiler. Type mex-setup in the MatLAB command window, and select Microsoft Visual C++ 6.0 as the compiler for co-simulation. In this way, the effective use of the control system is realized, and the operation effect and function of the system are improved.

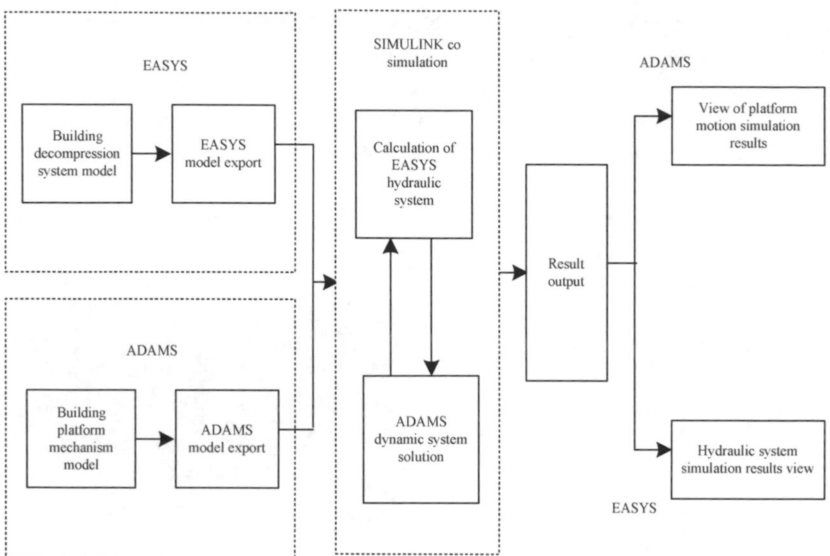

Fig. 10. Optimization of system operation steps

3 Analysis of Results

Use the STK simulation buoy to communicate with the ARGOS satellite to generate the data needed for the experiment (Figs. 11 and 12).

Through STK scene simulation, under different radiation source coordinates, simulation data such as the speed and position of the satellite and the received Doppler frequency can be obtained at different times.

An electromechanical measurement and control system for marine mobile buoys based on neuron PID is proposed. In order to verify the function of the system, a comparative test was performed compared with the traditional system. From the conclusion of the simulation analysis, it can be seen that the passive system with accumulator has a better compensation effect on high frequency waves, but the compensation effect for low frequency is poor, and the compensation rate is even negative. That is, the accumulator has a negative effect on high frequency waves. Hydraulic shock has a better absorption effect, and for low-frequency pulsation, because the process of storing and releasing energy is faster than the process of system pressure change, it will be difficult to play a good role, and sometimes even a negative effect. Refer to the explanations of some inherent parameters of the hydraulic system in the product description of the hydraulic drive system of the deep-sea mining ship motion simulation platform, and set the parameters of the mathematical model of the valve-controlled non-stack symmetrical system. The above-mentioned valve-controlled cylinder system is used as the construction For the model object, the parameter estimation table of the nonlinear mathematical model is as follows (Table 3):

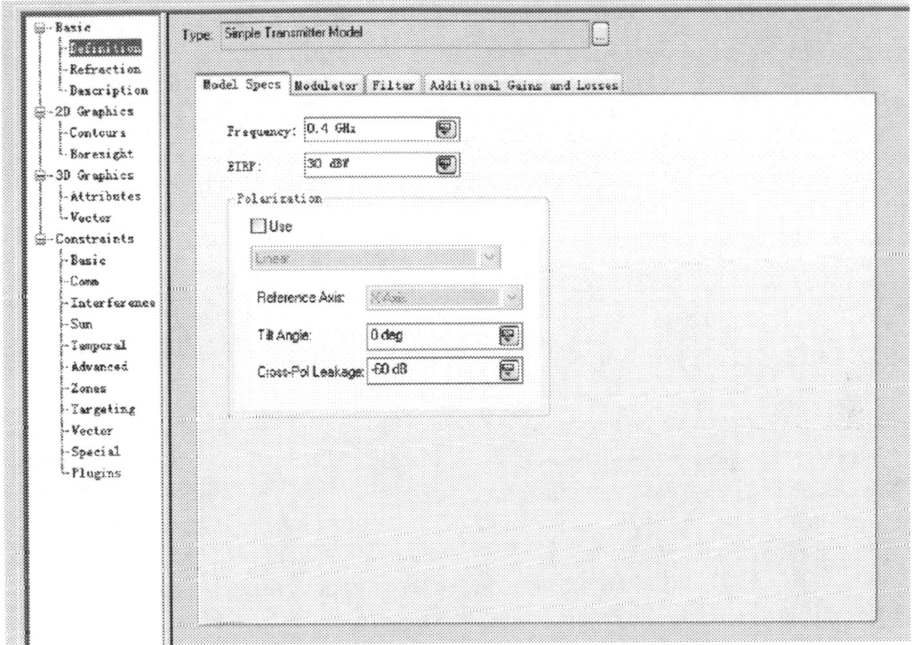

Fig. 11. STK radiation source setting diagram

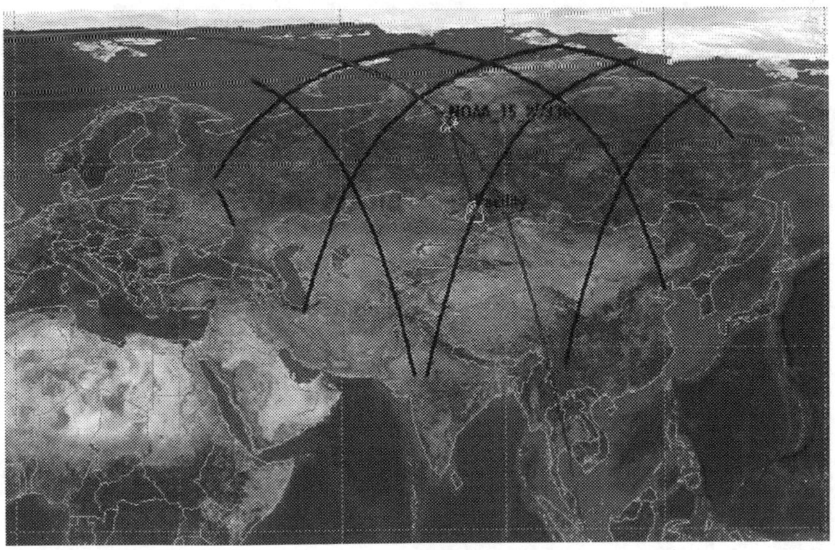

Fig. 12. Schematic diagram of STK two-dimensional simulation

Table 3. Non-linear mathematical model parameter estimation table

Parameter name	Parameter value (unit)	Parameter name	Parameter value (unit)
C_d	0.67	V10	3.216e−4 (m^3)
A_1	5.026e−3 (m^2)	V20	2.110e−1 (m^3)
A_2	2.513e−3 (m^2)	P_S	11 MPa
B	850 (N·s/m)	C_e	7.654e−13 (m/s·Pa)
β_e	750 (MPa)	C_1	2.863e−13 (m/s·Pa)
W_2	1.88e−2 (m)	M	100 (Kg)
W_1	3.76e−2 (m)	P	0.875e3 (kg/m^3)

Through Matlab/Simulink toolbox, the nonlinear mathematical model of the valve-controlled cylinder system is established, and the S-function is introduced to describe the nonlinear model. Substitute the estimated values of model parameters in the table to simulate the nonlinear model. The communication function of the marine mobile buoy mainly sends control commands through the upper computer control center to control the movement of the mobile buoy. This test reflects the communication function through the test of wireless communication performance. Mainly to test the wireless module program of the mobile buoy, the wireless drive module can send and receive data correctly and timely. The lower computer and the upper computer of the marine mobile buoy communicate by means of digital radio, and the maximum transmission distance is 20 km. During the test, the motorized buoy was placed in an open lake, and at the same time, a computer installed with the upper computer monitoring center software sent control commands to the marine mobile buoy. First turn on the power supply, turn on the main control board, initialize the wireless module, then turn on the host computer monitoring center, after the setting is completed, the wireless connection is performed, and finally the start control command is sent. At this time, the marine mobile buoy can move in the lake. At the same time, the host computer monitoring center can receive the neuron PID and other information sent back by the marine mobile buoy in real time. The marine mobile buoy can move according to the control instructions issued by the upper computer monitoring center, and the upper computer monitoring center can also receive the data fed back by the marine mobile buoy. In order to ensure that the identified model has higher reliability, the system input signal must meet the following conditions: during the test period, the input signal can continuously stimulate the system to produce a response, and can excite the modes within the working range of the system. The working condition of the simulation platform is required to reflect the motion curve within the frequency 0.5. In order to be able to build a model closer to the actual working condition, select the same control quantity input to obtain the comparison between the displacement output of the nonlinear model and the displacement output measured by the experiment as the picture shows (Figs. 13 and 14):

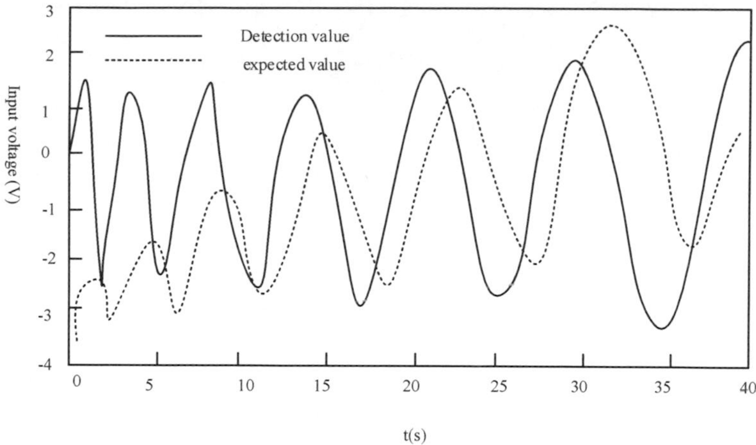

Fig. 13. Traditional method detection results

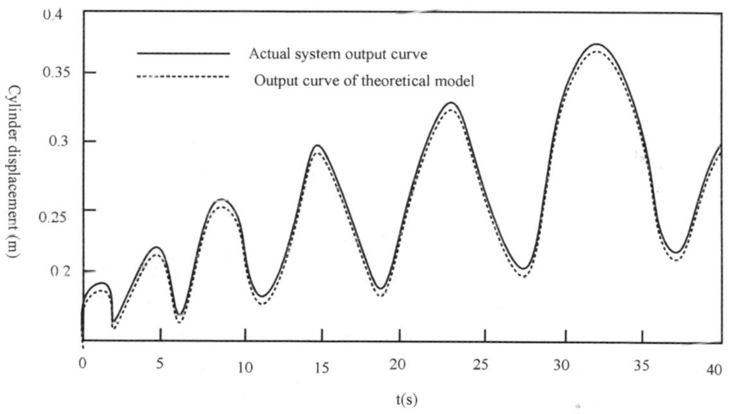

Fig. 14. Test results of the method in this paper

The upper computer control center sends the set target position to the marine mobile buoy in the form of coordinates. Then the marine mobile buoy will perform corresponding movement according to the program setting after receiving the start control command. The test result is shown in the Fig. 15.

In the actual test, the movement of the marine mobile buoy can be controlled through the upper computer monitoring center. After setting the target position of the movement in the monitoring center of the upper computer, the main control board of the lower computer controls the motor and the propeller drives the movement of the marine mobile buoy. At the same time, it controls the steering gear to adjust the direction of the marine mobile buoy.

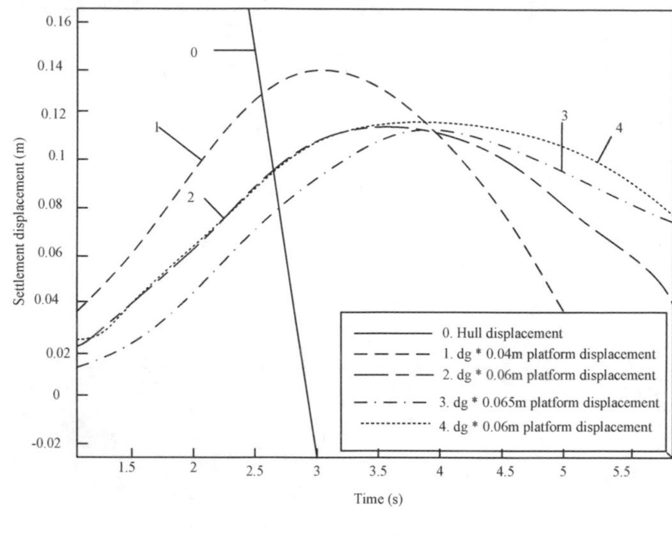

Partial enlarged drawing

Fig. 15. The influence of different pipe diameters on the heave compensation effect

It can be seen from the figure that the theoretical model output response curve of the nonlinear mathematical model established in this paper is basically consistent with the actual system output response curve trend, reflecting the effectiveness of the nonlinear mathematical model of the above-mentioned motion simulation platform, and it can be used as a valve-controlled cylinder A theoretical research platform for performance analysis in all aspects, to conduct research and analysis on control strategies and other aspects. The results of the experiment show that the motorized buoy can realize its motion control, positioning, data transmission and other functions through remote control. The movement of the motorized buoy in the lake can be accurately represented on the map, with high resolution and strong reliability.

4 Concluding Remarks

This paper develops a neuron PID-based electromechanical measurement and control system for marine mobile buoys that can monitor a wide range of marine environment, completes the mechanical structure design and measurement and control system design of marine mobile buoys, and compiles the embedded control board and the various modules. Program design. The system realizes the functions of remote control, positioning, information transmission and display of the mobile buoy, and can be equipped with various monitoring instruments for marine environment monitoring, and has strong practicability.

On the basis of this system, more comprehensive monitoring content can be added to facilitate testing of newly designed buoys, reducing experiment costs and experiment time.

References

1. Knight, P.J., Cai, O.B., Sinclair, A., et al.: A low-cost GNSS buoy platform for measuring coastal sea levels. Ocean Eng. **203**(9), 107198 (2020)
2. Palm, J., Eskilsson, C.: Mooring systems with submerged buoys: influence of buoy geometry and modelling fidelity. Appl. Ocean Res. **102**(10), 102302 (2020)
3. Ghiasi, M., Soares, C.G.: Influence of the shape of a buoy on the efficiency of its dual-motion wave energy conversion. Energy **214**(118998) (2020)
4. Ma, Y., Mao, Z.Y., Qin, J., et al.: A quick deployment method for sonar buoy detection under the overview situation of underwater cluster targets. IEEE Access **PP**(99), 1 (2019)
5. Wei, X., Liu, N., Dong, T., et al.: A preliminary assessment of an innovative air-launched wave measurement buoy. Appl. Ocean Res. **106**(1–2), 102458 (2020)
6. Numerical Study on the Optimization Design of the Conical Bottom Heaving Buoy Convertor. Ocean Engineering **173**(FEB.1), 235–243 (2019)
7. Xi, F., Pang, Y., Liu, G., et al.: Self-powered intelligent buoy system by water wave energy for sustainable and autonomous wireless sensing and data transmission. Nano Energy **61**(8), 1–9 (2019)
8. Lerch, M., De-Prada-Gil, M., Molins, C.: The influence of different wind and wave conditions on the energy yield and downtime of a Spar-buoy floating wind turbine. Renewable Energy **136**, 1–14 (2019)
9. Li, M., Wu, B.J., Jiang, C.Y., et al.: Effect of reciprocating and unidirectional airflow on primary conversion of a pentagonal Backward Bent Duct Buoy. Appl. Ocean Res. **89**, 85–95 (2019)
10. Tna, B., Rk, B., Rn, A., et al.: Numerical analysis of boundary conditions in a Lagrangian particle model for vertical mixing, transport and surfacing of buoyant particles in the water column. Ocean Model. **136**, 107–119 (2019)
11. Jullien-Corrigan, A., Ahmadi, K.: Measurement of high-frequency milling forces using piezoelectric dynamometers with dynamic compensation. Precis. Eng. **66**(9), 1–9 (2020)
12. Silva-Saravia, H.D., Pulgar, H.A., Tolbert, L.M., et al.: Enabling utility-scale solar PV plants for electromechanical oscillation damping. IEEE Trans. Sustainable Energy **PP**(99), 1 (2020)
13. Wang, R., Gao, J., Li, S., et al.: Condition-based dynamic supportability mechanism for the performance quality of large-scale electromechanical systems. IEEE Access **PP**(99), 1 (2020)
14. Yang, D., Zhang, T., Cai, G., et al.: Synchrophasor-based dominant electromechanical oscillation modes extraction using OpDMD considering measurement noise. IEEE Syst. J. **13**(3), 3185–3193 (2019)
15. Luo, Q., Guo, Z., Huang, H., et al.: Nanoelectromechanical switches by controlled switchable cracking. IEEE Electron Device Lett. **40**(99), 1209–1212 (2019)
16. Bordone, A., Ciuffardi, T., Raiteri, G., et al.: Improved current estimates from spar buoy-mounted ADCP measurement station: a case study in the Ligurian sea. J. Marine Sci. Eng. **9**(5), 466 (2021)
17. Hu, Z., Wu, T.: Research on multidimensional information collection algorithm of marine buoy wireless communication network. Microprocessors Microsyst. **80**(2), 103582 (2021)
18. Gish, L.A.: Concept design of a deployable marine energy testing system. Marine Technol. Soc. J. **54**(6), 84–90 (2020)

Software Architecture Evolution and Technology Research

Ruiqi Zeng(✉), Yiru Niu, Yue Zhao, and Haiyang Peng

Science and Technology on Communication Security Laboratory, No. 30 Research Institute of China Electronics Technology Group Corporation, China Electronics Technology Cyber Security Co., Ltd., Chengdu, China
zengruiqi@sina.com, physea@mail.ustc.edu.cn

Abstract. In order to solve the problem of the complexity of software system and the efficiency of large-scale collaboration, software architecture design methods have gradually developed. In the process of software architecture evolution, users and concurrency are increasing day by day, and software technology is constantly evolving and improving accordingly. Although there are many researches and technical discussions on software architecture methodology, there are few papers that fully discuss the development of technology in terms of the evolution of the architecture. This paper takes the evolution of software architecture as the research object, discusses problems, solutions and related technologies involved in the evolution of software architecture, and comprehensively explains how the system gradually evolves from a single architecture to a complex high-concurrency architecture.

Keywords: Architecture design · Architecture evolution · High concurrency

1 Introduction

With the rapid development of computer science and programming, software design can be applied to everything, from daily life to aerospace. The development method has also evolved from a single-person development model to a large-scale collaborative engineering model.

Various uncertainties in management and technology have increased explosively with the explosive growth of large-scale collaboration efficiency and software complexity. As a result, the quality of software development cannot be effectively guaranteed, and the cycle and cost cannot be effectively controlled.

People have been seeking to find solutions to these problems. However, in the software engineering bible "The Mythical Man-Month" published in 1975, Fred Brooks said that there is no silver bullet that can solve all problems1.

Since then, people have developed project research and development process management to control the uncertainty of management activities, and also developed software architecture design methods to control technical uncertainty. At the same time,

S. Liu and X. Ma (Eds.): ADHIP 2021, LNICST 416, pp. 708–720, 2022.
https://doi.org/10.1007/978-3-030-94551-0_54

they have continuously summarized and improved in practice, to effectively guide and to the greatest extent guarantee the quality, cycle and cost of software development.

In terms of software architecture design, software architecture is often evolving. With the increasing number of users and concurrency, the actual software system will also experience a process from simple in the early stage to complex in the later stage. In the process of software architecture expansion, various problems will arise, and corresponding technologies to solve these problems will emerge accordingly. The emergence of these technologies ensures the availability, stability, scalability and concurrency of the software during the evolution of the system2.

This paper explains problems that may be encountered in software evolution, such as how to provide high concurrent access when resources are limited, how to quickly query database tables, how to reduce user access latency, how to improve analysis efficiency for unstructured data, how to carry out rapid expansion of the system, how to carry out rapid upgrade and deployment of the system, how to carry out all-round monitoring of the system and so on. For these problems, discussed how to solve them, the related technologies used, and explained the effects of these technologies in solving practical problems.

The rest of this paper is organized as follows. Section 2 introduces the evolution of architecture and technologies involved. Section 3 gives a general overview of a complex architecture. Section 4 is a summary.

2 Architecture Evolution Technology

When the initial user and concurrency were very small, the simplest software architecture was often used. With the increase of users and concurrency, system performance encountered bottlenecks. At this time, various technologies need to be used to expand and enhance various parts of the architecture. Then the system became more and more complex. This is the software architecture evolution.

2.1 Stand Alone Deployment

In the early stage of system construction, there were few concurrent visits and there were no requirements for performance. At this time, the application system and the database are often placed on the same server. The entire deployment structure is a stand-alone deployment, as shown in (Fig. 1).

The problem with stand-alone deployment is that the application system and the database of the same server will compete for the resources. When the number of visits increases, the server will become a performance bottleneck, so architecture evolution is required.

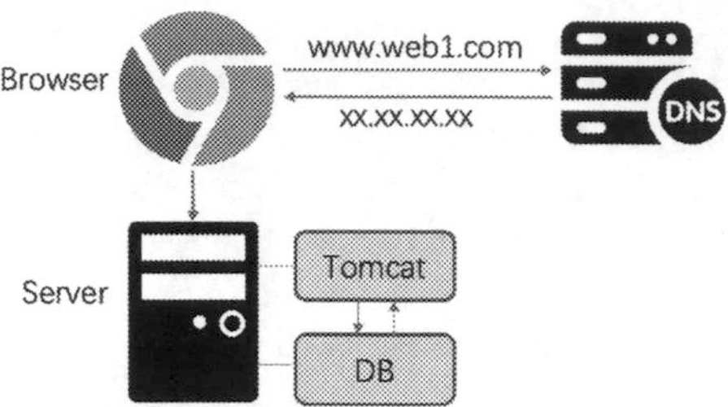

Fig. 1. Stand alone deployment

2.2 Separation of Application and Database

In order to cope with the lack of performance of the single-machine deployment server, the application system and the database are usually placed on two different servers, which can significantly improve their performance. The separated architecture is shown in (Fig. 2).

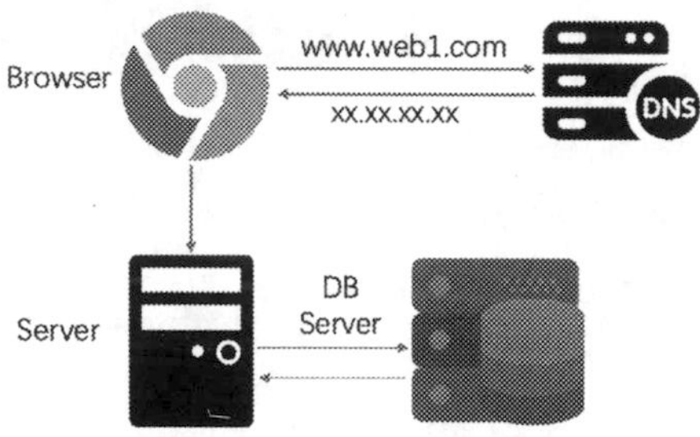

Fig. 2. Separation of application and database

When concurrency continues to grow, the read and write capabilities of the database will become a bottleneck, requiring continued architectural adjustments.

2.3 Introducing Local Cache and Distributed Cache

In order to solve database performance problems, caching technology is introduced, which includes adding a local cache on the application server side and adding a distributed cache externally3. The cache is mainly used to store popular html pages or frequently used data. Through caching technology, most requests can be intercepted before they reach the database, which greatly improves efficiency. The architecture at this time is shown in (Fig. 3).

Although cache can handle most of the requests, with the growth of user access, the stand-alone server, where the application is deployed, will under increasing pressure of concurrency, resulting in slower response, so the architecture will continue to evolve.

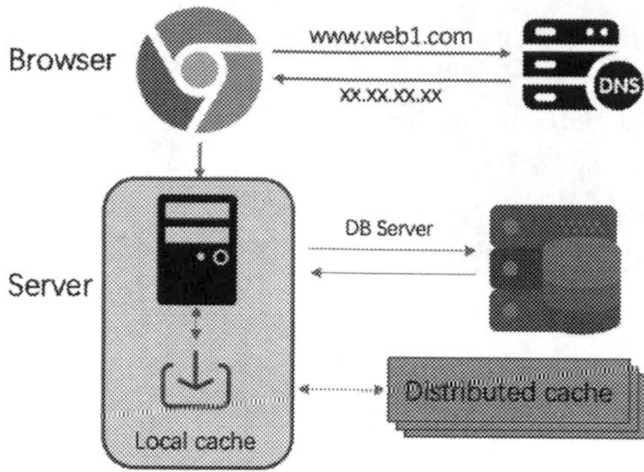

Fig. 3. Introducing local cache and distributed cache

2.4 Introducing Reverse Proxy to Realize Load Balancing

When the application pressure is too great, multiple instances of the application can be deployed, and reverse proxy software can be used for load balancing to reduce the pressure on each server and improve the comprehensive concurrency capabilities4. Generally, you can use Nginx as a load balancer. The evolved architecture is shown in (Fig. 4).

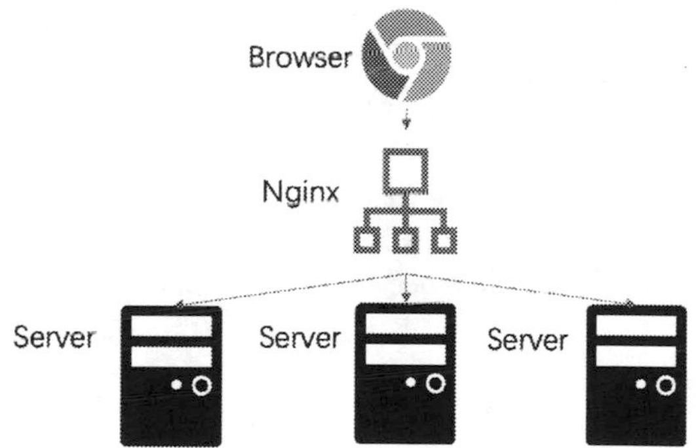

Fig. 4. Introducing reverse proxy to realize load balancing

Although the load balancer greatly increases the amount of concurrency, more concurrency means more database access, and the database deployed on a single machine becomes the bottleneck of system performance.

2.5 Database Read/Write Separation

In order to reduce the pressure on the database, the database is divided and multiple databases are used for business processing. It is mainly divided into reading DB and writing DB. The reading DB is used when reading data, and the writing DB is used when writing data5. There can be multiple reading DB, and synchronization mechanism is used to synchronize data from writing DB to reading DB. The latest written data that needs to be queried can be written to the cache, and obtained through the cache. The architecture is shown in (Fig. 5).

There are some specialized middleware technologies that can support the separation of database read and write, and the client can access the lower-level database through the middleware. After reading and writing are separated, problems such as data synchronization need to be solved.

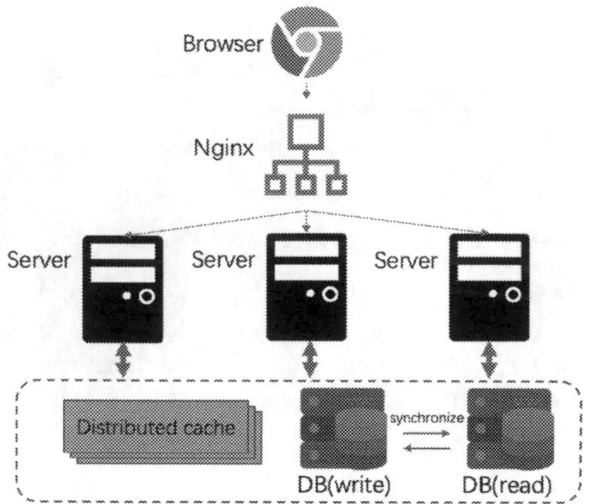

Fig. 5. Database read write separation

When the business continues to increase gradually, there will be a big gap in the number of visits between different businesses. Different businesses directly compete for the database, which will affect each other's performance, so the architecture needs to be upgraded again.

2.6 Database Divided by Business

In order to solve the problem of competition between different businesses on database, database is divided by business, so that data of different businesses are stored in different databases. Resource usage is isolated according to business, and competition for resources is reduced. For businesses with high traffic, more servers can be deployed to support them. The architecture is shown in 0. This architecture needs to solve problem of correlation analysis across DB (Fig. 6).

Under this architecture, as number of users grows, the stand-alone writing DB will gradually reach performance bottleneck.

2.7 Split a Large Table into Small Tables

As the amount of data in written database becomes larger and larger, the query efficiency will become lower and lower. In order to improve the efficiency of a single table, it is

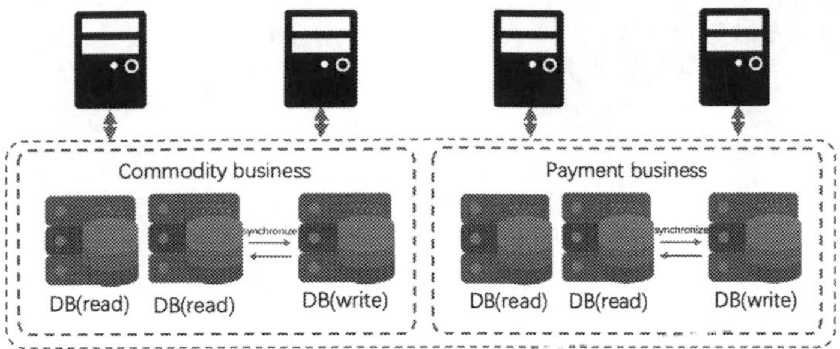

Fig. 6. Database is divided by business

necessary to split the tables. Sub-table is to disperse data of one table into multiple tables to increase the efficiency of data query and operation efficiency.

As long as the amount of table data for real-time operation is small enough and request can be distributed to these small tables on multiple database servers as evenly as possible, the database performance can be improved through horizontal expansion. At this time, the system architecture is shown in (Fig. 7).

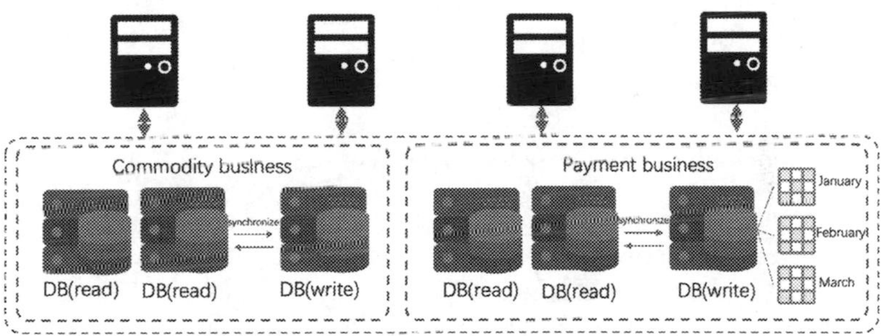

Fig. 7. Split a large table into small tables

Under this architecture, both the database and application system can be scaled horizontally, and the supportable concurrency is greatly improved. However, as the number of users continues to grow, a single-machine load balancer will eventually become a performance bottleneck.

2.8 Use LVS or F5

Since the bottleneck is in Nginx, it is impossible to achieve performance improvement by adding multiple layers of Nginx, so a more powerful load balancing technology is required6.

LVS and F5 are load balancing solutions that work at the fourth layer of network. LVS is software, running in kernel state of operating system, and can forward TCP requests or higher-level network protocol requests, so it supports more protocols, and its performance is also much higher than Nginx. F5 is a load balancing hardware, similar to the capabilities provided by LVS, with higher performance than LVS, but it is expensive. The enhanced system architecture is shown in (Fig. 8).

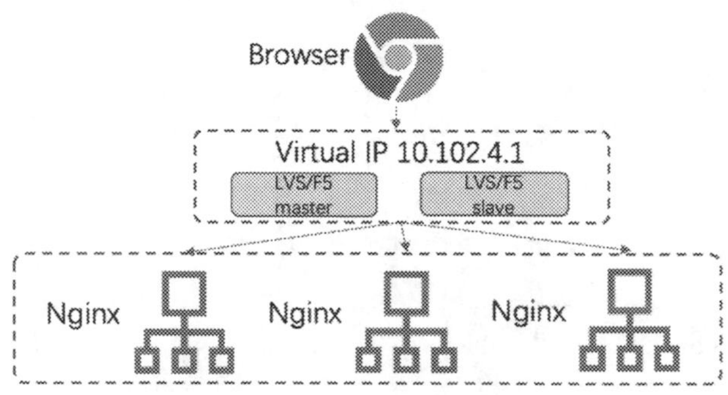

Fig. 8. Use LVS or F5

The number of users continues to grow, reaching tens of millions or even hundreds of millions. Because users are distributed in different regions and their distance from server room is different, the delay for users to access server will also be significantly different, which causing the response to some users is too slow, so the architecture still needs to evolve.

2.9 Load Balancing of Data Centers by DNS Polling

By distributing business requests to different data centers or computer rooms, you can continue to increase concurrent visits.

In DNS server, one domain name can be configured to multiple IP addresses, and each IP address corresponds to a virtual IP in different computer rooms. This method can realize load balance between computer rooms, the system can achieve horizontal expansion at computer room level, and concurrency of tens of millions to hundreds of millions can be solved by adding computer room7. The architecture is shown in (Fig. 9).

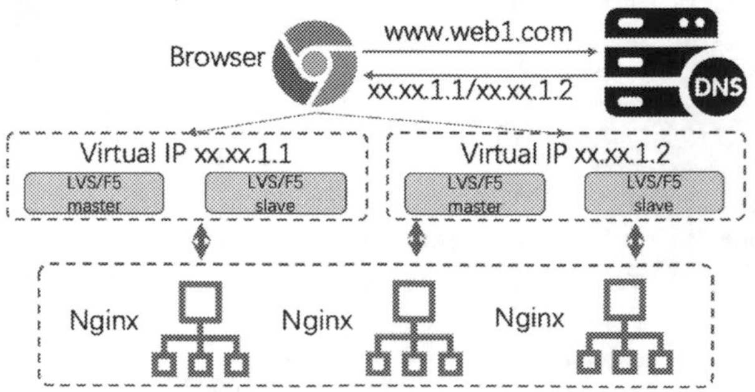

Fig. 9. Load balancing of data centers by DNS polling

With enrichment of data and development of business, the requirements for retrieval and analysis become more and more abundant, new needs require new technologies.

2.10 Introduce NoSQL Database and Search Engine Technology

For massive file storage, it can be solved by distributed file system HDFS. For key value data, it can be solved by HBase and Redis8. For full-text retrieval scenarios, it can be solved by search engines such as ElasticSearch. For multi-dimensional analysis scenarios, it can be solved by Kylin or Druid9. The enhanced architecture is shown in (Fig. 10).

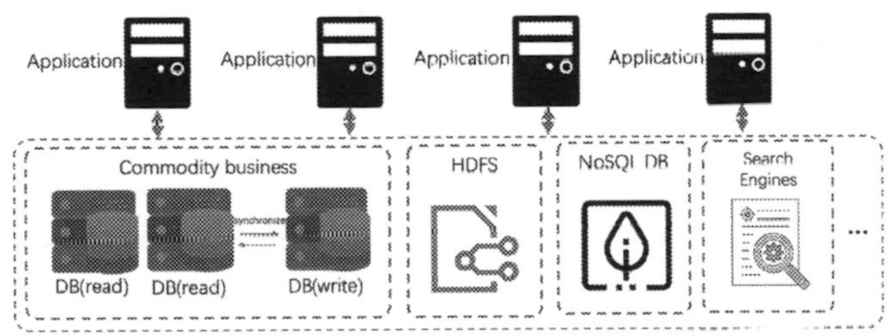

Fig. 10. Introduce NoSQL database and search engine technology

Abundant technologies have solved a wealth of needs, and the business dimensions can be greatly expanded. As a result, one application contains too much business code, so its business upgrade and iteration become difficult.

2.11 Big Application Split into Small Application

In order to solve the difficult upgrade and iteration problem caused by a large application, application code is divided according to business sectors, so that the responsibilities of a single application are clearer, the maintenance cost is lower, and each module can be upgraded and iterated independently10. The architecture after application split is shown in (Fig. 11).

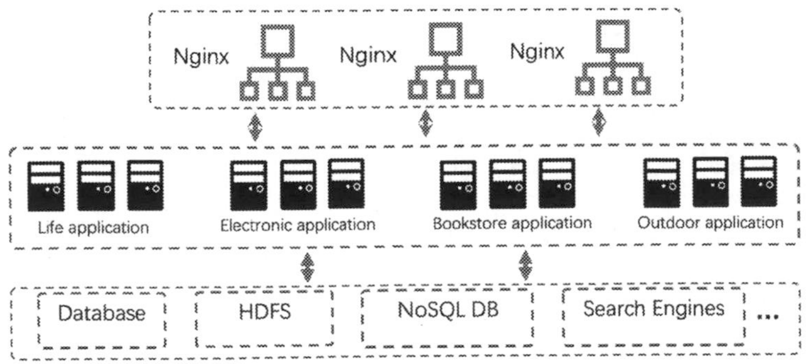

Fig. 11. Big application split into small application

After application is split, there are shared modules between different applications, and separate management of each application will cause the same code to be redundant, which lead to that when a public function need upgraded, all application code need upgraded. To improve reusability, the common code needs to be stripped out.

2.12 Separating Public Functions into Micro Services

In order to solve the problem of reuse and efficiency of public functions, these services need to be deployed separately. For functions that are used by multiple applications such as user management, orders, payment, and authentication, their codes are extracted to form a single service, which is made into a microservice model1. Applications and services access public services through multiple methods such as HTTP, TCP, or RPC requests, and each individual service can be managed by a separate team (Fig. 12).

Fig. 12. Separating public functions into micro services

Although microservices improve service stability and availability, when applications and services access each other, their call chains will become very complicated and logic will become chaotic, so new technologies are needed to solve this problem.

2.13 Introduce ESB to Shield Access Difference of Service Interface

In order to improve efficiency of call chain between services, ESB is used to perform unified access protocol conversion, applications and services access and call each other through ESB, so as to reduce degree of system coupling12. The ESB technical architecture is shown in (Fig. 13).

When the number of applications and services continues to increase, the deployment of applications and services becomes complicated, and the number of services deployed on the same server increases sharply, and service operating environment is prone to conflicts. In addition, for scenarios that require dynamic expansion and contraction, the performance of service needs to be scaled horizontally, which makes operation and maintenance very difficult.

Fig. 13. Introduce ESB to shield access difference of service interface

2.14 Introduction of Containerization Technology

In order to solve operation, maintenance and deployment problems of large-scale microservice systems, containerization technology is needed.

Containerization technology can make application deployment fast and efficient, and the deployment environment can be reused13. At the same time, container orchestration technology provides a set of mechanisms for deploying, maintaining, expanding and repairing containerized microservices, making container management very efficient. At present, the most popular containerization technology is Docker, and the most popular container management tool is Kubernetes14. Applications/services can be packaged as Docker images, and images can be dynamically distributed and deployed through Kubernetes15. The containerized deployment architecture is shown in (Fig. 14).

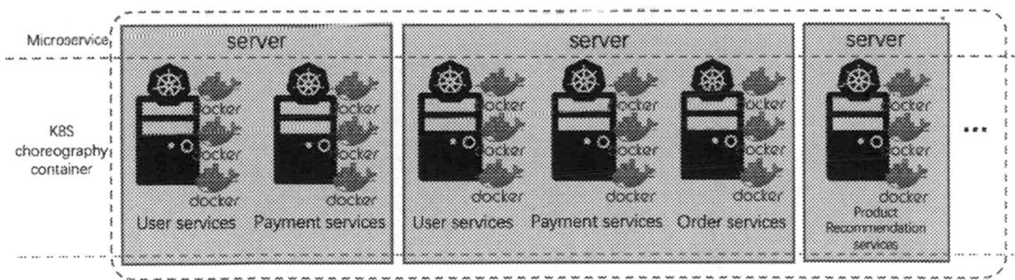

Fig. 14. Introduction of containerization Technology

Containerized deployment solves the problem of service dynamic expansion and contraction, and greatly improves operation and maintenance efficiency. However, hardware resources need to be managed, the costs of machine itself, operation and maintenance are extremely high, and resource utilization is low.

2.15 Using Cloud Platform System

In order to solve the problem of high cost of physical resources and dynamic expansion of hardware resource, the system can be deployed on a public cloud to use massive public machine resources.

On cloud platform, hardware resources (such as CPU, memory, network, etc.) can be dynamically applied for on demand, a general operating system, common technical components (such as Hadoop technology stack, MPP database, etc.) and even well-developed applications are provided for users.16. The cloud platform architecture is shown in (Fig. 15).

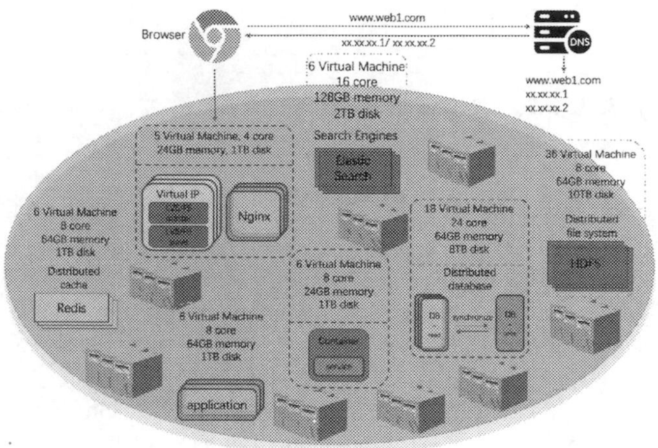

Fig. 15. Using cloud platform system

Since then, the entire technical architecture has solutions from the problem of high concurrent access to service architecture and system implementation.

3 Global View

0is a global view of architecture evolution. In actual application, the architecture design of a system with clear performance indicators needs to be able to support the performance of system, while leaving interfaces for architecture extending.

In our system design, we adopted front-end and back-end separation technology, local cache technology, business database separation, database read-write separation technology, load balancing technology, NoSQL database, microservice-related technology, and containerization technology. Through the use of these technologies, the high concurrency requirements of our system have been met, and our system can provide 10,000 concurrent access for outside system (Fig. 16).

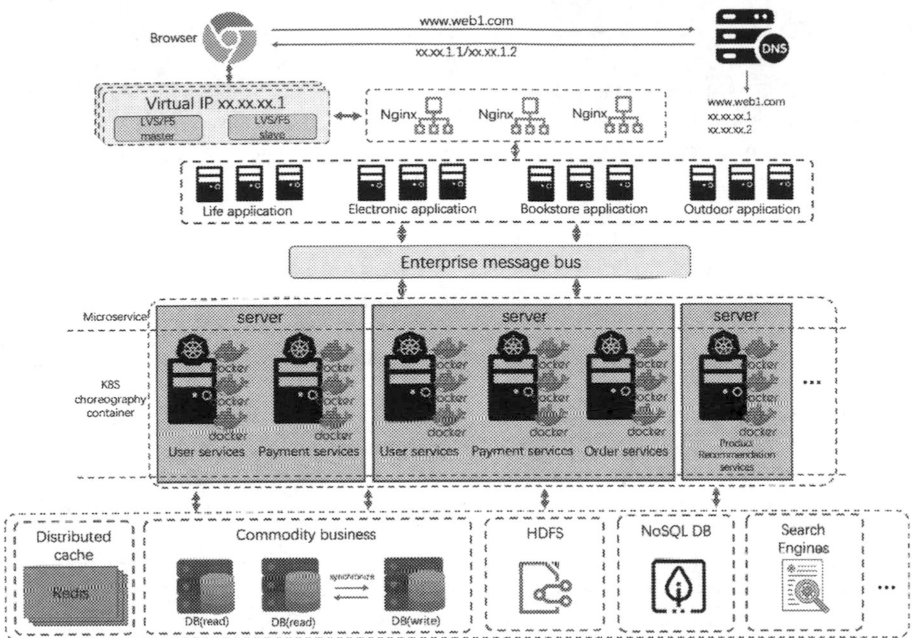

Fig. 16. Global view

4 Conclusion and Outlook

This paper discusses the technologies involved in the evolution of the architecture. Firstly, we explain why the software architecture methodology emerged, then we illustrate problems, methods, and technologies involved in software architecture evolution by describing the relatively complete process of architectural evolution. Finally, a comprehensive display and summary of software architecture evolution are presented. This paper covers all aspects of architecture evolution, discusses the usage scenarios and effects of various technologies, and provides a comprehensive reference for high-concurrency software design.

In future work, we will continue to pay attention to the development of related technologies. For example, the evolution of microservice technology to service mesh, the containerized deployment of heterogeneous systems, etc., and continue to discuss the application scenarios of related technologies and their promotion of architecture evolution.

Acknowledgment. Foundation Item: Supported by Sichuan Science and Technology Program (No. 2020YFG0292).

References

1. Brooks, F.: The Mythical Man-Month, 20th Anniversary Edition (1995)
2. Abbas, G., Imran, M., Hafeez, Y., et al.: Improving software architecture design decision by selecting set of solutions. In: 2020 3rd International Conference on Computing, Mathematics and Engineering Technologies (iCoMET) (2020)
3. Jing, Z., Wu, G., Hu, X., et al.: A distributed cache for hadoop distributed file system in real-time cloud services. IEEE (2012)
4. Moharir, M., Shobha, G., Oppiliappan, A., et al.: A study and comparison of various types of load balancers. In: 2020 5th IEEE International Conference on Recent Advances and Innovations in Engineering (ICRAIE). IEEE (2020)
5. Xue-jun, LI, Qian-jun, et al.: Application and Research of Read-write Separation Technology in Active-active Database
6. Randhawa, N.S., Dhami, M., Singh, P.: Enhanced load balancer with multi-layer processing architecture for heavy load over cloud network. Department of Electronics and Communication Engineering, Swami Vivekanand Institute of Engineering & Technology, Banur, India; Department of Information Technology, Chandigarh Engineering College, Landran, Mohali, India; Department of Information Technolog
7. Hong, Y.S., No, J.H., Kim, S.Y.: DNS-based load balancing in distributed web-server systems. In: IEEE Workshop on Software Technologies for Future Embedded & Ubiquitous Systems, & the Second International Workshop on Collaborative Computing, Integration. IEEE (2006)
8. Cattell, R.: Scalable SQL and NoSQL data stores. ACM SIGMOD Rec. **39**(4), 12–27 (2010)
9. Daniel, B.K.: Big Data and data science: a critical review of issues for educational research. British J. Educ. Technol. **50**(1) (2019)
10. Bernstein, D.: Containers and cloud: from LXC to docker to kubernetes. IEEE Cloud Comput **1**(3), 81–84 (2014)
11. Nadareishvili, I., Mitra, R., Mclarty, M., et al.: Microservice Architecture (2016)
12. Hua, J.X., Xia, Q.U.: Application Research on the Enterprise Service Bus Technology in Application System Integration of Universities. Journal of Xi'an University (Natural Science Edition) (2019)
13. Develop with Docker [EB/OL] [2020 - 09 - 10]. https://docs.docker.com/develop/
14. Orzechowski, M., Balis, B., Pawlik, K., Pawlik, M., Malawski, M.: Transparent deployment of scientific workflows across clouds - kubernetes approach. In: 2018 IEEE/ACM International Conference on Utility and Cloud Computing Companion (UCC Companion), Zurich, 2018, pp. 9–10 (2018). https://doi.org/10.1109/UCC-Companion.2018.00020
15. Ruiqi, Z.Y., Zhao, H., et al.: A novel construction technology of enterprise business deployment architecture based on containerized microservices (2020)
16. Sugam, S., Chang, V., Sunday, T.U., et al.: Cloud and IoT-based emerging services systems. Cluster Comput. **22**, 1–21 (2019)

Design of Fuzzy PID Control System of Indoor Intelligent Temperature Based on PLC Technology

Fa-yue Zheng(✉)

Beijing Polytechnic, Mechanical and Electronic Engineering School, Beijing 100176, China
fayue766@sohu.com

Abstract. In order to meet the temperature requirements of residents and improve the accuracy of indoor temperature control, PLC technology is used to design an indoor intelligent temperature fuzzy PID control system. The hardware includes a sensor selection unit, a PLC hardware selection unit, a fuzzy PID controller unit and a power circuit design unit, and the software includes a PLC software module, a fuzzy PID control program module and a serial communication program module.Through the design of hardware unit and software module, the indoor intelligent temperature fuzzy PID control system is realized. The experimental data show that: compared with the existing system, the temperature control accuracy of the design system is higher, which fully proves the effectiveness and feasibility of the design system, and provides a more comfortable living environment for residents.

Keywords: PLC technology · Indoor · Temperature · Intelligence · Fuzziness · PID control

1 Introduction

With the rapid development of China's economy and the remarkable improvement of people's quality of life, the number and scale of Chinese buildings have achieved rapid development. Especially in recent decades, intelligent buildings have developed rapidly, and people have higher requirements for indoor comfort, especially temperature. In order to meet the needs of residents, a large number of new buildings adopt central air conditioning to provide a comfortable environment or meet the technological requirements in the room. But the high energy consumption of central air conditioning is also a problem which can not be ignored today, the situation of high energy consumption of central air conditioning and how to improve the utilization rate of energy is also an important topic of domestic and foreign scholars, which is directly related to the sustainable development of the country [1].

My country is a large resource-consuming country, and the energy consumption of output value is very high compared to developed countries. This requires us to save energy and reduce energy consumption in the use of energy. This is my country's

S. Liu and X. Ma (Eds.): ADHIP 2021, LNICST 416, pp. 721–732, 2022.
https://doi.org/10.1007/978-3-030-94551-0_55

basic national policy. The energy consumption of buildings in my country accounts for 30% of the total energy consumption. In buildings that use central air-conditioning, the energy consumption of central air-conditioning accounts for about 50% of the total building energy consumption, and this data is continuously increasing every year, so we can See the importance of energy saving in central air conditioning. According to the data of Guangzhou and Shenzhen, the power consumption of central air conditioning has accounted for 24% and 29% of the total power consumption of the city, which undoubtedly brings huge pressure to the city's power grid.

At present, in the domestic central air-conditioning chillers, the capacity of water pumps and fans is too large. The actual indoor load refers to the amount of cooling or heat required by the terminal equipment of the central air conditioner due to changes in indoor temperature. The indoor load will change with the seasons, environmental climate, equipment operation, and the size of the flow of people, this is a variable amount [2]. However, in order to meet the requirements of indoor load, central air conditioners generally operate under partial load conditions 90% of the time. In fact, the cooling or heat required by the equipment at the end of the air conditioner is often less than the set flow rate. It can be seen that if we adopt a constant flow operation mode, it will bring huge energy consumption problems.From the current situation, the energy saving research of central air conditioning, especially the water system, is the key task of energy saving of central air conditioning. In view of the defects of large energy consumption and long working time in partial load state, the control of water system is unreasonable. The lack of control and regulation device in water system will make the water system run at constant flow state. When the load changes and deviates from the design value, the cooling capacity of the system cannot change with the load, which will cause huge waste; The response capacity of the system is poor, and the water delivery speed is behind the load change speed, which will cause the parameter instability of the control system and affect the control effect.

In order to meet the temperature needs of today's residents, reduce energy consumption, and reduce the pressure on the power grid, how to control indoor temperature has become one of the national key research topics, so the indoor intelligent temperature fuzzy PID control system based on PLC technology is designed. Design system hardware through sensor selection unit, PLC hardware selection unit, fuzzy PID controller unit and power supply circuit design unit, design system software through PLC software module, fuzzy PID control program module and serial communication program module, according to system hardware and software The design realizes the design of the indoor intelligent temperature fuzzy PID control system, and the validity of the designed system in this paper is verified through simulation experiments.

2 Hardware Design of Indoor Intelligent Temperature Fuzzy PID Control System

Hardware is the basis and premise of the stable operation of the design system. According to the requirements of indoor temperature control, the hardware of the system is divided into sensor selection unit, PLC hardware selection unit, fuzzy PID controller unit and power circuit design unit.

2.1 Sensor Selection Unit

The design system mainly selects four kinds of sensors: temperature sensor, humidity sensor, carbon dioxide concentration sensor and light intensity sensor, and controls the temperature through the sensor data [3]. We installed six temperature and humidity sensors, one light intensity sensor and one carbon dioxide concentration sensor in the room. Outdoor installation of a temperature and humidity sensor, as well as a light intensity sensor, the transmission mode of these sensors are analog transmission [4].

The sensor performance index is shown in Table 1.

Table 1. Sensor performance index table

Name	Model	Range
Indoor temperature sensor	LT/M	0–50 °C
Outdoor temperature sensor	LT/M	−40–60 °C
Humidity Sensor	LT/S	0–100% RH
Light intensity sensor	LT/G	0–200000 Lx
Carbon dioxide sensor	LT/CO_2	0–3000 ppm
Name	Output	Precision
Indoor temperature sensor	Rs485	±1 °C
Outdoor temperature sensor	Rs485	±1 °C
Humidity Sensor	Rs485	±7% RH
Light intensity sensor	Rs485	±5000 Lx
Carbon dioxide sensor	4–20 mA	±(20 + reading%) ppm

2.2 PLC Hardware Selection Unit

PLC is a high performance programmable logic controller. PLC takes microprocessor as the core, and integrates computer technology, automatic control technology and network communication technology [5]. It is a widely used industrial automation device with strong control ability, high reliability and flexible configuration.

Siemens S7-300 series PLC is a medium sized programmable logic controller, which is widely used in industrial automation and power industries. Modularization, easy implementation of distributed configuration, easy entry, fast processing speed, and strong communication capabilities make Siemens S7-300 series PLC the main control PLC for many complex control systems. Siemens S7-300 series PLC adopts modular structure design, and each individual module can be combined and expanded, mainly composed of rack, power module, CPU module, communication module, interface module, signal module and function module. The PLC structure module is shown in Fig. 1.

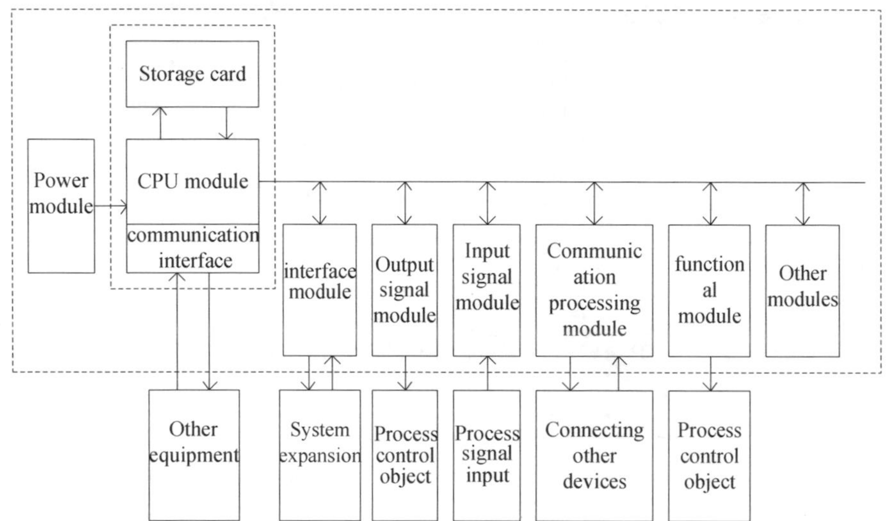

Fig. 1. Schematic diagram of PLC structure module

The purpose of each module of S7-300 series PLC is as follows:

Rack (RACK): a guide rail used to install and fix each PLC module.

Power supply module (PS): used to convert 120/240 V AC to 24 V or SV DC for use by PLC modules.

CPU module: used to execute user program and support plug-in memory card to expand memory. With RS485 and MPI communication interface, some models of CPU with PROFIBUS-DP communication interface.

Interface module (IM): used to extend the PLC rack, connecting the main rack and the extended rack, the distance is generally no more than 10 m.

Signal module (SM): used for input and output data. According to different data types, there are digital input/output modules and analog input and output modules.

Communication module (CP): used for point-to-point network connection between PLC and other devices with communication functions (PLC, computer, etc.).

Function module (FM): used to control positioning and high-speed counting operation in open-loop or closed-loop system.

CPU is a key component in PLC hardware, and its technical parameters are shown in Table 2.

Table 2. Technical parameters

Project	Parameter
Communication Interface	X1:MPI X2:DP/PN
Working memory capacity	256 kB

(*continued*)

Table 2. (*continued*)

Project	Parameter
Carrying memory capacity	MMC max 8 MB
DB/FC/FB/OBSHUL Quantity/Capacity	1024/256 kB
Nonvolatile memory capacity	128 kB

2.3 Fuzzy PID Controller Unit

The design of fuzzy PID controller should first determine the basic structure of fuzzy PID controller according to the controlled object. According to the controlled object and the number of input and output variables, the fuzzy PID controller is divided into single input single output and multiple input multiple output from the structure [6].

Fuzzy PID controllers are divided into one-dimensional controllers, two-dimensional controllers and three-dimensional controllers according to the structure. The schematic diagrams are shown in Fig. 2, which correspond to Figures (a), (b), and (c) in Fig. 2.

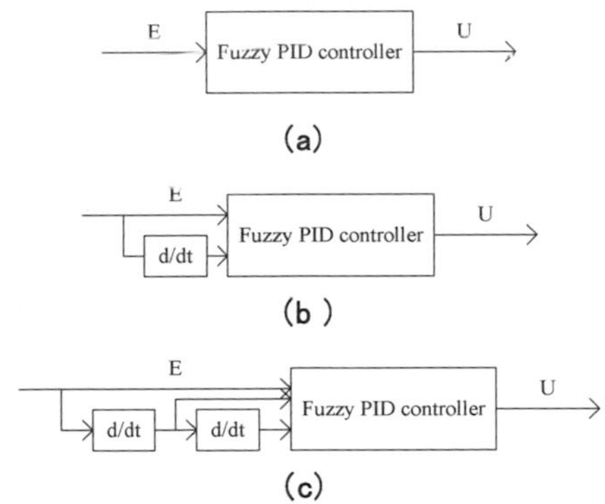

Fig. 2. Structure diagram of fuzzy PID controller

Theoretically speaking, the control effect of a fuzzy PID controller is related to its dimensionality, but as the dimensionality increases, the control law corresponding to the fuzzy PID controller's control object becomes more complicated, and the control algorithm has a higher level of achievement. The difficulty. However, the dimensionality is too low, and the deviation collection input points of the fuzzy PID controller are few, which makes it difficult to reflect the dynamic characteristics of the control system. Generally, a two-dimensional fuzzy PID controller is used, which can better reflect the dynamic quality of the control process, and is relatively not very complicated in the realization of the control algorithm. Therefore, the design system selects a two-dimensional fuzzy PID controller.

2.4 Power Circuit Design Unit

The design system adopts three-phase three wire power supply mode, M1 is the main circuit of water pump motor, M2 is fan motor, and the fan motor has 27 fan motors in total. The I/O distribution of the system slave is shown in Table 3.

Table 3. Input and output I/O allocation table

Output		Enter	
Q0.0	No. 1 fan unit 1	A1W0	PH 1
Q0.1	No. 1 fan unit 2	A1W2	No. 1 EC
Q0.2	No. 2 fan unit 1	A1W4	Pressure
Q0.3	No. 2 fan unit 2	A1W8	PH 2
Q0.4	No. 3 indoor fan	A1W10	EC 2
Q0.5	No. 4 fan unit 1		
Q0.6	No. 4 fan unit 2		
Q0.7	No. 1 indoor window		
Q0.8	No. 2 indoor window		
Q0.9	No. 3 indoor window		

The manual control circuit and the automatic control circuit share the relay and contactor. The toggle switch is used to directly control the relay KA, and the contactor is controlled by KA. The actions of the actuator are all controlled manually.

The hardware of the above design can not realize indoor temperature control, so based on hardware, the software module of the design system is designed.

3 Software Design of Indoor Intelligent Temperature Fuzzy PID Control System

The design system software includes PLC software module, fuzzy PID control program module and serial communication program module. The specific design process is as follows:

3.1 PLC Software Module

STEP7-Micro/WIN programming software is an extremely powerful software specially developed by Siemens for its S7-200 series PLC. It is an indispensable R&D tool for 200PLC users. It has the characteristics of simple, easy to learn, practical and efficient, it can solve the complicated automation control requirements of the process, save the user's programming time, and make the operation of domestic users more convenient after the Chinese version is launched. STEP7-Micro/WIN is based on the WINDOWS platform, but its applicability to the system is relatively strong.It can write, compile, debug and configure the program in offline mode. In this process, the program and parameters are saved in the computer. After connecting with PLC, the communication of program and parameter data written by users is completed by uploading and downloading, and various operation functions of PLC are realized. In the process of programming, the software itself has a simple and effective syntax error detection function. With the help of this function, programmers can find syntax and data errors in advance. The software can also manage and encrypt the user program, which protects the programmer's intellectual property rights [7].

The software uses the WINDOWS program interface, which makes it easy for programmers to master. The programming window includes a browse bar, instruction tree, cross reference window, data block/data window, main menu bar, and tool bar. Switching between various windows is very convenient, enabling users to easily find the required data or symbols and their usage, which provides convenience for programming and debugging. Compared with the same type of Small PLC software, the main program, subroutine and interrupt program can be written in different windows, and the program written by the software can be edited into a library file. When the programmer writes the same type of program, he can directly call his own library file and edit it as his own recognizable name, so as to avoid the need of repeated work, At the same time, the design of sub window enables users to accurately find the subprogram where the problem is and make changes during on-site debugging, which is a function that many programming software do not have [8].

The design of this article uses the CPU226CN AC/DC/relay type with the highest CPU model among 200PLCs as the controller. Its performance advantages mainly include: first, more input and output points, strong expansion capability, 24 input points, 16 output points, and expansion The capacity is 7 I/O modules; second, the input voltage of the input power is 85–264 VAC, the digital input is 24 V DC, and the rated output voltage of the digital is 24 V DC or 250 V AC; third, it has two RS-485 Communication interface, PPI communication baud rate supports up to 187.5 kbaud, and free port communication baud rate is 1.2 KBaud to 115.2 KBaud.

3.2 Fuzzy PID Control Program Module

Through the method of fuzzy mathematics, the long-term experience of human beings is summed up as the rules of fuzzy reasoning, and then the fuzzy rules are input into the microcomputer. The deviation signal and deviation change rate signal produced by comparing the measured value with the known set value are sent to the fuzzy controller. After the completion of fuzzification, fuzzy reasoning and anti fuzzification, the corresponding three correction quantities ΔK_p, ΔK_i and ΔK_d are obtained, By inputting these three corrections into the PID controller, the PID parameters can be adjusted to the best state online, so as to realize the effective control of the controlled object.

This article uses a two-dimensional fuzzy controller. The two variables input to the fuzzy PID controller include the deviation of the return water temperature of the cold water system compared with the given value and the rate of change of the deviation. The output of the fuzzy controller is Parameters ΔK_p, ΔK_i and ΔK_d required by conventional PID controllers.

In practical application, the control effect depends on the selection of appropriate fuzzy rules, the quantization factor of input variables and the scale factor of output variables [9]. The given value of water temperature is set as 12oc. Considering the actual weather conditions, we set the range of deviation as [−14, 14] and the range of deviation change as [−18, 18]; According to the above expert experience and experimental trial and error method, the parameter variation range of controller ΔK_p, ΔK_i and ΔK_d is $[-0.07, 0.07]$, $[-0.001, 0.001]$ and $[-0.02, 0.02]$ in turn. The fuzzy domain of each variable is set to $\{-6, -5, -4, -3, -2, -1, 0, 1, 2, 3, 4, 5, 6\}$; the input and output variables select 7 fuzzy subsets: $\{NB, NM, NS, Z, PS, PM, PB\}$, which correspond to negative large, negative medium, negative small, negative, positive small, positive medium, and positive large. The quantization factors $K_e = 0.5$ and $K_{ec} = 0.4$ of the deviation and deviation change, and the scale factor of the output variable are set to 0.015, 0.0002 and 0.065 respectively.

When establishing the fuzzy rule table, the setting requirements of K_p, K_i and K_d of PID controller are different when the value of deviation $|e|$ and deviation rate $|ec|$ changes.

(1) When the value of $|e|$ is relatively large, if the control system is required to have good tracking performance and speed up the system response, then a relatively large K_p and a relatively small K_d should be selected. In this process, in order to prevent the system from overshooting Excessive adjustment will cause integral saturation, and the effect of integral should be limited. Generally, $K_i = 0$ is selected;
(2) When the values of $|e|$ and $|ec|$ are medium, if the system is required to have a small overshoot and a certain response speed, then a smaller K_p and an appropriate K_i should be selected. In this process, K_d has a greater impact on the system, and an appropriate K_d value should be selected;
(3) When the value of $|e|$ is relatively small, it means that the output value is close to the set value. If the system is required to have better stability, the values of K_p and T should be increased at this time. In order to avoid shocks during this process, If $|ec|$ is relatively large, K_d should have a medium value; if $|ec|$ is relatively small, the value of K_d should be relatively large to make the system's anti-interference ability stronger.

For any input e and ec of the measured variable, its output u is required, and the fuzzy reasoning relation set R is needed first. The fuzzy inference relation set R of the system temperature is expressed as follows:

$$R = \underset{i,j}{U}\left(E_j \times EC_j \times U_{ij}\right) \quad (1)$$

In formula (1), E_j, EC_j and U_{ij} represent deviation, deviation change and the fuzzy state of output u; the value range of i and j is.

According to the above, the fuzzy relation corresponding to rule R_i can be obtained

$$\underset{\sim}{R_1} = NB_e \times (NB_{ec} \cup NS_{ec} \cup Z_{ec}) \times PB_u \quad (2)$$

According to formula (2), the fuzzy relationship matrix $\underset{\sim}{R}$ is inferred, and each element of the universe E and EC is used as input, and the corresponding fuzzy amount of the control variable is obtained through the fuzzy inference synthesis operation:

$$\underset{\sim}{u}(k) = \left(\underset{\sim}{E}(k) \times \underset{\sim}{EC}(k)\right) \circ \underset{\sim}{R} = (NBe \times PSec) \circ R \tag{3}$$

In formula (3), $(NBe \times PSec)$ represents the input fuzzy state, and its membership function vector must be combined with the subsequent $\underset{\sim}{R}$. Therefore, the operation result of $(NBe \times PSec)$ is expanded into a row vector, and then combined with it.

In the design system, the minimum rule is used in the fuzzy operation, and the maximum rule is used in the fuzzy synthesis rule; In the process of anti fuzziness, we use the maximum membership method. By synthesizing the assignment table of different membership degrees of known fuzzy subsets and the fuzzy control model with different parameters, we can get the accurate values of ΔK_p, ΔK_i D and ΔK_d to adjust the PID parameters.

3.3 Serial Communication Program Module

The communication port of S7-200 CPU is the standard RS-485 half duplex serial communication port. The format of serial character communication can include: a start bit, 7 or 8 characters (data bytes), a parity bit, or no parity bit, and a stop bit. The baud rate of communication can be set to 9600 bit/s, 19200 bit/s, 115200 bit/s. All serial communication devices that conform to these formats can communicate with S7-200 CPU. With the help of free port communication mode, S7-200 CPU can communicate with other devices and controllers disclosed by many communication protocols.

When sending XMT and receiving RCV commands as the core command of Freeport communication, note that multiple commands cannot be valid at the same time. The sending and receiving of data needs to pass through the data buffer. The sending command is to send the data in the buffer through the communication port and accept the command. The data is received from the communication port into the buffer. When calling and sending XMT and receiving RCV commands, only the starting byte address of the communication port and the data buffer is required.Sending XMT and receiving RCV instructions have nothing to do with the address of the communication object on the network, but only operate on the local communication port. If there are multiple devices on the network, the message must contain address information, which is the processing object of sending XMT and receiving RCV instructions [10].

Through the design of the above hardware unit and software module, the operation of the indoor intelligent temperature fuzzy PID control system is realized, which provides effective system support for indoor temperature control and provides a good living environment for residents.

4 Experiment and Result Analysis

4.1 Selection of Experimental Software

With the development of computer technology, simulation technology has been paid more and more attention, which makes people get better design reliability and research prospect in scientific research and production application, and saves the cost of scientific research and production. System simulation has a wide range of applications, it is not only widely used in the field of natural science, but also in the management and other fields has been successfully applied. There are many kinds of software used for simulation. Here, the simulation of PMV value control is MATLAB software. Therefore, it is necessary to introduce the application of this typical system simulation software in the control field.

The MATLAB software was launched by Math Works in the United States. The software provides a wealth of functions such as numerical analysis, matrix calculations, graphics rendering, data processing, image processing, etc. In addition, MATLAB has also launched more than 30 functions for different application disciplines. Application toolboxes, such as control system toolbox, fuzzy control toolbox, neural network toolbox, optimization toolbox, statistics toolbox and symbolic mathematics toolbox, etc. At present, it has become one of the most widely popular languages in the international control community, and is known as the "universal" tool for engineers, scientists, educators, and professionals from all walks of life all over the world.

SIMIJLINK toolbox is an extension of MATLAB software. It is mainly used for modeling, analysis and simulation of dynamic systems. It provides graphical user interfaces with most of the dynamic system structure modules required to establish system models. After entering the MATLAB environment, just type the SIMULINK command to open the module library. The user can select the required module according to his own system, drag it to his own system model with the mouse, and then use the mouse to draw a line to connect it. It constitutes the SIMULINK description of the system. After the model of the system is built, the user can set or change the parameters of the module according to the different needs of the system, then open the simulation menu, set the simulation parameters, and start the simulation process. After the simulation, the user can observe the simulation output of the system through the output oscilloscope or plot function。

4.2 Analysis of Experimental Results

According to the above selected experimental software, the simulation experiment is carried out for Indoor Intelligent Temperature Fuzzy PID control, and the application performance of the system is displayed by the control accuracy. The analysis process of the specific experimental results is as follows.

The control precision obtained by experiment is shown in Table 4.

Table 4. Control accuracy data sheet

Number of experiments	Design system	Existing system
20	78.56%	56.45%
40	70.12%	45.12%

(continued)

Table 4. (*continued*)

Number of experiments	Design system	Existing system
60	71.23%	48.79%
80	80.11%	41.20%
100	75.45%	45.18%
120	76.69%	49.51%
140	78.52%	40.00%
160	81.32%	58.49%

As shown in Table 4, the design system temperature control accuracy range is 70.12%–81.32%, and the existing system temperature control accuracy range is 40.00%–58.49%. The comparison of the above data shows that the temperature control accuracy of the design system is higher, which fully proves the effectiveness and feasibility of the design system.

5 Conclusion

This research uses PLC technology to design a new indoor intelligent temperature fuzzy PID control system, through the sensor selection unit, PLC hardware selection unit, fuzzy PID controller unit and power supply circuit design unit design system hardware, based on the hardware design through PLC Software module, fuzzy PID control program module and serial communication program module design system software, so as to realize the design of indoor intelligent temperature fuzzy PID control system. The design system in this paper greatly improves the temperature control accuracy and can better control the indoor temperature for residents. Provide a more comfortable environment.

Fund Projects. Research on intelligent temperature control system based on PLC (Project No.: YZK2015 050).

References

1. Li, X., Han, Z., Zhao, T., et al.: Online model for indoor temperature control based on building thermal process of air conditioning system. J. Build. Eng. **39**(2), 102270 (2021)
2. Zhong, X., Ridley, I.A.: Verification of behavioural models of window opening: the accuracy of window-use pattern, indoor temperature and indoor PM 2.5 concentration prediction. Build. Simul. **13**(3), 527–542 (2020)
3. Jiang, D., Wang, L.: Constant temperature control system for indoor environment of buildings based on internet of things. Int. J. Internet Protoc. Technol. **13**(2), 1 (2020)
4. Matta, A., Reddy, M.S.: Numerical study on indoor temperature in multi-storey buildings. Int. J. Control Autom. **13**(1), 245–249 (2020)
5. Kwon, O.I., Kim, Y.I., Kim, S.H.: Prediction model of indoor temperature distribution for optimal control of building energy. Korean J. Air-Conditioning Refrigeration Eng. **33**(3), 130–141 (2021)

6. Hussein, S., Sharifmuddin, N.B., Rabeei, A., et al.: Black box modelling and simulating the dynamic indoor air temperature of a laboratory using (ARMA) model. Indonesian J. Electr. Eng. Comput. Sci. **21**(2), 791–800 (2021)
7. Fontes F , R Anto, Mota A , et al. Improving the ambient temperature control performance in smart homes and buildings. Sensors **21**(2), 423–423 (2021)
8. Liu, S., Sun, G., Fu, W. (eds.): eLEOT. LNICST, vol. 339. Springer, Cham (2020). https://doi.org/10.1007/978-3-030-63952-5
9. Liu, S., Liu, X., Wang, S., Muhammad, K.: Fuzzy-aided solution for out-of-view challenge in visual tracking under IoT assisted complex environment. Neural Comput. Appl. **33**(4), 1055–1065 (2021)
10. Liu, S., Li, Z., Zhang, Y., Cheng, X.: Introduction of key problems in long-distance learning and training. Mob. Networks Appl. **24**(1), 1–4 (2018). https://doi.org/10.1007/s11036-018-1136-6

Author Index

Printed in the United States
by Baker & Taylor Publisher Services